Modelling Spatial and Spatial-Temporal Data

A Bayesian Approach

Chapman & Hall/CRC Statistics in the Social and Behavioural Sciences

Series Editors

Jeff Gill, *Washington University, USA*
Steven Heeringa, *University of Michigan, USA*
Wim J. van der Linden
Tom Snijders, *Oxford University, UK, University of Groningen, NL*

Recently Published Titles

Multilevel Modelling Using Mplus
Holmes Finch and Jocelyn Bolin

Applied Survey Data Analysis, Second Edition
Steven G. Heering, Brady T. West, and Patricia A. Berglund

Adaptive Survey Design
Barry Schouten, Andy Peytchev, and James Wagner

Handbook of Item Response Theory, Volume *One*: Models
Wim J. van der Linden

Handbook of Item Response Theory, Volume *Two*: Statistical Tools
Wim J. van der Linden

Handbook of Item Response Theory, Volume *Three*: Applications
Wim J. van der Linden

Bayesian Demographic Estimation and Forecasting
John Bryant and Junni L. Zhang

Applied Multivariate Analysis in the Behavioral Sciences, Second Edition
Kimmo Vehkalahti and Brian S. Everitt

Analysis of Integrated Data
Edited by Li-Chun Zhang and Raymond L. Chambers

Multilevel Modeling Using R
W. Holmes Finch, Joselyn E. Bolin, and Ken Kelley

Modelling Spatial and Spatial-Temporal Data: A Bayesian Approach
Robert Haining and Guangquan Li

For more information about this series, please visit: https://www.crcpress.com/go/ssbs

Modelling Spatial and Spatial-Temporal Data

A Bayesian Approach

By
Robert Haining and Guangquan Li

CRC Press
Taylor & Francis Group
Boca Raton London New York

CRC Press is an imprint of the
Taylor & Francis Group, an **informa** business

A CHAPMAN & HALL BOOK

CRC Press
Taylor & Francis Group
6000 Broken Sound Parkway NW, Suite 300
Boca Raton, FL 33487-2742

© 2020 by Taylor & Francis Group, LLC
CRC Press is an imprint of Taylor & Francis Group, an Informa business

No claim to original U.S. Government works

Printed on acid-free paper

International Standard Book Number-13: 978-1-4822-3742-9 (Hardback)

Visit the Taylor & Francis Web site at
http://www.taylorandfrancis.com

and the CRC Press Web site at
http://www.crcpress.com

I dedicate the book to my wife, Rachel.
Robert

I dedicate the book to my parents, Yaosen and Shaoying, my

wife, Hui, and our boys, Jamie and Freddie
Guangquan

Contents

Preface xix
Acknowledgements xxv

Part I Fundamentals for Modelling Spatial and Spatial-Temporal Data

1 Challenges and Opportunities Analysing Spatial and Spatial-Temporal Data 3
   1.1   Introduction 3
   1.2   Four Main Challenges When Analysing Spatial and
       Spatial-Temporal Data 4
       1.2.1   Dependency 4
       1.2.2   Heterogeneity 8
       1.2.3   Data Sparsity 10
       1.2.4   Uncertainty 12
               1.2.4.1   Data Uncertainty 12
               1.2.4.2   Model (or Process) Uncertainty 13
               1.2.4.3   Parameter Uncertainty 13
   1.3   Opportunities Arising from Modelling Spatial and Spatial-Temporal Data 14
       1.3.1   Improving Statistical Precision 14
       1.3.2   Explaining Variation in Space and Time 18
               1.3.2.1   Example 1: Modelling Exposure-Outcome Relationships 18
               1.3.2.2   Example 2: Testing a Conceptual Model at the
                       Small Area Level 20
               1.3.2.3   Example 3: Testing for Spatial Spillover (Local
                       Competition) Effects 22
               1.3.2.4   Example 4: Assessing the Effects of an Intervention 24
       1.3.3   Investigating Space-Time Dynamics 27
   1.4   Spatial and Spatial-Temporal Models: Bridging between Challenges
       and Opportunities 27
       1.4.1   Statistical Thinking in Analysing Spatial and Spatial-Temporal
               Data: The Big Picture 27
       1.4.2   Bayesian Thinking in a Statistical Analysis 30
       1.4.3   Bayesian Hierarchical Models 33
               1.4.3.1   Thinking Hierarchically 33
               1.4.3.2   Incorporating Spatial and Spatial-Temporal Dependence
                       Structures in a Bayesian Hierarchical Model Using
                       Random Effects 36
               1.4.3.3   Information Sharing in a Bayesian Hierarchical Model
                       through Random Effects 37
       1.4.4   Bayesian Spatial Econometrics 39
   1.5   Concluding Remarks 40
   1.6   The Datasets Used in the Book 41
   1.7   Exercises 45

2   **Concepts for Modelling Spatial and Spatial-Temporal Data: An Introduction**
    to "Spatial Thinking" 47
    2.1   Introduction 47
    2.2   Mapping Data and Why It Matters 48
    2.3   Thinking Spatially 52
           2.3.1   Explaining Spatial Variation 52
           2.3.2   Spatial Interpolation and Small Area Estimation 54
    2.4   Thinking Spatially and Temporally 56
           2.4.1   Explaining Space-Time Variation 56
           2.4.2   Estimating Parameters for Spatial-Temporal Units 59
    2.5   Concluding Remarks 59
    2.6   Exercises 60
           Appendix: Geographic Information Systems 61

3   **The Nature of Spatial and Spatial-Temporal Attribute Data** 63
    3.1   Introduction 63
    3.2   Data Collection Processes in the Social Sciences 63
           3.2.1   Natural Experiments 64
           3.2.2   Quasi-Experiments 67
           3.2.3   Non-Experimental Observational Studies 68
    3.3   Spatial and Spatial-Temporal Data: Properties 71
           3.3.1   From Geographical Reality to the Spatial Database 71
           3.3.2   Fundamental Properties of Spatial and Spatial-Temporal Data 74
                    3.3.2.1   Spatial and Temporal Dependence 74
                    3.3.2.2   Spatial and Temporal Heterogeneity 75
           3.3.3   Properties Induced by Representational Choices 76
           3.3.4   Properties Induced by Measurement Processes 83
    3.4   Concluding Remarks 84
    3.5   Exercises 84

4   **Specifying Spatial Relationships on the Map: The Weights Matrix** 87
    4.1   Introduction 87
    4.2   Specifying Weights Based on Contiguity 88
    4.3   Specifying Weights Based on Geographical Distance 90
    4.4   Specifying Weights Based on the Graph Structure Associated with a Set
           of Points 90
    4.5   Specifying Weights Based on Attribute Values 92
    4.6   Specifying Weights Based on Evidence about Interactions 92
    4.7   Row Standardisation 93
    4.8   Higher Order Weights Matrices 95
    4.9   Choice of W and Statistical Implications 97
           4.9.1   Implications for Small Area Estimation 97
           4.9.2   Implications for Spatial Econometric Modelling 102
           4.9.3   Implications for Estimating the Effects of Observable Covariates
                    on the Outcome 104
    4.10   Estimating the W Matrix 105
    4.11   Concluding Remarks 106
    4.12   Exercises 106
    4.13   Appendices 107

Appendix 4.13.1 Building a Geodatabase in R 107
Appendix 4.13.2 Constructing the *W* Matrix and Accessing Data
               Stored in a Shapefile 110

**5 Introduction to the Bayesian Approach to Regression Modelling with Spatial
and Spatial-Temporal Data** 115
  5.1   Introduction 115
  5.2   Introducing Bayesian Analysis 116
        5.2.1   Prior, Likelihood and Posterior: What Do These Terms Refer To? 116
        5.2.2   Example: Modelling High-Intensity Crime Areas 120
  5.3   Bayesian Computation 121
        5.3.1   Summarising the Posterior Distribution 121
        5.3.2   Integration and Monte Carlo Integration 123
        5.3.3   Markov Chain Monte Carlo with Gibbs Sampling 127
        5.3.4   Introduction to WinBUGS 129
        5.3.5   Practical Considerations when Fitting Models in WinBUGS 133
                5.3.5.1   Setting the Initial Values 133
                5.3.5.2   Checking Convergence 134
                5.3.5.3   Checking Efficiency 136
  5.4   Bayesian Regression Models 137
        5.4.1   Example I: Modelling Household-Level Income 138
        5.4.2   Example II: Modelling Annual Burglary Rates in Small Areas 143
  5.5   Bayesian Model Comparison and Model Evaluation 147
  5.6   Prior Specifications 148
        5.6.1   When We Have Little Prior Information 148
        5.6.2   Towards More Informative Priors for Modelling Spatial and
               Spatial-Temporal Data 150
  5.7   Concluding Remarks 151
  5.8   Exercises 153

Part II   Modelling Spatial Data

6 Exploratory Analysis of Spatial Data 159
  6.1   Introduction 159
  6.2   Techniques for the Exploratory Analysis of Univariate Spatial Data 160
        6.2.1   Mapping 161
        6.2.2   Checking for Spatial Trend 165
        6.2.3   Checking for Spatial Heterogeneity in the Mean 168
                6.2.3.1   Count Data 168
                6.2.3.2   A Monte Carlo Test 169
                6.2.3.3   Continuous-Valued Data 170
        6.2.4   Checking for Global Spatial Dependence (Spatial Autocorrelation) 172
                6.2.4.1   The Moran Scatterplot 173
                6.2.4.2   The Global Moran's **I** Statistic 174
                6.2.4.3   Other Tests for Assessing Global Spatial Autocorrelation 177
                6.2.4.4   The Global Moran's **I** Applied to Regression Residuals 178
                6.2.4.5   The Join-Count Test for Categorical Data 179

6.2.5   Checking for Spatial Heterogeneity in the Spatial Dependence
        Structure: Detecting Local Spatial Clusters 181
        6.2.5.1   The Local Moran's *I*.. 182
        6.2.5.2   The Multiple Testing Problem When Using Local Moran's *I*....185
        6.2.5.3   Kulldorff's Spatial Scan Statistic 186
6.3   Exploring Relationships between Variables 191
        6.3.1   Scatterplots and the Bivariate Moran Scatterplot.................... 191
        6.3.2   Quantifying Bivariate Association.. 193
        6.3.2.1   The Clifford-Richardson Test of Bivariate Correlation in
                  the Presence of Spatial Autocorrelation..................................... 193
        6.3.2.2   Testing for Association "At a Distance" and the Global
                  Bivariate Moran's *I*.. 195
        6.3.3   Checking for Spatial Heterogeneity in the Outcome-Covariate
                Relationship: Geographically Weighted Regression (GWR) 195
6.4   Overdispersion and Zero-Inflation in Spatial Count Data 203
        6.4.1   Testing for Overdispersion... 204
        6.4.2   Testing for Zero-Inflation 206
6.5   Concluding Remarks.. 207
6.6   Exercises 209
Appendix. An R Function to Perform the Zero-Inflation Test by
        van Den Broek (1995).. 210

7   **Bayesian Models for Spatial Data I: Non-Hierarchical and Exchangeable
    Hierarchical Models** 213
7.1   Introduction 213
7.2   Estimating Small Area Income: A Motivating Example and Different
        Modelling Strategies... 214
        7.2.1   Modelling the 109 Parameters Non-Hierarchically 216
        7.2.2   Modelling the 109 Parameters Hierarchically 217
7.3   Modelling the Newcastle Income Data Using Non-Hierarchical Models 218
        7.3.1   An Identical Parameter Model Based on Strategy 1 218
        7.3.2   An Independent Parameters Model Based on Strategy 2 220
7.4   An Exchangeable Hierarchical Model Based on Strategy 3.............................. 223
        7.4.1   The Logic of Information Borrowing and Shrinkage............................ 224
        7.4.2   Explaining the Nature of Global Smoothing Due to
                Exchangeability.. 225
        7.4.3   The Variance Partition Coefficient (VPC).. 226
        7.4.4   Applying an Exchangeable Hierarchical Model to the Newcastle
                Income Data.. 228
7.5   Concluding Remarks.. 230
7.6   Exercises 230
7.7   Appendix: Obtaining the Simulated Household Income Data......................... 231

8   **Bayesian Models for Spatial Data II: Hierarchical Models with Spatial
    Dependence** 233
8.1   Introduction... 233
8.2   The Intrinsic Conditional Autoregressive (ICAR) Model 234
        8.2.1   The ICAR Model Using a Spatial Weights Matrix
                with Binary Entries... 234

8.2.1.1   The WinBUGS Implementation of the ICAR Model 236
8.2.1.2   Applying the ICAR Model Using Spatial Contiguity to the Newcastle Income Data 238
8.2.1.3   Results 240
8.2.1.4   A Summary of the Properties of the ICAR Model Using a Binary Spatial Weights Matrix 244
8.2.2   The ICAR Model with a General Weights Matrix 245
8.2.2.1   Expressing the ICAR Model as a Joint Distribution and the Implied Restriction on W 245
8.2.2.2   The Sum-to-Zero Constraint 246
8.2.2.3   Applying the ICAR Model Using General Weights to the Newcastle Income Data 247
8.2.2.4   Results 248
8.3   The Proper CAR (pCAR) Model 249
8.3.1   Prior Choice for ρ 251
8.3.2   ICAR or pCAR? 251
8.3.3   Applying the pCAR Model to the Newcastle Income Data 253
8.3.4   Results 253
8.4   Locally Adaptive Models 256
8.4.1   Choosing an Optimal W Matrix from All Possible Specifications 258
8.4.2   Modelling the Elements in the W Matrix 258
8.4.3   Applying Some of the Locally Adaptive Spatial Models to a Subset of the Newcastle Income Data 262
8.5   The Besag, York and Mollié (BYM) Model 266
8.5.1   Two Remarks on Applying the BYM Model in Practice 268
8.5.2   Applying the BYM Model to the Newcastle Income Data 269
8.6   Comparing the Fits of Different Bayesian Spatial Models 274
8.6.1   DIC Comparison 274
8.6.2   Model Comparison Based on the Quality of the MSOA-Level Average Income Estimates 276
8.7   Concluding Remarks 277
8.8   Exercises 279

9 **Bayesian Hierarchical Models for Spatial Data: Applications** 281
9.1   Introduction 281
9.2   Application 1: Modelling the Distribution of High Intensity Crime Areas in a City 282
9.2.1   Background 282
9.2.2   Data and Exploratory Analysis 283
9.2.3   Methods Discussed in Haining and Law (2007) to Combine the PHIA and EHIA Maps 285
9.2.4   A Joint Analysis of the PHIA and EHIA Data Using the MVCAR Model 286
9.2.5   Results 290
9.2.6   Another Specification of the MVCAR Model and a Limitation of the MVCAR Approach 293
9.2.7   Conclusion and Discussion 293
9.3   Application 2: Modelling the Association Between Air Pollution and Stroke Mortality 296

9.3.1   Background and Data 296
9.3.2   Modelling..300
9.3.3   Interpreting the Statistical Results 302
9.3.4   Conclusion and Discussion 305
9.4   Application 3: Modelling the Village-Level Incidence of Malaria in a Small Region of India 308
9.4.1   Background..308
9.4.2   Data and Exploratory Analysis.......................................308
9.4.3   Model I: A Poisson Regression Model with Random Effects...............310
9.4.4   Model II: A Two-Component Poisson Mixture Model......................311
9.4.5   Model III: A Two-Component Poisson Mixture Model with Zero-Inflation 313
9.4.6   Results 314
9.4.7   Conclusion and Model Extensions.................................318
9.5   Application 4: Modelling the Small Area Count of Cases of Rape in Stockholm, Sweden..321
9.5.1   Background and Data 321
9.5.2   Modelling..322
9.5.2.1   A "Whole-Map" Analysis Using Poisson Regression.............322
9.5.2.2   A "Localised" Analysis Using Bayesian Profile Regression...323
9.5.3   Results 326
9.5.3.1   "Whole Map" Associations for the Risk Factors.....................326
9.5.3.2   "Local" Associations for the Risk Factors 326
9.5.4   Conclusions 329
9.6   Exercises 330

10 Spatial Econometric Models...333
10.1   Introduction 333
10.2   Spatial Econometric Models..334
10.2.1   Three Forms of Spatial Spillover 334
10.2.2   The Spatial Lag Model (SLM)..335
10.2.2.1   Formulating the Model....................................335
10.2.2.2   An Example of the SLM 336
10.2.2.3   The Reduced Form of the SLM and the Constraint on δ........337
10.2.2.4   Specification of the Spatial Weights Matrix..........................338
10.2.2.5   Issues with Model Fitting and Interpreting Coefficients 339
10.2.3   The Spatially-Lagged Covariates Model (SLX)...............339
10.2.3.1   Formulating the Model....................................339
10.2.3.2   An Example of the SLX Model 340
10.2.4   The Spatial Error Model (SEM)..340
10.2.5   The Spatial Durbin Model (SDM).....................................341
10.2.5.1   Formulating the Model....................................341
10.2.5.2   Relating the SDM Model to the Other Three Spatial Econometric Models 342
10.2.6   Prior Specifications 342
10.2.7   An Example: Modelling Cigarette Sales in 46 US States......................343
10.2.7.1   Data Description, Exploratory Analysis and Model Specifications 343
10.2.7.2   Results...345

10.3 Interpreting Covariate Effects 346
    10.3.1 Definitions of the Direct, Indirect and Total Effects of a Covariate 346
    10.3.2 Measuring Direct and Indirect Effects without the SAR Structure
         on the Outcome Variables 347
        10.3.2.1 For the LM and SEM Models 347
        10.3.2.2 For the SLX Model 347
    10.3.3 Measuring Direct and Indirect Effects When the Outcome
         Variables are Modelled by the SAR Structure 350
        10.3.3.1 Understanding Direct and Indirect Effects in the
                Presence of Spatial Feedback 350
        10.3.3.2 Calculating the Direct and Indirect Effects in the
                Presence of Spatial Feedback 351
        10.3.3.3 Some Properties of Direct and Indirect Effects 351
        10.3.3.4 A Property (Limitation) of the Average Direct and
                Average Indirect Effects Under the SLM Model 354
        10.3.3.5 Summary 354
    10.3.4 The Estimated Effects from the Cigarette Sales Data 355
10.4 Model Fitting in WinBUGS 356
    10.4.1 Derivation of the Likelihood Function 357
    10.4.2 Simplifications to the Likelihood Computation 361
    10.4.3 The Zeros-Trick in WinBUGS 361
    10.4.4 Calculating the Covariate Effects in WinBUGS 362
10.5 Concluding Remarks 365
    10.5.1 Other Spatial Econometric Models and the Two Problems of
         Identifiability 365
    10.5.2 Comparing the Hierarchical Modelling Approach and the Spatial
         Econometric Approach: A Summary 367
10.6 Exercises 370

11 Spatial Econometric Modelling: Applications 373
11.1 Application 1: Modelling the Voting Outcomes at the Local Authority
     District Level in England from the 2016 EU Referendum 373
    11.1.1 Introduction 373
    11.1.2 Data 374
    11.1.3 Exploratory Data Analysis 375
    11.1.4 Modelling Using Spatial Econometric Models 375
    11.1.5 Results 378
    11.1.6 Conclusion and Discussion 381
11.2 Application 2: Modelling Price Competition Between Petrol Retail
     Outlets in a Large City 382
    11.2.1 Introduction 382
    11.2.2 Data 383
    11.2.3 Exploratory Data Analysis 383
    11.2.4 Spatial Econometric Modelling and Results 384
    11.2.5 A Spatial Hierarchical Model with t_4 Likelihood 385
    11.2.6 Conclusion and Discussion 388
11.3 Final Remarks on Spatial Econometric Modelling of Spatial Data 388
11.4 Exercises 389
Appendix: Petrol Retail Price Data 390

Part III   Modelling Spatial-Temporal Data

12 **Modelling Spatial-Temporal Data: An Introduction**.. 395
   12.1   Introduction 395
   12.2   Modelling Annual Counts of Burglary Cases at the Small Area Level: A
          Motivating Example and Frameworks for Modelling
          Spatial-Temporal Data 398
   12.3   Modelling Small Area Temporal Data 401
          12.3.1 Issues to Consider When Modelling Temporal Patterns in the
                  Small Area Setting.. 403
                  12.3.1.1 Issues Relating to Temporal Dependence.............................. 403
                  12.3.1.2 Issues Relating to Temporal Heterogeneity and Spatial
                            Heterogeneity in Modelling Small Area Temporal Patterns........ 404
                  12.3.1.3 Issues Relating to Flexibility of a Temporal Model................ 404
          12.3.2 Modelling Small Area Temporal Patterns: Setting the Scene 406
          12.3.3 A Linear Time Trend Model... 407
                  12.3.3.1 Model Formulations.. 407
                  12.3.3.2 Modelling Trends in the Peterborough Burglary Data........... 411
          12.3.4 Random Walk Models... 416
                  12.3.4.1 Model Formulations.. 417
                  12.3.4.2 The RW1 Model: Its Formulation Via the Full
                            Conditionals and Its Properties 418
                  12.3.4.3 WinBUGS Implementation of the RW1 Model......................... 421
                  12.3.4.4 Example: Modelling Burglary Trends Using the
                            Peterborough Data 421
                  12.3.4.5 The Random Walk Model of Order 2 424
          12.3.5 Interrupted Time Series (ITS) Models... 426
                  12.3.5.1 Quasi-Experimental Designs and the Purpose of ITS
                            Modelling 426
                  12.3.5.2 Model Formulations.. 428
                  12.3.5.3 WinBUGS Implementation... 429
                  12.3.5.4 Results.. 432
   12.4   Concluding Remarks.. 434
   12.5   Exercises 435
   Appendix: Three Different Forms for Specifying the Impact Function f................... 436

13 **Exploratory Analysis of Spatial-Temporal Data**.. 439
   13.1   Introduction 439
   13.2   Patterns of Spatial-Temporal Data 440
   13.3   Visualising Spatial-Temporal Data 443
   13.4   Tests of Space-Time Interaction 448
          13.4.1 The Knox Test 449
                  13.4.1.1 An Instructive Example of the Knox Test and Different
                            Methods to Derive a p-Value 451
                  13.4.1.2 Applying the Knox Test to the Malaria Data 453
          13.4.2 Kulldorff's Space-Time Scan Statistic... 454
                  13.4.2.1 Application: The Simulated Small Area COPD
                            Mortality Data.. 456
          13.4.3 Assessing Space-Time Interaction in the Form of Varying Local
                  Time Trend Patterns 459

13.4.3.1 Exploratory Analysis of the Local Trends in the
Peterborough Burglary Data 460
13.4.3.2 Exploratory Analysis of the Local Time Trends in the
England COPD Mortality Data 460
13.5 Concluding Remarks 463
13.6 Exercises 464

**14 Bayesian Hierarchical Models for Spatial-Temporal
Data I: Space-Time Separable Models** 465
14.1 Introduction 465
14.2 Estimating Small Area Burglary Rates Over Time: Setting the Scene 465
14.3 The Space-Time Separable Modelling Framework 467
14.3.1 Model Formulations 467
14.3.2 Do We Combine the Space and Time Components Additively or
Multiplicatively? 469
14.3.3 Analysing the Peterborough Burglary Data Using a Space-Time
Separable Model 470
14.3.4 Results 474
14.4 Concluding Remarks 478
14.5 Exercises 479

**15 Bayesian Hierarchical Models for Spatial-Temporal Data II: Space-Time
Inseparable Models** 481
15.1 Introduction 481
15.2 From Space-Time Separability to Space-Time Inseparability: The Big Picture 481
15.3 Type I Space-Time Interaction 484
15.3.1 Example: A Space-Time Model with Type I Space-Time Interaction 484
15.3.2 WinBUGS Implementation 486
15.4 Type II Space-Time Interaction 486
15.4.1 Example: Two Space-Time Models with Type II Space-Time
Interaction 489
15.4.2 WinBUGS Implementation 490
15.5 Type III Space-Time Interaction 490
15.5.1 Example: A Space-Time Model with Type III
Space-Time Interaction 491
15.5.2 WinBUGS Implementation 492
15.6 Results from Analysing the Peterborough Burglary Data 493
15.7 Type IV Space-Time Interaction 498
15.7.1 Strategy 1: Extending Type II to Type IV 499
15.7.2 Strategy 2: Extending Type III to Type IV 501
15.7.2.1 Examples of Strategy 2 502
15.7.3 Strategy 3: Clayton's Rule 504
15.7.3.1 Structure Matrices and Gaussian Markov
Random Fields 505
15.7.3.2 Taking the Kronecker Product 506
15.7.3.3 Exploring the Induced Space-Time Dependence Structure
via the Full Conditionals 508
15.7.4 Summary on Type IV Space-Time Interaction 512
15.8 Concluding Remarks 513
15.9 Exercises 515

16   Applications in Modelling Spatial-Temporal Data 517
   16.1   Introduction 517
   16.2   Application 1: Evaluating a Targeted Crime Reduction Intervention 518
       16.2.1 Background and Data 518
       16.2.2 Constructing Different Control Groups 520
       16.2.3 Evaluation Using ITS.. 521
       16.2.4 WinBUGS Implementation 523
       16.2.5 Results 523
       16.2.6 Some Remarks 526
   16.3   Application 2: Assessing the Stability of Risk in Space and Time.................... 527
       16.3.1 Studying the Temporal Dynamics of Crime Hotspots and
               Coldspots: Background, Data and the Modelling Idea 527
       16.3.2 Model Formulations 530
       16.3.3 Classification of Areas.. 531
       16.3.4 Model Implementation and Area Classification 532
       16.3.5 Interpreting the Statistical Results 535
   16.4   Application 3: Detecting Unusual Local Time Patterns in Small Area Data 539
       16.4.1 Small Area Disease Surveillance: Background and Modelling Idea........ 539
       16.4.2 Model Formulation 540
       16.4.3 Detecting Unusual Areas with a Control of the False
               Discovery Rate... 543
       16.4.4 Fitting BaySTDetect in WinBUGS 543
       16.4.5 A Simulated Dataset to Illustrate the Use of BaySTDetect 546
       16.4.6 Results from the Simulated Dataset... 546
       16.4.7 General Results from Li et al. (2012) and an Extension of
               BaySTDetect.. 548
   16.5   Application 4: Investigating the Presence of Spatial-Temporal
       Spillover Effects on Village-Level Malaria Risk in Kalaburagi,
       Karnataka, India... 550
       16.5.1 Background and Study Objective.. 550
       16.5.2 Data 551
       16.5.3 Modelling.. 552
       16.5.4 Results 553
       16.5.5 Concluding Remarks... 556
   16.6   Conclusions... 558
   16.7   Exercises 560

Part IV   Addendum

17   Modelling Spatial and Spatial-Temporal Data: Future Agendas?............................ 565
   17.1   Topic 1: Modelling Multiple Related Outcomes Over Space and Time............ 565
   17.2   Topic 2: Joint Modelling of Georeferenced Longitudinal and
       Time-to-Event Data 567
   17.3   Topic 3: Multiscale Modelling 567
   17.4   Topic 4: Using Survey Data for Small Area Estimation.................................. 568
   17.5   Topic 5: Combining Data at Both Aggregate and Individual Levels to
       Improve Ecological Inference 571

17.6   Topic 6: Geostatistical Modelling..572
          17.6.1   Spatial Dependence 573
          17.6.2   Mapping to Reduce Visual Bias..574
          17.6.3   Modelling Scale Effects..574
17.7   Topic 7: Modelling Count Data in Spatial Econometrics...................575
17.8   Topic 8: Computation..575

References 577

Index...597

Preface

Large volumes of fine-grained spatial and spatial-temporal data, some routinely collected by public and private organizations, are available to students, researchers and policymakers, as well as the general public. Advances in information technology over the last few decades have made possible a revolution in the collection, storage, retrieval and presentation of social, economic, health and environmental data, enabling us to observe and study our world in ever finer geographical detail. With the increasing temporal frequency with which such data are collected, we are also in a better position to observe, measure and evaluate changes taking place over sometimes short time spans. Such *spatial* and *spatial-temporal precision* in the data we have access to should help us to better understand our changing world and provide us with information that can be used to inform policymaking and to evaluate the effectiveness of any intervention.

But to reap the full benefits of such precision in our data we need to be able to obtain reliable estimates of attribute properties for these small spatial and spatial-temporal units. Estimates for small geographical areas suffer from many problems, including data sparsity and data uncertainty that undermine *statistical precision*. Unreliable estimation of variability associated with small area estimates undermines the benefits that can be derived from such data. The big question that lies behind this book is: can we have the best of both worlds – spatial and spatial-temporal precision *and* statistical precision? Here, the answer to that question follows a path that leads through data modelling.

What Are the Aims of the Book?

- To provide students and researchers whose interests lie in the analysis of social, economic, political and public health data with an introduction to modelling fine-grained spatial and spatial-temporal data. This includes data collected for small areas such as census tracts, where populations may be numbered in the low hundreds, and data collected at point sites. Throughout the book, we consider situations where the small areas partitioning the study region are fixed in their locations, sizes and shapes, or, when point sites are of interest, they are fixed in their location. Random variation is associated with the attributes (i.e. the data we observe), *not* the spatial objects to which the attribute values are attached.

- To present spatial and spatial-temporal modelling at a level that is accessible to students and researchers from all areas of the social, political, economic and public health sciences who have a grounding in statistical theory up to and including the simple linear regression model and who have a basic understanding of matrices and matrix manipulation. This means that we believe parts of the book should be accessible to social science students who are following a quantitative path through their undergraduate degree; and fully accessible to postgraduate students and those researchers further along in their careers whose work involves analysing quantitative data. For students of statistics whose interests have been aroused by the sorts of questions social scientists habitually wrestle with, we hope this book will provide a bridge from what they already know about modelling data to the challenges that lie in wait when working with spatial or spatial-temporal data. Some (and increasingly more) research in the social sciences involves collaboration between social scientists and statisticians. For that to be effective there needs to be a shared understanding of statistical methods, and our hope is that this book will help to promote informed conversations between both groups on the relevance of the important methodologies contained herein.

- To highlight *the challenges* encountered when modelling spatial and spatial-temporal data and to present *the opportunities* that analyses of spatial and spatial-temporal data offer. Bridging between challenges and opportunities is central to any kind of statistical modelling, and modelling spatial and spatial-temporal data raises some very interesting issues in this regard.

We take the view that presenting methodology through a combination of theory and application is essential to understanding. Models are constructed to answer questions. So, to better understand the process of modelling, it helps to see how one pair of researchers (at least) have tackled certain substantive questions given the data to hand. It means that our approach includes exposing how a model is constructed in order to tackle a specific question and why that particular model is better suited over others in the given situation. We look at how to fit the given model to observed data, explain why the model estimates behave in the way they do, how the model estimates can be used to address the question(s) of interest, how to compare different models given the data that we observe, how to assess the fit of the model of choice to the observed data and how to modify/extend the model of choice. These are questions that are directly relevant to the modelling of spatial and spatial-temporal data, and they are the questions that help the analyst to advance the process of data modelling.

What Are the Key Features of the Book?

- We cover both hierarchical and spatial econometric modelling – two approaches to modelling spatial and spatial-temporal data that are widely used. In hierarchical modelling, dependency, which as we shall see is a fundamental property of spatial and spatial-temporal data, is modelled in what is called the data generating process (or process model) so that, conditional on the process model, data values are assumed to be independent. This, as we shall explain, brings a number of benefits to the analysis of spatial and spatial-temporal data. By contrast, in spatial econometric modelling, dependency is modelled in the likelihood (or data model). In the course of this book we shall explore the differences between these two modelling paradigms and their implications for statistical analysis and explain why each has a role to play in spatial and spatial-temporal data modelling.

- This book adopts a Bayesian approach to inference. Much has been written contrasting the Bayesian and frequentist approaches to inference. We take the view that Bayesian inference (with its emphasis on revising prior beliefs, vague as they sometimes are, in the light of new data) has particular appeal in the observational sciences not only because of the non-experimental, non-replicable nature of the data social scientists work with but also because of the types of questions often asked of that type of data. When modelling spatial and spatial-temporal data, adopting the Bayesian approach to inference gives rise to two practical advantages. First is its *flexibility*, which allows us: to construct complex models in order to adequately describe the observed data; to acknowledge the presence of different sources of uncertainty identified within a given study; to straightforwardly assess potential impacts of various plausible modelling assumptions on the findings of a study; to provide a summary that is relevant to the question(s) in hand. Second is its *efficiency* in utilising all the available information (from data as well as from prior knowledge) so that parameters involved in the modelling can be estimated reliably. These two intertwining features of the Bayesian approach are central to tackling the challenges presented in spatial and spatial-temporal data. Throughout this book, we shall demonstrate how Bayesian hierarchical modelling and Bayesian spatial econometric modelling can help to address substantive questions arising in the social, economic, political and health sciences.

- Parameter estimation is an integral part of the process of modelling. For model fitting in the Bayesian framework, we draw heavily on WinBUGS, an open-source software that is extremely flexible for fitting complex spatial and spatial-temporal models. In addition to its flexibility, implementing such complex models in WinBUGS, as we shall see, is reasonably straightforward. We provide the WinBUGS code with detailed explanations throughout the book. In cases where WinBUGS may not be an efficient choice for model fitting, we point the reader to (more) appropriate computational methods/software. In addition to WinBUGS, we make use of R, also an open-source software, in different places throughout the book, primarily at the exploratory stage.

- We provide the datasets and the associated codes in R and WinBUGS used in the examples and applications in an online repository. Making this material publicly available not only allows readers to replicate the results presented in the book but can also be used as a template for analysing one's own data. One of the important aspects of modelling spatial and spatial-temporal data is visualisation, for which mapping is often required. We will illustrate how we can produce maps in R and how we can use the resulting maps to explore features of the data observed over space and/or time and to evaluate the models that we develop against the observed data.

- We include applications from many social science fields: from criminology, sociology, economics and political science to epidemiology, public health and geography. The common thread in these different applications is that they all involve the analysis and modelling of spatial or spatial-temporal data. We believe that diversity promotes the cross-fertilization of ideas.

The Structure of the Book

The book is divided into three main parts. In Part I we introduce some of the basic ideas and principles involved in spatial and spatial-temporal modelling. It is divided into five chapters. Chapter 1 discusses the main methodological challenges in analysing spatial and spatial-temporal data and the opportunities that spatial and spatial-temporal data offer and how one can address these challenges through the use of Bayesian hierarchical models and Bayesian spatial econometric models. It offers a non-technical overview, a map if you will, of the book. Chapter 2 introduces some of the key ideas associated with what it means to "think spatially". Taken together, Chapters 1 and 2 attempt a fusion of spatial thinking and statistical thinking. Chapter 3 provides an overview of important characteristics of spatial and spatial-temporal attribute data, whilst Chapter 4 overviews the different ways in which the relationships between spatial units can be captured/quantified and the implications for different types of statistical modelling. Chapter 5 provides an introduction to Bayesian modelling and the use of WinBUGS for fitting (non-spatial) models to spatial data.

Part II examines approaches to spatial modelling where data have been collected on a set of attributes (variables) at one instant, or for one period, of time, and each data value is georeferenced to one of the spatial units (e.g. census tracts or fixed point sites such as retail outlets) in the study region. In Chapter 6 we describe a number of selected exploratory data analysis methods that provide useful tools to start the modelling process for such spatial data. Chapters 7 to 9 deal with Bayesian hierarchical modelling. Chapter 7 first presents non-hierarchical models (as a starting point for obtaining small area parameter estimates) then looks at exchangeable hierarchical models that borrow information from the whole map for the purpose of small area estimation. Chapter 8 introduces hierarchical models with spatial dependence where spatial dependence is modelled through different forms of the so-called conditional autoregressive (CAR) structure. These spatial models include the intrinsic CAR model, the proper CAR model and the Besag, York and Mollié (BYM) model. Locally adaptive spatial smoothing models will also be discussed. In the case of these models, information is borrowed locally rather than globally from the whole map. Through a number of applications, Chapter 9 illustrates the use of Bayesian hierarchical modelling to tackling a range of different substantive questions. Chapter 10 describes the spatial econometric modelling approach, and in Chapter 11 we provide two examples of their application. Spatial econometric models are important when the goal of an analysis is to estimate what are called spatial spillover and feedback effects, and they do so by extending the class of standard normal linear regression models.

Part III deals with modelling spatial-temporal data – each data value is observed at one of the instants of time (or during one of the time windows) within the study period and is also georeferenced to one of the spatial units in the study region. In Chapter 12 we discuss the special challenges that arise when moving from spatial to spatial-temporal data. To complement the material of Part II we discuss various time series models in Chapter 12, paying particular attention to the considerations we need to reflect on when using these time series models to analyse *spatial* time series data. These temporal models in combination with the spatial models of Part II provide the building blocks (the "Lego blocks") with which we construct what are known as space-time separable models for spatial-temporal data. Space-time separable models do not provide a satisfactory fit to spatial-temporal data when space-time interaction is present, as is often the case, in spatial-temporal observed outcome values. One form of space-time interaction is associated with concentrations of events at particular times in particular places (such as localized hotspots of crime or disease or traffic accidents). In Chapter 13 we first visually describe what the structure of a space-time dataset looks like in the presence and in the absence of space-time interaction. We then describe exploratory methods for assessing the presence (or absence) of space-time interaction in a given space-time dataset and methods for studying local properties of space-time interaction. Chapter 14 deals with space-time separable models, and Chapter 15 extends the space-time separable models to space-time inseparable models, where we need to embed models that capture space-time interaction. In Chapter 16 we look at some examples of spatial-temporal modelling with relevance to the field of policymaking:

evaluating whether a geographically targeted police initiative has been effective or not and setting up a surveillance system for detecting sudden increases in disease rates in particular places. We also include an application that examines the stability in space and time of crime hotspots and another application that shares common ground with the spatial econometric models of Part II, where we are looking for evidence of spillover effects in disease transmission over space and/or time, but where now we have spatial-temporal data to work with. We conclude the book, Part IV, with Chapter 17, which briefly suggests some future directions for spatial and spatial-temporal data modelling that we believe to be of importance.

The contents of some sections of this book took shape or began to take shape in the BIAS project (Bayesian methods for combining multiple Individual and Aggregate data Sources in observational studies: www.bias-project.org.uk/), and the authors acknowledge a particular debt to Nicky Best (director) and Sylvia Richardson. The BIAS project was sponsored by the National Centre for Research Methods (NCRM) through the Economic and Social Research Council (ESCR), UK. NCRM also funded short courses at the University of Cambridge, open to social science postgraduates and researchers from the UK and abroad, where the material for this book continued to evolve. Some material in Chapter 5 originates from several short courses on Bayesian analysis using WinBUGS (with the course material originally developed by Nicky and Sylvia) given at the Cathie Marsh Institute (CMIST) at the University of Manchester. Some of the material was also trialled in short courses taught at universities in several different parts of the world. We are grateful to the students who attended these courses and gave us useful feedback on the presentation of some of our material. The book that has materialised has been written so that it might serve both as a reference book as well as a hands-on learning and teaching text. The latter claim is based partly on the fact that we have included modelling code and exercises that students might use to develop their understanding and technical skills.

When using this book as a basis for a first one-semester-long postgraduate level course, the five chapters in Part I provide a general overview of spatial data analysis. Chapter 5 introduces the Bayesian approach to inference and outlines instructions for fitting a model in WinBUGS, so knowledge in those areas is not a prerequisite to this first course. We recommend covering selected material in Chapter 6 (e.g. mapping in R and carrying out global tests of spatial heterogeneity and spatial dependency) and extending the non-spatial models discussed in Chapter 5 to the non-hierarchical models (in Chapter 7) as well as some selected hierarchical spatial models (e.g. the exchangeable model, the ICAR and the BYM models) introduced in Chapters 7 and 8. Selected applications presented in Chapter 9 can be included to demonstrate the application of some of the spatial models. A second one-semester-long more advanced course will primarily focus on spatial econometric modelling (Chapters 10 and 11) and modelling spatial-temporal data. When covering the topic of spatial econometric modelling, it is important to draw students' attention to the comparison between the hierarchical modelling approach and the spatial econometric approach as discussed in Section 10.5.2. For the discussion on modelling spatial-temporal data, we recommend including material from Chapter 12 to Chapter 15, with selected applications from Chapter 16. As discussed in Chapter 15, some of the methodological advances (in particular those associated with constructing a space-time model with Type IV space-time interaction (Section 15.7)) and/or some of the computational challenges (e.g. Table 15.4) can provide a basis for a postgraduate-level research project. The same recommendation applies to the topics outlined in Chapter 17. Greater synergy between the methodologies of this book and the methodology of geostatistics (Topic 6 in Chapter 17) has long been advocated.

Acknowledgements

As a teaching text we acknowledge that the applications are biased (no pun intended) towards UK examples. This reflects the origins of the work and the fact that this is first and foremost a statistical methodology text. However, not all the datasets are from the UK – we have datasets from Sweden and India that reflect non-BIAS collaborations we have both been involved in. We are grateful to Vania Ceccato and to Sulochana Shekhar and Ashfaq Ahmed for the opportunity to work with them on the Stockholm rape study (funded by the Swedish Research Council, FORMAS) and the Karnataka malaria project (funded by the British Council's UK-India Education and Research Initiative) respectively, and for permission to include analyses of their data in our book. The work on the Sheffield stroke data was a collaborative project funded by the NHS Executive Trent Research Scheme and we wish to acknowledge Ravi Maheswaren, Steve Wise, Tim Pearson and Paul Brindley. Early work on the high intensity crime data by RPH was undertaken with Max Craglia and Paul Wiles with funding from the UK Government's Home Office and later with Jane Law (with funding from the Isaac Newton Trust, University of Cambridge). It was when working with Jane that RPH first encountered Bayesian hierarchical modelling. A big thank you to all these people – and many others whose work has shaped the content of this book – and to the funders of the above-mentioned projects.

This book is the product of 11 years of collaboration between RPH and GL on various projects and papers. We think it is fair to say we have both learned a great deal from each other, and the order of authors is purely alphabetical.

GL would like to thank Mark Little and Paolo Vineis, his PhD supervisors, who introduced him to the statistical world, and more generally, the world of scientific research. He would like to thank Nicky and Sylvia for their generous sharing of their knowledge and experience when working within the BIAS project. It was during the BIAS project that he was exposed to Bayesian modelling, spatial statistics, hierarchical modelling and their flexibility in tackling a variety of real-world problems. Thanks to Majid Ezzati for his advice and stimulating discussions on modelling small area health data. Over the past few years, GL has had the privilege to work with a number of colleagues, so thanks also to Lea Fortunato, Marta Blangiardo, Virgilio Gómez-Rubio and James Bennett for their insightful knowledge on modelling spatial and spatial-temporal data; Anna Hansell and Ismaïl Ahmed for shaping the small area disease surveillance work; Pete Philipson for introducing GL to the joint modelling of longitudinal and survival data. Thanks also to Mark Hancock and Matthew Quick, as well as the undergraduate and postgraduate project students at Northumbria, who have, to varying degrees, tried out parts of the book.

We both wish to thank the Cartography Unit (Phil Stickler) and the Geomatics Unit (Gabriel Amable) in the Department of Geography at the University of Cambridge for their assistance with some of the figures. We would like to thank John Kimmel for his help and support (and patience!) over the process of publishing this book. Thanks also to the anonymous referees for their comments and suggestions to shape this into a better book.

GL wishes to thank his parents, Yaosen and Shaoying, his wife, Hui, and their twin boys, Jamie and Freddie, for their love, understanding and support – he can now start fulfilling the many promises to the boys: "When the book is done, Daddy will …" RPH wishes to thank his wife, Rachel, who has been wonderfully understanding of his version of "retirement".

   We have set up a website where the datasets and codes can be accessed – and where any errors detected by or reported to us will be flagged and corrected. For those of our readers who are willing to make their datasets available to add to the learning materials, we plan to post web links there to their datasets. The link to the book's website is www.sptmbook.com.

Part I

Fundamentals for Modelling Spatial and Spatial-Temporal Data

1

Challenges and Opportunities Analysing Spatial and Spatial-Temporal Data

1.1 Introduction

When setting out to build a statistical model for spatial or spatial-temporal data, the analyst typically encounters a number of key methodological *challenges*. In the past it has been rather common, especially in the social sciences, to dwell on two of these challenges. The existence of spatial dependency invalidates conventional (or traditional) statistics, which assumes independent observations, whilst spatial heterogeneity sometimes requires us to consider how to estimate large numbers of parameters, perhaps more than the number of data points we have. In Section 1.2, we begin by describing *four* challenges that need to be met when analysing spatial or spatial-temporal data, thereby placing the challenges presented by the existence of spatial dependency and spatial heterogeneity within a wider set of methodological issues. In doing so, we discover that one of these challenges, spatial dependency, presents us with an *opportunity* when meeting another of the challenges, data sparsity. In this chapter, and where there is no ambiguity, for short we shall just refer to "spatial data" when referring to both spatial and spatial-temporal data.

In Section 1.3 we give examples of the *opportunities for analysis* that the methods of this book open up for us. Then, in Section 1.4, we overview the two main approaches to spatial and spatial-temporal modelling that constitute the core of this book. These two approaches enable us to address the types of questions specified in Section 1.3 whilst engaging, in different ways, with the challenges described in Section 1.2. These two approaches are: Bayesian hierarchical models and Bayesian spatial econometric models. Both types of model, hierarchical and econometric, are widely found in the social science literature, although not always in their Bayesian forms. We will touch on the reasons for adopting a Bayesian approach to inference, leaving more detailed discussion until Chapter 5. At the end of this chapter the reader will find a list of the datasets used in this book.

Particularly for the researcher unfamiliar with analysing spatial data, effort also needs to be directed at what it means to "think spatially". Explaining what this phrase implies we leave until Chapter 2.

1.2 Four Main Challenges When Analysing Spatial and Spatial-Temporal Data

1.2.1 Dependency

For spatial and spatial-temporal data, values that are close together in space and/or in time are unlikely to be *independent*. Dependence (or lack of independence) is a fundamental property of spatial and spatial-temporal data. A data value observed for an area during some interval of time typically contains some information about data values for the same variable at other (nearby) areas within the same (or nearby) time window. For example, a close inspection of Figure 1.1 shows that, although there are exceptions (see for example the areas marked x and +), the raw burglary rates in the areas that are adjacent generally tend to be more alike than those from areas that are far apart. We can also observe some local clusters of areas with high rates of crime events. The raw burglary rate in an area is calculated as the ratio of the number of burglary cases reported in the area to the number of houses at risk in that area. We refer to these as the *raw* burglary rates, as the calculation is based only on the data without sharing information through modelling (see Section 1.3.1 later).

In the case of spatial-temporal data, the sequence of maps in Figure 1.2 illustrates a similar, although more complex, space-time dependence structure – in part because of the use of smaller spatial units. The value observed in a spatial unit tends to be similar not only to the data values of its spatial neighbours but also to those of its temporal neighbours and to those of each spatial neighbour's temporal neighbours. Here, two areas are considered to be neighbours of each other if they share a common boundary. For time, the temporal neighbours of time t are considered to be $t - 1$ and $t + 1$. We will return to the topic of defining spatial neighbourhood structures in Chapter 4, and temporal and spatial-temporal neighbourhood structures in Chapters 12 and 15, respectively. Both examples

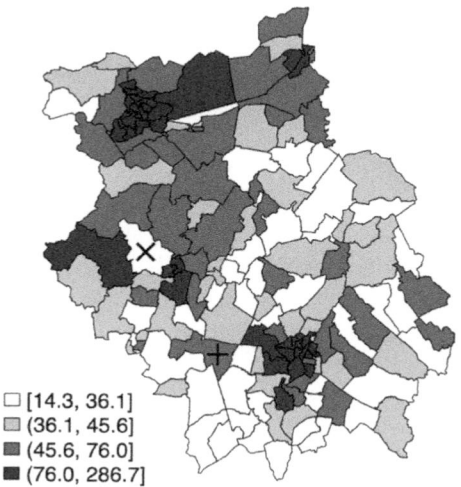

□ [14.3, 36.1]
▨ (36.1, 45.6]
▪ (45.6, 76.0]
■ (76.0, 286.7]

FIGURE 1.1

The raw burglary rates – number of burglary cases per 1000 houses in each area – across the 157 electoral wards in Cambridgeshire, England in 2004. The raw burglary rate in an area is calculated by dividing the number of reported burglary cases in that area during 2004 by the number of houses at risk in that area then multiplied by 1000. For scale: Cambridgeshire is approximately 60km (east-west) and 70km (north-south).

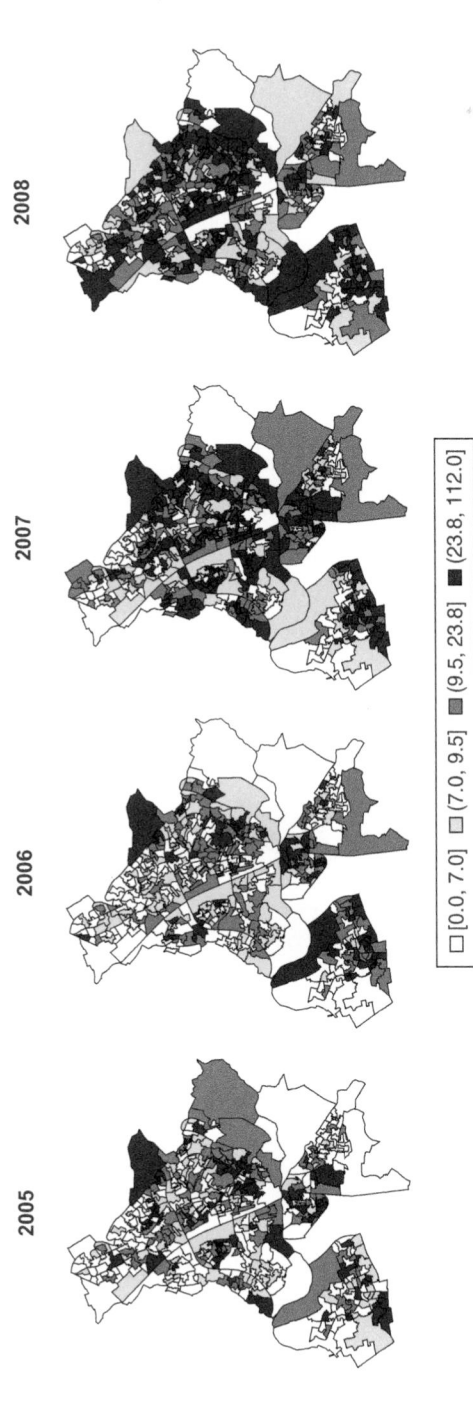

FIGURE 1.2
Maps of the raw annual burglary rates per 1000 houses for Peterborough, England 2005–08 at the Census Output Area (COA) level. For scale: in 2010 Peterborough's population was approximately 180,000.

show evidence of *positive autocorrelation*,[1] a property often encountered in practice. The presence of positive autocorrelation in spatial and spatial-temporal data implies that values close together in space and/or in time tend to be similar. However, as we shall discuss in Chapter 3, this property is dependent on the scale of the spatial and/or spatial-temporal units used to report the data. We shall consider ways to subject these ideas to testing in Chapter 6.

But dependency is not only a property of the *data values* that we observe, it can also be present in model *parameters* that represent a statistical property that varies spatially and/or temporally. Consider fitting, separately, a simple linear trend model to the time sequence of burglary counts reported for each of the 452 Census Output Areas (COAs) in Peterborough. The spatial distribution of the slope estimates is shown in Figure 1.3, where it appears that there is a tendency for areas close together to show similar increasing (decreasing) trends over time.

The presence of dependency in data presents a challenge to standard regression modelling where model errors are assumed to be independent. Although the use of covariates can account for some of the dependence structure in outcome data (for instance, in the context of analysing disease risks, the similarity in risk levels of respiratory disease in nearby neighbourhoods may be partly due to, and hence accounted for by, their proximity to a main road), it is often the case that the residuals from a model are still found to be spatially and/or temporally autocorrelated. This can be for a variety of reasons, but a common reason is that the model omits important covariates that are themselves spatially and/or temporally autocorrelated. Thus this property of the missing covariates is inherited by the

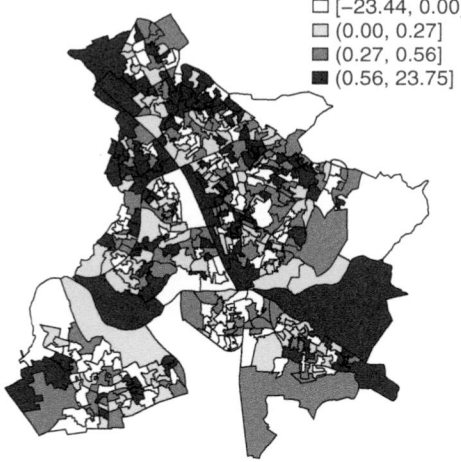

□ [−23.44, 0.00]
□ (0.00, 0.27]
■ (0.27, 0.56]
■ (0.56, 23.75]

FIGURE 1.3
A map of the estimates of the local slopes using the COA-level burglary data in Peterborough between 2005 and 2008.

---

[1] Correlation measures the strength of the association between two sets of values from two different variables. For example, a correlation of 0.6 between age and level of income suggests a higher income is associated with a higher age and vice versa. In the case of spatial and spatial-temporal data, we are interested in the strength of association between a set of values from the *same* variable, hence the prefix "auto". Spatial autocorrelation measures the strength of the correlation between the value of a variable (say income level) at a place with values of the same variable from other places within the study region separated, for example, by a specified distance. A more detailed discussion will be given in Chapter 6.

model's residuals, which represent the unexplained part of the variation in the observed values of the outcome variable.

When positive spatial autocorrelation is present, applying a model that assumes independent errors leads to *an underestimation of parameter uncertainty* (while this uncertainty is overestimated if negative spatial autocorrelation is present). We take a moment to expand this point further, the discussion of which leads onto some important implications to the modelling of spatial and spatial-temporal data.

Consider the following simple example of estimating the unknown mean, μ, of a normal distribution using a set of N observations $x_1,...,x_N$, which are sampled from the normal distribution in question. Assume for simplicity that the variance of this normal distribution, σ^2, is known. An unbiased point estimate of μ is given by the sample mean $\bar{x}$ of the observations. To derive the uncertainty interval of the point estimate, we need to estimate the error variance (denoted as EV, so $\sqrt{EV}$ gives the standard error), a measure of variability of the estimate of μ. If the observations are assumed to be independent, then the error variance is estimated by $\widehat{EV}_{ind} = \dfrac{\sigma^2}{N}$. The hat symbol $(\hat{\ })$ means that $\widehat{EV}_{ind}$ is a point estimate of the unknown parameter EV_{ind}, and the subscript *ind* emphasises the independent assumption about the observations. However, if the observations are in fact *dependent*, the estimations given above are based on a misspecified model. While the sample mean is still an unbiased point estimate of μ, the error variance is now $\widehat{EV}_{dep} = \dfrac{\sigma^2}{N} + \hat{\gamma}$, where $\hat{\gamma}$ is the sample autocovariance that quantifies the dependence amongst all the observations. If there is positive spatial dependence amongst the observations, $\hat{\gamma} > 0$ and hence $\widehat{EV}_{ind}$ *underestimates* the uncertainty of our estimate of μ. The resulting uncertainty interval becomes too narrow. In the case of hypothesis testing, we can see that the risk of committing a Type I error (for example, concluding the mean is not equal to zero when the conclusion is not justified at the chosen level of significance) is increased. In fact, the situation will usually be rather worse because typically σ^2 is not known, so it has to be estimated. The usual estimator of σ^2, in the case of independent observations, is $s^2 = \dfrac{1}{N-1}\sum_{i=1}^{N}(x_i - \bar{x})^2$.

However, when observations are dependent, s^2 underestimates σ^2 (see Haining, 1988, p.579), thus further increasing the risk of a Type I error (see Exercise 1.8).

If we now rephrase our argument about μ, replacing it with a set of covariates and their associated regression coefficients, then we can see that the same problem carries over to hypothesis tests on the contribution that the different covariates make to accounting for the variation in the observed outcome values. The examples in Section 1.3.2.1 below provide some illustrations of this point.

The above discussion provides justification for accounting for dependence in spatial data. Underlying the results described is the fact that "less information" about the parameter of interest (μ in the above illustration) is contained in a dataset where observations are positively autocorrelated. This can be compared to the situation where observations are independent. The term "effective" sample size has been coined and used to measure the information content in a set of autocorrelated data. Compared with the case of N independent observations, if we have N positively autocorrelated data values then the effective sample size is less than N (how much less depends on how strongly autocorrelated data values are). The effective sample size can be thought of as the *equivalent number of independent observations available for estimating the parameter* – see Cressie (1991, p.14–15) for a short worked example. We will return to the effective sample size in Chapter 6. It is this reduction in the information content of the data that increases the uncertainty of the parameter estimate, as reflected by $\widehat{EV}_{dep} > \widehat{EV}_{ind}$. This same data property is responsible for giving rise to inefficient estimators of the regression parameters on which we base our conclusions about the contribution covariates make to explaining variation in the data values that we observe on the outcome.

The presence of positive dependence in *observations* means that neighbours tend to have similar observed values, a property that poses a *challenge* to the sort of *likelihood-focused modelling* described in the previous paragraph. The information loss in a set of positively autocorrelated *observations* arises because each observation contains what we might call "overlapping" or "duplicate" information about other observations (if we know the data value at one location it tells us something about data values on the same variable at nearby locations). However, there is another angle from which to look at spatial dependence which, as it turns out, offers us a way to address the challenge described above as well as some of the other problems that we encounter when modelling spatial and spatial-temporal data. Because of the presence of dependency, the information that we have about an area regarding some *characteristic*, say a parameter such as its burglary rate, tells us something about the same characteristic in other neighbouring areas. As we shall see, the existence of spatial and spatial-temporal dependency turns out to provide an *opportunity* to "share information" across space and/or time in estimating these characteristics. This idea is central to much of the *process-focused modelling* described in this book. We shall have much to say about what "information sharing" (or "borrowing strength") means in practice, but for now we turn to the second of the four challenges we must face when analysing spatial data.

1.2.2 Heterogeneity

Things seldom stay the same across space or through time. Income levels vary from one part of a city to another. While different areas may experience different risk levels of a disease, the temporal dynamics of the disease risk may vary locally. For example, the risks in some areas may remain high while those in others show evidence of decreasing over time. A dataset where all subsets have the same statistical properties is said to be homogeneous. When homogeneity does not hold, the dataset is said to be heterogeneous. Heterogeneity is sometimes referred to as the second fundamental property of spatial data after spatial dependency, although its presence may depend on the size (or extent) of the geographical area under study rather than an inherent property of an attribute in geographic space.

Spatial heterogeneity can take different forms. It can be present in the *mean level* of a set of data. In studying burglary risk, different areas may present different risk-reward trade-offs for would-be burglars so that the *risk* of a burglary being committed varies from area to area. As a consequence, data on *observed* household burglary rates will show geographical variation. Figure 1.4 illustrates the uneven spatial distribution in the (raw) burglary rates in Cambridgeshire, England, in 2004. There appears to be a marked difference between the two urban areas of Peterborough (circled in the northwest) and Cambridge (circled in the south) on the one hand and rural areas (much of the rest of the county) on the other. The burglary rates are higher (with a darker colour) in the urban areas. The explanation for this difference could be that although the same risk factors are responsible in both urban and rural areas and the impact of each risk factor is the same, *levels* of the risk factors are higher in the case of urban areas than rural areas. Such heterogeneity in the mean would be captured by fitting a standard regression model – differences in the

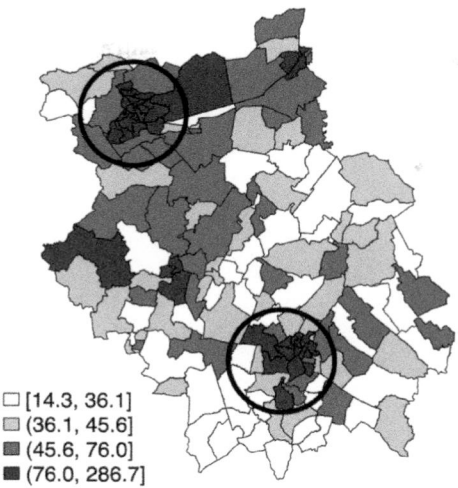

FIGURE 1.4
The raw rates of household burglary (per 1000 houses) at the ward level, Cambridgeshire, England, 2004. This is the same map as shown in Figure 1.1.

mean brought about by different levels of the risk factors in different areas (higher levels of poverty, less social cohesion, for example).

Another explanation for the rural-urban difference could be that although the same risk factors may be at work in both urban and rural areas, there could be differences between the two types of area in terms of the *effects* that these risk factors have on rates of household burglary. In other words, heterogeneity may be present in the *relationship* between the outcome (Y) and the risk factors (X). The relationship between X and Y may be dependent on location. The hypothesis of a heterogeneous regression relationship typically leads to the fitting of some form of "local" regression model. Here we might anticipate different parameter values depending on whether a spatial unit is in a rural or urban area. In the case of some local regression models, such as geographically weighted regression (see Section 6.3.3), each area has its own parameter value (see Lloyd, 2011, Chapter 5). But the point to emphasize here is that to deal with heterogeneity, we may need to construct models with many unknown parameters. And, in the case of some models, there are a lot more parameters to estimate than there are data points. This clearly presents a *challenge* to estimation.[2]

In the case of analysing spatial-temporal data, heterogeneity can occur in local temporal evolutions. Figure 1.5 demonstrates a diverse range of trend patterns in annual raw burglary rates across 452 COAs in Peterborough, UK from 2005 to 2008. In particular, the highlighted local trends display patterns that appear to be very different from the citywide increasing trend. That said, a word of caution is needed. While Figure 1.5 is useful

---

[2] To deal with an heterogeneous outcome-covariate relationship, a local regression model of the form $y_i = a + b_i x_i + e_i$ allows the regression coefficient to be area-specific, i.e., using b_i (one for each area) as opposed to using just a scalar parameter b. The specification of this model poses a challenge in estimation. While in the case of the standard simple regression model there are only three parameters to estimate (the intercept coefficient (a), a single regression coefficient (b) and the error variance), the use of N area-specific regression coefficients results in the need to estimate $N + 2$ parameters. There are more unknown parameters than the N pairs of (x_i, y_i) data points available. This estimation problem becomes even more severe if, for example, we allow the intercept to be area-specific too. In that case, there are $2N + 2$ parameters to estimate.

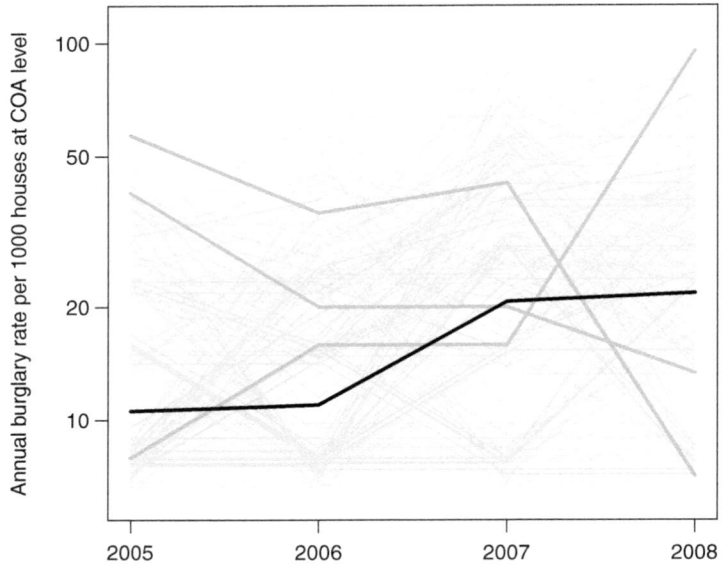

FIGURE 1.5
COA-level time trends of the raw annual burglary rates (numbers of burglary cases per 1000 houses) in Peterborough from 2005 to 2008. Each annual raw rate is calculated as the ratio of the number of burglary cases in each year to number of houses at risk. The thicker grey lines show some examples of the local trends that differ from the overall trend for the whole of Peterborough, represented by the black curve.

as a tool for exploratory data analysis, a topic that we will return to in Chapters 6 and 13, uncertainty associated with these risk estimates also needs to be accounted for before drawing any conclusions.

The challenge introduced by heterogeneity is that the modelling needed must be flexible to accommodate a statistical property of the system that may vary spatially and/or temporally. As we will explain and illustrate, this estimation problem is tackled by imposing appropriate *spatial structure* on the area-specific parameters so that these parameters are assumed to be *dependent*, as opposed to being independent.[3] In other words, the regression coefficient in one area is assumed to be *dependent on* the regression coefficients from other, perhaps nearby, areas. Assuming such a dependence structure effectively reduces the number of parameters in the model to be estimated. More importantly, this dependence assumption provides the basis for *information sharing*, a powerful modelling concept that allows us to provide reliable estimates for area-specific parameters. In addition to tackling the estimation issue, the idea of information sharing also helps address the issue of data sparsity, the third main challenge when analysing spatial data that we turn to next.

1.2.3 Data Sparsity

The availability of spatial and spatial-temporal data is increasing at a rapid pace. Particularly in more developed countries, such data are being made available at fine spatial

---

[3] In the space-time context, statistical independence in respect of parameters means that if a parameter from area i at time t (say the level of burglary risk associated with that space-time cell) takes on a particular value, this value does not provide any information about the values that same parameter takes on in other areas at other times.

and/or temporal scales. For example, since late 2010, police forces in England, Wales and Northern Ireland have been publishing monthly data on crime and policing at the street level (https://www.police.uk). Population data are published down to the block (USA) and output area (UK) levels. Health data for geographical areas are available through national and regional centres for health statistics. Such fine-scale spatial and spatial-temporal data offer unique opportunities to explore and reveal characteristics of the system of interest in ever finer detail in both space and time. But, *spatial* and *spatial-temporal precision* may not be matched by *statistical precision*. The lack of statistical precision arises because of insufficient data at the chosen space-time scale, the issue often known as the *data sparsity problem*.

Data sparsity can arise from two scenarios. The first is when the spatial and spatial-temporal data we analyse come from small populations or when the event itself is uncommon. This scenario is usually associated with registry-based data such as crime events in police recorded databases or disease counts from regional or national health registries. When observed from small populations over relatively short periods of time, event counts are often too low to provide reliable risk estimates. These estimates are often highly variable, with wide uncertainty intervals. As an example, Figure 1.6 shows the point and interval estimates of burglary rates in Peterborough in 2005 at the COA-level. A COA typically contains only around 100 houses. In 2005 79% of the 452 COAs in Peterborough experienced two or fewer burglary cases. Using only these sparse data, the resulting uncertainty intervals are very wide, hindering the use of these data to address practical questions. In the case of Figure 1.6, all the uncertainty intervals overlap with each other, suggesting that the underlying burglary risk for all these COAs may not differ. However, for the COA on the extreme right edge of the plot, although its uncertainty interval overlaps with all the others, its point estimate is considerably larger than the others. This observation may raise the question of whether the underlying burglary risk of that COA is much higher than the rest. Because of data sparsity, the area-by-area estimates presented in Figure 1.6 do not allow us to answer that question. The approach that this book is taking to try to answer the question is through modelling.

The second scenario that leads to data sparsity is when we need to analyse data at a spatial or space-time resolution that is finer than that at which the data were initially collected.

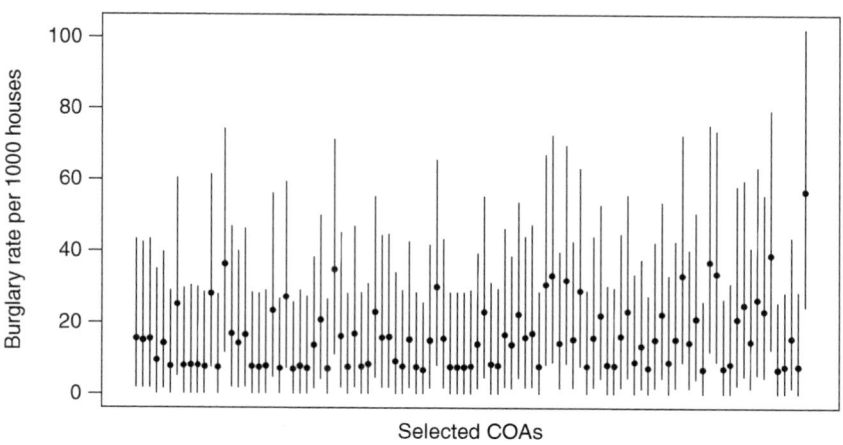

FIGURE 1.6
The raw burglary rates (the solid dots) in 2005 per 1000 houses with their associated uncertainty intervals (the vertical bars) for a selection of COAs in Peterborough.

TABLE 1.1

Distribution of Data Points per MSOA in Newcastle

Number of households included (inclusive)	0	1–5	6–9	10–13
Number of MSOAs	10	5	75	19

Available from the National Survey.

This second scenario typically arises from carrying out a subnational analysis using national level survey data. In contrast to the first scenario, the subnational populations considered here may be large, but in the absence of any spatial stratification in the national level survey, there is a risk that the number of data points in at least some subnational units may be too small to directly provide reliable estimates for these populations. In Chapters 7 and 8, we will consider a study that estimates the level of average weekly household income at the middle super output area (MSOA) level in Newcastle, England. The data used in this study were extracted from a national survey. While the data volume is sufficiently large to provide a good national estimate, when disaggregated to the MSOA level (just under 6800 MSOAs in England according to the 2011 UK census definition), household data per MSOA become scarce. Table 1.1 shows that most MSOAs in Newcastle only have a small number of households from the national survey and some have none. This scarcity of data at the chosen analysis level presents a serious problem to inference. For example, using only the data in each MSOA, the resulting interval estimates for each MSOA tend to be either too wide or too narrow because the widths of these intervals depend solely on the spread of the handful of data values in each MSOA. No estimates can be provided for MSOAs with no data values and no interval estimates can be provided for MSOAs with only one data value.

1.2.4 Uncertainty

In any statistical analysis, there are three main sources of uncertainty we need to account for: data uncertainty, model (or process) uncertainty and parameter uncertainty. We now go through each in turn.

1.2.4.1 Data Uncertainty

Data uncertainty derives from the inevitable errors and uncertainties associated with the collected data. Typically we use a probability distribution (called the likelihood; see Chapter 5) to describe the randomness associated with, say, the number of recorded cases of a disease or voting preferences or road traffic accidents in an area during an interval of time. The probability model captures our uncertainty about the observed data values, recognizing that some of the variation we observe is associated with the many, often complex, processes that lie behind the data values we have for our analysis. Some of these processes are associated with the data collection activity itself (e.g. sampling processes, measurement processes, recording processes), some with the inherently stochastic or complex nature of the underlying processes generating the outcome values that we observe.

Uncertainty in data also arises from missing data values and data incompatibilities. The latter is a particular problem for spatial and spatial-temporal data analysis when different attributes are reported on different spatial frameworks and/or on different time scales – the incompatible data problem (see for example Gotway and Young (2002)). In later sections of the book we will meet with case studies where data values are missing in certain spatial units (see the Newcastle income data analyses in Chapters 7 and 8) and incompatible data problems (see Section 9.2, where we model the relationship between stroke mortality and air pollution where health and demographic data had been collected on census tracts whilst air pollution data were reported on a grid). Data uncertainty also arises from the fact that much of the spatial data we analyse in the health and social sciences comes from small irregularly shaped administrative areas that vary in size both physically and in terms of population size.

1.2.4.2 Model (or Process) Uncertainty

There is uncertainty associated with all forms of model building in part because of the inherent simplification of the system under study that modelling involves. While no one model can represent the "truth", different models possess different degrees of usefulness. Even the simplest model provides information towards the next step in model building. As George Box commented on several occasions: all models are wrong, but some are useful (see Box, 1976, for apparently the first instance). One of the goals of statistical analysis is to measure how useful different models are.

Our understanding of health and social-economic phenomena is often highly contested. However, where different theories/understandings can be expressed by different statistical models, these models may be subsequently tested against data in order to quantify how well each model, and by implication the theory behind it, is supported by data. Haining (2003, p.356–358), for example, cites two different models to explain spatial patterns of retail pricing for petrol. One model emphasizes the supply side and stresses the role of local competition effects between retailers together with site effects (range of other services offered; quality of site) and locational effects (whether the retailer is on a main road or not) to explain price variation. The other model emphasizes the demand side and in particular consumer behaviour, focusing on how accessible sites are in relation to flows of consumers within an urban area.

The observational nature of much social and public health science makes rigorously identifying associations between variables, let alone establishing causality, highly uncertain. Which covariates to include in a model present challenges – perhaps for practical reasons (data are not available), perhaps because of ignorance. We shall discuss the issues arising in drawing valid conclusions in observational science further in Chapter 3. At this point we note that one of the coping strategies we will encounter at many points in the book will be the inclusion of so-called "random effects" terms in the process model. We shall encounter two types of random effects – spatially structured random effects and spatially unstructured random effects. Both types are often included to cope with the effects of missing covariates in model specification, some of which may show spatial dependence, others not. A similar idea will be seen when modelling temporal data in Chapter 12 and when modelling spatial-temporal data in Chapters 14 to 16.

1.2.4.3 Parameter Uncertainty

In a statistical analysis, there are quantities whose values are not known. We call these quantities *parameters*, which include not only regression coefficients and the variance of the errors but also any missing data values. We use data to provide estimates for these parameters. However, we need to go further. *The uncertainty associated with the estimate of any parameter ought to be propagated throughout the estimation process.* In other words, the estimation of a parameter ought to take into account the uncertainty associated with the *other* parameters in the model. For example, suppose we are trying to evaluate the effectiveness of an intervention. To do this, we need to compare the outcome trajectory (over time) in the "treated" areas against the outcome trajectory in areas which have not been treated. Any difference between the two trajectories would be indicative of an impact from the intervention. In making such a comparison we need to model the outcome trajectory in the non-treated areas, and this will require the estimation of model parameters. There will be uncertainty associated with these estimates. When estimating the parameters of the model for the treated areas, in order to establish whether the treated areas have followed a different outcome trajectory since the intervention, we should recognize and allow for that uncertainty. We will see an illustration of this in Section 16.2 in Chapter 16, when we consider the problem of evaluating a targeted small area policy introduced by the Cambridgeshire Police Constabulary. Exercise 1.7 illustrates how the uncertainty in estimating the variance σ^2 of a normal distribution affects the interval estimate of the unknown mean, μ, of that normal distribution. In Section 1.4.2 we will discuss *how* parameter uncertainty is incorporated. This constitutes a fundamental difference between Bayesian and frequentist approaches to inference, the two statistical frameworks that are commonly used for the analysis of spatial and spatial-temporal data.

1.3 Opportunities Arising from Modelling Spatial and Spatial-Temporal Data

Analyses of spatial data can address questions relating to the "here and now" of the system of interest. Such questions include estimating the subnational income levels in the current year for resource allocation to address local poverty; the identification of neighbourhoods that have experienced high risks of crime or disease in a given year in order to formulate intervention strategies. Analyses of spatial-temporal data offer a dynamic view of the system of interest. With spatial-temporal data, there is the possibility of studying the process and even, potentially, establishing casual relationships. In this section, we will illustrate some practical applications that can be undertaken using the modelling strategies discussed in this book.

1.3.1 Improving Statistical Precision

Estimation is an essential element in the analysis of spatial data. Because of the data sparsity problem (Section 1.2.3), we seldom draw conclusions based on estimates (say of risk or average household income) for an area that are calculated solely using the available data *from that area*. Instead, through model building, we exploit the dependence structure present in data in order to carry out *information sharing*, an idea that is central to the estimation process. As noted previously, positive autocorrelation implies the overlapping of information amongst the observations, or, expressed differently: the data value in one area contains *some information* about the data values of its neighbours. Analogous to the case of positively autocorrelated *observations*, by imposing a positive dependence structure onto a set of *area-specific parameters*, parameters in neighbouring areas are assumed to be similar. For example, if the average household income levels (i.e. the area-specific parameters) can be assumed to be positively autocorrelated, and if the data values from an area's neighbours suggest that these neighbouring areas tend to have high income levels, then this information can be used for the estimation of the average income level of that area, in addition to using the available (but typically scarce) data from that area. As a result, information can be shared when estimating these area-specific parameters, improving the accuracy of the estimates. This information sharing is achieved using Bayesian hierarchical models.

To demonstrate the effect of information sharing on estimation, we take a look at a dataset that we shall revisit several times in Part II of the book. This dataset contains the weekly income (in £s) at the household level extracted from a national survey, and each household is georeferenced to one of the 109 MSOAs in Newcastle, England (see Table 1.1 for a summary of the data). With the aim to provide the MSOA-level average weekly income per household, Figure 1.7 shows a map of these MSOA averages, each calculated using only the household income values available in a given MSOA. Amongst the 109 MSOAs, there are 10, each highlighted with a "x" in Figure 1.7, with no data values. Clearly, if only the local data are used, we cannot produce any estimates for these 10 MSOAs. Besides these 10 MSOAs, also evident in Figure 1.7 is that MSOAs that are close in space tend to have similar levels of average income, a sign of positive autocorrelation. When estimating the parameters of interest, which, in this case, are the average weekly household income levels across the MSOAs, the presence of such a dependence structure in data provides the basis for information sharing. Figure 1.8 compares the MSOA-level estimates obtained by using only the local data as discussed above (Panel A) with the estimates obtained from two Bayesian hierarchical models (Panels B and C). While the detail of the two Bayesian hierarchical models is withheld till Chapters 7 and 8, both models feature information sharing but to different extents. The estimates shown in Panel B come from a model that shares information *globally* so that the estimation of the income level of an MSOA uses not only information from the data in that MSOA but also information from all the other MSOAs across Newcastle. Global information sharing is achieved based on the assumption that

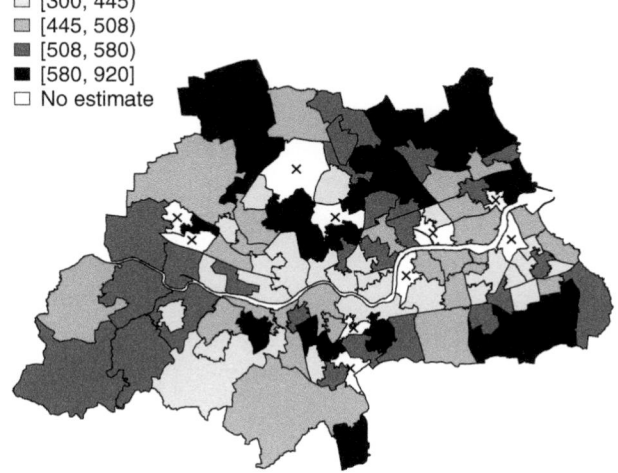

FIGURE 1.7
Average weekly household income (2007/08) at the middle super output area (MSOA) level for Newcastle, England obtained from survey data. The elongated white line is the River Tyne (and hence Newcastle is also known as Newcastle upon Tyne). For scale: Newcastle is approximately 40 km (east-west) and 22km (north-south) and in 2010 Newcastle's population was about 290,000.

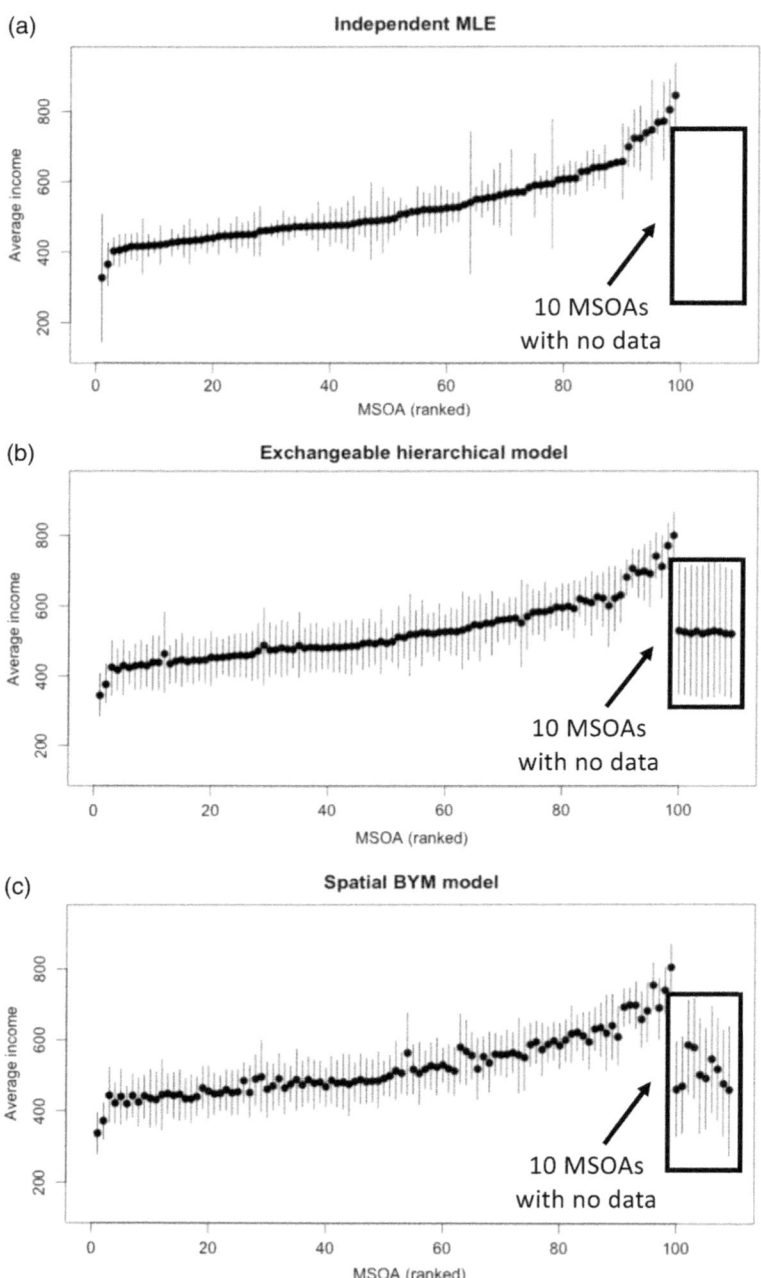

FIGURE 1.8
Estimates of the average weekly income per household over the 109 middle super output areas (MSOAs) in Newcastle, England. In all the plots, the solid dots represent the point estimates and the vertical lines indicate the 95% uncertainty intervals. Each estimate in panel A is obtained using only the survey data available in a given MSOA, while those in Panels B and C are from two Bayesian hierarchical models that borrow information only-globally and both globally and locally across the MSOAs respectively (see main text for detail). When available, the estimates of the 10 MSOAs without data are presented on the right-hand-side of each panel, enclosed by a rectangular box.

the average income level of each MSOA is similar to the (global) Newcastle average. In addition to global information sharing, the model associated with Panel C also borrows information *locally*, whereby information is shared amongst neighbours. The assumption that underlies local information sharing is that income levels of nearby MSOAs tend to be similar, which is evident in Figure 1.7.

An inspection of the three panels in Figure 1.8 immediately reveals that whether information is shared or not, and if it is shared, *how* it is shared, make a noticeable difference to the estimates for the 10 MSOAs with no data. As discussed above, the strategy of producing an estimate using only data available in an area does not yield any estimates for these 10 MSOAs, hence no estimates in the rectangular box on the right-hand end of Panel A. On the other hand, because of information sharing, both Bayesian hierarchical models can produce estimates for these 10 MSOAs, but the estimates from these two models are different. In Panel B, the estimates for these 10 MSOAs are virtually the same. This is because global information sharing assumes the income level of each MSOA is similar to the Newcastle average. For an MSOA without any data, the estimate of its income level basically takes the estimate of the Newcastle average. By contrast, the model associated with Panel C also shares information locally, in addition to globally. Thus, for each of these 10 MSOAs, this local information sharing modifies the estimate based on the Newcastle average according to the income levels of that MSOA's neighbours. For example, the income estimate of an MSOA becomes higher (lower) than the Newcastle average if its neighbours tend to have higher (lower) income levels than the Newcastle average. It is sharing information locally that results in the different estimates across these 10 MSOAs.

Another, perhaps subtler, difference across the three panels in Figure 1.8 is the lengths of the uncertainty intervals (represented by the vertical lines) associated with the 99 MSOAs with data. The uncertainty intervals in Panel A, obtained using only the available intra-MSOA data, vary a great deal in length, some being very wide and some very narrow. Both extremes are unrealistic. By contrast, information sharing in both the Bayesian hierarchical models has had a substantial impact on the interval estimates, which become more consistent in length. What has happened is that because of the gain of information, borrowed from other MSOAs, the very wide intervals are shortened, and uncertainty of the income estimates has been reduced. At the same time, because of information sharing, the variation associated with that shared information widens the unrealistically narrow intervals, which now represent (perhaps more appropriately) greater uncertainty in the associated income estimates.

Common to the above two Bayesian hierarchical models is the assumption that the income level of an area is statistically related to the income levels of other, possibly nearby, areas within the study region, an assumption that has its origins in Tobler's First Law of Geography: "everything is related to everything else, but near things are more related than distant things" (1970, p.236) and is also captured, more formally, in what Banerjee et al. (2004, p.39) refer to as "the first law of geostatistics".[4] We shall discuss the above two Bayesian hierarchical models in Chapters 7 and 8, where we will explore in detail how information borrowing works in a Bayesian hierarchical model.

---

[4] Spatial and spatial-temporal data dependency is exploited in many spatial interpolation methods including inverse distance weighting and kriging (see for example Isaaks and Srivastava, 1989). It is also exploited when estimating missing values in spatial and spatial-temporal databases (see for example Haining, 2003, p.164–174).

1.3.2 Explaining Variation in Space and Time

Explaining, or accounting for, the variability observed over space and/or time underlies most analyses of spatial and spatial-temporal data in the public health and social sciences. Having estimated the spatial or spatial-temporal patterns, the next inferential task is to explain these patterns using covariates that are available to us. Regression models can be constructed to address various practical problems such as quantifying covariate-outcome relationships, evaluating theories, testing for the presence or absence of spillover effects in space and/or time[5] and assessing policy performance, amongst other areas of application. We will provide four examples to illustrate the types of questions that can be tackled by modelling spatial and spatial-temporal variation using the Bayesian hierarchical modelling approach or the Bayesian spatial econometrics approach that we will explore in this book. We will briefly describe the models that enable us to tackle these types of problems as well as some practical considerations. Readers for whom this type of material is unfamiliar may wish to focus, at this stage, just on the broad aims and objectives of these examples since all will figure again later in the book.

1.3.2.1 Example 1: Modelling Exposure-Outcome Relationships

In environmental/geographical epidemiology, analysing aggregate data is viewed as the weakest form of analysis for establishing environmental exposure-disease relationships at the individual level. However, such forms of analysis, known as ecological inference, are useful when there are problems associated with obtaining reliable measures of exposure for individuals. In situations where health data are only made available at the aggregate level but not at the individual level due to confidentiality, ecological inference becomes "the only hope of making progress" (King et al., 2004, p.1). Armstrong (2001) offers a defence of ecological analyses. Such analyses are typically undertaken to generate hypotheses that may then be followed up by other methods such as a cohort study. Richardson (1992, p.199–200) provides a list of good practice in the design of geographical studies. King et al. (2004) surveys various strategies to address problems and issues arising from ecological analyses across a number of research fields. Here, we present an example of ecological inference in a public health context. However, this form of inference is not problem-free. We defer detailed discussion till Chapters 9 and 17.

To provide an example of this type of modelling we consider Maheswaren et al. (2006), who studied the effect of exposure to atmospheric nitrogen oxides (NO_x; the exposure variable) on stroke deaths (the health outcome of interest) at the enumeration district (ED) scale in Sheffield, England. Each enumeration district (a reporting unit last used for the 2001 census of England and Wales) had on average about 200 households. Figure 1.9 shows a map of the stroke deaths by ED and the data on NO_x exposure by grid square. One of the challenges of this research was to transfer all the data to a common spatial framework and to try to make allowance for the uncertainty associated with such a transference (Brindley et al., 2004).

In addition to the covariate on NO_x exposure, data were also available to account for two ED-level confounding variables,[6] deprivation and population smoking prevalence. A fourth ED-level covariate was also included, where each value of this covariate measured

---

[5] The term "spillover" refers to the situation where events (including outcomes) in one spatial or space-time unit impact on outcomes in other spatial or space-time units.

[6] A confounding variable is a variable that influences both the dependent variable (the health outcome in this case) and the independent variable (the environmental exposure in this case), thus undermining any measure of association between the two unless the confounder is controlled for (i.e. included in the regression model).

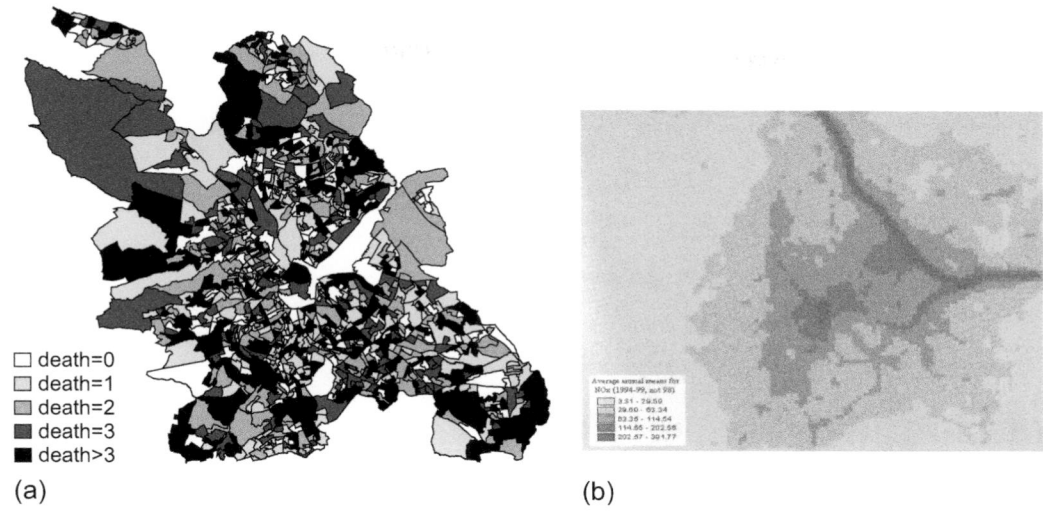

(a)   (b)

FIGURE 1.9
Maps showing (a) Number of stroke deaths by enumeration district in Sheffield 1994–1999; (b) Average annual mean NO_x levels 1994–1999 (excluding 1998) by grid square. For scale: in 2010 Sheffield's population was about 550,000.

the average level of NO_x in the neighbouring EDs that lay within a specified distance of each ED but excluding its own NO_x level. This fourth covariate – often called a spatially-lagged covariate – was added to account for the fact that populations are not static but in the course of their daily lives, for work and for leisure, move around the city. This spatially-lagged covariate is included to capture an aspect of the NO_x exposure associated with people's movement (see Chapter 4 for the construction of a spatially-lagged variable). Another challenge for this research (and in fact for ecological analyses in general) was to account for the effects of the covariates that were thought to be associated with the outcome of interest but were not included in the analysis due to, for example, data unavailability or where we (the analysts) just did not realize their importance. To represent the effects of such omitted covariates (also referred to as unmeasured/unobserved covariates), *random effects* (or *random effect terms*) are included in a Bayesian hierarchical model. As opposed to a fixed effect, which is the coefficient associated with a covariate whose values are available to the analysts (e.g. the fixed effect of deprivation or the fixed effect of smoking prevalence), these terms are called *random effects* because they represent the effects of a set of covariates whose values are not observed or available to the analyst (hence, "random"). We shall have more to say about these random effects in Section 1.4: how to model them and why Bayesian hierarchical modelling plays an important part in modelling these random effects.

Amongst the models considered in Maheswaren et al. (2006, p.508–509), Table 1.2 summarizes the estimated effects of NO_x exposure on the risk of stroke death from two Bayesian models, one that includes random effects while the other does not. Here, we focus on the following two points. First, compared to the one without random effects, the model with random effects is a better model, according to the Deviance Information Criterion (DIC). As we shall discuss in Chapter 5, DIC is a measure that allows us to compare different models (hierarchical or otherwise). In general, a model with a smaller DIC value is better supported by the data. The model with random effects is better because these random

TABLE 1.2

Comparison of the Estimated Relative Risks (RRs) Associated with
Different Levels of NO_x Exposure between Two Bayesian Models Where
One Includes Random Effects Whilst the Other Does Not

NO_x Exposure Category	Estimates of Exposure Effect on Stroke Mortality	
	From a Bayesian Model without Random Effects	**From a Bayesian Model with Random Effects**
5 (the highest exposure)	1.48 (1.31–1.67)	1.27 (1.05–1.53)
4	1.26 (1.12–1.42)	1.16 (0.96–1.39)
3	1.10 (0.98–1.24)	1.04 (0.86–1.24)
2	1.13 (1.00–1.26)	1.08 (0.89–1.29)
1 (reference category)	1.00	1.00
DIC	4871.57	3929.11

Each cell reports the posterior mean (as a point estimate of the exposure effect) and
the 95% credible interval (a measure of uncertainty in the Bayesian framework). For
both models, NO_x levels are measured by category, in which 5 is the highest level of
exposure, 1 (the reference category) the lowest level of exposure, and both models
include all four covariates as discussed in the main text.

effects enable the model to capture part of the variability that is beyond what can be captured by the covariates included in the model.

The second point is that when random effects are included, it is only at the highest (5th)
quintile of exposure that the point estimate of the relative risk exceeds 1.00 (the value
indicating no effect on the stroke risk) and the associated 95% credible interval excludes
1.00. In other words, only the highest level of exposure is found to be associated with an
elevated stroke mortality. However, when random effects are excluded, both the 4th and
5th quintiles of exposure are found to be associated with an increased stroke mortality.
In fact, when random effects are included, all the uncertainty intervals are wider. This
increase in the uncertainty in the estimates is the result of recognising the fact that beyond
the four covariates included in the model, there are still other unobserved/unmeasured
factors at work. One of the reasons to include random effects into a regression model is
to account for such effects. We will present a fuller analysis in Chapter 9, where we will
illustrate ways to address some of the issues often associated with spatial epidemiology,
including spatial misalignment of datasets, spatial units with missing data, uncertainty
associated with measuring a covariate because of the small number of cases of smoking
and ecological (or aggregation) bias arising from within-area variation in exposure.

1.3.2.2 Example 2: Testing a Conceptual Model at the Small Area Level

The second example arises often in the social sciences where we start from some con-
ceptual model which organizes our current understanding of the key dimensions that
influence the outcome under investigation. Selected covariates are included in the model
because they are deemed to provide a satisfactory representation of these key dimensions,
and data are available. The statistical model containing these covariates is then fitted to
the observed outcome values to assess the contribution of the various elements of the con-
ceptual model to variation in the outcome. The example considered here is in the context
of small area crime analysis.

Analysing crime data at the small area level is frequently justified by reference to the two-stage rational choice theory of offender behaviour. At the first stage, the would-be offender selects an area within which they consider committing their crime. At the second stage they choose their victim within the selected area.[7]

Figure 1.10 shows the distribution of sexual assaults in public spaces during 2008–09 in the city of Stockholm, Sweden by small area census tracts called *basområdes*. Ceccato et al. (2019) identified three classes of risk factor to explain small area variation in the number of sexual assaults. These took the form of three hypotheses:

- *Opportunity hypothesis*: High counts of rape are associated with areas offering opportunity to the offender (e.g. relatively large numbers of vulnerable female targets perhaps because of the location of particular amenities or institutions);

- *Meso- and micro-scale anonymity hypothesis*: Areas with poor natural surveillance and/or poor social control where residents have a high fear of crime provide *meso-scale anonymity* to the offender. Neighbourhoods with buildings and street layouts with inadequate lighting or with many quiet alleyways and secluded areas provide areas for concealment. Areas with many such places may provide a number of suitable "micro-spaces" for the offence to be committed, thereby providing *micro-scale anonymity*;

- *Accessibility hypothesis*: Accessible areas may be attractive to offenders because they bring potential targets into an area and offer a quick and easy escape route from the crime scene.

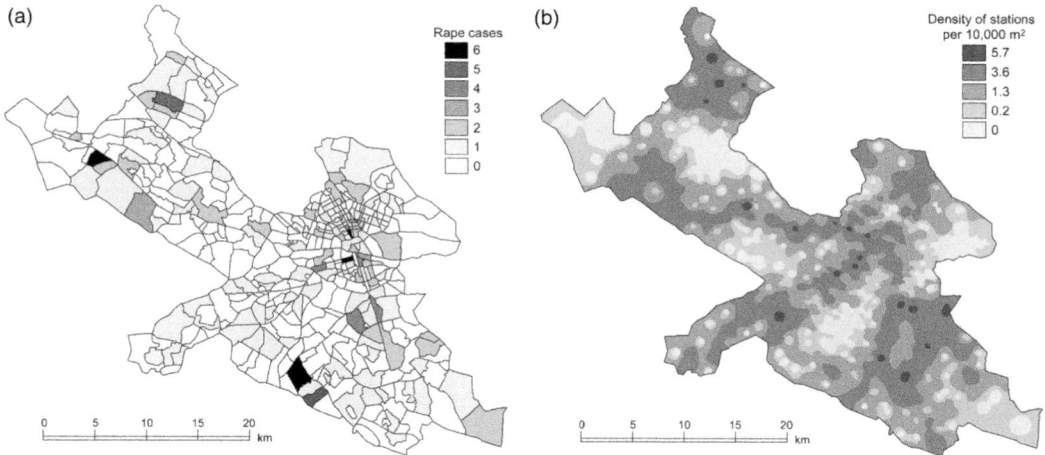

FIGURE 1.10
(a) Numbers of sexual assaults at the *basområde*-level reported between 2008 and 2009 in Stockholm, Sweden. (b) Density of transit stations as evidence of a possible link with area accessibility. The city centre (the most accessible area in the city) is identifiable from the large number of very small *basområdes* visible in the northeast corner in (a). After Ceccato et al., 2019.

---

[7] Bernasco and Luykx (2003, p.984–988) provide a discussion of two-stage rational choice theory in the context of household burglary. They specify three relevant sets of area-level risk factors that determine the burglary risk in a neighbourhood. Offenders take into account the *attractiveness* of the neighbourhood in terms of the likely value of the goods that can be stolen; the *opportunities* offered by the neighbourhood in terms of how likely it is that the offender will be able to successfully carry out the crime without being caught and finally *accessibility* in terms of how familiar a neighbourhood is and how close it is to where burglars live. Which particular house is targeted within a neighbourhood is explained by other factors.

To explain spatial variability of the observed numbers of rape cases, Ceccato et al. (2019) identified 11 covariates (risk factors), each associated with one, in some cases two, of the hypotheses as previously discussed. The modelling task was then to estimate the effects of these risk factors in order to test their importance, and in turn the hypothesis (or hypotheses) that they represent. However, the quantification of the risk factor effects raised the following three modelling issues:

- The issue of multicollinearity whereby some of the risk factors are found to be correlated
- The issue of interactions amongst the risk factors where the effect of a risk factor on the outcome depends on the value of another risk factor (or those of several other risk factors)
- The issue of spatially-varying risk factor effects where the effect of a risk factor can vary from one area to another, as opposed to remaining constant across the study region – an example of parameter heterogeneity as discussed in Section 1.2.2

With these issues in mind, Ceccato et al. (2019) analysed the dataset through a two-stage approach in which a "global" (or whole map) model was first fitted to assess the overall effects of the risk factors. To complement the global model, a "local" analysis was then carried out using Bayesian profile regression. In addition to addressing the aforementioned three methodological issues, the second-stage local analysis revealed, amongst other results, two groups of high-risk *basområdes*, each of which possesses distinctively different characteristics – one group located in the city centre (as evident in Figure 1.10(a)) with a large number of alcohol outlets, high population turnover and large counts of robbery cases, a second group consisting of poor suburban areas characterised by large female residential populations with subway stations and schools located within them and residents expressing a high fear of crime. The results of analysis suggest the importance of the dimensions of accessibility, opportunity and anonymity. But more crucially, this study highlights the importance of undertaking crime analysis at the small area level in order to better understand the geography of crime. Such an opportunity only arises because of the availability of the recorded crime data at a fine spatial scale and a set of statistical models that are appropriately constructed. We shall return to this example in Chapter 9.

1.3.2.3 Example 3: Testing for Spatial Spillover (Local Competition) Effects

Consider a group of petrol retailers distributed across a city. Price posting is in operation (in the form of large, elevated display boards at the site) so consumers driving past can gain price information easily and at no cost in the course of their various intra-city journeys. In the case of a more or less homogeneous good, such as petrol, a retailer at one site might give thought to how the price set by a neighbouring petrol station might affect the level of demand at their site. By the same token, other petrol retailers might consider how the first retailer's price affects the level of demand at their sites. Whilst there are likely to be "site effects" associated with each retailer (for example the number of pumps which might help reduce waiting times for customers; other services offered) which would need to be controlled for, there remains the question as to whether there is a "competition" effect associated with the prices offered by the set of competitors of each site. This is one form of "spatial spillover", which we will discuss further in Chapter 2.

Figure 1.11 is a map of 63 petrol retail outlets in Sheffield and the prices they charged on a single day in March 1982. Haining (1990) used a similar dataset to test for local price competition across the retail outlets.

The following is a spatial econometric model that expresses these ideas in the form of a regression model with a spatially lagged dependent variable term included as a covariate. It is an example of a so-called spatial lag model (SLM):

$$Y_i = \alpha + \delta \sum_{j=1}^{N} w_{ij}^* Y_j + X_i \beta + e_i \tag{1.1}$$

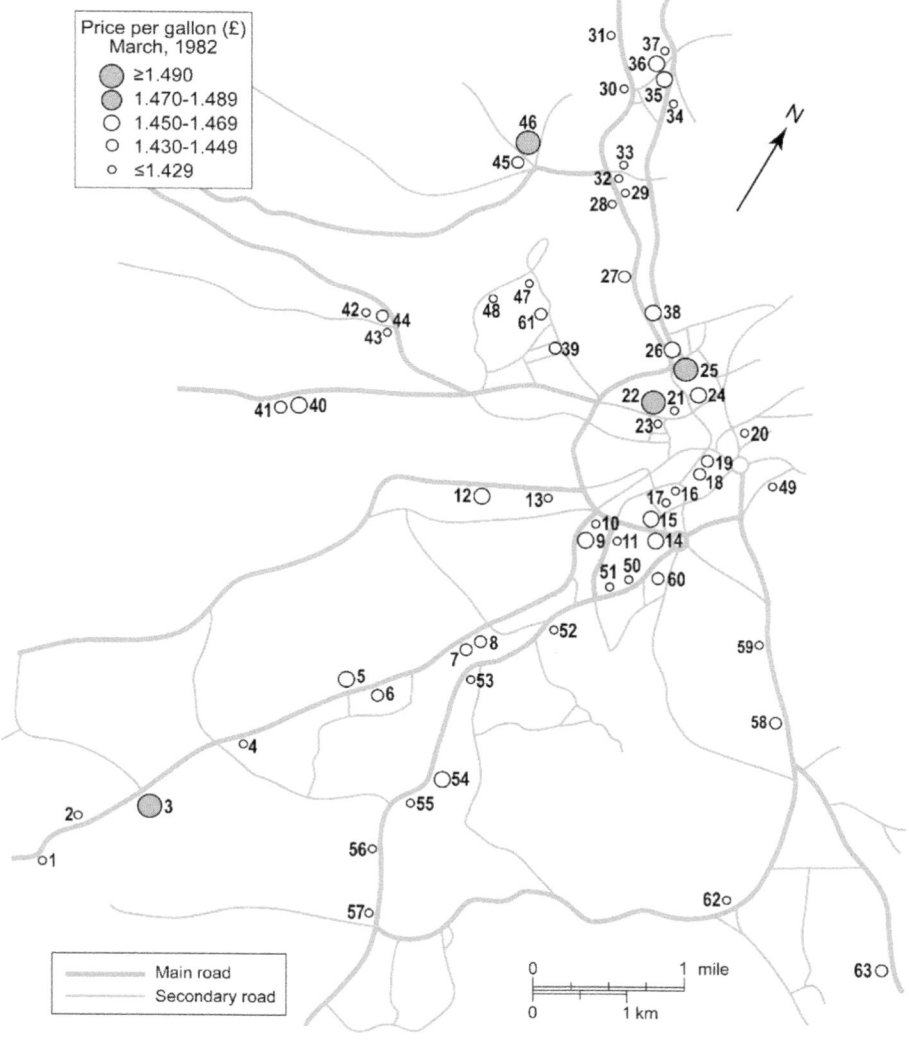

FIGURE 1.11
Map of petrol prices at retail outlets in Sheffield, March 1982.

Here Y_i is a variable representing the petrol price at retail site i, one of the N retail sites within the study region, and α and $X_i\beta$ are, respectively, the intercept and the effects of a set of covariate values (X_i) at site i on its petrol price (i.e. the "site effects"). e_i is a normally distributed error term. At first glance, Eq. 1.1 resembles the form of a multiple linear regression model. But the additional term $\delta \sum_{j=1}^{N} w_{ij}^* Y_j$ makes it interesting, and the definition of this term relates to a number of important quantities that we encounter in modelling spatial and spatial-temporal data. First of all, w_{ij}^* is an element from the so-called *spatial weights matrix*, a matrix that specifies a neighbourhood structure amongst the N spatial locations (or spatial units). While we will discuss various ways to define the spatial weights matrix in Chapter 4, for the discussion here, we set $w_{ij}^* = 0$ if site j is not considered to be (or known to be) one of the competitive neighbours of site i, and w_{ij}^* takes a positive value if j is assumed to be (or known to be) a competitive neighbour. Since a site cannot compete with itself, $w_{ii}^* = 0$. So, $\sum_{j=1}^{N} w_{ij}^* Y_j$ is basically a weighted sum of the petrol prices of site i's competitive neighbours. In other words, Eq. 1.1 models the petrol price at site i as a function of its "site effects" and the petrol prices at its competitive neighbours. The inclusion of the latter enables us to examine for the presence or absence of price competition effects. This is done through the estimation of the parameter δ. For example, if δ is estimated to be positive and away from zero (say, its 95% credible interval does not cover zero), this suggests that there may be interaction (competitive) effects in price setting between adjacent retail sites – low (high) prices at one site will tend to be associated with low (high) prices in the set of neighbouring sites. If, for example, δ is estimated to be close to zero, this suggests any spatially-defined competition effect may be weak or not present.

Although the formulation in Eq. 1.1 has many of the characteristics of a standard multiple linear regression, as we shall see, complications in terms of both model fitting and parameter interpretation arise due to the feature that the set of outcome variables, $Y_1,...,Y_N$, appears on both sides of the regression equation. We will investigate spatial econometric modelling in Chapter 10 and return to the petrol price competition example in Chapter 11.

1.3.2.4 Example 4: Assessing the Effects of an Intervention

This example is taken from Li et al. (2013) where statistical modelling was undertaken to investigate whether some locally targeted policy implemented by police, where some areas within a city were "treated" with the policy and the rest not, had had a (positive) impact on crime rates in the treated areas of the city. This scenario can be formulated as a case-control study whereby the census tracts that had been treated contribute to the case group, whilst those that had not been treated form the control group. One of the ways of evaluating the effectiveness of the policy is to see if there has been a reduction in household burglary in the treated (case) areas compared to the untreated (control) areas. Li et al. (2013) focuses on the evaluation of a so-called "no cold calling" (NCC) initiative, a scheme implemented at a selection of areas where various policing activities took place to deter cold callings – unsolicited visits to households to sell products or services.

Figure 1.12(a) shows the locations of the NCC areas and their dates of implementation. The NCC areas are scattered across Peterborough, England. The accompanying Figure 1.12(b) shows the (raw) annual burglary rates for the individual NCC areas and for the aggregations of the NCC and the non-NCC areas. Figure 1.12(b) highlights the large amount of variability present across the individual NCC areas. However, the two solid

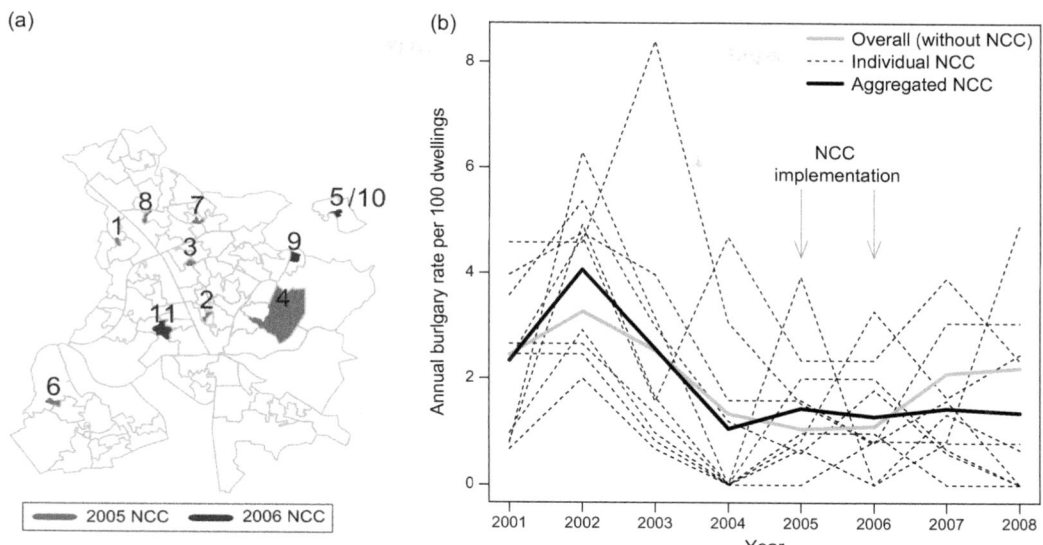

FIGURE 1.12
(a) Map showing the locations of the no cold calling (NCC) areas in Peterborough with the year in which the NCC scheme was implemented. (b) The graph showing the variation in the (raw) annual household burglary rates by NCC area – each raw burglary rate is calculated by dividing the number of burglary cases by the number of houses at risk in that area then multiplied by 100. The black dotted lines are the rates for the individual areas, whilst the dark solid line is the overall NCC trend obtained by aggregating all the 10 NCC areas. Similarly, the light solid line is the overall non-NCC trend obtained by aggregating all the other non-NCC areas in Peterborough.

lines provide the first indication that there might have been some reduction in household burglary rates in the NCC areas relative to the rest of Peterborough. But to properly assess the impact of the policy, we need to undertake modelling to address the large variability caused by data sparsity. Cases of burglary when reported annually by census output area (COA), the spatial level at which the evaluation was carried out, are generally few in number.

The evaluation method developed in Li et al. (2013) employs interrupted time series modelling. First proposed by Campbell and Stanley (1963), an interrupted time series model comprises two (sub) models: *the baseline model* that describes the time trend of the control group (i.e. the light solid line in Figure 1.12(b)); whilst *the assessment model* quantifies the departure of the NCC trend (i.e. the dark solid line in Figure 1.12(b)) from the estimated control trend on and after the NCC implementation. Estimating post-NCC departures helps answer the question of whether the NCC scheme had a measurable impact on the burglary rates in the NCC areas compared to the non-NCC areas. The evaluation method was placed within the Bayesian framework, an important element in the evaluation. This is because the uncertainty in the estimated control trend is fully propagated (or accounted for) in the estimation of the policy's impact – a particularly desirable feature of Bayesian inference in this context.

Although there were a number of challenges in the evaluation, which we shall discuss in Chapter 16, the availability of the police-recorded crime data at a fine spatial-temporal resolution offers the opportunity to evaluate not only the overall (city-wide) impact of the NCC scheme but also *the local impacts* associated with the individual NCC areas.

Extracted from Li et al. (2013, p.2021), Figure 1.13 summarises the estimated overall impact as well as the impact associated with each of the 10 NCC areas. The modelling result supports our initial observations (in Figure 1.12(b)) that there had been some reduction in household burglary rates in the NCC areas taken as a whole relative to all the other non-NCC areas in Peterborough. But also evident in Figure 1.13 is that the NCC impact varies locally. Some NCC areas appear to have benefited from the NCC scheme whilst others show the policy has had little impact. Li et al. (2013) took the further step of considering whether such local variability was due to how the NCC scheme was implemented in different areas as well as variation in local characteristics including deprivation. The results from this further modelling have implications for the future implementation of the policy. As Mason (2009) pointed out, neighbourhood policing programmes, such as the NCC initiative, are best evaluated with a "local focus". The importance of undertaking analysis at

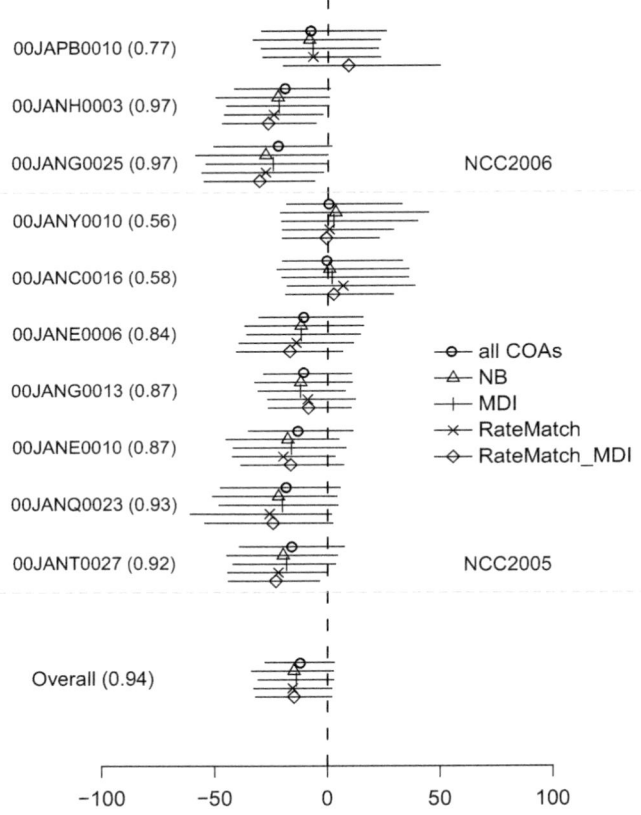

Percentage change in burglary rate compared with controls

FIGURE 1.13
The estimated percentage change in burglary rates in the NCC areas as a group and the estimated percentage change in burglary rate for each of the 10 NCC areas after the first year of implementation. Five different control groups were used (see Table 16.2 in Chapter 16). Each point symbol represents the posterior mean with the horizontal bar showing the 95% credible interval. A negative value corresponds to a "success", i.e., a reduction in burglary rate for all the NCC areas as a group or as an individual NCC area relative to the respective control group. The number in brackets is the posterior probability of success (based on using all non-NCC areas in Peterborough as control). This value is the proportion of the posterior distribution lying to the left of the vertical dotted line, which indicates no NCC impact.

both the "global" and the "local" levels was also recognized in the Stockholm rape study described in Section 1.3.2.2. With data available over space and time at a fine scale, we have the opportunity to better understand various forms of spatial heterogeneity. We will return to the NCC evaluation in Chapter 16 in Part III of the book.

1.3.3 Investigating Space-Time Dynamics

In addition to addressing questions regarding the "here and now" of a set of relationships between variables, the time dimension in spatial-temporal data provides information on how these relationships have developed over time. In the health and social sciences, one application of spatial-temporal data is change detection, identifying areas with "unusual" time trends. Change detection can be performed under two different scenarios: policy evaluation and surveillance. Under policy evaluation, a policy or an intervention has been implemented in a known subset of areas at a particular time, and we wish to evaluate whether this has had a measurable impact on the event rate in those areas. The NCC evaluation in Section 1.3.2.4 provides an example of this type of analysis.

For surveillance, there is no subset of areas of interest *a priori*. The goal of surveillance is to identify areas whose event rates differ markedly from the overall time trend of the study region. Unusual local trends may be due to the emergence of localised predictors, risk factors or the impact of a policy/intervention that is not known prior to the surveillance analysis. To provide an example, Li et al. (2012) developed a space-time detection model BaySTDetect that aims to identify areas with time trends that differ from the general trend pattern. BaySTDetect was applied to a set of district-level chronic obstructive pulmonary disease (COPD) mortality data in England and Wales between 1990 and 1997. Figure 1.14 shows the locations of the five districts detected and their corresponding trend patterns. The interesting finding is that one district detected by the method started from 2008 to commission various local health services to tackle high levels of COPD mortality. The authors suggested that this rising trend might have been recognised earlier, in the 1990s, using this type of surveillance technique. We will explore this application in detail in Chapter 16.

1.4 Spatial and Spatial-Temporal Models: Bridging between Challenges and Opportunities

1.4.1 Statistical Thinking in Analysing Spatial and Spatial-Temporal Data: The Big Picture

Statistical thinking concerns the reasoning processes behind the application of *a set of statistical techniques* that are appropriate to model the *observations* in hand with the purpose of revealing and understanding the *characteristics of a system or relationships amongst a set of variables*.[8] Figure 1.15 identifies some key stages in the process of statistical thinking (adapted from Waller (2014), who in turn adapted Box et al. (1978)). In practice, a statistical analysis may start with a hypothesis grounded in domain-specific theory (so-called *theory-driven analysis*). The underlying theory informs model specification, including which covariates to

---

[8] Throughout the book, "data" and "observations" are used interchangeably.

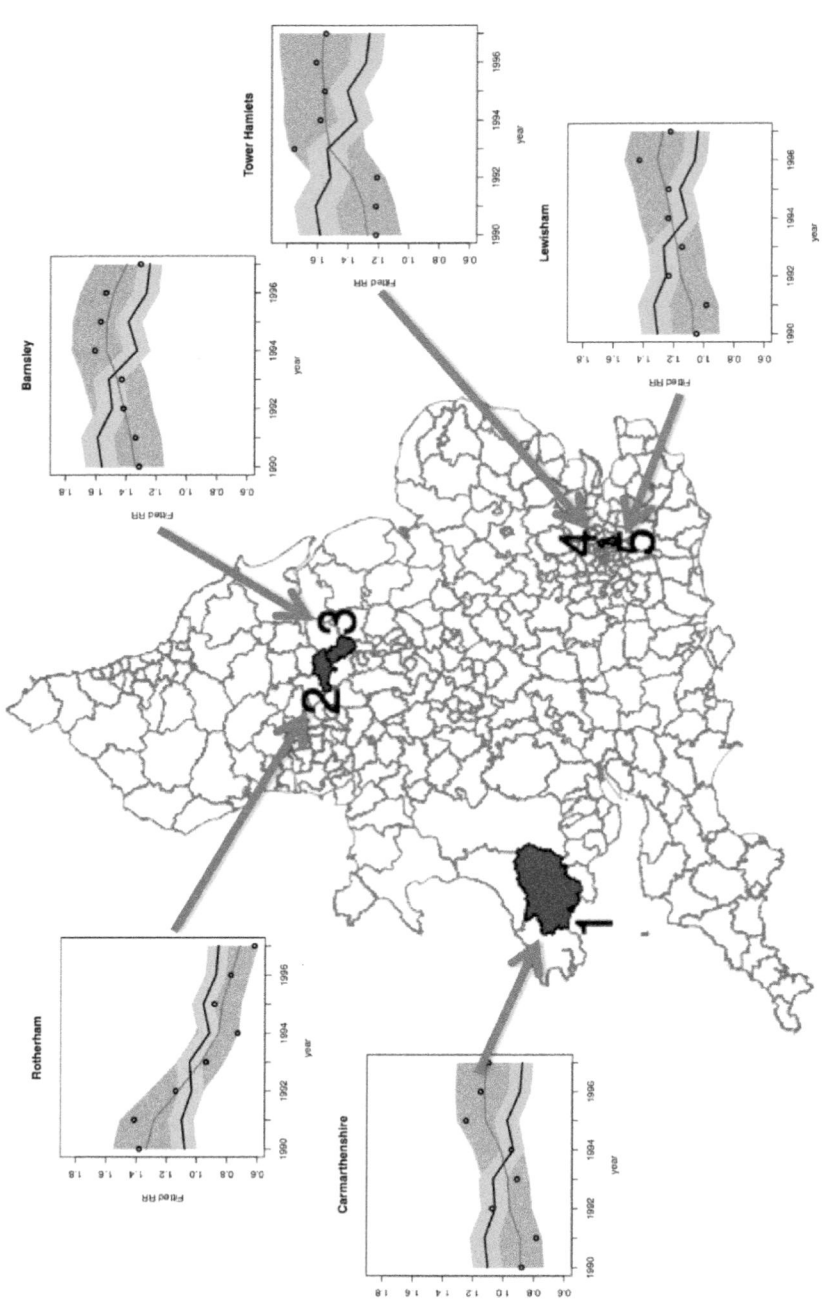

FIGURE 1.14

Highlighting unusual local (district-level) trends of COPD mortality in England and Wales. Results were obtained from BaySTDetect, a space-time detection method (Li et al., 2012). Each inserted plot shows the estimated overall trend of COPD mortality (the dark black line showing the posterior means and the light grey band representing the 95% uncertainty intervals), the estimated local trend (the light grey line showing the posterior means and the darker band showing the uncertainty) and the (raw) relative risks (shown by the black dots) calculated using only the mortality data without modelling.

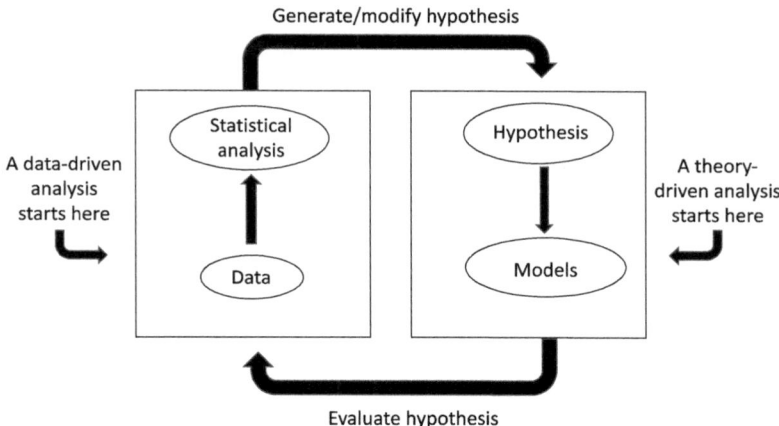

FIGURE 1.15
The process of statistical thinking.

include in the model and hence what data to collect. Statistical analysis may also begin from a set of observations with the aim of generating new insights or new hypotheses (so-called *data-driven analysis*). The reason to mention these two types of statistical analyses is that in contrast to the experimental design setting discussed in Box et al. (1978), spatial analyses in the health and social sciences typically arise from observational settings. We will explain what characterizes observational science in Section 3.2. Given a set of observational spatial data, a statistical analysis can engage in activities related to both theory-driven (e.g. testing theory) and data-driven (e.g. pattern discovery in the form of surveillance) forms of analysis. Hypotheses generated from the findings of (or patterns obtained from) one or more data-driven analyses sometimes serve as a starting point for subsequent theory-driven analysis using a new set of data. These two types of statistical analyses largely overlap in terms of the reasoning process involved. They only differ in terms of their starting point, which in turn reflects our current state of knowledge about what we are studying.

The statistical analysis of data includes model building, parameter estimation, model evaluation, model comparison and result interpretation. For a theory-driven analysis – confronting theoretical ideas with empirical data – our aim is to evaluate our initial hypotheses, and the theory from which they were derived, against data. For a data-driven analysis, the patterns discovered may contribute to the generation of new research hypotheses that, together with a new set of data, may initiate a theory-driven analysis. But in the case of both data-driven and theory-driven forms of analysis we need to construct probabilistic models to account for various forms of uncertainty (see Section 1.2.4) and to estimate parameters through which we attempt to reveal and understand the characteristics of the system that we are studying. The types of probabilistic modelling we will see in this book are Bayesian hierarchical modelling and Bayesian spatial econometric modelling.

However, it is important to stress the following two points when undertaking *any* form of statistical analysis. First, the theoretical understanding of the system is an integral part of the statistical reasoning process. A statistical model cannot be separated from its subject area because, for example, parameter interpretation relies heavily on subject knowledge. Second, a statistical analysis is an *iterative* process, and each iteration reveals new insights into the system under study. Model evaluation and model comparison are two crucial elements in this process. Therefore, in Part II and Part III of the book, we have included chapters with various applications in order to (a) illustrate the interplay between statistical reasoning and subject knowledge and (b) demonstrate how to navigate through the iterative process of a statistical analysis.

1.4.2 Bayesian Thinking in a Statistical Analysis

Parameter inference (i.e. parameter estimation) can be undertaken using either the frequentist approach or the Bayesian approach (see, for example, Bayarri and Berger (2004) for a thorough discussion between these two inferential approaches). These two inferential approaches differ fundamentally in terms of *how the unknown parameters are defined*, regardless of which of the two modelling approaches is taken, either the hierarchical modelling approach or the spatial econometric approach (see Sections 1.4.3 and 1.4.4 for more detail on the two modelling approaches).

The frequentist approach views parameters as fixed but unknown quantities. The term "fixed" means that there is no uncertainty associated with any of these unknown quantities. In other words, each parameter is assumed to take on a single value, although this value is not known. Parameter estimation is based on the likelihood (also known as the data model), which is a probability distribution (or a set of probability distributions) specified by the analyst to model the observed outcome values. For example, to evaluate the overall impact of the NCC scheme (Section 1.3.2.4), the frequentist approach would start by constructing a data model to describe the annual burglary counts observed in each of the eight years in both the NCC group and the non-NCC group. Through the data model, the post-policy difference in burglary risk between the NCC and the non-NCC groups can then be measured. The differences in risk between the two groups, during the post-policy period, are the parameters of interest, and the point estimates of these parameters are obtained by maximising the likelihood, i.e., by maximising the probability of the count data observed for the two groups under the statistical model constructed. The construction of a data model is not unique to the frequentist approach. The Bayesian approach takes the same step (see Section 1.4.3.1 below). However, since the frequentist paradigm treats any unknown quantities as fixed quantities with no inherent uncertainty, the derivation of a *confidence interval*, the frequentist way to express the uncertainty of an estimate in relation to an unknown parameter, is based on the idea of "infinite hypothetical replications of the situation under consideration" (Wakefield, 2013, p.23). Whilst the idea of "infinite hypothetical replications" is appropriate to experimental science where experiments can be repeated, it rarely sits comfortably within the health and social sciences where, because of the observational nature of the science, there is usually no opportunity for repeatable experimentation. In the case of the NCC evaluation, we only have a single realization of the system, and it is impossible to carry out the same implementation under the same conditions over many times, the foundation on which the frequentist approach rests.

In contrast to the frequentist idea of investigating the long run frequency of an event of interest over an infinitely large set of repeated experiments, Bayesian inference relies on the idea of updating our prior belief about an event of interest using the data we observe. In the case of the NCC evaluation, before seeing the data, we may state that "we do not know whether the NCC scheme has had an impact on the burglary risk of the NCC group". In other words, the burglary risk of the NCC group may be lower, not different from or even higher than that of the non-NCC group. This expresses our prior belief, where "prior" means that the statement above is our opinion on the potential policy impact *prior to* seeing any data. Now to investigate the policy's impact, we build a Bayesian model to combine our prior belief with the data that we observe (i.e., the burglary counts observed in both the NCC and the non-NCC groups) to form an updated belief about the policy's impact. Bayesian inference does not invoke the concept of repeated experiments but rather focuses on the extent to which the observed data leads us to revise our prior beliefs regarding the quantities of interest (e.g. the policy's impact on burglary risk).

Bayesian inference, unlike the frequentist approach, treats all unknown parameters as random variables, each associated with a probability distribution. When constructing a statistical model, the key difference between the Bayesian and the frequentist approaches is that the Bayesian approach requires us to assign prior probability distributions (a mathematical way to reflect our prior belief) to all the unknown parameters, a step that is not needed under the frequentist approach. However, this "extra" step under the Bayesian approach gives rise to several advantages when analysing spatial and spatial-temporal data, impacting on every aspect in a statistical analysis from model building, parameter estimation and interpretation to model evaluation and model comparison.

The assignment of prior distributions, a distinctive requirement in Bayesian inference, offers a natural way to model a large number of (potentially correlated) parameters. This is particularly relevant when analysing a set of spatial (or spatial-temporal) data because the number of unknown parameters involved in the modelling is typically large. For example, in the case of the income example (Section 1.3.1), there are 109 parameters, each associated with the unknown average income level of one of the 109 MSOAs in Newcastle. In the case of evaluating the local impacts of the NCC scheme (Section 1.3.2.4), we have a set of spatial-temporal parameters, each of them measuring the difference in burglary risk between an NCC (treated) area and the non-NCC (control) group in one of the post-policy years. Such a large number of parameters is needed in order to deal with the issue of heterogeneity. We could reduce all the 109 parameters in the income example down to just one, the Newcastle average, but it is rather unrealistic to assume that all the MSOAs in Newcastle have the same average income level. The same argument applies in the policy evaluation example. The assumption that all NCC areas experience the same impact from the policy is questionable. In addition, we wish to assess the policy's impact not only at the overall level but also at the local level. The problem is how to model these unknown parameters. To complicate the problem further, these parameters may be correlated rather than independent of each other. For example, the average income level in one MSOA may be similar to the average income levels of other (possibly nearby) MSOAs. The local NCC impacts may vary from one treated area to another, but might it not be reasonable to expect that these local impacts are similar based on the observation that the scheme was implemented by the same police force and all treated areas are urban areas within the same city of Peterborough? Of course, one needs to validate the appropriateness of such an assumption against the observed data, a topic that we shall come back to next.

So how should we account for such a dependence structure in these parameters? The Bayesian treatment is to formulate a multivariate prior probability distribution, often in the form of a multivariate normal distribution, for the unknown parameters (e.g. the 109 parameters for the MSOA average income and the parameters measuring the post-policy risk differences between each NCC area and the non-NCC group). Through these multivariate prior distributions, we can impose a dependence structure (through the variance-covariance matrix in the case of a multivariate normal distribution) on these parameters, allowing them to be correlated, which in turn addresses the dependence property in the observed data. Another important implication is that the imposed dependence structure on these parameters induces information sharing when we come to estimating these parameters. The estimation of the average income level in one MSOA can borrow information from other MSOAs, in addition to using the available data (if there is any) in that MSOA.

Modelling parameters jointly through various multivariate prior distributions helps to estimate such a large number of parameters through information sharing and to capture the dependence structure in data. However, one should bear in mind that the dependence structure imposed on the parameters is based on *our* assumption (e.g. all MSOAs have similar income levels or all NCC areas experience similar policy impacts). This leads on to a set of questions: "what would be a reasonable dependence structure for a given set of parameters and data?"; "how should we borrow information appropriately – should we borrow information *globally* in the sense that when estimating the parameter in one area we borrow information from all other areas regardless of where those areas are, or should we borrow information *locally* so that an area's spatial (or spatial-temporal) neighbours have a greater influence on the estimation of the parameter in that area compared to those that are far away, or some combination of both?"; "what should we do about the 'outliers' that do not fit well into the whole picture (e.g. the two areas marked with x and + in Figure 1.1 – in that case, we may not want to borrow information locally but rather to treat them differently from the rest)?"; and "how would the findings of the analysis (e.g. the efficacy of the NCC scheme) change if another plausible dependence structure on the parameters was used?" Tackling these questions requires us to have a solid understanding of both the problem and the data in hand. The Bayesian approach offers a natural way to handle many parameters, but we, the modellers, need to think (harder) about what we are trying to achieve. It is also worth noting that a Bayesian model incorporates our assumptions about the dependence structure of the parameters through the prior distribution. Putting it slightly differently, under the Bayesian paradigm, we can incorporate the geographical configuration of a set of areas as a form of prior information (or *spatial knowledge* as we introduce in Section 5.6.2), supplementing the information coming from the more conventional form of data (e.g. the burglary counts or the household income values). In this regard, the frequentist approach might be said to be inefficient because inference is based only on the observed data (the data are everything) disregarding any other forms of information (including spatial knowledge).

Another advantage with adopting the Bayesian approach in parameter estimation, in addition to information sharing, is the propagation of uncertainty. Under the Bayesian approach, parameter inference is based on the posterior distribution – a probability distribution that combines the information from both the data (through the specification of the likelihood) and any prior information that we might have about the parameters before seeing the data (through the specification of prior distributions). The posterior distribution encapsulates not only the available information but also any possible sources of uncertainty: uncertainty associated with the data, uncertainty associated with the process and/ or uncertainty associated with the parameters. The advantage of putting all the uncertainty that we can identify into the posterior distribution is that the estimation of a particular parameter in the model takes into account the uncertainty associated with other parameters in the model. For example, under the Bayesian approach, the estimation of the NCC impacts takes full account of the uncertainty associated with the estimated burglary risks of the non-NCC group. The frequentist approach, by contrast, ignores that uncertainty.

In terms of parameter interpretation, since we have the entire posterior distribution, in addition to providing conventional point and uncertainty estimates (e.g. in the form of posterior means and 95% credible intervals; see Section 5.3.1), we can derive probabilistic statements that are directly relevant to the problem in hand. For example, we can calculate the posterior probability that the NCC scheme has had a positive impact in each of the 10 NCC areas. We can also construct and make inference about new quantities that are some (possibly nonlinear) transformation of the parameters in the model. For example, the overall and the local percentage changes in burglary rates shown in Figure 1.13 are not parameters in the model but are derived based on a nonlinear transformation of some model parameters (see Section 16.2.3). The estimation of such transformed parameters is generally difficult under the frequentist approach.

For model evaluation, in addition to analysing model residuals, the Bayesian approach also allows us to investigate how well the model under consideration describes the observed data as well as certain characteristics of the data using posterior predictive checks. Such model checking helps highlight any discrepancies between the observed data and the model, informing the next step(s) in order to extend or improve the modelling.

In summary, compared to frequentist methods, Spiegelhalter et al. (2004) argue that the Bayesian approach is more flexible in adapting to each unique situation, more efficient in using all available evidence and more useful in providing relevant quantitative summaries. Carrying out an analysis using Bayesian inference is often regarded as updating prior beliefs in the light of data. This is arguably a less controversial basis for inference in an observational science than that which underlies frequentist inference. As a result, Bayesian inference has been widely applied in a variety of spatial and spatial-temporal contexts including policy evaluation (Li et al., 2013), disease surveillance (Shekhar et al., 2017) and spatial epidemiology (Maheswaran et al., 2006; Blangiado et al., 2016; and Haining et al., 2010). In Chapter 5, we will discuss in more detail the specifications and computations necessary for Bayesian inference. At this point, however, we turn in Sections 1.4.3 (Bayesian hierarchical models) and 1.4.4 (Bayesian spatial econometric models) to the two modelling approaches that we will explore in this book.

1.4.3 Bayesian Hierarchical Models

The challenges presented in Section 1.2 reflect the complex nature of spatial and spatial-temporal data and the processes that are thought to generate these data. As a result, the analysis of spatial and spatial-temporal data requires a flexible modelling framework in order to incorporate complex structures both in the data and in the model parameters while dealing with data sparsity and various sources of uncertainty. The Bayesian hierarchical modelling approach offers the flexibility required and has a wide range of applications in the social, political and health sciences (see for example Gelman and Hill, 2006; Jackman, 2009; Lawson et al., 2016; Lawson, 2018). Books by Banerjee et al. (2004), Cressie and Wikle (2011) and Blangiardo and Cameletti (2015) have illustrated the applications of hierarchical models to spatial and spatial-temporal data. In what follows we sketch some of the key features of Bayesian hierarchical modelling whilst leaving details until Part II and Part III of the book.

1.4.3.1 Thinking Hierarchically

We noted in Section 1.2 the two forms of complexity often associated with spatial and spatial-temporal data: dependence and heterogeneity. If we are to analyse such data then we need modelling frameworks that will be flexible enough to accommodate these properties. Hierarchical models provide that framework. Because observations are not independent, inferences made about one spatial unit affect inferences about another. For example, inferences about crime rates or disease rates or voting preferences in one census tract affect inferences about those rates or preferences in other census tracts. Interventions to influence crime rates, as in Section 1.3.2.4, or voting behaviour (that we will come to in Section 11.1 in Chapter 11), might have larger (or smaller) effects in some census tracts than in others, and we want to be able to consider such possibilities. Hierarchical models provide us with different ways of handling the complexities associated with spatial data so that our inferences are more soundly based.

The Bayesian hierarchical modelling framework formalises a statistical model into three components (or levels, hence the name "hierarchical"): a *data model*, a *process model* and a *parameter* model. Each of these three models is specified to deal with the three sources of uncertainty associated with the data, the process and the parameters (see Section 1.2.4). Our aim of building a statistical model is to learn about the process, as represented by the process model, and the unknown parameters in the parameter model *given*, or stated more formally *conditional on*, the observed data, in the presence of uncertainty. To achieve that goal, we translate the above statement into the following probability statement:

$$\Pr\left(process,\ parameters \mid data\right) \qquad\qquad (1.2)$$

The above expression defines a *conditional probability*. The basic idea of a conditional probability is that for two events, A and B, $\Pr(A \mid B)$ asks the question: "given that event B has already happened, what is the chance of event A happening?" Expressing the above differently, the power of conditional probability allows us to learn about something that is unknown to us (e.g. whether A would happen or not) from a related thing (e.g. B) that we have observed (B has already happened). Instead of defining through events, Eq. 1.2 utilises exactly the same idea but on the process and the parameters – the things that are not known but are of interest to us – given the data – the things we have observed.

In a Bayesian analysis, $\Pr\left(process,\ parameters \mid data\right)$ is referred to as the joint posterior probability distribution (see Chapter 5). As discussed in Section 1.2.4, there is uncertainty associated with the process and the parameters. To express this uncertainty, we use probability. As we shall soon see, we also use probability to deal with data uncertainty.

Given the complexity of the problems that we are dealing with, building a probability model for the process and the parameters is rather difficult. This is where Bayes' theorem plays a vital role: it allows us to break a complex problem into smaller, more manageable "chunks"; namely, the three components in the Bayesian hierarchical modelling framework. We will state the decomposition of Eq. 1.2 here (in Eq. 1.3 next) but leave the proof to Chapter 5 after Bayes' theorem has been introduced. Using Bayes' theorem, Eq. 1.2 becomes

$$\Pr\left(process,\ parameters \mid data\right) \ \propto\ \Pr\left(data \mid process,\ parameters\right)$$

$$\times\ \Pr\left(process \mid parameters\right) \qquad (1.3)$$

$$\times\ \Pr\left(parameters\right)$$

Eq. 1.3 defines the structure of *a Bayesian hierarchical model*, allowing us to formulate the complex conditional probability of interest, $\Pr\left(process, parameters \mid data\right)$, as a product of three conditional probabilities:

- the data model $\rightarrow \Pr\left(data \mid process, parameters\right)$,
- the process model $\rightarrow \Pr\left(process \mid parameters\right)$,
- the parameter model $\rightarrow \Pr\left(parameters\right)$.

The symbol $\propto$ in Eq. 1.3 is the proportionality symbol, meaning that the probability on the left-hand side of the symbol is equal to the product on the right-hand side multiplied by a constant. So $\propto$ allows us to ignore the constant, thereby simplifying the expression on the right-hand side.

Highlighted by Eq. 1.3 is why the Bayesian hierarchical modelling framework is not only flexible but also suitable for modelling spatial and spatial-temporal data. The reasons are:

1. The decomposition into the three model components allows us to build complex models in a modular fashion – the three model components can be built separately as opposed to considering all the parts simultaneously, a near impossible task in general.

2. The ability to build models in a modular fashion allows us to investigate various different specifications for one (or more) of the three model components by "swapping" the current specification for another one – this enables us to extend models more straightforwardly and to examine how the findings of an analysis might be affected by other plausible modelling assumptions.

3. The probabilistic nature of the three model components enables us to deal with the uncertainty associated with the data, the process and the parameters.

4. The use of conditional probability brings the three individually-built model components together to form a *joint* inference about the process and the parameters, meaning that various sources of uncertainty are properly accounted for.

In addition to the above, there are also features of the Bayesian hierarchical modelling framework that specifically address the properties of dependency and heterogeneity in spatial and spatial-temporal data (see Section 1.4.3.2) and additional reasons as to why we have taken the Bayesian approach (as opposed to the frequentist approach) to parameter estimation (as discussed in 1.4.2). But first we look at the role that each of the three model components plays.

1.4.3.1.1 The Data Model

Under the data model (the likelihood), each of the observed outcome values is associated with an *outcome variable*. To reflect possible uncertainty associated with the observed outcome values, each outcome variable is associated with a probability distribution (so each outcome variable is a random variable). Under a Bayesian hierarchical model, these outcome values, thus the outcome variables that they are associated with, are modelled independently *conditional on* the underlying spatial or spatial-temporal process and all the unknown parameters. When modelling spatial and spatial-temporal data, this conditional independence assumption on modelling the outcome values has two important implications for model specification and parameter estimation. First, when specifying the data model, we do not need to take the dependence structure of the data into account, but instead each outcome variable is treated independently (conditional on the process and the parameters) of one another. Thus we can use a wide variety of *univariate* probability distributions to model the observed outcome values. For example, one can specify a Poisson distribution for describing the number of burglary cases reported in an area or a normal distribution for describing the weekly income level of a household. The second implication is that the spatial or spatial-temporal dependence structure in the observed outcome values is incorporated in the process model, the component we look at next.

1.4.3.1.2 The Process Model

The data generating process is modelled through the process model. It is in the process model where spatial and spatial-temporal dependence, present in the observed outcome values, is modelled. We attempt to explain the dependence structure in the observed outcome data through the inclusion of *covariates* that can be spatially and/or temporally autocorrelated. We specify *random effects* to capture the part of the variability in the observed outcome data that is unexplained by the covariates we have included. Very often, a process model includes both of the above (see the stroke example in Section 1.3.2.1). Because of the conditional specification of the data model, embedding dependence in the process model induces dependence in the set of outcome variables and, in turn, in the outcome values that we observe. This feature of the hierarchical modelling framework contrasts with the situation where the dependence structure in the observed outcome values is modelled through the set of outcome variables themselves, a feature that characterises the spatial econometric modelling approach which we describe later in this chapter.

1.4.3.1.3 The Parameter Model

The parameter model describes the properties of the unknown parameters that appear in both the data model and the process model. Throughout the book, we use the terms "unknown parameters", "unknown quantities" and "unknown variables" interchangeably. These unknown parameters represent a wide range of quantities, including not only statistical parameters, such as regression coefficients and residual variance, but also parameters involved in the modelling of the random effects (e.g. random effect variance and/or a spatial autocorrelation parameter – these are terms that we will define later on in the book). In some cases, these unknown parameters also include missing values in a covariate (see, for example, the stroke example in Section 9.2 in Chapter 9 where some areas did not have data on smoking prevalence, a covariate in that study; see also Section 1.3.2.1). It is essential to propagate the uncertainty associated with these unknown quantities throughout all parts of a model so that the estimation of a parameter of interest (say a policy's impact on an area's burglary risk or the association of NO_x exposure with the risk of stroke death) fully acknowledges all possible sources of uncertainty present in the modelling. This is a key feature of the Bayesian hierarchical modelling framework. As we have discussed in Section 1.4.2, the treatment of these unknown quantities is where Bayesian inference and frequentist inference, the two most commonly-used inferential frameworks, differ.

1.4.3.2 Incorporating Spatial and Spatial-Temporal Dependence Structures in a Bayesian Hierarchical Model Using Random Effects

In a Bayesian hierarchical model, the dependence structure present in the spatial or spatial-temporal data is captured in the process model. Quite often, the variability in the observed outcome data cannot be fully accounted for using the available covariates alone because there may well be a set of unobserved/unmeasured (omitted) covariates at work. As a result, in the context of spatial modelling, a map of the residual values after accounting for the effects of the observable covariates may display a spatial pattern. Failing to account for the autocorrelation structure in the residual values results in biased estimates of the regression coefficients and inflated Type I errors when the residuals are found to be positively autocorrelated (see Section 1.2.1). To account for the effects of such omitted covariates, random effects are included, acting as surrogate measures of these unobserved/unmeasured covariates. What are these random effects? A set of random effects are a set of random variables, one for each spatial unit in the context of spatial modelling or one for each space-time unit in the context of spatial-temporal modelling. These random variables are modelled jointly through a multivariate (prior) probability distribution. It is through the specification of such a multivariate probability distribution we structure these random effects spatially or spatially-temporally.

To impose spatial dependence onto a set of area-specific random effects, we shall introduce two specifications, the conditional autoregressive (CAR) structure (in Chapter 8) and the simultaneous autoregressive (SAR) structure (in Chapter 10) and show their application in Chapters 9 and 11 in Part II of the book. In Part III of the book, we will introduce model specifications for incorporating temporal dependency (in Chapter 12) and spatial-temporal dependency (in Chapters 14 and 15). Although these model specifications differ, they have one important feature in common – all depend on the specification of a weights matrix (**W**) in order to define the spatial, the temporal or the spatial-temporal neighbourhood structure. Expressed slightly differently, this matrix defines which spatial units (temporal units or space-time units) are to be treated as "neighbours" of one another and, in some circumstances, the "strength" of that neighbouring relationship. It is through the spatial (spatial-temporal) weights matrix that we define the set of inter-unit relationships on a single map or on a sequence of maps. We consider the specification of a spatial weights matrix in Chapter 4, a temporal weights matrix in Chapter 12 and that associated with the modelling of spatial-temporal data in Chapters 14 to 16.

It should be noted that because these random effects represent the joint effects of omitted covariates, the dependence structure of these random effects, by definition, is *unknown*. To consider this point further, suppose we were to have the values for all these omitted covariates and we were to produce maps that show their spatial variation. It is likely that whilst some of the maps might show a pattern of random variation over the study region, others might display a spatial pattern. In practice, these covariates are absent from the analysis and so are their spatial (or spatial-temporal) dependence structures. Therefore, *how* we model the random effects reflects our *assumption* on whether the unobserved covariates in the application are spatially-unstructured, spatially-structured or some combination of both. To represent these three assumptions respectively, in Chapters 7 and 8, we will introduce three modelling options for the area-specific random effects: an exchangeable model where all random effects are assumed to be similar and come from a common probability distribution, a spatially-structured model where a local spatial dependence structure is imposed on the random effects and a convolution model that encapsulates both the exchangeable model and the spatially-structured model. Similarly, in Chapter 15 we will present four different space-time dependence structures for modelling random effects defined over space and time. It is essential to investigate the sensitivity of results to the assumptions we make about the random effects. Different dependence structures imposed on the random effects can lead to different parameter estimates because another role of random effects in a Bayesian hierarchical model is to carry out information sharing, a topic that we now turn to.

1.4.3.3 Information Sharing in a Bayesian Hierarchical Model through Random Effects

Hierarchical models for spatial and spatial-temporal data contain many parameters, all needing to be estimated reliably from data. This creates a challenge for estimation, particularly when data are sparse. It is important to point out that a reliable estimator for a parameter requires not only that the point estimate is close to the "true" value but also that the uncertainty of the estimator is "not too large". In other words, it is undesirable to have either an unbiased estimator that has very large uncertainty or a very biased estimator with little uncertainty. In some situations, the uncertainty associated with an unbiased estimator can be too small, resulting in an uncertainty interval that is "unrealistically" narrow (see Section 1.3.1 and also Section 7.3.2 in Chapter 7). A reliable estimator aims to strike a balance between bias and variance (uncertainty). As represented graphically in Figure 1.16, a reliable estimator can be slightly biased (hence in Figure 1.16, the cross representing the true value is slightly off the centre of the distribution of the reliable estimator), providing the uncertainty of that estimator is such that the true value does not fall on either of the extreme tails of the associated posterior distribution.

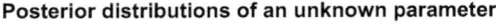

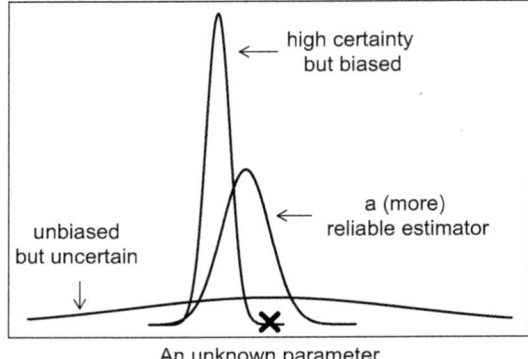

An unknown parameter

FIGURE 1.16
Illustrations of two undesirable posterior distributions, one that is unbiased but highly uncertain and the other that is highly certain but biased, and a posterior distribution that provides a more reliable estimate for the unknown parameter. The cross at the bottom indicates the true value of the parameter.

Placed within the context of modelling spatial data, each parameter is typically associated with an area (or with a space-time cell in the case of modelling spatial-temporal data). Particularly when data are sparse, information sharing plays a critical role in stabilizing the uncertainty associated with parameter estimates so that the uncertainty intervals are neither too wide nor too narrow (see Panels B and C in Figure 1.8). At the same time, a certain degree of bias may be introduced, because information sharing involves smoothing (formally known as shrinkage), whereby the point estimate of an area's parameter is pulled towards the overall mean in the case of global information sharing (smoothing) or towards the mean of an area's neighbours in the case of local information sharing (smoothing). The amount of smoothing (shrinkage) depends on a number of factors, some of which we shall explore, briefly, next. We shall provide more detail in Chapters 7 and 8 in Part II of the book. Appreciating the smoothing property of information sharing also highlights the importance of model evaluation because we need to ensure that the estimates obtained from information sharing describe the observed data adequately (i.e. the estimates are not "too biased"). We shall look at various ways to carry out model checking and model evaluation in Part II and Part III of the book, but we look next at the role that random effects play in information sharing and what we mean by shrinkage.

To explain the role of random effects in the process of information sharing, let us return to the Newcastle income data first introduced in Section 1.2.3 and then discussed again in 1.3.1. It is through the specification of the random effects (in the process model) that information is shared with the aim of improving the precision of our estimates. The problem, as we have seen, is that we do not have enough data points to arrive at reliable estimates for all the MSOAs. There were no sample data for some of the MSOAs, whilst several others had only one or two sample data points (see Table 1.1). Eq. 1.4 presents a Bayesian hierarchical model with the purpose of sharing information globally (across all the MSOAs):

$y_{ij} \sim N\left(\mu_i, \sigma_y^2\right)$	The data model
$\mu_i \sim N\left(\alpha, \sigma_\mu^2\right)$	The process model
$\Pr\left(\sigma_y^2, \alpha, \sigma_\mu^2\right)$	The parameter model

(1.4)

where y_{ij} is the income value for the jth household in MSOA i and μ_i is the unknown average income for MSOA i. It is the latter that we wish to provide an estimate for. All the MSOA-level means are assumed to follow the common distribution $N(\alpha, \sigma_\mu^2)$, implying that the average income levels are different across MSOAs, but they are all distributed around α, the Newcastle average. σ_μ^2 represents the between-MSOA variability. In this model, the set of MSOA-level averages, $\mu_1, ..., \mu_{109}$, across the 109 MSOAs are modelled as a set of random effects. For the purpose of illustration, we fix α and σ_μ^2 to the mean ($\bar{y}_{all}$) and the variance (s_y^2) of all the household-level data in the survey, respectively.[9] A result from Chapter 7 shows that the point estimate of each μ_i is given by:

$$\frac{w_i}{w_i + v_i} \bar{y}_i + \frac{v_i}{w_i + v_i} \bar{y}_{all}$$

where $w_i = \dfrac{n_i}{s_y^2}$, $v_i = \dfrac{1}{\sigma_\mu^2}$ and n_i is the size of the sample in MSOA i.

This expression highlights the essence of information borrowing under a hierarchical model: the point estimate of μ_i is a weighted average of the sample mean income of this MSOA ($\bar{y}_i$) and the overall mean of Newcastle ($\bar{y}_{all}$). Stating this more generally: the estimate from a hierarchical model combines information from its own unit and information from other units in the study region. The expression above also illustrates the following two points regarding the nature of information borrowing. First, the weights are functions of the sample size, n_i. The smaller the n_i (i.e. the fewer data points we have within MSOA i), the more information we borrow from the Newcastle average. Hence the estimate is *pulled* more towards the overall mean – this is referred to as the *shrinkage* (or *smoothing*) that results from implementing a hierarchical model. In the extreme, the point estimate for an MSOA without data (i.e. $n_i = 0$) will be equal to the Newcastle average, $\bar{y}_{all}$, resulting in a complete shrinkage. On the other hand, if the size of the data within an MSOA is sufficiently large, little information will be borrowed from the common distribution, and hence there will be little shrinkage of the point estimate towards the Newcastle average for this MSOA. Second, this hierarchical model imposes global information borrowing (hence a global smoothing) because the estimate of each MSOA borrows information from the global distribution, which is estimated using all the data in Newcastle. As we will see in Chapter 8, the CAR model implies a *local* smoothing, where information is borrowed based on the neighbourhood structure imposed by the chosen spatial weights matrix. As we shall also see in Chapter 8, this local smoothing idea can be extended further to perform *adaptive* local smoothing so that information from other spatial units is borrowed "appropriately". The smoothing idea will be further explored in the context of modelling temporal data (Chapter 12) and spatial-temporal data (Chapters 14 to 16).

1.4.4 Bayesian Spatial Econometrics

One of the primary interests amongst users of spatial econometric models is to quantify how the outcome in area (or location) i affects outcomes in its spatial neighbours; how changes in a covariate in one area may impact on outcomes in other areas. These are often

---

[9] To fully specify a Bayesian hierarchical model we would also need to assign priors to the unknown parameters σ_y^2, α and σ_μ^2; that is, specify the prior probability distributions to each in the parameter model $\Pr(\sigma_y^2, \alpha, \sigma_\mu^2)$ in Eq. 1.4, but we can set this to one side for the purpose of this sketch.

referred to by spatial econometricians as "spatial spillover effects", which may also give rise to "feedback effects". More generally, however, spatial econometrics has developed to address the statistical issues that arise when applying the standard normal regression model to spatial data, recognising: (a) the need to handle spatial dependence; (b) the importance of explanatory factors located in other spaces ("space-distant explanatory factors"); (c) the need to model space explicitly when analysing spatial data (Paelinck and Klaassen, 1979, p.5–11). To this historic list we should add: (d) the need to model various forms of spatial heterogeneity (Anselin, 1988). Spatial econometrics has evolved as a branch of standard regression modelling accommodating the special requirements of spatial data. In so doing, it takes a particular interest in certain types of quite distinctive *spatial regression models*, examples of which we shall see in Chapter 10. As we have indicated, the spatial econometrics approach uses several ways to capture the spatial dependence structure in an outcome variable, dependence that is embedded in the likelihood function for the data rather than in a separately specified *data generating process*. This does lead to some challenges for model fitting and parameter interpretation, as we shall see in Chapter 10.

Consider again the model presented in Eq. 1.1 (Section 1.3.2.3), which is an example of the spatial lag model (SLM), one of several formulations in the family of spatial econometric models. For ease of discussion, we present the model again. The SLM directly specifies a series of N simultaneous equations (one equation for each spatial unit) on the outcome variables in the following form:

$$Y_i = \alpha + \delta \sum_{j=1}^{N} w_{ij}^* Y_j + \mathbf{X}_i \boldsymbol{\beta} + e_i \qquad (1.5)$$

The above formulation reflects an emphasis on estimating the *interactions* taking place amongst the outcomes in different spatial units: the outcome in one spatial unit affecting the outcomes in other spatial units – the notion of spatial spillover. One of the consequences of this type of model specification is that the set of outcome variables, $Y_1, \ldots, Y_N$, are jointly modelled through a *multivariate* normal distribution. In other words, the SLM reflects the dependence structure of the outcome values through the *data model* (i.e. the likelihood), a feature of a *likelihood-based* model. This contrasts with the hierarchical modelling approach, where the dependence structure is incorporated through the process model. The implications of modelling spatial dependence through the likelihood are: (a) a multivariate probability distribution needs to be used for modelling the observed outcome values; (b) care needs to be taken when interpreting the effects of the covariates; and (c) specific algorithms need to be derived for parameter estimation. While, as we shall see in Chapter 10, the joint modelling of the outcome values through the likelihood poses difficulties in model specification and estimation, it is this joint modelling feature of spatial econometric models that enables us to tackle a distinctive set of research questions relating to spillover and feedback effects. These points will be discussed in detail in Chapter 10, where we shall encounter other models in the family of spatial econometric models.

1.5 Concluding Remarks

The statistical modelling of spatial and spatial-temporal data presents four significant challenges: dependence, heterogeneity, data sparsity and uncertainty (data, process and parameter uncertainty). These *challenges* need to be met if we are to realise the *opportunities* that are provided by spatial and spatial-temporal data in the social, economic and public health sciences. These opportunities include: improving small area statistical precision; modelling covariate-outcome relationships; testing social science theories; testing for spatial interaction effects; assessing the effects of a geographically targeted policy intervention; implementing effective geographical surveillance.

A Bayesian hierarchical model, which formalizes a statistical model into a data model, a process model and a parameter model, provides a methodology for meeting all these challenges, enabling us to realise most (but not all) of the opportunities we have identified. Bayesian spatial econometric modelling is another important methodology that has been developed principally to address the challenges presented by spatial dependency and spatial heterogeneity in "standard" regression modelling. Different from a Bayesian hierarchical model, a spatial econometric model addresses the properties of spatial dependency and spatial heterogeneity through the data model (i.e. the likelihood function). In a Bayesian hierarchical model, we can embed both these properties in the process model so that data values can be modelled independently *given* the process model and the parameters. Although the spatial econometric approach does give rise to some unfortunate complications due to the joint modelling of the outcome values, it does offer two important benefits. First, it provides a simple and rather natural way of extending the regression model for normally distributed outcomes, widely used by social scientists, in order to address inference problems arising when either or both spatial dependence and spatial heterogeneity are present. Second, it allows us to test for spatial spillover and feedback effects arising when *outcomes* in different spaces interact with one another.

When estimating the parameters in a model, the approach taken in this book is to use Bayesian inference rather than frequentist inference. Our reasons are the following. First, Bayesian models allow us to incorporate various sources of uncertainty, leading to more reliable inference. Second, Bayesian inference is based on updating prior beliefs (however vague they might be) in the light of the observed data through the use of Bayes' theorem. This approach to inference is arguably much more relevant to social science research than frequentist inference. The idea of invoking a very large number of replications of the data generating "experiment" is, to say the least, challenging in the context of the social sciences. In many areas of social science, research experimentation is impossible, sometimes unethical. Finally, however intellectually appealing Bayesian inference might be, the case would be moot if the means did not exist to fit "realistic" spatial models to actual data. Whereas that was the situation only a few decades ago, today there is a wide variety of software that enables us to fit various Bayesian spatial and spatial-temporal models. In this book we demonstrate the use of WinBUGS (version 1.4.3), an open-source software for fitting Bayesian models, and R (R Core team, 2019), another open-source software for more general statistical analyses and visualisation, for modelling spatial and spatial-temporal data.

1.6 The Datasets Used in the Book

In the course of this book we shall discuss and report results from the analysis of certain datasets. These datasets will also be made available to the reader on the book's website. Below is a list of these datasets together with some information about them.

	Discipline	Name of dataset	Space	Time	Data collection process	Type of response	Used in chapters
Spatial data	Economics	Household level income data in Newcastle, England	Household level income data georeferenced to one of the 109 Middle Super Output Areas (MSOAs) in Newcastle, England	2007/08	Observational through survey (simulated)	Continuous-valued data	Chapters 6–8 to illustrate ESDA and various hierarchical and non-hierarchical spatial models
		Cigarette sales data in US states	Across 46 US states	Data collected in 1992	Observational	Continuous-valued data	Chapter 10 to illustrate the fitting of various spatial econometric models and to highlight issues related to the interpretation of covariates effects in spatial econometric modelling
	Criminology	Stockholm outdoor rape data, Sweden	Number of outdoor rape cases recorded in 407 *basområdes* in Stockholm, Sweden	Cases reported between 2008 and 2009	Observational (police recorded crime database)	Count data	Chapter 9 to illustrate the use of Bayesian profile regression to deal with issues such as multicollinearity, interactions and spatial heterogeneous effect of parameters
		High Intensity crime Area (HIA) data, Sheffield, England	337 Census Output Areas (COAs) in Sheffield	1998	Observational (police recorded crime database + data based on senior police officer perception)	Binary data	Chapters 5 and 9 on modelling binary response using univariate and bivariate Bayesian logistic regression models with and without random effects. Chapter 6 as an example of the join-count test

(Continued)

Discipline	Name of dataset	Space	Time	Data collection process	Type of response	Used in chapters
Public health	The Stroke-NO_x study in Sheffield, England	1030 Enumeration Districts (each with approximately 150 households) in Sheffield	Stroke mortality data between 1994 and 1998	Observational	Count data for mortality plus NO_x air pollution data from monitoring stations	Chapter 9 to demonstrate ecological modelling to examine disease risk and environmental exposure and to illustrate the use of hierarchical models for combining individual and area level data to strengthen ecological inference
Political science	Brexit – constituency-level voting outcomes from the EU referendum in England	326 Local Authority Districts (LADs) in England (English Administrative Districts 2011 definition)	Voting outcomes from the EU referendum in 2016	Observational	Continuous-valued data (proportions of votes for leaving the EU)	Chapter 11 as an illustration of spatial econometric modelling
Spatial-temporal data Economics	Petrol price data in Sheffield, England	Data collected on petrol price at 63 petrol stations.	Monthly data from January to March, 1982	Observational	Continuous-valued data	Chapter 11 to illustrate various spatial econometric models and a Bayesian hierarchical model with a t_4 likelihood
Public health	Malaria incidence data in Kalaburagi taluk, Karnataka, South India	Malaria cases in 139 villages in Kalaburagi taluk, Karnataka, South India	Monthly counts of new malaria cases over three years from 2012 to 2014	Observational	Count data	A subset of this dataset is used in Chapter 9 to illustrate the development of a spatial mixture model with zero-inflation. The full set of space-time data is visualized in Chapter 13, then analysed in Chapter 16 using a spatial-temporal mixture model with zero-inflation focusing particularly on the investigation of spillover effects

(Continued)

Discipline	Name of dataset	Space	Time	Data collection process	Type of response	Used in chapters
	The mortality data on chronic obstructive pulmonary disease (COPD)	Numbers of COPD deaths across the 354 Local Authority Districts (LADs) in England (English Administrative Districts 2001 definition)	Annual mortality counts from 1990 to 1997	Observational (simulated)	Count data	This simulated dataset is used in Chapter 13 to illustrate a number of exploratory techniques and is analysed in Chapter 16 for disease surveillance using BaySTDetect, a Bayesian space-time detection method
Criminology	Peterborough domestic burglary data, England	Counts of burglary events in each of the 452 Census Output Areas (COA) in Peterborough, with each COA containing approximately 100 houses.	Annual count over eight years from 2001 to 2008	Observational (police recorded crime database)	Count data	This dataset is introduced in Chapter 12 then analysed using various spatial-temporal models in Chapter 15. A subset of this dataset is modelled in Chapter 16 for studying temporal dynamics of crime hotspots and coldspots
	The "No Cold Calling" (NCC) study embedded in the Peterborough burglary data, England	10 COAs in Peterborough, England	NCC scheme implemented 2005/06	Quasi-experimental design	Count data on crime + binary indicator for policy implementation	Chapter 16 on policy evaluation

1.7 Exercises

Exercise 1.1. The R script, `map_ward_burglary_cambridgeshire.R`, on the book's website produces the map of the ward-level raw burglary rates in Cambridgeshire in 2004 presented in Figure 1.1. You may not be familiar with the R commands in the script, but we shall discuss them in detail in Chapter 6. See the comments provided in the script, outlining what each line is doing. Run the script in R to produce the map.

Exercise 1.2. With reference to an area of research with which you are familiar, consider the extent to which the four challenges described in Section 1.2 need to be recognized and why.

Exercise 1.3. Again, with reference to an area of research with which you are familiar, specify a problem for which a Bayesian hierarchical model might provide the appropriate methodology and a problem for which a Bayesian spatial econometric model might provide the appropriate methodology. Justify your choice.

Exercise 1.4. Assess the relative merits of Bayesian and frequentist inference in your field of study.

Exercise 1.5. Recall the model from 1.4.3.3 where:

$y_{ij} \sim N\left(\mu_i, \sigma_y^2\right)$	The data model
$\mu_i = \alpha + U_i$ $U_i \sim N\left(0, \sigma_\mu^2\right)$ (or equivalently $\mu_i \sim N\left(\alpha, \sigma_\mu^2\right)$)	The process model
$\Pr\left(\sigma_y^2, \alpha, \sigma_\mu^2\right)$	The parameter model

We remarked that it can be shown that the point estimate of μ_i is given by:

$$\frac{w_i}{w_i + v_i} \bar{y}_i + \frac{v_i}{w_i + v_i} \bar{y}_{all},$$

where $w_i = \dfrac{n_i}{s_y^2}$, $v_i = \dfrac{1}{\sigma_\mu^2}$ and n_i is the size of the sample in MSOA i. Experiment with different values of the parameters n_i, s_y^2 and σ_μ^2 to see the effects on the relative weighting attached to $\bar{y}_i$ and $\bar{y}_{all}$ (you may fix the values of $\bar{y}_i$ and $\bar{y}_{all}$ to say 400 and 600, respectively).

Exercise 1.6. In Section 1.2.1 we noted that an unbiased estimator for the mean μ is the sample mean of the observations $X_1, \dots X_N$ given by $\bar{X} = \dfrac{\sum_{i=1}^N X_i}{N}$. When the observed values can be assumed independent and all have come from a normal distribution with mean μ and variance σ^2, the error variance for the estimator of μ is $\widehat{EV}_{ind} = E\left[(\bar{X} - \mu)(\bar{X} - \mu)\right] = \sigma^2/N$. Here, σ^2 is assumed to be known. However, when sample values are dependent, $\widehat{EV}_{dep} = \dfrac{\sigma^2}{N} + \hat{\gamma}$. Show that $\hat{\gamma} = \dfrac{2}{N^2} \sum\sum_{i<j} Cov(X_i, X_j)$, where $Cov\left(X_i, X_j\right)$ is the

autocovariance between X_i and X_j. (Hint: we want to calculate the variance of $\bar{X}$, i.e.

$$Var(\bar{X}) = Var\left(\frac{\sum_{i=1}^{N} X_i}{N}\right) = \frac{1}{N^2}\left(\sum_{i=1}^{N}\sum_{j=1}^{N} Cov(X_i, X_j)\right)).$$

Exercise 1.7. Consider again the situation in Section 1.2.1 where a set of N observations are used to estimate the unknown mean μ. Assuming that these N observed values are drawn independently from the normal distribution in question and σ^2, the variance of the normal distribution, is known, the calculation of the 95% confidence interval for μ is $\left(\bar{X} - 1.96 \times \frac{\sigma}{\sqrt{N}}, \bar{X} + 1.96 \times \frac{\sigma}{\sqrt{N}}\right)$. When σ^2 is not known, comment on the differences between the following two versions of a 95% confidence interval for μ where (a) the value of the sample standard deviation, s, replaces σ in the calculation; and (b) the uncertainty in estimating σ through s is taken into account in the interval calculation. Calculate the two versions using a dataset with the following summary statistics: $\bar{x} = 6.3$, $s = 7.8$ and $N = 10$. Carry out the same calculation using the same values but changing N to 100. How does the comparison of the two intervals change when N is large or when N is small?

Exercise 1.8. As in Exercise 1.7, when σ^2 is not known and if the observations, $X_1, \ldots X_N$, can be assumed to be independent, show that the sample variance s^2 is an unbiased estimator of σ^2. When the observations are positively autocorrelated, show that s^2 underestimates σ^2 (hint: see Haining 1988, p.579).

2

Concepts for Modelling Spatial and Spatial-Temporal Data: An Introduction to "Spatial Thinking"

2.1 Introduction

A "spatial" database or a "spatially-referenced" database is one where each individual datum in that database is attached to a location. The implication is that we know the places on the earth's surface to which the data refer. When spatial data are also recorded for a sequence of times (instants in time or time periods, for example over a week, a month or a year), we refer to a "spatial-temporal" database. Knowledge of the spatial or spatial-temporal location of each datum is part of the information content of that database and is important both for the statistical analysis of that data and the interpretation of results.

Many areas of the social, economic and public health sciences analyse spatial and spatial-temporal data. Through organizational and technological advances it has become possible to access ever larger volumes of such data and to be ever more precise in the geographical assignment process and ever more prompt in the temporal updating process. In this chapter we describe the key concepts that inform the analysis of spatial and spatial temporal data – sometimes referred to as "spatial and spatial-temporal thinking" or "spatial thinking" for short.

Core geographical concepts that are fundamental to this mode of thinking include: the location, distribution and patterning of objects and events (attributes) in geographical space and space-time; places and regions and their particular attributes; the interdependencies that exist between places and regions and the various types of spatial-temporal interaction processes between places at different times; the geographical extent of an area or domain of study; and the scale (or level of generalization) at which we observe that domain. Clifford et al. (2009) discusses core geographical concepts which ground the academic discipline of geography, and these concepts are also important to spatial thinking, which is about developing a geographical "habit of mind" when scientifically approaching certain types of problems and issues.[1] We note here that although this book does not deal with Geographical Information Systems (GISs), such systems have an important contribution to make to the modelling activities that this book deals with. So in the appendix to this chapter we briefly discuss them.

Attributes are observed and measured, and our aims include one or more of: providing a *description* of the spatial or spatial-temporal variation of attributes; *explaining* the observed variation in one or more attributes by identifying spatial associations and developing theory; and making *predictions* or *forecasts* about future spatial patterns with the aim of

---

[1] See: https://blogs.esri.com/esri/gisedcom/2013/05/24/a-working-definition-of-spatial-thinking/.

47

informing *policy*. Our focus is the study of patterns and processes in geographical space, arriving at conclusions based on scientific principles. In this book, to achieve these aims, we apply statistical theory and method. In the previous chapter we considered some of the special challenges and opportunities that the statistical modelling of spatial and spatial-temporal data gives rise to. In this chapter we focus on the geographical and spatial concepts that are frequently called upon and which inform how statistical method is applied. Taken together, Chapters 1 and 2 are designed to promote a *fusing of two reasoning processes* – the spatial thinking and the statistical thinking. Both are essential for the effective implementation, by social, economic and public health scientists, amongst others, of statistical techniques applied to spatial and spatial-temporal data (Haining, 2014 and Waller, 2014).

2.2 Mapping Data and Why It Matters

Knowing and recording where objects are located or when and where events happen in geographical space-time is important for research in many academic disciplines as well as for those concerned with more immediate, practical concerns such as the design and delivery of public or private sector policies. We use the term "object", in the geographical information science sense, to refer to an entity (a house, a road, a city) that endures through some extended, though not necessarily long, period of time (Warboys and Hornsby, 2004). In the short to medium term at least, the location of an object is fixed, for example the location of a business, a retail outlet, a landfill site, a hospital or clinic, an archaeological artefact, a transit terminal, a line of communication (road, railway, river). Recording the geographical location of objects in particular classes at a point in time means such data can be organized and visualized in map form. Mapping becomes a starting point for interrogating the data both in terms of characterising and making comparisons between places.

We use the term "event", again drawing on the geographical information science literature, to refer to an entity that happens and is then gone (Warboys and Hornsby, 2004).[2] For example, an event can be a street robbery or a burglary that occurred on a particular day at a particular location within a city; a new case of malaria in a region; or an individual diagnosed as suffering from a chronic respiratory condition or some limiting long-term illness. Recording and mapping the space-time location of events (new cases of a disease, instances of household burglary) may provide an important starting point for developing a better understanding of the underlying processes responsible for the observed outcomes. Combining real-time event data, such as data on disease occurrences, and current object data, such as the location of clinics and other support services, as well as areas with a known increased risk of an outbreak, provides a starting point for designing response strategies in the case of a new outbreak. Combining event data on street crime with object data on the location of transport routes, transit terminals and certain types of retail outlets may suggest strategies for crime prevention and/or responding to local crime hotspots.

The study of the attributes that attach to objects or events and which describe their properties is an aspect, to a greater or lesser degree, of many academic fields within the arts, humanities and the social, economic and public health sciences. In the environmental sciences, mapping the physical, environmental and human activity of an

---

[2] Warboys and Hornsby (2004) refer to objects as "continuant" entities and events as "occurrent" entities. Events and processes, they argue, essentially "speak to the same idea".

area helps in meeting environmental challenges and threats. The study of particular places involves not just cataloguing the attributes of different geographically defined areas, it also involves developing an awareness of the role different processes, operating at different geographical scales, have in shaping the characteristics of each place and how those characteristics may in turn impact on or modify larger scale processes. Thinking spatially means thinking through the implications of where objects and events are located, why objects are located where they are, why events happen where they do and why the changes we observe over time (perhaps in response to a common stimulus such as a national economic recession, a regional crime wave or a threat to health) may not be the same everywhere. In the fields of policy-making and policy evaluation, to think spatially means to think about how the impacts associated with (say) a national policy vary between areas as well as thinking about the need for geographic targeting (distinguishing between areas in terms of whether or not they should receive financial support) or geographic tailoring (making adjustments to policies at the local level to reflect local conditions). Thinking spatially, as opposed to non-spatially, usually involves presenting data in particular ways, asking different kinds of questions and using different kinds of methods to answer those questions. For a wide range of examples, see Lawson et al., 2016; Beck et al., 2006; Clarke, 1997; Cromley and McLafferty, 2012; Cuzick and Elliott, 1992; Gamarnikow and Green, 1999; Martin et al., 2015; Tita and Radi, 2010; and Weisburd et al., 2012.

The map is an essential presentational tool for any science that works with data with a spatial or geographical reference (or "geo-reference"). The geo-reference of a single piece of data identifies the location on the earth's surface to which the datum refers. The location may be a point co-ordinate that defines the position of the datum on east-west (x) and north-south (y) axes, it may be a line such as a transport route or it may be an area/polygon such as a census tract or region. The top left-hand diagram in Figure 2.1 shows a dot map recording the location of some event – a household burglary for example or a case of street robbery. All the events have been *binned* (aggregated) into a single time interval. Such data may be spatially binned, that is aggregated into small spatial units as shown on the top right-hand diagram in Figure 2.1, where the greyscale goes from white (0 cases) to black (5 or more cases). Much of the economic, social and demographic data that social scientists work with (and which we will work with in this book) are reported through irregular spatial bins, such as census tracts. On the other hand, much environmental data, particularly remotely sensed data, are reported using pixels as shown on the top right-hand diagram in Figure 2.1. How much spatial detail is lost by binning depends on the size of the bins (map scale or resolution).

These data can be refined by examining the time (t; a point in time or interval of time) when each event occurred. Such data can be represented by a space-time plot in which the location of any event is captured by x and y co-ordinates (two horizontal axes) and time, t, is captured by a third (vertical) axis. The data can then be aggregated or "binned" into space-time units as shown in the bottom diagram in Figure 2.1, where a time has been attached to each of the events shown in the top diagram in Figure 2.1. Each bin records the number of cases in that space-time bin. In the process of binning, one should also bear in mind that in some situations, the exact time of events may not be known. For example, residential burglaries often occur when the occupiers are away for a period of time. A patient may be diagnosed sometime after they first contracted a disease. The significance of this uncertainty depends on the interval of time used for the purposes of data analysis and the length of time of the uncertainty. With the demand for greater temporal precision to meet the operational needs of, say, police forces or health services, some adjustment to

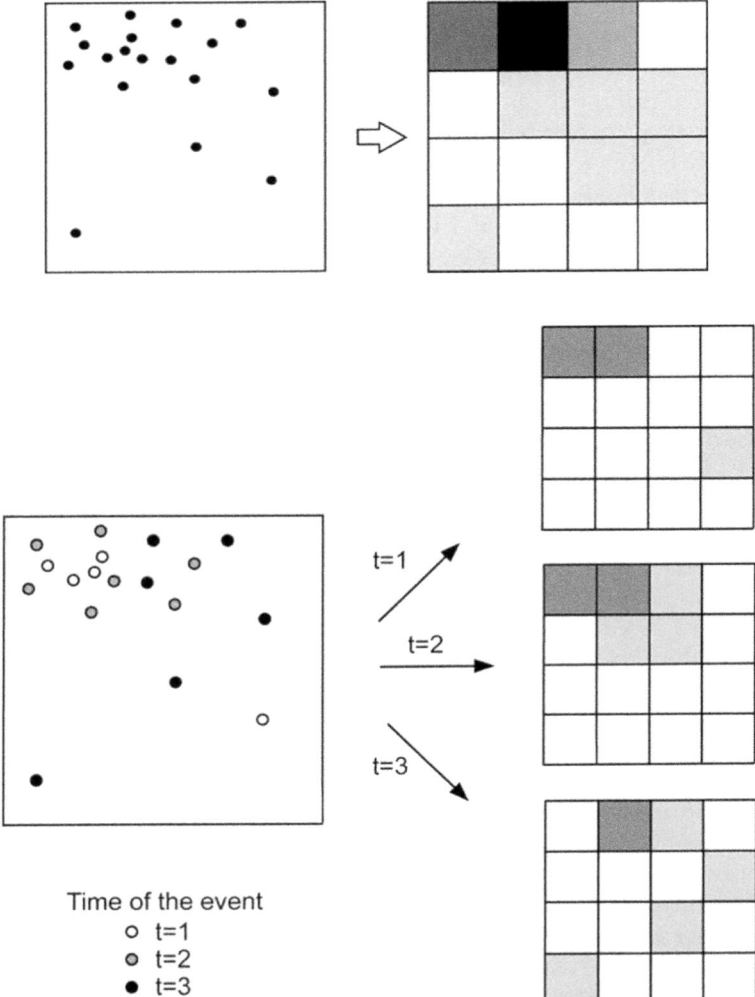

FIGURE 2.1
Top: event data showing location binned into small, regular spatial units. Bottom: if the time of the event is recorded then event data can be binned into space-time bins.

the binning process to allow for this form of uncertainty may be necessary. In the case of crime recording, see for example Ratcliffe and McCullagh, 1998 and Ratcliffe, 2002.

A purely spatial analysis examines each one of the (horizontal) slices through the space-time data. A space-time analysis, considering all the space-time data together, allows us to show change over time and to suggest the presence of space-time interaction, where events occurring at a particular space-time bin are correlated with events occurring in nearby space-time bins – a topic that we shall examine closely in Part III of the book. The bottom diagram in Figure 2.1 might suggest some localized diffusion process associated with the event, an observation that would not be evident from the top diagram in Figure 2.1 where the binning of events in time is too coarse.

A space-time multivariate dataset, where many attributes are recorded, can be thought of as a "space-time-attribute" data cube in which one axis identifies location ("where"), the

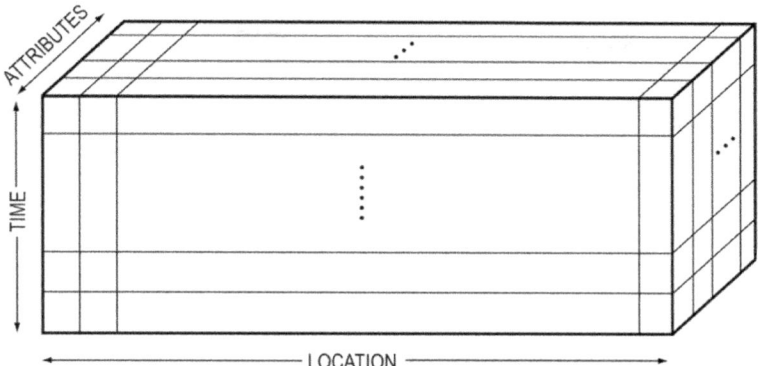

FIGURE 2.2
The space-time-attribute data cube. Each small "box" inside the cube identifies an attribute value associated with a single area at a point in, or during an interval of, time.

second axis identifies time ("when") and the third axis specifies the set of attributes that are recorded for each space-time unit of observation ("what"). Each individual bin records the value of a specified attribute in a specified space-time interval, such as the number of cases of a particular crime or disease occurring in a specified area during some interval of time (see Figure 2.2). The representations in Figure 2.1 retain the spatial and temporal relationships between the data points for the purpose of visualising the data. By contrast, the representation given by Figure 2.2 does not retain the spatial relationships between the spatial bins, and if these are important, as they usually are, we need another vehicle for storing this information. In Chapter 4 we discuss the connectivity or weights matrix, W, used for this purpose amongst others.

Maps are central to the methods of this book. A map can be defined as a symbolic depiction of the arrangement or distribution of entities (objects or events) and their attribute values across some area. The emphasis of a map is the spatial relationships between those entities and their attribute values. A map may be printed or it may be held in digital form as in a Geographical Information System (GIS), where the data are held as a series of layers where each layer can be independently switched "on" or "off", giving rise to the possibility of multiple different views. If values are recorded over time so that each layer refers to a different time window, displaying such data in time order can be used to show, for example, urban expansion (or contraction), network evolution or the spread of some infectious disease.

From the perspective of statistical analysis, we can think of a map as both a presentation and a scientific visualization tool. As a presentation tool the map is used to provide the reader with a geographical summary or display, either of the original data or of important results arising from some analysis of that data. As a visualization tool the map is used in exploratory data analysis (see Chapters 6 and 13) in which the data analyst displays and experiments with their data in map form in order to gain insights into data characteristics and data relationships. A single map referring to a defined interval of time or snapshot gives us a picture of the "here and now" of what is happening and can suggest associations between attributes. Maps over time reveal how that "here and now" changes and may indicate the presence of space-time interaction. Maps that show movement between nodes, for example if the data refer to flows of goods or people or information between urban areas over a period of time, indicate how places interact with one another economically or socially.

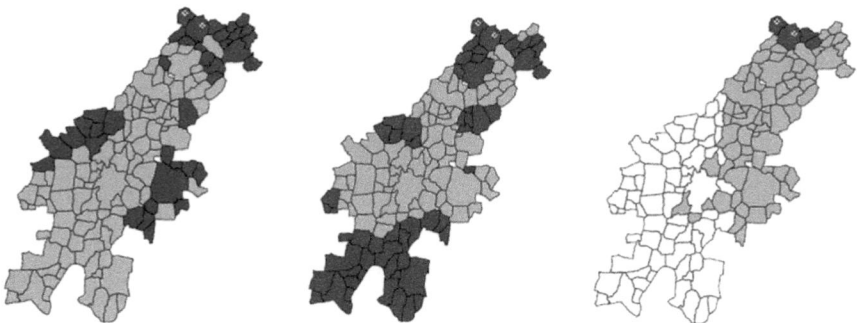

FIGURE 2.3
New cases of malaria reported at the village level for Karnataka, India. From left to right: August–September 2014 (warm-wet season); October–November 2014 (warm-dry season); December 2014–January 2015 (cool-dry season). On each map, shading goes from dark (high risk) to light (low risk). Villages identified as malaria "hotspots" are marked (⊗) (Shekhar et al., 2017). For scale: From North East to South West is approximately 80km.

A sequence of maps covering a series of time points or intervals allows us the possibility of investigating process. If geographical data are available over time then not only can we observe changes over time, we can observe the way a process unfolds in space and time – for example, whether events occurring close together in space also occur close together in time – and vice versa. This is important in many areas of research, including studying the spread of an infectious disease (see Figure 2.3); how the local geography of crime risk varies over time; the geography of the take-up over time of an innovation or new technology; the impact over space and time of a new targeted policy (for illustration see: Bowers and Johnson, 2005; Gao et al., 2013; Grubesic and Mack, 2008; Hagerstrand, 1967; Newton and Felson, 2015; and Qi and Du, 2013).

2.3 Thinking Spatially

2.3.1 Explaining Spatial Variation

To "think spatially" is not only to be concerned with the study of places. Maps reveal more than that the numbers of events or the presence/absence of objects vary from place to place. Our attention may be drawn to spatial or geographical structures and relationships in one or more attributes, which may be important in terms of the causes and consequences of the spatial variation that we observe.[3]

(i) *Variation in attribute values relative to significant fixed points or areas.* Census tracts within a city will vary in terms of their distance from the city centre. Levels of anti-social behaviour by census tract might show a declining trend with increasing distance from the centre. Street robberies may be observed to cluster around bars. Vandalism hot spots are often found at or near public transit stops. Certain health indicators for

---

[3] Maps of several attributes may draw attention to relationships between different attributes. Scatterplots and added variable plots are also useful, but such tools may need to be adapted if we want to explore the relationship between attributes across a set of spatial units (see Chapter 3 and Chapter 6).

a group of settlements might correlate with their distance and direction from possible sources of environmental risk. Archaeological artefacts may be found in specific locations, such as close to and with views of coastal areas that imply the sea may have some symbolic or practical significance. For a range of examples see: Ceccato, 2013; Elliot et al., 2009; Fisher et al., 1997; Gallup et al., 1999; and Venables, 1999.

(ii) *Spatial gradients in attribute values between adjacent areas.* The close juxtaposition of neighbourhoods that show a marked contrast in terms of their affluence or social cohesion may have implications for a range of social, economic and health-related outcomes. Consider the following thought experiment. Two communities are identical in terms of their internal or "place" characteristics, but they differ in terms of the characteristics of their immediate neighbouring communities. Outcomes might differ between the two areas because of interaction effects taking place across their respective boundaries – interaction that is triggered by the attribute gradient. There is evidence, for example, that communities that present very similar socio-economic characteristics may differ in terms of health outcomes or crime rates depending on the social and economic characteristics of adjacent communities. For some examples see: Block, 1979; Bowers and Hirschfield, 1999; Dow et al., 1982; Gatrell, 1997; and MacLeod et al., 1999.

(iii) *Macro-spatial configuration.* The spatial structure of a society and its economy may have implications for social and economic outcomes. The presence of significant income inequality in a society may have implications that are influenced by the geography of that inequality. Consider the following thought experiment. Imagine two cities identical in all important respects and in particular with equivalent numbers of people living in different income bands. In one of the two cities a significant majority of the population living in poverty live in one large ghetto, whilst in the other city the same population is more dispersed, perhaps as a consequence of the city's public housing policy. Both the city level health profiles and crime rates in those two cities as well as their micro-geographies may be influenced by the difference in the macro-spatial configuration of the population living in poverty. Recent literature in economics has drawn attention to the benefits of spatial clustering of economic activity. Geographical proximity may confer benefits on firms that cluster in the form of positive externalities that derive from their geographical context. These externalities include benefits in terms of factor conditions for firms within the cluster (e.g. labour availability), demand conditions (e.g. access to markets), firm strategy (e.g. knowledge acquisition) and the presence of related and supporting industries (e.g. proximity to supply industries). For some examples see: Kahn et al., 2006; Krugman, 1996, 1998; Porter, 1998; Sampson et al., 1997; Snow and Moss, 2014; Sparks et al., 2009; Szwarcwald et al., 2000; and Wilson, 1997.

Recognizing spatial relationships in the study of events is also important. That is because the processes that underlie the events we observe are not constrained by the spatial units in terms of which the data are reported. People move about in their daily lives and over their life course, and they have information about what is happening elsewhere – never more so than in an age of social media. Environmental influences can be carried from place to place by movements of air and water. The spread of influences from one place to another are sometimes referred to as "spatial spill-over (or spillover) effects" (see Sections 1.3.2.3 and 1.4.4). When considering data reported over a set of well-defined spatial units, for example local government areas or health districts, spillover effects are important because decisions made (e.g. resources allocated and outcomes arising) in one area may have consequences for other areas. But spillover effects also need to be considered when the spatial units are artificial constructions, such as census tracts. The outcomes we observe in one place may in part be a consequence of circumstances elsewhere. How a spillover effect impacts on a study depends on the relationship between the scale at which the spillover process is operating in the real world and the size or scale of the spatial units through which outcomes are observed. If spatial spillover processes in the real world are highly localized but the spatial units are large, then the contribution of any spillover effect may result in an internal multiplier effect with numbers of cases (where cases are present) inflated by the localized spillover effect. So a map of case counts might show large numbers of areas with zero or small counts, whilst there are other areas with large counts. If spillover processes operate at a scale that is larger than the spatial units of analysis, this may then lead to spatial correlation effects between adjacent units. As a result, areas with high (low) case counts tend to be located close to other areas also with high (low) case counts. Spillover effects may operate at many scales, so that both characteristics might be evident in a dataset. For some examples, see Brett and Pinkse, 2000; Case et al., 1993; Brueckner, 2003; Hanes, 2002; Hassett and Mathur, 2015; and Revelli, 2002.

The differences that are observed from place to place will be a consequence of many different factors. Differences, for example in health outcomes and crime rates, reflect differences in the composition of the population. With all other things being equal, a place with a larger proportion of older people will tend to show worse morbidity statistics than places with a younger population, but an area with a larger proportion of younger people (especially young males) may have higher rates of certain types of offending. Because people move around in the course of a day, a week, a year and over the course of their lives, morbidity differences in the case of chronic conditions reported by small area are likely to depend on the life histories of the population living in the area. In the case of infectious diseases, on the other hand, morbidity differences will depend on recent movement behaviours in the population (for work or for leisure).

Differences between places will reflect the presence or absence of other attributes – the presence of subway stations or bars in a neighbourhood on criminal assaults and antisocial behaviour. Differences between places may also reflect group or ecological scale effects associated with the population. For example, places with higher levels of social cohesion tend to experience lower rates of crime. Differences in population health profiles may reflect differences in area level environmental characteristics, such as the amount and location of green space in an urban area, levels of air pollution or natural radiation.

To illustrate some of the ideas described above, the reader is referred again to the first three examples in Section 1.3.2 where regression is used to model the number of deaths due to stroke in Sheffield, the number of cases of sexual assault by small areas in Stockholm and the spatial spillover effects in a retail price setting. In each case, spatial thinking, as defined here, has either played a role in model specification (including a spatially-lagged covariate to measure the spatial aspects of NO_x exposure in modelling stroke deaths; including covariates to measure accessibility in order to explain variation in the number of cases of sexual assault) or underpinning the purpose of the analysis (testing for price competition effects between retail sites).

2.3.2 Spatial Interpolation and Small Area Estimation

The previous examples of spatial thinking refer to how this mode of thinking may help us to better *understand or explain* what we observe in a response or outcome variable. There is a further dimension to what we mean by the phrase "spatial thinking" that stems from the observation that data values near together in geographical space tend to be more alike than data values that are further apart – the property of spatial dependency we have already discussed in Chapter 1. This property of spatial data is the basis for a number of statistical methods, including spatial interpolation and small area estimation, which broadly speaking are methods that enable us to tackle specific types of data challenges.

Spatial interpolation methods are used to estimate (or predict) *data values* at unsampled locations on a continuous surface from which a map of spatial variation of the attribute can be constructed. These methods can also be used to estimate data values which are missing in the case of an area partitioned into small spatial units such as census tracts. Spatial interpolation methods exploit the property of spatial dependence for the purpose of estimation (or prediction).

By way of illustration, Figure 2.4 shows six point sites on a continuous surface where sample values of the attribute y have been measured. These observed attribute values are labelled $y_1,...,y_6$. The problem is to estimate the unknown value at s (y_s) and attach a measure of error to the estimate. It is assumed that because values in geographical space are correlated, if the sample points are close enough to s, then as we have discussed in Chapter 1, their observed values carry information about the value at s. The closer a sample point is to s, the more information its observed value is expected to carry about y_s.

Let $\hat{y}_s$ denote the estimated value of y_s. Let $\psi_1,...,\psi_6$ denote six weights assigned to the six sample point values $y_1,...,y_6$. Then, for this sample, let:

$$\hat{y}_s = \sum_{i=1}^{6} \psi_i y_i \qquad \sum_{i=1}^{6} \psi_i = 1.0$$

There are many ways in which the weights can be determined. In geostatistics, the weights, $\psi_1,...,\psi_6$, are estimated using a valid spatial correlation function or variogram that describes the spatial dependency in the set of sample values $y_1,...,y_6$, particularly over short distances. The weights also adjust downwards to lower the contribution of the observed values that come from a spatial cluster of sample points – for example, locations 4, 5 and 6 in Figure 2.4 (Isaaks and Srivastava, 1989). This is referred to as cluster downweighting. This sketch is an example of a spatial interpolation method, known as kriging, grounded in spatial thinking. Spatial interpolation is carried out using the locational information contained in the database. It is also an example of how spatial dependence can be exploited to borrow information, in this case, to estimate an unknown data value.

Suppose we now replace the sample points by a set of polygons that partition the area. That is, the data now refer to attribute values for a set of polygons such as census tracts. Some polygons have data values whilst others, for no special reason (perhaps the data were just lost, or never recorded) do not.[4] Subject to finding a satisfactory set of weights, this is a similar problem to that illustrated in Figure 2.4. So, spatial interpolation methods can be used to obtain estimates of the "missing values" (see for example Haining, 2003, p.164–174). Whilst we will not be covering geostatistics in this book, we shall return in Chapter 17 to look at some recent developments in geostatistics that are of interest when tackling problems where data values are attached to small, irregularly shaped polygons.

Small area estimation methods share common ground with spatial interpolation methods. These methods are employed to improve the precision of each small area parameter

---

[4] To be precise, we say the missing data values are "missing at random" as opposed to missing because they were deleted, perhaps for confidentiality reasons, or suppressed for some reason, perhaps because their values were large or small.

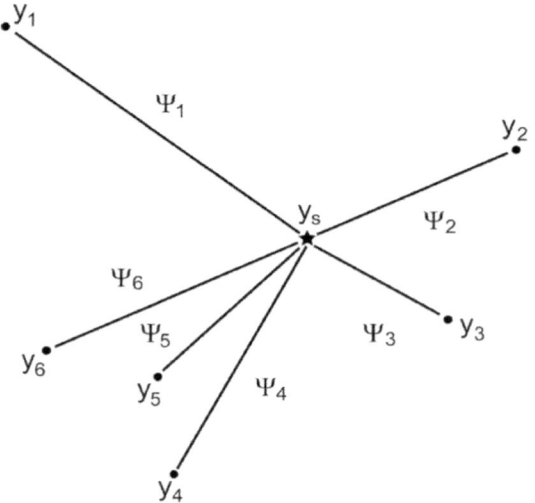

FIGURE 2.4
The weights, $\psi_1, \ldots, \psi_6$, are used to estimate, y_s, the data value at location s. Each of the observed values at locations 1 to 6, $(y_1, \ldots, y_6)$, are weighted by their corresponding $\psi_i \geq 0$, with the weights reducing in size as distance to location s increases.

estimate (e.g. the underlying disease or crime risk in each area) and/or the predicted outcome values for areas that have no data. We gave an example of this in Section 1.3.1 in Chapter 1. If an estimate for an area is produced using only the data points falling within that area, and if that area only has a small number of data points, the resulting estimate is likely to be highly unreliable (e.g. with a very wide or very narrow uncertainty interval). This can arise, for example, if the sampling scheme is not spatially stratified, so some tracts end up with few or no samples in them. In the extreme case where there are no sample values in a census tract, no parameter estimate can be obtained. It can also arise if the event is rare (e.g. a rare disease) and/or the at-risk population in each area is small. The underlying problem here is data sparsity (Section 1.2.3).

As we have seen in Sections 1.4.2 and 1.4.3.3, Bayesian hierarchical models tackle the above problem by embedding the spatial thinking that parameters are *spatially dependent*. Such a modelling assumption allows Bayesian hierarchical models to utilise ("borrow") the data from an area's neighbours in order to improve the estimation of its own parameter. In Chapter 8 we shall see the spatial models used for this borrowing process. In addition to improving the precision of small area parameter estimates, Bayesian hierarchical modelling provides a robust methodology for fitting models to small area data with the goal of explaining spatial variation (see Section 1.4.3).

2.4 Thinking Spatially and Temporally

2.4.1 Explaining Space-Time Variation

By the term "spatial-temporal process" we mean a process where spatial-temporal relationships, between people or places, for example, enter explicitly into the process. All such processes involve some form of interaction between entities (people, businesses, political and other institutions) in both space *and* time. The interaction process itself may be highly complex, and those which involve interactions between individuals might not be observed (or even observable), although with increased usage of location-aware technologies such as mobile phones, the possibilities for studying interaction effects and identifying network linkages amongst individuals, using big data algorithms, are increasing. The term "spatial-temporal interaction process" describes a broad class of dynamic flows from one location to another along a network and where the flows might comprise physical items (goods), people, information or ideas. The network's character might be physical and fixed like a transport link or trade route. Or its character may be various, intangible and fluid like connections between people as they communicate with each other or move around in the course of their lives. These interactions in space and time are likely to underlie the variation in attribute values that we observe in our data.[5] In Part III we look closely at how to model spatial-temporal data and how to handle the presence of what is termed "space-time interaction" in a dataset. We briefly describe some generic examples of space-time processes that may underlie this data property, which appears as clusters of cases in particular places at particular times.

(i) *Diffusion processes.* These are processes where the attribute of interest, such as information, a rumour, a disease or a new technology, is adopted, or taken up, or otherwise spreads through a population of individuals (e.g. firms or people) who for the purposes of analysis or because of data limitations are assumed to have fixed locations. By "fixed" we mean, for example, the map of adoption identifies the residential location of the adopting (and non-adopting) individuals, or the location of the adopting (and non-adopting) businesses. At any point in time it can be specified which individuals have the attribute and which do not. The spread of an infectious disease is modelled as a diffusion process where infected individuals mix with individuals from the susceptible population who, dependent on the infection risk, may then become infected. The spatial distribution of the individuals that are susceptible and the extent of mixing can have important implications for the spread of the disease (speed, spatial extent, size of the epidemic). In the case of an outbreak of rioting or civil unrest, new participants may be drawn in by copycat behaviour through inter-personal contact or first-hand experience. The spread of cultural traits may be the product of a diffusion process or it may be the product of a dispersal process, a type of process that we now turn to.

(ii) *Dispersal processes.* In contrast to a diffusion process, a dispersal process involves the re-location or spatial distribution of individuals, including people, animals and plants. The geographical spread of cultural or physical traits may be the consequence of individuals dispersing over a region carrying those traits with them and bringing them to the newly settled area. Other examples include: seed dispersal from parent plants where the spatial distribution of the attribute (the number of new plants in an area) depends on the scale and intensity of the dispersal

---

[5] Depending on the scale of the study, the spatial locations of the interactions might differ significantly from the spatial locations of where attributes are recorded. In the case of chronic diseases with long latency periods, individuals exposed to an environmental risk at one time in their lives might have moved to a new residential location before they are diagnosed with the condition. Similar problems arise when mapping cases of an infectious disease, although on a shorter time scale. Other examples of where the critical interactions may have happened at a different location from where an outcome is measured include voting preferences, crime victimization, social attitudes and life expectancy.

process as well as environmental attributes, such as the capacity of an area to support new arrivals; the spread of a human population over a region as a result of a migration process, with people populating a previously uninhabited or sparsely populated area; or enforced migration of an ethnically defined sub-population due to war or persecution. Various "push" and "pull" factors may determine the migrant distribution, including the distance travelled in the course of migration. Factors also include the existence of any "intervening opportunities", as well as any resistance encountered in newly settled areas.

(iii) *Interaction processes involving exchange and transfer between entities with a fixed location.* Urban and regional economies are bound together by processes of commodity exchange and income transfer. Income earned in one place may be spent elsewhere, thereby binding together the economic fortunes of different places. These transfers will be reinforced by intra-area and inter-area economic multiplier processes. Such spillover effects may be reflected in the spatial structuring of social and economic attributes such as per capita income or economic growth rates.

(iv) *Interaction processes involving action and reaction between agents.* In a market economy, the determination of prices at a set of retail outlets may involve a process of action and reaction amongst retailers. For example, a price adjustment by one retailer may, depending on the effect that price change has on levels of demand at other retail sites, lead to a price response by other retailers. The adjustment may depend on spatial proximity, with retailers nearby responding more rapidly to meet the price change than those further away. A pattern of prices may develop across the set of retail sites that reflect these competitive interactions as retailers seek to defend their market share and profitability.

(v) *Interaction processes involving spatial convergence.* The necessary condition for the occurrence of an offence, following the arguments of Routine Activity Theory, is the convergence in space and time of a motivated offender, a victim or "suitable target" (a person or a house) in the absence of a capable guardian. The underlying interactions involve potentially three groups of individuals, since the definition of a capable guardian includes the presence of other people who by their mere presence discourage the criminal act.[6] Different crimes, since they may involve different individual elements (offender, victims, capable guardians), have different space-time interaction signatures. In some cases we observe intense concentrations of crime events known as crime "hotspots".

Many different types of spatial-temporal processes have been studied by social, political and economic scientists over many years, and the interested reader can use the previously cited literature in this chapter together with the following small sample to explore this literature and other related areas of research. Archaeology (Hodder and Orton, 1976); anthropology (Alexander and Maschner, 1996; Relethford, 2008); behavioural ecology (Fayet et al., 2014); criminology (Felson and Cohen, 1980; Skogan, 1990; Bursik and Grasmick, 1993; Braga and Weisburd, 2010); economics (Besley and Case, 1995; Besley et al., 1997; Brown and Rork, 2005; Florkowski and Sarmiento, 2005; Garrett and Marsh, 2002; Haining, 1984; Kalnins, 2003; Patton and McErlean, 2003); epidemiology and public health (Bailey, 1967; Gorman et al., 2005); veterinary epidemiology (Ward and Carpenter, 2000); geography (Haining, 1987;

---

[6] This idea underlies forms of urban design that encourage "eyes on the street" as a way of reducing urban crime. See for example Jacobs (1961).

Sheppard et al., 1992); political science (Bailey and Rom, 2004; Cho, 2003; Gleditsch and Ward, 2006; Salehyan and Gleditsch, 2006; Williams and Whitten, 2014); sociology (Land et al., 1991; Morenoff, 2003; Papachristos et al., 2013; Tolnay, 1995; Tolnay et al., 1996); and transport research (Eckley and Curtin, 2013).

Some processes evolve over time to a final state, and it is this we may be principally interested in (for example, as in the case of the outcome of a political election by constituency). In other situations, we are interested in the way processes develop or evolve over time as well as the final state (as in the case of the spread of an infectious disease or the emergence of cultural or physical traits in a population). However, many of the outcomes that are observed and analysed in the social, economic and health sciences are the result of processes that are in a constant state of flux involving a wide range of different types of interactions between people and places and with policy shifts originating in both the public and private sectors, adding further complexity.

2.4.2 Estimating Parameters for Spatial-Temporal Units

When data are available in time as well as in space, the presence of positive temporal autocorrelation in the observed values provides a further basis for information sharing when estimating the parameters associated with each space-time unit. For example, when estimating (say) the disease risk for each small area, the numbers of disease cases observed at time point t may give us some information about the risk of disease at time point $t - 1$ or $t + 1$. We will see eventually that "temporal neighbours" of a time point t include time points that are both before *and* after t (assuming both are in the study period), that is $t - 1$ *and* $t + 1$. This definition of temporal neighbours raises an important point: instead of our usual thinking that time is "unidirectional" (time goes forward), for *information borrowing*, it is "bi-directional", meaning that we borrow information both from the "past" (e.g. $t - 1$) and from the "future" (e.g. $t + 1$). If we are willing to assume that the levels of disease risk between year $t - 1$ and year t are similar and, likewise, the levels of disease risk between year t and year $t + 1$ are similar, then when estimating the disease risk in year t, why not borrow information from both $t - 1$ and $t + 1$? We shall discuss these ideas further in Chapter 12 (Section 12.3). When modelling spatial-temporal data, the idea of spatial information sharing (Section 2.3.2) and the idea of temporal information sharing can be combined, so that when estimating the level of disease risk at one space-time unit we can borrow information from its *temporal neighbours*, its *spatial neighbours*, as well as its *temporal neighbours' spatial neighbours*. Chapter 15 provides more detail.

2.5 Concluding Remarks

"Spatial thinking" means recognizing that where entities (objects and events) and their attributes are located in space, and when they occur in time, are not just a set of co-ordinates but, rather, are important elements of the information contained in a spatial database. This information can be used both to address certain types of questions (small area estimation, spatial interpolation), to study patterns, associations and processes in space and time and to make forecasts and support policy-making. In the next chapter we consider the nature of the data in a spatial or spatial-temporal database and the implications for the application of statistical methodology.

2.6 Exercises

Exercise 2.1. With reference to any area of study with which you are familiar, describe a research question where interest focuses on:

- Spatial variation in an attribute relative to some significant fixed point(s) or area(s) (2.3.1(i)).
- How spatial gradients in attribute values between adjacent areas might influence outcomes (2.3.1(ii)).
- How the macro-spatial configuration of a society might impact on certain outcomes (2.3.1(iii)).

Give your reasons for thinking these associations might exist.

Exercise 2.2. The dataset, `malaria_individual_cases.csv`, contains individual level data on new cases of malaria reported across 139 villages in Karnataka, India. Each case is georeferenced to one of the villages and is allocated to one of three time windows: August–September 2014, October–November 2014 and December 2014–January 2015. Use a software of your choice to aggregate the individual cases to a case count (i.e. number of malaria cases) for each village-time-window. In Section 13.3, we will consider methods to visualise the resulting set of spatial-temporal count data.

Exercise 2.3. Give examples, from your field of study, where (a) spatial interpolation (applied to point or area data) and (b) small area estimation techniques would be useful.

Exercise 2.4. From your field of study, give an example of the following:

- A diffusion process (2.4.1(i))
- A dispersal process (2.4.1(ii))
- An interaction process (between places) involving some physical exchange or transfer (2.4.1(iii))
- An interaction process involving action and reaction amongst agents at different locations or in different areas (2.4.1(iv))
- An interaction process involving the convergence in geographical space of different actors or agents (2.4.1(v))

Exercise 2.5. Suppose the values observed at the six locations in Figure 2.4 are $y_1 = 20$, $y_2 = 2$, $y_3 = 4$, $y_4 = 10$, $y_5 = 6$ and $y_6 = 5$. The coordinates of these six locations are (–10, 10), (6, 4), (3, 3), (–3, –6), (–4, –4) and (–8,–4) respectively. Explore the impacts on y_s, the estimated value for location s, which is located at (0, 0) from the following three different ways to define the weights, $\psi_1, \dots, \psi_6$:

(a) Equal weights assigned to all six points.
(b) Equal weights to the nearest four neighbours defined by Euclidean distance.
(c) Weights are defined using an exponential decay function, $\exp(-0.5 \cdot d_i)$ where d_i is the Euclidean distance between point i and point s. Note that one needs to modify the values from the exponential decay function to satisfy the condition that $\sum_{i=1}^{6} \psi_i = 1$. What is the implication if $\sum_{i=1}^{6} \psi_i \neq 1$?

Compute y_s using the above three sets of weights and explain why the resulting values for y_s are different.

Exercise 2.6. From Exercise 2.5, consider again defining the weights using the exponential decay function, which, as we shall see in Section 4.3, can be written more generally as $\exp(-\lambda \cdot d_{ij})$, where d_{ij} is a distance measure between two points i and j and λ is a positive-valued parameter. In Exercise 2.5, λ was fixed at 0.5, but in practice the value of λ is unknown but is typically estimated using data. Would it be reasonable to just calculate a value for y_s by fixing this unknown parameter to a single value (say 0.5, a point estimate from some software)? Or, would it be more reasonable to take both the point estimate (0.5) and the associated interval estimate (often reported as a 95% confidence interval between, say, 0.2 and 0.8) into the estimation of y_s? Explain why the latter is more appropriate than the former. How can we incorporate the uncertainty associated with the point estimate (as quantified by the interval estimate) into the estimation of y_s? (Hint: see Section 5.2 in terms of how to express uncertainty of a parameter using a suitable probability distribution; once a plausible probability distribution is formed based on the point and interval estimates for λ, we can perform a Monte Carlo simulation to examine the impact on y_s – see Exercise 8.7 in Chapter 8 and Exercise 16.8 in Chapter 16 for the idea of Monte Carlo simulation.)

Exercise 2.7. Review, descriptively and comparatively, the principles behind information borrowing in space and in space-time.

Exercise 2.8. With reference to some specific scientific research question within your field of study, sketch how you think the functionality of a GIS might support your work (see Appendix).

Appendix: Geographic Information Systems

Scientific reasoning depends on the existence of good quality data; clearly formulated hypotheses that can be empirically tested; a rigorous methodology which allows data-based conclusions to be drawn about these hypotheses; and an enabling technology that allows practical and precise implementation of the methodology. The scientific approach to spatial thinking requires for its development (Goodchild and Haining, 2004):

- The existence of good-quality, spatially-referenced, geocoded data
- Spatial questions that are clearly formulated as spatial hypotheses
- A rigorous methodology to enable us to compare observed data with the expectations derived from our spatial hypotheses
- An enabling technology that allows us to handle practically and precisely spatially referenced data and implement the methodology

A Geographic Information System (GIS) is an important part of the enabling technology for spatial thinking. There are several definitions of what is meant by a GIS, each highlighting the different ways in which a GIS can support working with spatially referenced data:

- GIS as a toolbox: "a powerful set of tools for collecting, storing, retrieving at will, transforming and displaying spatial data from the real world for a particular set of purposes" (Burrough and McDonnell, 2000, p.11)

- GIS as a database management system: "any… computer based set of procedures used to store and manipulate geographically referenced data" (Burrough and McDonnell, 2000, p.11)
- GIS as a spatial decision support system: "a decision support system involving the integration of spatially referenced data in a problem solving environment" (Cowen, 1988)

A GIS is part of a "constellation of computer technologies for capturing and processing geographic data" which includes the global positioning system (GPS), satellite data collection systems and digital scanners (Cromley and McLafferty, 2012, p.16). Physically, a GIS comprises computer hardware (e.g. networked computers; large format scanners and printers), software (e.g. that enables data input and output, storage and database management) and an organizational structure that includes skilled people who are able to operate the system. But growth of the internet means users of GIS do not need to have their own in-house, physical GIS. Distributed GIS services have made it possible for many more users to take advantage of GIS capability. Books by Maheswaren and Craglia (2004) and Cromley and McLafferty (2012) discuss how GIS has contributed to research and practice in public health.

GIS functionality falls into the following broad categories (Cromley and McLafferty, 2012, p.30):

(i) Measurement (e.g. distance, length, perimeter, area, centroid, buffering, volume, shape)

(ii) Topology (e.g. adjacency, polygon overlay, point and line in polygon, dissolve, merge)

(iii) Network and location analysis (e.g. connectivity, shortest path, routing, service areas, location-allocation modelling, accessibility modelling)

(iv) Surface analysis (e.g. slope, aspect, filtering, line of sight, viewsheds, contours, watersheds)

(v) Statistical analysis (e.g. spatial sampling, spatial weights, exploratory data analysis, nearest neighbour analysis, spatial autocorrelation, spatial interpolation, geostatistics, trend surface analysis)

At the time of writing there is an entry in Wikipedia (last edited 28th December 2018), "List of geographic information systems software", which gives both open-source and commercial GIS products. Notable in the former category are GRASS GIS and QGIS. Notable in the latter category are ERDAS IMAGINE, ESRI (which includes ArcMap, ArcGIS, ArcIMS), Intergraph and Mapinfo. However, there are many more – see https://en.wikipedia.org/wiki/List_of_geographic_information_systems_software.

3

The Nature of Spatial and Spatial-Temporal Attribute Data

3.1 Introduction

In Chapter 1 we identified four important and commonly encountered challenges that need to be addressed when modelling spatial data: the presence of spatial dependence, various forms of spatial heterogeneity, different forms of uncertainty (data, model and parameter uncertainty) and data sparsity. In this chapter we consider those challenges, and in particular the presence of spatial dependence and spatial heterogeneity, in the context of the way that data are collected and, ultimately, the relationship between the spatial database and the geographical "reality" that the database represents. We shall find that the structure of spatial dependence and some aspects of spatial heterogeneity can be a consequence of how the database is constructed. We shall explore these issues in Section 3.3 and in the process identify other properties of spatial data that are important in spatial data modelling. However, before that, in Section 3.2, we look at the types of data collection processes that generate spatial and spatial-temporal data. Understanding the form of the "experiment" (if any) that lies behind the data we analyse has implications for how we can interpret the results from any statistical analysis.

3.2 Data Collection Processes in the Social Sciences

The randomized controlled experiment is widely considered the gold standard in experimental social and public health science for the purpose of establishing causality and avoiding bias (DiNardo, 2008). Whilst one group of subjects or individuals receives a treatment, another comparable group does not. The subjects within each of the two groups are referred to as cases (the treated) and controls (the non-treated), respectively. Three criteria define the randomized controlled experiment (Dunning, 2012, p.15):

(1) Comparison is made between responses by those subjects that have received one or more treatments against those in one or more control groups.

(2) The assignment of subjects to groups (case or control) is done at random.

(3) The manipulation of the treatment (the intervention) is under the control of the researcher.

TABLE 3.1

Principal Types of Data Collection Processes (* denotes an "observational" study)

Type of Study	Treatment: Cases and Controls	Randomization (to Reduce Confounding Effects)	Treatment Controlled by Researcher	Replication
Randomized controlled experiment	Yes	Yes	Yes	Yes
Natural experiment*	Yes (due to some "natural" process)	Yes (perhaps after data manipulation)	No	No
Quasi-experimental study*	Yes	No	No	No
Non-experimental study*	No	No	No	No

The use of randomized controlled experiments in the social sciences can be found in the fields of education (Bradshaw et al., 2009), criminology (Farrington and Welsh, 2005) and international development studies (Karlan and Zinman, 2010), whilst Melia (2014) provides a critical review of their value in transport research. In many areas of social science research, however, the researcher controlling which subject does and which subject does not receive any particular treatment (criterion (3) above) is not always possible. Randomization may be objected to on ethical or moral grounds or it may not be practical. As a consequence, it may be particularly difficult to control for confounding effects. DiNardo (2008) observes that if neither control nor randomization are possible, then the term "experiment" is not appropriate to describe the study.

Table 3.1 provides a summary of the four main types of study and what distinguishes an observational study from a randomized controlled experiment. All the examples in this book fall into the category of observational studies. We now discuss some examples of observational data collection processes involving spatial and spatial-temporal data and discuss implications.

3.2.1 Natural Experiments

A *natural experiment* is where a clearly defined sub-population is exposed to some treatment so that any comparison between the exposed and the non-exposed groups may be reasonably attributed to the exposure. However, which subjects are exposed and which are not is the outcome of some natural or other process, and the allocation is not under the control of the researcher, thereby failing to meet criterion (3). However, if the assignment of subjects to the two groups can be said to be "at random" or "as-if at random", so that the data collection or subsequent data manipulation process means that the data can be argued to meet criterion (2), then the researcher can be said to be analysing data from a natural experiment (Dunning, 2012, p.16). Studies based on natural experiments, where criteria (1) and (2) for a randomized controlled experiment are satisfied, but not (3), have increased greatly since the 1990s, especially in political science and economics. We now see some examples.

Changes in transport networks or boundary changes bring about changes in spatial relationships and can be used to explore the role of spatial interactions on outcomes (see for example Redding and Sturm, 2008; Machin and Salvanes, 2010). Bursztyn and Cantoni (2016) test whether exposure to foreign media affects the economic behaviour of households living under totalitarian regimes. They do this by exploiting a natural experiment, which resulted in differential access to West German television broadcasting in East Germany (the German Democratic Republic or GDR) during 30 years under a communist regime, and using a regression discontinuity design. This design depends on being able to identify a known threshold to distinguish between those who are cases and those who are controls. Some parts of the GDR, notably areas in the northeast and southeast, were not reached by West German broadcasts either because of distance or the effects of terrain, so that signal strength and hence reception was below threshold. Those areas not reached by a sufficiently strong signal are defined as the control areas, and the areas that were reached by a sufficiently strong signal are defined as the treatment areas. The interested reader is referred to the original paper for discussion of the threshold and how it was determined. Households are allocated either to the treatment group (cases) or to the control group (controls), as determined by their location (see Figure 3.1). Bursztyn and Cantoni study household consumption behaviour immediately following reunification of Germany, focusing on products with a high intensity of advertisements on West German television in the 10 years prior to reunification.

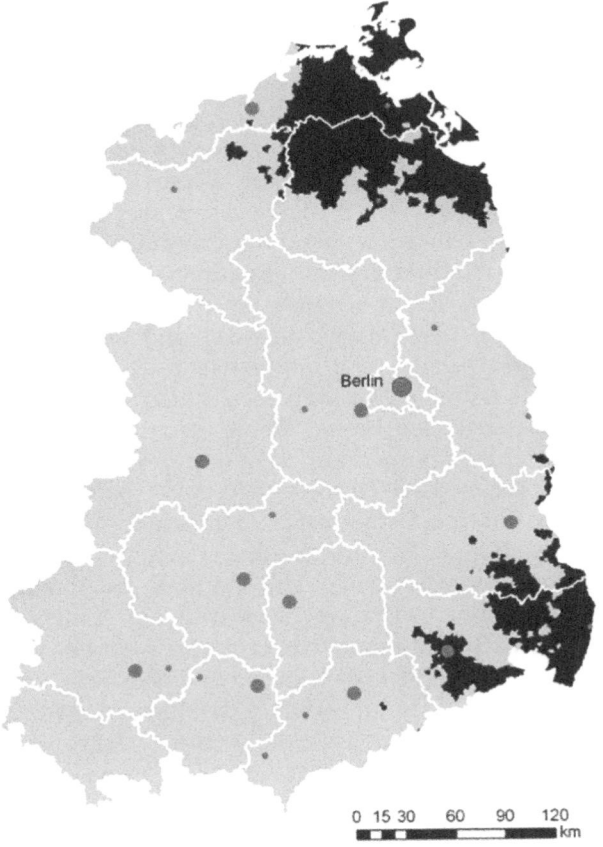

FIGURE 3.1
The former East Germany. The dark shaded areas define the areas within which households in the control group were selected, the grey shaded areas are the treatment areas. The dots denote major cities and the white lines denote district borders (Bezirke) (after Bursztyn and Cantoni, 2016).

The comparability principle underpins the selection of cases and controls because it is the group not receiving the treatment that provides the benchmark against which to test for treatment effects. Controls are selected so as to reduce or eliminate selection bias, confounding bias and information bias (see Wacholder et al., 1991a and 1991b). This is achieved through ensuring that cases and controls are drawn from the same "base experience"; that confounding effects are minimized either through the selection process or subsequent analysis; that comparable and satisfactory levels of accuracy can be achieved in measuring properties of the two groups. Bursztyn and Cantoni (2016) argue that the households in the treatment and control areas in their study are comparable for the following reasons: (a) there was no selective spatial sorting across the region (i.e. individuals more susceptible to Western advertising did not move into areas with better reception); (b) people in the GDR who could watch West German television did watch it; (c) treatment effects were driven by demand differences and not supply differences. The reader is referred to the original paper for more detail. One of the models they fitted regressed the natural log (to the base e) of expenditure by household i on good j on the following covariates: whether the household i was located in a "treated" area and hence exposed to West German television (1) or not (0); the average number of minutes of television advertising per day devoted to good j in 1980–89; a vector of household demographic and economic characteristics; the minimum driving distance from the municipality where household i lives to the West German border (excluding West Berlin). The inclusion of the distance covariate controls for the confounding effect of distance from West Germany. It is possible that some Western products are less likely to be available in more remote areas (where the control areas are located). It is the parameter associated with the exposure covariate (whether the household was located in the treated area multiplied by the average number of minutes of television advertising per day) that is of interest. It is expected to be positive and significant if individuals exposed to Western television spend more heavily on the advertised goods. The interested reader is referred to their paper for the findings of the study.

Using geographical boundaries in a regression discontinuity design presents a number of distinctive challenges, as discussed in Keele and Titiunik (2015). They study the effects of political advertisements on voter turnouts in US presidential campaigns in the West Windsor-Plainsboro school district, an area which straddles two media markets – the Philadelphia media market (southwest), where the volume of political advertisements was high, and the New York City media market (northeast), where the volume of advertisements was zero (see Figure 3.2).[1] They identify three special features that distinguish a geographic regression discontinuity design from a non-geographic two-dimensional regression discontinuity design. The first feature is *compound treatments*: there might be multiple treatments that may affect outcomes, and these treatments may originate from political or administrative sources that operate within the same geographical boundary frameworks, making it difficult to separate their effects. They recommend focusing attention on areas where overlap does not occur. The second feature is *naive distance*: that simple distance to the border does not take into account distance *along* the border when measuring distance between members of the case and control subgroups. The third feature is *spatial treatment effects*: since we are dealing with a two-dimensional boundary between cases and controls, it is theoretically possible that treatment effects could be spatially heterogeneous along the boundary.

Keele and Titiunik (2015) also point out that spatial sorting of individuals near geographical boundaries (in terms of income, for example) may undermine the "as-if at random"

---

[1] They also discuss the conditions for identifiability in a geographic regression discontinuity (GRD) design.

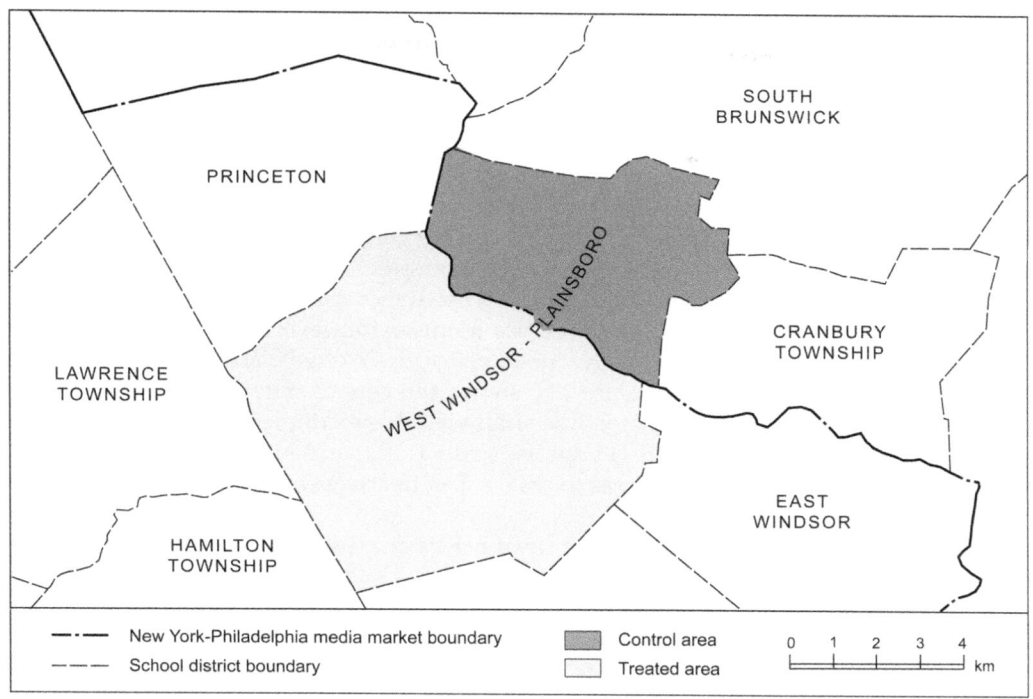

FIGURE 3.2
Boundary between New York City and Philadelphia media markets (Keele and Titiunik, 2015).

requirement. In assessing media impacts, spatial spillovers across boundaries linked to people's friendship networks may undermine the allocation of individuals to either the treatment or the control group. Treatment spillovers have been identified in the case of neighbourhood crime prevention measures, making it unclear exactly what aspects of the treatment might be significant. The authors note that "considerable substantive knowledge is needed to credibly exploit geographic boundaries as regression discontinuity designs" (Keele and Titiunik, 2015 p.128). Natural experiments have been exploited in several contexts in the social and economic sciences. We refer the interested reader to Rosenzweig and Wolpin (2000), DiNardo and Lee (2004) and Hearst et al (1986) for more examples.

3.2.2 Quasi-Experiments

Quasi-experiments are distinguished from natural experiments by virtue of a non-random assignment of the treatment so that in this case neither criteria (2) nor (3) for a randomized controlled experiment are met, only criterion (1). Most of quasi-experimental studies are of the "before and after" kind, typically involving a "difference in differences" approach in seeking to establish a relationship. That means computing the difference in the treated group before and after; the difference in the control group before and after; and then testing the difference between these two differences. Campbell and Ross (1970) studied the introduction of speeding restrictions by law in Connecticut and asked whether the introduction of the law had had the effect of cutting traffic fatalities. The crackdown on speeding came at a time when fatalities were unusually high. This in turn prompts the concern that any reduction detected by the study would just be a regression to the mean effect, not a consequence of the new law. DiNardo and Lemieux (2001) made a comparison of the change between 1980 and 1990 in the fraction of 18– to 21-year-olds smoking marijuana between those US states who already had in place in 1980 a minimum alcohol drinking age of 21 and those which, as a result of federal pressure, raised the legal age of alcohol consumption from 18 to 21 after 1980. Li et al. (2013) used a quasi-experimental design to test the impact of a no-cold calling (NCC) intervention by Cambridgeshire Constabulary on burglary rates. The scheme, designed to deter doorstep cold calling, which can be a smoke-screen for criminal activity, including distraction burglary and selling substandard goods or services at inflated prices, was targeted at specific streets in Peterborough, a large urban area in England. In their analysis, Li et al. (2013) constructed five different types of control groups based on similar local characteristics (deprivation levels; burglary rates; spatial proximity) to address regression to the mean concerns. We have already seen this application (see Section 1.3.2.4), and Figure 1.12 shows the census output areas within which the NCC "treated areas" were located. We shall visit this example in detail in Chapter 12, then again in Chapter 16. On a larger spatial scale, Kolko and Neumark (2010) compared policy treated areas with control areas to assess the effectiveness of enterprise zones in job creation in California.

A variant on the quasi-experimental design was used in the evaluation of the UK's reassurance policing programme (Tuffin et al., 2006). The programme was trialled across sixteen wards in England, and the results from six treated wards were compared with the results from six matched untreated (control) wards. Matching was undertaken in terms of: population density, ethnicity and proportion in managerial employment, but similarity in crime levels was also taken into account. Control wards were selected from the same police force area and where no other interventions were in place that might introduce an element of treatment confounding. Matched pairs analysis compares before and after attitudes of residents in each treated area with before and after attitudes in the corresponding matched control area (Tuffin et al., 2006). One should bear in mind that establishing causal connections based on any form of quasi-experimental design is problematic because of possible unmeasured confounding effects as well as selection or allocation bias.

3.2.3 Non-Experimental Observational Studies

The third and final type of observational study is the *non-experimental observational study*, where data are collected on a population in order to study associations and relationships typically between an outcome of interest and a set of independent variables. In this case, there has been no explicit or well-defined exposure of some of the units to one or more treatments with a consequent partitioning of the units into those that have been treated and those that have not (the control group). In such studies, none of the three criteria for a randomized controlled experiment are met. In the absence of a well specified experiment underlying the data (in addition to the presence of potential data shortcomings), modelling of non-experimental spatial data often needs to confront the problem of spatially correlated omitted variables, as evidenced by spatially correlated regression residuals. Gibbons and Overman (2012) point to the role that carefully chosen instrumental variables can play in addressing this challenge, as well as the challenge arising from correlation between an explanatory variable and the error term which occurs when fitting some spatial econometric models (see Chapter 10). For some examples cited by Gibbons and Overman in the analysis of spatial problems, see Cutler and Glaeser (1997), Hoxby (2000), Michaels (2008) and Luechinger (2009).

The arguments above might lead the reader to the conclusion that observational studies are of little value – particularly non-experimental observational studies that dominate the examples in this book. That would be a mistake. Observational studies analyse real world data, providing direct evidence about the world, evaluating real policies and yielding insights into associations and correlations that otherwise would not be possible. Subject matter theory becomes essential in order to interpret findings calling for a continuous and subtle interplay between data, methodology and theory in the advancement of any discipline that depends on observational data. The results of observational studies may in some circumstances suggest hypotheses that can be followed up by other, sometimes more expensive, methods. For example, in spatial epidemiology, the effects of different environmental exposures on health are usually difficult to assess except by observational methods, but results so obtained can be followed up by more detailed fieldwork (e.g. see Section 1.3.2.1 in the case of assessing health impacts due to exposure to atmospheric pollutants).

To turn the tables, the problem with allowing only experimental data is that it would place excessive constraints on what social scientists could study and the amount of evidence available for policymakers to draw upon (Melia, 2014). The social world does not always divide neatly into those "treated" and those who are "not treated". Using appropriate statistical methodology, analysis of observational data allows us to test for the effects of varying many independent variables simultaneously on the dependent (outcome) variable, providing that the independent variables are not strongly correlated. However, social scientists and policymakers who draw on research findings from observational data must be cautious in what they claim to know as a result of their analysis. The lack of real control means that research findings always have weak internal validity: a weak understanding of cause and effect – at best an understanding of *associations* between variables. It also means that results have weak external validity: considerable uncertainty about the extent to which research findings can be generalized.

What of the social scientific study of relationships and processes using spatial and spatial-temporal datasets? Places observed at a point in time or over a period of time do not allow control (of the levels of different variables to see what the effect is of varying the level of another variable), randomization (in selecting which places to treat and which to use as controls) or replication. In controlled experiments, replication may take a number of forms, including: replication of the same experiment (with the same levels of the different variables) to help assess measurement error; replication of the complete experiment (involving all treatment combinations) in order to confirm findings.[2] In trying to assess the effects of a geographically targeted policy, it may be difficult to introduce randomized control (indeed, it may not be ethical), raising concerns about allocation bias when evaluating that policy. Through social and other forms of interaction, geographical areas have permeable boundaries, so that it may not be clear the extent to which treatment effects carry over into other supposedly non-treated areas through spillover or spread effects. As Melia (2014) remarks in the context of geographical transport studies, the researcher cannot ensure that any intervention is the only relevant change which affects the treated group differently from the control group. Nor are "treatments" all necessarily identical, as noted by DiNardo and Lee (2004) in their study into the effects of unionisation on firm survival rates, because not all unions operate in the same way. Geographically targeted policies may be implemented in different places by different groups of people, which could

---

[2] Replication is distinct from the reproducibility of an experiment. Whilst replication involves the repetition of treatment combinations within an experiment to assess variability, reproducibility refers to the ability of one or more independent researchers to duplicate any given experiment.

affect its delivery. In social sciences, treatments are "much more fragile objects" when compared to those in the physical or medical sciences (DiNardo, 2008).

There are further problems when working with aggregated spatial data in establishing the external validity of findings by reproducing a given experiment. When analysing relationships between attributes recorded for geographical areas, results are conditional on the size of areas (the "scale effect"), and at any given scale, the choice of area boundaries (the "partition effect").

As remarked by Holt et al. (1996, p.181): "if a statistic is calculated for two different sets of areal units which cover the same population, or sample, a difference will usually be observed even though the same basic data have been used in both analyses. This difference is cited as evidence of the modifiable areal units problem." The values of ecological variables will be dependent on how the spatial units have been defined. Some of the differences in findings between two studies in two different areas could be linked to differences in the way the areal units have been constructed relative to the underlying (continuous space) geography.

There is an element of uniqueness to any study where data are collected for particular places during particular periods of time – uniqueness in terms of each area's specific combination of place-attributes as well as in terms of their spatial relationships. Gibbons and Overman (2012) ask if this rules out any form of causal analysis on spatial data using experimental methods because of the impossibility of finding valid control areas against which to compare the treated areas. They argue that it is not necessary to find control and treated places that are identical in every way and that it is sufficient to ensure the control and the treated places are "comparable along the dimensions that influence the outcome being studied" (p.186). However, the problem of "place-uniqueness" may be more severe when seeking to establish external validity by reproducing studies in other places or at other times. Replication and reproducibility may present special challenges for those working with spatial and spatial-temporal data.

There will be other complications to consider: unconsidered or unobserved confounding variables influencing outcomes; a high degree of correlation amongst the independent variables that are observed (often referred to as the problem of multicollinearity);[3] sometimes there is only small variation in the levels of at least some of the independent variables, making it difficult to be sure about their effects on outcomes. We may be interested in the relationship between the outcome and some combinations of values of some of the independent variables. But complications can arise if such combinations of values of the independent variables are not present in the observed data. This can arise in non-experimental, observational settings, because we are not able to control the values of independent variables – typically we have to "take what we are given". The social scientist working with observational spatial and spatial-temporal data encounters considerable uncertainty in terms of what is being observed and what conclusions can be drawn. There seems little doubt that the challenges presented in demonstrating the internal and external validity of research findings are particularly acute when analysing spatial and spatial-temporal data.

---

[3] The problem of multicollinearity arises when using the regression model to test relationships between a response (outcome) variable and a set of independent variables (covariates). Material deprivation, population turnover and social cohesion measure different facets of a community. But such covariates tend to be correlated (areas with high population turnover often have high levels of material deprivation and low levels of social cohesion), presenting the problem of multicollinearity. It will be possible to make inference on their joint effect, but care needs to be taken in interpreting the effect estimates of the individual correlated covariates (see Hill and Paynich, 2014, Chapter 2).

3.3 Spatial and Spatial-Temporal Data: Properties

3.3.1 From Geographical Reality to the Spatial Database

To better understand and classify the properties of spatial and spatial-temporal data, we need to understand the relationship between the multivariate space-time data cube introduced in Section 2.2 (see Figure 2.2) and the "real world" from which such data have been taken. In order to carry out either spatial or spatial-temporal data analysis, the complexity of the real world must be captured in a finite number of bits of data. As described in the geographical information science literature, this is achieved through the processes of conceptualization and representation (Goodchild, 1989; Longley et al., 2001). We shall focus here on these two processes as they apply to capturing objects in space and time whilst acknowledging that the measurement of attributes is also preceded by similar processes of abstraction.[4]

For the purposes of storing geographical data in a database, there are two conceptualisations or ways of viewing the real world – the field and the object views. The field view conceptualizes space as covered by surfaces with the attribute varying continuously or nearly continuously across space. The object view, by contrast, conceptualizes space as essentially empty but populated by well-defined, indivisible objects that form points, lines or polygons on the earth's surface. Whilst the field view is appropriate for many types of environmental and physical attributes on the earth's surface (rainfall, temperature, land cover, air quality), the object view is appropriate for many types of social, economic and political attributes of populations (unemployment; election results; cases of disease).[5]

These two views constitute models of the real world. In order to reduce a field to a finite number of bits of data, either point sampling is undertaken or the surface is partitioned. The partitioning may take the form of a grid of regular polygons referred to as a raster. The grid is laid down independently of the underlying field and its surface variation. This is called an intrinsic partition. Another type of partitioning is to parcel up an area into subareas which have broadly uniform characteristics. This is called a non-intrinsic partition and is analogous to the process of constructing geographical regions. The resulting partition will necessarily be irregular. How closely any field is captured (represented) will depend on the density of the sample points or the size of the raster in relation to surface variability. There is a substantial literature, both theoretical and empirical, on the efficiencies of different sampling designs – for example, random, systematic and stratified random sampling – given the nature of surface variation (see for example Cressie, 1991; Ripley, 1981; and Haining, 2003). This process of representing the surface in terms of a finite number of discrete bits of data involves a loss of information about surface variability at distances shorter than the sampling interval of points or at scales smaller than the size of the raster.

The same loss of detail arises selecting a representation based on the object view. A city comprises many households, but in a national census, household information is aggregated to census tracts for confidentiality reasons. However, the recent trend in economically developed countries has been to improve the reporting of small area

---

[4] Familiar social science attributes such as "material deprivation" and "social cohesion" require a decision on what such terms mean (their conceptualisation) and then how they should be captured or represented so that they can be measured.

[5] Some attributes can be conceptualised by both views – population distributions at a given point or interval in time are sometimes conceptualised as density surfaces (fields), sometimes as points (objects). If we add transport costs to the price a consumer pays for a good, then prices for commodities sold at discrete locations produce a price field.

census data whilst trying to maintain continuity with the past for comparative purposes. Since the 2001 UK census (1981 in Scotland), Census Output Areas (COAs) have been used as the smallest spatial units for which household data are reported. COAs were built using clusters of adjacent unit postcodes with the aim of having broadly similar populations.[6] In England and Wales, 80% of the over 175,000 COAs have between 250 and 320 residents (approximately 125 households). They were designed (except in Scotland) to be as socially homogeneous as possible (unlike the previously used enumeration district (ED) framework) in terms of dwelling type and household tenure. COAs are in turn aggregated to what are termed lower and above them middle super output areas (LSOAs and MSOAs), which have average resident populations of around 1,500 and 7,500, respectively. There are almost 35,000 LSOAs and around 7,200 MSOAs in England and Wales.[7] In the US, the smallest geographic census unit is the block, reporting data collected for all households. There were over 11 million blocks in the US and Puerto Rico in 2010 (almost half of which had 0 population in 2010). These are typically grouped into block groups (containing between 600 and 3,000 people), which are in turn grouped into census tracts which have approximately 4,000 residents on average. Tract boundaries generally follow visible and identifiable features, but in some US states they follow legal boundaries to allow linkage with government defined spatial units. They have been defined to be relatively homogeneous in terms of population characteristics, economic data and living conditions.[8] In Europe the move to a common framework for the European Union led to the creation of Nomenclature of Territorial Units for Statistics (NUTS) regions at three levels, although even at the finest scale (NUTS 3), these regions are still by comparison with the above census geography very coarse (for example, there are just 174 NUTS 3 units in the UK).

Any aggregation involves a loss of information about variability at scales below that of the aggregation. There may be a further loss of information in capturing the polygon if, for example, a representative point, such as a population weighted centroid, is used to locate the polygon in the database. Choice of graph structure for the purpose of defining the spatial relationships between such points or polygons may involve another type of information loss. This will be discussed further in Chapter 4.

The conceptualization of a geographic space either as a field or as an object is largely dictated by the attribute. The representation, the process by which information about the real world is made finite using geometric constructs (points and polygons and in some cases lines), involves making choices. These choices include the size and configuration of polygons, the location and density of sample points. For further discussion of these issues see Martin (1998; 1999).

Figure 3.3 provides a schematic summary of the two processes involved in moving from the complexity of geographical or spatial "reality" to the model for the spatial database. Figure 3.4 provides a schematic summary through to the spatial database itself populated with the collected data. This diagram distinguishes between model quality and data quality. As noted, these processes apply not only to how space is captured in the spatial database but also how time is captured too. The principles are the same. Time may be conceptualised by the equivalent object and field views. The field view of time conceptualizes events in time as occurring continuously (e.g. stock market prices), whilst the object view

---

[6] Unit postcodes include 15 households (approximately 40 residents) on average, although in practice the number can vary from 1 to 100 households.
[7] For details see the ONS website at http://www.ons.gov.uk/ons/guide-method/geography/beginner-s-guide/census/output-area--oas-/index.html.
[8] For details see https://www.census.gov/geo/reference/gtc/gtc_ct.html.

From geographical reality to a model for the Spatial Data Matrix (SDM)

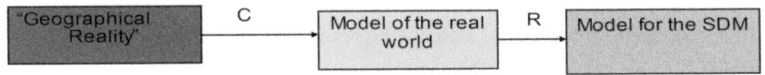

C: Conceptualisation (object and field views)

R: Representation (points, lines and polygons)

FIGURE 3.3
From geographical reality to a model for the spatial data matrix (SDM).

From geographical reality to the spatial data matrix

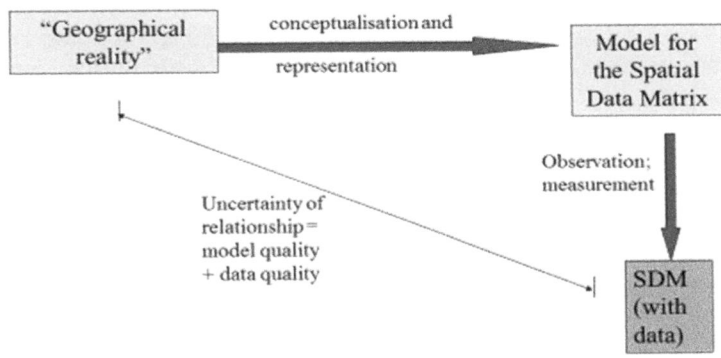

FIGURE 3.4
From geographical reality to the spatial data matrix (with data).

of time imagines time as empty but populated by discrete events which might be treated as points in time (e.g. the price of a house at the point of sale) or as intervals of time (e.g. the period from the start to the end of a war or period of civil unrest; the accumulated number of car thefts in an area during some window of time). Time introduces fewer complications, but as with space, transformational processes are implemented for the purpose of creating finite quantities of such data.

The purpose of the preceding section has been to distinguish between the various stages in the construction of the spatial and spatial-temporal database. We are now in a position to classify data properties into three groups: fundamental properties; properties arising from representational choices; and properties due to measurement processes. Fundamental properties are properties that are inherent in the nature of things as they exist in space and time. Representational properties are properties that are *acquired* by the data because of the choices made in making the data available (by the data provider). Measurement properties are properties acquired as a result of the processes involved in collecting data values. We now discuss each of these in turn.

3.3.2 Fundamental Properties of Spatial and Spatial-Temporal Data

3.3.2.1 *Spatial and Temporal Dependence*

Lack of independence, at least over short distances or periods of time, is a fundamental property of attributes in space and time. Objects and events in space and time display continuity, and the world would be a strange place if it were not so. Imagine (if you can!) a world where a field variable changes suddenly and randomly as we move from one geographical location to another immediately nearby, or as we travel forwards or look backwards in very small increments of time. If we know the level (value) of an attribute at one point in time, or at one location, we can usually make an informed estimate of the same attribute's value soon after or just before, or at another nearby location. *Information about the level of an attribute at a given location can be exploited to estimate the level of the same attribute at nearby locations*, as we discussed in Section 2.3.2. The same remark applies in the case of events in time for the purpose of estimating the value of the same attribute at nearby points in time, either before or after, as discussed in Section 2.4.2. But as distance between locations increases or as separation in time lengthens, then the association between levels of the attribute decreases, as does the usefulness of the information they carry for the purpose of estimation.[9]

Awareness of this property has a long pedigree in both environmental and social sciences. Fisher (1935) wrote in the context of agricultural field trials: "patches in close proximity are commonly more alike, as judged by yield of crops, than those which are further apart" (p.66). Webster (1985) remarked that such dependence could be traced back to soil properties that were a consequence of processes operating at many different spatial scales. Neprash (1934) made an early observation about the similarity of socio-economic characteristics across adjacent areas, and Stephan (1934) remarked: "data of geographic units are tied together … we know by virtue of their very social character, persons, groups and their characteristics are inter-related and not independent" (p.165). Tobler (1971) invoked the idea that nearby attribute values were more similar than those further apart when setting up his simulation model of Detroit. This observation later became known to geographers and geographical information scientists as Tobler's First Law of Geography. Whilst it may not meet the standards required to be called a "law", it has certainly proved to be a useful aphorism in many areas of spatial analysis, and as we shall see shares common ground with the choice of informative prior distributions used in most forms of spatial and spatial-temporal data modelling for the purpose of capturing spatially structured random effects (see Sections 1.4.2.3 and 5.6.2).[10]

These properties of spatial and spatial-temporal dependence follow partly from the space and space-time occupying nature of objects (a city, a labour market area), but they also arise from the outcomes of various types of processes that unfold over space and time. We refer the reader to Sections 2.3 and 2.4 for discussion of different types of "spatial-temporal processes".

From the perspective of studying process or making forecasts, there are important differences between the dependence found in spatial as opposed to temporal data. Dependence in time typically flows forward with events at time $t-1$ influencing events at time t. In time, the present is dependent on the past, but not vice versa – time has a natural

---

[9] Whilst temporal dependence may in general decay with increasing separation, there can be cyclical (as in the case of business cycles involving several years, purchasing behaviour during the course of a week or month) as well as seasonal components to the variation.

[10] For continuous data, Banerjee et al. (2004, p.39) referred to a more formal expression of the same idea as the "first law of geostatistics".

uni-directional flow, and this influences how we model *processes* or make *forecasts* using time series data (this contrasts with when we are seeking to improve the statistical precision of estimates from a database: data values at *both $t + 1$ and $t - 1$* can be used to inform estimates at time t; see Section 2.4.2). Space, by contrast, usually has no such ordering, and the two-dimensional nature of space means that dependency structures can vary not only with distance but also direction.[11] Processes unfolding in space and time may involve a hybrid of these two structures, recognizing that attributes at location i at time t may be a function of the same attribute *at* location i at times prior to t and the same attribute at locations *near to i* at times prior to t.

Modelling dependency is a key element in the specification of spatial and spatial-temporal models. How such dependency is captured, how the presence of spatial and spatial-temporal dependency is exploited, how information is shared, depends on the scientific problem and whether the goal is to improve statistical precision, explain variation in space and time or to investigate space-time dynamics (see Section 1.3).

3.3.2.2 Spatial and Temporal Heterogeneity

Homogeneity is present in the distribution of attributes in space and time when all subsets in space and time have the same statistical properties. By contrast, heterogeneity is present when statistical properties are not the same across all subsets. Heterogeneity can exist in any part of a statistical distribution, but in practice we tend to be most concerned with heterogeneity in the mean, in the variance and in the autocorrelation structure and with heterogeneity in relationships between attributes.[12]

When a process in time is interrupted, there can be a mean shift (to a higher or lower mean value). This shift may be permanent or temporary, and the onset of the shift can be abrupt or gradual. Examples include an interest rate rise or the start of a labour dispute and the consequences for economic output; the onset of civil unrest arising from a contested election result; a fall in vehicle-related crime rates (car theft or breaking into a car) in an area following the introduction of a targeted police intervention. If there is a geography to the causal factor (labour disputes are restricted to certain regions; civil unrest is more pronounced in some regions than others due to differences in the strength of political affiliations; police interventions vary by police force areas), then there may be different mean shifts in different geographical areas.

Spatial heterogeneity in the mean of an attribute may arise because of different *levels* of the causal variables in different sub-regions of the study area. In Section 1.2.2 we discussed the case of household burglary rates in Cambridgeshire. The principal urban areas of Cambridgeshire, Cambridge and Peterborough, experience higher rates of household burglary in comparison to the rural parts of the county. It is possible that whilst the same risk factors affect outcomes in the same way across Cambridgeshire, the levels of the risk factors in these two places are greater than they are elsewhere in the county, thus accounting for the observed differences. We also considered the possibility that the *nature* of the causal relationship might be different between rural and urban areas of Cambridgeshire – *heterogeneity in the relationship* between burglary rates and the set of causal variables. This latter form of heterogeneity is observed in hedonic house pricing models. The values

---

[11] Such dependency structures are called anisotropic, with dependency different in the east-west direction compared to the north-south direction. An infectious disease spread by environmental factors (wind, for example) may display anisotropy reflecting the direction of the prevailing wind.

[12] In stochastic process modelling, homogeneity corresponds to the property of stationarity; heterogeneity corresponds to the property of non-stationarity.

attached to particular housing features (a garage, a basement, a garden and so on) can vary over time but can also vary geographically due to climate differences, cultural preferences or market differences (for example, between affluent and poorer neighbourhoods).

Spatial heterogeneity can also be present in *variance-covariance* properties. The presence of isolated clusters of sub areas on a map with high or low levels of an attribute are indicative of heterogeneity. Such clusters could be associated with the levels of independent variables, as in the preceding example. But when clusters occur because of highly localized processes that may additionally involve contagious or spillover processes within and between areas, then another option is to view such a map property as heterogeneity in the variance-covariance part of the distribution, with the variances and covariances depending on location. In the case of a locally contagious process, the existence of early cases of adoption or infection (say) in one area triggers further cases in the same and adjacent areas. There may be nothing special, in terms of adoption or risk factors, to explain why the process started where it did, but having done so, a sequence of events is triggered which leads to a cluster of cases much greater than the numbers of cases observed elsewhere.

3.3.3 Properties Induced by Representational Choices

The extent to which dependency and heterogeneity are present in a database depends not only on the fundamental nature of events in space and time but also on the chosen representation. If sample points on a continuous surface are far enough apart, then attribute values may be uncorrelated.[13] Assessing the effect of aggregation on the spatial dependence properties of object data can be quite complicated to unravel or to generalise about. For example, there is likely to be some level of spatial dependence in household income levels at the scale of individual households in a city. But average household income levels for small spatial groupings of households may display a higher level of spatial dependence because aggregation introduces local smoothing. Larger spatial groupings introduce further smoothing, although at some scale the data may start to capture large-scale differentiation in the socio-economic structure of the city. The level of spatial dependence may then become weaker compared to other, finer, scales.

Much social science data, especially national census data, are made available in the form of aggregate counts, rates or proportions by small area. Area-specific proportions or rates may be calculated with reference to the resident population or number of households, which acts as the denominator or with reference to some other property such as physical size. Such aggregation is often for confidentiality reasons. Irrespective of whether the data refer to a continuously varying phenomenon (field data) or aggregations of individuals such as households (object data), spatial aggregation has the effect of smoothing spatial variation. Within-area variation (spatial heterogeneity) is lost the larger the areas used and when the range of values is compressed. Figure 3.5 shows the effects of reporting 2011 unemployment rates for Cambridgeshire (excluding Peterborough) by census output area (of which there are 1937) and by lower and middle super output areas (of which there are 375 and 76, respectively).[14] Note, for example, the loss of spatial variation in the northern part of the map when data are presented by middle super output areas compared with those presented by census output areas. Note also that at the census output area scale the

---

[13] In geostatistics, the term "range" is used to define the distance such that any pair of sample points separated by that distance or greater are uncorrelated.

[14] The number of census tracts vary from census to census as a result of population increase or decrease. In the case of Cambridgeshire, the number of census output areas in 2001 was 1820.

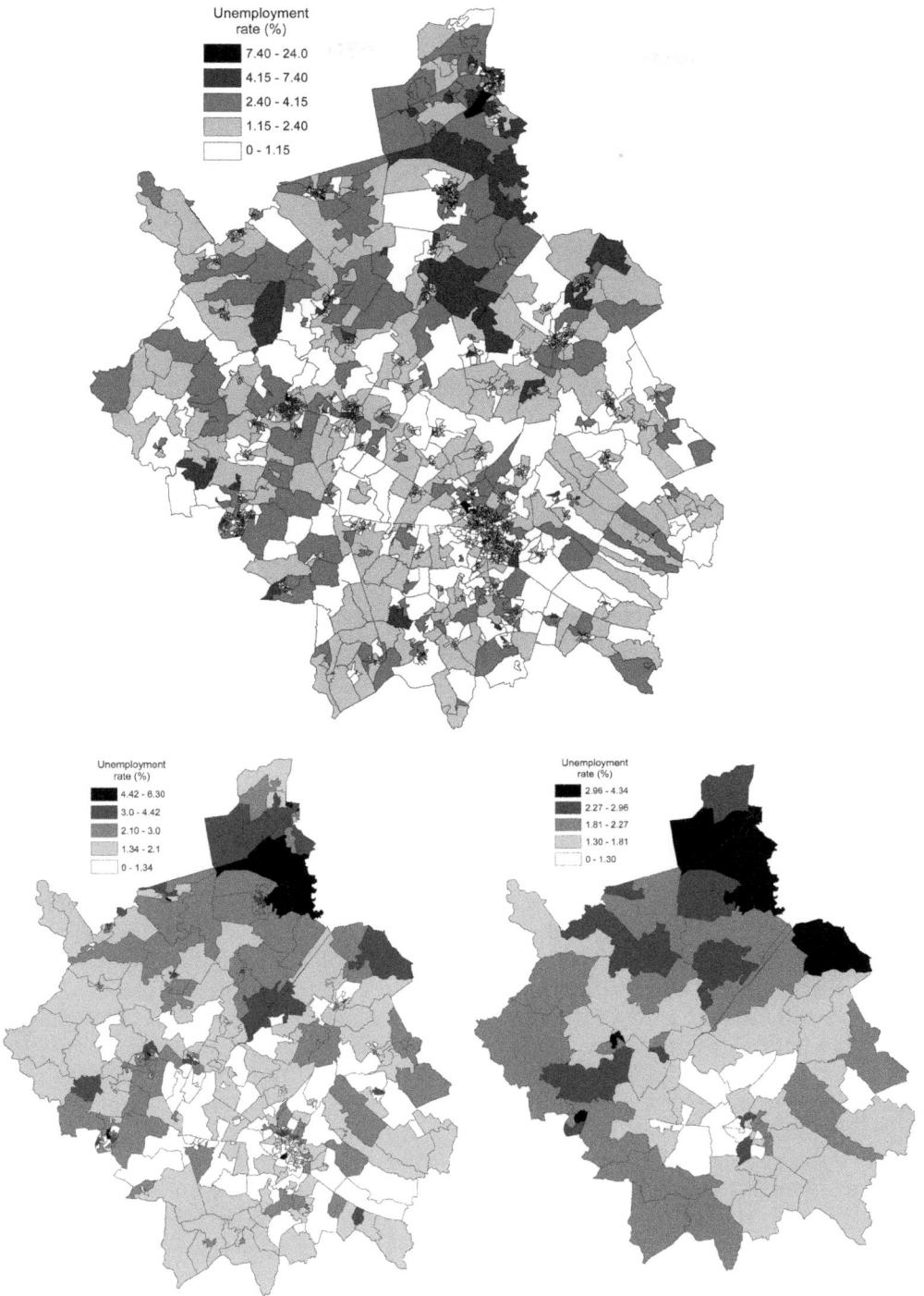

FIGURE 3.5
Unemployment rate for Cambridgeshire (excluding Peterborough) in 2001 per 100 of working-age residents. Data shown by: Top: Census Output Areas (COAs); Bottom left: lower super output areas (LSOAs); Bottom right: middle super output areas (MSOAs).

maximum unemployment rate is 24.0, whilst at the middle super output area scale the maximum unemployment rate is 4.34 (see also Exercise 3.5).

This loss of detail is usually less pronounced in the case of a non-intrinsic partition compared with an intrinsic partition because of the way the two types of partitions are constructed – census tracts are constructed to be relatively homogeneous in terms of certain socio-economic characteristics and thus, at least with respect to those characteristics, less within-area detail is lost. In the case of an intrinsic partition (such as pixelated remotely sensed data), the degree of smoothing and hence loss of detail on land use type will depend on the size of the pixels. Where attributes representing means are recorded by areal aggregates such as census tracts (for example average household income), the analyst may have no information on variability around the mean figure. If an ecological attribute is recorded, such as area-level material deprivation or social cohesion, the calculation will be conditional on the chosen partition in terms of the size (scale) of the individual spatial units *and* where the boundaries are drawn between them.[15] This is an aspect of the modifiable areal units problem, discussed in the previous section, which in certain circumstances can obscure important features in the geography of the data. Figure 3.6 shows the effect of presenting a dot map of cases of a disease (based on Snow's map of cholera cases in London in 1854) in terms of three different areal aggregations, the result of which is to suggest the presence of two, one or no clusters of cases depending on where the areal unit boundaries are drawn. This map also illustrates the importance of the size (or extent) of a study area in the detection of clusters, for were the study area to be larger with no significant numbers of cases of cholera elsewhere, then the concentration of cases would be even more evident. Chou (1991) studies the effect of map resolution on spatial autocorrelation statistics.

It is often noted that data collected for a group of small geographical areas tends to have a large variance-to-mean ratio, indeed were spatial resolution to become finer still, the ratio would tend to increase still further. The reason is that spatial data obtained at a fine geographical scale give us more fine-grained or detailed information about the map pattern, and it is this that leads to an increased variance (see Figure 3.5). By contrast, aggregation reduces between-area variability – as we note when applying map smoothing (see Section 6.2.2). There are several possible causes of a large variance-to-mean ratio. It may be due to the effect of localized cluster processes, such as that described above in the case of a point source of pollution or where the disease is infectious. Some areas have large counts (where the disease starts and spreads locally); others have much smaller or even zero counts (where the disease has not yet occurred in the population). There may be heterogeneity in the mean, which can also produce a large variance-to-mean ratio. Modelling data with a large variance-to-mean ratio does not present a problem when using the normal model (as mean and variance are represented by two different parameters in a normal distribution) but does have consequences when modelling count data using the Poisson model. The Poisson model is often proposed as the probability model for small area count data, and the reader will encounter it many times in this book. But the Poisson distribution assumes the mean and variance are the same. When the variance in a dataset is much greater than the mean and we propose to use the Poisson distribution to model the data, we say that the data shows greater variability than predicted by the Poisson model, or equivalently that the data display *overdispersion*. A large variance-to-mean ratio signals overdispersion, and any model for the data needs to accommodate this property (see

---

[15] An ecological or contextual attribute is an attribute that is defined, and only has meaning, at the group (e.g. area) level.

FIGURE 3.6
Effect of areal aggregation on the detection of clusters of cases. Snow's map of cholera cases in London in 1854. Adapted from Monmonier, 1996. In (B), the area(s) in black denotes a cluster of cases.

Section 6.4 for more detail). Including covariates in the Poisson model to form a Poisson regression model may account for any apparent overdispersion when the cause is heterogeneity in the mean. But often in practice, including covariates may not be sufficient, and if that is the case, other solutions such as including random effects in the case of a hierarchical model are needed. Various statistics have been proposed to test for overdispersion in order to anticipate the problem (Dean and Lundy, 2016). We shall discuss such tests in Chapter 6.[16]

Aggregation has implications for the types of inference that are possible. Ecological inference is the process whereby aggregated data are used to infer individual level relationships – usually because individual level data are not available for confidentiality reasons. It can also arise because individual level data would not be reliable. For example, it may be difficult and/or too expensive to get reliable measures of exposure to environmental or community risk factors at the individual level. Ecological (or aggregation) bias is the difference between the estimates of relationships obtained using grouped data and those estimates obtained using individual level data. The analyst who uses estimates obtained

---

[16] Overdispersion may arise when using other models for count data – such as the binomial, where although the mean and the variance are not required to be equal, they are still related (one is a function of the other).

from aggregated data to infer individual level relationships without specifying the conditions under which the estimates are reasonable is guilty of committing an example of the ecological fallacy. The converse, using individual level estimates uncritically to infer group level relationships, thereby ignoring the possibility of group level or contextual effects, is referred to as the atomistic (or individualistic) fallacy. Whilst we shall return to the latter in Chapter 17, in Section 9.3 we shall see how aggregation bias can be reduced in a study of the relationship, at the small area level, between stroke mortality and exposure to air pollution (NO_x).

It might appear from the foregoing that it is better to have smaller areal aggregates rather than larger ones, and on grounds of within-area homogeneity that is generally true. But spatial precision may not be matched by statistical precision. Data errors or small random fluctuations in the numbers of events (e.g. numbers of household burglaries) can have a big effect on rate calculations (number of events per household in an area) when populations are small, and the event of interest is uncommon. Consider the case where the unknown burglary risk of a Census Output Area (COA) is estimated using the burglary rate: the observed number of burglary cases divided by the number of houses in that area. A COA typically has around 125 houses, and household burglary is an uncommon event – typically around two to three cases per year at the COA level based on the Peterborough burglary data that we will see in Chapter 12. The burglary rate of a COA with two cases of burglary will rise or fall by 50% with the addition or subtraction of a single case (due to, say, recording error). This has big impacts for cluster detection. Consider two COAs i and j and both have the same number of houses. If three and two cases of household burglary are reported in COA i and COA j respectively during one year, the burglary rate in i is 50% higher than that in j. However, given the small populations, the uncertainty associated with these two rate estimates is likely to be large, as we shall soon see. More generally, rates and ratios calculated for areas with large populations tend to be more robust (their statistics less affected by small perturbations in the data that may or may not be due to any errors in the database) than those for areas with small populations, but larger areas conceal higher levels of internal (intra-area) heterogeneity, as we have noted (e.g. in Figure 3.5). As a result, a small cluster in a large spatial unit will be diluted and hence may not to be detected.

Suppose the areas within the study region vary in population size. Consider the effect of this on the burglary rate taking the form of the number of burglary cases divided by the number of houses at risk. When analysing crime counts, this population adjustment is undertaken to control for differences in the size of the at-risk population. If areas vary in terms of the size of the at-risk population, then sampling error, in addition to the burglary rate, will not be the same across the map. To see this, let O_i and n_i denote the observed count of burglaries and the number of at-risk houses in area i respectively. So, the burglary rate (per house) in i is O_i / n_i. If burglaries are uncommon events, we can assume O_i is Poisson distributed with parameter $n_i r_i$ (i.e. $O_i \sim Poisson(n_i r_i)$), where r_i is the true but unknown burglary risk in area i. Under these conditions, O_i / n_i is the maximum likelihood estimator of r_i. It follows that the sampling variance (Var) of O_i / n_i is

$$Var\left(\frac{O_i}{n_i}\right) = \left(\frac{1}{n_i}\right)^2 Var(O_i) = \left(\frac{1}{n_i}\right)^2 (n_i r_i) = \left(\frac{1}{n_i}\right)^2 O_i.$$

So the sampling error (SE), the square root of the sampling variance, is $O_i^{1/2} / n_i$. Since the denominator is the at-risk population (n_i), the sampling error decreases as the size of the at-risk population increases. Conversely, areas with smaller n_i will have larger variances associated with their rate estimates, and so are more likely to give rise to extreme values. This is referred to as the *small number problem* (see Exercise 3.7). Furthermore, any apparent inter-area spatial variation in rates could be an artefact of whatever spatial structure may exist in the distribution of the population sizes (for an example see Gelman and Price, 1999). Note also that the number of observed cases, O_i, also affects the variance of the burglary rate, but we will defer the discussion till Chapter 12 (Section 12.3.3.2.1).

The standardised ratio, O_i / E_i, where E_i is the expected count for area i, is the maximum likelihood estimator for the unknown area-specific relative risk of events of interest (e.g. the occurrence of a household burglary or a case of a particular disease) in area i.[17] If the interest lies in determining the statistical significance of counts (whether there are more or fewer cases than would be expected in an area), the small number problem again plays a role. Under the Poisson assumption and assuming E_i is known (a fixed number), the standard error is given by $O_i^{1/2} / E_i$ and is inversely related to the expected number of cases. When studying uncommon events over small populations, the expected number of cases tends to be small, resulting in standardised ratios that are (extremely) large or small. Although such extreme ratio values are likely to be far away from 1, they are typically not considered to be statistically significant because of the large standard errors they are associated with. As Mollie (1996) points out, standardised ratios that are declared to be significantly different from 1 tend to be associated with areas with large populations. Increases in *spatial precision*, the growing availability of small area data linked to the growing use of GPS and related technologies, gives rise to the statistical challenge of how to ensure the *statistical precision* of quantities estimated using such data. It should also be noted that the small area problem is not solely associated with the modelling of small area count data. The same challenge can arise when each area's population size is large but a limited amount of data within each area are available to make inference about the area's characteristics (Sections 1.2.3 and 1.3.1). In that case, the area is not small in terms of its population size, but the number of available data points is small. But we will defer the discussion of that situation till Chapters 7 and 8 in Part II of the book.

If small area populations, and hence sampling errors, vary significantly across the map, it may be advisable, at least in a first pass through the data, to reflect this in the way maps and scatterplots are presented. Thus, attention is drawn to those small area estimates that have smaller sampling error and are therefore the most reliable in the set. When exploring associations, it can be informative to display the scatterplot in such a way that the size of each point on the scatterplot reflects the reliability of the underlying data value. Figure 3.7 shows an example of two scatterplots of incidence rates of malaria against rainfall levels for an area in Karnataka, South India. We shall visit this application in Chapters 9 and 16.

A cartogram is a useful tool for visualizing variability in map form. The physical size of each spatial unit is adjusted in order to capture some feature of the data, such as the population size of each area. Cartograms are particularly useful when physical size and population size do not correlate, so that in order to see all the census tracts on a regional map (from small high-density tracts near city centres to large lower-density rural tracts),

---

[17] An expected count for an area within a study region is the number of cases to be expected if the events of interest (e.g. household burglaries) occurred at random across the entire study region, thus depending only on the distribution of the population at risk (see Section 6.2.3.1). In spatial epidemiology, the calculation of the expected counts typically accounts for the effects of age and gender (and possibly other factors, such as deprivation) on the disease outcome (see Section 9.3.2).

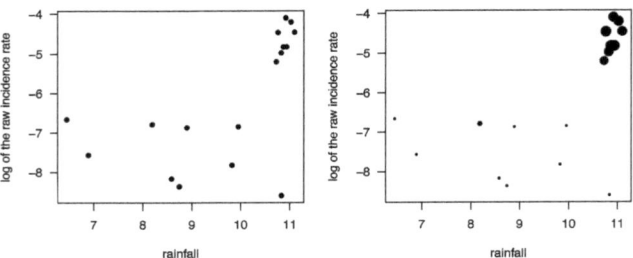

FIGURE 3.7
A scatterplot of the log incidence rates of malaria against rainfall levels across 139 villages: (a) a standard scatterplot and (b) the size of each point is inversely related to the variance (hence uncertainty) of the incidence rate.

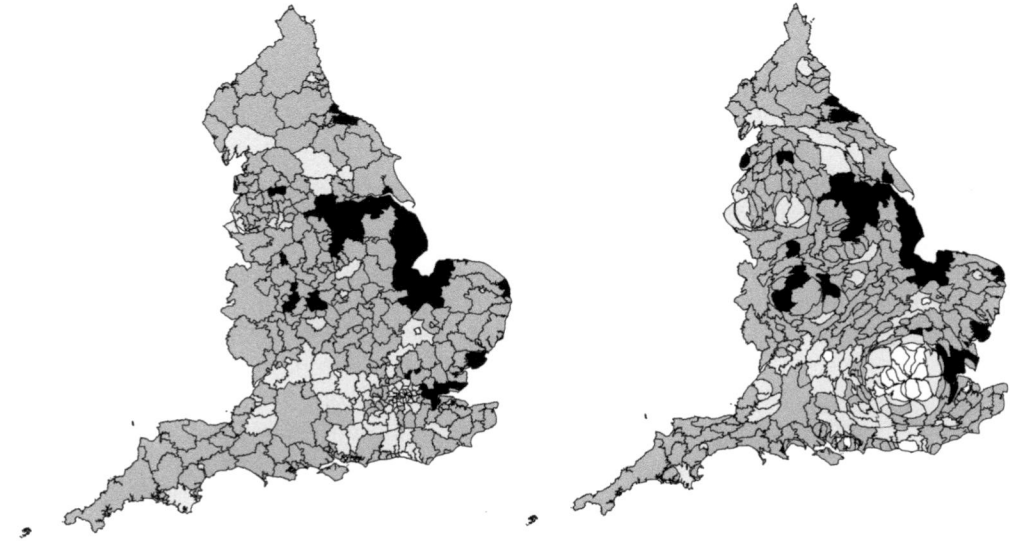

FIGURE 3.8
A conventional map (a) and a cartogram (b) showing the district-level leave vote proportions from the EU referendum in England. In the cartogram, the polygons (districts) have been resized according to their population sizes from the 2011 UK census. Darker shading denotes higher proportion of leave votes.

the physical size of each tract is made a function of its population size. Figure 3.8 compares a conventional map with a cartogram of voting outcomes in the EU referendum by local authority districts – one of the applications in Chapter 11. See Exercise 6.16 in Chapter 6 for instructions on how to draw a cartogram in R.

Particularly when analysing small area data (as opposed to larger, more familiar national level data), the use of cartograms may call for creativity in displaying such maps in ways that enable the viewer to retain their understanding of where places are. Linked windows software (such as GeoDa) provides one way of achieving this. The viewer can click on an area on the cartogram and see where it lies on the "real" map. When analysing national level data, cartograms can be constructed that retain the shape and the location of individual nation states. See for example https://gcaptain.com/who-has-the-oil-a-map-of-world-oil-reserves/, which shows the distribution of world oil reserves, by country, in 2010 (Konrad, 2010). Whether drawing a cartogram or a conventional map, different methods for choosing class intervals also affect the visual impression of spatial variability on a choropleth map. The reader has the opportunity to explore this issue in one of the exercises at the end of this chapter (Exercise 3.12). We shall have more to say about mapping and data visualisation in Chapter 6 for spatial data and Chapter 13 for spatial-temporal data.

3.3.4 Properties Induced by Measurement Processes

We assess the quality of spatial and spatial temporal data against four characteristics: accuracy, completeness, consistency and resolution (Guptill and Morrison, 1995; Batini and Scannapieco, 2006).

Accuracy. A common assumption in error analysis is that attribute errors are independent. But small errors in geocoding may mean events are assigned from their correct area to a nearby area, resulting in over-counts and under-counts in adjacent areas. Incorrect geocoding or in the timing of events may mean that localized clusters of cases are not detected. The time when a crime event such as a household burglary occurred may be subject to considerable uncertainty (Ratcliffe and McCullagh, 1998; Ratcliffe, 2002). The timing of the event may be known only to within some window of time (a weekend, a week), and there may be a geography to this uncertainty. Neighbourhoods where residents take a close interest in the welfare of others may be better informed about timing than other neighbourhoods.

In the case of remotely sensed data, the values recorded for any pixel are not in one-to-one relationship with an area of land on the ground because of the effects of light scattering, so that errors in adjacent pixels are positively spatially correlated.[18]

Error effects may propagate through a database when different arithmetic or cartographic operations are carried out on the data. Source errors are compounded and transformed by such operations (see Haining, 2003, p.124–127 for a short overview and references).

Completeness. Data incompleteness refers to situations where there are missing data values (perhaps due to non-recording as a result of industrial action or where a national survey has been carried out, but because there has been no spatial stratification, some areas have no data points) or events are under-reported. Missing cases in a point pattern of events may mean local clusters of cases are not detected. Not all cases of domestic burglary are reported. Whilst break-ins to suburban residences are usually well reported by the homeowner for insurance purposes, those taking place in inner city neighbourhoods may not be for fear of reprisals against the householder or because there is no household insurance to provide motivation. In the latter case, the incompleteness has a geography which will be problematic for any form of spatial analysis. The 1991 UK census was thought to have undercounted the population by up to 2% because of fears that the data would be used to enforce the new local "poll tax". Inner city areas show higher levels of undercounting in national censuses than suburban areas, where populations are easier to track.

The absence of data in even a small number of cells of a spatial or spatial-temporal database can potentially have serious consequences for analysis if whole rows and/or columns of the database have to be discarded because they are incomplete. Effective imputation methods are needed to estimate the missing values and to attach measures of uncertainty to those estimates. Examples of coping with missing values will be encountered in Chapter 9.

Consistency. Data inconsistency arises where two different databases provide mutually contradictory data – for example, cases of unemployment in neighbourhoods with no resident population or average house price values for neighbourhoods where there has

---

[18] The point spread function quantifies this overlap. The error is analogous to a weak spatial filter passed over the surface. See Forster (1980).

been extensive re-development. Usually inconsistency arises because the data on different attributes were not collected at the same time. National censuses are collected every 10 years, but some areas of cities, especially inner-city areas, may experience rapid population mobility and re-development, which may render counts out of date. As data collection methods improve, with more frequent and better updating, these sorts of inconsistency might become less of an issue.

Resolution. Resolution refers to the spatial and temporal scales at which data are collected. This introduces some of the issues raised in earlier sections of this chapter about how data values are smoothed as a function of the scale of aggregation. Data for a project may draw on databases that use different areal frameworks, some of which may be at different scales of resolution. If areal interpolation methods are used to fuse the data onto a common framework, this introduces further levels and patterns of error and uncertainty into the database (Cockings et al., 1997).

3.4 Concluding Remarks

In Chapter 1 we drew attention to four key challenges in the statistical analysis of spatial data. In this chapter we have seen how some of these challenges are tied to fundamental properties of spatial data (spatial dependence and spatial heterogeneity), but that the form the challenges take is also affected by how spatial data are made available (representation choices) as well as measurement processes. The fact that much of the spatial data we are concerned with in this book is small area aggregate data raises a number of incidental challenges in addition to the challenge of data sparsity. They are not incidental in the sense of not being important, but incidental in the sense of adding further challenges to the previous four that will impact on model specification (e.g. overdispersion, large numbers of zero counts and the small number problem) and how model results should be interpreted or understood. Various forms of uncertainty run deeply through spatial data modelling, not least because, as we have seen, the source of our data often falls far short of the ideal "experiment".

3.5 Exercises

Exercise 3.1. With reference to one or more areas of study with which you are familiar, give an example of each of the four principal types of data collection processes (see Table 3.1). What are the main challenges that need to be faced when trying to implement each of these data collection processes?

Exercise 3.2. With reference to any area of study with which you are familiar, sketch the transformational processes that lead from "reality" to a spatial-temporal database with quantitative data. Consider the question in terms of how geographical space and time are captured, as well as how attributes are represented in the database (the "space-time data cube" of Chapter 2).

Exercise 3.3. Review your understanding of the following terms in this chapter, giving examples from your area of study:

a. Modifiable areal units problem

b. Ecological (or aggregation) bias

c. Ecological fallacy

d. Atomistic fallacy

e. Small number problem

f. Confounding bias (see also Chapter 1, footnote 6)

g. Selection bias and information bias

Exercise 3.4. What is meant by the terms "spatial precision" and "statistical precision"? Why is there a tendency for each to move in opposite directions (i.e. as spatial precision increases, statistical precision decreases, and vice versa)?

Exercise 3.5. In the UK, data on many social and economic variables are reported by the national census at three spatial scales: census output area level and at the lower and the middle super output area levels (see, for example, Figure 3.5). For a variable and a region (city) of your choice, explore how different levels of spatial aggregation affect the properties of the data. Census data at various spatial scales can be obtained from https://www.ons.gov.uk/census/2011census or https://www.nomisweb.co.uk/ for the UK and https://www.census.gov/data.html for the US. Appendix 4.13.1 provides web links to obtain shapefiles of UK and US census geography, as well as illustrating how to combine a spatial dataset with a shapefile in R.

Exercise 3.6. With reference to an area of data analysis with which you are familiar, list the main data quality challenges that need to be faced relating to data: accuracy; completeness; consistency; and resolution. Why, in your chosen area, do they present problems?

Exercise 3.7. Using a set of burglary count data at the census output area level in Peterborough (available online), calculate the variability associated with a rate estimate (the number of observed cases divided by the size of the population) in order to verify the small number problem arising from small area count data (Section 3.3.3).

Exercise 3.8. What is meant by a "regression to the mean effect"? How might concerns about such an effect be addressed?

Exercise 3.9. What is meant by the term "overdispersion" with respect to using the Poisson model, and why is it commonly encountered when analysing small area data?

Exercise 3.10. Sketch some of the special challenges that arise when visualizing spatial data and relationships between spatial variables. How might these challenges be met?

Exercise 3.11. Review the main challenges encountered when looking for clusters in small area data.

Exercise 3.12. Different methods for choosing class intervals affect the visual impression of spatial variability on a choropleth map. Illustrate this using a data set of your choice. Consider the following methods for defining class intervals: equal interval; quartiles; quintiles; natural breaks. (Hint: you may want to go through the material in Section 6.2.1 first, then produce a choropleth map using the Peterborough burglary data introduced there.)

4

Specifying Spatial Relationships on the Map: The Weights Matrix

4.1 Introduction

In this chapter we describe various ways of representing spatial relationships between areas (or fixed points, such as the locations of petrol stations or the centroids of census tracts) on a map using a spatial weights matrix. This matrix is often referred to as the W matrix. When modelling spatial data, the spatial weights matrix, W, plays an important role because it provides a way of summarising the spatial relationships between the areas (or points) to which the observed data are georeferenced.[1] As we shall discuss in this chapter, the spatial relationships amongst areas constitute an important data element, and the construction of the W matrix allows us to turn these relationships into numbers. However, different from conventional data (such as the observed numbers of crime or disease cases), this form of data is often based on our assumptions about how the areas are connected to each other. A purpose of this chapter, then, is to introduce different ways of defining connections, as well as the strength of those connections, amongst pairs of areas. Some of these methods are based on the geographical configuration of the areas, and some methods are attribute-based, in which distance between any two areas is defined by how similar (or dissimilar) the two areas are in terms of a chosen attribute.

When modelling a spatial dataset that comprises N areas (or fixed points), a spatial weights matrix W is an $N \times N$ (N rows by N columns) matrix that defines how we choose to specify the spatial relationships between the areas (or the points). In this chapter (also throughout this book), we use w_{ij} to denote the element of W on the ith row and jth column ($i = 1, \ldots, N$ and $j = 1, \ldots, N$). Often the elements of W take the value either 0 or 1. If $w_{ij} = 1$, then we say that the two areas (or points) i and j are *neighbours* of each other. If $w_{ij} = 0$, then the two areas (or points) i and j are not neighbours. Each diagonal element w_{ii} in W is set to 0 (for all i), meaning that an area (or point) is not allowed to be a neighbour of itself. However, in some situations where, for example, we are applying some spatial smoothing operation to the data, we may want to set w_{ii} to 1 (for all i) so that the value in area i, together with the values in the neighbours of i, are used in the calculation. We discuss this option in more detail later in the chapter. The elements in W can also be non-binary. That is, they can be real numbers (not necessarily integers) greater than or equal to zero. This option allows us to vary the strength of "neighbourliness" from one pair of neighbouring areas to another, for example as a function of how close two areas (or points) are geographically. The focus

---

[1] The areas (or points) are the nodes, whilst the elements of W specify the edges that connect the nodes. We use the term "matrix" here for explanatory purposes and because formulae typically use matrix notation. In computer software, spatial relationships are usually presented as lists of areas with their defined "neighbours".

of this chapter is on defining spatial weights matrices and discussing the important consequences of any definition. We defer the discussion of weights matrices for modelling temporal and spatial-temporal datasets until Chapters 12 and 15, respectively.

Later in the chapter we discuss the roles that the W matrix plays in various forms of statistical analyses of spatial data and the statistical implications arising from any choice made for the form of this matrix. As we shall discuss, the specification of W has been a particular subject of study in spatial econometrics, a collection of models for spatial data that we will discuss in depth in Chapter 10. But one of the principal goals in spatial econometric modelling is to obtain reliable estimates of spatial spillover and feedback effects – for example, how economic decisions or activity in one area impact on outcomes in neighbouring areas, both directly and indirectly (i.e. through third-party areas), as well as in the originating area. Thus, an important aspect of that modelling is how the spatial relationships between areas are defined through the W matrix. As we shall also see, there are implications for other areas of spatial modelling, including small area estimation and the estimation of covariate effects.

In the first five sections in this chapter (Sections 4.2–4.6), we describe five main methods for defining the elements in the W matrix: contiguity (Section 4.2), geographical distance (4.3), graph-based methods (4.4), attribute-based methods (4.5) and interaction-based methods (4.6). The first three methods, contiguity, geographical distance and graph-based methods, construct W based on purely geometrical or topological relationships amongst the set of areas (or points). The other two methods, attribute-based and interaction-based, define W based on data (observed values) that describe certain characteristics of the areas where such characteristics are thought to underlie the interconnectedness of these areas. In Section 4.7, we discuss row standardisation of the W matrix, and in Section 4.8 we describe higher order weights matrices (such as W^2, W^3 and so on). In Section 4.9 we draw the reader's attention to the roles of the W matrix in the statistical modelling of spatial data and some of the statistical consequences arising from the way that W is specified.

Although different ways of defining the W matrix exist, the reader should note that in much statistical modelling to date it is the contiguity and geographical distance criteria that have been most frequently employed. Whilst Tobler's First Law (see Section 3.3.2.1) is often invoked to justify a choice of W based on geometrical or topological relationships between the areas, or the choice of W is based just on what software makes possible, we argue that the specification of W really ought to be placed in the context of the problem in hand. In some situations, as we shall describe in Section 4.9, we may even need to model the elements in the W matrix, as opposed to treating them as known quantities. Therefore, in Section 4.10 we briefly discuss approaches to estimating the weights matrix. Some of these methods will be discussed in more detail in Part II of the book. The two appendices at the end of this chapter deal with some practical problems: how to combine a spatial dataset with a shapefile; how to create various spatial weight matrices from a shapefile; and how to manipulate the data values stored in a shapefile.

4.2 Specifying Weights Based on Contiguity

There are two methods that lie within this category. Under the first method, if areas i and j share a common border, then $w_{ij} \neq 0$. In other words, these two areas are defined as neighbours. Typically, $w_{ij} = 1$. If two areas do not share a common boundary, then $w_{ij} = 0$ and they are not neighbours. By analogy with a chess board, this first method is sometimes referred to as the "rook's move" definition for contiguity.

The second method defines two areas i and j to be neighbours, i.e. $w_{ij}=1$, if they share either a common border *or* a common vertex (two areas touch at just one point). Otherwise they are considered to be non-neighbours ($w_{ij}=0$). By analogy again with a chess board, this is sometimes referred to as the "queen's move" definition for contiguity.

In both methods, $w_{ij}=w_{ji}$ so that W is a symmetric matrix. This distinction between rook's move and queen's move is important in the case of data reported on square pixels. Compared to rook's move contiguity, queen's move contiguity increases the number of neighbours for each pixel on the map. In particular, for the pixels that are not at the boundary of the map, the increase is from four to eight.

In the case of data reported on census units, which are typically irregular in shape, this distinction is usually of much lesser significance. Figure 4.1 shows a binary weights matrix for a small set of spatial units using queen's move contiguity as the criterion. Note that areas 2 and 6 are treated as neighbours ($w_{26}=w_{62}=1$) and so are areas 3 and 5 ($w_{35}=w_{53}=1$) under queen's move, but they would not be under rook's move.

Constructing the weights matrix via spatial contiguity can be refined by specifying each weight (the strength of the "neighbourliness") as a function of the length of the common border as a proportion of the total border of i. That is,

$$w_{ij} = \left(\frac{l_{ij}}{l_i}\right)^a \tag{4.1}$$

where l_i is the length of the border of spatial unit i and l_{ij} is the length of the shared border between i and j. The power (exponent) a is usually specified by the user rather than estimated. Applying this weight specification to an irregular spatial partition, generally $w_{ij} \neq w_{ji}$, and thus W is no longer a symmetric matrix.

A question that may need resolving is what to do with islands or if the spatial system comprises blocks of areas that are non-contiguous. If spatial unit i is an island, then the ith row of the W matrix under spatial contiguity would consist entirely of zeros. When using the weights matrix as part of a spatial model, this is not permitted, and a typical solution is to "join" islands to the spatial units on the mainland that are nearest (see an example in Section 11.1.2 in Chapter 11).[2]

	A1	A2	A3	A4	A5	A6	A7	A8	A9	A10
A1	0	1	0	1	1	0	0	0	0	0
A2	1	0	1	0	1	1	0	0	0	0
A3	0	1	0	0	1	1	0	0	0	0
A4	1	0	0	0	1	0	1	0	0	0
A5	1	1	1	1	0	1	1	0	0	0
A6	0	1	1	0	1	0	1	1	1	0
A7	0	0	0	1	1	1	0	0	1	1
A8	0	0	0	0	0	1	0	0	1	0
A9	0	0	0	0	0	1	1	1	0	1
A10	0	0	0	0	0	0	1	0	1	0

FIGURE 4.1
A map of spatial units and an associated binary (0/1) weights matrix based on queen's move contiguity.

---

[2] This is not a problem if the weights matrix is only being used for smoothing purposes, although the user might wish to smooth the island value with one or more of the spatial units close by on the mainland.

4.3 Specifying Weights Based on Geographical Distance

This method of defining the weights matrix W can be applied quite naturally to the case where data values are attached to fixed point sites. But any collection of spatial units in the form of polygons can also be represented by a collection of points. For example, we could take the geometric centroid of each polygon or, in other circumstances, we might use the population weighted centroid where the location of the centroid depends on the distribution of population within the polygon. This might be particularly appropriate if spatial interaction is defined through resident populations (in each spatial unit) and these populations are not uniformly distributed. In any of these cases, we can specify spatial relationships in terms of Euclidean (straight-line) distance. Other distance metrics could be used, and in small area analyses of urban data the Manhattan metric is sometimes chosen, which measures distance by following the street network.

Let d_{ij} be the distance (Euclidean, Manhattan or other distance measure) between points i and j with $i \neq j$. Listed below are three different functions to define the weight between those two points:

$$(i) \rightarrow \text{Inverse distance}: \ w_{ij} = d_{ij}^{-1}$$

$$(ii) \rightarrow \text{Inverse distance raised to the power } \gamma > 0: \ w_{ij} = d_{ij}^{-\gamma} \qquad (4.2)$$

$$(iii) \rightarrow \text{Negative exponential with } \lambda > 0: \ w_{ij} = \exp\left(-\lambda \cdot d_{ij}\right)$$

The diagonal elements in W are all equal to zero. In both (ii) and (iii), the two parameters γ and λ are often specified by the analyst as opposed to being estimated using data. All the above three functions have the same properties that (a) as distance increases the weight decreases, and thus neighbourliness gets weaker; and (b) the resulting weights are symmetric ($w_{ij} = w_{ji}$ for all i and j). All the off-diagonal elements in W will be non-zero (albeit declining close to zero as distance increases) unless we impose a threshold. Two commonly applied thresholds are:

(i) Absolute distance threshold: if $d_{ij} \geq x$ then $w_{ij} = 0$.
(ii) Based on k nearest neighbours: allowing only the k nearest neighbours of spatial unit i to be non-zero. In this case, weights will not be symmetric.

The distance and common border definitions of neighbourliness can be combined, for example, using the following form:

$$w_{ij} = \left(\frac{l_{ij}}{l_i}\right)^a \cdot d_{ij}^{-\gamma} \qquad (4.3)$$

As before, both a and γ are positive, and their values are specified by the analyst.

4.4 Specifying Weights Based on the Graph Structure Associated with a Set of Points

Where the spatial units on which the data have been collected are, or can be treated as, a set of fixed points (such as a set of retail outlets in a city, or a set of urban places across a large region), then we can impose a graph on the set of points and use that as a basis for defining a set of weights. In the case of the Gabriel graph (Matula and Sokal, 1980), as illustrated in Figure 4.2, any two points A and B are neighbours if and only if all other points are outside the circle on whose circumference A and B lie and whose centre lies on the line segment joining A and B.

Another method of defining neighbours for a set of fixed points based on graph structure is through a Dirichlet tessellation (also referred to as a Voronoi partition or Voronoi decomposition). To illustrate the method, Figure 4.3(a) shows a set of points which are called fixed seed points, and Figure 4.3(b) shows a Dirichlet partition defined on those fixed seed points. A Dirichlet partition for a set of fixed seed points is a partitioning of the space into regions with each region containing only one of those fixed seed points. These regions (called Voronoi polygons, Thiessen polygons or Dirichlet cells) are formed such that if you are standing at any position inside a region, you are always closer to the seed point in that region than to any of the other seed points. This partitioning can then be used to provide a set of contiguity-based weights so that, for example, in Figure 4.3(b), the two fixed points A and C are neighbours, but A and F are not. Alternatively, a Delaunay triangulation, as illustrated in Figure 4.3(c), can be used to connect each pair of fixed points if the respective regions that these two points are in share a common border. Then a set of distance-based weights (combined with contiguity because of the use of the Delaunay triangulation) can be defined.

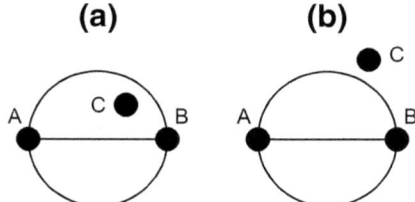

FIGURE 4.2
Defining neighbours on a Gabriel graph. In (a), the two points A and B are not defined to be neighbours, because point C is inside the circle, whereas in (b), A and B are defined to be neighbours since C is outside the circle.

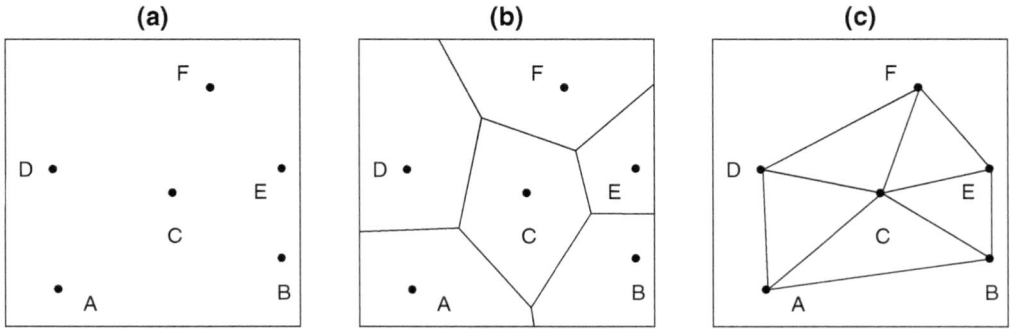

FIGURE 4.3
(a) A set of point sites; (b) a Dirichlet tessellation defined on the point sites, where the line that separates each pair of points dissects perpendicularly the line segment that joins those two points (for example, the nearly horizontal line that separates points B and E perpendicularly dissects the line joining B and E); and (c) the Delaunay triangulation based on the Dirichlet tessellation in (b).

4.5 Specifying Weights Based on Attribute Values

Neighbourliness can be defined in terms of how similar two areas are based on some attribute or set of attributes. Case et al. (1993), in a study of budget spillovers amongst US states, defined w_{ij} based on how similar two states were in terms of their socio-economic composition. Their definition of the weight between states i and j is given by

$$w_{ij} = \frac{1}{|x_i - x_j|} \tag{4.4}$$

where x_i and x_j are the values of a selected socio-economic variable in the two states and $|x_i - x_j|$ is the absolute value of the difference. Law (2016), in a study of fall injuries amongst senior citizens by small area in a part of Eastern Canada, specified the weight between two areas as

$$w_{ij} = E_i \cdot E_j \tag{4.5}$$

where E_i and E_j are the expected numbers of fall injuries in areas i and j, respectively, taking into account the age and sex composition of the areas (see Section 9.3.2 in Chapter 9). This formulation of weights implies that the neighbourliness of two areas is strong if the expected numbers of fall injuries in both areas are large. These weights can be further modified to include spatial separation of the two areas by, for example, dividing by d_{ij}. We will see some examples of the use of attribute values to define the W matrix in Chapter 8 (in particular, Sections 8.3.3 and 8.4.2).

4.6 Specifying Weights Based on Evidence about Interactions

Interaction data such as data on flows of goods or people or communications can be used to specify weights. The underlying idea is that real flows may be indicative of the strength of neighbourliness (contact) that may exist between different places and which may therefore help in explaining spatial variation.

Bavaud (1998) proposed two different ways to define interaction weights:

$$\text{Export} - \text{based weights}: w_{ij} = f_{i \to j} / f_{i \to \cdot}$$
$$\text{Import} - \text{based weight}: w_{ij} = f_{j \to i} / f_{\cdot \to i} \tag{4.6}$$

where

- $f_{i \to j}$ measures the export from area i to area j.
- $f_{j \to i}$ measures the import from area j to area i.
- $f_{i \to \cdot}$ is the total export from i to all other areas in the study region.
- $f_{\cdot \to i}$ measures the total import from all other areas to i.

In general, weights defined through interaction between areas will not be symmetric, i.e. $w_{ij} \neq w_{ji}$.

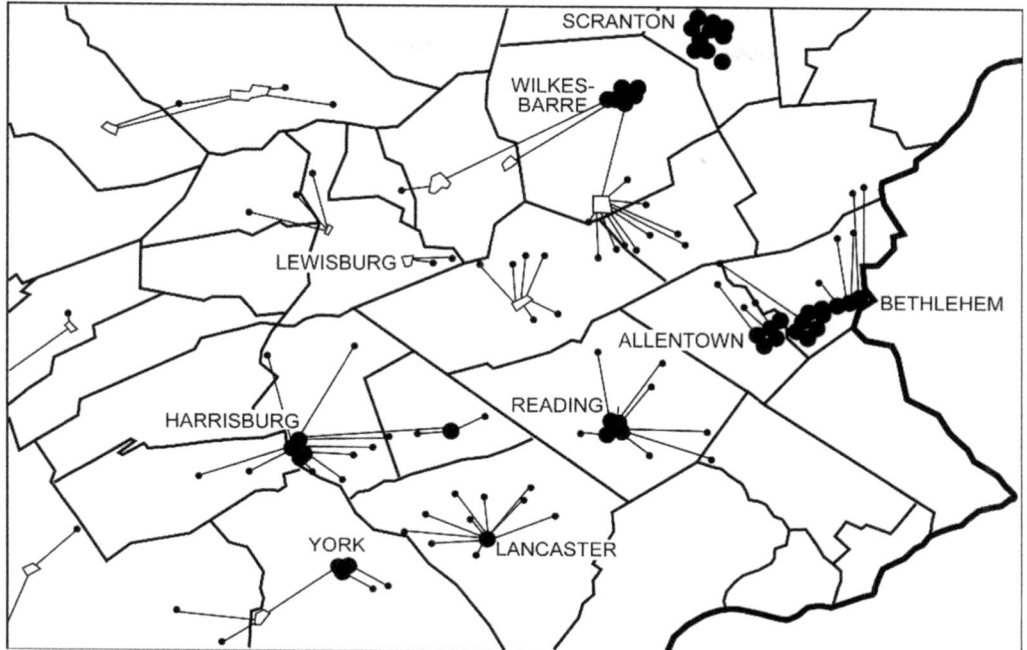

FIGURE 4.4
Urban places and their economic links based on interaction data: Pennsylvania, USA (after Haining, 1987).

Figures 4.4 and 4.5 illustrate two other examples of defining weights based on interaction. In Figure 4.4, urban places in Pennsylvania are linked together based on a set of hierarchical relationships drawn from Central Place Theory (Losch, 1957). Smaller urban places close to larger urban places are assumed to be economically dependent on those larger places (through wage transfers and expenditure patterns), but larger places are not dependent on smaller places. In the case of Figure 4.5, managers of petrol retail outlets in a city were asked to name the sites they believed were their principal competitors in the urban market. Linkages were then drawn using that information.

Conway and Rork (2004) used migration data to construct weights in their study of US state taxes. Davies and Voget (2008), in their study of tax competition within the European Union, constructed weights based on relative market potential, a theoretical concept from the new economic geography literature (see Baldwin and Krugman, 2004; Krugman, 1996, 1998). Their paper includes a review of other approaches to specifying spatial weights (including distance and levels of GDP) in studies of tax rates, some of which can be classified under the methods described in Sections 4.3 and 4.4.

4.7 Row Standardisation

Where polygons have different numbers of neighbours that we wish to control for, then the weights matrix needs to be row standardised. In some areas of spatial modelling, working with a row-standardised matrix helps in the interpretation of model parameters. For

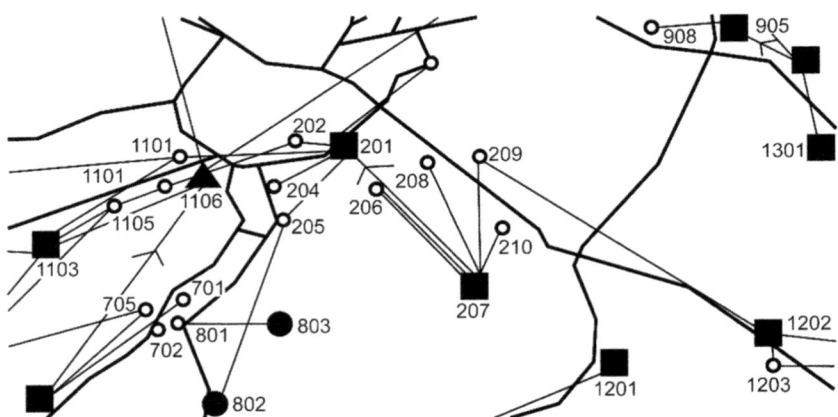

FIGURE 4.5
Competitive links between petrol retail sites based on proximity and evidence from a questionnaire survey
of site managers in Sheffield, England. Each number refers to a retail site, the shapes refer to their corporate
type (solid square = "Major" retailer; solid circle = "Minor" retailer; solid triangle = supermarket retailer; open
circle = a reference station for the retailer, to which it is joined by a line. If no arrow, both sites consider the other
a reference station in price setting; arrow head indicates direction of referencing). From Ning and Haining,
2003, Figure 4.

example, in the context of spatial smoothing, we may assume that the attribute value of a
spatial unit is similar to an average (as opposed to the sum) of the attribute values of its
neighbouring units.

To row standardise the W matrix, each element on a row is divided by the corresponding row sum. So the sum of all elements on a row after row standardisation equals 1, and
the sum of all values in the row-standardised matrix, often referred to as W^*, equals N,
the number of spatial units in the study region. It follows that row-standardised weights
matrices are no longer symmetric, as shown in Figure 4.6. For the conditional autoregressive (CAR) models, the asymmetry of the row-standardised weights matrix has several
implications, which will be discussed in Chapter 8.

As part of exploratory spatial data analysis, row-standardised matrices are used to
obtain spatially smoothed maps in order to iron out the noises (the sudden "jumps") in the

	A1	A2	A3	A4	A5	A6	A7	A8	A9	A10	Sum
A1	0	1	0	1	1	0	0	0	0	0	3
A2	1	0	1	0	1	1	0	0	0	0	4
A3	0	1	0	0	1	1	0	0	0	0	3
A4	1	0	0	0	1	0	1	0	0	0	3
A5	1	1	1	1	0	1	1	0	0	0	6
A6	0	1	1	0	1	0	1	1	1	0	6
A7	0	0	0	1	1	1	0	0	1	1	5
A8	0	0	0	0	0	1	0	0	1	0	2
A9	0	0	0	0	0	1	1	1	0	1	4
A10	0	0	0	0	0	0	1	0	1	0	2

	A1	A2	A3	A4	A5	A6	A7	A8	A9	A10	Sum
A1	0	0.33	0	0.33	0.33	0	0	0	0	0	1
A2	0.25	0	0.25	0	0.25	0.25	0	0	0	0	1
A3	0	0.33	0	0	0.33	0.33	0	0	0	0	1
A4	0.33	0	0	0	0.33	0	0.33	0	0	0	1
A5	0.17	0.17	0.17	0.17	0	0.17	0.17	0	0	0	1
A6	0	0.17	0.17	0	0.17	0	0.17	0.17	0.17	0	1
A7	0	0	0	0.20	0.20	0.20	0	0	0.20	0.20	1
A8	0	0	0	0	0	0.50	0	0	0.50	0	1
A9	0	0	0	0	0	0.25	0.25	0.25	0	0.25	1
A10	0	0	0	0	0	0	0.50	0	0.50	0	1

FIGURE 4.6
An original, unstandardised, contiguity W matrix (left; as in Figure 4.1) and its row-standardised form (right).

map so that spatial trends and patterns in the data can be easier to identify (Section 6.2.2). Let z denote a column vector of length N containing the data values to be smoothed. Let W denote an $N \times N$ binary (0/1) weights matrix, and let I_N denote the identity matrix of size N. When the purpose of spatial smoothing is to remove noise on the map, we want to include the value from area i in calculating the spatial average for area i. That can be done by adding the identity matrix to W, i.e. $(I_N + W)$. So, denoting $(I_N + W)^*$ as the row-standardised version of $(I_N + W)$, and $(I_N + W)^*_i$ as the ith row of $(I_N + W)^*$, then the spatial average for area i is given by

$$z_i^{(SL1)} = (I_N + W)^*_i \, z = \sum_{j=1}^{N} h_{ij}^* z_j \tag{4.7}$$

where h_{ij}^* is the element on row i, column j of the row-standardised matrix $(I_N + W)^*$. We use the superscript (SL1) in $z_i^{(SL1)}$ to signify that it is a spatially smoothed (also referred to as a *spatially-lagged*) value. Then $z^{(SL1)}$ is a spatially-smoothed version of z. Eq. 4.7 is one of many ways of spatial smoothing, and thus we label it as *SL1*.

Another way to calculate a spatially-lagged value for area i is to exclude its own value in the calculation. For example, in spatial epidemiology, we may wish to distinguish between the impacts on health from exposure to some environmental hazard (atmospheric pollution, for example) in an area from exposure arising in neighbouring areas. The latter aims to capture the residents' exposure arising from short distance movements (for example, for commuting and shopping). See Section 1.3.2.1. Then the first kind of effect could be quantified via the inclusion of the exposure covariate x, whilst the second kind of effects could be investigated through a spatially-lagged version of the exposure covariate:

$$x_i^{(SL2)} = W_i^* x = \sum_{j=1}^{N} w_{ij}^* x_j \tag{4.8}$$

Different from Eq. 4.7, this version of spatial smoothing (labelled as *SL2*) excludes the exposure level in area i so that $x_i^{(SL2)}$ represents an average exposure level of the neighbours of i. In Eq. 4.8, W_i^* denotes the ith row of the row-standardised W. This way of calculating a spatially-lagged version of an observable covariate is also used in the spatially-lagged covariates (SLX) model in spatial econometric modelling (see Section 10.2.3 for more detail).

4.8 Higher Order Weights Matrices

The binary (0/1) matrix W based on contiguity is sometimes referred to as the first order contiguity weights matrix because it defines the immediate or first order neighbours. Spatial weights matrices can be defined for second, third and higher orders of contiguity. As the order increases, the spatial separation between an area and its neighbours at the corresponding order increases. Here, spatial separation between two areas is measured by the number of steps along the shortest path. As illustrated in Figure 4.7, the second order neighbours of area 6 are generally further away from area 6 than the first order neighbours. For this map, there are no third order neighbours for area 6. A purpose in defining

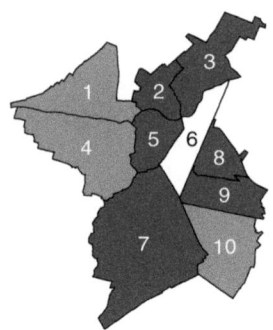

FIGURE 4.7
The first order (the polygons in a darker grey colour) and the second order neighbours (the polygons in a lighter grey colour) of area 6 based on queen's move contiguity. Note that the two-colour shadings do not represent the strength of relationships; they are only used to distinguish the first and the second order neighbours.

higher order weights matrices is to assess how similarity (measured by the strength of autocorrelation) decays with increasing spatial separation (see Chapter 6 and Section 6.2.4 in particular). Spatial smoothing can be carried out by including higher order neighbours (e.g. averaging values from both the first and the second order neighbours) so that the resulting map becomes smoother than if only first order neighbours are included in the smoothing procedure.

Different orders of contiguity can be obtained by raising the first order contiguity binary (0/1) matrix, W (e.g. as in Figure 4.1), to a positive integer power e.g. $W^2 = W \times W$ for the second order contiguity and $W^3 = W^2 \times W = W \times W \times W$ for the third order and so on. To look at these powered W matrices in more detail, let $w_{ij}^{(k)}$ denote the element in row i and column j of the W matrix of order k. The value of $w_{ij}^{(k)}$ equals the *total* number of paths going from i to j in k steps. For example, using queen's move contiguity, we can go from area 6 in Figure 4.7 to any one of the following areas, 2, 3, 5, 7, 8 and 9, in one step, and there is only one pathway leading to one of those areas from area 6. Thus, $w_{6j} = 1$, with $j = 2,3,5,7,8$ and 9. Note that the superscript (k) in $w_{ij}^{(k)}$ is removed when $k = 1$. However, since we cannot go from area 6 to the other three areas (1, 4 and 10) in one step, $w_{6j} = 0$, with $j = 1,4$ and 10. The diagonal element $w_{66} = 0$, since by taking one step, we have to move away from area 6. Therefore, we can write down all the first order neighbours of an area directly from the binary W matrix.

However, as far as obtaining a higher order contiguity matrix of, say, second (or third) order is concerned, the problem with raising W to the power two (or three) is that the resulting matrix will count *all* the two- (or *all* the three-) step paths from i to j. Consider again area 6 in Figure 4.7. There are two ways of stepping from area 6 to area 1 in two steps: one way is via area 2, and the other way is via area 5. Thus, $w_{61} = 2$. But if all we want to know is whether area 2 is a second order neighbour of area 6, the actual *number* of ways of going from 6 to 1 in two steps is redundant information.[3] Then there is the problem of the

---

[3] There are circumstances where we want to know the number of pathways between two areas, for example if we are interested in constructing a measure of influence of areas on each other. Deriving the number of pathways becomes important when we come to explain the properties of direct and indirect effects in spatial econometric modelling (see Section 10.3.3.3 for more detail). Here, however, we are interested in constructing a way of identifying second and higher order neighbours for the purpose of calculating a second or higher order spatial autocorrelation statistic, for example.

pathways themselves and whether they involve backtracking or circularity (or following a "scenic" route) – all of which would be undesirable if we are seeking a proper measure of second or third, or higher order neighbourliness. To take an example, suppose we calculate W^2 (see Exercise 4.4). We will see that w_{66} is not zero but six, meaning that area 6 is a "neighbour" of itself in two steps. The six two-step pathways involve going from area 6 to one of its six first order neighbours, then back to itself. However, area 6 should not really be considered a second order neighbour of itself. It is not a "proper" second order neighbour since it involves a form of backtracking. The possibilities for taking "scenic" pathways between areas tend to increase as order increases. When raising W to order three, there are many pathways linking area 6 and area 1 in three steps, but we have already defined area 1 as a second order neighbour of area 6, so we do not also want to define it as a third order neighbour too. Anselin and Smirnov (1996) develop an efficient algorithm for sweeping out redundant, backtracking, "scenic" and circular paths for the purpose of constructing the proper second and higher order contiguity matrix from the powered W matrix.

It is also worth noting that raising a symmetric W matrix to any positive integer power yields a symmetric matrix. However, since the row-standardised W is not symmetric, $(W^*)^2$, $(W^*)^3$ and so on are also not symmetric. But when $w_{ij} \geq 0$ for all i and j, and if $w_{ij}^{(k)} = 0$, then $\left(w_{ij}^*\right)^{(k)} = 0$, where $\left(w_{ij}^*\right)^{(k)}$ is the element in row i and column j of $(W^*)^k$ (see Exercise 4.7).

All the above observations on the elements of powered W matrices have significance beyond the calculation of higher order contiguity. As we shall see in the next section, these powered W matrices give rise to complex ripple effects when using the W matrix for local information sharing and assessing the effects of spatial spillover and feedback in spatial econometric modelling.

4.9 Choice of W and Statistical Implications

An important role for the W matrix is to turn spatial relationships amongst the N areas (or points) into "data". Through modelling, this form of data supplements the conventional form of data (e.g. data on crime/disease case counts observed across the census tracts in a city, or data on levels of income amongst a set of geo-referenced households within a survey), allowing us to strengthen the estimation of quantities at the small area level, to study how the characteristics of one area might affect those in other areas (i.e. the notion of spatial spillovers and feedbacks), and to obtain a more accurate estimate of the effect of an observable covariate on an outcome. The W matrix plays an integral part in any form of spatial analysis in this book. Whilst we defer the more technical details of the modelling to Part II, here we discuss the implications of the choice of W on specific statistical analyses and problems.

4.9.1 Implications for Small Area Estimation

"Borrowing strength" is a methodology for improving the precision of estimates of small area characteristics such as crime/disease rates or average income levels. If we represent the unknown quantities of interest, say the unknown crime rates in two areas i and j as random variables, and if these two random variables are *positively correlated*, then the information that we have on the crime rate in one area tells us something about the crime rate in the other area. For example, if area i has a low crime rate, as suggested by its small number of reported crime events, then this information can be used (or borrowed/shared) when estimating the crime rate in area j, in addition to using the count of reported crime events in j and vice versa. This is the idea behind *local information sharing*. The stronger the correlation between two random variables, the more information is shared locally between the two areas. Note that negative correlation is also allowed but not often used.

Now, what is the role that the W matrix plays? In Chapter 8, to operationalise local information sharing, we will introduce a class of spatial models based on the conditional autoregressive (CAR) structure, a modelling structure that allows us to impose a set of spatial relationships onto the random variables that represent the quantities of interest (e.g. the small area crime rates). These random variables are also called random effects. The so-called proper CAR (pCAR) model structures these random effects such that the variance-covariance matrix of these random effects takes the following form,

$$\Sigma = \sigma^2 D_w^{-1}\left(I_N - \rho W^*\right)^{-1}, \tag{4.9}$$

where we now see the appearance of W^*, the row-standardised W matrix. In Eq. 4.9, σ^2 is a variance parameter, D_w^{-1} is the inverse of D_w, an $N \times N$ diagonal matrix (all off-diagonal elements equal to zero) with the diagonal elements equal to the row sums of the W matrix, and ρ is called the spatial autocorrelation parameter. The variance-covariance matrix Σ, a function of W, determines the properties of local information sharing. There are two ways to study the form of Σ: one is through an infinite regular array (see Besag and Kooperberg, 1995 and Künsch, 1987) and the other is by using a torus (Held and Rue, 2010, p.205–207). Here we describe the latter in more detail whilst referring the reader to the reference provided for details about the former. A torus illustrated in Figure 4.8 is formed, as Held and Rue (2010, p.206) instruct, by wrapping a regular lattice into a "sausage" then joining the two ends of the sausage together. Whilst the geometric shape of a torus hardly ever appears in any social science applications, it has the benefit that when using contiguity (Section 4.2) to construct the W matrix, every pixel on the torus has the same number of neighbours (four in the case of the rook's move or eight in the case of the queen's move). As a result, every single pixel has the same set of correlations with all the other pixels on

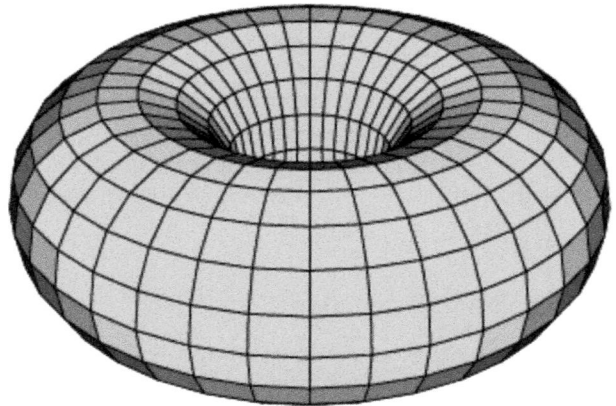

FIGURE 4.8
An illustration of a torus.

the torus. So it is sufficient to study the spatial correlation structure of just one pixel. In the setting considered by Held and Rue (2010), their torus is formed using a 29×29 lattice so the W matrix is of size 841×841. Their W matrix is defined as a binary matrix based on first order rook's contiguity. Figure 4.9 shows the correlation coefficients of pixel (15, 15) with all other 840 pixels on the torus (see Held and Rue, 2010, p.205–207 for further details).

Three observations can be drawn from Figure 4.9. First, all correlations are non-zero and positive. This means that local information sharing is not restricted to between each pixel and its immediate neighbours defined by W, but instead this process spans the entire map, mimicking a "ripple effect". Second, although the random effect in one area is positively correlated with the random effects in all other areas, the strongest correlations are with its immediate neighbours defined by the binary W matrix. For example, in Figure 4.9, the strongest correlation is between pixel (15, 15) and its immediate neighbours (15, 14), (14, 15), (16, 15) and (15, 16). When a non-binary weights matrix is used, those neighbours are the ones that have the largest values within the W matrix. Finally, Figure 4.9 clearly shows the tailing off in spatial correlation with increasing distance separation. In terms of information sharing, the information that area i borrows from area j $(i \neq j)$ becomes less as j moves away from i.

These three observations can also be explained mathematically by expanding the matrix inversion, $(I_N - \rho W^*)^{-1}$, in Eq. 4.9. Note that both σ^2, a scalar parameter, and D_w^{-1}, a diagonal matrix with positive, non-zero diagonal elements, do not affect the discussion of the off-diagonal elements in Σ (the focus of our discussion here). That is because if an off-diagonal element in $(I_N - \rho W^*)^{-1}$ is non-zero and positive, the element at the corresponding position in the variance-covariance matrix is also non-zero and positive.

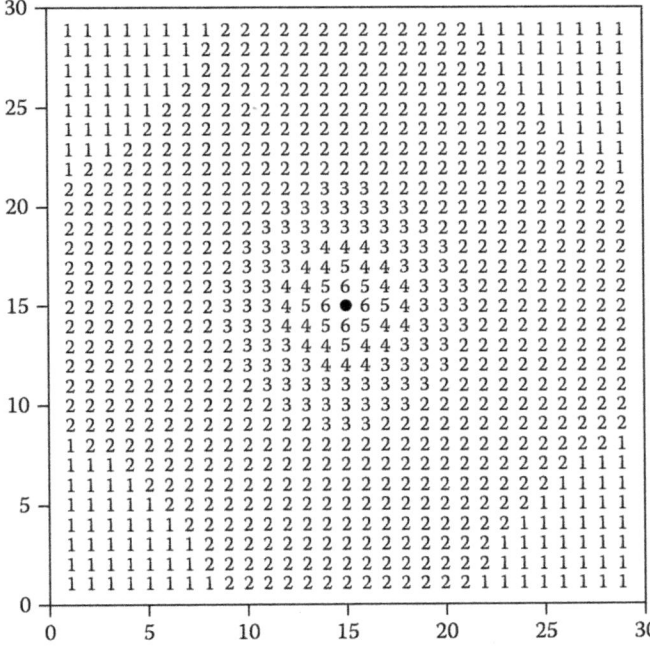

FIGURE 4.9
Plot of the correlation between pixel (15, 15) and all other pixels (i, j) (with $i \neq 15$ and $j \neq 15$) for a first order proper CAR model with parameter $\rho = 0.2496$ on a torus formed by wrapping a 29×29 lattice. Shown are the spatial correlations (x10) truncated to an integer (Held and Rue, 2010, p.207).

Providing ρ lies between -1 and 1, i.e. $|\rho| < 1$, a constraint that we will return to in Section 10.3.3.3, $(I_N - \rho W^*)^{-1}$ can be written as the sum of an infinite series:

$$\left(I_N - \rho W^*\right)^{-1} = I_N + \rho \cdot W^* + \rho^2 \cdot \left(W^*\right)^2 + \rho^3 \cdot \left(W^*\right)^3 + \cdots \qquad (4.10)$$

From Eq. 4.10, it becomes clear that not only is information borrowed from the first order neighbours it is also borrowed from higher order neighbours due to the terms involving higher orders of the row-standardised W matrix (see discussion in Section 4.8). However, the amount of information borrowing reduces as the order increases, since ρ^k gets smaller as k increases. Therefore, the key points obtained from the expansion are: (a) under the conditional autoregressive structure, information is not only borrowed from the immediate (first order) neighbours as defined by the chosen W but ripples across higher order neighbours and (b) the amount of borrowing reduces as order increases. Dubin (2009) discusses the spatial correlation structures associated with different spatial models and different forms of W.

However, the situation becomes more complicated when W refers to the spatial relationships amongst a set of irregular polygons, as Wall (2004) demonstrates using the states of the USA. Table 4.1 shows the correlations between two selected states (Missouri and Tennessee) and their first order contiguous neighbours derived from the proper CAR model (for specification of the model see Wall (2004, p.316)). Although both states have eight neighbours, the correlation varies within each state. For example, the correlation between Missouri and Kansas is stronger compared to that between Missouri and Kentucky. Similarly, Tennessee is more correlated with Alabama than it is with Arkansas. Since both states have the same number of first order neighbours, such differences do not reflect the influence of differences in the numbers of first order neighbours. We should ask the same question that Wall (2004, p.318) poses: "is this reasonable?" Wall also remarks that the variability in these correlations shows no systematic structure and also varies with the size of ρ, the spatial autocorrelation parameter in Eq. 4.9.

The specification of the spatial weights matrix, W, determines how information is shared locally. However, we seldom know the *true* spatial relationships of the areas, and thus we need to question the appropriateness of the choice of W. For example, if the W matrix is specified based on geographical proximity, are the adjacent areas the most appropriate to

TABLE 4.1

Implied Correlations Between Tennessee and Missouri and their First-Order Neighbours Under the Proper CAR Model

Missouri First Order Neighbours		Tennessee First Order Neighbours	
Arkansas	0.238	Alabama	0.324
Illinois	0.247	Arkansas	0.257
Iowa	0.244	Georgia	0.327
Kansas	0.263	Kentucky	0.229
Kentucky	0.223	Mississippi	0.300
Nebraska	0.248	Missouri	0.216
Oklahoma	0.251	N. Carolina	0.312
Tennessee	0.216	Virginia	0.265

Source: Wall, 2004, p.319.

borrow information from? If we are seeking to improve the precision of small area estimates of social or economic attributes, and if geographically nearby areas, at least on some parts of the map, differ markedly in terms of social or economic characteristics, a better basis for "information borrowing" might be to define the set of neighbours of any area i with reference to those attributes that suggest the areas will have similar characteristics. This is the argument for defining the W matrix using attributes of areas (Section 4.5). For example, Law (2016) defines "distance" between areas by the expected number of fall-induced injuries. The larger the expected number of falls in both areas i *and* j, the "closer" they are considered to be in terms of spatial epidemiological "distance". The larger the weighting (w_{ij}), the stronger the information borrowing from those areas as opposed to other areas, even including those that are geographically adjacent. We shall see some other examples of such attribute-based definitions in Chapter 8.

A consequence of local information sharing is that the resulting map is made smoother looking. However, where pronounced differences exist between adjacent areas, which can be seen by mapping the observed data (Chapter 6), sharing information locally runs the risk of oversmoothing. Figure 4.10 provides an illustration where, if information

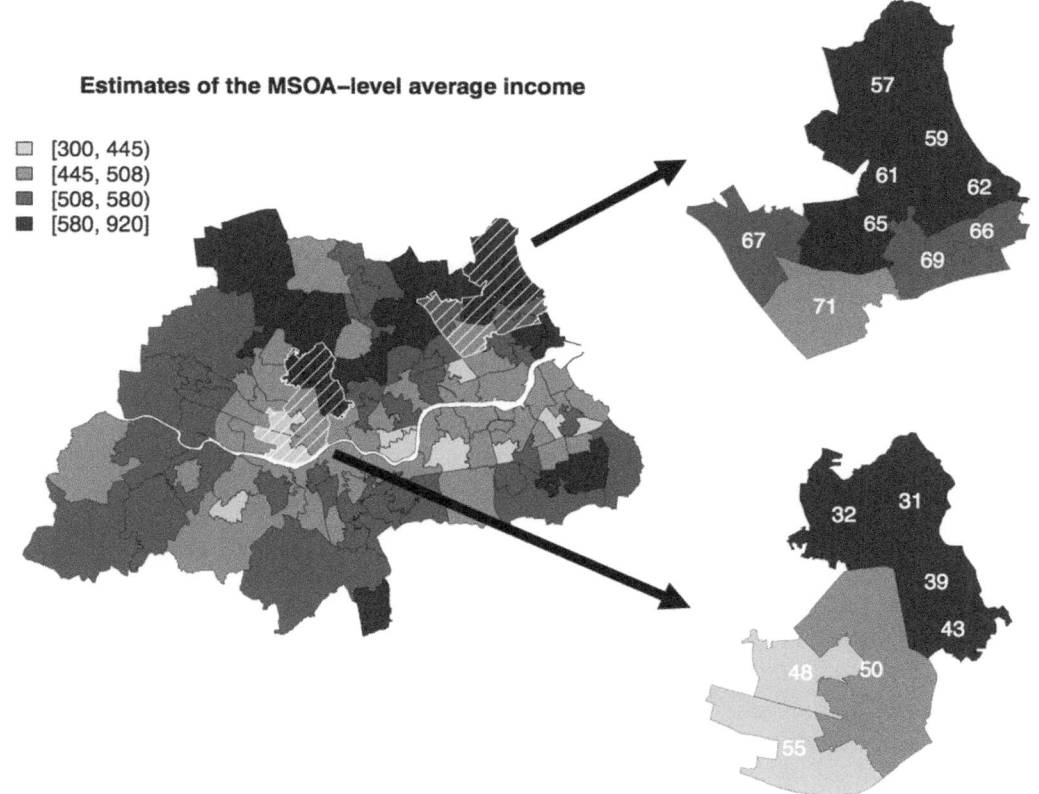

FIGURE 4.10
A map of the estimated average income at the middle super output area (MSOA) level in Newcastle. The income estimates are obtained from a version of the conditional autoregressive model. Whilst the resulting map is generally smooth-looking, there is evidence of the presence of pronounced differences amongst some subsets of spatially-contiguous areas. See Section 8.4 for further detail.

sharing is applied to these data, in some areas of the map (highlighted) there is a risk of masking real differences between adjacent areas. Other forms of bias are possible too, including undersmoothing. Various methods for adaptive spatial smoothing have been proposed to address these problems. Common to these adaptive smoothers is the modelling of the W matrix as opposed to treating it as fixed (or given). We will discuss the modelling of the W matrix in Section 4.10 and will detail some of the spatial adaptive models in Section 8.4.

When borrowing information in estimating a set of spatial-temporal random effects (parameters), the same principles apply. For a dataset with N spatial units and T time periods, if the aim is to improve the precision of the estimates associated with the parameter in each of the $N \times T$ space-time cells (say the crime/disease rate in area i at time t), then it is natural to consider not only how information should be borrowed spatially, but also how to borrow information temporally across the T time periods. As we shall see, borrowing information in time is *bi-directional*, meaning that the estimate at time t will borrow information from *both* $t-1$ *and* $t+1$ – that is, from both the past and the future time points with respect to t. This definition of temporal neighbours is different from the situation where we are modelling the underlying space-time *process*. In that case, we might reasonably assume the present time point t depends on the past time point $t-1$ but not the future time point $t+1$. However, in the context of small area estimation, bi-directional information sharing over time is justifiable and appropriate because we are exploiting the temporal dependence structure present in the dataset in order to improve the estimation of these $N \times T$ parameters. We can extend such an idea further to borrow information *both* spatially *and* temporally. We shall discuss these topics in detail in Part III of the book.

4.9.2 Implications for Spatial Econometric Modelling

An important reason for fitting spatial econometric models is to empirically estimate spatial spillovers and feedbacks (see Section 2.3.1). One of the models in spatial econometrics is the so-called spatial lag (SLM) model, and the mathematical form of this model helps clarify the fundamental role played by the W matrix. Suppose we are modelling a set of random variables, $Y_1, \ldots, Y_N$. Each Y_i represents the outcome of interest in area i within a study region that has N areas. The SLM model is formulated as

$$Y_i = \alpha + \sum_{j=1}^{N} b_{ij} Y_j + \beta \cdot x_i + e_i \tag{4.11}$$

Eq. 4.11 essentially defines a regression model with a single observable covariate x_i (although the SLM model can include more than one covariate), and e_i representing an independent error term. The distinctive feature of the SLM model lies within the summation term, $\sum_{j=1}^{N} b_{ij} Y_j$. The inclusion of that term allows the outcome variable of each area i to be affected by the outcome variables of other areas. The parameter b_{ij} measures the effect of the outcome variable in area j on the outcome variable in area i. Typically, $b_{ii} = 0$, for all $i = 1, \ldots, N$, so that the outcome of each area is not allowed to affect itself. It is the estimation of these $N \cdot (N-1)$ parameters (i.e. b_{ij} for $i \neq j$) that enables us to study the spatial interdependence of these areas. However, that poses a challenge: there are more parameters to estimate than available observations. One way to solve the problem is to replace b_{ij} by $\delta \cdot w_{ij}^*$, where w_{ij}^* is the element from the row-standardised version of a chosen W matrix.

The number of unknown parameters is reduced dramatically from $N \cdot (N-1)$ to just one, namely, δ. The SLM model then becomes

$$Y_i = \alpha + \delta \sum_{j=1}^{N} w_{ij}^* Y_j + \beta \cdot x_i + e_i \qquad (4.12)$$

As we shall see in Section 10.2.2.3, the SLM model in Eq. 4.12 can be rewritten into the so-called reduced form[4]:

$$Y = \alpha \left(I_N - \delta W^* \right)^{-1} + \left(I_N - \delta W^* \right)^{-1} \beta x + \left(I_N - \delta W^* \right)^{-1} e \qquad (4.13)$$

Whilst we shall describe each of the terms in Eq. 4.13 in Section 10.2.2.3, we can immediately see that the matrix inverse $(I_N - \delta W^*)^{-1}$, the same term as we discussed in Section 4.9.1, comes into play. At this point, we shall highlight some of the properties of this model induced by the W matrix.

First, in Eq. 4.13, the independent error terms (represented by e) are multiplied by $(I_N - \delta W^*)^{-1}$ so that the error terms in the SLM model are no longer independent. In fact, some spatial econometric models (the SLM model is an example) incorporate spatial dependence in the observed outcome values through the likelihood function. This contrasts with hierarchical modelling where spatial dependence in data is modelled through the process model. But we shall defer further discussion of these two approaches to Chapters 7, 8 and 10.

Second, in contrast to a standard regression model, the covariate effect, β, is multiplied by $(I_N - \delta W^*)^{-1}$. As a result, a change to a covariate's level in area i not only leads to changes to the outcome in the same area (the "direct impact" of LeSage and Pace, 2009), it also leads to changes to the outcomes in all the other areas due to the ripple effect – as we illustrated in Figure 4.9. However, the closer (further away) an area is to the originating area i, the stronger (weaker) is the spillover effect it experiences. This can be seen through the power expansion of $(I_N - \delta W^*)^{-1}$ in Eq. 4.10.

Finally, the interrelatedness of the outcome variables, as in Eq. 4.12, gives rise to a feedback mechanism due to backtracking. For example, a change in a covariate's level in area i can lead to changes in the outcomes in other neighbouring areas, which in turn affect the outcome in the originating area i (the "indirect impact" of LeSage and Pace, 2009).

Quantifying the direct and indirect impacts of an observable covariate is a topic of importance in certain spatial econometric models. We shall investigate their calculation in Chapter 10 (Section 10.3). However, in identifying these direct and indirect impacts, the above arguments have demonstrated the crucial role played by the (assumed) W matrix.

It is also worth highlighting two identification problems,[5] as listed below, in spatial econometric modelling, both of which are associated with the W matrix:

1. Whilst the specification of the W matrix eases the estimation problem, it cannot be said to test directly which regions interact with one another nor the strength of their interactions (Harris, 2011, p.263). The *true* W matrix is unknown – however

---

[4] The reduced form of a spatial econometric model is the form of the model in which the outcome variables are expressed as a function of the covariate(s) and the error terms.

[5] An identification problem is the inability in principle to identify the best estimate of the values of one or more parameters in a regression model. Put differently, more than one set of parameters (including those associated with spillover effects) can generate the same distribution of observations.

reasonable it might seem to base its construction on contiguity or distance criteria or interaction data, for example. This has led some authors to stress the importance of theory for specifying **W** (Davies and Voget, 2008) or at least to assess the robustness of model results to alternative and equally plausible specifications of **W**. There is uncertainty associated with specifying **W**.

2. Even if the model and the **W** matrix are correctly specified, a second identification problem arises when the unknown parameters of the model cannot be uniquely recovered from the reduced form of the model. Vega and Elhorst (2015, p.341) observe that this type of identification problem can arise with what they call the general nesting spatial model (Section 10.5.1) that contains spatial interaction terms in the outcome variables, the explanatory variables and the errors. An identification problem can also arise if the **W** matrix is specified using an economic variable (using for example the attribute-based methods discussed in Sections 4.5 and 4.6), and the variation of that economic variable can be explained by any of the variables (covariates or the outcome variable) in the model. This last form of identification problem is avoided if, for example, contiguity or distance criteria can legitimately be used (on economic grounds) to specify **W**.

Vega and Elhorst (2015, p.341) conclude: "the basic identification problem in spatial econometrics is the difficulty to distinguish different models and different specifications of **W** from each other without reference to specific economic theories."

4.9.3 Implications for Estimating the Effects of Observable Covariates on the Outcome

As we have already noted in Chapter 1 (see examples 1 and 3 in Section 1.3.2), regression modelling can be employed to provide estimates of fixed effect parameters, i.e. the regression coefficients that quantify the effects of observable covariates on the outcome. But suppose, after fitting the model, spatial dependence is still encountered in the model's residuals. Remedial steps include adding a set of spatially structured random effects in the case of hierarchical modelling or specifying a spatially structured error term in the case of spatial econometric modelling. This is often done to adjust the fixed effect estimates for the (unobserved) presence of spatially structured missing covariates that are impacting the outcome. Suppose the fixed effect of interest is the association between an environmental exposure (e.g. air pollution) and stroke deaths, as in Section 1.3.2, Example 1. Suppose, as is often the case, this environmental exposure is itself spatially structured. Having accounted for the spatial dependence structure in the model's residuals by including spatially-structured random effects (or error terms), the estimated exposure-outcome fixed effects relationship will change. Yet the choice of the spatial structure for the residuals, through the use of the **W** matrix, is typically a modelling assumption, as opposed to something determined from the data. If the spatially structured missing covariates are confounders in the relationship, it will be difficult to assess the true impact of the environmental exposure on the health outcome. This is known as the *problem of spatial confounding* (see Hodges and Reich, 2010; Wakefield, 2007; Paciorek, 2010). Currently, a response is to assess the sensitivity of findings to different forms of the spatial model for the errors, including the choice of the **W** matrix (Haining et al., 2010).

4.10 Estimating the *W* Matrix

Section 4.9 discusses the importance of the *W* matrix in spatial modelling. Yet its true form is unknown to us, and hence there is uncertainty associated with its specification. Instead of treating the *W* matrix as a matrix of fixed numbers, various methods, in the literature of both hierarchical modelling and spatial econometrics, have been proposed to estimate the *W* matrix from data.

Methods for estimating *W* have been described in the spatial econometrics literature in the context of model selection. One strategy proposed is to fit the model in hand with the *W* matrix specified by a set of plausible configurations and then select the best configuration through model comparisons (see, for example, Seya et al., 2013 and LeSage and Pace, 2009). When the *W* matrix is specified using distance-based methods (such as those in Sections 4.2 and 4.3), the parameters, for example, a in Eq. 4.1 and λ and γ in Eq. 4.2, could be estimated rather than simply prespecified by the analyst. Such estimation methods have been implemented by and within a wider class of geostatistical models by Diggle et al. (1998). See also Vega and Elhorst (2015), who specify the elements in *W* using method (ii) in Eq. 4.2, then estimate the power γ from data.

Methods for estimating the *W* matrix have also been considered in the spatial epidemiology literature in the context of obtaining small area estimates of disease rates in order to reduce the risk of oversmoothing. Lee and Mitchell (2013) describe a method using locally adaptive spatial smoothing. Their method seeks to take into account spatial heterogeneity in disease risk, where for some parts of the map a smooth transition (from one area to another) is evident between adjacent neighbours, whilst in other parts abrupt step changes (or boundaries) are observed.[6] In the former case, it is appropriate to borrow information from the adjacent neighbour; in the latter case, not. Identifying step changes may also be helpful in defining the location of disease clusters as well as the aetiological factors contributing to the variation in risk. The basis of their approach is to treat the non-zero elements in *W* as random variables (these are the pairs of areas that are contiguous) whilst the zero elements in *W* remain fixed at zero. If $w_{ij}=1$, this implies that areas i and j are correlated (see Section 4.9.1) and the random effects for both areas will be locally smoothed according to the neighbourhood structure. If $w_{ij}=0$ then areas i and j are conditionally independent. The approach of Lee and Mitchell (2013) is to define the *W* matrix as a matrix containing random elements, then to construct a decision rule based on the marginal posterior distributions of the corresponding random effects for deciding whether each non-zero weight element should be reset to zero. We will present details of their modelling in Section 8.4. But it is important to realise that resetting $w_{ij}=0$ (indeed under any of the methods described where $w_{ij}=0$) does *not* mean there will be no local smoothing (information sharing) between i and j, but rather the smoothing will be much reduced and will depend on, for example, the geography of step changes involving the neighbours of the neighbours. As Section 4.9.1 describes, there is a "ripple effect" associated with the smoothing process when using a spatial model such as the proper CAR model (see Chapter 8 for more detail).

---

[6] The detection of step changes on a map is referred to as "wombling", after Womble (1951).

4.11 Concluding Remarks

The W matrix summarizes our modelling assumptions about the spatial relationships between the areas (or fixed points or sites or polygons) to which data values refer in the database. Defining relationships between areas in geographical space raise many more challenges than the equivalent problem in time series analysis. Time series data are usually recorded at regular intervals, with time imposing a natural ordering. By contrast, as we have seen, there are many ways of defining spatial relationships, from those based purely on geometric properties to those that reflect, in some sense, real similarities in terms of attribute values or real interactions between places. We have also presented approaches based on estimating the non-zero elements in the W matrix. Many of these methods for defining W will appear again in Part II of the book.

Constructing the W matrix carries significant implications for map smoothing, small area estimation, spatial econometric modelling, as well as regression modelling more generally. Those implications differ between hierarchical modelling and spatial econometric modelling because of the way spatial dependence is specified in the two classes of models – in the process model in the first case, in the likelihood function in the second. As a consequence, the "rippling" effect associated with spatial models, such as the pCAR model, and which depend on how W is specified, have different consequences in the two methodologies. We shall have more to say about the two modelling approaches in Part II, especially in Section 10.5.2.

4.12 Exercises

Exercise 4.1. What is the "identifiability problem" associated with the weights matrix (W) in spatial econometrics?

Exercise 4.2. Explain and give an example of the "problem of spatial confounding" in regression modelling.

Exercise 4.3. Describe circumstances where it is better to work with the row-standardised form of the weights matrix rather than the unstandardised form.

Exercise 4.4. Obtain W^2, W^3 and W^4 for the map shown in Figure 4.1 based on binary weights under queen's move contiguity. Inspect a selection of the elements in each matrix and identify all the different pathways that link the corresponding pairs of polygons.

Exercise 4.5. For W^2 and W^3, obtained from Exercise 4.4, identify the second and third order neighbours of each polygon (if any), that would be used to compute second and third order measures of spatial autocorrelation.

Exercise 4.6. We presented a form of the spatial lag (SLM) model in Eq. 4.12. The model, without any covariates, has the form ($i = 1,...,N$):

$$Y_i = \alpha + \delta \sum_{j=1}^{N} w_{ij}^* Y_j + e_i.$$

(i) Rewrite the model in matrix notation (using the W matrix).

(ii) Specify the model on the map given by Figure 4.1 and for the associated binary queen's contiguity matrix shown in Figure 4.1. Trace the spillover and feedback effects (using both the mathematical form of the SLM model and the map) that spread from area 6 following a one unit increase in the value of Y_6.

(iii) Consider the full version of the SLM model, where a set of k observable covariates (exogenous variables), $X_i = (x_{i1}, \dots, x_{ik})$, are included:

$$Y_i = \alpha + \delta \sum_{j=1}^{N} w_{ij}^* Y_j + X_i \beta + e_i.$$

Write this model in matrix notation and re-arrange terms to give the reduced form of the model (in the form of Eq. 4.13).

Exercise 4.7. Let $w_{ij}^{(k)}$ and $\left(w_{ij}^*\right)^{(k)}$ be the elements on row i column j of W^k and $\left(W^*\right)^k$, respectively. Show that if $w_{ij}^{(k)} = 0$, then $\left(w_{ij}^*\right)^{(k)} = 0$ when $w_{ij} \geq 0$ for all i and j.

4.13 Appendices

Appendix 4.13.1 Building a Geodatabase in R

This appendix introduces some of the aspects associated with constructing a geodatabase in R. Figure 4.11 illustrates the particular geodatabase that we are focusing on here. Each row in a dataset is linked to the corresponding geographical features in a shapefile. A shapefile is a commonly used file format for storing the geometric and attribute information of some geographical features. These geographical features can represent areas (such as census tracts) or points (such as the locations of petrol stations or retail units in a city). In the case of areas, the geometric information contains multiple sets of coordinates that form the boundaries of the areas, whereas in the case of points, the geometric information stored in a shapefile contains the coordinates of the points. A shapefile also contains some attribute information about the geographical features, for example, the names of the areas (or the petrol stations). There are several online resources from which shapefiles can

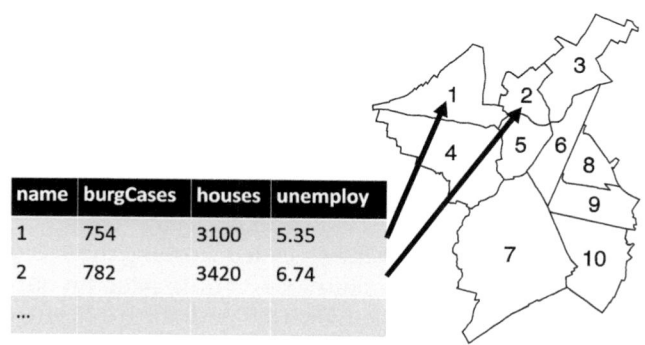

name	burgCases	houses	unemploy
1	754	3100	5.35
2	782	3420	6.74
...			

FIGURE 4.11
Linking a (external spatial or spatial-temporal) dataset to a map.

be obtained.[7] For example, the UK Data Service provides digitised boundary data for different layers within the census geography (https://census.ukdataservice.ac.uk/use-data/guides/boundary-data). The U.S. Census Bureau also provides a similar service (https://www.census.gov/geo/maps-data/data/tiger-cart-boundary.html).

In addition to the shapefile, we also have a dataset that contains our spatial or spatial-temporal data (e.g. the number of burglary events or disease cases observed in each area within the study region over a number of years), where the rows in this dataset are georeferenced to the areas (or points) in the given shapefile. However, quite often, such a dataset is given to us in a file that is separate from the shapefile. So our first task is to use R to link this dataset to the shapefile.

Solving this task involves the following four steps:

1. Read the shapefile into R.
2. Read the spatial (or spatial-temporal) dataset into R.
3. Link the shapefile with the spatial (or spatial-temporal) dataset.
4. Save the resulting shapefile.

To illustrate the above four steps, we use the example given in Figure 4.11. There are two files that we need to download from the book's website. The first file, TenAreas.zip, contains the relevant files to create the map in Figure 4.11. The second file, data_for_TenAreas.csv, is a spatial dataset where each row is georeferenced to each of the 10 areas in the above map. There are four variables (attributes) in this spatial dataset:

- name: the names of the areas
- burgCases: the number of burglary cases reported in each area
- houses: the number of houses at-risk in each area
- unemploy: the percentage of unemployed residents in each area

To proceed, create a folder called sptmbook on the C drive of your computer. Download the two files as described above and save them to the folder that you just created. Unzip the file TenAreas.zip so that three additional files appear in the folder: TenAreas.shp, TenAreas.dbf and TenAreas.shx.[8] The R code below reads the shapefile into R:

```
1   setwd('c:/sptmbook/') # specify the working directory
2   library(maptools) # load the maptools library in
3   shp <- readShapePoly('TenAreas.shp') # read the shapefile into R
4   plot(shp) # produce a simple map
5   str(shp@data) # show structure of the attribute data in the shapefile
```

Line 1 uses the setwd function to specify the so-called working directory where the shapefile and the data file are stored. This simplifies the coding later, because when we read a file into R, we only need to point R to the file name without specifying the full directory

---

[7] There may be circumstances where you need to construct your own shapefile. Constructing shapefiles is beyond the scope of this book. For advice see, for example: http://gis.yohman.com/up206a/how-tos/how-to-create-your-own-shapefile/ or https://docs.qgis.org/2.8/en/docs/training_manual/create_vector_data/create_new_vector.html.

[8] A shapefile consists of three mandatory files with the file extensions of .shp, .shx and .dbf. These three files must be in the same folder when reading a shapefile into R.

every time. For example, without setting the working directory, Line 3 would become shp <- readShapePoly('c:/sptmbook/TenAreas.shp'), where we have to specify the full directory in the function's argument. Line 2 loads the R package maptools, a package in R containing a collection of R functions for handling spatial objects. If this is the first time you are using this package (or indeed using R), you need to install the package by typing the command install.packages('maptools') prior to running the R code above. Line 3 uses the readShapePoly function (from the maptools package) to read the shapefile into R. Then Line 4 displays the map as shown in Figure 4.11 (without the labelling), and Line 5 shows the structure of the attribute data stored in that shapefile. In particular, the syntax shp@data accesses the attribute data, and the R function str displays the structure of the data. R returns the following output, showing that there is one variable called areaName stored in the shapefile, and this variable consists of the names of the areas (or polygons). The first polygon is called A1, the second polygon is called A2 and so on.

```
'data.frame': 10 obs. of  1 variable:
 $ areaName: chr  "A1" "A2" "A3" "A4" ...
```

It is important to make a note of the name of this variable, as it will be used as the variable to merge the shapefile with the other dataset. In other words, this variable is the *link* between the shapefile and the spatial dataset that we will look at next.

To read the data file, data_for_TenAreas.csv, into R, we run the following R code:

```
6   # read the data into R
7   crimeData <- read.csv('data_for_TenAreas.csv',header=TRUE)
8   crimeData # display the dataset on screen; you can also use
9    # str(crimeData) to show the structure of the dataset
```

Here, the spatial dataset is in a csv (comma-separated-values) file, so we use the R function read.csv to read the data in. The second argument header in the read.csv function is set to TRUE (all capital letters), because the first line in the data_for_TenAreas.csv file contains the column names. So, in practice, it is useful to inspect the data file before importing it into R. When the dataset is provided as an Excel file, one can export the dataset to a csv file (using the Save As ... option in Excel). Then run the R commands as given here. Another important point is that, in the spatial dataset, the column containing the names of the areas is labelled as name. However, the name of this column has to match the name of the column in the shapefile (which is labelled as areaName there). The following line of R code solves the problem by simply changing the name of the first column in crimeData to areaName:

```
10  colnames(crimeData)[1] <- 'areaName'
```

We are now in position to merge the shapefile with the dataset crimeData. This is done using the merge function. Note that the setting of the third argument, by='areaName', tells R to match the rows in crimeData with the polygons in shp using the variable areaName so that the row for A1 in crimeData is assigned to the polygon called A1 and the row for A2 in crimeData is assigned to the polygon called A2 and so on.

```
11  new.shp <- merge(shp,crimeData,by='areaName')
12  new.shp@data
```

The R output below from running the code on Line 12 confirms the correct matching.

```
areaName SP_ID burgCases houses unemploy
      A1     A1       754   3100     5.35
      A2     A2       782   3420     6.74
      A3     A3       700   4480     8.29
      A4     A4       470   1930     5.09
      A5     A5       557   2150     7.91
      A6     A6       896   4020     5.50
      A7     A7       305   3120     6.27
      A8     A8       274   3320     5.71
      A9     A9       182   3490     6.56
     A10    A10       287   3360     4.06
```

Now, run the following in R to save the resulting shapefile to the sptmbook folder:

```
13   writePolyShape(new.shp,'TenAreas_with_data.shp')
```

This completes our task of constructing a geodatabase.

Appendix 4.13.2 Constructing the *W* Matrix and Accessing Data Stored in a Shapefile

This appendix focuses on the following two tasks. The first task is to derive a spatial weights matrix *W* using a shapefile. The second task is to access some relevant columns of data stored in the shapefile in order to carry out some calculations and/or to export the data for WinBUGS modelling. Using the shapefile constructed in Appendix 4.13.1, we illustrate how to accomplish the first task.

First, we read the modified shapefile into R. Go through the material in Appendix 4.13.1 if you have not already done so.

```
1   setwd('c:/sptmbook/') # specify the working directory
2   library(maptools)     # load the maptools library
3   # read the modified shapefile into R
4   new.shp <- readShapePoly('TenAreas_with_data.shp')
```

The function poly2nb in the spdep package derives the neighbourhood structure from a shapefile. So Line 5 in the R code below loads the required package into R, then Line 6 obtains the first order neighbours of each area using queen's move contiguity. If queen=FALSE in the polyg2nb function, rook's move contiguity is then used. Line 7 (str(nb)) displays the structure of the resulting R object, nb.

```
5   library(spdep) # load the spdep library
6   nb <- poly2nb(new.shp,queen=TRUE)
7   str(nb)
```

As shown below, nb is a list with 10 elements, and each element is an array containing the first order neighbours of an area. For example, the first array in the list nb tells us that the first order neighbours of area 1 are areas 2, 4 and 5. The second array tells us that the first order neighbours of area 2 are areas 1, 3, 5 and 6.

```
List of 10
 $ : int [1:3] 2 4 5
 $ : int [1:4] 1 3 5 6
 $ : int [1:3] 2 5 6
 $ : int [1:3] 1 5 7
 $ : int [1:6] 1 2 3 4 6 7
 $ : int [1:6] 2 3 5 7 8 9
 $ : int [1:5] 4 5 6 9 10
 $ : int [1:2] 6 9
 $ : int [1:4] 6 7 8 10
 $ : int [1:2] 7 9
 - attr(*, "class")= chr "nb"
 - attr(*, "region.id")= chr [1:10] "0" "1" "2" "3" ...
 - attr(*, "call")= language poly2nb(pl = new.shp, queen = TRUE)
 - attr(*, "type")= chr "queen"
 - attr(*, "sym")= logi TRUE
```

Using the neighbourhood structure in nb, we can now derive various versions of the **W** matrix via the nb2mat function:

```
8    W <- nb2mat(nb, style='B')   # binary (0/1) weights
9    W                            # display W on screen
10   std.W <- nb2mat(nb, style='W') # row-standardised W
11   std.W                        # display std.W on screen
```

In the above code, W is the **W** matrix shown in Figure 4.1, and std.W is the row-standardised version given in Figure 4.6. The second argument style in the nb2mat function determines whether a set of binary (0/1) weights are used (by setting style='B') or the weights are row-standardised (by setting style='W'). Higher order neighbours can be obtained using the nblag function. For example, the R code, higher.order.neighbours <- nblag(nb,2), finds the neighbours of each area up to and including the second order. The second argument of the nblag function determines the maximum order of the neighbourhood structure to be constructed. Type str(higher.order.neighbours) into R to show these neighbours.

We can also calculate the **W** matrix raised to some power in R. For example, W.squared <- W%*%W calculates W^2 and W.power.three <- W%*%W%*%W calculates W^3 and so on. Note that the R syntax %*% performs a matrix multiplication.

As we shall discuss in Chapter 8, the intrinsic conditional autoregressive (ICAR) model and the proper conditional autoregressive (pCAR) model are two spatial models that allow us to impose a spatial structure on a set of unit-specific parameters, one parameter for each spatial unit. Whilst we will discuss the statistical properties of these spatial models in Chapter 8, the implementation of these models in WinBUGS requires us to input the chosen spatial weights matrix as data. In particular, when fitting the ICAR model, the car.normal function in WinBUGS is used and the **W** matrix enters the car.normal function via three data arrays, namely, num[], adj[] and weights[]. The array num[] is of length N, which is the number of areas in the study region. The ith element in the array num[] shows the number of neighbours that area i has. The IDs of the neighbours and their associated weights are stored in adj[] and weights[], respectively. Both adj[] and weights[] have the same length, which is the total number of neighbours across all areas.

These three arrays can be obtained in R using the nb2WB function. Running the following two lines of R code, we have

```
12   spatial.data.for.WinBUGS <- nb2WB(nb)
13   spatial.data.for.WinBUGS # display the results on screen
```

```
$adj
 [1]  2  4  5  1  3  5  6  2  5  6  1  5  7  1  2  3  4  6  7  2  3  5  7  8  9  4
[27]  5  6  9 10  6  9  6  7  8 10  7  9
$weights
 [1] 1 1 1 1 1 1 1 1 1 1 1 1 1 1 1 1 1 1 1 1 1 1 1 1 1 1 1 1 1 1 1 1 1 1 1 1 1 1
$num
 [1] 3 4 3 3 6 6 5 2 4 2
```

The resulting R object, spatial.data.for.WinBUGS, is a list containing the three arrays that WinBUGS requires. We shall defer the detailed explanation of the values in each array to Section 8.2.1.2.

We now turn our attention to the second task: accessing data values stored in a shapefile. Suppose we want to produce a choropleth map of a variable (or attribute) that is some function of a variable (or several variables) stored in a shapefile. To do that, we first need to extract the relevant columns of data from the shapefile, then perform the calculation. Take the example of calculating the burglary rates – the burglary count in each area divided by the corresponding number of houses – across the 10 areas in Figure 4.11. The three lines of R code below carry out this calculation:

```
14   y <- new.shp@data$burgCases   # extract the burglary counts
15   pop <- new.shp@data$houses     # extract the numbers of houses
16   burgRate <- y/pop              # calculate the burglary rates
```

Again, the R syntax, new.shp@data, accesses the dataset (the attribute values) stored in the shapefile, and new.shp@data$burgCases accesses the column labelled as burgCases in that dataset. Line 16 then calculates the burglary rates as required. We will come back to the above procedure in Section 6.2.1, where we show how to create a choropleth map in R.

Another reason for accessing the data values in a shapefile is to export them from R to WinBUGS. Suppose we want to fit a model in WinBUGS taking the burglary counts, the numbers of houses and the spatial neighbourhood structure as input data. We need to put these data values into a list in R, then write the list of data to a file. The following lines of R code accomplish that:

```
17   # use the length function to obtain the number of areas
18   # in the study region
19   nareas <- length(new.shp@data$burgCases)
20   data.for.WinBUGS <- list(N=nareas,cases=new.shp@data$burgCases
21    ,pop=new.shp@data$houses)
22   data.for.WinBUGS <- c(data.for.WinBUGS,spatial.data.for.WinBUGS)
```

In particular, Lines 20 to 21 create a list with the burglary case counts, the numbers of houses and the number of areas. Line 22 combines the list data.for.WinBUGS with the list spatial.data.for.WinBUGS, where the latter contains the neighbourhood structure.

To write the resulting data list to a file, we use

```
23   library(R2WinBUGS)
24   write.datafile <- R2WinBUGS:::write.datafile
25   write.datafile(data.for.WinBUGS,towhere='WinBUGSdata.txt')
```

Line 25 writes the data list (data.for.WinBUGS) to a file called WinBUGSdata.txt in the working directory (which is c:/sptmbook/). The R function used there is write.datafile. However, that function is a hidden function (a non-visible function) in the R package R2WinBUGS. Before the hidden function can be used, Line 23 loads the package, then Line 24 accesses this hidden function from the package (R2WinBUGS:::write.datafile) and makes it visible. We can then use it on Line 25.

5

Introduction to the Bayesian Approach to Regression Modelling with Spatial and Spatial-Temporal Data

5.1 Introduction

This chapter provides an important bridge between the material that has been covered so far in the book and the analytical material that follows in Parts II and III. We have three key objectives. The first is to provide the reader with an introduction to Bayesian inference from a theoretical perspective (what do we mean by "Bayesian inference"), from a model building perspective (how do we construct Bayesian models to tackle the problem in hand) and from a computational perspective (how do we implement/fit a Bayesian model using WinBUGS). The second objective is to demonstrate how to apply Bayesian inference to analyse spatial data of the type that we illustrated in the examples in Chapter 1 (Section 1.3.2). Finally, to help fix ideas, the third objective is to provide some illustrative examples using spatial data.

There are three key features of the Bayesian approach that offers itself as an efficient and pragmatic way to statistical inference (i.e. learning about the unknown parameters that we are interested in). First, the Bayesian approach allows us to utilise all sources of information that are available to us (e.g. information from the data that we observe; information from expert opinion; and/or information from previous studies). Second, the Bayesian approach represents various sources of information using probability distributions. Third, the application of Bayes' theorem combines all the available information for learning (or making inference) about the underlying process and the associated parameters. Bayes' theorem is a simple theorem concerning conditional probabilities of the form $\Pr(A \mid B)$– the probability of an event A occurring given that an event B has occurred. Yet, as we shall see in this chapter, Bayes' theorem allows us to form a probability distribution (called the posterior distribution) that encapsulates all the available information about the underlying process and the parameters.

This chapter is structured as follows. In Section 5.2 we will, through examples, introduce the three key components of any Bayesian model: the prior distribution, the likelihood function and the posterior distribution. In a Bayesian model, prior distributions need to be assigned to all unknown model parameters. A prior distribution is a probability distribution that represents our knowledge about an unknown parameter before the data are analysed. The data we observe contain another source of information about the unknown parameters, and this information is represented through the likelihood, a probability distribution for the data. We then use Bayes' theorem to combine the prior information with the likelihood to form the posterior distribution. The posterior distribution can be viewed as the updated knowledge about the unknown parameters in light of the data. Inference about an unknown parameter is based on its posterior distribution. In Section 5.3, we will discuss some issues associated with summarising the posterior distribution, and we will introduce WinBUGS, a flexible statistical programme to implement Bayesian models. We will discuss the Bayesian implementation of regression models to spatial data in Section 5.4. These regression models will lay the foundation for the more complex spatial and spatial-temporal models that will be described in subsequent chapters. Section 5.5 discusses model comparison and model evaluation, two important elements in any statistical modelling within the Bayesian framework. As we shall see, prior specification is a key element in Bayesian inference and it is particularly so in modelling spatial and spatial-temporal data. In Section 5.6 we look at different prior specifications and in particular "non-informative" and "informative" priors and see how informative priors can be used to express geographical-substantive as well as spatial knowledge (we will draw a distinction), and the nature as well as the source of that knowledge.

5.2 Introducing Bayesian Analysis

5.2.1 Prior, Likelihood and Posterior: What Do These Terms Refer To?

To introduce these three terms and the relationship between them, consider the simple example of learning about the probability of obtaining a head when a coin is flipped once. The quantity of interest is the unknown probability of getting a head from a flip of a given coin and is denoted as θ (the Greek letter theta). To learn about this unknown probability, θ, an obvious way forward is to conduct an experiment by flipping this coin several times to see how many heads we observe. However, *before* carrying out any experiment, we may be able to say something about θ based on either our own belief or knowledge. For example, if the experimenter is a coin expert, then after noting that the coin is a genuine British one pound coin, she forms the belief that the chance of getting a head should be quite close to 50%, say, between 45–55%. However, suppose close inspection of the coin reveals scratches on both sides of the coin which may affect its rotation in the air. The coin expert may now revise her belief, asserting that "θ is most likely to be 0.5 and is unlikely to be outside the range 0.4 to 0.6." By contrast, suppose the experimenter knows absolutely nothing about coins, has never seen a British one pound coin before and is sceptical about whether it is genuine. This experimenter may say "θ will take any value between 0 and 1." Why 0 and 1? It is because θ is a probability. Both of those two statements about θ are valid,[1] and they both represent beliefs/knowledge about the unknown parameter θ *prior to* seeing the results from the experiment (i.e. the data). Clearly, the first statement implies stronger prior belief (or less prior uncertainty) about θ than the second one.

Prior information can also come from knowledge generated from previous experiments. For example, if other experimenters have flipped coins that were produced from the same batch as the one in hand, such results, say 48 heads out of 100 flips, can be used in the current analysis as a form of prior information. We will say more about this in Section 5.6.

In a Bayesian model, the analyst specifies a *prior distribution*, a probability distribution to represent the prior information for an unknown parameter. Hereafter, we will use "a

---

[1] If you say "θ will likely be between –1 and 9", then this is not a valid statement about your prior knowledge about θ because θ clearly cannot be negative and cannot go beyond 1!

TABLE 5.1

The Probability Density Functions for the Beta Distribution and the Uniform Distribution with Some Examples

$\theta\sim Beta(a,b)$	$\theta\sim Uniform(c,d)$
Probability Density Function	Probability Density Function
$Pr(\theta\mid a,b)=\dfrac{\Gamma(a+b)}{\Gamma(a)\cdot\Gamma(b)}\cdot\theta^{a-1}\cdot(1-\theta)^{b-1}$, with θ defined between 0 and 1.	$Pr(\theta\mid c,d)=\dfrac{1}{d-c}$, with θ defined between c and d.

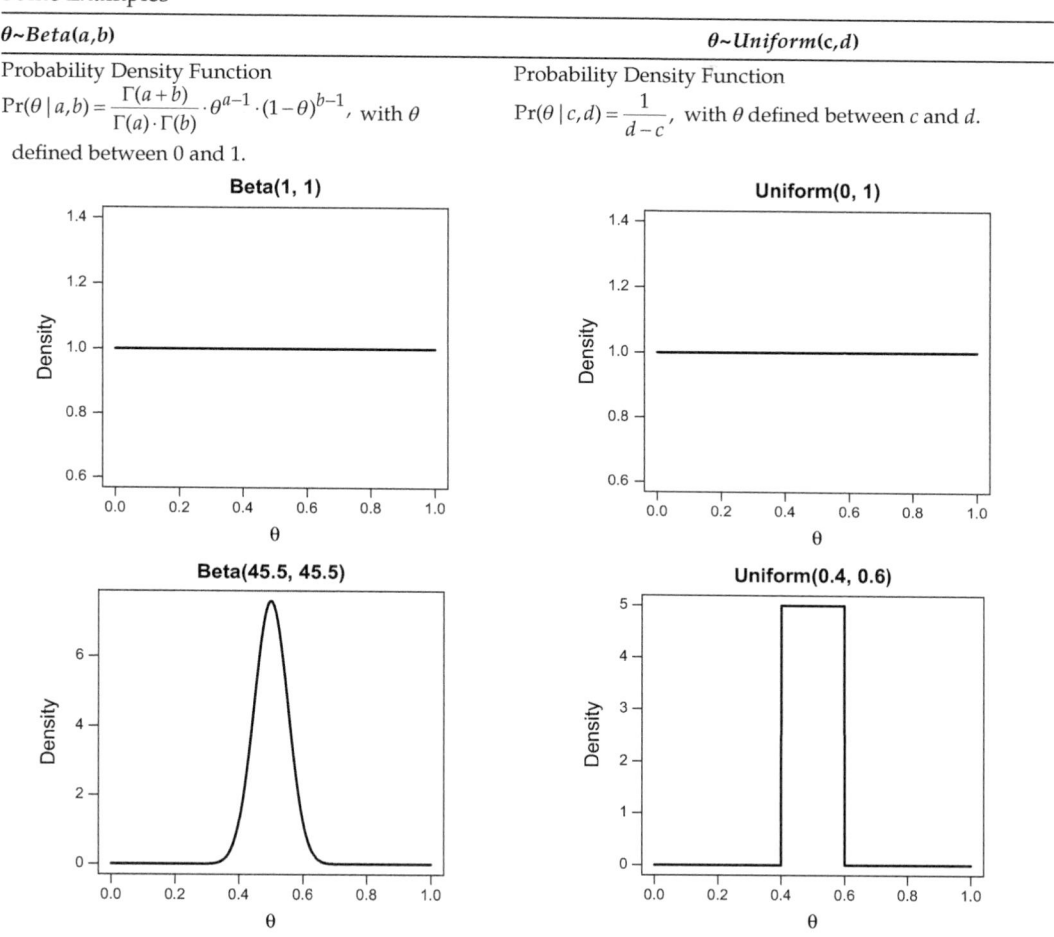

prior distribution" and "a prior" interchangeably. For example, to represent the non-expert belief, we can assign either the uniform distribution *Uniform*(0,1) or the Beta distribution *Beta*(1,1) as a prior for θ. Mathematically, we use the notation $\theta \sim Beta(1,1)$ to mean that θ follows the *Beta*(1,1) distribution. Note that the *Beta*(1,1) distribution is equivalent to *Uniform*(0,1) in the sense that both distributions present the same prior information, that is, all values between 0 and 1 are assumed to be equally likely for the unknown parameter θ before seeing any data.

Table 5.1 shows the shapes of both distributions as well as their probability density functions. The expert opinion – the parameter θ is most likely to be 0.5 but unlikely to be outside the interval between 0.4 and 0.6 – can be represented using either the *Beta*(45.5, 45.5) distribution or the *Uniform*(0.4, 0.6) distribution.[2] As opposed to a flat distribution from the

---

[2] The former may capture the expert's opinion more closely, as the latter implies (a) all values between 0.4 and 0.6 are equally likely and (b) it is impossible for θ to go outside the interval defined by this distribution.

non-expert belief, these two prior distributions are more *informative* about where θ lies: it is more likely to lie between 0.4 and 0.6 than anywhere else (Table 5.1). The two distributions arising from the expert's belief are called *informative priors*, whereas the two distributions arising from the non-expert's belief are called *vague* or *weakly-informative priors*. We will have more to say about prior specification in Section 5.6. Exercise 5.1 shows how to derive *Beta*(45.5, 45.5) based on the expert's opinion.

The other piece of information about θ comes from data. Suppose an experiment has been carried out in which this coin was flipped 10 times and six heads were observed. This is data. In a Bayesian model, data are considered to be fixed and non-random, but how many heads we observe is governed by the process of flipping this particular coin 10 times. This process itself depends upon the unknown parameter, θ. As a result, data are considered as *realizations* of the process, hence providing some information about the parameter that we want to learn about. In order to link the data to the process and thus to the parameter θ, a *likelihood* function – a probability distribution for describing the data – is specified. For the coin data, a natural choice for the likelihood function is the binomial distribution, which describes the chance of observing y successes ($y = 0,1,...,n$) out of n trials with the outcome of each trial being either a success or a failure. So, the observation $y = 6$ heads is a realization from a binomial distribution *Binomial*(θ,n) where θ is the unknown parameter and $n = 10$, the number of flips in the experiment. We write the likelihood as $y \sim Binomial(\theta,n)$, and the probability density function of a binomial distribution is:

$$\Pr(y \mid \theta, n) = \binom{n}{y} \theta^{y} (1-\theta)^{n-y} \tag{5.1}$$

where $\binom{n}{y}$ is the binomial coefficient that computes the number of ways of having y successes out of n trials.

Now we need to combine the two sources of information, the prior information and the information from the experimental data, to form *the posterior distribution*, a probability distribution that contains our updated/current knowledge about the unknown parameter θ. This combination is done using Bayes' theorem. Proposed by Thomas Bayes in 1763 (Bayes, 1763), Bayes' theorem is defined through two events, A and B:

$$\Pr(A \mid B) = \frac{\Pr(B \mid A) \times \Pr(A)}{\Pr(B)} \tag{5.2}$$

In the above expression, Pr($A \mid B$) is a conditional probability of event A given event B. The conditional probability Pr($A \mid B$) can be seen as the probability of something that we are interested in, i.e. event A occurring (or not) given the information that something else, i.e. event B, has happened. Bayes' theorem allows us to express that conditional probability in the form given on the right-hand side of Eq. 5.2. When dealing with parameters and data, we can consider parameters as event A and data as event B in the sense that parameters are the "something" that we wish to learn about whilst data are the "something" that have been observed. Applying Bayes' theorem, we have

$$\Pr(parameters \mid data) = \frac{\Pr(data \mid parameters) \times \Pr(parameters)}{\Pr(data)} \tag{5.3}$$

By "ignoring" the normalizing constant Pr(*data*) in the denominator,[3] Eq. 5.3 simplifies to:

$$\Pr\left(parameters \mid data\right) \propto \Pr\left(data \mid parameters\right) \times \Pr\left(parameters\right), \tag{5.4}$$

where "$\propto$" means "is proportional to", due to the removal of the normalizing constant. $\Pr\left(parameters \mid data\right)$ is the posterior distribution and is given by the product of $\Pr\left(data \mid parameters\right)$, known as the likelihood function, and Pr(*parameters*), the prior distribution. Hence, Eq. 5.4 is often written as

$$posterior \propto likelihood \times prior \tag{5.5}$$

In words, the posterior distribution is proportional to the product of the likelihood and the prior.

Return now to the coin flipping example. Using the prior distribution *Beta (a,b)*, where *a* and *b* can be replaced by the corresponding numbers (Table 5.1) depending on whether a vague prior or an informative prior is used, the posterior distribution for θ, the probability of getting a head from flipping the coin in question once, is given by

$$\Pr\left(\theta \mid y,n,a,b\right) \propto \left[\binom{n}{y}\theta^{y}(1-\theta)^{n-y}\right] \times \left[\frac{\Gamma(a+b)}{\Gamma(a)\cdot\Gamma(b)}\cdot\theta^{a-1}(1-\theta)^{b-1}\right] \tag{5.6}$$

The two pairs of square brackets in Eq. 5.6 enclose the binomial likelihood and the Beta prior, respectively. Combining like terms, we have:

$$\Pr\left(\theta \mid y,n,a,b\right) \propto \frac{\Gamma(a+b)}{\Gamma(a)\cdot\Gamma(b)}\cdot\binom{n}{y}\cdot\theta^{y+a-1}(1-\theta)^{(n-y)+(b-1)} \tag{5.7}$$

Since the four quantities, *a*, *b*, *n* and *y*, are known numbers, the term $\frac{\Gamma(a+b)}{\Gamma(a)\cdot\Gamma(b)}\cdot\binom{n}{y}$ is simply a multiplicative constant that does not involve θ. Thus, the posterior distribution for θ can be simplified to

$$\Pr\left(\theta \mid y,n,a,b\right) \propto \theta^{y+a-1}(1-\theta)^{(n-y)+(b-1)} \tag{5.8}$$

Comparing Eq. 5.8 to the probability density function of the Beta distribution in Table 5.1 (ignoring the multiplicative constant $\frac{\Gamma(a+b)}{\Gamma(a)\cdot\Gamma(b)}$), the posterior distribution for θ is just another Beta distribution with parameters $(y + a)$ and $(n - y + b)$.

When the posterior distribution and the prior distribution are of the same probability distribution (but with different values for the parameters), then the prior distribution is called a *conjugate* prior for the chosen likelihood. Thus, the Beta distribution is a conjugate prior for the binomial likelihood since the resulting posterior distribution is also a Beta distribution. The Uniform distribution, on the other hand, is not a conjugate prior (thus

---

[3] The normalizing constant Pr(*data*) ensures the resulting posterior distribution is a proper distribution, namely, integrates to one. This multiplicative constant can be removed since (a) the shape of the posterior distribution is unaffected and (b) WinBUGS, the software that carries out all the required computation of the posterior distribution, does not need to know this constant (WinBUGS will be introduced in Section 5.3).

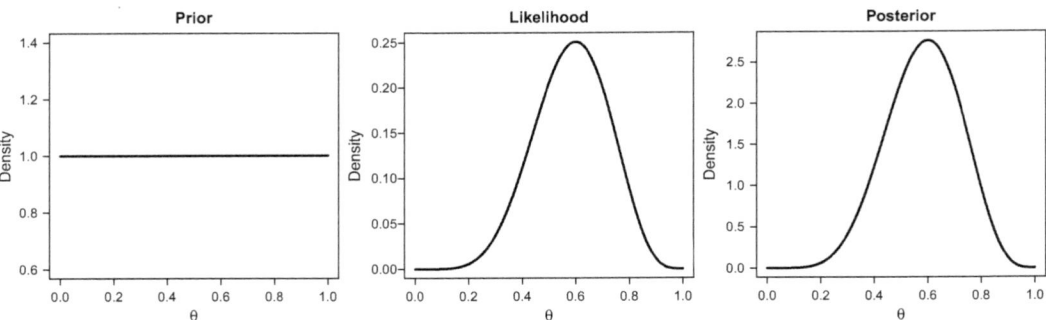

FIGURE 5.1
The prior, the likelihood and the posterior for the coin flip example.

called a *non-conjugate* prior) for the binomial likelihood. A benefit of using a conjugate prior is that the posterior distribution is a known distribution. There are closed-form expressions for various summaries (such as the mean and different percentiles) of the posterior distribution. Using a conjugate prior, as we will see in Chapters 7 and 8, also allows us to gain a better understanding of some spatial models. However, the availability of numerical methods (e.g. Markov chain Monte Carlo) and software (e.g. WinBUGS) allows us to use various forms of prior distributions, conjugate or not, that are appropriate for dealing with the problem in hand. We defer the discussion of Bayesian computation to Section 5.3.

Using the vague prior $Beta(1,1)$ (i.e. $a = 1$ and $b = 1$), and with the data $y = 6$ and $n = 10$, Figure 5.1 shows the prior, the binomial likelihood and the resulting posterior distribution $\theta \mid y = 6, n = 10, a = 1, b = 1 \sim Beta(7,5)$. The likelihood and the posterior distribution are very similar in shape. That is because, compared to the likelihood, the $Beta(1,1)$ prior contains very little information about θ. As a result, the posterior distribution is largely dominated by the likelihood. In general, when vague priors are used, the results from a Bayesian analysis are similar to those from a frequentist analysis.

To summarise, in Bayesian modelling, we need to specify a likelihood function to link the observed data to a probability model and a set of prior distributions for all the unknown parameters in that model. Using Bayes' theorem, we then combine the likelihood with the priors to form the posterior distribution, a probability distribution for the parameters. Encapsulating information from both the data and the priors, the posterior distribution allows us to learn about the unknown parameters. Before we talk about how to summarise the posterior distribution, we will look at an example of Bayesian regression modelling.

5.2.2 Example: Modelling High-Intensity Crime Areas

This example analyses a set of binary 0/1 outcome values across 337 census output areas (COAs) in Sheffield. These binary outcome values indicate whether a COA was considered as a high-intensity crime area (= 1; labelled as a PHIA) or not (= 0; labelled as a non-PHIA) based on police perceptions. Here we want to investigate whether ethnic heterogeneity affects the likelihood of being considered as a PHIA, hence a regression type of analysis.

Again, we need to specify a likelihood function for the binary outcome values and a prior distribution for each of the unknown parameters involved. For COA i where $i = 1,\ldots,337$, let y_i be the binary value indicating whether COA i is a PHIA ($y_i = 1$) or not ($y_i = 0$). Each of these binary outcome values can be modelled using a Bernoulli distribution. If a random variable X follows a Bernoulli distribution with parameter π, then X takes the value 1 with probability π and takes the value 0 with probability $1 - \pi$. So, the likelihood is written as $y_i \sim Bern(\pi_i)$, where π_i is the probability of COA i being considered as a PHIA. For each COA, we have an index, labelled as $x_{i,ethnic}$ that quantifies the level of ethnic heterogeneity. This COA-level index of ethnic heterogeneity takes a value between 0 and 1, and the larger the value the greater the ethnic mix in that COA. We want to assess whether this observable covariate on ethnic heterogeneity can be used to explain the COA-level PHIA probabilities. To do that, a logistic regression model can be used, and it specifies a regression relationship as follows:

$$logit(\pi_i) = \alpha + \beta \cdot x_{i,ethnic} \tag{5.9}$$

In Eq. 5.9, the logit transformed π_i (i.e. $logit(\pi_i) = \log\left(\dfrac{\pi_i}{1-\pi_i}\right)$) is expressed as a function of the covariate on ethnic heterogeneity. The logit transformation is used to ensure $\pi_i = \dfrac{\exp(\alpha + \beta \cdot x_{i,ethnic})}{1 + \exp(\alpha + \beta \cdot x_{i,ethnic})}$ always lies between 0 and 1. In this model, there are two unknown parameters: α, the intercept, and β, the regression coefficient. It is β that we are interested in as it measures the effect of the covariate $x_{i,ethnic}$ on the outcome. As we do not have any prior information on these two parameters, a typical choice of a vague prior for the intercept and regression coefficient(s) is a normal distribution with mean 0 and a large variance, say, 1000000 (see Section 5.6 for more detail). Therefore, for the prior specification, $\alpha \sim N(0,1000000)$ and $\beta \sim N(0,1000000)$. Combining the likelihood with the priors, the (joint) posterior distribution for α and β is

$$Pr(\alpha, \beta \,|\, data) \propto \left[\prod_{i=1}^{337} \pi_i^{y_i} \times (1-\pi_i)^{1-y_i}\right] \times \left[e^{-\frac{\alpha^2}{2 \times 1000000}}\right] \times \left[e^{-\frac{\beta^2}{2 \times 1000000}}\right] \tag{5.10}$$

On the right-hand side of Eq. 5.10, the product in the first pair of square brackets gives the likelihood and the following two pairs of square brackets are the priors for the two parameters. Putting the mathematical detail aside, the posterior distribution in Eq. 5.10 is more complicated than that from the coin flip example (i.e. Eq. 5.8) for two reasons. First, Eq. 5.10 is a multivariate probability distribution over multiple parameters. This is often the case in spatial and spatial-temporal models where there are many unknown parameters and hence the posterior distribution is higher dimensional. Second, the posterior distribution in Eq. 5.10 is not a known probability distribution, so we have no known formula to work out, for example, the posterior mean of the regression coefficient, β. In Bayesian regression modelling, determining the posterior summary of unknown parameters is done numerically via Markov chain Monte Carlo. So, to learn about β, we need to discuss the topic of Bayesian computation.

5.3 Bayesian Computation

5.3.1 Summarising the Posterior Distribution

In Section 5.2, we established that all knowledge about an unknown parameter is contained in the posterior distribution. An advantage of the Bayesian approach is, in addition to reporting a point and an interval estimate of a parameter, we can also construct any probability statements regarding the unknown parameter. We can also provide posterior estimates about any transformation of that parameter. All can be done easily using the posterior distribution.

Typically, we report the posterior mean (i.e. the mean of the posterior distribution) as a point estimate of an unknown parameter and use the 2.5 percentile and the 97.5 percentile of the posterior distribution to form the 95% *credible interval*. For example, based on the posterior distribution *Beta*(7,5) from the coin example, the posterior mean of θ, the probability of getting a head from a single coin flip, is about 0.58 and the 95% credible interval of θ is (0.31,0.83). These values can be calculated using either known formulae for the Beta distribution (see Exercise 5.2) or through Monte Carlo integration (see Section 5.3.2 later). A 95% credible interval of the form (a,b) corresponds to a probability statement, meaning that there is a 0.95 probability that the unknown parameter falls between the two numbers a and b. This is different from the interpretation of a 95% confidence interval from the frequentist approach. A 95% confidence interval (a,b) implies that if we obtain a large number of datasets through repeating the same experiment many times, then 95% of these datasets would yield values of the unknown parameter that go between the two numbers a and b. A confidence interval is based on the idea of repeated experiments (Section 1.4.2).

Since we have the entire probability distribution, we can also report probability statements such as "what is the chance that the given coin is unbiased if a coin is considered to be unbiased when its probability of getting a head from a single flip is highly likely to be between 0.45 and 0.55" or "what is the chance that the covariate on ethnic heterogeneity plays a role in explaining the observed binary values on whether a COA is considered to be a PHIA or not". These two questions can be answered by calculating the following two *posterior probabilities*, $Pr(0.45 < \theta < 0.55 | data)$ and $Pr(\beta > 0 | data))$, respectively. Figure 5.2 represents these two posterior probabilities graphically.

For the coin flip example, $Pr(0.45 < \theta < 0.55 | data) = 0.22$ (see Exercise 5.3 for the calculation). In other words, the posterior probability that θ lies between 0.45 and 0.55 is 0.22, a value that is perhaps too low to be considered as "highly likely" whilst not small enough to say the coin is biased. Basically, we cannot draw any firm conclusion based on the limited amount of data (e.g. small number of flips) and the vagueness of the prior

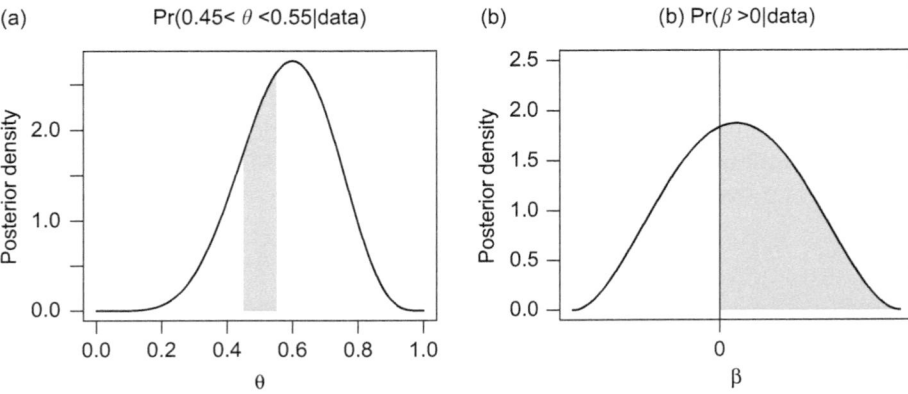

FIGURE 5.2
Graphical representation of the two posterior probabilities, $Pr(0.45 < \theta < 0.55 | data)$ in panel (a) and $Pr(\beta > 0 | data)$ in panel (b). The shaded area in each plot represents the corresponding posterior probability.

information – a situation that we often encounter in modelling spatial and spatial-temporal data (see Chapter 7). Getting more data (e.g. doing more coin flips) or incorporating additional information through the prior (e.g. expert opinion and/or previous experiments) are ways to provide more information. We will see how we can achieve the latter through modelling in Part II and Part III of this book.

For the HIA example, although we do not yet have the computational tools to calculate the posterior probability $\Pr(\beta > 0 \mid data)$, we can still make a few comments. In regression modelling, we want to assess whether a covariate is associated with the outcome or not. The frequentist approach would answer that by testing whether the corresponding regression coefficient β is equal to 0 – the value 0 indicates no effect. But would it be meaningful to calculate the posterior probability: $\Pr(\beta = 0 \mid data)$? The answer is no because that probability is always equal to 0 for a continuous-valued parameter. Instead we calculate the posterior probability that the regression coefficient β is greater than 0. If the resulting posterior probability is very large, say larger than 0.95, then this means that the majority of the posterior distribution of β lies above 0. That would imply a COA with a larger value of the ethnic heterogeneity index (i.e. with a greater ethnic mix) tends to be more likely to be considered as a PHIA. If, on the other hand, the resulting posterior probability is very small, say less than 0.05, then the majority of the posterior distribution is lying below 0. The result would then indicate that a COA with a greater ethnic mix would be less likely to be considered as a PHIA. If the posterior probability is not far away from 0.5, then β is not different from 0, suggesting that ethnic heterogeneity does not play a role in explaining the observed outcome.

In some situations, we are interested in some transformation of a parameter, in addition to the parameter itself. For example, we may be interested in (predicting) the number of heads observed if we flip the coin in question 100 times more. When using a logistic regression model, as in the HIA example, we are interested in the exponentiated coefficient β, i.e. $OR = e^{\beta}$, which is interpreted as the odds ratio (OR). From the odds ratio, we can calculate the change in odds (e.g. the probability of being a PHIA divided by the probability of being a non-PHIA) for a one-unit change in the corresponding covariate. So, in the HIA example, if the posterior mean of OR is 1.003 then a one-unit increase in the index of ethnic heterogeneity would be associated with a 0.3% $=(1.003 - 1) \times 100$ increase in the odds of being a PHIA. We can also derive a 95% credible interval for such a change because β has a posterior probability and any transformation of β also has a posterior probability. However, when the transformation is nonlinear, e.g. the exponential transformation, it must be carried out within the model fitting – we will see how to do that in Sections 5.3.2 and 5.4.2. This is because the exponential transformation of the posterior mean of β is not equal to the posterior mean of e^{β}. The reader is encouraged to demonstrate that.

5.3.2 Integration and Monte Carlo Integration

Mathematically, summaries of the posterior distribution can be written in the form of a definite integral. Consider the simple situation where we only have a single parameter, θ, the form of the integral is given by

$$\int_{H} g(\theta) \cdot \Pr(\theta \mid data) d\theta \tag{5.11}$$

where $\Pr(\theta \mid data)$ is the posterior distribution of θ, $g(\theta)$ is any function of θ and H denotes the interval over which the integral is evaluated. Eq. 5.11 gives various posterior summaries of θ through specifying $g(\theta)$ and H accordingly. For example, setting $g(\theta) = \theta$ and the interval H to $\pm\infty$, the posterior mean of θ is given by

$$\mathbb{E}(\theta \mid data) = \int_{-\infty}^{+\infty} \theta \cdot \Pr(\theta \mid data)d\theta \qquad (5.12)$$

The posterior probability of θ greater than some threshold C is calculated by

$$\Pr(\theta > C \mid data) = \int_{C}^{+\infty} \Pr(\theta \mid data)d\theta \qquad (5.13)$$

where $g(\theta) = 1$ and the integral is evaluated between C and $+\infty$.

Through Eq. 5.11, we can also compute posterior summaries of some transformation of θ. For example, the posterior mean of e^{θ} and the posterior probability of e^{θ} greater than a given threshold Q can be calculated respectively by

$$\mathbb{E}\left(e^{\theta} \mid data\right) = \int_{-\infty}^{+\infty} e^{\theta} \cdot \Pr(\theta \mid data)d\theta \qquad (5.14)$$

and

$$\Pr\left(e^{\theta} > Q \mid data\right) = \int_{Q}^{+\infty} \Pr(\theta \mid data)d\theta \qquad (5.15)$$

In general, however, there are no closed-form solutions for these integrals (except for conjugate models). These integrals need to be computed numerically via Monte Carlo integration.

The idea of Monte Carlo integration goes as follows. Suppose we can sample M random values independently from the posterior distribution. Then the empirical distribution (e.g. the histogram) of these sampled values approximates the posterior distribution and the integral in Eq. 5.11 can be approximated by the average of the function $g(\theta)$ calculated using the sampled values within the interval H. That is, denoting these values as $\theta^{(1)}, \ldots, \theta^{(M)}$,

$$\int_{H} g(\theta) \cdot \Pr(\theta \mid data)d\theta \approx \frac{1}{M} \sum_{m=1}^{M} g\left(\theta^{(m)}\right) I\left(\theta^{(m)} \in H\right) \qquad (5.16)$$

where $I\left(\theta^{(m)} \in H\right)$ is an indicator function that returns 1 if the sampled value $\theta^{(m)}$ falls within the interval H and returns 0 otherwise. The larger the M, the better the approximation becomes. Therefore, we can approximate the posterior mean and the 95% credible interval of θ by the mean of the sampled values $\theta^{(1)}, \ldots, \theta^{(M)}$ and the 2.5 and 97.5 percentiles of the sampled values. Similarly, using the exponentiated sampled values $e^{\theta^{(1)}}, \ldots, e^{\theta^{(M)}}$, the same idea can be applied to approximate the posterior mean and the 95% credible interval of e^{θ}. The posterior probability of the form $\Pr(\theta > C \mid data)$ is approximated by the proportion of the sampled values that are greater than the given threshold C.

To illustrate, consider again the coin example, where θ is the unknown probability of getting a head when flipping the given coin once. The R code given in Figure 5.3 first samples M random values independently from the posterior distribution $\theta \mid data \sim Beta(7,5)$, then uses these sampled values to approximate the posterior distribution, the posterior mean, the 95% credible interval and the posterior probability, $\Pr(0.45 < \theta < 0.55 \mid data)$. Lines 53 to 64 in Figure 5.3 calculate the above posterior summaries using the closed-form solutions

```
1   ##############################################################
2   #  specify number of random values to be sampled
3   #  independently from the posterior distribution
4   ##############################################################
5   M <- 50
6
7   ##############################################################
8   #  sample M random values independently from the posterior
9   #  distribution Beta(7,5)
10  ##############################################################
11  sampled.values <- rbeta(M,7,5)
12
13  ##############################################################
14  #  the histogram of the sampled values
15  #  Note that in the hist function, the argument xlim sets the
16  #  minimum and maximum values when plotting the histogram and
17  #  freq=FALSE tells R to plot the probability density (so
18  #  that we can superimpose the beta(7,5) density)
19  ##############################################################
20  hist(sampled.values,xlim=c(0,1),freq=FALSE)
21
22  ##############################################################
23  #  superimpose the posterior distribution Beta(7,5)
24  ##############################################################
25  #  first generate a sequence of values between 0 and 1 for
26  #  calculating the Beta density
27  x <- seq(0,1,length.out=1000)
28  #  calculate the density at each of the 1000 values generated
29  #  from the previous step
30  beta.density <- dbeta(x,7,5)
31  #  superimpose the curve of the posterior distribution
32  lines(x,beta.density)
33
34  ##############################################################
35  #  using the sampled values to approximate the posterior
36  #  mean, the 95% credible interval and the posterior
37  #  probability Pr(0.45<theta<0.55|data)
38  ##############################################################
39  mean(sampled.values)   #  posterior mean
40  #  95% credible interval using the quantile function
41  quantile(sampled.values,c(0.025,0.975))
42  #  select the sampled values within the required range
43  v <- which(sampled.values>0.45 & sampled.values<0.55)
44  #  calculate Pr(0.45<theta<0.55|data); the R function length
45  #  counts the number of values in the object v
46  length(v)/M
47
48  ##############################################################
49  #  calculating the above summaries using closed-form
50  #  solutions (some via built-in functions in R)
51  ##############################################################
52  #  the mean of Beta(a,b) is a/(a+b)
53  7/(7+5)
54  qbeta(0.025,7,5)    #  For a random variable X~Beta(a,b), the x
55                      #  solutions function returns x such that
56                      #  Pr(X<x) is equals the 1st argument of
57                      #  the function; the 2nd and 3rd
58                      #  arguments specify the parameter values
59                      #  of the Beta distribution, giving the
60                      #  lower bound of the 95% credible interval
61  qbeta(0.975,7,5)    #  the upper bound of the 95% CI
62  #  the pbeta function (below) returns the prob. of X less
63  #  than the value given by the 1st argument
64  pbeta(0.55,7,5) - pbeta(0.45,7,5)
```

FIGURE 5.3
R code to carry out a Monte Carlo integration and calculate various summary statistics from the posterior
distribution $\theta \,|\, data \sim Beta(7,5)$ using closed-form solutions (via built-in R functions).

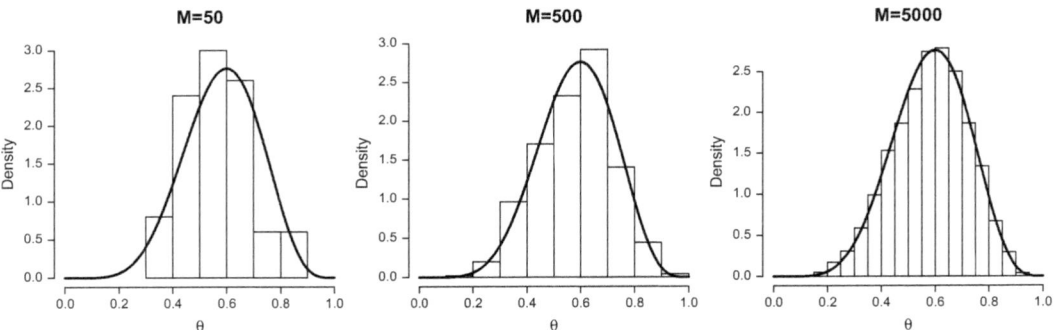

FIGURE 5.4
Comparing the approximation to the posterior distribution $\theta \,|\, data \sim Beta(7,5)$ through M random values independently drawn from the posterior distribution. See Figure 5.3 for detail.

via some built-in functions in R. Figure 5.4 compares the true posterior distribution to the histogram of the sampled values across different sample sizes. As the size of the sample M increases, the approximation becomes better. This is also evident in the numerical summaries tabulated in Table 5.2: the approximated values through Monte Carlo integration are getting close to the true values calculated from $Beta(7,5)$ as M increases.

It should be noted that we are able to compare the approximated distribution to the true posterior distribution and various approximated summary statistics to their true values because the posterior distribution in this particular example is a known probability distribution, a direct consequence of using a conjugate prior. When using non-conjugate priors, as is often the case in practice, the true values of various posterior summaries cannot be calculated analytically. Thus, Monte Carlo integration becomes essential. Figure 5.4 also highlights the importance of having a large enough set of sampled values in order to approximate the posterior distribution well. This relates to the topic of efficiency that we will return to in Section 5.3.5.3.

More generally, when modelling spatial and spatial-temporal data, there are multiple parameters. The same Monte Carlo integration idea applies, although the discussion becomes more complicated because instead of having just one single parameter θ, we have a vector of k parameters, $\boldsymbol{\theta} = (\theta_1,\ldots,\theta_k)$. We use a boldface letter to denote a vector of parameters. The idea proceeds as follows. Suppose we can sample M sets of values independently from the (multivariate) posterior distribution $\Pr(\boldsymbol{\theta}|\,data)$ and denote the ith set as $\boldsymbol{\theta}^{(i)}= \left(\theta_1^{(i)},\ldots,\theta_k^{(i)}\right)$, with $i = 1,\ldots,M$. The posterior distribution $\Pr(\boldsymbol{\theta}|\,data)$ is known as the joint posterior distribution, as it is defined for all the parameters jointly. Then, using the

TABLE 5.2

Posterior Summaries for θ Using Either the Closed-Form Solutions for the Beta Distribution or through Monte Carlo Integration with Different Sizes of Sampled Values

		Posterior Mean	95% Credible Interval	$\Pr(0.45 < \theta < 0.55 \,|\, data)$
From closed-form solutions		0.58	(0.31, 0.83)	0.22
MC integration	$M = 50$	0.57	(0.36, 0.85)	0.30
	$M = 500$	0.58	(0.30, 0.83)	0.21
	$M = 5000$	0.58	(0.30, 0.83)	0.21

sampled values associated with a particular parameter, say θ_j (where the index j takes an integer value between 1 and k), i.e. $\theta_j^{(1)}, \ldots, \theta_j^{(M)}$, we can approximate the *marginal posterior distribution* of this parameter $\Pr(\theta_j \mid data)$ and any posterior summaries as in the single parameter case.

The prerequisite for Monte Carlo integration is that we can sample values (or sets of values) from a (multidimensional) posterior distribution that is not of a standard form (e.g. not a Beta distribution or a univariate/multivariate normal distribution and so on). As a result, these samples cannot be drawn directly using, say, the `rbeta` (or `rnorm` or `rmvnorm`) function in R. One way to deal with this problem is through the use of Markov chain sampling, a topic that we discuss next.

5.3.3 Markov Chain Monte Carlo with Gibbs Sampling

Monte Carlo integration, as we have observed, is a technique for numerical integration using values sampled from a given posterior distribution. Markov chain sampling is a powerful way to provide the values required to carry out that integration technique. Sampling using Markov chains allows us to sample values from any given posterior distribution, and that distribution can be of high dimension (i.e. with many parameters). There are many Markov chain sampling methods, but we are focusing on Gibbs sampling here because it is the sampling algorithm that WinBUGS uses. Before that, we will first define what a Markov chain is and describe the general idea of sampling using Markov chains.

A sequence of random variables, $X_1, X_2, \ldots$, forms a Markov chain if the conditional distribution of X_{t+1} given $X_1, \ldots, X_t$ only depends on X_t. When applied to sampling, the Markovian property tells us to (a) explore the posterior distribution iteratively and (b) sample the next value from the distribution based on the current value. This iterative sampling idea is illustrated graphically in Figure 5.5. When this iterative procedure is carried out for long enough (over a sufficient number of iterative steps), we are able to examine the entire posterior distribution. Using these sampled values, Monte Carlo integration can then be used to provide various summaries of the posterior distribution.

Markov chain Monte Carlo (MCMC) is a collection of computational methods (or algorithms) that combine both Markov chain sampling and Monte Carlo integration to provide posterior summaries of unknown parameters. There are a large number of theoretical results on MCMC, but the discussion of these results is well beyond the scope of this book. We refer interested readers to van Ravenzwaaij et al. (2018) for an introduction and Gilks

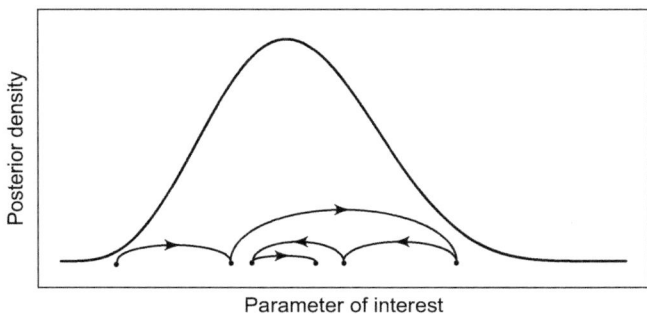

FIGURE 5.5
Illustrating the iterative nature and the Markovian property of Markov chain sampling.

et al. (1996) for more technical details and applications. We now turn our attention to Gibbs sampling.

Let $\boldsymbol{\theta} = (\theta_1,\ldots,\theta_k)$ denote a vector of k unknown parameters. Gibbs sampling generates a Markov chain by sampling from the *full conditional distributions*. For a parameter θ_j in the vector $\boldsymbol{\theta}$, its full conditional distribution is given by

$$\Pr\left(\theta_j \mid \theta_1, \theta_2, \ldots, \theta_{j-1}, \theta_{j+1}, \ldots, \theta_k, data\right),$$

which is the conditional distribution of θ_j given all other parameters and the observed data. Gibbs sampling draws values from the posterior distribution through the following three steps.

Step 1: Choose a set of initial (or starting) values for every single parameter in $\boldsymbol{\theta}$. We denote the set of initial values as $\boldsymbol{\theta}^{(0)} = \left(\theta_1^{(0)}, \ldots, \theta_k^{(0)}\right)$, where the superscript of each element in $\boldsymbol{\theta}^{(0)}$ denotes the iteration number and the subscript is the parameter index.

Step 2: Update the values of the parameters in turn using the full conditional distributions. This updating step goes as follows. Starting with θ_1, we sample a new value $\theta_1^{(1)}$ from its full conditional distribution $\Pr\left(\theta_1 \mid \theta_2 = \theta_2^{(0)}, \ldots, \theta_k = \theta_k^{(0)}, data\right)$, where all other parameters are fixed at their initial values. Similarly, for θ_2, we sample a new value $\theta_2^{(1)}$ from its full conditional distribution $\Pr\left(\theta_2 \mid \theta_1 = \theta_1^{(1)}, \theta_3 = \theta_3^{(0)}, \ldots, \theta_k = \theta_k^{(0)}, data\right)$ but, in this case, θ_1 is fixed at the updated value $\theta_1^{(1)}$, whilst all other parameters are still fixed at their initial values. We then follow the same procedure to update θ_3, θ_4 and so on until all k parameters have been updated. Then we have finished one MCMC iteration and $\boldsymbol{\theta}^{(1)} = \left(\theta_1^{(1)}, \ldots, \theta_k^{(1)}\right)$ is the set of values sampled at Iteration 1.

Step 3: Repeat the updating in Step 2 thousands of times.

The many sets of sampled values are then used to produce posterior summaries of the unknown parameters as described in Section 5.3.2, providing that we have satisfied the checks that we will come to in Section 5.3.5.

For a simpler situation where there are just two parameters, $\boldsymbol{\theta} = (\theta_1, \theta_2)$, Figure 5.6 illustrates various stages in the sampling of a (bivariate) posterior distribution using Gibbs sampling. Specifically, Panel (a) of Figure 5.6 shows that two MCMC chains are used to explore the posterior distribution, and the two MCMC chains have different initial values: one with $\theta_1^{(0)} = 4$ and $\theta_2^{(0)} = -4$, whilst the other with $\theta_1^{(0)} = -4$ and $\theta_2^{(0)} = 4$. Each solid dot represents the set of values sampled at each iteration, and the zig-zag shape of the two chains is the result of the conditional updating at Step 2. For example, from its starting position, each chain first gets a new value for θ_1 whilst keeping θ_2 at its initial value so the chain moves horizontally. A new value for θ_2 is then drawn whilst keeping θ_1 at its current position so that the chain moves vertically. After many iterations, the two MCMC chains appear to come together, sampling values within the same region (Figure 5.6(b)). There are, however, a number of checks that need to be carried out before we use the sampled values to compute posterior summaries of the two parameters. We will return to these checks after we have introduced WinBUGS, the software that performs Bayesian inference through the use of Gibbs sampling.

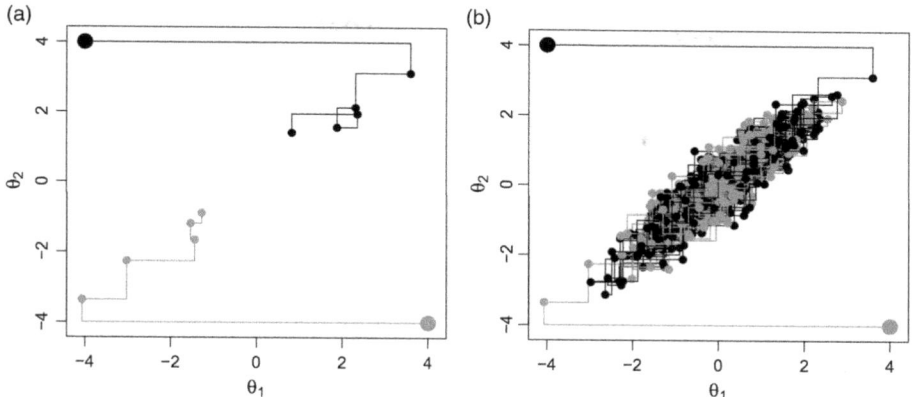

FIGURE 5.6
A graphical illustration of Gibbs sampling with two parameters. Two MCMC chains are shown. Panel (a) shows the first few iterations and panel (b) shows many hundred iterations, including those in panel (a).

5.3.4 Introduction to WinBUGS

WinBUGS is a flexible programme for carrying out Bayesian inference. WinBUGS is based on the BUGS language where the acronym BUGS stands for *Bayesian inference Using Gibbs Sampling*. In WinBUGS, the analyst specifies a likelihood function for the observed data and priors for the unknown parameters. WinBUGS then automatically constructs the resulting posterior distribution, draws samples from the posterior distribution via Gibbs sampling and subsequently produces posterior summaries of the parameters. WinBUGS is an open-source software that is freely available online. Throughout this book, Version 1.4.3 is used.[4]

We consider the simple coin example discussed in Section 5.2.1 to illustrate how to fit a model in WinBUGS. Figure 5.7 shows the full model in its mathematical form and the corresponding implementation in WinBUGS syntax.

To fit a Bayesian model in WinBUGS, we first need a model file, a file that contains the specifications of the likelihood function, the priors and possibly other quantities of interest. Every model file starts with the keyword `model`, and the model description is enclosed within a pair of curly brackets {...} (see Lines 1 and 11 in Figure 5.7). Lines 2 and 4 specify the binomial likelihood for the observed data and the Beta prior for θ, respectively. The two quantities y and `theta` are called *stochastic nodes*, and each stochastic node is associated with a probability distribution via the tilde sign ~. Typically, a stochastic node is used to specify the likelihood (e.g. Line 2) or a prior distribution for an unknown parameter (e.g. Line 4). The WinBUGS syntax for a probability distribution starts with the letter d followed by the abbreviation for that distribution, so `dbin` and `dbeta` are the WinBUGS syntax for the binomial distribution and the Beta distribution, respectively.

There is another type of node in WinBUGS called the *logical node*. For example, `diff` and `pGT.7` on Lines 7 and 10 (Figure 5.7) are logical nodes. A logical node stores the result from a set of deterministic calculations such as arithmetic operations (e.g. Line 7) and logical operations (e.g. Line 10). The left arrow sign, <-, links a logical node to its calculation. Specifically, on Line 7, the logical node `diff` stores the difference between a sampled

---

[4] OpenBUGS is another version of the BUGS language. Apart from some exceptions (see for example Section 8.2.1.2), the syntax between WinBUGS and OpenBUGS is similar. All material discussed in this chapter applies to both packages.

	WinBUGS code to implement the coin model with the Beta(1,1) prior for	Model in its mathematical form
1	`model{`	$y \sim Binomial(\theta, n)$
2	`    y      ~ dbin(theta,n) #  the binomial likelihood`	
3	`                          #  for y heads in n flips`	
4	`    theta ~ dbeta(a,b) #  the Beta prior on theta,`	$\theta \sim Beta(a,b)$
5	`                       #  the probability of head`	
6	`    # compute the difference between theta and 0.7`	
7	`    diff   <- theta - 0.7`	
8	`    # compute the posterior probability that theta`	
9	`    # is greater than or equal to 0.7`	
10	`    pGT.7 <- step(diff)`	$Pr(\theta \geq 0.7 | data)$
11	`}`	

FIGURE 5.7
The WinBUGS implementation of the coin example with the *Beta*(1,1) vague prior. Note that the line numbers are included for reference and they are not part of the WinBUGS code. Neither are the equations listed in the last column. The mathematical form is provided for explanation only.

value for `theta` and a fixed value 0.7 at each MCMC iteration. The difference then enters the `step` function on Line 10, where the `step` function returns 1 if the argument `diff` is non-negative and 0 otherwise. Therefore, the posterior mean of `pGT.7`, i.e. the proportion of the MCMC iterations where θ is greater than or equal to 0.7, gives the required posterior probability, i.e. $Pr(\theta \geq 0.7 | data)$. As we will see in later examples, the `step` function is often used to calculate posterior probabilities. We calculate the posterior probability $Pr(\theta \geq 0.7 | data)$ to simplify the discussion here. The calculation of $Pr(0.45 < \theta < 0.55 | data)$ also uses the `step` function but is slightly more complicated – see Exercise 5.4.

Two more features of the WinBUGS syntax from this simple example should be noted. First, WinBUGS is case-sensitive so, for example, `pGT.7` and `pgt.7` refer to two different quantities. Second, the hash sign, #, is the annotation symbol in WinBUGS, so comments after the # sign, on that line, are ignored. It is always good practice to annotate code.

We now need to enter the observed data and the values for the two parameters in the Beta prior. This is done using a data list as follows:

1	`list(y=6, n=10        # observed data: y=6 heads in n=10 flips`
2	`    ,a=1, b=1)        # defines the Beta(1,1) vague prior`

Data entry starts with the keyword `list` and the pair of parentheses encloses the data values.

Both the model file and the data list fully define the model. What we need now are the initial values to initiate the MCMC chains. Typically, two MCMC chains are run so that we can assess convergence of the chains (see Section 5.3.5.2). The idea of checking convergence is to ensure that the MCMC samples used to obtain summaries are from the posterior distribution. Below are two lists with initial values for the two chains:

1	`# initial value for theta for MCMC chain 1`
2	`list(theta=0.1)`

1	`# initial value for theta for MCMC chain 2`
2	`list(theta=0.9)`

The model file, the data file and the two lists of initial values are typeset and saved in separate plain text files with a .txt extension. Figure 5.8 shows all four files open in WinBUGS (to arrange the files, go to Window then choose Tile Horizontal or Tile Vertical).

To run a model in WinBUGS, follow the steps listed below (adapted from Lunn et al., 2012, p.17–20). We now go through these steps in turn.

Step 1. Open all the files containing model, data and initial values in WinBUGS.

Step 2. Open the Specification Tool (Figure 5.9) from Model -> Specification …

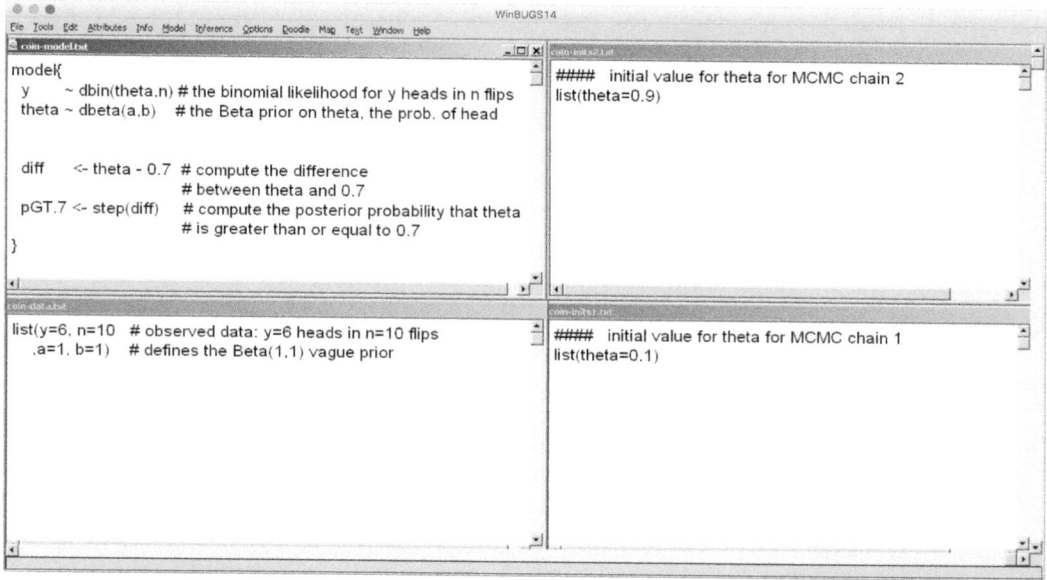

FIGURE 5.8
The four text files (the model file, the data file and the two lists of initial values) in WinBUGS in a tiled format.

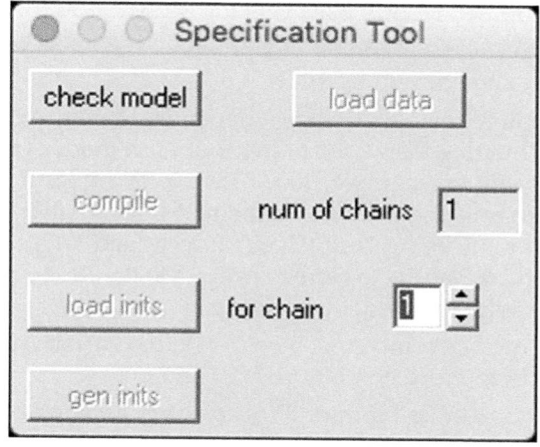

FIGURE 5.9
The specification Tool window.

Step 3. Activate the window containing the model code by clicking the banner of the model file (when a window is activated, the banner with the file name is in blue), then click on check model in the Specification Tool; WinBUGS displays any message at the bottom left-hand corner. If the model is correctly specified, it will display "model is syntactically correct".

Step 4. Activate the window with the data and click on load data in the Specification Tool; the bottom left-hand corner will read "data loaded".

Step 5. Type 2 in the text box labelled num of chains in the Specification Tool to specify running two MCMC chains.

Step 6. Now click compile in the Specification Tool and you will see the message "model compiled".

Step 7. Activate the window with the initial value for chain 1, then click on load inits in the Specification Tool. At this point, two things happen: (a) the message shows "chain initialized but other chain(s) contain uninitialized variables" and (b) the box next to the label "for chain" in the Specification Tool automatically changes to 2.

Step 8. Activate the window with the initial value for chain 2, then click on load inits in the Specification Tool and the display message will read "model is initialized".

At this point, WinBUGS knows that we are using the binomial likelihood and the Beta prior and internally formulates the (Beta) posterior distribution for theta. We have instructed WinBUGS to form two MCMC chains to sample from the posterior distribution with the first chain starting at theta = 0.1 and the second one at theta = 0.9. If you make a mistake at any point during this process (steps 1 to 8), you need to go through the steps again *from the beginning*.

Before updating the MCMC chains, Step 9 below informs WinBUGS of the quantities for which the sample values are to be stored or monitored:

Step 9. Open the Sample Monitor Tool from Inference -> Samples ... and type theta into the text box labelled node and then click set. This sets the monitor for theta. Repeat for pGT.7.

With the monitors set, Step 10 performs the MCMC updating:

Step 10. Open Update Tool from Model -> Update ... and enter 10000 in the text box labelled updates and click on update.

The number of iterations (or updates) depends on convergence and efficiency (see Section 5.3.5), but in general, the more complex a model, the more updates are required. For this simple example, 10000 iterations are sufficient. As WinBUGS updates the two chains, the number in the text box labelled iteration in the Update Tool window is being "refreshed" at a regular frequency defined in the refresh text box (the default is to refresh every 100 iterations). This feature shows how quickly (or slowly) the model is running.

Once the updating has completed, Step 11 obtains various posterior summaries of the parameters of interest, namely, theta and pGT.7.

Step 11. Type * in the Sample Monitor Tool to select all quantities that have been monitored, then click density – to produce the marginal posterior distribution for each parameter approximated using the sampled values (Figure 5.10) – and stats – to obtain various posterior summaries of the parameters (Figure 5.11).

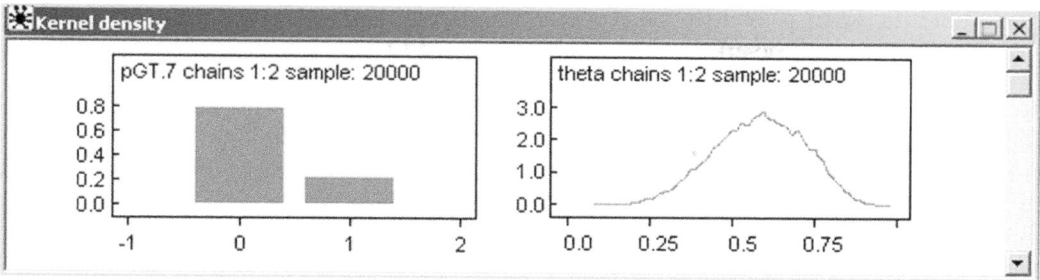

FIGURE 5.10
The marginal posterior distribution (or referred to as kernel density in WinBUGS) for each of the two parameters, theta and pGT.7.

node	mean	sd	MC error	2.5%	median	97.5%	start	sample
pGT.7	0.2154	0.4111	0.002982	0.0	0.0	1.0	1	20000
theta	0.5832	0.1372	0.001069	0.3056	0.588	0.8307	1	20000

FIGURE 5.11
Various numerical summaries for both parameters from the posterior distribution.

Figure 5.11 shows that the posterior mean for θ is 0.58 and the 95% credible interval for θ is (0.31, 0.83). The latter two values are the 2.5th and the 97.5th percentiles from the posterior distribution for θ (the values from the 5th and the 7th column in Figure 5.11). The numerical summaries based on the 20000 sampled values (MCMC samples) from two MCMC chains are in agreement with those from the closed-form solutions given in Table 5.2. The posterior probability $Pr(\theta \geq 0.7 \mid data)$ is estimated to be about 0.22, the posterior mean of pGT.7. The form of the posterior density for pGT.7 (Figure 5.10) reflects the fact that pGT.7 is obtained from the step function that returns either 1 or 0. As we shall see in Parts II and III of this book, WinBUGS can deal with more complex and realistic models for analysing spatial and spatial-temporal data. The posterior distributions for those models are no longer in closed form. However, the 11 steps described above are generic in running any model in WinBUGS.

5.3.5 Practical Considerations when Fitting Models in WinBUGS

In this section, we will discuss a number of issues that arise when fitting a model in WinBUGS. Readers are also referred to the paper by Brooks (1998) for a more general discussion. We start with the issue of how to set the initial values.

5.3.5.1 Setting the Initial Values

Initial values are required for all parameters that are given prior distributions. These initial values provide a set of starting values for the Gibbs sampler (see Section 5.3.3). There is no right or wrong initial value for a parameter, but the general advice is to choose the initial values sensibly! If a chosen initial value is near where the posterior distribution is, the MCMC chain may approach the posterior distribution faster compared to a chosen initial value that is far away. A badly chosen initial value may sometimes lead to slow or non-convergence. In practice, the analyst can explore the data to arrive at an "educated" guess for the initial values. For instance, if the unknown parameter is the population mean, one can use the sample mean from the data as an initial value. For regression coefficients, initial values can be taken as some small values, say ±0.001, or estimates from fitting the model in the frequentist framework (e.g. using the lm or glm function in R). If the parameter is a variance, then the sample variance can be used. To assess convergence reliably, Gelman and Rubin (1992) suggest choosing overdispersed initial values for different chains. Here, overdispersed means that the initial values need to be very different whilst still being sensible. Intuitively, we would be more confident of convergence if our two MCMC chains had started from very different initial positions but come together to sample from the same parameter space.

5.3.5.2 Checking Convergence

Convergence checking is concerned with ensuring the MCMC chains are sampling from the target posterior distribution. Here we will focus on the convergence diagnostics available in WinBUGS whilst we refer the reader to Brooks (1998, p.75–77) for a more general discussion of the topic. Convergence can be checked visually using the history plot. A history plot shows the sampled values against iteration numbers. Once a chain has converged, the history plot should show the sampled values scattering randomly around a stable mean value (Figure 5.12(c)). The converse, however, may not be true. Imagine two chains have run and the history of both chains shows a random scatter but about two different mean values. In that case, convergence has not been reached because these two chains may well have become trapped in some local modes of a multimodal posterior distribution, or there may be issues with the model itself, e.g. parameters are not individually identifiable.[5] For these reasons, it is important to run two or more MCMC chains (typically we take the minimum number of two) to ensure convergence is reliably assessed and the posterior distribution is explored fully.

In addition to the visual inspection of the history plot, the Brooks-Gelman-Rubin (BGR) diagnostic (Gelman and Rubin, 1992 and Brooks and Gelman, 1997), a formal statistic for detecting non-convergence, is implemented in WinBUGS: the bgr diag button in the Sample Monitor Tool. Running multiple chains is required to carry out the BGR diagnostic, and running two chains meets the requirement. The basic idea is that the values sampled from multiple chains may have come from the same underlying distribution if, for a given value, we can no longer tell which chain this value is from. To formalise this, the BGR diagnostic calculates the ratio of the overall variability (by pooling all sampled values together) to the averaged within-chain variability. Once chains have converged, the BGR diagnostic should be close to 1 (Gelman et al., 2014, p.285). In practice, convergence is achieved if the BGR diagnostic is below 1.05 for *all* parameters (Lunn et al., 2012, p.75). Figure 5.12(b) shows an example of the BGR plot from WinBUGS. In the BGR plot, we are looking for the kth iteration, beyond which the blue and the green lines (representing the overall variability and the within-chain variability, respectively) are stable and the red line (the BGR diagnostic) is stable and close to the horizontal dashed line at 1. So, the BGR plot in Figure 5.12(b) suggests that the two chains have reached convergence after around

---

[5] Two parameters are not individually identifiable if we only have information on some combination of the two. For example, a and b cannot be estimated individually if we only know the value of $(a + b)$.

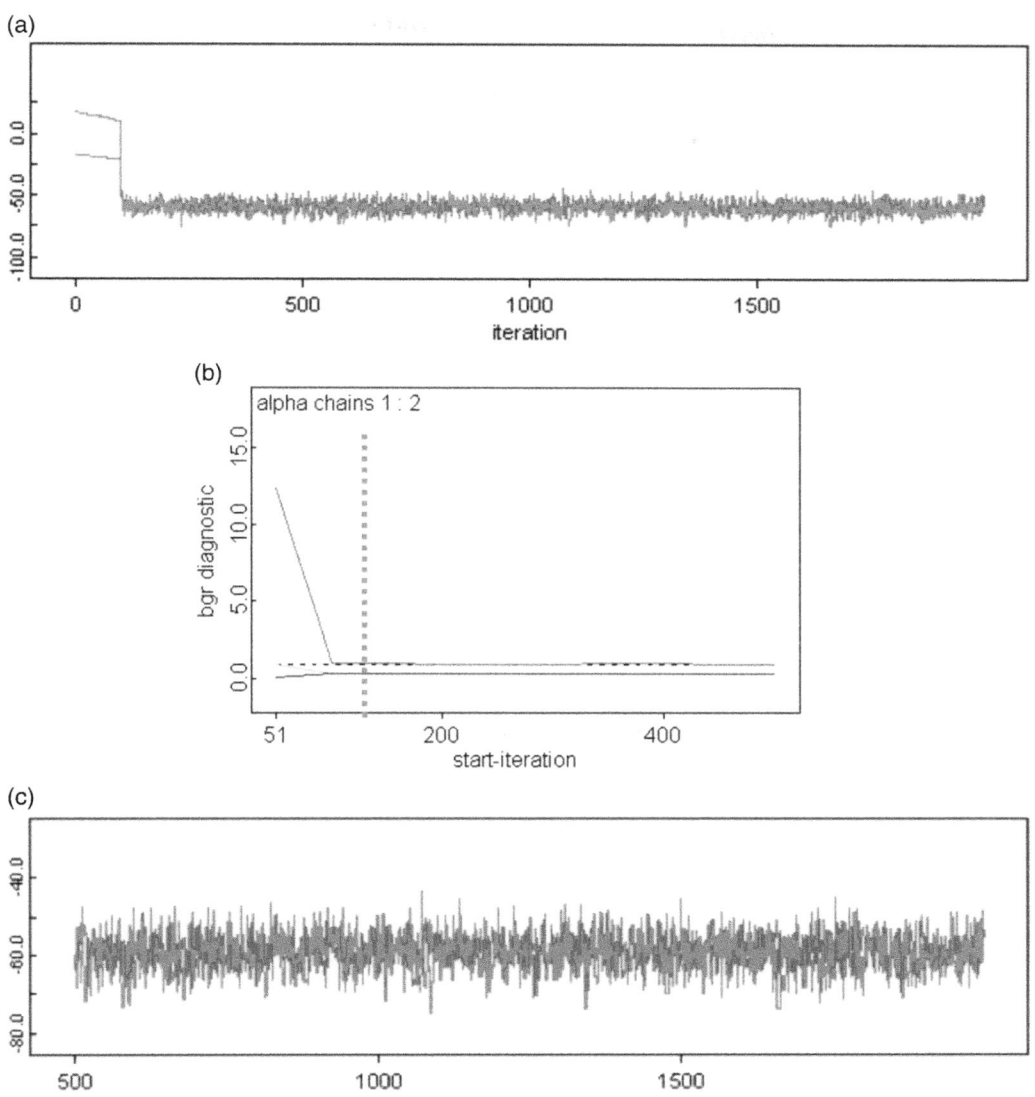

FIGURE 5.12
(See colour insert.) (a) A history plot of the 2000 iterations from two MCMC chains where the beginning of the two chains is clearly not from the target posterior distribution; (b) the BGR plot from WinBUGS showing that after the 150th iteration all three lines are stable and the red line is close to 1 (the grey vertical line is super-imposed for ease of interpretation and is not part of the BGR plot); (c) after discarding the first 500 iterations the resulting history plot has the required form (more iterations than suggested by the BGR plot have been discarded to be on the safe side).

150 iterations, after which the two chains start to come together and settle around a stable mean thereafter as shown in the history plot in Figure 5.12(a).

  Iterations before convergence are known as "burn-in", and they should be removed (or discarded) before making posterior summaries. To discard the burn-in, enter the beginning iteration for any posterior summary into the text box labelled beg in the Sample Monitoring Tool. For example, to discard the first 500 iterations in Figure 5.12(a), set the beginning iteration to 501 so that all iterations from (and including) the 501st iteration will be used to calculate the posterior summary for this parameter.

5.3.5.3 Checking Efficiency

After convergence, we now need to obtain samples (from the posterior distribution) to compute posterior summaries. The more iterations we have, the better the posterior distribution is approximated (see the discussion of the sample size M in Section 5.3.2). But we cannot and do not want to run the model indefinitely. So, we need some criterion to tell us to stop sampling once the set of sampled values becomes sufficiently large to provide a good approximation to the posterior distribution. This is the check of efficiency. The Monte Carlo (MC) error is a standard output from WinBUGS for this purpose. The MC error is the standard error of the mean based on the MCMC samples as an estimate of the true posterior mean. The MC error reduces with increasing sample size because, if we assume the MCMC samples are independent, the formula for the MC error would be $s / \sqrt{n}$, where s and n are, respectively, the sample standard deviation and the number of MCMC samples. Unfortunately, MCMC samples are not independent, but rather they are autocorrelated (because of the Markovian property; also see below). So, the MC error based on the autocorrelated MCMC samples is larger than $s / \sqrt{n}$, and how much larger depends on the strength of the autocorrelation – see Section 6.3.2.1 for the definition of the effective sample size in the context of modelling spatial data. Nevertheless, we can see that the MC error is inversely related to the sample size. We want the mean of the MCMC samples to be as close to the true mean as possible, hence we need the MC error to be small. As suggested by Lunn et al. (2012, p.78–89), we can stop updating when the MC error goes below 5% of the corresponding posterior standard deviation. More iterations are required if we are interested in estimating the tail probability of the posterior distribution. In such cases, it is recommended to run the model till the MC error is less than 1.5% of the posterior standard deviation (Raftery and Lewis, 1992). For the output in Figure 5.13, a sample of 1000 MCMC iterations (second row in the output) is sufficiently large to meet the 5% criterion if we are interested in estimating the parameter beta. But if the interest is in estimating the upper tail probability of beta greater than 0.4 (roughly three standard deviations away from the mean), then we would need about 9000 iterations to meet the more stringent criterion of 1.5%.

As mentioned above, MCMC samples are autocorrelated. The greater the autocorrelation, the more samples are required to meet the below 5% (or below 1.5%) efficiency criterion. The auto cor button in the Sample Monitor Tool calculates the autocorrelation at various lags. The two MCMC chains in the top panel of Figure 5.14 are highly autocorrelated (showing a "snake-like" behaviour), and the mixing for these two chains is poor. In general, high-autocorrelation and poor mixing are not problems, and running the two chains longer often helps solve any issue (see the bottom panel of Figure 5.14). In addition,

node	mean	sd	MC error	2.5%	median	97.5%	start	sample
beta	0.1329	0.09812	0.01029	-0.01505	0.1201	0.3514	501	100
beta	0.124	0.09138	0.002902	-0.05298	0.1232	0.3067	501	1000
beta	0.13	0.09625	0.00105	-0.05274	0.1278	0.3241	501	9000

FIGURE 5.13
Posterior summary of a parameter with various number of MCMC iterations. The MC error (4th column) reduces as the number of MCMC samples increases (the last column).

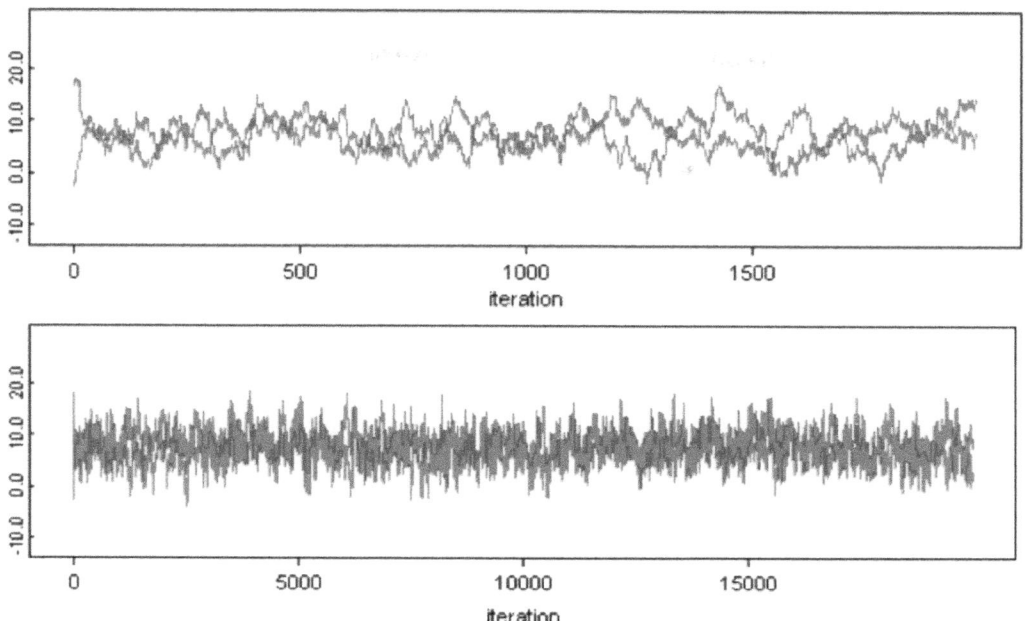

FIGURE 5.14
Two MCMC chains with high autocorrelation (top panel), and the same two chains but run for longer (20000 iterations), showing the required form (bottom panel).

the option of thinning, i.e. using every k^{th} iteration instead of every single iteration to produce the posterior summary, also helps reduce autocorrelation. To use every 10th iteration, for example, we enter 10 into the text box labelled `thin` in the Sample Monitor Tool.

5.4 Bayesian Regression Models

In this section, we discuss the Bayesian implementation of a regression model. Regression modelling is a technique to study the relationship between the outcome values that we have observed and a set of explanatory variables. Models considered in this book are mostly from the family of generalized linear models (GLM), a class of models developed by Nelder and Wedderburn (1972) and McCullagh and Nelder (1989) to deal with both continuous- and discrete-valued (including count and binary) outcome data.

Regression modelling, as we shall see, forms the basis of many analyses that we will discuss in Parts II and III of this book. When analysing spatial and spatial-temporal data, building a Bayesian regression model involves specifying *a likelihood function* for describing the observed outcome values; *a process model* that specifies a function of the explanatory variables together with spatial and spatial-temporal random effects; and finally, *a set of prior distributions* for all the unknown parameters (see Section 1.4.3.1).

In general, specification of the likelihood function depends on the type of outcome values (continuous- or discrete-valued) and the statistical properties of the outcome values (for example, a roughly symmetric, bell-shaped histogram of the response values suggests a normal likelihood function). In some situations, the context of the application and/or data availability provides hints as to which likelihood might be suitable. For example, when modelling counts of events with rare occurrence, such as annual counts of domestic burglary or street assaults involving violence over a set of small areas, either the Poisson or the binomial distribution may seem to be a reasonable choice. However, using the binomial distribution as the likelihood function for modelling the number of street assaults involving violence across a city may be difficult. This is because the binomial distribution requires us to specify the quantity n_i, the population at-risk in sub-area i of the city, which is not well-defined in that situation (see Section 9.5). For domestic burglary counts, on the other hand, the at-risk population can be defined as the number of houses in each spatial unit, so both the Poisson and the binomial distributions could be used and, for rare events, both would produce comparable results.

The process model considered in this section only includes explanatory variables (also known as covariates or predictors, and these terms will be used interchangeably hereafter). It specifies a relationship between the outcome and the observable covariates. The choice of covariates to be included in a regression analysis is typically based on theoretical grounds and data availability. In practice, when modelling spatial and spatial-temporal data, we typically include spatial and spatial-temporal random effects to account for the effects of unobserved/unmeasured covariates, in addition to allowing for information sharing spatially and/or temporally. We defer the descriptions of these random effect models till Parts II and III.

Finally, in the illustrative examples below, vague priors are assigned to parameters so that posterior inferences are primarily based on the observed data. Specifically, a normal distribution with mean 0 and a large variance 1000000 (i.e. 10^6) is used as a vague prior for the intercept and each of the regression coefficients. When applicable, a vague Gamma prior, Gamma(0.001,0.001), is assigned to the data precision, which is the inverse of the data variance. The support of a Gamma distribution is on the positive real line, a feature that satisfies the requirement of a variance parameter, which must be strictly positive. For a reader who is unfamiliar with the Gamma distribution, see Exercise 5.7 for some more detail. We will return to the discussion of prior specification in Section 5.6.

We have seen an example of the logistic regression model in Section 5.2.2. In the next two sections, we look at two examples. The first example illustrates the specification of a Bayesian linear regression with a normal likelihood. We pay particular attention to the implementation of this model in WinBUGS as well as the practical issues discussed in Section 5.3.5 when performing Markov chain Monte Carlo in WinBUGS. The second example deals with a set of small-area burglary counts. This example gives an illustration of Bayesian Poisson regression modelling. We will look at how to make posterior inference on some transformation of parameters in WinBUGS and talk about model evaluation using posterior predictive checks.

5.4.1 Example I: Modelling Household-Level Income

The dataset used in this example is a subset of the survey data on household-level weekly total gross income in Newcastle introduced in Section 1.3.1. We extracted data on 200 households with the following three variables:

y_i:	Weekly total gross income (in £s) for household i
age_i:	Age of the head of household
r_i:	Rurality index, a categorical variable indicating whether household i is in a rural area (= 1), an urban area (= 2) or a mixed rural-urban area (= 3)

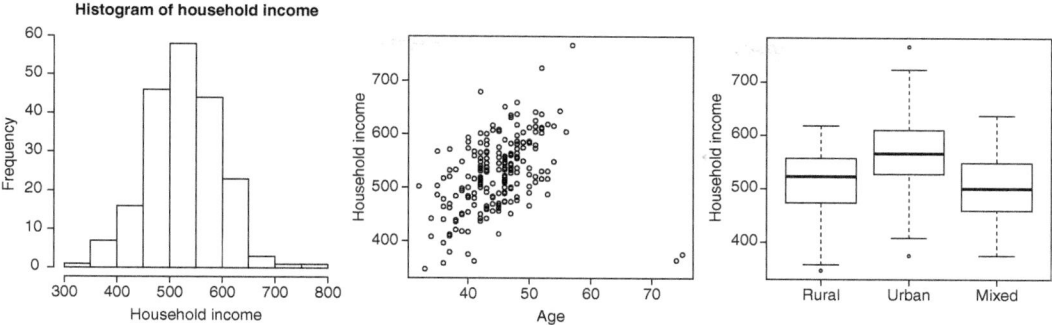

FIGURE 5.15
Exploring the household-level income data in Newcastle: the histogram of the household-level income values; a scatterplot of the income values against age; and a boxplot for assessing the difference in mean income across the three area types: rural, urban and mixed.

The subscript i indicates household with $i = 1,\ldots,200$. The aim of the analysis here is to assess the effects of the two covariates, age_i and r_i, on household income.

Figure 5.15 explores the data graphically. The bell-shaped and symmetric distribution of the income values shown in Figure 5.15(a) suggests that a normal likelihood is reasonable. Figure 5.15(b) indicates that household income tends to increase as the head of household gets older, with the exception of the two data points lying in the bottom right corner. These two outlier points may pull the regression line downwards, so we need to bear this feature in mind when building the model. For the association between rurality and income, Figure 5.15(c) shows that urban households tend to have higher levels of income compared to rural and mixed areas, while the income levels of the latter two appear to be similar.

Based on the observations from the exploratory plots, a regression model is given as follows. For $i = 1,\ldots,200$,

$$y_i \sim N\left(\mu_i, \sigma^2\right)$$

$$\mu_i = \alpha + \beta \cdot age_i + \eta_{r_i}$$

(5.17)

In the above model, the normal distribution is used as the likelihood function. The regression relationship (the process model) expresses household income as a linear combination of the intercept, α, and the effects from the two covariates, $\beta \cdot age_i$ and η_{r_i}. Specifically, the regression coefficients β and η (η is a Greek letter pronounced eta) quantify the age effect and the rurality effect on income, respectively. Note that the three-level categorical variable r_i enters the model through the subscript on η. As a result, η represents three elements, η_1, η_2 and η_3. The corresponding element is then selected depending on the rurality of a household. Here, η_1 is fixed to 0 (i.e. setting the rural category as the reference category) so that the intercept α can be estimated. Thus, η_2 and η_3 are measuring the income difference between urban and rural areas and between mixed and rural areas, respectively. Readers are encouraged to make the distinction between the two formulations of the rurality covariate, η_{r_i} or $r_i \cdot \eta$.[6] Finally, σ^2 is the residual variance, measuring the between household variability that is not explained by the two covariates in the model.

---

[6] See, for example, Lunn et al. (2012, p.104–106) and Gelman and Hill (2007, p.66–68) for more detail on how to handle categorical covariates in regression modelling in general.

To complete the model specification, priors need to be assigned to the five unknown parameters: α, β, η_2, η_3 and σ^2. A vague normal prior with mean 0 and a large variance 10^6 is assigned to the intercept α and the regression coefficients β, η_2 and η_3. For the residual variance σ^2, we use a vague Gamma prior with parameters 0.001 and 0.001 on the precision τ, which is $1/\sigma^2$. Figure 5.16 shows the WinBUGS model, the structure of the data and the initial values for two separate MCMC chains.

```
1    ################################################################
2    #   Block 1:
3    #   The WinBUGS code for specifying the normal regression
4    #   model for the income data
5    ################################################################
6    model{
7       #   looping through all households in the data
8       for (i in 1:N) {
9          y[i] ~ dnorm(mu[i],tau)   #  normal likelihood for income
10         #    specifying the regression relationship
11         mu[i] <- alpha + beta*age[i] + eta[r[i]]
12      }
13      #   setting the rural category as the reference
14      eta[1] <- 0
15      #   specifying priors on the intercept and the regression
16      #   coefficients
17      alpha ~ dnorm(0,0.000001)
18      beta  ~ dnorm(0,0.000001)
19      eta[2] ~ dnorm(0,0.000001)
20      eta[3] ~ dnorm(0,0.000001)
21      #   prior on residual precision (=1/variance)
22      tau ~ dgamma(0.001,0.001)
23      #   calculate the residual variance from precision
24      residual.variance <- pow(tau,-1)
25      #   calculate the residual standard deviation from variance
26      residual.SD <- pow(residual.variance,0.5)
27      #   quantifying income difference between mixed and urban
28      eta[4] <- eta[3] - eta[2]
29   }
30   ################################################################
31   #   Block 2:
32   #   the dataset on household-level income with two
33   #   household-level covariates in the WinBUGS format
34   ################################################################
35   list(N    = 200
36          ,y   = c(604, 500, 467, 542, ... , 466, 565, 379, 556)
37          ,age = c(51, 46, 41, 47, ... , 50, 43, 37, 45)
38          ,r   = c(1, 1, 1, 1, ..., 1, 1, 1, 2, 2, ..., 2, 2, 2, 3
39               ,3, ... , 3, 3)
40   )
41   ################################################################
42   #   Block 3:
43   #   two sets of initial values
44   ################################################################
45   #   initial values for MCMC chain 1
46   list(alpha=100, beta=1, eta=c(NA,10,10,NA),tau=0.001)
47   #   initial values for Chain 2
48   list(alpha=700, beta=-1, eta=c(NA,5,-10,NA),tau=0.01)
```

FIGURE 5.16
The WinBUGS code for specifying the regression model (Block 1), the income data (Block 2) and two sets of initial values for running two MCMC chains (Block 3).

In Figure 5.16, `eta` (η) contains four elements. The first three elements, `eta[1]`, `eta[2]` and `eta[3]`, correspond to η_1, η_2 and η_3, as defined in Eq. 5.17. The fourth element, `eta[4]`, the difference between `eta[3]` and `eta[2]` (Line 28 in Figure 5.16), quantifies the income difference between mixed and urban areas. Both `eta[1]` and `eta[4]` are logical nodes – both are on the right-hand side of the <- symbol (Section 5.3.4) – and do not need initial values, so `NA` is placed in their positions in the two sets of initial values (Lines 46 and 48 in Figure 5.16). But initial values are required for the two stochastic nodes `eta[2]` and `eta[3]`, as they both have priors (Lines 19 and 20 in Figure 5.16). Another point to note is that `dnorm`, the WinBUGS syntax to define a normal distribution (see Line 9), is parameterised using mean (the first argument in `dnorm`) and *precision* (the second argument in `dnorm`), where the precision is the inverse of a variance. The vague normal prior, $N(0,1000000)$, is therefore implemented as `dnorm(0,0.000001)` on Lines 17–20. Save the three blocks to three separate files then follow the 11 steps described in Section 5.3.4 to run the model. The results below are based on running the two MCMC chains over 10000 iterations.

First, we need to check convergence. Figure 5.17 shows the history plots and the BGR diagnostic plots for two chains. After a few hundred iterations, all parameters, apart from `residual.SD`, have reached convergence: all three lines in each BGR plot remain stable, and the red line virtually overlaps the dotted line 1; each history plot shows a random scatter of the sampled values around a stable mean; and the two MCMC chains overlap nicely, as seen in the history plot, showing a good mixing. For `residual.SD`, both the blue and green lines only become stable after around 5000 iterations. Hence, we will discard the first half of the 10000 iterations in each chain as burn-in by setting the `beg` in the Sample Monitor Tool to 5001. The remaining iterations are used for posterior summary.

We next check efficiency. Table 5.3 shows that all MC errors (the values in the fourth column) are less than 5% of the corresponding posterior standard deviations (the values in the third column), meaning that the total 10000 iterations (5000 from each of the two MCMC chains) are sufficient to provide a good approximation to the posterior distribution. At this point, we can proceed to interpret the estimates of the parameters tabulated in Table 5.3.

The regression coefficient β (`beta`), measuring the age effect, is estimated to be 4.00 (the posterior mean), with a 95% credible interval of 2.57–5.43, suggesting a strong effect of age on income. A £4.00 increase in weekly household income is estimated for each additional year for the head of the household, whilst the 95% credible interval tells us that this increase can be as small as £2.57 or can be as large as £5.43. These results agree with what we have observed in Figure 5.15(b).

Compared to rural areas, households in urban areas have considerably higher income, and the income difference, quantified by η_2 (`eta[2]`), is about £48.5 (95% credible interval: 28.5–68.4). While there appears to be no income difference between rural and mixed areas (the posterior mean of η_3 (`eta[3]`) is –13.4 with a 95% credible interval of –33.3–7.2, thus including the value 0), the income level in mixed areas tends to be much lower than that in urban areas (η_4 (`eta[4]`) estimated to be –61.8 with a 95% credible interval from –83.1 to –40.5; excluding the value 0).

As a form of model checking, Figure 5.18 plots observed household-level income, the fitted values (i.e. the posterior means of μ_i in Eq. 5.17) from the regression model, against the age of the head of household. Although the fitted line captures the overall pattern that income increases with age, it is perhaps pulled down by the two outliers in the bottom right corner. To minimize the influence of the outliers, one might either remove them from the data or modify the model. Unless we know that those two observations are errors created by, for example, mistakes in data entry, modifying the model is better. In Exercise 5.9,

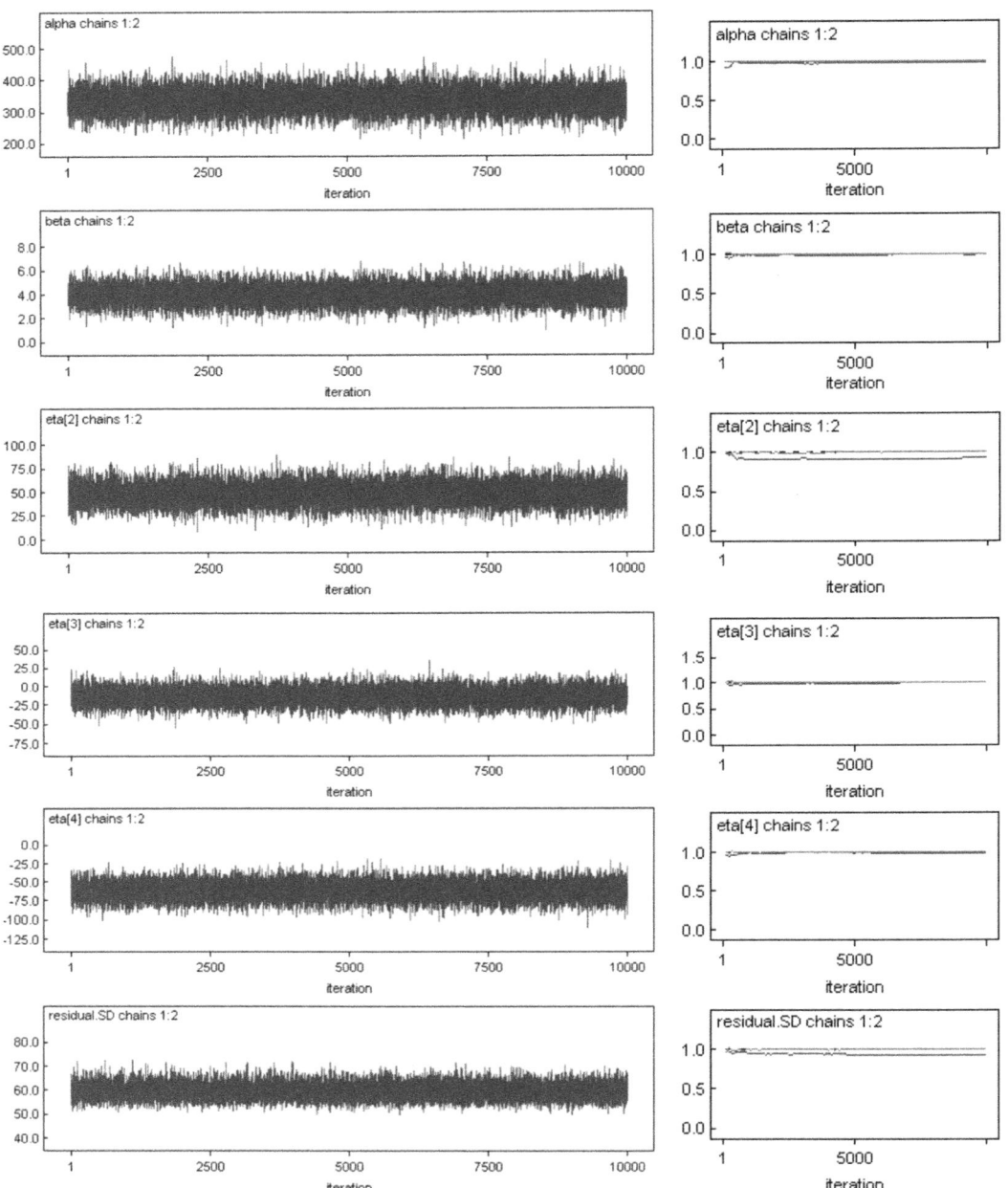

FIGURE 5.17
(See colour insert.) Checking convergence: history plots and the BGR diagnostic plots.

the normal likelihood in Eq. 5.17 is replaced by a Student-t distribution with four degrees of freedom to make the regression more robust to (i.e. less affected by) outliers.

As illustrated here, model checking plays a key role in regression modelling (in fact, any kind of statistical modelling): it helps to identify potential issues with the model in hand and, in turn, inform possible model improvements.

TABLE 5.3

Summary Statistics of the Posterior Distributions

node	mean	sd	MC error	2.50%	median	97.50%	start	sample
alpha	336.6	33.69	0.3135	271.6	336.5	401.5	5001	10000
beta	3.996	0.7387	0.007304	2.569	3.992	5.429	5001	10000
eta[2]	48.47	10.24	0.1052	28.49	48.38	68.43	5001	10000
eta[3]	−13.36	10.27	0.1155	−33.3	−13.41	7.232	5001	10000
eta[4]	−61.83	10.83	0.1071	−83.06	−61.77	−40.48	5001	10000
residual.SD	59.65	3.019	0.03227	54.09	59.54	66.0	5001	10000

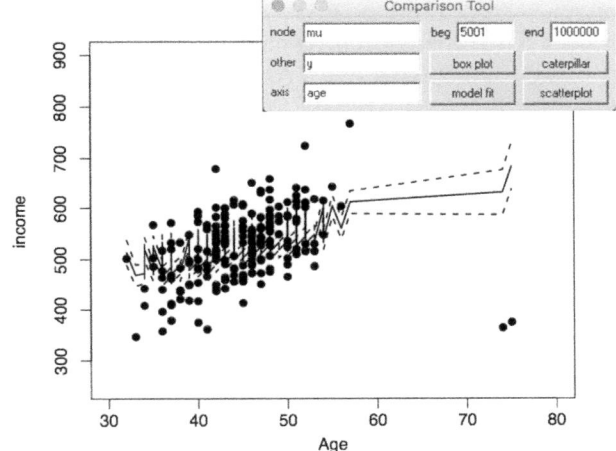

FIGURE 5.18
Model checking: A scatterplot of the household-level income values against age of the households with the fitted regression line (with uncertainty) superimposed. The black dots are the actual observations, the solid line is the fitted regression line (the posterior means of μ_i in Eq. 5.17) and the two dashed lines represent the 95% uncertainty band (i.e. the 95% credible intervals of the fitted values). The window insert shows how this plot is produced in WinBUGS.

The plot in Figure 5.18 is produced within WinBUGS using The Comparison Tool (open through Inference -> Compare...). The inserted window shows the detail.

5.4.2 Example II: Modelling Annual Burglary Rates in Small Areas

This example uses a dataset containing the numbers of burglary events recorded in each of the 452 census output areas (COAs) in Peterborough, UK, in 2005 (see Section 1.6 in Chapter 1). The aim here is to construct a Poisson regression model that estimates the effects of three selected covariates on small area burglary rates. For each COA i ($i = 1,...,452$), the dataset contains the following information:

O_i :	The number of recorded burglary events during 2005
n_i:	The number of at-risk houses
$x_{i,own}$:	The percentage of owner-occupied houses

$x_{i,det}$:	The percentage of detached/semi-detached houses
$x_{i,ethnic}$:	An index measuring ethic heterogeneity; this index ranges between 0 and 1, and the larger the value the greater the ethnic mix

Here we specify the Poisson distribution as the likelihood function to model the variation in burglary counts. For each COA i, the observed burglary count is modelled as

$$O_i \sim Poisson(\mu_i)$$

$$\mu_i = n_i \cdot \theta_i$$

(5.18)

where n_i is the number of at-risk houses and θ_i is the underlying burglary rate for that COA. Note that Eq. 5.18 is expressed according to a syntax restriction in WinBUGS where any calculation of the Poisson mean cannot be done within dpois, the WinBUGS function to define a Poisson distribution. In other words, we cannot do O[i] ~ dpois(n[i]*theta[i]) in WinBUGS, but instead we have to rewrite it as O[i] ~ dpois(mu[i]), then define the Poisson mean on a separate line, e.g. mu[i] <- n[i] * theta[i]. See Lines 9 and 10 in Figure 5.19 for the implementation. Otherwise, Eq. 5.18 is equivalent to $O_i \sim Poisson(n_i \cdot \theta_i)$.

Our interest is to see how the COA-level burglary rates are affected by the three COA-level covariates. To do that, the following regression relationship (the process model) is formulated:

$$\log(\theta_i) = \alpha + \beta_1 \cdot x_{i,own} + \beta_2 \cdot x_{i,det} + \beta_3 \cdot x_{i,ethnic}$$

(5.19)

Eq. 5.19 models the log-transformed burglary rate in each COA as a function of the intercept and the effects of the three COA-level covariates. The log transformation (known as the log link function (McCullagh and Nelder, 1989)) is used to ensure the burglary rate, θ_i, is a positive quantity. For prior specification, each of the four unknown parameters in the model, α and β_k (k = 1,2,3), has a vague normal prior with mean 0 and variance 10^6. Apart from the use of the Poisson likelihood, the model specification is largely the same as that for the income example in Section 5.4.1. But we want to focus on the following two points, (a) making posterior inference on some transformation of parameters and (b) model checking using posterior predictive probabilities. We now discuss the two in turn.

For a Poisson regression, in addition to the regression coefficients β_k, we are also interested in the exponentiated β_k, i.e. e^{β_k}, which are interpreted as the rate ratios (*RRs*). For example, $RR_1 = e^{\beta_1}$ measures the change in the burglary rate due to a one percent increase in owner-occupied housing. An estimate of the rate ratio that is much higher (lower) than 1 would suggest an increase (decrease) in the burglary rate with a unit increase in the corresponding covariate. If the rate ratio is estimated to be close to 1, then the corresponding covariate does not play an important role in explaining the variation in burglary rates at the COA level. To obtain the posterior estimates of the rate ratios, the nonlinear exponential transformation needs to be carried out within WinBUGS. We will comment on how this is done in Figure 5.19.

Posterior predictive check is a useful tool to check how consistent a model is with the observed data. As stated by Gelman et al. (2014, p.143), "If the model fits, then replicated data generated under the model should look similar to observed data. To put it another way, the observed data should look plausible under the posterior predictive distribution." In this example, we want to see how well the Poisson regression model defined in Eqs. 5.18 and 5.19 describes the observed burglary counts. To do that, for each COA, we simulate

```
1    ##############################################################
2    #   The WinBUGS code for specifying the Poisson regression
3    #   model for the burglary data
4    ##############################################################
5    model {
6      #  looping through the 452 COAs  (N=453)
7      for (i in 1:N) {
8        # the Poisson likelihood for burglary counts
9        O[i]   ~ dpois(mu[i])
10       mu[i] <- n[i] * theta[i]    #  defining the Poisson mean
11       # the regression relationship
12       log(theta[i])    <- alpha
13                         + beta[1]*(x_own[i]-mean(x_own[]))
14                         + beta[2]*(x_det[i]-mean(x_det[]))
15                         + beta[3]*(x_ethnic[i]-mean(x_ethnic[]))
16     }
17     #   vague priors on the intercept and the regression
18     #   coefficients
19     alpha ~ dnorm(0,0.000001)
20     for (k in 1:3) {
21       beta[k] ~ dnorm(0,0.000001)
22     }
23     #   compute the rate ratios (RRs) associated with the
24     #   covariates
25     RR_own <- exp(beta[1])
26     RR_det <- exp(beta[2])
27     RR_ethnic <- exp(beta[3])
28     #   posterior predictive check against the observed
29     #   COA-level case counts
30     for (i in 1:N) {    #  go through each COA
31       O.pred[i] ~ dpois(mu[i])    #  predict the burglary
32                                   #  count for each COA
33       #   calculate the difference between the predicted
34       #   and the observed counts
35       diff[i] <- O.pred[i] - O[i]
36       # calculating ppp[i]; see main text for detail
37       ppp[i] <- equals(O[i],0)*equals(O.pred[i],O[i])
38             + (1- equals(O[i],0))*step(diff[i])
39     }
40   }
41   ##############################################################
42   #   two sets of initial values
43   #   (specified only partially; see the main text)
44   ##############################################################
45   #   initial values for MCMC chain 1
46   list(alpha=-2, beta=c(0.01,0.01,0.01))
47   #   initial values for Chain 2
48   list(alpha=-4, beta=c(-0.01,-0.01,-0.01))
```

FIGURE 5.19
The WinBUGS code to fit the Poisson regression model defined in Eq. 5.18 and Eq. 5.19 for the COA-level burglary counts in Peterborough reported during 2005.

O_i^{pred}, the number of burglary cases predicted from the fitted model, then compare the posterior distribution of O_i^{pred} to the actual observed case count in that COA. Note that O_i^{pred} has a posterior distribution since it is a function of the unknown parameters as formulated in Eq. 5.19. Any discrepancy between the model and the observed value can be highlighted via the so-called posterior predictive p-value:

$$\Pr\left(O_i^{pred} \geq O_i \mid data\right) \text{ if } O_i > 0$$

$$\Pr\left(O_i^{pred} = O_i \mid data\right) \text{ if } O_i = 0$$

(5.20)

An extremely large (or small) posterior predictive p-value would indicate that the model is producing too many (or too few) burglary cases – the reader is encouraged to visualize this interpretation (hint: using Figure 5.2). In addition to checking against the observed outcome values, posterior predictive check can also be used to check against some particular characteristics of the dataset (see for example Sections 9.1 and 9.3 in Chapter 9 and Section 16.5 in Chapter 16).

Figure 5.19 shows the WinBUGS implementation of the Poisson regression model for the burglary data, featuring both the calculation of the rate ratios and the posterior predictive check. There are a few points to comment on. First, on Lines 13 to 15, each of the three covariates is mean-centred, subtracting each covariate value from its mean with with `mean(x_own[])` being the syntax to calculate the mean of the covariate `x_own`. Mean-centring the covariates sometimes helps speed up convergence and results in better mixing. It also gives a more meaningful interpretation for the intercept. For example, in this implementation, alpha is interpreted as the log burglary rate when all covariates are set at their mean values.

Second, Lines 25 to 27 calculate the rate ratios. For each of the three regression coefficients, `beta[1]`, `beta[2]` and `beta[3]`, a value is sampled from the posterior distribution at each MCMC iteration. This sampled value is then exponentiated to give the value for the corresponding rate ratio. Posterior summaries for the rate ratios are then obtained using these exponentiated values over the MCMC samples.

Third, Line 31 predicts the number of burglary cases for each COA. Then Lines 37 and 38 calculate the posterior predictive p-value. Since there is no `if` statement in WinBUGS, the calculation may seem convoluted. The `equals` function in WinBUGS returns 1 when its two arguments are equal and returns 0 otherwise. Therefore, `equals(O[i],0)*equals(O.pred[i],O[i])` contributes to the posterior predictive p-value of a COA where zero cases are observed (i.e. the second line in Eq. 5.20). For a COA with at least one case, `equals(O[i],0)*equals(O.pred[i],O[i])` is 0, whilst the part on Line 38, `(1- equals(O[i],0))*step(diff[i])`, contributes towards the calculation of the first line in Eq. 5.20.

Finally, the two sets of initial values are only partially specified because we still need to assign initial values for each `O.pred[i]`, the predicted number of burglary events. There are two options to accomplish this. The first option is to explicitly specify a list of 453 integer values for each chain, e.g. `O.pred=c(1,1,1,...,1)` for chain 1 and `O.pred=c(3,3,3,...,3)` for chain 2. The second option is to use the `gen inits` button in the Specification Tool. Proceed from Step 1 to Step 8 as described in Section 5.3.4. After loading each set of initial values in (i.e. Step 7 and Step 8), WinBUGS displays the message "This chain contains uninitialized variables" in the left-hand bottom corner. Once both sets of initial values have loaded in (i.e. after Step 8), press the `gen inits` button in the Specification

TABLE 5.4

Posterior Summaries of the Estimated Rate Ratios Associated with the Three COA-Level Covariates

Rate Ratios	Posterior Mean (95% credible interval)
RR_{own}	0.988 (0.983,0.992)
RR_{det}	0.996 (0.993,1.000)
RR_{ethnic}	0.997 (0.991,1.004)

Tool so that initial values are generated by WinBUGS internally for all the parameters that do not have starting values (see also the WinBUGS manual). WinBUGS will display "initial values generated, model initialized". We can then proceed onto the subsequent steps. We will discuss some of the results below, deferring the details of the WinBUGS implementation to Exercise 5.10.

Table 5.4 summarises the posterior estimates of the rate ratios across the three covariates. The posterior means of all three rate ratios are estimated to be below 1, suggesting that a one-unit increase in each of the covariates would be associated with a decrease in burglary rate. However, only the covariate on the percentage of owner-occupied houses appears to be statistically significant in the sense that the 95% credible interval does not include 1. The change in the burglary rate can also be quantified. A one percent increase in owner-occupied housing (roughly one or two houses more in a COA occupied by the owners) is associated with a reduction in the burglary rate of 1.2% (= $(0.988 - 1) \times 100$, with the negative value indicating a reduction). The same calculation yields the 95% credible interval of that reduction: (0.8%,1.7%). Note that the above calculation is a linear transformation of the posterior summaries of the rate ratio and hence not required to be carried out within the WinBUGS fit. There is weak evidence of an effect associated with the percentage of detached/semi-detached houses since the upper bound of the 95% credible interval is 1. Ethnic composition does not appear to be a strong factor for explaining variability in small area burglary rates.

How well does this model fit the data? The estimated posterior predictive p-values for 423 COAs fall in a reasonable range between 0.05 and 0.95, a criterion proposed by Gelman et al. (2014, p.150–151) to assess discrepancy. These estimates suggest that the regression model considered here fits the observed burglary counts in these 423 COAs adequately. For the remaining 29 COAs, however, the model predicts either too many (two COAs) or too few (27 COAs) cases of burglary, indicating a lack of fit. Exercise 5.10 investigates these 29 COAs in more detail.

We finish this example by noting that we will see many aspects of the modelling presented here again in Parts II and III. The regression model considered in this example is certainly much simpler than the spatial and spatial-temporal models that we will discuss later in the book, but it is always advisable to start simple. A simple model often offers some insights into the dataset in hand, e.g. what the likely effects are from the covariates and which parts of the data the model fits well and which parts it fits poorly. Some of the information helps us to refine the model to provide a better description of the data.

5.5 Bayesian Model Comparison and Model Evaluation

For any given dataset, we might often come up with several different models. We might, for example, be undecided as to which of several models should be preferred on theoretical grounds; we might progress in our analysis from fitting a rather simple model to more complex models. In these circumstances, a question arises: "how do I compare these models and decide between them?"

In the Bayesian framework, models can be compared using the Deviance Information Criterion (DIC; Spiegelhalter et al., 2002), which takes the following form:

$$DIC = \bar{D} + pD \tag{5.21}$$

In Eq. 5.21, $\bar{D}$ is the posterior mean of the deviance, measuring the goodness of fit (of the model to the data), and pD is the effective number of parameters, measuring model complexity. A smaller $\bar{D}$ value indicates a better fit to the data. A model with more "effective parameters" is a more complicated model. For non-hierarchical models (all the models considered in this chapter are non-hierarchical), pD is very close to the number of actual unknown parameters in the model. However, for hierarchical models, to which most of the spatial and spatial-temporal models that we will discuss in Parts II and III of this book belong, a better way to measure model complexity is through pD, rather than counting the number of actual unknown parameters. This is because some of these unknown parameters (e.g. random effects) are not independent but correlated and hence the *effective* number of parameters in the model may be fewer than the actual number.

For comparing models invoking Occam's razor, our preference is usually for simple rather than complicated models, providing the simpler model does not result in a poor fit to the data. DIC is the Bayesian equivalent of Akaike's Information Criterion (AIC), which is often used in the frequentist approach for model comparison. As in the case of AIC, a model with a smaller DIC value is better supported by the data. When comparing two models, a rule of thumb is that where there is a difference between two DIC values greater than 5, the model with the smaller DIC value is preferred. A difference in DIC less than 5 suggests the two models are indistinguishable (based on DIC) (Lunn et al., 2012, p.166–167).

It is important to note that whilst the DIC comparison helps us to decide which model or set of models best describes the data, DIC itself does not tell us how *well* the chosen model(s) describes the observed data. The best model amongst the models considered can still be a poor model. For this reason, it is important to perform appropriate model checks. We have seen some forms of model checking in Sections 5.4.1 and 5.4.2 by comparing the fitted values to the observed values. We shall see other forms of model checking in subsequent chapters (see for example Chapter 9, Section 11.2 and Section 16.5).

5.6 Prior Specifications

The choice of prior is, in principle, subjective. It reflects our knowledge about the parameter before seeing the observed data in hand (it expresses our "prior" knowledge). We divide our discussion on the prior into two situations depending on how much information we have about the parameter(s) before seeing the data.

5.6.1 When We Have Little Prior Information

If the analyst feels she knows very little about the possible value, then a vague prior should be chosen. Such a prior is also referred to as a diffuse or weakly-informative prior.

Throughout this book, we will refer to a prior as "vague", "diffuse" and "weakly-informative" interchangeably. The rationale for using a vague prior "is often said to be 'to let the data speak for themselves' so that inferences are unaffected by information external to the current data" (Gelman et al., 2014, p.51). In these circumstances a prior is selected which is vague with respect to the likelihood, meaning that the prior distribution spreads diffusely over the range of values for the parameter that are supported by the likelihood. If we use a vague prior, estimates from Bayesian analysis tend to be similar to maximum likelihood estimates because the posterior distribution is dominated by the likelihood (see the use of the vague $Beta(1,1)$ prior in the coin example in Section 5.2.1). Depending on what parameter we are dealing with, there are some typical probability distributions that we can use for specifying a vague prior.

For a location parameter, θ (for example a mean or a regression parameter), the uniform distribution on the whole real line from $-\infty$ to $+\infty$, denoted as $Uniform(-\infty,+\infty)$, looks to be a good candidate for being a vague prior because all possible values of θ are assumed to be equally likely. Posterior inference is then based on the likelihood, $\Pr(\text{data}|\theta)$. However, $Uniform(-\infty,+\infty)$ is an improper probability distribution because it does not integrate to 1, a requirement for any proper probability distribution. Using an improper distribution as a prior should usually be avoided, but there are exceptions (see Section 8.2.1.1 when specifying the intrinsic conditional autoregressive model for spatial data). There are options to make an approximation to this improper distribution. One is to use $Uniform(a,b)$ as a prior for θ where a and b are chosen to reflect a "wide" range (say $a = -1000$ and $b = 1000$). Another possible option is to use the normal distribution with mean 0 and a "large" standard deviation. For example, $\theta \sim N(0,\sigma^2)$ with σ set to say 1000. It is easier to conceptualize a normal distribution in terms of standard deviation because of the so-called three-sigma rule: almost all values covered by a normal distribution $N(0,\sigma^2)$ are within $\pm3\sigma$ (Pukelsheim 1994), giving us a rough idea of the range of a normal distribution.

But how "wide" is wide and how "large" is large? One principle we can adopt is to write down what we think the range of values that θ should take, then stretch the range out. For example, if we expect θ to lie in the range between 1 and 10, then $Uniform(-1000,1000)$ or $N(0,100^2)$ represent vague priors for θ. On the other hand, if we expect θ to lie in the range between 500 and 1000, then neither of the two priors before are vague (they are in fact very informative – too informative!). In that case, better choices (giving a more spread out range) would be $Uniform(-100000,100000)$ or $N(0,10000^2)$. WinBUGS parameterizes the normal distribution in terms of mean and *precision*, which is 1/variance. So the WinBUGS code to implement the prior specification $\theta \sim N(0,10000^2)$ is `theta ~ dnorm(0, 0.00000001)`.

For a variance parameter, a prior needs to have a positive support. Typical choices as a vague prior are

- A $Gamma(\varepsilon,\varepsilon)$ distribution with ε set to be small, say 0.001, on the precision τ (see Exercise 5.7 for more information about the Gamma distribution). For example, the WinBUGS implementation of $\tau \sim Gamma(0.001,0.001)$ is `tau ~ dgamma(0.001,0.001)`.

- A uniform distribution, $Uniform(a,b)$, on the standard deviation σ where a is typically chosen to be a small value close to but greater than 0, say 0.0001, and b is set to be a large value, say 1000.

- A half normal distribution with mean 0 and a large variance on the standard deviation σ (Gelman 2006). "Half" means that the normal distribution is bounded below by 0 such that σ is strictly positive. For example, $\sigma \sim N_{+\infty}(0,10^2)$, where $N_{+\infty}$ denotes a half normal distribution bounded below by 0. In WinBUGS: sigma ~ dnorm(0,0.01)I(0,), where I(0,) restricts the distribution to be defined only on the positive real line.

Similar to the discussion for a location parameter, the range of a prior distribution for a variance parameter should be sufficiently wide to be vague. When dealing with hierarchical models, which we shall introduce in Parts II and III, we may want to consider assigning moderately informative priors (e.g. using a smaller standard deviation on the half normal prior or a narrower range for the uniform prior) to help the estimation. We will see some examples later (see for example Section 16.4).

Whatever strategy the analyst adopts to specify priors, sensitivity analysis should form part of any model fitting process. Different priors should be tested to ensure that an apparently innocuous uniform prior is not introducing substantial information into the fitting. The analyst needs to consider, and try out, different specifications of a vague prior and widen the range of a uniform prior (or increase the variance of a normal prior).

5.6.2 Towards More Informative Priors for Modelling Spatial and Spatial-Temporal Data

Bayesian inference becomes particularly interesting in those cases where we believe, with justification, that we have some prior knowledge about one or more of the model's parameters. Bayesian inference allows us to make use of that knowledge rather than simply discarding it and proceeding as if it is only the present dataset that matters for the problem at hand. The reader is reminded that in Section 5.2.1 we provided some rather general examples where, in the case of a coin flipping experiment, we might move away from a vague prior towards a more informative prior. The examples we provided fell into two categories: (i) where we are able to argue about the probable range of the parameter from first principles (about coins and how they behave when flipped); (ii) where we are able to call on previous relevant experimentation (with coins like the one under study) or other relevant empirical research experience. When the analyst feels that they are an expert on the subject (based on either (i) or (ii) or some combination of both) and feels that they have a great deal of prior knowledge, they are likely to choose an informative prior – one that introduces a source of information, in addition to that from data, into the estimation of the parameter. Between the two extremes of "I know nothing" and "I am an expert", we can select priors that are to some degree informative. That degree of informativeness becomes stronger as we move to the latter state. Before we can do that we must have a clear understanding of what the parameter means (its interpretation). This will depend on the problem we are studying.

So what kind of prior knowledge is likely to be relevant when analysing spatial and spatial-temporal data? We suggest two types of prior knowledge, which we refer to as geographical-substantive knowledge and spatial knowledge.

By geographical-substantive knowledge we mean the kind of knowledge that derives from previous work in the substantive field – studying areas of a city prone to high levels of crime (Section 5.2.2); studying variation in household income (Section 5.4.1); studying why some areas of a city have higher levels of household burglary than others (Section 5.4.2). Typically, such substantive knowledge is used to specify the covariates that are included in the process model rather than proscribing, in some sense, the values of one or more of the regression parameters. So this type of prior knowledge, at least in many areas of the social sciences, informs *the specification of the process model* rather than the inclusion of an *informative prior distribution* on one or more of the parameters.[7]

The second type of prior knowledge, what we have termed spatial knowledge, is interesting in a different way. In Section 3.3.2 we discussed dependence and heterogeneity and saw that they constituted two of the fundamental properties of spatial and spatial-temporal data. Data values close together in space (and in space-time) tend to be more alike than data values that are further apart. If we are willing to impose that observation about data onto a set of parameters, then we can assume that parameters close together in space (and space-time) tend to have values that are more alike than parameters that are further apart. This second observation provides the basis for specifying a type of prior distribution (or prior model) on parameters that leads to information borrowing amongst spatial or space-time units that are close together. We shall see examples of spatial prior models in Chapter 8 and spatial-temporal prior models in Chapter 15 that enable such information borrowing and, in the process, how the data sparsity challenge is addressed. As a form of "expert" knowledge, spatial knowledge, as we have defined here, is a hybrid between types (i) and (ii) described above – combining knowledge that expresses how things vary in space and space-time from first principles with knowledge that has been observed in many empirical contexts about how social science and environmental phenomena vary in space and space-time.

It is worth noting that there can be a link here with geographical-substantive knowledge. Consider a situation where previous work suggests the need to include a particular covariate in the process model but data on this covariate is either unavailable or incomplete. One way to approach this situation is to introduce random effects into the process model. To reflect our spatial knowledge about the covariate for which data are missing or incomplete, appropriate prior distributions can be specified so that the random effects are spatially structured and/or spatially unstructured (see for example Sections 1.3.2.1 and 9.2).

5.7 Concluding Remarks

One of the aims of this chapter has been to introduce Bayesian inference from a theoretical perspective, a model building perspective and a computational perspective. As we have seen, the Bayesian approach provides the joint probability distribution for all the parameters. This feature brings several benefits that we briefly summarize here. We will illustrate these benefits in many different spatial and spatial-temporal modelling contexts in the course of Parts II and III.

First, Bayesian inference incorporates information not only from data (via the data model) but also from other sources external to the observed data (via the prior distributions). For the analysis of spatial and spatial-temporal data, this latter feature becomes particularly useful, because as we have noted in 5.6.2 and will describe in depth in Chapters 8 and 15, prior distributions provide a way of capturing dependence structures across space and space-time. Second, the joint posterior distribution for all parameters

---

[7] See Section 3.2 for reasons as to why our knowledge about the association is often quite limited.

allows various sources of uncertainty to propagate throughout the statistical analysis. The estimation of regression coefficients takes into account the uncertainty arising from, for example, the presence of missing data and/or the errors in measuring the outcome data. Third, since parameter inference is based on the entire posterior distribution, in addition to the conventional point and interval estimates, probability statements can be readily obtained in order to directly address the question(s) in hand. For example, what is the probability that the average household income level in an area is below a given poverty threshold, or how likely is it that the overall burglary rate in the treatment group is less than in the control group? Fourth, in addition to the parameters in the model, inference on interpretable quantities that are some transformation of the model parameters can be obtained easily via Markov chain Monte Carlo (MCMC). We saw examples of this in Sections 5.2.2 and 5.4.2 (where the parameter of interest in each case is a nonlinear transformation of the regression coefficient as opposed to the coefficient itself) and Section 1.3.2.4 (where we were interested in posterior inference on $[\exp(b)-1]\times100$, the percentage change in the burglary rate compared to the controls, rather than b; see Section 12.3.5.4 for more detail). Finally, as we shall see many times, the Bayesian approach offers a convenient way to construct complex models – a feature that enables the analyst to construct a range of models to examine, for example, the robustness of the findings or to explore different features of the data.

Much of the rest of this book focuses on Bayesian regression modelling, so it is appropriate at this point to draw attention to a few points that the modeller of spatial and spatial-temporal data ought to be aware of and that the reader will encounter at various points in Parts II and III. First, the choice of a likelihood is certainly not unique. Certain features of the outcome values may require us to consider alternatives. For example, the Student's t distribution with v degrees of freedom is often used as an alternative to a normal likelihood in robust regression when outliers are present in response values (see Section 5.4.3 and Exercise 5.9; see also Gelman and Hill, 2007, p.124–125). When dealing with count data, the issue of overdispersion, where the variance exceeds the mean in the data, is prevalent in the case of spatial and spatial-temporal count data (see Section 3.3.3). The Poisson likelihood does not allow overdispersion in data since both the mean and the variance are assumed to be equal under a Poisson distribution. The negative binomial distribution offers an alternative (e.g. Hilbe, 2011), while including random effects within the Poisson likelihood is also used (e.g. Gschlößl and Czado, 2008). In application three in Chapter 9, we will discuss some choices of likelihood to tackle the issue of zero-inflation (where the chosen likelihood cannot accommodate the large number of 0 values in the observed data) when analysing count data in a spatial setting.

With a selection of likelihood choices to hand, which one should we use first? We recommend "starting simple". Some likelihoods (e.g. normal, binomial and Poisson) are considered to be "standard" while others (e.g. the Student's t and the negative binomial) are typically used once you are more familiar with the data.[8] Start with the standard one, then consider other alternatives. Placing the regression analysis in the Bayesian framework makes it easy to examine different likelihoods, and the change of the likelihood function only requires you to modify a few lines of WinBUGS code. This gives you the flexibility to fully explore the features in the data. Over the course of model building, model checking and model comparison are vital to a statistical analysis.

---

[8] The t_v and negative binomial distributions are not in the exponential family, so the corresponding regression models are not in the class of GLMs.

The other main aim of this chapter has been to show how Bayesian inference is applied to model spatial data. We have approached this through the three examples in this chapter – the HIA example in Section 5.2.2 and the income and the burglary examples in Section 5.4. Those three examples provide a starting point where a Bayesian analysis has been implemented on spatial data. Beyond specification of the process model, the modelling that we have seen so far is not spatial in any distinctive or "special" way. However, in Section 5.6 and Section 5.6.2 in particular, we have started to explore how that most characteristic feature of Bayesian inference, embedding prior knowledge into data analysis, might proceed in the case of spatial data.

5.8 Exercises

Exercise 5.1. For the coin example in Section 5.2.1, justify the use of a Beta distribution, $Beta(a,b)$, as a prior distribution for θ, the probability of obtaining a head when a coin is flipped once. The expert's opinion that θ is most likely to be 0.5 but unlikely to be outside the interval between 0.4 and 0.6 can be expressed mathematically as the mean of the Beta distribution is 0.5 and the standard deviation of the Beta distribution is 0.05 (why 0.05? Hint: the so-called three sigma rule). We can form two simultaneous equations with two unknown quantities, a and b, then solve for a and b. (Hint: if $\theta \sim Beta(a,b)$, then the mean of θ is $\dfrac{a}{a+b}$ and the variance of θ is $\dfrac{ab}{(a+b)^2(a+b+1)}$).

Exercise 5.2. Again, for the coin flip example, derive the posterior distribution for θ using the $Beta(a,b)$ distribution as a prior for θ. Calculate the posterior mean and the posterior variance for θ when using (a) the vague $Beta(1,1)$ prior and (b) the informative $Beta(45.5, 45.5)$ prior. Compare and comment on your results.

Exercise 5.3. Use R to simulate a set of 10000 random values from the distribution $Beta(7,5)$, the posterior distribution for θ when using the vague prior $Beta(1,1)$. (Hint: to simulate n values randomly from the distribution $Beta(a,b)$ in R, use the function `rbeta(n,a,b)`, then use the function length to see how many of those simulated values fall between 0.45 and 0.55; see also the R code given in Figure 5.3).

Exercise 5.4. Follow the steps outlined in Section 5.3.4 to carry out the analysis for the coin flip data. First, run the model exactly as given in Figure 5.7 and make sure you can obtain the results as presented in Figure 5.10 and Figure 5.11. Then modify the WinBUGS code in Figure 5.7 to calculate the posterior probability $Pr(0.45 < \theta < 0.55 \,|\, data)$. (Hint: you may need to multiply the output from two step functions, i.e. `step(diff1)*step(diff2)`, where `diff1 <- theta - 0.45` and `diff2 <- 0.55 - theta`).

Exercise 5.5. For the coin example, suppose we are interested in the probability of observing more than 20 heads if the same coin is flipped a further 30 times. Calculate this probability under the following two scenarios: (a) fixing θ (the probability of getting a head if this coin is flipped once) to the mean of the posterior distribution $Beta(7,5)$ when using the vague $Beta(1,1)$ prior; and (b) taking the entire posterior distribution of θ (i.e. $Beta(7,5)$) into account when calculating the above probability. Comment on the results and explain why the resulting probabilities are different. Hint: for (a), one can use the R command 1-pbinom(20,30,0.58) to calculate the required probability, whilst for (b), one needs to add the following six lines of WinBUGS code to Figure 5.7 (between Line 10 and Line 11 there) to do the prediction then calculate the required probability:

```
# make a prediction for the future 30 flips
y.pred <- dbin(theta,30)
# why subtract y.pred from 20.7 (not 20)?
d <- y.pred - 20.7
# Pr(number of heads over 30 flips > 20 | data))
prob.gt.20 <- step(d)
```

Exercise 5.6. Comment on the similarity between Exercise 5.5 above and Exercise 2.6 in Chapter 2.

Exercise 5.7. Use R to explore the properties (the support, the distributional shape, the mean and the variance) of the following Gamma distributions: (a) *Gamma*(0.01,0.01); (b) *Gamma*(1,1); (c) *Gamma*(10,10); and (d) *Gamma*(10,5). If $X \sim Gamma(a,b)$, then the mean of X is a/b and the variance of X is a/b^2. Some useful R functions are

1) rgamma(n,a,b) simulates n values randomly from a *Gamma*(a,b) distribution.

2) dgamma(x,a,b) calculates the density function for a *Gamma*(a,b) distribution at a given value x.

3) pgamma(x,a,b) calculates the probability $\Pr(X \leq x)$ for a given value of x and $X \sim Gamma(a,b)$.

4) qgamma(p,a,b) returns a value x such that $\Pr(X \leq x) = p$ for a given probability of p.

Exercise 5.8. Carry out the analysis of the income data in WinBUGS using the model presented in Eq. 5.17; pay particular attention to checks of convergence and efficiency and the form of model checking given in Figure 5.18. Monitor the DIC value.

Exercise 5.9. Repeat the modelling of the income data but replace the normal distribution in the likelihood by the t_4 distribution. Compare the results (e.g. the posterior estimates of the model parameters and the fit to the data) with those obtained from the model with the normal likelihood (see Exercise 11.2 in Chapter 11 for more detail).

Exercise 5.10. Analyse the burglary count data at the census output area (COA) level in Peterborough in 2005 using the Poisson regression model presented in Eq. 5.18 and Eq. 5.19. Investigate the goodness of fit to the data from the model via the posterior predictive check (Eq. 5.20). Comment on which parts of the data the model fits well and which parts of the data the model fits poorly. (Hint: for an area where the model fits poorly, compare the posterior predictive distribution of the case count with the actual observed count to understand whether the model predicts too many or too few cases.)

Exercise 5.11. The dataset PHIA_with_covariates.csv, on the book's website, contains the binary 0/1 outcome values that we discussed in Section 5.2.2. Also contained within that dataset are four COA-level covariates:

1) ethnic: index of ethnic heterogeneity that varies between 0 and 1, where the larger the value the greater the ethnic mix in a COA

2) carvan: percentage of households in a COA without a car or van

3) turnover: percentage of population living in a COA who in the previous year were living at another address either in the same COA or elsewhere

4) lonepar: percentage of households in a COA classified as a single-parent family

Extend the logistic regression model presented in Section 5.2.2 to include all four covariates and fit that model to the PHIA data. Comment on the estimated covariate effects. (Hint: the WinBUGS syntax `y[i]~dbern(pi[i])` specifies the Bernoulli likelihood for modelling the 0/1 binary outcome value `y[i]` and `logit(pi[i]) <- alpha + beta*ethnic[i]` specifies the regression relationship on the logit scale in Eq. 5.9.)

Exercise 5.12. Repeat the analysis in Exercise 5.11 but using the dataset `EHIA_with_covariates.csv`, which contains another set of COA-level binary outcome values that indicate whether a COA was considered as an HIA based on recorded crime data (the letter E in EHIA represents "empirical"). We will return to this dataset with more detail in Section 9.2.

Exercise 5.13. Suppose the values in each of the following three sets of data are independently drawn from a common normal distribution with an unknown mean and a known variance:

Set 1: 18, 4, 16, 1, −11, 0, −14, −32, 16, −19

Set 2: 842, 1166, 1346, 1091, 767, 1073, 1331, 1017, 996, 731, 816, 1219, 984, 797, 955, 916, 694, 919, 883, 937

Set 3: 19899, 33191, 13677, 33565, 20708, 24951, 26319, 50313, 40523, 34858, 37719, 35145, 40998, 33051, 34282, 17261, 22102, 7501, 41055, 27525

For each set of data, determine if each of the probability distributions listed below is a vague prior, a moderately informative prior, an informative prior or a prior distribution that may not be appropriate for the unknown mean:

(a) $N(0,10^2)$; (b) $N(0,100^2)$; (c) $N(0,10000^2)$; (d) $N(0,100000^2)$;

(e) $Uniform(0,100)$; (f) $Uniform(0,10000)$; and (g) $Uniform(0,1000000)$.

Exercise 5.14. In Chapter 1, we stated the structure of a Bayesian hierarchical model in the following form:

$$\Pr\left(process, parameters \mid data\right) \; \propto \; \Pr\left(data \mid process, parameters\right)$$

$$\times \; \Pr\left(process \mid parameters\right)$$

$$\times \; \Pr\left(parameters\right)$$

with the three components, $\Pr(data, process \mid parameters)$, $\Pr\left(process \mid parameters\right)$ and $\Pr(parameters)$, denoting the data model, the process model and the parameter model, respectively. Give a proof of the decomposition of $\Pr(process, parameters \mid data)$ using Bayes' theorem.

Part II

Modelling Spatial Data

6

Exploratory Analysis of Spatial Data

6.1 Introduction

In this chapter we describe methods for analysing a new set of spatial data. Exploratory data analysis (EDA) is a collection of numerical and graphical techniques for summarizing data properties, identifying unusual or interesting features in data, detecting data errors and suggesting relationships that might be incorporated into a model for the purpose of inference.[1] The careful application of EDA methods can help to narrow down the set of possible models. These same techniques might also be used in a later stage of data analysis, for example to examine the results of model fitting; provide evidence on model fit and whether model assumptions have been met; or assess the influence of particular data points on model fit. Resistant numerical methods are often recommended (the median rather than the mean, the inter-quartile range rather than the standard deviation) because they are not much affected by a small number of extreme, possibly "rogue" values in the dataset (Tukey, 1977). EDA methods help the analyst to arrive at some early understanding of a part of the variability in a dataset that it might be possible to account for, or "explain", and another part that it might not be. These broadly correspond, respectively, to Tukey's reference to the "smooth" and the "rough" parts of a dataset.

We assume the reader is familiar with the basic principles behind, and techniques of, EDA (see Tukey, 1977; Hoaglin et al., 1983; and Pearson, 2018). Andrienko and Andrienko (2006) provide extensive coverage of exploratory methods for spatial and temporal data. Here we concern ourselves with some of the specialist EDA techniques that are relevant to exploring the properties of small area data and which support spatial data modelling. The term *Exploratory Spatial Data Analysis* (ESDA) is frequently used to refer to this branch of EDA. For the reader interested in studying the wider aspects of ESDA, not necessarily concerned with data modelling as such, the software GeoDa developed by Anselin (2005) provides several of the ESDA methods discussed in this chapter as well as others not covered here, including computer-based graphical techniques that involve interaction (between researchers and their data) and multiple linkages between a displayed set of graphs and maps (see also Anselin et al., 2006). Haining et al. (1998) discuss ESDA in a Geographic Information Systems environment whilst Haining (2003, Chapter 6) provides an overview of ESDA.[2]

---

[1] Some caution is needed here. Using the same dataset to test a hypothesis that was formulated from that dataset biases the test result in favour of accepting the hypothesis. Good practice requires that more data are collected in order to test rigorously a hypothesis that has been formulated in this way.

[2] Different ESDA techniques are needed for other types of spatial data. For geostatistical data, sample data points on a continuous surface, see Cressie (1991, Chapter 2), who discusses a number of resistant techniques.

In the next section, Section 6.2, we scope the types of exploratory techniques needed for univariate small area data and discuss the role of mapping as a useful data visualization tool (6.2.1). We then describe techniques for checking for spatial trends (6.2.2), spatial heterogeneity in the mean (6.2.3), for global spatial dependence (6.2.4) and finally for spatial heterogeneity in the spatial dependence structure, which is concerned with detecting event clusters (6.2.5). In Section 6.3 we look at exploratory techniques for examining relationships between two or more variables. This raises interesting challenges because an association between two variables need not be constrained to the same area (for example, a covariate's level in one area could affect outcomes not only in that area but also in other areas) nor be constant across the study region (for example, an outcome-covariate relationship may vary from one area to another). Finally, in Section 6.4, we describe tests for overdispersion and zero-inflation in count data because these are two challenges commonly encountered when modelling counts in small geographical areas.

6.2 Techniques for the Exploratory Analysis of Univariate Spatial Data

Tukey (1977) argued that the "smooth" part of a single variable is the regular or predictable part of the data, whilst the "rough" part is the irregular or unpredictable component. Each data value that we observe on the variable contains these two elements:

$$DATA = SMOOTH + ROUGH$$

To take a simple univariate example, a boxplot can be used to suggest the rough and smooth elements in the distribution of a single variable where the box indicates the smooth part of the data whilst the whiskers and the values that lie beyond the ends of the whiskers reflect the increasingly rough elements of the data.

Tukey's decomposition of the observed data values is a useful starting point for our discussion of exploratory methods for univariate spatial data:

$$SPATIAL\ DATA = SPATIAL\ SMOOTH + SPATIAL\ ROUGH$$

But what are these two elements and, to make a start, what do we mean by "SPATIAL SMOOTH"? We suggest the following:

a. Spatial trend. In the case of a *linear* trend, data values might decrease with constant gradient in a certain direction across the map (e.g. from north to south or southwest to northeast). Higher order trends (such as quadratic, cubic or quartic trends) have maximum and minimum inflexion points with different gradients on different parts of the map.

b. Spatial heterogeneity. Variation in data values may not necessarily vary as smoothly as implied by some order of trend. We need ways of checking for differences in data values across a map that we may subsequently be able to explain using a regression model.

c. Global spatial dependence. This refers to the situation where there is a propensity for similar values to be found close together (positive spatial autocorrelation) irrespective of position on the map. This is sometimes referred to as "whole-map" or "global" *clustering* of similar data values. Suppose we take a pair of areas that are separated by a distance d, and another pair of areas separated by a distance k, where k is greater than d. In general, and providing the distances d and k are not large, the observed values of a variable from the pair of areas separated by distance d will be more alike, whilst the observed values of the same variable from the pair of areas separated by distance k tend to be less similar. Global spatial dependence implies that *this is the case anywhere on the map* (thus the term "*global*").

d. Spatial heterogeneity in the spatial dependence structure. Local spatial dependence refers to localized patterns of dependence with different structures of spatial autocorrelation on different parts of the map, for example, taking the form of small scattered *clusters* of high values (hotspots) or low values (coldspots).

What do we mean by "SPATIAL ROUGH"? This might take the form of an independent (spatially uncorrelated) white noise process. We might also single out individual extreme values or outliers. In a spatial dataset, there are two possible types of extreme value. The first type are data values that are different from most of the other values in the dataset. An outlier of the first type is often referred to as a *distributional outlier*, since it is different in relation to the overall distribution of data values. The second type of outlier is sometimes called a *spatial outlier*, where the data value observed in a spatial unit is very different from the values observed in the neighbouring spatial units. Although a distributional outlier will also often occur as a spatial outlier, a spatial outlier need not necessarily be a distributional outlier.

Two of the elements that we have referred to as "SPATIAL SMOOTH" are amongst the four data challenges presented at the outset in Chapter 1: the presence of spatial dependency and spatial heterogeneity. Many of the techniques in this chapter are designed to identify these properties of spatial data as essential contributions to data modelling. As we shall see, the need to check for spatial heterogeneity gives rise to what are termed "local" statistics, which generate as many values of the statistic as there are areas. This is in contrast to "global" statistics, which provide a single statistic for the whole map. However, before discussing these types of statistics and their role in exploratory data analysis, we start with considering mapping – an essential graphical technique in any exploratory analysis of spatial data.

6.2.1 Mapping

A map provides a visualisation of the data and is particularly useful at the exploratory stage to help reveal interesting features of the data (e.g. spatial trends, global and local spatial dependence structures as well as spatial outliers). Such findings help inform model building. Choropleth maps are the most commonly used type of map for visualizing areal data and are used extensively throughout this book for various purposes (such as initial exploration of the data and examining residuals for model assessment). In a choropleth map, each polygon (or spatial unit) is shaded according to the value of the variable that we are mapping. So, for example, in a greyscale choropleth map, a spatial unit with a high (low) attribute value is typically assigned a darker grey (a lighter grey) colour.

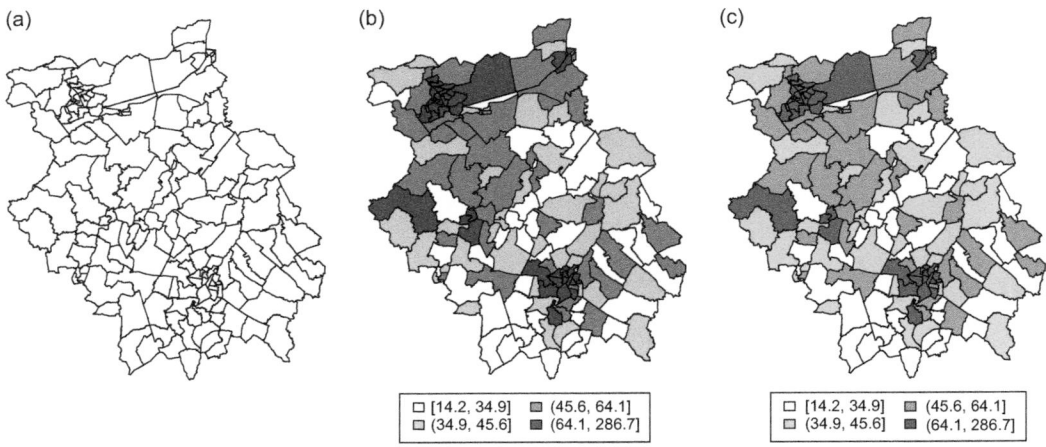

FIGURE 6.1
(See colour insert.) (a) A map of the 157 wards in Cambridgeshire, England; (b) and (c) are the choropleth maps of the burglary rates per 1000 houses in greyscale (b) and in shades of blue (c).

In R, a simple yet flexible way to produce a choropleth map is to use the generic plot function.[3] Typing the following commands into R, we obtain a map of the 157 wards in Cambridgeshire, England contained in the shapefile of the Cambridgeshire burglary data (Figure 6.1(a)).

```
library(maptools)   #  load the maptools package in for reading a shape
                    #  file into R

#  read in the shapefile of the Cambridgeshire
cambridgeshire <- readShapePoly('cambridgeshire.shp')

#  burglary data
plot(cambridgeshire)  # draw the 157 wards (or polygons) in Cambridgeshire
```

We now want to display information on a variable of interest, for example the burglary rates per 1000 houses across these wards as illustrated in Figure 6.1(b) in greyscale or in Figure 6.1(c) in shades of blue. Creating a choropleth map involves five basic steps:

Step 1: Extract the variable to be mapped from the shapefile (or create this variable if it is defined based on variables in the shapefile; for example, disease rates can be computed if the observed numbers of cases and the at-risk population are given in the shapefile).

---

[3] There are many other ways to create choropleth maps in R, for example, using the spplot function in the sp package and through the ggplot2 package. These methods do provide additional functionalities, but they require more knowledge to operate. There is also the choropleth function in the GISTools package.

Step 2: Define a set of cut points to convert the variable of interest into a categorical variable (for example, using quartiles as the cut points, each value is assigned to one of the four categories).

Step 3: Define a shading scheme (e.g. in greyscale or in shades of blue).

Step 4: Create the choropleth map using the `plot` function.

Step 5: Add a legend defining the shadings (or categories) using the `legend` function.

Following the above five steps, we now go through the R codes for creating the greyscale choropleth map in Figure 6.1(b).

The burglary rates are not in the shapefile, but we can compute the rates using the numbers of reported burglary cases and the numbers of at-risk houses. This calculation is done using the lines below.

```
####################################################################
#   Step 1: calculate the burglary rates per 1000 houses
#   (note: the syntax cambridgeshire@data allows us to access the
#   dataset in the shapefile; see Appendix 4.13.2 in Chapter 4 for more
#   detail)
####################################################################
O <- cambridgeshire@data$burgCases
pop <- cambridgeshire@data$houses
burgRate <- O/pop * 1000   #   burglary rates per 1000 houses
```

We use the quartiles to convert the burglary rates into four categories, each containing an equal number of areas (or data values). This step is done using the `quantile` function in R, which by default returns the minimum, the first, second and third quartiles and the maximum of the data values.

```
####################################################################
#   Step 2(a): define the cut points
####################################################################
cutpoints <- quantile(burgRate) # return 5 values: min, Q1, Q2, Q3 and max
```

Using the resulting cut points, each rate is assigned to one of the four categories (or levels) using the `cut` function:

```
####################################################################
#   Step 2(b): assign each burglary rate to one of the four categories
#   based on quartiles
####################################################################
rate.level <- cut(burgRate,cutpoints,labels=1:4,include.lowest=TRUE)
```

The first argument, `burgRate`, in the above command specifies the variable to convert, and the second argument, `cutpoints`, enters the set of cut points on which the categories are defined. We label the four categories, 1, 2, 3 and 4, using the third argument so that, for example, a burglary rate that is greater than the minimum but smaller than the first quartile is labelled 1 and a burglary rate between the third quartile and the maximum is labelled 4. The last argument, `include.lowest`, is set to TRUE so that the first category also includes the smallest value (see what would happen if you set `include.lowest=FALSE`).

For step 3, we define four shadings on the greyscale, one for each category.

```
#########################################################################
# step 3: define shadings
#########################################################################
shadings <- grey(c(1,0.7,0.4,0.2)) # grey(1) = white and grey(0.2)=(very)
                                    # dark grey (and grey(0)=black)
```

If blue is used (i.e. Figure 6.1(c)), replace the R command above by

```
#########################################################################
#  the brewer.pal function (from the RColorBrewer package) produces
#  4 shades of blue as defined in the 1st and 2nd arguments,
#  respectively
#########################################################################
library(RColorBrewer)  #  load the RColorBrewer package
shadings <- brewer.pal(4, 'Blues')
```

We are now in position to create the choropleth map:

```
#########################################################################
# step 4: create the choropleth map using the plot function
#########################################################################
plot(cambridgeshire,col=shadings[rate.level])
```

The command above is the same as that for producing the "plain" map in Figure 6.1(a), but for each spatial unit, the second argument assigns one of the four shadings according to which rate category this spatial unit lies within. For example, an area in rate category 1, the lowest rate category, has the white shading (`grey(1)`, which corresponds to the colour name "#FFFFFF"), whereas an area in category 4, the highest rate category, has a dark grey shading (`grey(0.2)` with the corresponding colour name "#333333") – type the three objects, `rate.level`, `shadings` and `shadings[rate.level]`, in turn into R and match their entries to see how this colour assignment works.

Finally, it is essential to add a legend describing the rate categories:

```
#########################################################################
#  step 5: add a legend of category definitions at the top left
#  corner of the figure
#########################################################################
```

```
intervals <- c('[14.2, 34.9]','(34.9, 45.6]','(45.6, 64.1]','(64.1,
286.7]')
#  the first argument in the legend function defines the position of the
#  legend and the last argument assigns the shadings to the intervals
legend('topleft',legend=intervals,fill=shadings)
```

The cut points define the intervals. The right (or left) square bracket,] (or [), means that the interval includes values equal to the upper (or lower) bound of the interval, while the left parenthesis, (, means the interval excludes values equal to the lower bound of the interval.

We turn the focus now to the resulting choropleth map itself. The burglary rate varies greatly across the 157 wards in Cambridgeshire. Two clusters of wards, Peterborough (the cluster in the north of Cambridgeshire) and Cambridge (the cluster in the south), are shown to have higher burglary rates compared to other wards in the study region. Such observations often lead us to generate a series of questions: "do the underlying burglary risks differ from one ward to another?"; "do wards that are geographically close tend to have similar burglary risks?"; "are there any (other) high-risk clusters?" and "can the spatial distribution of risk be explained using various socio-economic and demographic characteristics of the areas?" And these questions in turn lead us to perform subsequent analysis/modelling of the data.

It should be emphasised, however, that useful though a map is for visualizing this data, we cannot draw any firm conclusions based on a map derived using only the observed data. The main reason is that various sources of uncertainty (as discussed in Chapter 1) are not properly incorporated and quantified. To do this we require modelling.

6.2.2 Checking for Spatial Trend

Prior to mapping, the variable of interest can be smoothed spatially to better reveal any overall spatial pattern in the data. A first-order smoothed value for area i can be obtained by calculating a weighted average of the value in area i together with those in the areas that are neighbouring or adjacent to i. Similarly, a second-order smoothing can be carried out by averaging the value in i, the values of the neighbours of i and the values associated with the neighbours of the neighbours. Spatial smoothing at a third order or higher can be computed along similar lines. This operation smooths the variation in the data, which may help reveal any overall global patterns. Smoothing also helps to alleviate the data sparsity problem (see Section 1.2.3) through data borrowing, an operation that increases the size of the population on which each spatially-smoothed value is calculated.

To carry out a first-order spatial smoothing, the augmented row-standardised weights matrix, $(I+W)^*$, is multiplied by x, the set of values attached to the spatial units (see Section 4.7). To carry out a second-order smoothing around area i, we need to add in the neighbours of the neighbours before row standardising (see Section 4.8 and Appendix 4.13.2 on how to identify higher order neighbourhoods using powers of the W matrix). When calculating the spatial average for area i, the same weight is often assigned to i and to all its spatial neighbours, irrespective of the locations of the neighbours relative to i. To address this, unequal weights can be used instead. For example, a neighbour whose centroid is closer to the centroid of i could be given a bigger weighting than another neighbour whose centroid is further away (see Chapter 4 for ways to define W matrices meeting this requirement).[4]

---

[4] This is similar to kernel smoothing, where higher weights are given to points that are closer. See, for example, Hastie et al. (2009, Chapter 6).

While useful for revealing large-scale trends, the effect of smoothing is to *increase* spatial dependence in the resulting map, as the smoothed values are obtained using spatially-overlapping data subsets. As a consequence, map smoothing does not provide a suitable basis for assessing spatial autocorrelation. In addition, spatial smoothing tends to "iron out" small-scale variation (e.g. discrete jumps on the map due to very large or very small data values), making it unsuitable for detecting spatial anomalies (e.g. identifying crime or disease hotspots).

The two maps in Figure 6.2 show the effect of applying a first-order smoothing operation to the burglary data for Cambridgeshire. Using the spatially-smoothed burglary rates (Figure 6.2(b)), the higher rates in the two urban areas of the county, Cambridge (in the south) and Peterborough (north), become more apparent, as there is a general tendency for rates to decline from north to south in Cambridgeshire.

A careful comparison of the two maps in Figure 6.2 reveals evidence of oversmoothing. For example, before smoothing, the rate for the polygon with a cross sign is in the lowest rate category. However, after averaging over the (much higher) rates of its neighbours, the spatially smoothed rate for this area becomes much higher. The same happens to the polygon with a plus sign but in the "opposite direction", where the burglary rate in this area is high but the smoothed burglary rate is low, due to the much lower burglary rates from its neighbours.

The spatially-smoothed burglary rate of each area is calculated as the ratio of the sum of the cases in i (O_i) and its neighbours to the sum of the populations at risk in i (n_i) and its neighbours. That is, $r_i^{\text{smoothed}} = \sum_{j \in \Delta i} O_j / \sum_{j \in \Delta i} n_j$, where Δi denotes the spatial neighbours of i and area i itself – what would be the difference if r_i^{smoothed} is calculated as $\sum_{j \in \Delta i} r_j$, where r_j is the burglary rate in area j?

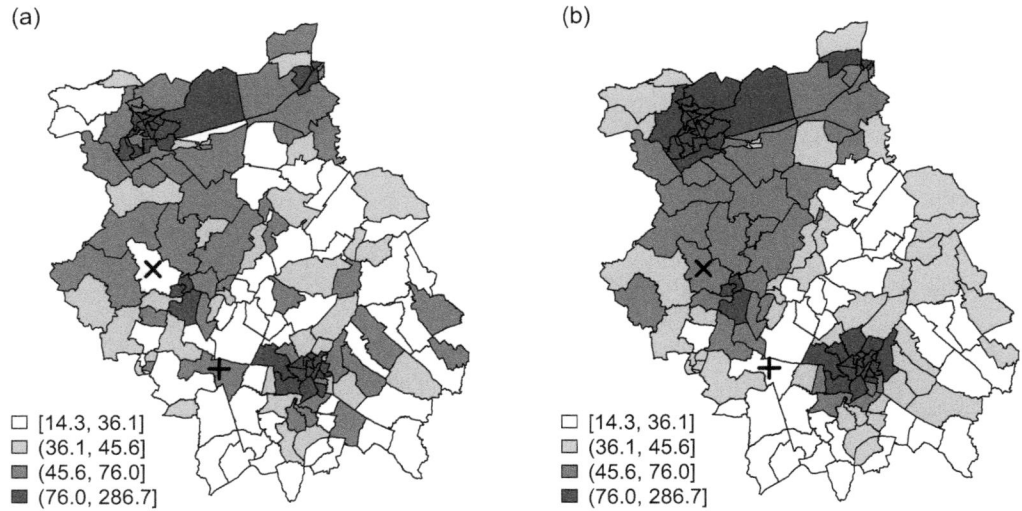

(a)

(b)

☐ [14.3, 36.1]
☐ (36.1, 45.6]
▣ (45.6, 76.0]
■ (76.0, 286.7]

☐ [14.3, 36.1]
☐ (36.1, 45.6]
▣ (45.6, 76.0]
■ (76.0, 286.7]

FIGURE 6.2
(a) a map showing the burglary rates per 1000 houses in Cambridgeshire and (b) a map of the spatially-smoothed burglary rates. In the spatially-smoothed map, two highlighted polygons (one with a plus sign "+" and one with a cross sign "x") illustrate evidence of oversmoothing. Note that the same set of cut points is used for both maps to ensure the two maps are comparable.

The following R commands carry out the spatial smoothing. For ease of understanding, the script is organised in blocks, each achieving a specific (small) task.

```
###########################################################
#   block 1: obtain the augmented row-standardised spatial weights
#   matrix (I + W)ˆ
###########################################################
#   first obtain the spatial weights matrix based on shapefile
#   (see Chapter 4 for more detail)
library(spdep)     #   load the spdep package
nb <- poly2nb(cambridgeshire,queen=FALSE)   #   neighbouring structure based
                                #   on rook's contiguity
W <- nb2mat(nb,style='B')   #   the spatial weights matrix (setting style to
                          ·#   'B' gives 0/1 weight => 1=neighbours 0=not)
#   then create the augmented spatial weights matrix I+W
N <- nrow(W) #   obtain number of areas
I <- diag(N) #   use the diag function to create an N-by-N identity matrix
IW <- I + W   #   the augmented spatial weights matrix
#   finally standardise each row
for (i in 1:N) {
  IW[i,] <- IW[i,]/sum(IW[i,])
}
###########################################################
#   block 2: spatial smooth the numbers of cases and the population
#   separately
###########################################################
smooth.O <- IW %*% matrix(O,ncol=1)
smooth.pop <- IW %*% matrix(pop,ncol=1)
###########################################################
#   block 3: calculate the spatially-smoothed burglary rates per
#   1000 houses
###########################################################
smooth.rate <- smooth.O/smooth.pop * 1000
```

In block 2 of the R code, the syntax `%*%` performs a matrix multiplication, i.e. multiplying an $N \times N$ matrix (IW) by an $N \times 1$ matrix of the burglary cases (or population). The latter is created using the `matrix` function, and the argument ncol=1 tells R to create only one column in the resulting matrix. The procedures in Section 6.2.1 can then be used to map the smoothed rates (as well as the unsmoothed ones), as shown in Figure 6.2(b). Note that, to ensure the shadings are comparable between the two maps in Figure 6.2, the same cut points are used for both and are taken to be the quartiles calculated by combining the spatially-smoothed and the unsmoothed rates as follows:

```
cutpoints <- quantile(c(burgRate,smooth.rate))
```

6.2.3 Checking for Spatial Heterogeneity in the Mean

6.2.3.1 Count Data

In the context of modelling count data, for example numbers of crime events or disease cases occurring over a set of N areas, we are often interested in testing for any differences in crime or disease risk across the set of areas. Note that we are now referring to the *risk* of crime, say burglary (or disease), rather than the burglary *rate*. A rate is obtained from a data calculation: the count of a particular event (household burglary) divided by the number of units (households) at risk (from being burgled). We call the value from that calculation the *raw rate* of an area, because it is calculated using only the data available in that area (no information sharing involved; see Chapters 7 and 8). Risk, on the other hand, is a theoretical quantity, a *parameter* which has to be estimated. Denoting θ_i as the true but unknown risk in area i ($i = 1, \ldots, N$), we wish to test the null hypothesis that all areas have the same level of risk (i.e. $\theta_1 = \theta_2 = \ldots = \theta_N$) against an alternative hypothesis that there is at least one area with a different risk level (i.e. $\theta_i \neq \theta_k$ for some $i \neq k$).

One option is to perform a chi-square test. Consider the example of modelling small-area burglary risk. Let O_i be the number of burglary cases reported in area i and n_i be the number of at-risk households in that area. Now, O_i / n_i is the maximum likelihood estimate of the underlying true risk θ_i (see Section 3.3.3). The chi-square test statistic takes the form

$$T = \sum_{i=1}^{N} \left\{ \frac{(O_i - E_i)^2}{E_i} \right\}, \tag{6.1}$$

where $E_i = n_i \cdot \left(\sum_{i=1}^{N} O_i \big/ \sum_{i=1}^{N} n_i \right)$ is the expected number of cases in area i assuming that burglary cases arise at random across the study region so that any area's share of the total number of cases observed is proportional to the size of its own population at risk. This assumption implies a constant risk across the study region. The test statistic measures an overall deviation from the constant risk surface. So if, for some areas, the observed numbers are much different from the expected, then T would become large, suggesting a departure from the constant risk assumption, or in other words, the presence of spatial heterogeneity in the mean of the data. To assess statistical significance, when N (the number of areas) is large, the test statistic T follows a χ^2 distribution (called a chi-square distribution, hence the name of the test), with $N - 1$ degrees of freedom under the null hypothesis of a constant risk surface over the map. The p-value can then be obtained and the decision on whether to reject or retain the null hypothesis can be made (e.g. reject the null hypothesis if the p-value is less than the typically assumed 5% significance level).

If the null hypothesis is rejected in favour of the alternative, a subsequent task will be to try to build a statistical model to explain the observed variation. The chi-square test can be readily adapted for the case of standardised ratios. In that case, the expected number of cases given in the dataset are used directly in the calculation of the test statistic.

To investigate whether or not the underlying burglary risk is the same across Cambridgeshire, the R script in Figure 6.3 calculates the chi-square test statistic (blocks 1 and 2) and computes the p-value from the chi-square distribution (block 3). The large test statistic of 16616.42 results in a p-value that is very close to 0, suggesting strong evidence against the null hypothesis of a constant burglary risk across Cambridgeshire.

```
####################################################################
#  block 1: calculate the expected numbers of cases
####################################################################
all.O <- sum(O)           #  summing all the cases
all.pop <- sum(pop)       #  summing all the at- risk populations
E <- pop * all.O/all.pop  #  calculate the expected numbers of cases


####################################################################
#  block 2: compute the test statistic T
####################################################################
test.stat <- sum((O-E)^2/E)


####################################################################
#  block 3: obtain the p-value using the chi-square dist with N-1
#  degrees of freedom
####################################################################
N <- length(pop)     #   obtain N, the number of areas
1 - pchisq(test.stat,df=N-1)
```

FIGURE 6.3
The R code to calculate the chi-square test statistic and the p-value from the corresponding chi-square distribution using the Cambridgeshire burglary data.

6.2.3.2 A Monte Carlo Test

Statistical significance can also be assessed using a Monte Carlo (MC) test. An MC test approximates the sampling distribution of the test statistic under the null hypothesis through calculating the test statistic over a large number of datasets, each being generated under the null hypothesis. We will see several variations of an MC test in this chapter. For the chi-square test of spatial heterogeneity, the MC test is carried out through two steps:

Step 1: All the observed cases are re-distributed over the areas at random such that each case has a probability of falling into area i proportional to n_i, the size of the at-risk population (or proportional to E_i, the expected numbers of cases, in the case of standardised ratios) – this procedure follows from the null hypothesis of a constant risk across the areas.

Step 2: The test statistic T is calculated for the resulting dataset generated at Step 1.

The above two steps are repeated many times. The resulting set of T values forms an approximation to the distribution of the test statistic T under the null hypothesis. The test statistic calculated using the original observed data is then compared to the above distribution to obtain an empirical p-value, which tells us how extreme the test statistic of the observed data is in relation to the distribution. From Davison and Hinkley (1997, p.140–141), with an early reference to Hope (1968), the formula to calculate the empirical (or Monte Carlo) p-value is

$$\frac{k+1}{m+1}, \tag{6.2}$$

where m is the total number of simulated datasets and k is the number of these datasets that produce a test statistic greater than the test statistic calculated using the original data.

Consider again the Cambridgeshire burglary data. With 999 simulations, the histogram in Figure 6.4 shows the approximated distribution of the test statistic T under the null hypothesis of a constant burglary risk. The test statistic of 16616.42 sits well beyond the right-hand tail of the distribution (Figure 6.4), producing the empirical p-value of 0.001 ($=(0+1)/(999+1)$). Therefore, the MC test provides strong evidence to reject the null hypothesis in favour of the alternative hypothesis that burglary risks vary spatially. The distribution of the test statistic from the Monte Carlo simulation is close to the χ^2_{156} distribution (Figure 6.4), thus the two approaches come to the same conclusion.

The 999 Monte Carlo simulations are carried out using the R script in Figure 6.5.

6.2.3.3 Continuous-Valued Data

Analysis of variance (ANOVA) is a standard method for testing differences in group means. Take the example of the Newcastle income data introduced in Section 1.3.1, where the inferential goal is to provide reliable estimates of average income per household for the 109 MSOAs in Newcastle. While Figure 1.7 in Chapter 1 suggests some variation in the income level across the MSOAs, ANOVA is an initial attempt towards answering the question: "do MSOAs have the same average income or not?" As we shall see in Chapter 7, the above question plays a part in determining an appropriate strategy for modelling the data. Here we demonstrate the technique using the Newcastle income data but refer readers to, for example, Chapter 16 in Faraway (2002) regarding the technical details as well as its implementation in R.

An R function to carry out an ANOVA is `aov` as specified towards the end of the next page. The `income` and `msoa` variables in the data object `hh.data` contain, respectively, the household-level income values from the survey and the MSOA indicator identifying which MSOA a household resides in. The results of the ANOVA are stored in the object `fit`, which is then summarised using the `summary` function in the third line.

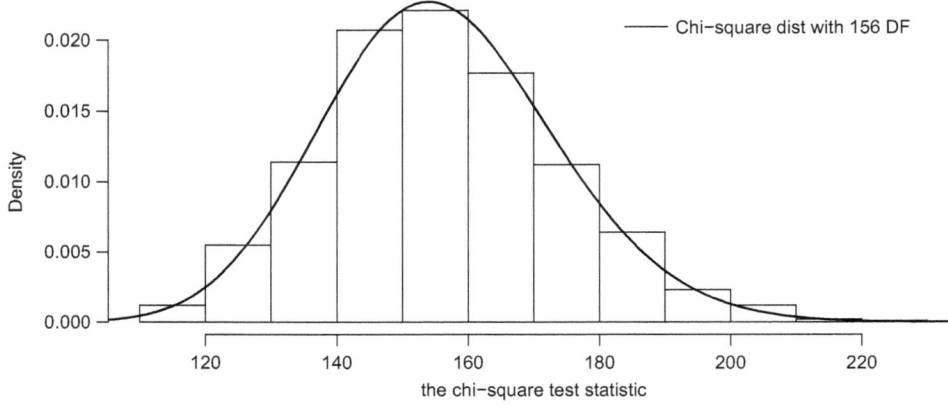

FIGURE 6.4
The approximated distribution of the chi-square test statistic based on 999 Monte Carlo simulations (the histogram). The distribution is well below 16616.42, the test statistic calculated from the original Cambridgeshire burglary data. The solid curve is the χ^2_{156} distribution, which matches with that from the Monte Carlo simulation.

```
########################################################################
#  block 4: specify number of MC simulations and create an array for storage
########################################################################
nsims <- 999          #  number of Monte Carlo simulations to be carried out
store <- rep(0,nsims) # create an array of size nsims to store the test stat
                      # calculated using the simulated data


########################################################################
#  block 5: calculate the probability of a case falling into each
#  area; these probabilities are proportional to the sizes of the
#  at-risk population so that if the population size of area i is
#  twice of that of area j, then a case is twice more likely to
#  fall into area i than into j
########################################################################
prob <- pop/sum(pop)


########################################################################
#  block 6: performing the 999 MC simulations using a for loop
########################################################################
for (isim in 1:nsims) {
  #  Step 1: using a multinomial distribution to redistribute
  #  all the (all.O=)25311 cases to the areas
  sim.O <- rmultinom(1,all.O,prob)
  #  Step 2: calculate and store the chi-square test statistic
  #  using the simulated data
  store[isim] <- sum((sim.O - E)^2/E)
}


########################################################################
#  block 7: calculate the empirical p-value: the proportion of the values in store
#           exceeds the test statistic calculated using the original observed data
########################################################################
k <- length(which(store>test.stat)) # counting how many values in the store array
                                     # that are greater than test.stat
(k+1)/(nsims+1) # computes the empirical p-value
```

FIGURE 6.5
The R code to perform the 999 MC simulations for the chi-square test applied to the Cambridgeshire burglary
data. The labelling of the blocks continues from that in Figure 6.3.

```
load('income.RData')  #  read in the household-level income data
fit <- aov(income ~ msoa, data=hh.data)
summary(fit)
```

R then returns the following:

```
            Df   Sum Sq Mean Sq F value Pr(>F)
msoa         1    88594   88594   3.301 0.0696 .
Residuals  758 20340834   26835
---
Signif. codes:  0 '***' 0.001 '**' 0.01 '*' 0.05 '.' 0.1 ' ' 1
```

The null hypothesis here is that all MSOAs have the same level of average income per household, while the alternative hypothesis is that at least one MSOA has an average income level that is different from the rest. The p-value reported is 0.0696, below the 10% significance level, suggesting some (weak) evidence of varying levels of average income. Therefore, when building a model, we need to allow the income level to be MSOA-specific, as opposed to restricting all MSOAs to having the same level. At the same time, we should also bear in mind that ANOVA, as well as the chi-square test discussed in Section 6.2.3.1, does not account for certain features of spatial data, for example, spatial autocorrelation, heterogeneous variance (also known as heteroscedasticity, i.e. the within-MSOA variance may not be constant across the MSOAs), as well as possible non-normality of the residuals. These issues are addressed through modelling (see later in Part II of this book).

6.2.4 Checking for Global Spatial Dependence (Spatial Autocorrelation)

The exploratory tools discussed in this section are concerned with assessing whether or not spatial autocorrelation is present in a set of spatial data. First we need to be clear what spatial autocorrelation is and get a sense of what it looks like in area-level data. Illustrated in Figure 6.6 are the three principal forms of spatial autocorrelation depicted on a square lattice of small areas where black pixels take the value 1 and white pixels take the value 0. Figure 6.6A is an extreme form of negative spatial autocorrelation (0s and 1s form an alternating, chessboard-like, pattern), whilst 6.6B represents a less extreme form of negative spatial autocorrelation. Figure 6.6E represents an extreme form of positive spatial autocorrelation (all the 0s cluster together and all the 1s cluster together), whilst Figure 6.6D represents a less extreme form of positive spatial autocorrelation. Positive spatial autocorrelation is the form of spatial autocorrelation most commonly encountered in practice. Figure 6.6C displays a random pattern of 0s and 1s. More generally (that is, where values are continuous rather than binary), the fundamental idea is that positive spatial autocorrelation describes the situation where, *at the scale of the spatial unit*, data values in nearby spatial units (areas) tend to be more alike than those from spatial units (areas) that are far apart. Areas with large (small) values tend to be neighbours to areas which also have large (small) values. To emphasize this point the reader should re-visit Figures 6.1(b, c) and 6.2(a, b) to see examples of this more general case. On the other hand, negative spatial autocorrelation, much less frequently encountered in practice, is where large (small) values tend to be adjacent to small (large) values. The absence of spatial autocorrelation (or spatial independence) is when there is no obvious pattern (or structure) in the spatial distribution of the data values. In other words, data values appear to be distributed randomly in space.

It is important to recognize that the property of spatial autocorrelation is scale dependent. To show this, consider Figure 6.6A. At the scale of the squares on the board, a chess board is negatively spatially autocorrelated (black next to white); but at scales smaller than the squares (e.g. points within a square), the chessboard shows positive spatial autocorrelation over short distances. Negative spatial autocorrelation cannot exist on a continuous surface as distance separation converges to zero because, with a few exceptions (e.g. geological fault lines), our geographical world is intrinsically continuous (and would be a strange and difficult place to inhabit were it not to be so).

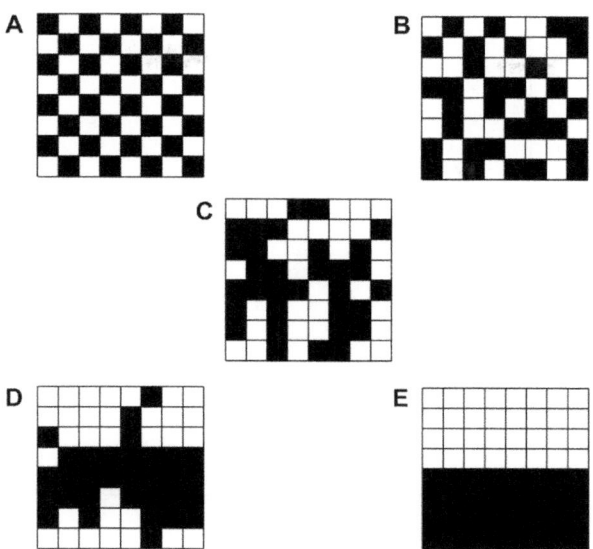

FIGURE 6.6
An illustration of three main types of spatial autocorrelation expressed in terms of a spatial arrangement of 0s and 1s on a square pixelated map. A: extreme negative spatial autocorrelation (chess-board); B: negative spatial autocorrelation; C: random pattern; D: positive spatial autocorrelation; E: extreme positive spatial autocorrelation (clustered map pattern).

In what follows, we will introduce methods that visually explore this feature of spatial data and methods that test the presence or absence of spatial autocorrelation.

6.2.4.1 The Moran Scatterplot

The Moran scatterplot is a graphical exploratory tool to check for spatial autocorrelation. The value of each spatial unit is plotted against a weighted average of the values from its spatial neighbours. Let x denote a column vector of size $N \times 1$ containing the values of interest over the N spatial units. A set of weighted averages is obtained by multiplying the row-standardised weights matrix W^* by x. The Moran scatterplot is a scatterplot between x and W^*x. It is conventional to put the original values on the horizontal axis and the spatially averaged values on the vertical axis. Both x and W^*x are standardised so that they have a mean of 0 and a variance of 1. This standardisation effectively partitions the Moran scatterplot into four quadrants, and the interpretation of each quadrant is in Figure 6.7. Note that the spatial averages W^*x are also known as spatially-lagged values (see more in Chapters 10 and 11), and the spatially-lagged value for spatial unit i takes the form $\sum_{j=1}^{N} w_{ij}^* x_j$, with $w_{ii}^* = 0$ (cf. the operation of spatial smoothing in Section 6.2.2).

Figure 6.8 shows the Moran scatterplot for the burglary rate map shown in Figure 6.2(a) with a line of best fit superimposed. The plot shows that in general, areas with large (small) values tend to be surrounded by areas with large (small) values, consistent with the visual

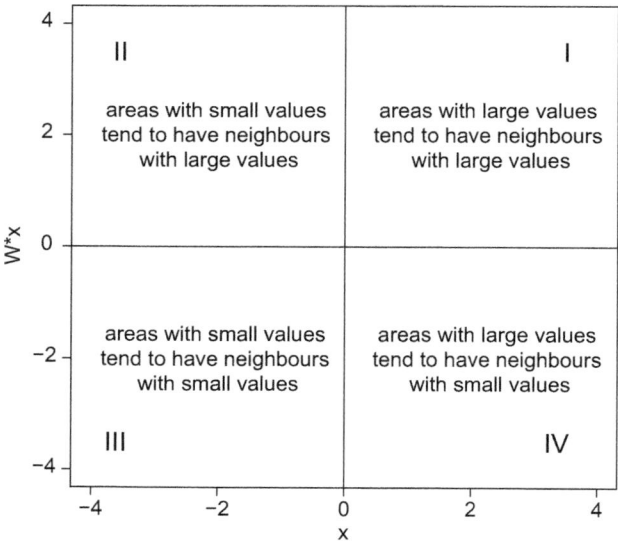

FIGURE 6.7
Interpretation of each of the four quadrants in a Moran scatterplot.

signs of positive spatial autocorrelation in the rate map itself. The plot can also be used to suggest the presence of spatial outliers. For example, the dot enclosed by a large circle identifies an area whose burglary rate is high but is surrounded by areas with, on average, lower burglary rates.

The Moran scatterplot is based on observing the similarity between values close together, where in the above case, "close together" means adjacent (first-order) neighbours. Such a scatterplot can also be constructed using spatial neighbours at the second order, third order and so on (see Section 4.8). Producing a sequence of such plots will allow us to observe how "similarity" weakens (assuming it does) as distance separation increases (see Exercise 6.3).

6.2.4.2 The Global Moran's I Statistic

The global Moran's autocorrelation statistic (or the global Moran's I; Moran, 1950) is the most commonly used statistic for testing for global spatial autocorrelation. It computes the cross products between data values, giving it a form that shows similarities to Pearson's bi-variate correlation coefficient:[5]

$$I = \frac{N \sum_{i=1}^{N} \sum_{j=1}^{N} w_{ij} d_i d_j}{S \sum_{i=1}^{N} d_i^2} \tag{6.3}$$

---

[5] Hence the term "auto" or "self" correlation. Unlike Pearson's statistic, Moran's I statistic does not range between −1 and +1.

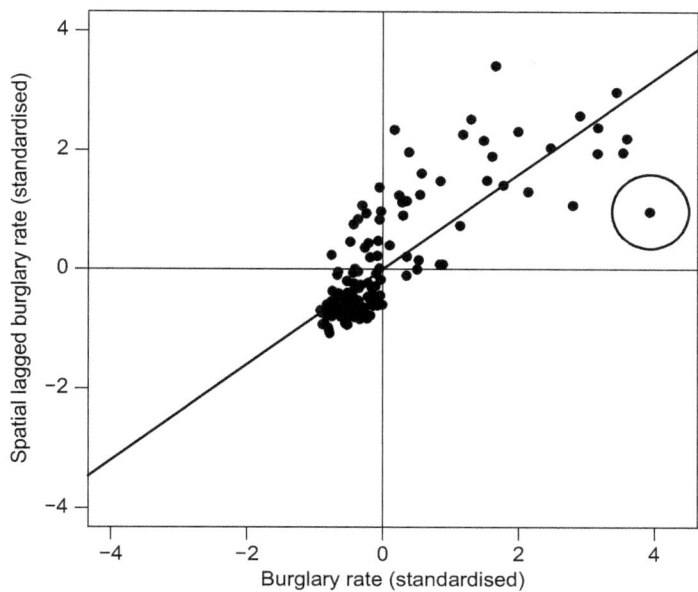

FIGURE 6.8
A Moran scatterplot based on the Cambridgeshire burglary data (see Exercise 6.2).

In Eq. 6.3, N denotes the number of areas; $d_i = x_i - \bar{x}$ measures the deviation of x_i from $\bar{x}$, the mean of all the observed values; w_{ij} is the entry of the ith row and the jth column in the spatial weights matrix W; and S is the sum of all the entries in W.

Intuitively, the formulation of I tells us that if most pairs of neighbouring areas have the same sign regarding their deviations from the mean, that is both areas are simultaneously above or are simultaneously below the mean, then the value of I will be positive and away from 0, indicative of positive spatial autocorrelation. On the other hand, if a majority of the pairs of neighbours have deviations from the mean in opposite directions, that is, one area is above while the other is below the mean, I will then be negative and away from 0. This suggests negative spatial autocorrelation is present. The global Moran's I becomes close to 0 when some pairs of neighbours have the same direction of deviation and others have deviations in the opposite direction and, when summing across all the pairs of neighbours, the products of deviations roughly cancel each other out. The global Moran's I then suggests the data values are spatially independent.

Cliff and Ord (1973 and 1981) have shown that under the null hypothesis of no spatial autocorrelation the expectation of I is close to 0. They have also derived the sampling distribution of I under the null hypothesis (Cliff and Ord, 1981, p.46–49), from which we can assess whether the value of I calculated from the observed data is sufficiently far away from 0 to establish evidence for the presence of spatial autocorrelation.

The R commands below compute the global Moran's I using the burglary rates from the Cambridgeshire data. The first line uses the `nb2listw` function (from the `spdep` package) to format the neighbourhood structure stored in `nb` (defined in Section 6.2.2) into the `listw` structure. For each area, the `listw` structure contains both the area's neighbours and the associated weights. The second argument of the `nb2listw` function specifies whether the binary (0/1) weights are used (`style='B'`) or the weights are taken from the row-standardised version of the spatial weights matrix (`style='W'`). The resulting `listw` object is then used in the `moran.test` function on the second line. The object `burgRate` contains the burglary rates as calculated in Section 6.2.1.

```
listw <- nb2listw(nb,style='W')
moran.test(burgRate,listw=listw)   #  the moran.test function is from the
                                   #  spdep package
```

The output is below:

```
Moran I test under randomisation
data:  burgRate
weights: listw
Moran I statistic standard deviate = 12.31, p-value < 2.2e-16
alternative hypothesis: greater
sample estimates:
Moran I Statistic      Expectation         Variance
      0.606409578    -0.006410256      0.002478418
```

The last line of the output reports the global Moran's *I* calculated for the Cambridgeshire burglary rates (=0.606), together with the expectation (=−0.006) and the variance (=0.002) of the sampling distribution of *I* under the null hypothesis. There is clear evidence suggesting the presence of positive spatial autocorrelation in the burglary rates, namely, areas with high burglary rates tend to be close together in space, whilst areas with low burglary rates tend to be close together.

A permutation test, a variant of the Monte Carlo procedure outlined in Section 6.2.3.2, is a common way to test for significance for the global Moran's *I*. A reason for its popularity is that a permutation test avoids the large sample assumption (the number of areas and/ or the total number of cases needs to be sufficiently large) in deriving the theoretical distribution of the test statistic under the null hypothesis. The permutation test for *I* involves repeating the following two steps many times:

Step 1: Generate a random permutation of the original data by reassigning the values randomly to the spatial units – this random reshuffling of the data follows from the null hypothesis of no spatial autocorrelation.

Step 2: Calculate the global Moran's *I* on the resulting dataset.

After many permutations, the values of *I* give an approximation to the distribution under the null hypothesis, which can then be used to obtain the empirical p-value using the formula given in Eq. 6.2.

The following R command performs a permutation test with 999 permutations for the Cambridgeshire burglary data using the `moran.mc` function in the `spdep` package. The first two arguments of the `moran.mc` function are the same as those in `moran.test`, while the third argument specifies how many permutations (or simulations) are to be carried out.

```
moran.mc(burgRate,listw=listw,nsim=999)  #  the moran.mc function is from
                                          #  the spdep package
```

The R output is

```
Monte-Carlo simulation of Moran I
data:   burgRate
weights: listw
number of simulations + 1: 1000
statistic = 0.60641, observed rank = 1000, p-value = 0.001
alternative hypothesis: greater
```

The output shows the empirical p-value is 0.001, meaning that none of the 999 permutations yielded a value of *I* greater than that calculated using the original burglary data. This can also be seen in Figure 6.9, where the *I* value based on the observed data (the vertical line) is compared to the approximated sampling distribution (the histogram) using the 999 permutations.

6.2.4.3 Other Tests for Assessing Global Spatial Autocorrelation

Other tests for examining global spatial autocorrelation have also been developed (see, for example, Cliff and Ord, 1973). The test statistics of these tests take a general form:

$$\sum_i \sum_j w_{ij} f(x_i, x_j),\qquad(6.4)$$

where $f(x_i, x_j)$ measures the similarity of two values, x_i and x_j, observed at two different spatial locations, i and j, and w_{ij} represents the connectivity of these two spatial locations (e.g. $w_{ij} = 1$ if they share a common boundary and $w_{ij} = 0$ otherwise). In the case of the global Moran's *I*, $f(x_i, x_j)$ is the product of the deviations from the mean. Geary's c (Geary, 1954) measures similarity of x_i and x_j by the squared difference, i.e. $f(x_i, x_j) = (x_i - x_j)^2$, which is also the basis of the estimator for the variogram used to measure spatial structure in geostatistics (see for example Cressie, 1991, p.40). The Getis-Ord general G statistic expresses pairwise similarity as $f(x_i, x_j) = x_i x_j$ (Getis and Ord, 1992, p.194).[6] The Moran test is most frequently used but needs to be interpreted with caution. Spurious positive spatial autocorrelation (rejecting the null hypothesis when it should be retained) may be the result of trends in the data or because of unequal variances across the set of data values – a particular problem if sample size varies spatially. For example, if we are analysing data across a number of census tracts, some of which are heavily populated urban tracts and some are more sparsely populated rural tracts, the extreme rates or ratios may tend

---

[6] The Getis-Ord general G statistic only applies to nonnegative data, so it cannot be used to assess spatial autocorrelation in residuals. Getis and Ord (1992, p.198): "It must be remembered that G(d) is based on a variable that is positive and has a natural origin. Thus, for example, it is inappropriate to use G(d) to study residuals from regression".

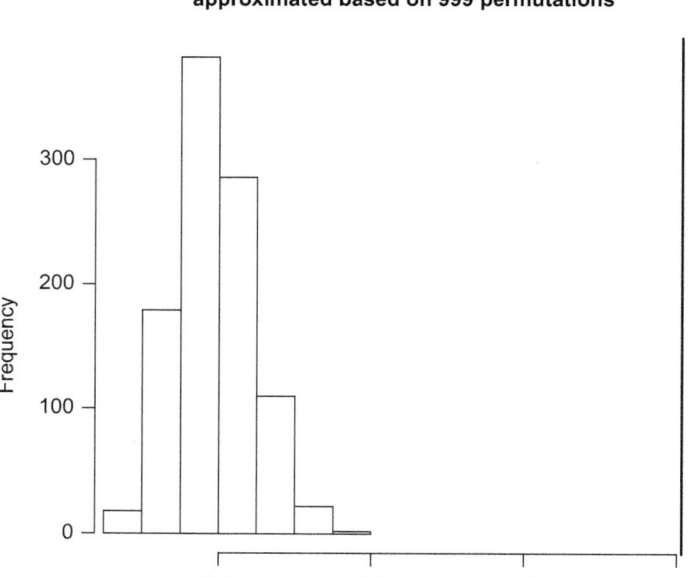

FIGURE 6.9

Assessing spatial autocorrelation in the Cambridgeshire burglary rates using the global Moran's *I*. The histogram is the approximated sampling distribution of Moran's *I* under the null hypothesis based on 999 permutations. The vertical line to the right is the Moran's *I* calculated using the original data (see Exercise 6.4).

to be found in the rural tracts (see Section 3.3.3). If unequal variance is a concern, other statistics have been proposed, such as Tango's test (Tango, 1995).[7]

6.2.4.4 *The Global Moran's* I *Applied to Regression Residuals*

One of the assumptions of ordinary least squares regression is that errors must be independent, and it is usually checked by calculating Moran's *I* statistic on the regression residuals (the estimates of the errors). Global spatial autocorrelation can arise in the residuals from a regression model if some of the unobserved/unmeasured independent variables are themselves spatially autocorrelated. In spatial econometrics, testing for residual spatial autocorrelation is often a first step before fitting one or another spatial regression model (see Chapters 10 and 11). At this point, two modifications to the discussion in Section 6.2.4.2 should be noted. First, $d_i = \hat{e}_i$, where $\hat{e}_i$ is the least squares residual for the *i*th observation. There is no need to subtract the residual mean, as residuals have a mean of zero by construction. Second, because the sum of the residuals is zero, the residuals themselves are not independent: if you are given $(N-1)$ residuals, knowing that the sum of the N residuals is zero, the *N*th residual can be calculated. For this reason, the mean and variance of the sampling distribution for the global Moran's *I* statistic under the null hypothesis of no residual

---

[7] Both Geary's c (the `geary.test` and `geary.mc` functions) and the Getis-Ord general G statistic (the `globalG.test` function) are implemented in the `spdep` package. The functions `tango.stat` and `tango.test` in the `DCluster` package perform Tango's test.

spatial autocorrelation is not the same as in the case under Eq. 6.3 (Cliff and Ord, 1973; Ripley, 1981, p.98–100). However, the previously described permutation test remains an option to derive an empirical p-value. Exercise 6.5 illustrates the use of the global Moran's *I* to examine the spatial autocorrelation structure in a set of spatial residuals.

6.2.4.5 The Join-Count Test for Categorical Data

When the data of interest are categorical, the join-count test (also known as the Krishna Iyer join-count test (Krishna Iyer, 1949)) can be used to assess spatial autocorrelation. The dataset on high intensity crime areas (HIA) introduced in Section 5.2.2 provides an example of such data. Consider the values based on the empirically-defined HIA (EHIA). For each census output area (COA) *i*, its EHIA status, x_i, is binary. That is, $x_i = 1$ if COA *i* is considered to be in an EHIA, or $x_i = 0$ if it is not in an EHIA. The join-count test allows us to ask whether the EHIA output areas are distributed randomly on the map or not.

Conventionally, the join-count statistic is formulated using two colours, black and white, so that area *i* is black if $x_i = 1$ and area *i* is white if $x_i = 0$. The statistic then counts up how many pairs of neighbouring areas (each pair is called a join) that are (a) both black (the BB joins); (b) both white (the WW join); and (c) of different colours (the BW or WB joins). Figure 6.10 illustrates how joins with the same colour and of different colours are counted. Eq. 6.5 provides the equations to carry out the calculation in general.

$$BB = \frac{1}{2}\sum_{i=1}^{N}\sum_{j=1}^{N}w_{ij}x_ix_j$$

$$WW = \frac{1}{2}\sum_{i=1}^{N}\sum_{j=1}^{N}w_{ij}\left(1-x_i\right)\left(1-x_j\right)$$

$$BW = \frac{1}{2}\sum_{i=1}^{N}\sum_{j=1}^{N}w_{ij}\left(x_i-x_j\right)^2 \tag{6.5}$$

The resulting numbers of joins can then be compared against their respective distribution under the null hypothesis of no spatial autocorrelation (Cliff and Ord, 1981, p.19–20). So, if there are considerably fewer BW joins compared to what would be expected, then the data exhibit positive spatial autocorrelation. One can further examine whether positive spatial autocorrelation is due to the clustering of the WW joins or the BB joins or both. On the

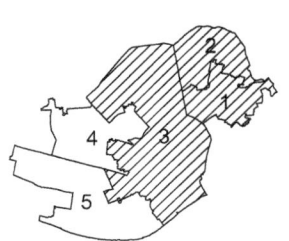

Area ID	x_i	For join count statistic
1	1 (B)	BB joins are (1,2), (2,3) and (1,3)
2	1 (B)	=> BB = 3
3	1 (B)	WW join is (4,5)
4	0 (W)	=> WW = 1
5	0 (W)	BW joins are (3,4) and (3,5)
		=> BW = 2
		(the WB joins (4,3) and (5,3) are
		not counted)

FIGURE 6.10
An illustration of counting the joins with the same colour or with different colours. Areas 1, 2 and 3 are black and areas 4 and 5 are white.

other hand, if the number of BW joins is not different from what would be expected, then there is little or no spatial autocorrelation. An excessive number of BW joins would indicate negative spatial autocorrelation.

In R, the join-count test is implemented as the `joincount.test` function in the `spdep` package. The R script below applies the test to the values on EHIA. Here we use a binary spatial weights matrix (block 3) so that the test statistics are calculated as illustrated in Figure 6.10. Exercise 6.7 performs the same test but using other specifications of the weights matrix.

```
########################################################################
#   block 1: load the Sheffield shape file containing the HIA data
########################################################################
sheffield <- readShapePoly('sheffield.shp')
########################################################################
#   block 2: extract the empirical defined HIA (EHIA)
########################################################################
EHIA <- sheffield@data$EHIA
EHIA <- as.factor(EHIA)  #  convert the binary data into a factor as
                         #  required by the joincount.test function
########################################################################
#   block 3: construct the neighbourhood structure based on queen's
#   contiguity
########################################################################
nb <- poly2nb(sheffield,queen=TRUE)
listw <- nb2listw(nb,style='B')   #  based on a binary W matrix (1 if two
                                  #  areas are neighbours but 0 otherwise)
########################################################################
#   block 4: carry out the join-count
#   test
########################################################################
joincount.test(EHIA,listw)
```

Below is the output from R:

```
     Join count test under nonfree sampling
data:  EHIA
weights: listw
Std. deviate for 0 = -1.0559, p-value = 0.8545
alternative hypothesis: greater
sample estimates:
Same Colour Statistic        Expectation         Variance
          686.0000            699.4529         162.3137

Join count test under nonfree sampling
data:  EHIA
weights: listw
Std. deviate for 1 = 5.3259, p-value = 5.023e-08
alternative hypothesis: greater
sample estimates:
Same Colour Statistic        Expectation         Variance
           40.00000           18.04340          16.99589
```

There are two parts to the R output, one testing spatial autocorrelation of the 0s (i.e. whether areas with 0 tend to be clustered together) and the other of the 1s (i.e. whether areas with 1 tend to be clustered together). For the EHIA data, there are 686 pairs of spatially-contiguous areas both with 0 (i.e. 686 WW joins). However, the corresponding p-value of 0.85 is large and well above the usual 5% significance level, suggesting the spatial distribution of the 0s appears to be random. In contrast, there are 40 pairs of spatially-contiguous areas with 1 (i.e. 40 BB joins), and the p-value is much smaller than the 5% significance level. Furthermore, the 40 BB joins are much higher than the expected number of 18.04. There is evidence suggesting the presence of positive spatial autocorrelation in the distribution of 1s, implying a spatial clustering of the EHIA census output areas. Note that the expectation and variance of the sampling distribution of the test statistic under the null distribution are derived based on the so-called non-free sampling assumption. This assumption is equivalent to sampling without replacement, whereby each area has the same probability of being 0 (or 1), but the number of areas with 0 (or 1) is constrained to be equal to that in the observed data (Cliff and Ord, 1981, p.12 and p.39).

6.2.5 Checking for Spatial Heterogeneity in the Spatial Dependence Structure: Detecting Local Spatial Clusters

The methods discussed in Section 6.2.4 aim to assess spatial dependence at the global scale, making an assumption that the same behaviour occurs uniformly across the study region. For example, the positive and significant global Moran's I calculated using the burglary rate data for Cambridgeshire implies that any two spatial neighbours tend to have similar burglary rates and that this is true everywhere on the map. Parts of the study region, however, may show spatial dependence structures that deviate from the above, suggesting spatial heterogeneity in the spatial structure. In this section, we focus on exploratory methods that are used to detect clusters of areas that exhibit localised patterns such as particularly high crime or disease risks (so-called crime/disease hotspots) or with low income levels. A cluster generally contains two or more spatially contiguous areas but, in some situations, a cluster may contain a single area in the form of a spatial outlier.

The Moran scatterplot is a graphical tool to assess whether there exist areas (or spatial outliers) that behave differently from their immediate neighbours (see Section 6.2.4.1 and Figure 6.8, for example). There are a large number of methods for cluster detection based on test statistics, and each can be broadly grouped into one of two classes. One class of cluster detection methods is based on calculating sets of local indices. Examples of this class include the local Moran's I (Anselin, 1995) and the Getis-Ord local G and G* statistics (Getis and Ord, 1992 and Ord and Getis, 1995).[8] The local Moran's I statistic (unlike either the local G or local G* statistics) is a local indicator of spatial association (or LISA), a term introduced by Anselin (1995) to refer to the fact that the *local* version of the statistic can be shown to be a decomposition of the corresponding *global* statistic. The second class of cluster detection methods is scan-based. A scan-based method involves passing a window over the study region. A window is reported as a cluster if the areas contained inside the window behave much differently compared to those outside the window in terms of, for example, the number of cases of a certain disease or of household burglary. Openshaw's geographical analysis machine (or GAM; Openshaw et al., 1987) and Kulldorff's spatial scan statistic and its variants (Kulldorff and Nagarwalla,

---

[8] The G^*_i statistic includes the value at location i; the G_i statistic excludes the value at location i.

1995 and Kulldorff, 1997) are scan-based statistics that are often used for cluster detection. We will introduce some of these methods in the next two sections while Waller and Gotway (2004, Chapter 7) and Haining (2003, p.237–263) both provide reviews of cluster detection methods.

6.2.5.1 The Local Moran's I

Anselin's local Moran's *I* statistic (Anselin, 1995) is a local decomposition of the global test for spatial autocorrelation. For each spatial unit i ($i = 1, \ldots, N$), the local Moran's *I* is defined as follows

$$I_i = \frac{N d_i \sum_{j=1}^{N} w_{ij} d_j}{\sum_{k=1}^{N} d_k^2}, \tag{6.6}$$

where $d_i = x_i - \bar{x}$, the deviation of the value in i from the overall mean, and w_{ij} is our usual notation for the element in the spatial weights matrix representing the connectivity between i and j. Anselin (1995, p.99) showed that the sum of the local Moran's *I* is proportional to the global Moran's *I* (hence is called a LISA):

$$I = \frac{1}{N} \sum_{i=1}^{N} I_i$$

The interpretation of I_i is the same as that of its global counterpart. A positive value of I_i implies local positive spatial autocorrelation where the value of area i is similar to those of its neighbours. A negative I_i, on the other hand, suggests negative spatial autocorrelation locally, meaning that the value of area i tends to be dissimilar to those of its neighbours.

The R code below defines a function called `local.moran.I` to compute I_i for each area i (see for example Section 10, in particular 10.1, in Venables et al. (2018) for detail on defining a function in R). As an example, the last three lines calculate I_i for area 14 using the burglary rates in the Cambridgeshire data.

```
###################################################################
#  Defining an R function (local.moran.I) to calculate Ii for each area
#  This function takes the following three arguments
#    1. x is an array of values over the N areas
#    2. area.id is an integer indicating which area Ii is calculated for
#    3. W is the NxN spatial weights matrix
#  This function returns
#     the Moran's I value for the area chosen
###################################################################
local.moran.I <- function(x,area.id,W) {
  N <- length(x)      #  number of areas
  d <- x - mean(x)    #  deviation from the mean for each area
```

```
      sum.d2 <- sum(d^2)  #  sum of the squared deviations
                          #  (denominator of Eq. 6.6)
      d.mat <- matrix(d,ncol=1)  #  convert the deviations into a matrix
                                 #  with size N×1
      # use matrix multiplication to calculate the spatial lagged values and
      #  select the corresponding lagged value for the chosen area
      x.sp.lag <- (W%*%d.mat)[area.id,1]
      #  calculate the local Moran's I (Eq. 6.6)
      loc <- N*d[area.id]*x.sp.lag/sum.d2
      return(loc)  #  return the result
}
######################################################################
#  an example: calculate the local Moran's I for area 14 on the
#  burglary rates
######################################################################
rate <- O/pop*1000  #  burglary rates per 1000 houses
nb <- poly2nb(cambridgeshire,queen=FALSE) #  neighbouring structure based
                                          #  on rook's contiguity
W <- nb2mat(nb,style='W')  #  the row-standardised W
area.id <- 14
local.moran.I(rate,area.id,W)  #  R returns the value 6.74, indicating
                               #  positive local autocorrelation
```

To test for significance, Anselin (1995) suggested the use of a conditional permutation test to obtain an empirical p-value for each local statistic. For I_i, each permutation generates a dataset in which the value of area *i* is *held fixed* while the data values of all other areas are redistributed randomly. Then I_i in Eq. 6.6 is calculated using the simulated dataset. Over a large set of simulations, we can obtain the empirical p-value (Eq. 6.2), which is used to test whether the value at *i* is autocorrelated with the values of its spatial neighbours or not. This conditional permutation test is carried out for each area separately, and the R codes to do that are given below.

```
######################################################################
#  block 1: calculate Ii for area 14 using the original observed values
######################################################################
area.id <- 14
test.stat <- local.moran.I(rate,area.id,W)
######################################################################
#  block 2: define number of permutations to carry out and create a
#  storing object
######################################################################
nsim <- 999
store <- rep(0,nsim)
```

```
############################################################
#   block 3: carry out the permutation test
#   Note that the code below uses the insert function from the R.utils
#   package which needs to be loaded in first
############################################################
library(R.utils)    #   load in the package so that we can use
                    #   the insert function later on
N <- 157            #   total number of areas in the study region
for (isim in 1:nsim) {
    #   ids contains the indices of all areas apart from the chosen
    #   one (i.e. 14 in this case)
    ids <- (1:N)[-c(area.id)]
    #   using the sample function to randomly permute the N-1
    #   indices (without replacement)
    s <- sample(ids,N-1)
    #   using the insert function to put the index 14 back to the 14th
    #   position (i.e. holding the value of area 14 fixed)
    s <- insert(s,area.id,area.id)
    #   use the permuted indices to get the simulated values
    new.x <- x[s]
    #   calculate the Moran's I based on the permuted values
    store[isim] <- local.moran.I(new.x,area.id,W)
}
############################################################
#   block 4: calculate the empirical p-value (Eq. 6.2)
############################################################
l <- length(which(store > test.stat))
(l + 1)/(nsim + 1)
```

The local Moran's *I* for area 14 is 6.74 and the empirical p-value is 0.001 over 999 permutations. This result presents strong evidence of positive local autocorrelation, suggesting that the value of area 14 tends to be similar to those of its spatial neighbours. Each local Moran's *I* can be linked to the Moran scatterplot in order to inform the nature of the local autocorrelation. For example, an inspection of the data tells us that the burglary rate for area 14 is 228.7 cases per 1000 houses, higher than the Cambridgeshire average (83.4 cases per 1000 houses) and the average burglary rate over the neighbours of area 14 is also high (187.5 cases per 1000 houses). So, the pair of values, the burglary rate of area 14 and the spatial lag burglary rate for area 14, lies in the first "high-high" quadrant on the Moran scatterplot (quadrant I in Figure 6.7). The same procedure can be applied to all other areas in the study region, and the results are presented in Figure 6.11.

Using the significance level of 0.01 (we defer the reason for using 0.01 to Section 6.2.5.2), there are 24 areas with positive and significant local Moran's *I*. But none are negative and significant (i.e. none of these local Moran's *I* are associated with an empirical p-value greater than 0.99). The empirical p-values of the remaining 133 areas lie between 0.01 and 0.99, offering little evidence to suggest the values of these areas are correlated with those of their neighbours. Figure 6.11(a) shows the locations of the identified areas and Figure 6.11(b)

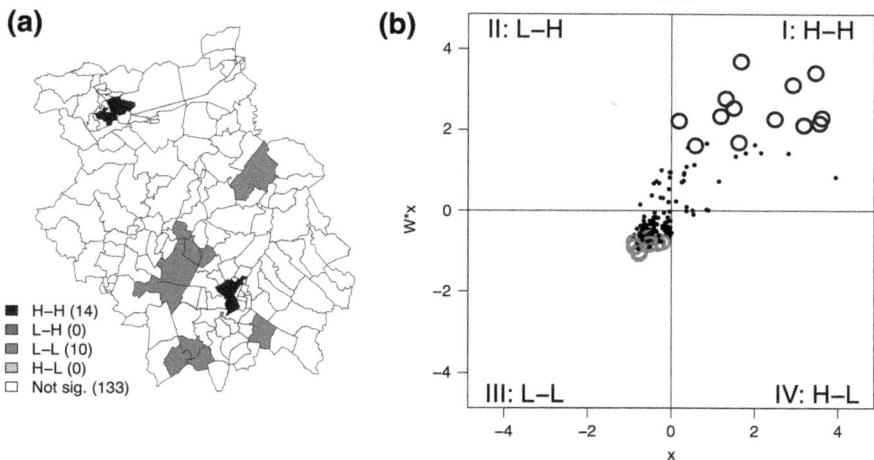

FIGURE 6.11

(a) Map of those areas identified with an empirical p-value either less than 0.01 (24 areas identified) or greater than 0.99 (0 areas identified). The polygons shaded dark grey correspond to the circles with the same shading in the first quadrant of the Moran scatterplot in (b). Similarly, the polygons in light grey in (a) correspond to the same shaded circles in the third quadrant in (b). The remaining 133 areas have empirical p-values lying between 0.01 and 0.99 (not significant) and they correspond to the small dots in (b).

links the areas to the corresponding points in the Moran scatterplot. Two high-risk clusters are identified, namely part of the city of Cambridge (in the south of Cambridgeshire) and part of the city of Peterborough (in the north of Cambridgeshire). There are also small pockets of areas identified to be low-risk clusters, and some of these contain only one area. Negative spatial autocorrelation is not detected locally. Exercise 6.8 carries out the same analysis but using the unemployment proportion as the variable of interest. Areas with significant negative spatial autocorrelation arise in that exercise.

6.2.5.2 The Multiple Testing Problem When Using Local Moran's I

When detecting clusters using the local Moran's I, there is the problem of multiple testing to consider because we are performing N tests, one for each area. There are several ways to address this issue. A Bonferroni adjustment can be used to reduce the risk of a type I error – rejecting the null hypothesis when it is true. However, the number of tests (or areas) is typically large, making the Bonferroni adjustment very conservative.[9] For the Cambridgeshire data, with a Bonferroni adjustment, the threshold for detecting significant positive autocorrelation would be 0.0003(=0.05/157) based on a significance level of 0.05, and the threshold becomes even smaller if the significance level is set at 0.01. Using the threshold of 0.0003, many fewer areas will be identified (see Exercise 6.9). Moreover, it requires 9999 permutations.[10] For spatial data, a Bonferroni adjustment may be particularly conservative because not only are the N tests not independent of each other because each

---

[9] A test is said to be conservative when, if the null hypothesis is true, you are less likely to commit a type I error than the significance value α (say, the usual 5% level) would suggest.

[10] Why would more permutations be needed in order for the threshold of 0.0003 to work? See for the interpretation of the empirical p-value: https://geodacenter.github.io/workbook/6a_local_auto/lab6a.html#interpretation-of-significance.

is based on overlapping subsets of data values, but in addition, data values may well be (positively) spatially autocorrelated (Anselin, 1995). Following Anselin (2018), we use the thresholds of 0.01 and 0.99 to detect local clusters (Figure 6.11). A more stringent thresholding criterion (say, 0.001 and 0.999) would result in fewer areas identified, while relaxing the thresholds to, say, 0.05 and 0.95, would lead to more areas being identified. At the exploratory stage, we prefer a less strict rule because the aim is to see whether there are any clusters present at all. In other words, we may not want to miss any "signals" in the data that are worth investigating via further modelling. But in general, we aim to employ a good decision rule that strikes a satisfactory balance between sensitivity (the power to reject the null hypothesis when it is false) and specificity (the ability to retain the null hypothesis when it is true), a topic that we shall return to in Chapter 16. Anselin (2018) and Brunsdon and Comber (2015, p.265–266) discuss another approach to address the multiple testing issue based on the false discovery rate, but again we defer that discussion to Chapter 16.

6.2.5.3 *Kulldorff's Spatial Scan Statistic*

A spatial scan statistic lays down windows of regular geometric shape (e.g. rectangular, circular or elliptical) over the study region to test for clusters of high (or low) concentrations of events. The scan test developed by Kulldorff and Nagarwalla (1995) and Kulldorff (1997) creates a set of circular windows of varying sizes over the study region.[11] For each of such windows, the rate of event occurrence *inside* the window is compared to that *outside* the window using the likelihood ratio. Intuitively, if the events are distributed randomly over the study region, the rate inside any one of the windows will be the same as that outside the corresponding window (subject to random noise). Then the likelihood ratio will be close to 1 across all the windows. However, if there exists a high concentration of events in one location, then the rate inside any window that contains this location (and hence the events in question) will yield a likelihood ratio greater than 1. The window that maximises the likelihood ratio is labelled as the most likely cluster.

The likelihood ratio serves as a test statistic for testing the null hypothesis that $p = q$ against the alternative hypothesis that $p > q$, where p and q denote the event rates inside and outside the window, respectively. A Monte Carlo procedure is used to obtain an empirical p-value to decide whether to reject or retain the null hypothesis. The Monte Carlo test is only applied once to the most likely cluster, thus avoiding the multiple testing problem.

For area-level data, each area is represented by a single co-ordinate (typically taken to be the area's centroid). An area falls within a window if its co-ordinate lies inside the window. The sets of windows considered vary in size, from the smallest sized windows, each containing only one area, to the largest ones that are usually restricted to contain a maximum of 50% of the total population at risk. Figure 6.12 illustrates the set of scanning windows for a map with five areas. Because of the 50% population restriction, no scanning window includes three or more areas in the illustrative example.

For each window, denote O_{in} and n_{in}, the number of events and the at-risk population *inside* the window respectively, O_{out} and n_{out}, the number of events and the at-risk population *outside* the window, respectively. Let $O = O_{in} + O_{out}$ be the total number of events and $M = n_{in} + n_{out}$ be the total at-risk population. The likelihood ratio – the ratio of the

---

[11] The spatial scan statistic (Kulldorff, 1997) generalizes scan statistics that were developed originally to scan for rare events over time by moving a time window and identifying the window with the maximum number of events.

likelihood of the data under the alternative hypothesis (L) to the likelihood of the data under the null hypothesis (L_0) – is given by

$$\lambda = \begin{cases} \dfrac{L}{L_0} & \text{if } \dfrac{O_{in}}{n_{in}} > \dfrac{O_{out}}{n_{out}} \\ 1 & \text{if } \dfrac{O_{in}}{n_{in}} \le \dfrac{O_{out}}{n_{out}} \end{cases},$$

where

$$L = \left(\frac{O_{in}}{n_{in}}\right)^{O_{in}} \left(\frac{n_{in}-O_{in}}{n_{in}}\right)^{n_{in}-O_{in}} \left(\frac{O_{out}}{n_{out}}\right)^{O_{out}} \left(\frac{n_{out}-O_{out}}{n_{out}}\right)^{n_{out}-O_{out}}$$

and

$$L_0 = \frac{O^O (M-O)^{M-O}}{M^M} \tag{6.7}$$

Note that the likelihood functions, L and L_0, assume that the events occur independently, so both are in the form of a binomial distribution. This independence assumption is appropriate when dealing with, for example, events comprising cases of a non-infectious disease. In addition, the derivation of L also assumes that the cases occurring inside the window are independent of those occurring outside the window. Prior to the use of the scan statistic, this assumption requires controlling for the effects of known risk factors that vary smoothly over space (see also the discussion at the end of this section).

As an illustration, Table 6.1 shows the calculation of the log likelihood ratio, $\log(\lambda)$, for each of the 10 windows constructed in Figure 6.12. The most likely cluster is the window containing area 1, whose log likelihood ratio is the maximum across all other windows. For the most likely cluster, the original data are randomly permuted many times (typically, over 999 or 9999 random permutations) to obtain an empirical p-value. The Monte Carlo

TABLE 6.1

Calculation of the Log Likelihood Ratio, $\log(\lambda)$, for Each of the 10 Scanning Windows Constructed in Figure 6.12

Window ID	Area(s) Inside	O_{in}	O_{out}	n_{in}	n_{out}	$\log(\lambda)$	Empirical p-value (999 sims)
1	1	20	8	100	400	18.65	0.001
2	1, 2	21	7	200	300	7.50	
3	2	1	27	100	400	3.46	
4	2, 1	21	7	200	300	7.50	
5	3	2	26	100	400	1.90	
6	3, 4	6	22	200	300	2.31	
7	4	4	24	100	400	0.33	
8	4, 3	6	22	200	300	2.31	
9	5	1	27	100	400	3.46	
10	5, 4	5	23	200	300	3.36	

Note: See main text for the definitions of O_{in}, O_{out}, n_{in} and O_{out}.

Area ID	Population	Events
1	100	20
2	100	1
3	100	2
4	100	4
5	100	1

FIGURE 6.12
An example illustrating the construction of the circular scanning windows, which are restricted to contain no more than 50% of the total population in the entire study region, resulting in 10 scanning windows. Five windows contain one area and the other five contain two areas. No window contains three or more areas because of the 50% population restriction. The dots represent the area centroids. Table 6.1 summarises the area(s) inside each window.

procedure is the same as that described in Section 6.2.3.2 and, for each dataset generated, λ in Eq. 6.7 is calculated for the detected window.

In R, the circular-version of the spatial scan statistic is implemented via the kulldorff function in the SpatialEpi package. The following R code applies the scan statistic to detect high-risk clusters in the Cambridgeshire burglary data.

```
#####################################################################
#  block 1: obtain the centroids of the spatial units
#####################################################################
coords <- coordinates(cambridgeshire)
#####################################################################
#  block 2: load in the R package
#####################################################################
library(SpatialEpi)
#####################################################################
#  block 3: carry out the circular spatial scan statistic using the
#  kulldorff function
#####################################################################
K <- kulldorff(geo=coords,     #  the centroids of the areas
               cases=O,        #  the numbers of burglary cases across
                               #  the spatial units
             population=pop,     #  the numbers of at-risk houses across
                                 #  the spatial units
             pop.upper.bound=0.5, #  the largest window contains max. 50%
                                  #  of all the houses in Cambridgeshire
             n.simulations=999,   #  number of permutations to carry out
                                  #  for testing the most likely cluster
             alpha.level=0.05     #  the significance level for the
                                  #  Monte Carlo test
)
```

```
#####################################################################
#     block 4: show contents in the result object K
#####################################################################
str(K)
```

Table 6.2 and Figure 6.13 summarise the results. The most likely cluster detected (labelled as Cluster 1 in Figure 6.13) covers a large part of Peterborough (in the north of Cambridgeshire). The risk associated with the cluster is 160.6 (=(9099/56650)×1000; Table 6.2) burglary cases per 1000 houses, almost twice the Cambridgeshire average (83.4 cases per 1000 houses). The empirical p-value from 999 permutations is 0.001, implying a considerably elevated risk of burglary in this group of areas compared to other parts of Cambridgeshire. There are three "secondary clusters" detected, one in the north of

TABLE 6.2

Summary of the Circular Spatial Scan Statistic Applied to the Cambridgeshire Burglary Data

	Most Likely Cluster	Secondary Clusters		
	Cluster 1	Cluster 2	Cluster 3	Cluster 4
Number of areas within	17	8	1	1
O_{in}	9099	5124	473	308
n_{in}	56650	25240	4190	2670
$\log(\lambda)$	2319	1967	21.9	16.2

Note: The locations of the detected clusters are in Fgiure 6.13.

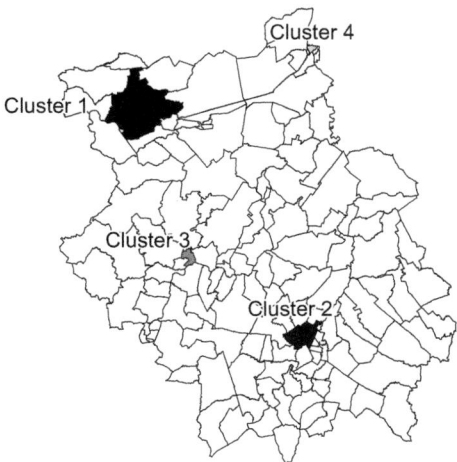

FIGURE 6.13

A map of the most likely cluster and the three secondary clusters detected using the circular spatial scan statistic.

Cambridge, while the other two each consist of a single area.[12] The burglary risks associated with these three clusters are also found to be higher than the Cambridgeshire average. It is worth noting that both Clusters 1 and 2 detected here to some extent match with the two "high-high" clusters detected using the local Moran's I (Figure 6.11). The circular window used makes an implicit assumption regarding the shape of the cluster (see below).

We end this section with a few remarks on Kulldorff spatial scan statistics in particular and on cluster detection in general. It should be noted that whether or not we detect a real cluster depends on the type of window used for the test, which may not match very well the shape of the "real" cluster "in the real world". To partially address this problem, Kulldorff et al. (2006) proposed a version of the spatial scan statistic using elliptical windows. But clusters can assume many shapes. Other tests for different shapes, based on the scan statistic, have been developed (Tango and Takahashi, 2005, 2012).

It is possible to miss a real cluster if, for example, the cases happen to be split across several subareas as a result of where the spatial boundaries of the subareas happen to fall (the partition problem; see Section 3.3.3). A "real" cluster can also be missed if there is dilution, for example the cases are geographically clustered but few in number in relation to the size of the population at risk in the subareas used for the analysis (the scale problem; see Section 3.3.3). Both these points illustrate how the way data are made available can impact on the results of a statistical analysis. Detection is sensitive to the geographical extent of the study area. For example, a random scatter of cases observed within a region may appear as a cluster if that region becomes a subarea of a larger study area.

Some caution is recommended when using cluster detection results to pin down the location of a cluster. Whilst the windows used are regular in shape (circular, elliptical), the region constructed based on the areas that lie inside the window that has yielded the most likely cluster may be irregular in shape. For example, the windows used for cluster detection in Figure 6.12 are all circular in shape, but window ID 10 (Table 6.1) comprises areas 4 and 5. Were window ID 10 to have been declared as the most likely cluster, the shape of the cluster on the map would be highly irregular. The same comment would apply if window ID 8 (containing areas 3 and 4) had been identified as the most likely cluster. As a final point, the spatial limits to an identified cluster should not be considered exact since the addition or subtraction of areas on the margins of a window may not greatly change p-values.

Factors such as the extent to which the cluster risk departs from the expected level as well as the population size within the cluster can both affect the power of detecting a real cluster. It is generally more difficult to detect clusters when the population size is small and/or the cluster risk is only elevated by a small amount. Finally, when available, one should include covariates in an attempt to explain why clusters are detected. In SaTScan, known risk factors can be adjusted for so that the cluster detected is unusual compared to "what would be expected".

Both the circular- and elliptical-versions of the spatial scan statistic are implemented in SaTScan™ (2018a), an open-source software developed by Kulldorff and colleagues. We refer the reader to the SaTScan™ User Guide (2018b) for details. In Chapter 13, we will use SaTScan™ in a space-time setting. In Section 16.3 in Chapter 16 we will provide an example of using observable covariates to explain the detected clusters.

---

[12] Secondary clusters are groups of areas that also have empirical p-values that imply an elevated risk. For further discussion of secondary clusters see Zhang et al. (2010).

6.3 Exploring Relationships between Variables

In a regression setting, we are interested in examining the association between a dependent (or outcome) variable and a set of independent variables (or covariates). Another situation is to evaluate the dependence structure amongst two (or more) dependent variables. For example, as will be discussed in Section 9.2, an inferential goal of analysing data on high intensity crime areas (HIAs) is to study the relationship between the HIAs defined using the recorded crime database (EHIA) and the HIAs identified by senior police officer assessments (PHIA), two sets of outcomes that are possibly related. Yet another situation is where we want to evaluate the dependence structure between two or more covariates in a regression model in order to anticipate multicollinearity problems.

In EDA, standard tools to examine relationships between variables include different forms of scatterplots (including added variable plots), correlation coefficients and curve fitting. Revisiting our earlier terminology, the smooth part of the relationship is described by the best fit regression line and quantified by the size of the correlation coefficient. The rough part of the relationship is captured by the scatter of points around the best fit line, the residuals and the variation in the dependent variable that cannot be accounted for by the available covariates.

When applying EDA techniques to spatial data, we need to acknowledge the properties discussed in Chapter 1. So, for example, since spatial units do not have "hard" boundaries, we need to recognize that relationships between variables can exist "at a distance", that is, a covariate can influence outcomes not only in the same spatial unit but also in others. Furthermore, the strength of relationships, whether in the same spatial unit or "at a distance", might vary spatially – spatial heterogeneity in covariate relationships calling for the use of local as well as global statistics. In addition, we need to acknowledge that the existence of spatial dependence in all or any of the variables might complicate, or indeed undermine, the insights obtained from an exploratory analysis. Figure 6.14 summarizes some of the types of association we are interested in as part of an exploratory analysis of bivariate spatial data and the interrelationships between them.

6.3.1 Scatterplots and the Bivariate Moran Scatterplot

A scatterplot visualises the relationship between two continuous-valued variables by plotting the values of a dependent variable (typically on the y-axis) against those of a covariate (on the x-axis). However, different points on a scatterplot may contribute to the overall

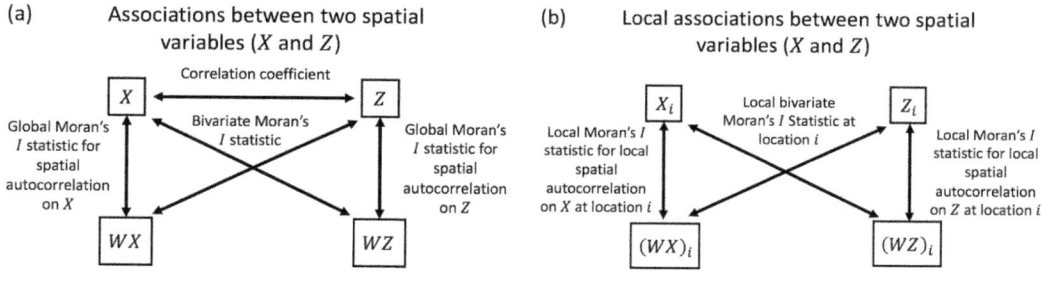

FIGURE 6.14
(a) Global measures of association; (b) local measures of association for a typical observation in area i.

relationship differently due to varying levels of uncertainty, as discussed in Chapter 3 (see in particular Section 3.3.3).

Consider the village-level malaria incidence data in Kalaburagi taluk, South India, which was first discussed in Chapter 2, Section 2.2. Figure 6.15(a) shows a scatterplot examining the relationship between the log incidence rates and the rainfall levels across the 139 villages. The scatterplot may initially suggest that rainfall level has a "piecewise linear" effect on the log rate: a clear positive linear relationship in a part of the plot where the log rates are larger than −6, while the rainfall level seems to have little effect on the risk of malaria for those areas with log rates lower than −6. However, this plot may be misleading because it ignores the uncertainty associated with the estimated rates. A more appropriate version is shown in Figure 6.15(b), where the size of each point is inversely related to the variance of the estimated incidence rate (Section 3.3.3). Visually, the points associated with high uncertainty assert less influence when interpreting the relationship.

While a standard scatterplot shows the association between an outcome and a covariate *at the same location*, a bivariate Moran scatterplot (Anselin et al., 2002) can be used to visualise how the outcome of an area is associated with the covariate values of *its spatial neighbours*. The latter represents a form of spatial spillover. A bivariate Moran scatterplot visualizes y, the outcome values, against W^*x, the spatially-lagged values of the covariate x, with W^* denoting the row-standardised spatial weights matrix where $w_{ii}^* = 0$ for all i. Both sets of values are standardized to have a mean of 0 and variance of 1, as in the case of the Moran scatterplot. However, the reason for using a bivariate Moran scatterplot is to assess the correlation between y and W^*x.

Using the Cambridgeshire burglary data as an example, Figure 6.16 shows the bivariate Moran scatterplot of the burglary rates (y) against the spatially-lagged values of the proportion unemployed (W^*x; i.e. the average unemployment proportions in the neighbours of area i but excluding that in i itself). A majority of the points in the bivariate Moran scatterplot lie in the first and the third quadrants, indicating that an area tends to have a high (low) burglary rate if the average unemployment proportions of its neighbours is

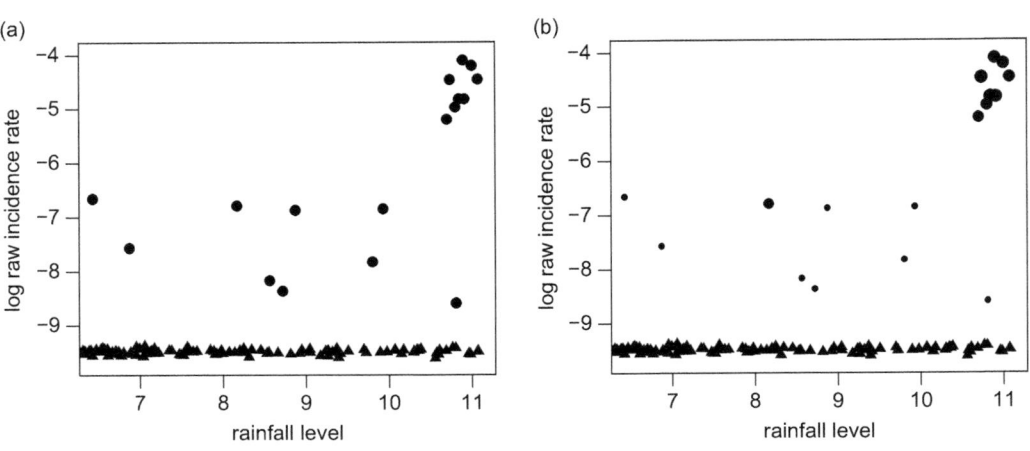

FIGURE 6.15
A scatterplot of the log incidence rates of malaria against rainfall levels across 139 villages: (a) a standard scatterplot and (b) the size of each point is inversely related to the variance (hence uncertainty) of the incidence rate. The triangles at the bottom of each plot correspond to the villages with zero reported cases of malaria (the triangles are jiggled slightly along the y-axis so that they are not lumped together). See Exercise 6.10 for the R code to produce the figures.

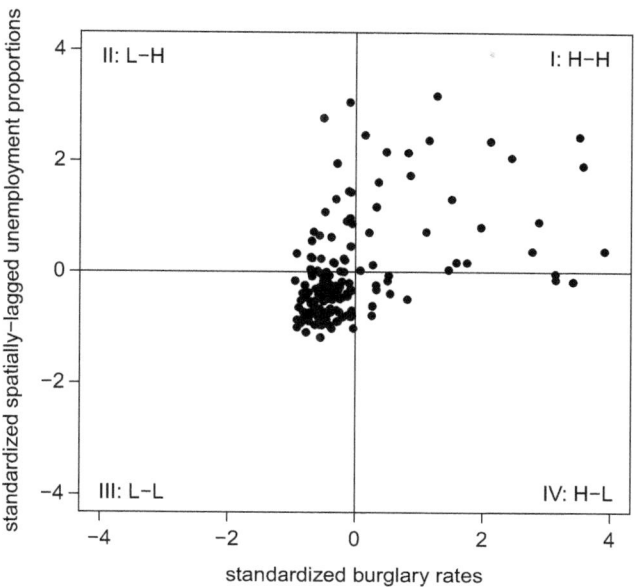

FIGURE 6.16
A bivariate Moran scatterplot between the standardized burglary rates (on the x-axis) and the standardized spatially-lagged values of the unemployment proportions (on the y-axis). The interpretations of the quadrants are different from those of a Moran scatterplot. For example, points in the 1st (3rd) quadrant correspond to areas with high (low) burglary rates, with their neighbours having on average high (low) levels of unemployment. Points in the 2nd (4th) quadrants are areas where the burglary rates are low (high), while their neighbouring areas tend to have high (low) levels of unemployment.

high (low). In other words, W^*x and y shows an overall positive correlation. However, the bivariate Moran scatterplot may over-interpret the spatial aspect of the correlation in the sense that the correlation observed may be largely due to the correlation between x and y. In other words, the correlation between W^*x and y may become weaker after we have controlled for the correlation between x and y. In regression modelling, this is termed omitted-variable bias, which arises when a regression model incorrectly excludes a covariate that is correlated with both the outcome variable and one (or more) of the covariates included in the model. The conclusion on the correlation between y and W^*x may be biased if variation in x is not controlled for. In Exercise 6.11, we use added variable plots (Haining, 1990a) to demonstrate this.

6.3.2 Quantifying Bivariate Association

6.3.2.1 The Clifford-Richardson Test of Bivariate Correlation in the Presence of Spatial Autocorrelation

Pearson's correlation coefficient, r, is a standard tool to quantify and test the correlation between two (continuous-valued) variables. When the observations on each variable are independent, the significance of the sample correlation coefficient can be tested by comparing the test statistic calculated using the observed data, $r\sqrt{(N-2)/(1-r^2)}$, against the t distribution, with $N-2$ degrees of freedom. The latter is the distribution of the test statistic under the null hypothesis that the two variables are uncorrelated. N

denotes the number of areas in the study region. However, when both sets of data values are shown to be spatially autocorrelated, although the calculation of the Pearson's correlation coefficient remains the same, the distribution of the test statistic under the null hypothesis is affected. In particular, the presence of positive spatial autocorrelation in both sets of values increases the variance of the distribution because of information loss – the positively autocorrelated data values contain less (or duplicated/overlapping) information (see Section 1.2.1; also Haining, 1988). It should also be noted that if only one of the two sets of values are spatially autocorrelated, the variance of the test statistic under the null hypothesis is unchanged (Haining, 1991), and in that case, the unmodified *t*-test can be used.

To address this problem, Clifford and Richardson (1985) modify the above *t*-test by replacing N, the *actual* sample size, by N', the *effective* sample size. For positive spatial autocorrelation, $N' < N$, reflecting the fact that data values that are positively spatially autocorrelated contain less information compared to the situation where these values are <u>independent</u> (not spatially autocorrelated). Thus, the modified test statistic becomes $r\sqrt{(N'-2)/(1-r^2)}$, and its value, calculated using the observed data, is compared to the *t* distribution with $N'-2$ degrees of freedom.

The R function `modified.ttest` in the `SpatialPack` package performs the Clifford-Richardson modified *t*-test. The R command below examines the correlation between areal burglary rates and the proportion unemployed in the presence of (positive) spatial autocorrelation (the reader encouraged to check the spatial autocorrelation structure in the unemployment data).

```
unemploy <- cambridgeshire@data$unemploy  #  get the unemploy column of
                                          #  values from the shapefile
coords <- coordinates(cambridgeshire)
modified.ttest(burgRate, unemploy, coords)
```

The modified *t*-test incorporates the spatial arrangement of the areas using the coordinates of the centroids. The function `coordinates` is used to obtain these coordinates of the centroids, which then enter the `modified.ttest` function through the third argument, `coords`. The output from R is as follows:[13]

```
Corrected Pearson's correlation for spatial autocorrelation
data: burgRate and unemploy; coordinates:
F-statistic: 15.3032 on 1 and 34.6178 DF, p-value: 4e-04
alternative hypothesis: true autocorrelation is not equal to 0
sample correlation: 0.5537
```

The sample correlation of 0.55 implies that areas with high (low) burglary rates are also areas with high (low) proportions of unemployed. The correlation is shown to be significant as the associated p-value is 0.0004, much lower than the 1% significance level. Note that instead of using the *t* distribution with $(N'-2)$ degrees of freedom, the function `modified.ttest` calculates the p-value using the F distribution with 1 and $(N'-2)$

---

[13] The R output presented is directly from the `modified.ttest` function. While it shows all the key numerical results, the following two modifications to the text would make it clearer: (1) Line 1 should read: "`Pearson's correlation corrected for spatial autocorrelation`" and (2) Line 4 should read: "`Alternative hypothesis: true Pearson's correlation is not equal to 0`".

as the degrees of freedom.[14] The output shows $N' - 2 = 34.62$, so the *effective* sample size is about 37. This implies that the information content for testing the bivariate correlation in the 157 pairs where the data values on each variable are spatially-autocorrelated is equivalent to that contained in 37 data pairs where the data values on each variable are not spatially autocorrelated. The unmodified test using $N = 157$ gives a p-value of 5×10^{-14}, suggesting even stronger evidence for rejecting the null hypothesis compared to the modified test. However, the unmodified test is clearly not appropriate for this example, as it overstates the amount of information we have in the data. Although in this example our conclusion remains the same whether we use N or N', it is clear that there will be occasions when we might incorrectly reject the null hypothesis at a given level of significance when we would have retained it had we (correctly) used N', making a Type I error. In Exercise 6.12, we will use the Clifford-Richardson modified *t*-test to examine the correlation between the two sets of binary indicators, EHIA and PHIA, in the HIA dataset.

6.3.2.2 Testing for Association "At a Distance" and the Global Bivariate Moran's I

Following the idea behind the bivariate Moran scatterplot, we might also want to test for the association between two variables "at a distance".[15] The global bivariate Moran's I examines the correlation between the value of a variable in area i with that of another variable in the neighbouring areas of i. Anselin et al. (2002) extended the global (univariate) Moran's I to a bivariate setting:

$$I_B = \frac{\sum_{i=1}^{N} \left(\sum_{j=1}^{N} w_{ij} x_j y_i \right)}{\sum_{i=1}^{N} y_i^2},$$

where x_i and y_i are the values of two variables observed in area i and w_{ij} is the element in the spatial weights matrix. This statistic suffers the same problem which we noted in the case of the bivariate Moran scatterplot, that is, it does not control for the correlation between x and y and thus may over-estimate the correlation between Wx and y. The bivariate spatial association measure L proposed by Lee (2001) is one way to address this problem. Another option is to use the spatial econometric approach, which we will return to in Chapters 10 and 11.

6.3.3 Checking for Spatial Heterogeneity in the Outcome-Covariate Relationship: Geographically Weighted Regression (GWR)

In addition to exploring a global pattern of correlation, in the case of spatial data the correlation between two variables can vary *locally*. An often-quoted example is the relationship between the price of a property and the number of bedrooms that the property has. In general, a property with more bedrooms tends to have a higher price. However, the added value

---

[14] A random variable X follows a *t* distribution with k degrees of freedom. If $Y = X^2$, then the random variable Y follows an F distribution with 1 and k degrees of freedom (Cacoullos, 1965). Use the R output to verify that the test statistic based on the F distribution is $r^2((N' - 2)/(1 - r^2))$.

[15] An early example of such a statistic was developed by Tjøstheim (1978) to test the association between air pollution and health outcomes, noting that the two sets of data may be collected on different spatial frameworks. This statistic measures the association between two variables based on the ranks of the observations and on their location coordinates. The statistic measures the spatial closeness of observations having the same rank on the two variables.

due to an additional bedroom depends on various factors, including characteristics of the property as well as characteristics of the neighbourhood in which the property resides. Since it is unlikely that we can identify all the factors and/or obtain data reliably representing these factors, the relationship is likely to vary spatially, showing *local deviations* from the *global, whole map effect*. In this section, we describe geographically weighted regression (GWR), an exploratory tool that allows us to check for spatial heterogeneity in regression parameters.[16]

A standard linear regression model in the form of Eq. 6.8 formulates the outcome values as a linear combination of K observable covariates and errors.

$$y_i = a_0 + \sum_{k=1}^{K} a_k x_{ik} + e_i \qquad (6.8)$$

Each of the regression coefficients, a_k, for $k = 1, \ldots, K$, quantifies the effect of the kth covariate on the outcome. Such an effect is assumed to be *global* in the sense that if, for example, an increase of the kth covariate is found to be associated with an increase in the outcome value (i.e. the point estimate of a_k is positive), this outcome-covariate relationship applies universally to all areas in the study region. However, in some situations, this assumption may not be realistic, leading to a *spatially-varying* outcome-covariate relationship (see Section 1.3.2.2 in Chapter 1).

Geographically weighted regression (GWR; Brunsdon et al., 1996), given in Eq. 6.9, extends the standard linear regression model in Eq. 6.8 by allowing each regression coefficient as well as the intercept to be area-specific. That is, a_k is replaced by a_{ki} for $k = 1, \ldots, K$ and likewise a_0 by a_{0i}. The estimates of these area-specific parameters allow us to visualise the spatially-varying relationship and to assess the extent to which the local relationship deviates from the global relationship.

$$y_i = a_{0i} + \sum_{k=1}^{K} a_{ki} x_{ik} + e_i \qquad (6.9)$$

GWR, however, is overparameterised. There are at least $(K+1)N$ parameters, e.g. $(a_{01}, \ldots, a_{0N})$, K sets of $(a_{k1}, \ldots, a_{kN})$ (for $k = 1, \ldots, K$), plus the error variance, to estimate using only N data points, one from each area. To estimate the area-specific coefficients, Brunsdon et al. (1996) proposed a weighted least squares method. The idea is to estimate the parameters of area i using the data from areas that are geographically close to i. The selection of the "geographically close" areas is done by specifying *weights*. As an example, Eq. 6.10 defines the exponential decay weight with d_{ij} measuring the Euclidian distance between the centroids of two areas i and j.

$$w_{ij} = \exp\left(-\frac{d_{ij}}{b}\right) \text{ with } b > 0 \qquad (6.10)$$

As the distance between any pair of areas increases, w_{ij} becomes smaller so that the data in area j contribute less towards the estimation of the parameters at i. The parameter b is called the bandwidth, which controls the rate of the exponential decay. As b tends to

---

[16] If instead, the focus of the analysis is just to examine the association between a variable in area i and a different variable from the neighbouring areas of i, the localised version of the bivariate Moran's I can be used (Anselin et al., 2006). However, we do not explore this further but refer the reader to the reference, and in particular to the GeoDa software that implements a number of univariate and multi-variate exploratory techniques for spatial data (Anselin 2003, 2005).

infinity, all the weights tend to 1, meaning that the data from all areas contribute equally towards the estimation of the parameters in each area. In that case, GWR reduces to the global regression model in Eq. 6.8. On the other hand, when b gets closer to 0, the exponential decay happens quickly as d_{ij} increases. The weights assigned to areas other than i effectively become 0, making no contribution towards estimating the parameters at i. As a result, the estimation of the parameters in area i uses data only from i. Although the estimated parameters will fit the observed outcome value well, they are likely to be highly uncertain because of the data sparsity problem (see Section 1.2.3). In practice, the bandwidth is estimated via a cross-validation procedure (see below).

Based on the weighted least squares approach, the (point) estimates of the intercept and the K regression coefficients, $(\hat{a}_{0i}, \hat{a}_{1i}, \ldots, \hat{a}_{Ki})$, for area i are given by the following expression,

$$\hat{a}_i = (x'D_i x)^{-1} x'D_i y, \tag{6.11}$$

where x is an $N \times (K+1)$ matrix containing the values of the K covariates, with the first column of x taking the value 1; x' denotes the transpose of x; and y is an $N \times 1$ matrix of the outcome values. D_i is an $N \times N$ diagonal matrix with the diagonal elements given by $(w_{i1}, \ldots, w_{ii}, \ldots, w_{iN})$, the weights between area i and all the areas including i itself. Note that the area-specific estimates from Eq. 6.11 reduce to the global estimates from the standard linear regression model in Eq. 6.8 when all the diagonal elements of D_i are set to 1 for all i. In that case, $\hat{a}_i$ becomes $(x'x)^{-1} x'y$, which are the parameter estimates using ordinary least squares.

Table 6.3 provides an illustration of the weights, the estimated intercepts and the regression coefficients based on a map of five areas. For each area, the weights become

TABLE 6.3

An Illustrative Example of the Calculation of the Exponential Decay Weights and the Local Intercepts and Regression Coefficients

Area ID	y_i	x_i	Weights with $b=953$	$\hat{a}_{0i}$	$\hat{a}_{1i}$
1	0.11	1.51	$(1.00, 0.40, 0.20, 0.06, 0.03)$	0.37	−0.19
2	0.02	0.39	$(0.40, 1.00, 0.14, 0.05, 0.02)$	0.23	−0.17
3	0.88	−0.62	$(0.20, 0.14, 1.00, 0.28, 0.14)$	0.61	−0.13
4	0.60	−2.21	$(0.06, 0.05, 0.28, 1.00, 0.25)$	0.62	−0.01
5	0.71	1.12	$(0.03, 0.02, 0.14, 0.25, 1.00)$	0.68	0.01

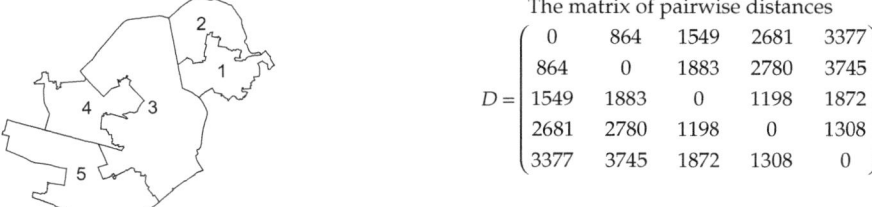

The matrix of pairwise distances

$$D = \begin{pmatrix} 0 & 864 & 1549 & 2681 & 3377 \\ 864 & 0 & 1883 & 2780 & 3745 \\ 1549 & 1883 & 0 & 1198 & 1872 \\ 2681 & 2780 & 1198 & 0 & 1308 \\ 3377 & 3745 & 1872 & 1308 & 0 \end{pmatrix}$$

Note: For this simple example, the coefficient estimates can be computed using Eq. 6.11 either by hand or in R (using the %*% syntax for matrix multiplication).

smaller as the pairwise distances increase. The estimated parameters, $\hat{a}_{0i}$ and $\hat{a}_{1i}$, are specific to each area because the set of weights is different from one area to another. The bandwidth $b = 953$ is estimated via a leave-one-out cross validation (LOOCV) method. The idea is to choose b in order to minimise the score of the form (Brundon et al., 1996; the subscript cv denotes that the score function is associated with the cross-validation procedure):

$$S_{cv}(b) = \sum_{i=1}^{N}\left(y_i - y_{\neq i}^*\right)^2, \tag{6.12}$$

where y_i is the observed outcome value of area i and $y_{\neq i}^*$ is the outcome value of area i predicted by the regression model fitted to all the data but excluding those from i. This LOOCV is repeated over all the areas in turn. Using the illustrative dataset in Table 6.3, Figure 6.17 plots the values of the score, $S_{cv}(b)$, against a range of values for b. The minimum appears when $b = 953$. The reader is encouraged to carry out the selection of b and the calculation of the estimated values for $\hat{a}_{0i}$ and $\hat{a}_{1i}$ in R (Exercise 6.13).

In R, the package GWmodel contains various functions to perform GWR. In particular, the bw.gwr function determines the bandwidth b using the cross-validation approach. The estimated bandwidth is then used in the gwr.basic function that estimates the area-specific regression coefficients. Using the Cambridgeshire burglary data, the R script below regresses the log burglary rates on the unemployment proportions using a GWR model. The burglary rates are log transformed so that the resulting distribution

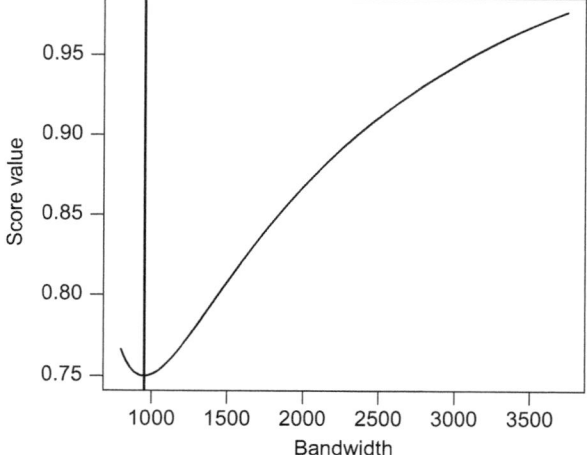

FIGURE 6.17
Selecting an optimal bandwidth using the leave one out cross validation approach. The vertical line indicates the selected bandwidth of $b = 953$, which yields the minimum score value.

is approximately normal. The reader is encouraged to produce a histogram of `logBur-gRate` to check.

```
#######################################################################
#  block 1: load in the R package
#######################################################################
library(GWmodel)
#######################################################################
#  block 2: log transform the burglary rates and assign the
#  transformed values to a new variable in the shapefile
#######################################################################
logBurgRate <- log(burgRate/1000)
cambridgeshire@data$logBurgRate <- logBurgRate
#######################################################################
#  block 3: extract the centroids/coordinates of the polygons and
#  calculate the pairwise (Euclidean) distances using the dist
#  function
#######################################################################
coords <- coordinates(cambridgeshire)
#  by default, the dist function (see the R command below) returns the
#  lower triangle of the symmetric distance matrix, the as.matrix function
#  fills the upper triangular part so that D is a full matrix
#  (e.g. the one shown in Table 6.3)
D <- as.matrix(dist(coords))
#######################################################################
#  block 4: select the bandwidth using cross-validation
#  The first part, logBurgRate ~ unemploy, defines the regression
#  relationship of the form y ~ x1 + x2 + ...
#  The object containing the entire shapefile enters via the data
#  argument
#  The last argument dMat takes on the full matrix of the pairwise
#  distances
#######################################################################
bw <- bw.gwr(logBurgRate ~ unemploy,data=cambridgeshire
            ,kernel='exponential',dMat=D)
bw  #  print out the estimated bandwidth ( b = 1902.346 )
#######################################################################
#   block 5: fit a GWR model
#######################################################################
gwr.fit <- gwr.basic(logBurgRate ~ unemploy,data=cambridgeshire
            ,kernel='exponential',bw=bw)
gwr.fit  #  display the model fits
```

There are two parts to the output of `gwr.fit`: (a) the fit of a global regression model and (b) the fit of a local GWR model. The global regression fit shown below suggests a positive significant effect of unemployment proportion on burglary rate as expected: an area with a high unemployment proportion tends to be associated with a high burglary rate.

```
*******************************************************************
*                     Results of Global Regression               *
*******************************************************************
Call:
 lm(formula = formula, data = data)
 Residuals:
    Min       1Q    Median       3Q       Max
 -1.31935 -0.36861 -0.03674  0.30700  1.68959
    Coefficients:
                Estimate Std. Error t value Pr(>|t|)
    (Intercept) -3.60976    0.08638 -41.792  < 2e-16 ***
    unemploy     0.09973    0.01171   8.518 1.33e-14 ***
```

The results from the GWR fit suggest evidence of spatial variability in the regression relation. First, the estimated bandwidth of 1902.346 lies between the minimum and the maximum pairwise distances, which are 754.54 and 79175.26, respectively. This suggests that the area-specific regression coefficients are not all equal to the estimates from the global regression model, as b is not estimated at the maximum distance (or equivalently, close to infinity). The parameter estimates for area i use more information from the data for areas that are geographically closer to i. This is visualised in Figure 6.18, which shows the exponential decay in weight against distance. The mean distances between the first, the second and the third order neighbours based on spatial contiguity are superimposed to help interpretation. The smooth decay in weight means that estimation uses information

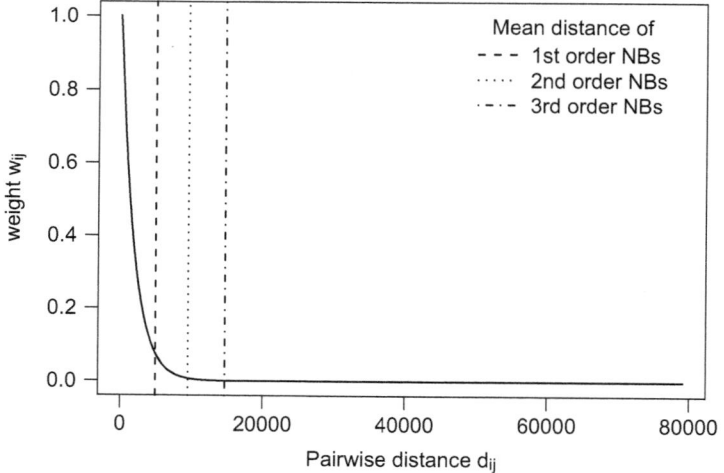

FIGURE 6.18
An exponential decay in weight against pairwise distance. The three vertical lines correspond to the average distances between the 1st, 2nd and 3rd order spatially contiguous neighbours (NBs).

from all areas, but the estimation of parameters in i uses less information from another area as their pairwise distance (or equivalently their neighbourhood ordering) increases. Based on the estimated bandwidth, there is virtually no information taken from neighbours at the third order and beyond.

```
*****************************************************************
*          Results of Geographically Weighted Regression       *
*****************************************************************
********************Model calibration information********************
Kernel function: exponential
Fixed bandwidth: 1902.346
Regression points: the same locations as observations are used.
Distance metric: Euclidean distance metric is used.
***************Summary of GWR coefficient estimates:***************
              Min.    1st Qu.    Median    3rd Qu.     Max.
Intercept  -6.135773 -3.972995 -3.605074 -3.177468 -1.7441
unemploy   -0.279303  0.070119  0.102461  0.148534  0.3686
```

The summary of the estimated regression coefficients for the unemployment effect suggests that unemployment is not associated with the burglary rate uniformly across Cambridgeshire. Figure 6.19 shows the spatial distribution of the point estimates of the local coefficients. For some areas the local estimates are positive but are either smaller or larger than 0.09973, the estimate from the global model. For some areas, however, the point estimates become negative, implying a "protective" effect of unemployment on burglary risk where a high unemployment proportion is associated with a low burglary rate. Of course, there may well be other local characteristics at work to "explain away" such a protective effect, but the results point us to further investigation of the dataset. In addition, the interpretation is only based on the point estimates, taking no account of the uncertainty associated with the point estimates. This leads us to the following discussion regarding the role that GWR plays.

Brunsdon et al. (1996, p.288) suggested two useful questions that may be examined using GWR:

- "Does the GWR model describe the data significantly better than a global regression model?"
- "Does the set of a_{ki} parameters exhibit significant spatial variation?"

To investigate both questions, they proposed a Monte Carlo permutation procedure where the observed data are redistributed to the areas at random and the value of b and the regression coefficients are estimated using each of the many simulated datasets. Empirical p-values (of the form in Eq. 6.2) can be computed to perform hypothesis testing. In particular, to tackle the first question, "[O]ne possible choice here is the weighting parameter obtained by the CV procedure which can be used to assess the difference of the GWR model from a global model" (Brunsdon et al., 1996, p.288), namely, see whether $1/b$ is close to 0 or not (or equivalently whether b tends to infinity). The second question is equivalent to assessing the spatial heterogeneity of the local coefficients compared to the global counterpart. While it is a similar idea to an ANOVA (Section 6.2.3.3), here the test is *specific to each area* because different areas have different neighbourhood structures, which in turn affect the sampling distribution of the test statistic under the null

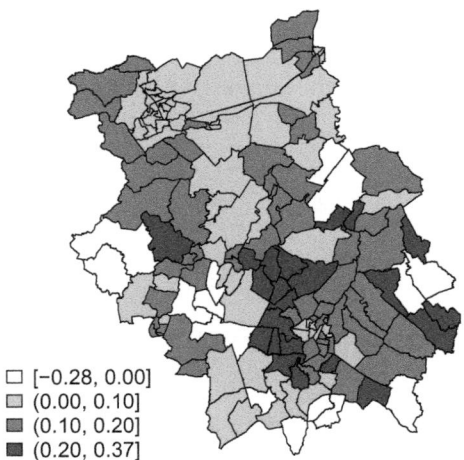

□ [−0.28, 0.00]
▨ (0.00, 0.10]
▩ (0.10, 0.20]
■ (0.20, 0.37]

FIGURE 6.19
A map of the estimated local regression coefficients measuring the effect of unemployment proportion on burglary rate. The effect estimate from the global regression model is 0.10.

hypothesis. This gives rise to the multiple testing problem (da Silva and Fotheringham, 2016; see also Section 6.2.5.2).

A GWR model can incorporate more than one covariate. However, care needs to be taken due to the issue of multicollinearity. In the context of a standard regression model, multicollinearity arises when some of the covariates are correlated, which may lead to difficulty in separating the effects of the correlated covariates (Fox, 1997, p.351). In GWR, Wheeler and Tiefelsdorf (2005) highlighted the issue that two sets of local regression coefficients can be correlated with each other even though the covariates that the coefficients represent raise no issue of multicollinearity in the corresponding global regression model. Wheeler (2007) subsequently extended the traditional diagnostic tools of multicollinearity such as the variance inflation factor (Neter et al., 1996) and the condition indexes based on singular value decomposition (Belsley, 1991) to GWR. When multicollinearity is suspected, one extension of GWR is to constrain the local parameters using ridge regression (Wheeler, 2007) or lasso regression (Wheeler, 2009). Alternatively, Brunsdon et al. (2012) proposed a locally-compensated *bandwidth* GWR in which the optimally-defined bandwidth is increased locally in areas where multicollinearity is detected. Another option that Brunsdon et al. (2012) proposed is a locally-compensated *ridge* GWR where, for the areas where multicollinearity is detected, a ridge regression model is used instead of a standard regression model. The paper by Fotheringham and Osham (2016) offers further insight into the issue of multicollinearity in GWR.

Under GWR, the estimation and the testing of the local coefficients are dependent upon the weighting. In addition to the exponential weight, there are other specifications, each imposing an assumption regarding how information is used in estimation. It is therefore important to make the assumption transparent so that its appropriateness to a given application can be discussed. For example, Brunsdon et al. (1996) considered a form of weighting in which only the data from areas within a certain distance of area *i* are used in the estimation of the parameters in *i* and data from other areas are not used. While this weighting "allows for efficient computation", they also pointed out a potential limitation that it "suffers the problem of discontinuity". The exponential weighting, in contrast, imposes spatial continuity, which may be desirable in certain applications. This relates back to the discussion of the importance of the weights matrix, which was the subject of Chapter 4. As we shall see, the specification of the spatial neighbourhood structure plays a crucial role in the hierarchical modelling of spatial data, as we shall observe in Chapter 8. A weighting function is called a kernel in the GWmodel package (as it serves the same functionality as a kernel function in kernel smoothing). Gollini et al. (2015) summarise various kernel functions implemented in GWmodel. Finally, in addition to the cross-validation approach, an optimal bandwidth can also be selected using a modified version of the Akaike Information Criterion (Gollini et al., 2015, p.19).

6.4 Overdispersion and Zero-Inflation in Spatial Count Data

The Poisson regression model is frequently used to model spatial data in the form of counts. A Poisson distribution assumes the mean is equal to the variance. However, when working with spatial data, this assumption can be violated, as the variance of the outcome values often exceeds the mean. This situation is referred to as overdispersion. Underdispersion, on the other hand, arises when the variance of the outcome values is smaller than the mean, but this situation is rarely encountered in practice.[17]

The presence of overdispersion can be due to any of a number of factors, but most often it arises as a result of heterogeneity in the mean, some form of clustering of events, or both. Heterogeneity in the mean might be addressed by the inclusion of covariates (or explanatory variables) in the regression model. Different *levels* of risk factors, for example, might explain the differences in disease rates amongst areas. However, most of the phenomena that we wish to study involve many risk factors, and it may be all but impossible to know about all of them or have data on them. In that case, there is a certain amount of variability in the observed outcome values that cannot be captured by the included variables. This uncaptured variability manifests itself as overdispersion.

Clustering, another cause of overdispersion, can occur at the intra-area scale due to, for example, some contagious, copy-cat or infectious process operating at a geographical scale smaller than the spatial units through which the data are reported. Such processes give rise to areas with large numbers of case counts, whilst other areas have no or only a very small number of cases. The geographical pattern of events we wish to model will be dependent on where such processes happen to start (e.g. where the first group of infected cases occur), with a cumulative effect producing an overdispersed pattern of counts. If the contagious process operates at a scale larger than the size of the spatial units, it is likely to give rise to global spatial autocorrelation in the count data or at least some localized clusters of events even if no global pattern of autocorrelation is present.

---

[17] Because the mean and variance of a normal distribution are independent parameters, this is not an issue when modelling normally distributed outcome values, although one needs to consider homoscedasticity (constant variance across areas) vs heteroscedasticity (varying variances across areas). See Section 8.6.1 in Chapter 8.

TABLE 6.4

A Summary of the Malaria Cases Occurring Between June and July 2012

	0	1	2	3–8	9–15	16–33
Number of new malaria cases						
Number of villages in the data	122	8	1	0	4	4
Number of villages predicted from a Poisson distribution with mean 1.12	45	51	28	15	0	0

Note: The frequencies in the second row are from the observed case counts, while those in the third row are predicted from a Poisson distribution with mean equal to 1.12, the mean number of malaria cases per village from the observed data.

For spatial count data, an initial assessment of overdispersion can be done using simple summary statistics (tabulating frequencies as well as calculating the mean and the variance of the count data) and mapping. To provide an example, Table 6.4 summarises the number of new malaria cases occurring between June and July 2012 across 139 villages in Kalaburagi taluk, South India (Section 1.6 in Chapter 1).

The variance of the observed numbers of cases, 21.4, is much higher than the mean of 1.12 cases per village. The large variability is caused by the existence of a few villages with large numbers of cases while a majority of the villages, about 88%, have no cases at all. As shown in Figure 2.3 in Chapter 2 for a period in 2014, typically most cases are highly concentrated in the north of the taluk, indicative of some clustering process. A Poisson distribution with mean 1.12 clearly does not describe the observed data well (see Table 6.4).

This dataset also highlights another problem that we often encounter when modelling spatial count data: zero-inflation. Zero-inflation refers to a situation where the outcome values that we observe contain many more 0 values than can be generated by the model under investigation. In the malaria data, for example, a simple model with a Poisson distribution with mean of 1.12 clearly suffers zero-inflation, as there are many more 0 values in the data than allowed by the model.

It is important to point out that overdispersion and zero-inflation are problems associated with the model, not with the data. If we could specify a model that fully describes the underlying data generating process, the model would be able to account for such features of the observed data. In that case, issues of overdispersion and zero-inflation would not arise. Therefore, tests for overdispersion and zero-inflation usually compare the fitted values from a Poisson regression model to the observed outcome values, thereby forming part of the model evaluation process.

Overdispersion and zero-inflation are generally addressed and modelled separately, although modelling the presence of an excessive number of zeros can, to some extent, help address the problem of overdispersion (see Section 6.4.2). In the following two sections, we describe exploratory tools that may be useful for detecting overdispersion and zero-inflation for a given Poisson regression model.

6.4.1 Testing for Overdispersion

Various tests have been developed to detect overdispersion in Poisson regression models, for example, Cameron and Trivedi (1986 and 1990), Dean and Lawless (1989), Dean (1992) and the Z and the Lagrange multiplier tests in Hilbe (2007, p.46–48). Here, we discuss the method proposed by Cameron and Trivedi (1990) as it is implemented in the R package AER.

A Poisson regression model describes each observed count, y_i (for $i = 1, \ldots, N$), as a random variable Y_i. Each outcome variable Y_i follows a Poisson distribution with mean μ_i, which is specified as a function of the available covariates, X. The use of a Poisson distribution means that $E[Y_i] = \text{var}(Y_i) = \mu_i$. The method proposed by Cameron and Trivedi (1990) tests the null hypothesis that $\text{var}(Y_i) = \mu_i$ against an alternative hypothesis that $\text{var}(Y_i) = \gamma \cdot \mu_i$. The parameter γ is called a dispersion parameter, and testing for overdispersion is equivalent to testing whether γ is greater than 1 or not. Note that this test can also be used to test for underdispersion when the sample estimate of γ is less than 1.[18]

In R, the function `dispersiontest` in the AER package implements the test. There are two steps involved. The first step fits a Poisson regression model to the count data using the `glm` function. Using the fitted values from the regression fit, the second step performs the overdispersion test. The R codes below apply the test to the malaria data.

```
###################################################################
#  block 1: read in the malaria data
###################################################################
load('malaria.RData')
###################################################################
#  block 2: fit a Poisson regression model with rainfall as a
#  covariate with the village-level population as an offset
###################################################################
poisson.fit <- glm(O ~ rainfall, offset=log(pop), family=poisson
                ,data=malaria)
###################################################################
#  block 3: carry out the overdispersion test by Cameron and
#  Trivedi (1990) using the dispersiontest function in the AER package
###################################################################
library(AER)   #  load in the AER package
dispersiontest(poisson.fit)
```

The R outputs are

```
Overdispersion test
data:  poisson.fit
z = 1.9972, p-value = 0.0229
alternative hypothesis: true dispersion is greater than 1
sample estimates:
dispersion
  3.987538
```

---

[18] Cameron and Trivedi (1990) specify a more general formulation of the alternative hypothesis: $\text{var}(Y_i) = \mu_i + \alpha \cdot g(\mu_i)$, where $g(\mu_i)$ defines a variance-mean relationship (Cameron and Trivedi, 1986). Common choices of $g(\mu_i)$ are: (a) $g(\mu_i) = \mu_i$, a linear variance function and (b) $g(\mu_i) = \mu_i^2$, a quadratic variance function. Using (a), the alternative hypothesis becomes $\text{var}(Y_i) = \mu_i + \alpha \cdot \mu_i = (1 + \alpha) \cdot \mu_i = \gamma \cdot \mu_i$, as given in the main text. Thus, testing whether or not $\alpha = 0$ as formulated in Cameron and Trivedi (1990) is the same as testing whether or not the dispersion parameter $\gamma = 1$. The quadratic variance function corresponds to Y_i, modelled by a negative binomial distribution (Cameron and Trivedi, 1986).

The dispersion parameter γ is estimated to be 3.99, a sign of overdispersion. The associated p-value is 0.0229, suggesting the rejection of the null hypothesis that $\gamma = 1$. The test result indicates that there is a considerable amount of variability in the observed count data that has not been captured by the fitted model. In Chapter 9, we will describe two modelling approaches to tackle overdispersion. The hierarchical modelling approach considers the uncaptured variability as a result of some unobserved risk factors on the response and uses random effects to account for such variability. Another approach is to use a (two-component) mixture model, which directly deals with the clustering nature of the malaria cases via a mixture of Poisson distributions.

6.4.2 Testing for Zero-Inflation

The problem of zero-inflation arises when a regression model fails to accommodate the number of zero values in the observed count data. The zero-inflated Poisson (ZIP) model (Lambert, 1992) and the hurdle model (Mullahy, 1986) are two commonly used methods to handle excessive zeros – we defer detailed discussion of both to Chapter 9. The focus of this section is to initially assess whether or not there are "too many" zero outcome values. For this, we consider the score test proposed by van Den Broek (1995), but we will first describe the zero-inflated Poisson (ZIP) distribution (Johnson et al., 1992, p.313) on which the score test is based.

Let Y be a discrete random variable that follows a ZIP distribution with two parameters μ and ϕ. We write $Y \sim \text{ZIP}(\mu, \phi)$, and the probability mass function is given by

$$\Pr(Y = y) = \begin{cases} \phi + (1-\phi) \cdot e^{-\mu} & \text{if} \quad y = 0 \\ (1-\phi) \cdot \dfrac{e^{-\mu} \mu^y}{y!} & \text{if} \quad y = 1,2,3\ldots \end{cases} \tag{6.13}$$

A ZIP distribution is essentially a mixture of a Poisson distribution with mean μ and a degenerate distribution at 0 (i.e. a distribution that takes the value of 0 with probability 1), with ϕ measuring the probability of an observed 0 coming from the degenerate distribution. $\phi = 0$ means that the Poisson distribution is capable of dealing with the observed number of zero values in the data. Using the ZIP distribution as the likelihood function for the observed values $y_1, \ldots, y_N$ over N areas, ϕ measures the proportion of areas where the chance of observing no event is 100%. In the case of the malaria data, these areas (or villages) are interpreted as "disease-free" (see Section 9.4). Exercise 6.14 visualises the ZIP distribution with $\mu = 1$ and different values of ϕ and compares it to the standard Poisson distribution with $\mu = 1$.

It is worth noting that if $Y \sim \text{ZIP}(\mu, \phi)$, $\text{var}(Y) \geq E[Y]$, a property that helps to address the overdispersion problem. To show the above mean-variance relation, the expectation and the variance of Y are $E[Y] = (1-\phi)\mu$ and $\text{var}(Y) = E[Y] + E[Y](\mu - E[Y])$ (Exercise 6.15). Substituting the expression of $E[Y]$ into $\text{var}(Y)$ and simplifying the resulting expression, we get $\text{var}(Y) = \mu - \mu\phi + \mu^2\phi - \mu^2\phi^2 = (1 + \mu\phi)(1-\phi)\mu = (1 + \mu\phi)E[Y]$. Because $(1 + \mu\phi)$ in the last expression is non-negative, $\text{var}(Y) \geq E[Y]$. When $\phi = 0$, $\text{var}(Y) = E[Y]$, and a ZIP distribution becomes the corresponding Poisson distribution.

To check whether a Poisson regression can accommodate the observed number of zeros, the score test of van Den Broek (1995) evaluates whether $\phi = 0$ or not. The score test is easy to implement because it only takes the fitted values from a standard Poisson regression as input and does not require the fitting of a ZIP model, an advantage pointed out by Lee et al. (2001). For a set of observed counts over N areas, the test statistic takes the form

$$S(\hat{\mu}) = \frac{\left\{ \sum_{i=1}^{N} \left(\frac{I(y_i = 0)}{\tilde{p}_{0i}} - 1 \right) \right\}^2}{\sum_{i=1}^{N} \left(\frac{1}{\tilde{p}_{0i}} - 1 \right) - N\overline{y}}, \tag{6.14}$$

where $\hat{\mu} = (\hat{\mu}_1, \ldots, \hat{\mu}_N)$, the fitted values from a Poisson regression model; $\tilde{p}_{0i} = \exp(-\hat{\mu}_i)$, the probability of observing a zero outcome value for area i under the fitted Poisson model; $I(y_i = 0) = 1$ if the observed outcome value of area i is 0 and $I(y_i = 0) = 0$ otherwise; and $\overline{y}$ is the mean of the observed outcome values. Under the null hypothesis that there is no zero-inflation, the test statistic follows an asymptotic chi-squared distribution with 1 degree of freedom.[19]

The R function ZI.test in the Appendix to this chapter takes the result object from fitting a Poisson regression model as the input and returns the test statistic, $S(\hat{\mu})$, and the p-value (i.e. the probability of exceeding $S(\hat{\mu})$ under the χ_1^2 distribution). Continuing from the overdispersion analysis of the malaria data, the R code below checks for zero-inflation.

```
####################################################################
#  block 4: carry out the zero-inflation test by van den Broek (1995)
#  poisson.fit is the result object after fitting a Poisson regression
#  model with the village rainfall levels as a covariate and the
#  village-level population as the offset
#  (see the R code in block 2 in Section 6.4.1)
####################################################################
ZI.test(poisson.fit)
```

The test statistic is 339.34 and the p-value is very close to 0. The results suggest that the fitted model cannot deal with the number of zero observations in the data, indicating the need to use a ZIP distribution as the likelihood function (see Section 9.4 in Chapter 9).

6.5 Concluding Remarks

Exploratory analyses of spatial data for the purpose of model specification raise a number of distinct challenges not necessarily encountered when analysing other sorts of (non-spatial) data. For this reason, we have described some special tools in this chapter.

---

[19] The term "asymptotic" means that the χ_1^2 distribution is only an approximation to the true distribution of the test statistic under the null hypothesis, but the approximation becomes closer when the size of the dataset N tends to be infinitely large. However, Waller (2014) criticises the idea of "N tends to infinity" in the spatial context where the closest equivalence to "statistical asymptotic" involves adding more areas to the study region. However, as he argues, this is more complicated than the "statistical thinking" which involves "adding more of the same type of observation" (Waller, 2014, p.11). Besag (1975, p.192) makes the same point, but using a more colourful example!

The simultaneous need to recognize the presence of *spatial dependence* and *spatial heterogeneity* (in both the mean and the dependence structure in the data) lie behind the development of several of the specialist tools described here (Sections 6.2 and 6.3). However, two of the other challenges associated with spatial data described in Chapter 1, namely the effects of *data uncertainty* (see for example Figure 6.15(b)) and, when modelling count data, *data sparsity*, which is often associated with overdispersion and zero inflation (see Section 6.4), have also played a role.

Exploratory methods can be divided into graphical techniques and numerical techniques and into those concerned with identifying the properties of individual variables and those concerned with identifying associations between variables. Both of these broad groups can be divided in turn into those that are "global" and those that are "local". Global techniques process all the small-area data in the form of a single plot (e.g. the Moran scatterplot), or a single numerical index (e.g. the global Moran's I statistic and the join-count statistic). Local techniques, by contrast, process spatially-defined subsets of the data (e.g. the local Moran's I and Kulldorff's scan statistic), and they are designed explicitly to expose the presence of spatial heterogeneity in the data. However, the application of local techniques raises problems when trying to establish which areas depart significantly from the global average. At the heart of these problems lie two issues: (i) tests associated with local index scores are not independent because the data values are typically spatially autocorrelated and the calculation of these local index scores uses overlapping sets of (autocorrelated) data values; (ii) the multiple testing problem arises because we perform at least as many tests as there are areas in the study region. Of the local tests discussed here, the local Moran's I suffers from the multiple testing problem (testing for significance is carried out on all areas), whereas Kulldorff's scan test for detecting local clusters avoids this problem.

The simultaneous presence of spatial heterogeneity and spatial autocorrelation gives rise to particular problems when exploring relationships between variables, whether using the conventional Pearson's correlation coefficient to test for a significant bivariate association or exploring "association at a distance". The latter, a challenge that is particularly relevant when analysing spatial data (but also temporal and spatial-temporal data, as we shall see in Part III of the book), refers to the challenge of trying to assess the association between two variables when effects can transfer across areas because of the permeability of many spatial boundaries (such as census tracts or administrative districts). We have shown how Geographically Weighted Regression (GWR) enables us to make a preliminary assessment as to whether there may be spatial heterogeneity in any outcome-covariate relationship.

Lastly, we have seen how data sparsity, manifest in the form of large numbers of areas with zero counts or a large variance-to-mean ratio across the set of areas, also calls for exploratory methods to help assess the extent to which particular extensions of Poisson models (the ZIP model or the hurdle model) may or may not be required.

Monte Carlo methods play an important role in testing significance, and we have seen several different forms of Monte Carlo simulation. Fortunately, with today's computing power such heavily computational methods do not present us with any significant problems – a comment that applies with even greater force when fitting Bayesian spatial models, the subject to which we now turn.

6.6 Exercises

Exercise 6.1. Draw choropleth maps of Cambridgeshire burglary rates as described in Section 6.2.1. Experiment with different cut points (Step 2) and shading schemes (Step 3).

Exercise 6.2. Write the R code to obtain a Moran scatterplot to produce Figure 6.8.

Exercise 6.3. Write the R code for a 2nd and 3rd order Moran scatterplot (see Section 6.2.4.1) (hint: the R function `nblag` in the `spdep` package creates higher order neighbours lists – see Appendix 4.13.2).

Exercise 6.4. Write an R script to obtain the sampling distribution for Moran's I statistic under the null hypothesis as shown in Figure 6.9.

Exercise 6.5. Use the `glm` function in R to fit a Poisson regression model to the Cambridgeshire burglary counts using the number of houses as an offset and the unemployment proportions as a covariate. Test spatial autocorrelation of the residuals using the global Moran's I.

Exercise 6.6. For each of the five maps in Figure 6.6, implement the join count test in R using a weights matrix derived from rook's move contiguity to check if each map is correctly classified as either positively or negatively spatially autocorrelated, or random. Repeat using a weights matrix derived from queen's move contiguity. Compare results.

Exercise 6.7. For the EHIA data (Section 6.2.4.5), implement the join count test in R using a spatial weights matrix derived using (a) queen's move contiguity and (b) rook's move contiguity. Compare results.

Exercise 6.8. Calculate the Local Moran's I for the Cambridgeshire unemployment data (Section 6.2.5.1).

Exercise 6.9. Explore the effect of varying the threshold, in order to adjust for multiple testing, when looking for significant clusters using the local Moran's I (see Section 6.2.5.2). The smaller the p-value threshold, the more permutations are required.

Exercise 6.10. Produce the two scatter plots in Figure 6.15 using the R script

`malaria_rainfall_scatter.R` (available online).

Exercise 6.11. Investigate the effects of the unemployment proportions and the spatially-lagged version of the unemployment proportions on burglary risk using the added-variable plots. (Hint: the `avPlots` function in R from the `car` package produces an added-variable plot.)

Exercise 6.12. Carry out the Clifford-Richardson modified t-test to examine the correlation between the two sets of binary indicators, EHIA and PHIA, in the HIA dataset (Section 6.3.2.1).

Exercise 6.13. Using Eq. 6.11, calculate, by hand, the estimated values of $\hat{a}_{0i}$ and $\hat{a}_{1i}$ using the data given in Table 6.3. Then, in R, carry out the leave-one-out cross validation to select the optimal bandwidth ($b=953$) based on the score function in Eq. 6.12.

Exercise 6.14. The R script, `visualise_ZIP.R`, online visualises the zero-inflated Poisson distribution in Eq. 6.13 with μ and ϕ. The standard Poisson distribution (without the zero-inflated part) is superimposed for ease of comparison. Experiment with different values for μ and ϕ.

Exercise 6.15. Prove that $E[Y] = (1-\phi)\mu$ and $\text{var}(Y) = E[Y] + E[Y](\mu - E[Y])$, where the random variable Y follows the zero-inflated Poisson distribution given in Eq. 6.13.

Exercise 6.16. As discussed in Section 3.3.3 (Chapter 3), a cartogram is a useful mapping tool when physical size and population size do not correlate (e.g. a polygon with a large (small) physical size tends to be associated with a small (large) population). If that is the case, a cartogram scales the physical size of each polygon according to its population size (see Figure 3.8 for an example). The R function `cartogram_cont` from the `cartogram` package creates a cartogram. This function requires the user to supply (a) the shapefile, a cartogram version of which is to be produced, and (b) the column name in the shapefile data that contains the population sizes (these are called `weights` in the `cartogram_cont` function). The output of this function is a shapefile containing the scaled polygons (by the chosen weights). Following the steps outlined in Section 6.2.1, that resulting shapefile can then be used to produce a cartogram with a chosen attribute (e.g. burglary rates). As an exercise, create a cartogram showing the ward-level burglary rates (per 1000 houses) in Cambridgeshire (as in Figure 6.1) but with the polygons resized according to their population sizes (the column `popSize` in the `cambridgeshire.shp` file contains the ward-level population).

Appendix. An R Function to Perform the Zero-Inflation Test by van Den Broek (1995). See more detail in Section 6.4.2.

```
ZI.test <- function(fits) {
############################################################
#  An R function to perform the zero-inflation test by
#  van Den Broek (1995) the argument fits is an object containing the
#  results from a Poisson fit to the observed count data (e.g. from
#  the glm function)
#
#  This function returns the test statistic and the p-value by
#  comparing the calculated test statistic to the chi-square
#  distribution with 1 degree of freedom
############################################################
   obs <- fits$y     #  extract the observed numbers of cases
                     #  from the result object
  N <- length(obs) #  number of areas
   mean.obs <- mean(obs)   #  mean number of cases
   fitted <- fits$fitted.values  #  extract the fitted values from
                                 #  the Poisson model
   p.zero.fitted <- exp(-fitted) #  calculate the probability of getting
                                 #  0 based on the fitted values
```

```
# calculate the numerator in Eq. 6.14
a <- sum((as.numeric(obs==0)-p.zero.fitted)/p.zero.fitted)^2
# calculate the denominator in Eq. 6.14
b <- sum((1-p.zero.fitted)/p.zero.fitted) - N*mean.obs
test.stat <- a/b  # calculate the test statistic (Eq. 6.14)
p.value <- 1 - pchisq(test.stat,1)
# putting the results together
res <- c(test.stat,p.value)
names(res) <- c('Test Statistic', 'p-value')
return(res)
}
```

7

Bayesian Models for Spatial Data I: Non-Hierarchical and Exchangeable Hierarchical Models

7.1 Introduction

In many analyses of spatial and spatial-temporal data, we wish to make inference about parameters that are specific to each area, θ_i (for $i = 1,\ldots,N$), or each space-time cell, θ_{it} (for $i = 1,\ldots,N$ and $t = 1,\ldots,T$). Even for a purely spatial analysis, this set of parameters can often be large because of the many small areas into which the study region is sub-divided. Estimating this large set of parameters reliably is the goal (or provides the basis for achieving the goal) of many small area analyses. Consider the income example introduced in Section 1.3.1 where the aim of the analysis is to provide reliable estimates of the 109 parameters, each representing the average weekly income level per household in one of the 109 middle super output areas (MSOAs) in Newcastle. In that example, we have 109 of these area-specific parameters, $\theta_1,\ldots,\theta_{109}$. The results from that analysis might later be used to construct a small area deprivation index or support policy at the small area level regarding, for example, resource allocation, infrastructure planning or providing support to small businesses.

In this chapter and the next (Chapter 8), we will present a family of Bayesian hierarchical models (BHMs) to deal with the estimation problem in the spatial context (and we will consider the Bayesian hierarchical modelling of spatial-temporal data in Part III). As we shall see, Bayesian hierarchical models encompass all the area-specific parameters within a common prior distribution, which itself has some unknown parameters (referred to as *hyperparameters*). This hierarchical structure on parameters gives rise to a distinctive feature known as *information borrowing* – the ability to borrow (or share) information across areas when estimating the area-specific parameters. This feature allows us to estimate the large set of parameters in a way that addresses the issues that arise from heterogeneity and data sparsity when analysing spatial data (see Chapter 1).

Prior to the action of borrowing information, there is a more fundamental question: *how* should information be borrowed? The basis for information borrowing is the concept of similarity. If two areas are considered to be similar in terms of their income levels, then any information about income in one area gives us some idea about the income level in the other. Because of their similarity, when we estimate the income levels for the two areas, the data observed in both areas can be pooled together to provide estimates for the two individual areas. In other words, the information obtained in one area can be borrowed (or shared) when estimating the parameter in the other area and vice versa. We shall introduce several ways to share information where each option is based on a (prior) assumption that we place on the dependence structure of the area-specific parameters.

More specifically, sharing information *globally* assumes the parameters of all areas are similar to one another in the sense that the spatial configuration of the areas plays no role in the process of information sharing. This is equivalent to assuming the income levels of all areas in Newcastle are similar to the Newcastle average. Another option is to borrow information *locally*. In that case, we assume a local spatial dependence structure – the values observed in nearby areas are more alike than those observed in areas that are far apart. Over the course of this chapter, we will develop Bayesian hierarchical models to allow information to be borrowed globally. In the next chapter we will consider models that allow information to be borrowed locally and extend the discussion to include models that combine the two options so that information can be shared both globally *and* locally. Through examples, we will explore the properties and implications of these modelling options.

In Section 7.2, we introduce the Newcastle income data, the illustrative example that will run through both Chapters 7 and 8. In Section 7.3, we fit non-hierarchical models to the income data in order to illustrate their problems. In Section 7.4, we describe a Bayesian hierarchical model with the so-called "exchangeable" structure on the area-specific parameters. This modelling structure allows information to be shared globally within the study region.

7.2 Estimating Small Area Income: A Motivating Example and Different Modelling Strategies

Obtaining data for all individuals in a study region is typically impractical. The data used for producing small area income estimates often come from surveys, where information on income is collected from a proportion of the population. To illustrate the modelling strategies discussed in this chapter and the next, we will be using the household-level data on weekly total gross income (in £) that were first introduced in Section 1.3.1. These household-level data are simulated to mimic the situation where data collected through a national survey are disaggregated to a subnational level, the middle super output area (MSOA) level in this application, to provide the average income estimates at the MSOA level. In particular, we are focusing on providing estimates for the 109 MSOAs in Newcastle, UK – there are just under 6800 MSOAs in England based on the definition of the 2011 UK Census Geography. Figure 7.1 shows a map of the 109 MSOAs. In Newcastle, each MSOA contains on average about 3300 households. The simulation procedure is detailed in the Appendix to this chapter. As discussed in Section 1.2.3, disaggregating data collected at the national level to a finer geographical level creates the issue of data sparsity. The size of the collected data is sufficiently large to provide an accurate estimate of the average income level per household at the national level. However, in the absence of spatial stratification at the MSOA level (or the spatial scale at which the average income estimates are needed) in the national survey, the number of households participating in the survey is likely to be small per MSOA, typically too few to be used directly to provide good estimates for the MSOA populations.

Table 7.1 provides a summary of the distribution of sample sizes per MSOA, highlighting the challenges arising due to the sparsity of the data. In particular, 10 of the MSOAs have no survey data at all due to the lack of spatial stratification at the MSOA level (see Appendix).

FIGURE 7.1

A map of the 109 MSOAs in Newcastle. The number shown in each polygon is the MSOA label.

TABLE 7.1

Distribution of Sample Size in the Newcastle Income Survey Data

Number of households included (inclusive)	0	1–5	6–9	10–13
Number of MSOAs	10	5	75	19

Table 7.2 summarises the income data, drawing particular attention to the considerable variability in the mean income level across MSOAs, a feature of the data that needs to be accounted for in the modelling. The challenge is then to obtain good quality estimates of the average income for each of the MSOAs in Newcastle using the survey data. To meet this challenge, how should we model the 109 unknown parameters, $\theta_1,...,\theta_{109}$, given this limited amount of data?

To tackle the problem, over the course of the next two chapters, four different strategies for modelling the area-specific parameters, $\theta_1,...,\theta_N$, are considered:

1. The parameters are identical.
2. The parameters are independent.
3. The parameters are similar globally.
4. The parameters are similar locally.

TABLE 7.2

Summary of the Newcastle Income Data (Weekly Total Gross Income in £) at the Household-Level (Using All the Income Values from the Survey) and at the MSOA-Level (Based on the Sample Means from the 99 MSOAs with Survey Data)

	Min	Q1	Mean	Median	Q3	Max	SD
Household-level	154	418	526	515	617	1244	164
MSOA-level	337	445	526	509	582	907	102

As illustrated in Figure 7.2, each modelling strategy assumes a *structure* on the area-specific parameters. We will now describe these strategies in turn and discuss their implications for the estimation of the parameters.

7.2.1 Modelling the 109 Parameters Non-Hierarchically

The first two strategies structure the unknown parameters *non-hierarchically*. Strategy 1 assumes that all areas are the same in terms of the variable of interest. For the income data, this modelling strategy states that all MSOAs in Newcastle have the same income level and, therefore, all the 109 unknown parameters are reduced to just one single unknown parameter, i.e., $\theta_i = \alpha$ for $i = 1,\ldots,109$. This assumption with identical parameters is illustrated

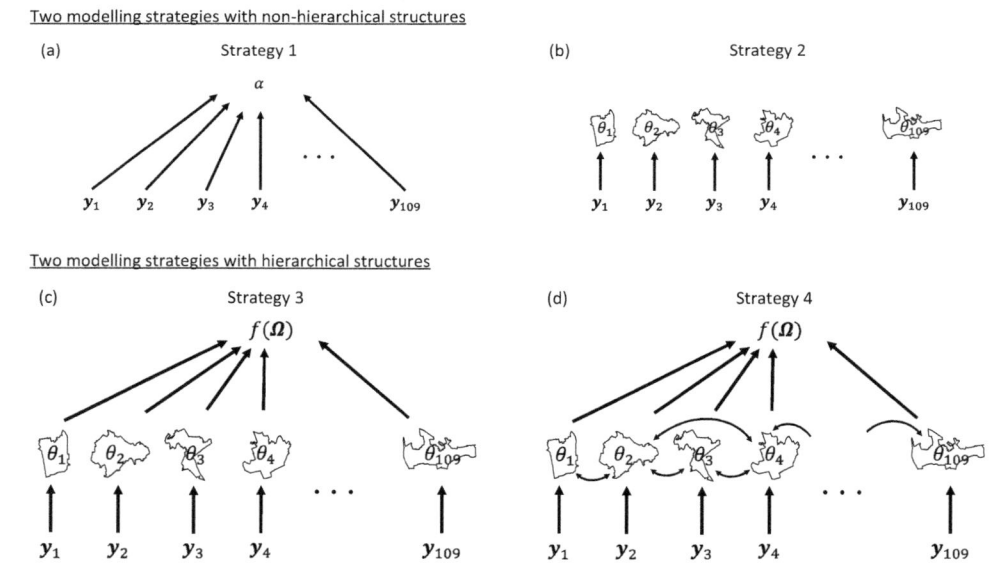

FIGURE 7.2

Schematic diagrams of the four modelling strategies: (a) identical parameters; (b) independent parameters; (c) parameters that are similar globally; and (d) parameters that are similar locally. We use y_i to denote outcome values – the household-level income values from the national survey – for MSOA i. In (b)–(d), the area-specific parameters, $\theta_1,\ldots,\theta_{109}$, are enclosed by the polygons of the respective MSOAs as shown in Figure 7.1 (not drawn to scale), and the labelling of the MSOAs is the same as that used in Figure 7.1. In (d), we use the connecting lines between MSOAs to emphasize the use of a local (neighbouring) structure of the MSOAs in information sharing. In this plot, two MSOAs are connected if they are adjacent to each other (e.g. MSOAs 1 and 2 are adjacent to each other; see Figure 7.1), and they are not connected otherwise (e.g. MSOAs 1 and 3; see Figure 7.1). This gives an example of a neighbourhood structure that provides a basis for local information sharing. See the main text for the description of each structure as well as the definitions of the mathematical symbols.

in Figure 7.2(a). Note that without the subscript i, α is simply a single unknown parameter representing the average weekly household income for the whole of Newcastle. All the data are essentially pulled together to estimate α, thus we can get a good estimate of this single parameter. This modelling strategy is often referred to as *complete pooling* (Gelman and Hill, 2007). However, such an identical parameter assumption is unrealistic. We do not expect all areas to have the same income level, and the data certainly do not suggest so (see Table 7.2). The main problem with this strategy is that it ignores the heterogeneity property of the spatial data.

To allow the θ_i's to be different across areas, Strategy 2 assumes that the areas are independent of each other regarding the variable of interest. For the income data, independence of areas implies that the income data obtained from one MSOA provide no information about the income levels of other MSOAs in Newcastle. As depicted in Figure 7.2(b), when modelling the θ_i's, this strategy estimates each θ_i only using the survey data obtained in MSOA i, and the data from other MSOAs do not affect the estimation of θ_i. Gelman and Hill (2007, p.256) label Strategy 2 a *no-pooling* strategy because the estimation of each θ_i gets no information from other areas. While this strategy retains the between-area variability, the estimate of each θ_i may be unreliable because of data sparsity. For most of the MSOAs, the inference on each θ_i is based on data from only a small number of households (see Table 7.1). Another problem with the income dataset is that no estimates can be produced for the 10 MSOAs with no data. For modelling spatial (as well as spatial-temporal) data, the inherent problem with this independence assumption is that it takes no account of any relationships that might exist between the areas (and/or across time points). Rather than a set of unrelated entities as assumed, some degree of similarity across areas is to be expected such that whilst income levels in nearby areas are likely to be *different*, they may well be *similar*. This is the rationale underlying the two hierarchical modelling strategies that we discuss next.

7.2.2 Modelling the 109 Parameters Hierarchically

The third and fourth modelling strategies are examples of the hierarchical modelling approach. As illustrated in Figures 7.2(c) and 7.2(d), both strategies impose *a hierarchical structure* on the unknown parameters, so that whilst the parameters are allowed to be different across the study region they are assumed to be similar and connected. Allowing the parameters to be different moves us away from the unrealistic assumption made under Strategy 1, acknowledging the heterogeneity property of spatial and spatial-temporal data. As opposed to being independent, as in Strategy 2, these parameters are assumed to be similar under the hierarchical modelling approach so that, as discussed in Section 7.1, information can be shared, addressing the data sparsity issue. In the context of the income example, assuming MSOAs to have similar income levels implies that the income data observed in MSOA i provide some *partial information* about the income levels in other MSOAs.

To achieve information borrowing, both of the hierarchical strategies use the same modelling concept for the area-specific parameters – all the θ_i's are embedded in a common (prior) probability distribution (denoted $f(\Omega)$ in Figures 7.2(c) and 7.2(d)), where Ω denotes a set of unknown parameters, known as hyperparameters, of this common distribution. One implication of information sharing is that the estimates of the area-specific parameters are smoothed – large values tend to get pulled down, while small values tend to get pulled up. As a result, a map of the point estimates of the θ_i's tends to be smoother than, for example, a map of estimates obtained under the independence assumption. For this reason, the Bayesian hierarchical models discussed in this chapter and the next are known as (spatial) smoothing models.

While all the above remarks apply to both Strategies 3 and 4, they differ by how information is pooled. Specifically, Strategy 3 assumes a *global* structure in which all areas are similar regardless of where they are on the map (e.g. the MSOA-level average income levels are all similar to the Newcastle average). This gives rise to global smoothing models. Strategy 4, on the other hand, imposes a *local* structure, so that the estimates for areas that are close together in space are more similar than for areas that are far apart (e.g. the average income levels in adjacent MSOAs are more alike than those from MSOAs that are far apart). Because of this local structure, the estimates from Strategy 4 are locally smooth.

In the next section, 7.3, using the Newcastle income data, we construct and fit statistical models derived from the two non-hierarchical modelling strategies. We will explore Strategy 3, the first of the two hierarchical modelling strategies in Section 7.4. Models based on Strategy 4 are discussed in Chapter 8.

7.3 Modelling the Newcastle Income Data Using Non-Hierarchical Models

7.3.1 An Identical Parameter Model Based on Strategy 1

To fix the notation, let y_{ij} be the income value of household j $(j = 1,...,n_i)$ in MSOA i $(i = 1,...,109)$, where n_i denotes the sample size in MSOA i. Here, we assume the household-level income follows a normal distribution so that

$$y_{ij} \sim N\left(\theta_i, \sigma^2\right) \tag{7.1}$$

In Eq. 7.1, θ_i is the unknown average income per household in MSOA i and σ^2 is the variance of the sampling distribution. Here, we use the normal distribution as the likelihood because it allows us to show some results in closed form (see Section 7.4), which is useful when exploring the properties of different modelling strategies. However, other probability distributions can be used as the likelihood function (e.g. either the log-normal distribution or the Gamma distribution can be considered to account for the positively-skewed nature often found in the distribution of income data).

Under the strategy of identical parameters, $\theta_i = \alpha$ for $i = 1,...,109$. Therefore, Eq. 7.1 becomes

$$y_{ij} \sim N(\alpha, \sigma^2) \tag{7.2}$$

To complete the model specification in the Bayesian framework, we assign vague priors independently to the two unknown parameters α and σ^2:

$$\alpha \sim N(0, 100000000)$$
$$\sigma \sim Uniform(0.0001, 1000) \tag{7.3}$$

Note that because σ, the sampling standard deviation, is strictly positive, the lower bound of the uniform prior is set to be a small value as opposed to 0. The data summary in Table 7.2 suggests the value for α, the Newcastle average, is likely to be in the hundreds. A (very) large variance of 100000000 is therefore used to ensure the prior for α is vague (or non-informative). A similar argument applies to the choice of the upper bound of the prior for σ. Would setting the variance of the prior for α to 1000000 be non-informative? Similarly, would setting the upper bound of the uniform prior for σ to 100 be non-informative? See Section 5.6 for the discussion of prior specifications.

Figure 7.3 shows the WinBUGS code for fitting this model, the format of the Newcastle income data and the two sets of initial values for running two MCMC chains. We will make a few comments regarding the WinBUGS implementation. The income data are in the so-called jagged array format, where the rows in the data matrix have different numbers of data values (see Figure 7.4). This is because the sample sizes are different across the MSOAs. To deal with this data format, an array called msoa of length 760, the same length as the array y containing the income values from all the households in the survey, is introduced in the data list (see Line 24 in Figure 7.3). When we go through each income value (Lines 4–7 in Figure 7.3), the corresponding element in msoa identifies which MSOA an income value is from. For example, the first five values in the msoa array are all 1, meaning that the first five income values in y, namely, 501, 616, 472, 816 and 637, are from MSOA 1. The subsequent two values in the msoa array are 2, so the corresponding values (i.e. in the 6th and the 7th positions) in y, namely, 500 and 506, are from MSOA 2. MSOA 3 is one of the 10 MSOAs that have no data, and hence the value 3 is not in the msoa array. Then the following five values in y are from MSOA 4 and so on. Line 10 assigns the

```
1   #  The WinBUGS code for specifying the identical-parameter model
2   model {
3      #  a for-loop to go through all the household level income data
4      for (j in 1:nhhs) {
5         #  defining the normal likelihood for each household
6         y[j] ~ dnorm(theta[msoa[j]],prec)
7      }
8      for (i in 1:nmsoas) {
9         #  all MSOAs have the same average income level
10        theta[i] <- alpha
11     }
12     #  vague priors for the Newcastle average and the sampling SD
13     alpha ~ dnorm(0,0.00000001)
14     sigma ~ dunif(0.0001,1000)
15     #  compute the sampling precision
16     prec <- pow(sigma,-2)
17  }
18
19  #  the household-level data in WinBUGS format
20  list(nhhs=760
21       ,nmsoas=109
22       ,y=c(501,616,472,816,637,500,506,560,542,447,644,522
23            ,487,275,...)
24       ,msoa=c(1,1,1,1,1,2,2,4,4,4,4,4,5,5,...))
25
26  #  initial values for chain 1
27  list(alpha=200,sigma=50)
28
29  #  initial values for chain 2
30  list(alpha=700,sigma=130)
```

FIGURE 7.3
The WinBUGS code, the data list and the two sets of initial values for fitting the identical-parameter model. The data list is shown in part for the purpose of illustration.

MSOA	Income values of the households in survey
1	501, 616, 472, 816, 637
2	500, 506
3	
4	560, 542, 447, 644, 522
5	487, 275, ...
...	

FIGURE 7.4
The jagged array format of the survey data. MSOA 3 is one of the 10 MSOAs with no survey data.

Newcastle average (`alpha`) to all the 109 MSOAs in Newcastle, including those 10 MSOAs without data.

Using the two sets of initial values given in Figure 7.3, two MCMC chains were run, each with 10000 iterations. The first half of each chain was discarded as burn-in and the second half was used (thus with 10000 iterations from both chains) to produce the posterior summary. For α, the posterior mean is 525.5, with a 95% credible interval of (514.0, 537.4), and this is also the estimate of the average income for each of the 109 MSOAs in Newcastle. There is little uncertainty about α due to the large amount of data used to estimate the two parameters in this model. In addition, because of the vague priors used, the posterior mean (=525.5) and the posterior standard deviation (=5.9) of α are very close to the sample mean (=525.6) and the standard error ($= s/\sqrt{N} = \dfrac{164.1}{\sqrt{760}} = 6.0$ with s and N being the sample standard deviation and the total number of households in the survey data, respectively). For this model, the Bayesian approach with vague priors and the frequentist approach give nearly the same results.

7.3.2 An Independent Parameters Model Based on Strategy 2

Instead of assuming all MSOAs have the same income level, the strategy with independent parameters estimates each θ_i using only the data from its own area. Effectively, each of the 109 MSOAs is treated as a stand-alone unit, the data from which are analysed using the (same) model given as follows:

$$y_{ij} \sim N\left(\theta_i, \sigma_i^2\right) \tag{7.4}$$

As in the case of the identical parameter model, the income values obtained in each MSOA are modelled using a normal distribution. For each MSOA, there are two unknown parameters to estimate, θ_i, the average income level, and σ_i^2, the within-MSOA variance. Given the small number of data points in each MSOA, we do not try to estimate σ_i^2 but instead fix it to the sample variance s_i^2. This is done for all those MSOAs that have at least two data points. For those MSOAs with only one or zero data points, s_i^2 and hence σ_i^2 is undefined, thus we fix s_i to 9999 (see the model fitting in Figure 7.5). No uncertainty measure for the income estimate θ_i can be produced for the MSOAs with one or zero data points. With σ_i^2 fixed, the only unknown parameter for each MSOA is θ_i, to which we assign the following vague prior

$$\theta_i \sim N\left(0, 100000000\right) \tag{7.5}$$

```
1    #  The WinBUGS code for fitting the model with independent parameters
2    model {
3       #  a for-loop to go through all the household level income data
4       for (j in 1:nhhs) {
5          #  defining the normal likelihood for each household
6          y[j] ~ dnorm(theta[msoa[j]],prec[msoa[j]])
7       }
8       for (i in 1:nmsoas) {
9          #  a vague prior independently assigned to each of the thetas
10         theta[i] ~ dnorm(0,0.00000001)
11         #  for each MSOA calculating the sampling precision using the
12         #  sample standard deviation (which is entered as data)
13         prec[i] <- pow(sigma[i],-2)
14      }
15   }
16
17   #  the household-level data in WinBUGS format (with the addition of
18   #  sigma, an array of length 109 containing the sampling standard
19   #  deviations of the MSOAs; for those with fewer than 1 data point,
20   #  the corresponding elements in sigma take the value of 9999)
21   list(nhhs=760
22       ,nmsoas=109
23       ,y=c(501,616,472,816,637,500,506,560,542,447,644,522
24          ,487,275,...)
25       ,msoa=c(1,1,1,1,1,2,2,4,4,4,4,5,5,...)
26       ,sigma=c(136.12, 4.24, 9999, 70.97,...))
27
28   #  initial values for the 109 thetas for chain 1
29   list(theta=c(100,100,...))
30
31   #  initial values for the 109 thetas for chain 2
32   list(theta=c(700,700,...))
```

FIGURE 7.5
The WinBUGS code, the data list and the two sets of initial values for fitting the model with independent parameters. The data list and the sets of initial values are shown in part for the purpose of illustration. The array sigma in Line 26 stores the sample standard deviations across the 109 MSOAs. The third element is 9999 because MSOA 3 does not have any survey data.

independently for $i = 1,...,109$. For the MSOAs with at least one data point, this model combines the normal likelihood with the vague prior distribution to form the posterior distribution for the MSOA-level average income. However, for the 10 MSOAs without data, there is no likelihood contribution to the posterior distribution, and hence the posterior is effectively just the vague prior that we assign. The WinBUGS code for fitting this model with the data list and initial values is shown in Figure 7.5.

The posterior distributions, summarised using the posterior means and the 95% credible intervals (denoted as the 95% CIs hereafter) of the MSOA-level averages, are shown in Figure 7.6(a), while Figure 7.6(b) maps the posterior means of the average income for the 99 MSOAs with data. Compared to the identical-parameter model, this independent-parameters model is able to show spatial variability in income at the MSOA-level. However, the estimates from this model rely too much on the limited amount of data available in each MSOA. For example, the estimate for MSOA 2 (indicated with an arrow in Figure 7.6(a)) uses only two data values, namely 500 and 506. The question is: how much confidence can we have in an estimate based on only two data points? – probably not much. Furthermore, these two observed values happen to be close to each other, resulting in a narrow 95% CI of (497.1, 509.0), much narrower compared to the 95% CIs for other MSOAs. By contrast, the first estimate at the extreme left edge of the plot is associated with a very wide 95% CI. This is a consequence of the two very different data values (186 and 488) in that MSOA (MSOA 10), again highlighting the unreliable estimate derived only using an area's own data. Another issue is with the MSOAs without data. In Figure 7.6(a), we leave the space blank for the estimates of these 10 MSOAs because this strategy cannot provide any estimates for them (or, as discussed before, their estimates merely take the vague prior that we assign to θ_i).

Despite the fact that these two non-hierarchical modelling strategies are rather unrealistic for tackling the problem in hand, through their application to the income data they offer some insights into the issues that we often encounter when modelling spatial (as well as spatial-temporal) data.

- The property of spatial heterogeneity requires parameters to vary over space, and hence the parameters representing the variable of interest need to be area-specific.

- The data sparsity problem hinders the independent estimation of the area-specific parameters, an assumption that also ignores the interrelatedness of the areas (i.e. they are all nested within the study region).

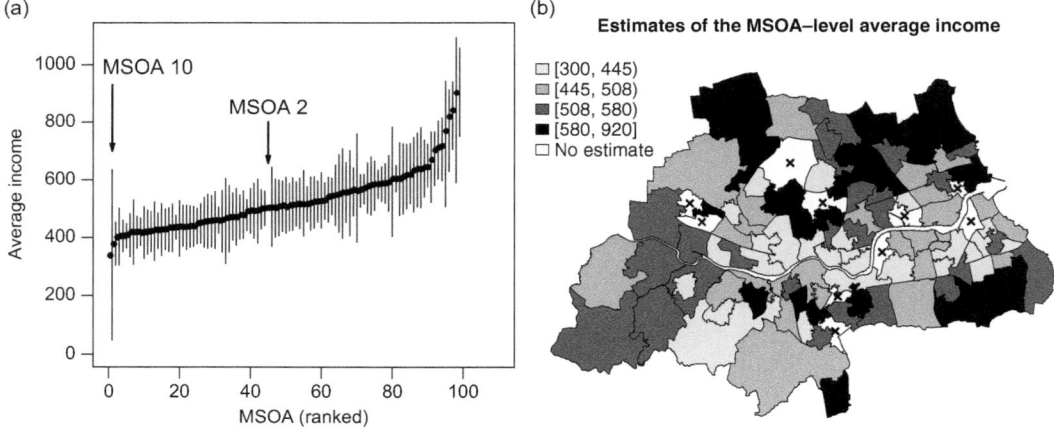

FIGURE 7.6
Summary of the results from the independent parameters model. Panel (a) shows the posterior means (the black dots) and the 95% credible intervals (the vertical lines) of the 99 MSOAs with survey data. For ease of interpretation, the estimates are in an ascending order (using the sample means). The blank space on the right is for the 10 MSOAs without data. Panel (b) is the map of the posterior means of the MSOA-level average income. The 10 MSOAs without data are located via the crosses. The elongated white space in the middle is the River Tyne.

In the next section, we start to see how the hierarchical modelling approach can help us to deal with these issues – an approach that is developed further in Chapter 8.

7.4 An Exchangeable Hierarchical Model Based on Strategy 3

Based on Strategy 3, we describe a Bayesian hierarchical model where parameters are specific to each area, and these area-specific parameters are modelled hierarchically based on the so-called exchangeable assumption. We will first present the model formulation, then explore the global smoothing nature of the model arising from the so-called exchangeable assumption.

To borrow information globally is to assume all areas within the study region are similar in terms of the variable of interest. When specifying a model, this assumption is equivalent to stating that each of the area-specific parameters, $\theta_1, \ldots, \theta_N$, follows a common prior probability distribution, but the parameters of this common prior distribution are unknown. The unknown parameters in a prior distribution are called hyperparameters, and the prior distributions assigned to the hyperparameters are called hyperpriors. Under the Bayesian framework, the hyperparameters (together with all the other parameters in the model, e.g. the area-specific parameters, $\theta_1, \ldots, \theta_N$) are estimated using the observed data. For the income data, Eq. 7.6 presents a Bayesian hierarchical model with global smoothing:

$$
\begin{aligned}
y_{ij} &\sim N\left(\theta_i, \sigma_y^2\right) \\
\theta_i &\sim N\left(\alpha, \sigma_\theta^2\right) \\
\alpha &\sim N(0,100000000) \qquad \text{for } i = 1, \ldots, 109 \\
\sigma_y &\sim Uniform(0.0001,1000) \\
\sigma_\theta &\sim Uniform(0.0001,1000)
\end{aligned}
\tag{7.6}
$$

Again, a normal likelihood is used to model the household-level income values from the survey. Each MSOA-level average income θ_i follows a common normal distribution $N\left(\alpha, \sigma_\theta^2\right)$, where α and σ_θ^2 are the two hyperparameters, representing the average income in Newcastle and the between-MSOA variability in average income, respectively. The two prior distributions, $N(0,100000000)$ for α and $Uniform(0.0001,1000)$ for σ_θ, are the hyperpriors. Note that we use the subscripts y and θ to differentiate the two variance parameters, σ_y^2 and σ_θ^2, where the former measures the household-level variability while the latter measures the MSOA-level variability. Since each θ_i follows a prior distribution, θ_i can be different from θ_j (for $i \neq j$). Yet θ_i and θ_j are not unrelated to each other because they are linked by the common probability distribution, $N\left(\alpha, \sigma_\theta^2\right)$. Conceptually, the rationale behind this model can be understood by considering the (hierarchical) structure of the data: the income values of the households in MSOA i are distributed around the average of that MSOA (thus y_{ij} is distributed as $N\left(\theta_i, \sigma_y^2\right)$ for each household in i), and the MSOA averages are distributed around (or similar to) the (global) Newcastle average (thus,

$\theta_i \sim N\left(\alpha, \sigma_\theta^2\right)$ for each MSOA in Newcastle). It is the data structure that forms the basis of this hierarchical model. Placing this model in the Bayesian framework acknowledges the uncertainty associated with the unknown parameters α, σ_y^2 and σ_θ^2.[1]

Here, we choose a normal distribution as the common prior distribution for the θ_i's because, together with the normal likelihood, we can write down an analytical form for the posterior means of θ_i. This helps clarify the logic behind information borrowing under this model (see below). Of course, in practice, similar to the likelihood, other distributions (e.g. the log-normal or the Gamma distribution) can be used. In what follows, we will first explain how information is borrowed under this hierarchical model then discuss why information is shared globally.

7.4.1 The Logic of Information Borrowing and Shrinkage

To see how information is borrowed, it is useful to write down the posterior mean of θ_i:

$$w_i \cdot \bar{y}_i + \left(1 - w_i\right) \cdot \alpha \tag{7.7}$$

where $\bar{y}_i$ denotes the sample mean in MSOA i, α is the Newcastle average and

$w_i = \dfrac{n_i \sigma_\theta^2}{n_i \sigma_\theta^2 + \sigma_y^2}$. The derivation of Eq. 7.7 can be found in Exercise 7.2. The form of Eq. 7.7 highlights the essence of information borrowing under a hierarchical model: the point estimate of θ_i is a weighted average of the sample mean, $\bar{y}_i$, and the global (Newcastle) mean, α, with the weights given respectively by w_i and $1 - w_i$. Stating this slightly differently: the estimate of θ_i uses information not only from its own unit (through $\bar{y}_i$) but also from other units through the estimation of α. The point estimate of each θ_i always lies somewhere between the sample mean for area i and the global mean.

How close θ_i is to the sample mean or the global mean depends on w_i, which is a function of the sample size in i (n_i), and the two variances, σ_y^2 and σ_θ^2. Here we will explore the role of n_i in information sharing and defer the discussion of σ_y^2 and σ_θ^2 to Section 7.4.4 when we consider the income example. Now considering σ_y^2 and σ_θ^2 fixed, Eq. 7.7 tells us that if the sample size of an area is very large (tends to infinity), w_i, the weight multiplying the sample mean, is close to 1, so the point estimate for that area is close to the sample mean. Or equivalently, since $1 - w_i$ is close to 0, the global mean, α, does not affect the point estimate, so information from other areas plays no role in the estimation of this area's parameter. Intuitively, if there is enough data in an area to estimate its own parameter, why should we want to borrow information from elsewhere? When the sample size of an

---

[1] The fully Bayes approach and the empirical Bayes approach are two ways to deal with the hyperparameter(s) in a hierarchical model. Both approaches are Bayesian in the sense that all parameters (apart from the hyperparameters) are unknown and, when combining the assigned priors with the likelihood, have posterior distributions. Under the fully Bayes approach, hyperpriors are assigned to the hyperparameters so that their uncertainty is captured through their posterior distributions. On the other hand, the empirical Bayes approach first estimates the values of the hyperparameters from the data, then estimates all other parameters in the model with the hyperparameters fixed at their estimated values (Clayton and Kaldor, 1987), effectively, ignoring the uncertainty associated with the hyperparameters. In this regard, the fully Bayes approach is better. See Bernardinelli and Montomoli (1992) for a comparison of the two approaches. Fitting the hierarchical model in Eq. 7.2 using the empirical Bayes approach would first estimate the three hyperparameters from the data: $\hat{\sigma}_y^2 = 164^2$, $\hat{\sigma}_\theta^2 = 102^2$ and $\alpha = 526$ (using the summary statistics in Table 7.2). The above values are then plugged into Eq. 7.6 to estimate the θ_is.

area is (sufficiently) large, the point estimate will be close to the sample mean, and information from other areas has little effect on this area's estimate.

However, because of data sparsity, having a very large sample of data in each and every one of the small areas does not happen in the applications considered in this book and is rarely the case, more generally, in modelling spatial and spatial-temporal data. The sample size, n_i, is rarely large. As the sample size of an area becomes smaller, the less information we will have from that area's own data. Then w_i will move towards 0 (or equivalently, 1 − w_i moves towards 1), so the point estimate of that area is pulled towards the global mean. In the extreme where there is no data point in an area (i.e. $n_i = 0$), w_i in Eq. 7.7 is 0, then the point estimate simply becomes the global mean α, resulting in a complete shrinkage. This is the effect of shrinkage (or smoothing) on the estimates of the area-specific parameters that results from implementing a hierarchical model. How much an area's estimate is shrunk towards the global mean depends on the sample size of that area. The smaller the sample size, the more shrinkage (smoothing) of its parameter estimate towards the global mean. In Section 7.4.4, we will illustrate the effect of information borrowing and shrinkage in the resulting estimates using the income data.

7.4.2 Explaining the Nature of Global Smoothing Due to Exchangeability

The global smoothing nature of the model defined in Eq. 7.6 comes from the so-called *exchangeable* assumption placed on the area-specific parameters. Stated formally, a set of parameters, $\theta_1,\ldots,\theta_N$, are assumed to be exchangeable if their joint probability distribution, $\Pr(\theta_1,\ldots,\theta_N)$, is invariant to any permutation of their indices. In other words, the labels of the parameters convey no information about the parameters. So if we model $\theta_1,\ldots,\theta_N$ as a set of exchangeable parameters, each θ_i will follow the same prior probability distribution but, in addition, we will assign prior distributions (hyperpriors) to the unknown parameters of that prior distribution. Therefore, in Eq. 7.6, $\theta_i \sim N\left(\alpha,\sigma_\theta^2\right)$ for each $i = 1,\ldots,109$, together with the hyperpriors assigned to the hyperparameters, α and σ_θ^2, specifies an exchangeable prior model on $\theta_1,\ldots,\theta_{109}$. By contrast, despite the same prior distribution being assigned to each θ_i, the specification $\theta_i \sim N(0,100000000)$ under the independent parameters model in Eq. 7.5 is *not* an exchangeable prior model. That is because (a) the parameters in that prior distribution are not unknown parameters but fixed numbers and (b) assigning a vague prior, $N(0,100000000)$, to each θ_i implies a prior assumption that each and every one of the θ_i's can be almost anywhere on the whole real line(!) – all the θ_i's are unrelated *a priori*. The exchangeable model in Eq. 7.6, on the other hand, implies a different prior structure on the parameters.

Exchangeability is a prior assumption that we place on the relationship amongst the area-specific parameters, and this prior assumption determines the nature of smoothing that these area-specific parameters receive. The exchangeable prior model, $\theta_i \sim N\left(\alpha,\sigma_\theta^2\right)$, over all MSOAs implies that, before seeing the survey data, the average income level of any MSOA is distributed around the Newcastle average (α), and the MSOA-level variance (σ_θ^2) gives us an idea of how different any MSOA-level average is from the Newcastle average. In other words, given the Newcastle average and the MSOA-level variability, all MSOAs in Newcastle are expected to have an income level that is similar to the Newcastle average. Thus, under an exchangeable prior model, all the area-specific parameters are pulled towards a global average *a priori*.

When combining this exchangeable prior model with the survey data, the Newcastle average will be estimated using all the available data across the MSOAs. Pulling the average income estimate of each MSOA towards the estimated Newcastle average means that the estimation of each MSOA's parameter borrows information from all the other MSOAs, resulting in global information sharing. As discussed in Section 7.4.1, the amount of pulling (or shrinkage) each MSOA receives depends on how much data that MSOA has: the fewer (more) data points an MSOA has, the more (less) shrinkage towards the estimated Newcastle average its parameter estimate receives. Regardless how much shrinkage each area receives, the shrinkage of any estimate is only towards the Newcastle average. Thus, an exchangeable prior model on the area-specific parameters leads to global shrinkage (or global smoothing). This global shrinkage (or global smoothing) is also evident in Eq. 7.7, where each point estimate is pulled towards the overall mean.[2]

In the context of modelling spatial data, another way of expressing exchangeability is that the estimation of θ_i receives *the same amount* of information from every other area in the study region regardless of how close (or far away) each area is to i. However, such an assumption may not be reasonable, as the spatial configuration of the areas may also offer some information in the sense that areas close together will be more alike than areas that are far apart. We refer to this as "local smoothing", which we shall look at in Chapter 8.

7.4.3 The Variance Partition Coefficient (VPC)

A hierarchical model partitions the total variability in the outcome data into different levels. For example, the model defined in Eq. 7.6 quantifies the variability at the household-level (or Level 1 in the language of multilevel modelling) and the variability at the MSOA-level (Level 2).[3] The variance partition coefficient (or VPC; also referred to as the intra-class correlation in multilevel modelling) is a useful summary for quantifying the percentage of the total variability explained by each level. The expression in Eq. 7.8 calculates the VPC measuring the percentage of the total variability attributable to the MSOA-level:

$$VPC = \frac{\sigma_\theta^2}{\sigma_y^2 + \sigma_\theta^2} \times 100 \tag{7.8}$$

A VPC that is close to zero means that only a small percentage of the total variability is explained by the MSOA-level averages, suggesting that the average income levels across MSOAs do not vary much compared to the variability at the household-level. Much of the variability is due to the variability at the household-level. A VPC that is close to 1, on the other hand, indicates that within-MSOA income is relatively homogeneous (i.e. the households within the same MSOA have similar incomes), but the differences in average income across MSOAs are large. A VPC of 0.5 suggests an equal partition of the total variability

---

[2] Another example of exchangeability is flipping a coin a few times and the probability of getting a head each time is similar to that of other flips. If the interest lies in modelling the unknown probability of heads across the coin flips, then it is reasonable to assume an exchangeable model where the unknown probabilities are independent and all follow a common distribution with the same mean and the same variance. However, if one can acquire the skill of getting a head through practice, then the sequence of probabilities of getting a head over all flips are no longer exchangeable. This is because the probability of getting a head for the 10th toss, for example, is systematically higher than that for the 1st toss due to the acquisition of that skill.

[3] Hierarchical models are also known as multilevel models because the data are structured into multiple levels (or layers) in a hierarchy. For example, the income data present a hierarchical or multilevel structure where the households are nested within different MSOAs. Another, more familiar, example is pupils nested within classrooms, which are nested within a school. In the language of multilevel modelling, the lowest level (Level 1) consists of the smallest unit in the data, so the household-level or the pupil level is Level 1. Then the MSOA-level or the classroom-level represents Level 2, and the school-level in the pupil example is Level 3 and so on.

between the two levels. The VPC is useful for informing the appropriate modelling strategy. For example, if the VPC is 0, i.e. no variability at the higher level, a model with the identical-parameters structure is appropriate. When the VPC is close to 1, getting only a few data points per area would be sufficient to make reliable inference about the area-specific parameters (because the outcome of interest is relatively homogeneous within each area). We will illustrate the calculation and interpretation of VPC using the income data next. We will also use VPC to assess the relative importance of different model components (see Chapters 8 and 15 for examples).

Figure 7.7 shows the WinBUGS implementation of the global smoothing model with exchangeable parameters. Lines 23–26 calculate the VPC defined in Eq. 7.8. Two MCMC

```
1    #  The WinBUGS code for fitting the model with exchangeable parameters
2    model {
3       #  a for loop to go through all the household level income data
4       for (j in 1:nhhs) {
5          #  defining the normal likelihood for each household
6          y[j] ~ dnorm(theta[msoa[j]],prec.y)
7       }
8       for (i in 1:nmsoas) {
9          #  each theta follows the same common normal distribution with
10         #  unknown mean and unknown variance
11         theta[i] ~ dnorm(alpha,prec.theta)
12      }
13      #  vague priors for the two hyperparameters
14      alpha ~ dnorm(0,0.00000001)
15      sigma.theta ~ dunif(0.0001,1000)
16      #  a vague prior for the sampling standard deviation
17      sigma.y ~ dunif(0.0001,1000)
18      #  calculate the two precisions
19      prec.theta <- pow(sigma.theta,-2)
20      prec.y <- pow(sigma.y,-2)
21
22      #  calculate the variances for the VPC calculation
23      var.y <- pow(sigma.y,2)
24      var.theta <- pow(sigma.theta,2)
25      #   calculate VPC defined in Eq. 7.8
26      vpc <- var.theta / (var.theta + var.y) * 100
27   }
28
29   #  household-level income data from survey
30   list(nhhs=760
31        ,nmsoas=109
32        ,y=c(501,616,472,816,637,500,506,560,542,447,644,522
33            ,487,275,...)
34        ,msoa=c(1,1,1,1,1,2,2,4,4,4,4,5,5,...))
35
36   #  initial values for chain 1
37   list(alpha=200,sigma.y=50,sigma.theta=20,theta=c(100,100,...))
38
39   #  initial values for chain 2
40   list(alpha=700,sigma.y=80,sigma.theta=50,theta=c(700,700,...))
```

FIGURE 7.7

The WinBUGS code, the data list and the two sets of initial values for fitting the global smoothing model with exchangeable parameters as defined in Eq. 7.6. The data list (same as that for the identical parameter model in Figure 7.3) and the sets of initial values are shown in part for the purpose of illustration.

chains, each with 10000 iterations, were run and the iterations from the second halves of both chains used to produce the results which we now consider.

7.4.4 Applying an Exchangeable Hierarchical Model to the Newcastle Income Data

The posterior distributions of the θ_i's are summarised in Figure 7.8(a) using the posterior mean and the 95% CI. For ease of comparison, the estimates from the model with independent parameters are shown in Figure 7.8(b). Compared to the estimates from the independent parameters model, the uncertainty intervals from the model with exchangeable parameters are more consistent in length, neither very wide nor very narrow. This is a consequence of information sharing. For example, for the estimates indicated with a solid arrow, the information borrowed from other MSOAs strengthens the highly-uncertain estimates that were obtained under the independent parameters model. Uncertainty has been reduced and the very wide uncertainty intervals shortened. At the same time, because of information sharing, very narrow intervals are widened (see, for example, those with a dotted line arrow), representing (perhaps more appropriately) greater uncertainty. Recall example area MSOA 2 mentioned in the results from the independent model. The 95% CI now is (384.6, 644.8), as opposed to the unrealistically narrow uncertainty interval of (497.1, 509.0) when only using its own two data points. Under the exchangeable model, the 10 MSOAs without data now have posterior distributions. The information for these MSOAs comes solely from the common prior distribution, $N\left(\alpha, \sigma_\theta^2\right)$, thus, as shown on the right-hand side of Figure 7.8(a), their posterior distributions are virtually identical.

Figure 7.9 demonstrates the effect of global shrinkage arising from the exchangeable assumption, which pulls the small area means from the independent parameters model towards the global Newcastle mean. However, a careful inspection reveals that different MSOA income levels receive different amounts of shrinkage. For example, the two MSOAs highlighted (MSOAs 27 and 57) have similar point estimates (i.e. sample means) under the independent parameters model, but, under the exchangeable model, the posterior mean of MSOA 57 is shrunk (slightly) more towards the Newcastle mean compared to that of MSOA 27. This is because there are 10 data points in MSOA 27 compared to six in

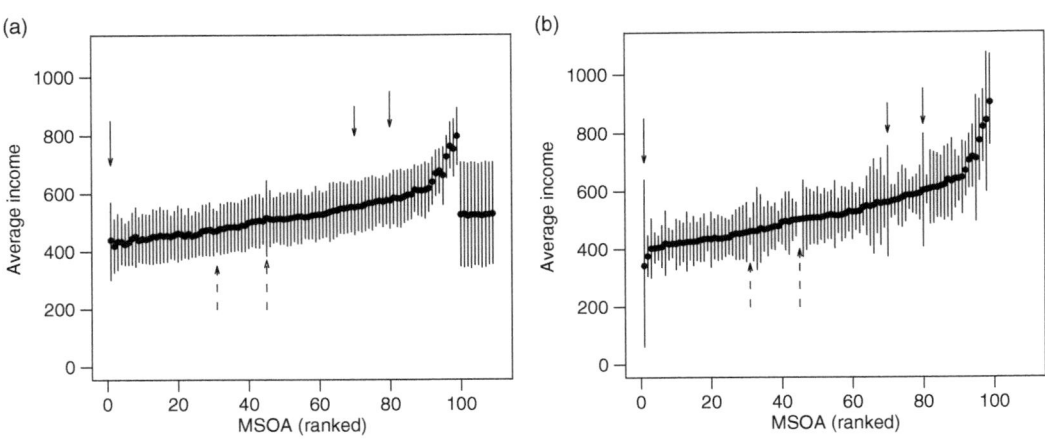

FIGURE 7.8
Summary of the posterior estimates for the MSOA-level average income based on (a) the hierarchical model with global smoothing and (b) the non-hierarchical model with independent parameters. In each plot, the black dots and the vertical lines respectively denote the posterior means and the 95% credible intervals for the θ_i's.

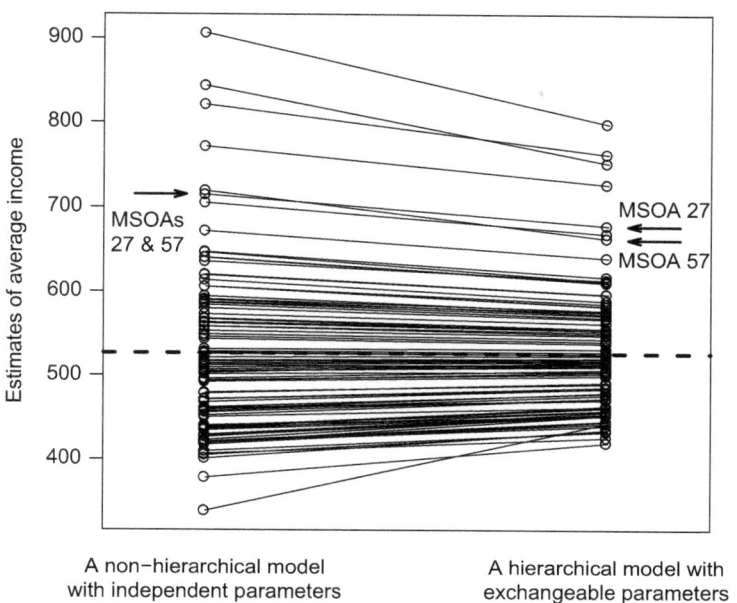

FIGURE 7.9
An illustration of shrinkage from a hierarchical model through the comparison of the posterior means of the MSOA-level average income from the independent parameters model and from the exchangeable parameters model. The horizontal dashed line shows the posterior mean of the Newcastle average from the exchangeable model. The estimates for the 10 MSOAs without data are not shown here.

MSOA 57. With more information available from the area itself, the estimate for MSOA 27 borrows less information from the common distribution and hence undergoes less shrinkage. For MSOA 57, on the other hand, there is less information available from the area to draw on (six sample points compared to 10), thus its estimate is pulled more towards the global mean. As is evident from Figure 7.9, different amounts of shrinkage result in some MSOAs becoming repositioned in the rank order of MSOAs by average income.

Recall Eq. 7.7; the amount of shrinkage towards the global mean is determined by the sample size, n_i, and the variances, σ_y^2 and σ_θ^2, at the household-level and at the MSOA-level, respectively. We carry out the following calculation to illustrate how these three parameters affect the amount of shrinkage. From the income data, the posterior means

of σ_y^2 and σ_θ^2 are 19127 and 8165, respectively. Using $w_i = \dfrac{n_i \sigma_\theta^2}{n_i \sigma_\theta^2 + \sigma_y^2}$ (defined in Eq. 7.7),

when n_i is about 2.3, an equal weight is assigned to both the small area mean and the global mean.[4] In other words, the point estimate of an MSOA with two or fewer data points is closer to the global mean than to the sample mean, whereas the point estimate of an MSOA with three or more data points remains closer to its sample mean. Following the same calculation, if $n_i > 231$ then $w_i = 0.99$, and the point estimate under this model for MSOA i will be almost identical to the sample mean of the data reported for MSOA i. In this dataset,

---

[4] We can rearrange $w_i = \dfrac{n_i \sigma_\theta^2}{n_i \sigma_\theta^2 + \sigma_y^2}$ to obtain $n_i = \dfrac{w_i \sigma_y^2}{\sigma_\theta^2 (1 - w_i)}$. When $w_i = 0.5$, $n_i = \dfrac{\sigma_y^2}{\sigma_\theta^2}$, which is the (inverse of the) variance ratio introduced in Gelman and Hill (2007, p.258) for interpreting the variances at different levels.

however, none of the MSOAs have such a large number of data points. All the estimates, to varying degrees, are shrunk towards the Newcastle average. Finally, the VPC, another way to interpret the estimated variances at different levels, is estimated to be 29.7%, with a 95% CI of (21.7%, 38.4%), suggesting there is more variability in the income data attributable to the household-level, compared to the MSOA-level.

7.5 Concluding Remarks

As we first discussed in Chapter 1, heterogeneity and data sparsity are two challenges that commonly arise when estimating a large number of small area parameters. In this chapter we have seen that it is not possible to meet both these challenges using non-hierarchical methods – Strategies 1 and 2 described at the beginning of Section 7.2. Strategy 1, complete pooling, whilst addressing the data sparsity issue, yields only a single global estimate (the Newcastle average) for all the small areas, thus not addressing the small area level heterogeneity property that is clearly present in the Newcastle data. Strategy 2, no-pooling, yields an estimate for each small area. Whilst addressing the heterogeneity issue, this strategy does not address the data sparsity issue because each area is treated independently of the rest, thus the estimate for each area only uses the data associated with that area. As a result, for areas that have no sample data, we end up with no estimate. For areas with only a few sample data points, the estimates are unreliable, as they rely too much on the limited data (see the discussion related to Figure 7.6).

We have taken our first steps to resolve these issues – hierarchical modelling in which, under Strategy 3, the estimate for any area is obtained by pooling the data from that area with the global average. Strategy 3 can be considered as a combination of Strategies 1 and 2. We obtain a small area estimate that is a weighted average of the estimates from Strategies 1 and 2 where the degree of weighting varies from small area to small area depending on data properties, thus going some way to meeting both the heterogeneity and data sparsity issues. In the next chapter we describe what more Strategy 4 offers us.

7.6 Exercises

Exercise 7.1. Fit the two non-hierarchical models to the Newcastle income data in WinBUGS. Produce the two plots in Figure 7.6 using the posterior estimates from the independent parameters model.

Exercise 7.2. Derive the expression in Eq. 7.7 for the posterior mean of each θ_i under the exchangeable model as formulated in Eq. 7.6.

Exercise 7.3. Fit the exchangeable model to the Newcastle income data in order to (a) verify the expression for the posterior mean in Eq. 7.7; (b) produce the plot in Figure 7.8(a); and (c) produce the shrinkage plot in Figure 7.9 together with the posterior estimates from the independent parameters model.

Exercise 7.4. Modify the WinBUGS code in Figure 7.7 to rank these 109 MSOAs based on their average income estimates. Based on the rank estimates, can you distinguish one MSOA from another? Now, suppose that an MSOA is considered to be income deprived if its average income level is 60% below the Newcastle average. Calculate the posterior probability of each MSOA being income deprived. How easy (or difficult) is it to identify the deprived areas using the resulting posterior probabilities?

Exercise 7.5. Discuss ways to compare the Bayesian models considered in this chapter given that the purpose of these models is to provide (reliable) estimates of average income across the small areas in a study region.

7.7 Appendix: Obtaining the Simulated Household Income Data

The household-level income data used in this chapter (and also in Chapter 8) are simulated based on a situation where a set of household-level data are collected through a national survey, and the survey data are then disaggregated to a small area level (the middle super output areas or MSOAs) to provide small area estimates of a characteristic of interest (average household income in each MSOA). There are two features that we wish to create in the simulation, and both are related to the data sparsity challenge that we discussed in Section 1.2.3. First, due to the lack of spatial stratification at the small area level, some areas do not have any survey data. So, in the simulation, we do not assign any income values to 10 selected MSOAs. These MSOAs are 3, 16, 28, 33, 35, 41, 78, 84, 88 and 96 in Figure 7.1. The second feature is the small number of data points (households) per MSOA. We set the number of households from the national survey for Newcastle to be 760, roughly 0.2% of the total number of households in Newcastle (based on the 2011 UK census). Each of these 760 households is equally likely to be assigned to any one of the 99 MSOAs in Newcastle. Table 7.1 summarises the sample sizes across the 109 MSOAs.

The income value for each of the 760 households is simulated based on the small area income estimates for 2007–08 produced by the Office for National Statistics (ONS; 2016). In the ONS database, each MSOA, i, in Newcastle (also available for all other MSOAs in England and Wales) is given a point estimate (denoted μ_i) for its average weekly household total gross income (in £) and an associated 95% confidence interval with the lower and the upper bounds of that interval denoted as l_i and u_i, respectively. Thus, for a household in MSOA i, its income value is drawn from a normal distribution with mean μ_i and standard deviation $\left[\dfrac{(l_i - u_i)}{4} \right] \times 3$. The quantity, $\left[\dfrac{(l_i - u_i)}{4} \right]$, is the (estimated) standard error, and the multiplicative factor 3 is to inflate the variability so that the within-area variability of household income is three times larger than the standard error. The database used to simulate the household-level data is available online.

8

Bayesian Models for Spatial Data II: Hierarchical Models with Spatial Dependence

8.1 Introduction

One of the fundamental properties of spatial data is the property of spatial dependence – data values close together in geographical space tend to be more alike than data values that are further apart.[1] For the purpose of what is to follow in this chapter, we can express this property in a slightly different way: an observation on a variable at location i carries some information about what is observed for the same variable in areas that are close to i.[2] Now, if this property is evident in spatial *data* as identified, for example, through an exploratory spatial autocorrelation analysis (see Chapter 6), is it not reasonable to assume that it will be a property of *parameters* too – so that, for example, average household income levels (i.e., the θ_is as in Chapter 7) will be more alike in MSOAs that are close together than in MSOAs that are far apart? We argue that this assumption can be invoked to further strengthen and improve small area estimates. The exchangeable, global, hierarchical model in Chapter 7 (Strategy 3) borrowed information from all other areas in the study region. Now, we propose to be more discriminating, but how can this spatial approach to borrowing strength be implemented? We turn to that in this chapter.

As discussed in Chapter 7, the options we consider for information sharing are based on prior assumptions that we place on the spatial dependence structure of the area-specific parameters. To express such spatial dependency so that we are able to implement these processes of information sharing, we need models. The discussion of such models forms a significant part of this chapter. We draw on the Newcastle income data to provide the illustrative example, and the reader is referred back to Section 7.2 for details of the dataset and Strategy 4 (see Figure 7.2 in Chapter 7).

We shall discuss various spatial models for localised information sharing, all of them involving some form of the conditional autoregressive (CAR) modelling structure – a modelling structure, like the simultaneous autoregressive (SAR) structure to be encountered in Chapter 10, commonly used to capture spatial dependence when analysing data associated with irregular polygons (e.g. outcomes attached to census tracts) or fixed points (such as a set of retail outlets).

The CAR structure imposes a neighbourhood structure on the spatial units using a spatial weights matrix (Chapter 4), and it is through that neighbourhood structure that local information sharing is achieved. We discuss the intrinsic conditional autoregressive

---

[1] As expressed informally in Tobler's First Law of Geography (see Chapter 3).
[2] This way of looking at spatial dependence underlies geostatistics and in particular methods for spatial interpolation including kriging (see Chapter 2).

(ICAR) model in Section 8.2, a version of the CAR modelling structure that is widely used in practice, and the proper CAR (pCAR) model in Section 8.3. We shall consider the key issues that arise when specifying binary and general weights matrices in the ICAR model (Sections 8.2.1 and 8.2.2). In Section 8.4, we look at adaptive ICAR models, which allow the elements in the spatial weights matrix to be estimated using data. In Section 8.5, we introduce the Besag-York-Mollie (BYM) model, which combines both an exchangeable model and the ICAR model, so that borrowing information is carried out both globally *and* locally. In section 8.6, we summarise all the modelling results and provide some insights into the application of these different modelling options.

8.2 The Intrinsic Conditional Autoregressive (ICAR) Model

The presence of positive spatial autocorrelation implies that the attribute values of areas that are geographically close together are more alike than those that are geographically widely separated. For the purpose of estimation, the presence of spatial dependence implies that we can borrow strength "locally" so that the parameter of area i will be similar to those from the *neighbours* of this area. The Intrinsic Conditional AutoRegressive (ICAR) model (Besag et al., 1991) is widely used to give a spatial structure to a set of area-specific parameters, $S_1, ..., S_N$, over N areas within a study region. Here, we denote the area-specific parameters as $S_1, ..., S_N$, as opposed to $\theta_1, ..., \theta_N$, to emphasise that these parameters are spatially structured. $S = (S_1, ..., S_N)$ denotes a collection of these area-specific parameters. The ICAR model is used as a *prior model* to impose a spatial dependence structure on S.

However, this does not entirely resolve the issue, because the ICAR model depends on the selected spatial weights matrix, W (see Chapter 4). The choice of W will have a significant impact on how information is borrowed amongst areas, and as we saw in Chapter 4, there are many ways to define a spatial weights matrix. In Section 8.2.1, we discuss the ICAR model that defines a neighbouring structure using a spatial weights matrix with binary (0 or 1) entries. In Section 8.2.2, we discuss the ICAR model using a general form of the spatial weights matrix, the entries of which are continuous-valued.

8.2.1 The ICAR Model Using a Spatial Weights Matrix with Binary Entries

In many statistical applications, the spatial structure of a map is often (but not always) defined through spatial contiguity. For example, recall from Section 4.2, a spatial weights matrix, W, defined based on rook's move contiguity consists of 0/1 entries, where $w_{ij} = 1$ if areas i and j share a common border (or edge) and $w_{ij} = 0$ otherwise. Also, $w_{ii} = 0$ for all $i = 1, ..., N$. Using the ICAR model with a rook's move W matrix induces a particular spatial dependence structure on the area-specific parameters. The choice of model (ICAR) and the choice of spatial weights matrix W, together, will have a strong influence on the geography of the spatial smoothing. Choosing a different spatial weights matrix, for example using the queen's move definition of contiguity, will result in a (slightly) different spatial smoothing.[3] Other neighbourhood definitions based on geographical distance, for example, $w_{ij} = 1$ if the distance between the centroids of areas i and j is less than a predefined

---

[3] In the case of irregular spatial units, there may be very few instances of spatial units joined at a vertex, so in practice these two definitions of the weights matrix typically lead to very similar if not identical smoothings.

threshold and $w_{ij}=0$ otherwise, will again yield a binary weights matrix but will produce yet another pattern of local smoothing.[4]

The ICAR model defines a *conditional* normal distribution for the parameter of each area i so that the specification of this distribution depends on the parameters of the neighbours of i. Based on a binary weights matrix, the conditional distribution for each S_i ($i=1, ..., N$) is given by

$$S_i \mid \boldsymbol{S}_{\{-i\}}, v, \boldsymbol{W} \sim N\left(\frac{\sum_{j \in \Delta i} S_j}{m_i}, \frac{v}{m_i}\right) \tag{8.1}$$

where

$\boldsymbol{S}_{\{-i\}}$ denotes the set of area-specific parameters excluding S_i, namely, $\boldsymbol{S}_{\{-i\}} = (S_1, ..., S_{i-1}, S_{i+1}, ..., S_N)$.

v is an unknown variance parameter whose role in this model will be discussed later

Δi denotes the set of neighbours of area i as defined in $\boldsymbol{W}$.

m_i is the number of neighbours that area i has. It equals the sum of the ith row of the $\boldsymbol{W}$ matrix and is also denoted as w_{i+}.

The conditional specification given in Eq. 8.1 implies a local dependence structure that leads to a local or neighbourhood smoothing of the area-specific parameters. Given the parameters of all other areas, the conditional mean of S_i, namely, $\dfrac{\sum_{j \in \Delta i} S_j}{m_i}$, only depends on the parameters of the neighbouring areas, implying a local dependence structure rather than a global dependence structure (in the latter case, the conditional mean of S_i would depend on all other S_j with $j \neq i$). In addition, the conditional mean is taken as the mean of the parameters of the neighbouring areas so that the parameters are similar locally. When applied to the income modelling, the ICAR model defined in Eq. 8.1 assumes that MSOAs that are close in space have similar average income levels. In effect, the ICAR model assumes positive spatial autocorrelation on the area-specific parameters, an assumption that should be tested at the exploratory stage of data analysis (see Chapter 6).

Positive spatial dependence induces local smoothing in the process of estimating the area-specific parameters. This local smoothing is evident in the conditional mean of S_i, which is set to be the mean of the parameters in the neighbouring areas. Instead of pulling towards the global mean as in the global smoothing model defined in Eq. 7.6 and Eq. 7.7 in Chapter 7, the ICAR model pulls the estimate of S_i towards the mean of the parameters in the neighbouring areas, thus smoothing the parameter estimates locally. For example, if we know (or are informed) that the average income levels of the three neighbours of area i are, say, 610, 570 and 620, then, prior to obtaining data for area i, we would expect the average income of area i to be around 600, the mean of its neighbours. Once we have obtained some data for this area, the posterior estimate for this area will lie between its sample mean and the average of its three neighbours. The question is, how close would the estimated income level be to the average of its neighbours? In other words, what are the factors that can affect the amount of local shrinkage under the ICAR model?

As in the case of global smoothing, the larger the sample size, n_i, the less local shrinkage a parameter will receive. Expressing this from a Bayesian standpoint, the ICAR model acts as a prior distribution for the area-specific parameters. If the dataset is sufficiently large,

---

[4] The ICAR model also requires the $\boldsymbol{W}$ matrix to be symmetric. Therefore $\boldsymbol{W}$ cannot be derived based on the k nearest neighbours or spatial interactions discussed in Chapter 4. This requirement will be discussed in Section 8.2.2.

the likelihood dominates the posterior distribution, and hence the prior distribution plays a minimal role. Another factor that can affect the amount of local shrinkage is m_i, the number of spatial neighbours of area i. Eq. 8.1 suggests that the more neighbours an area has, the more local smoothing it will receive. This is because, compared to an area that has only one or two neighbours, an area with a large number of neighbours (a large m_i), the conditional variance, $\frac{v}{m_i}$, as in Eq. 8.1, is small. Then its parameter, S_i, is more likely to be close to the average of the area's neighbours. In the extreme (although unrealistic) case where m_i tends to infinity (i.e. area i an infinite number of neighbours), the conditional variance tends to 0. This is equivalent to placing a strong prior that the estimate of S_i simply takes the value of the local neighbourhood average, even before obtaining data for this area. When strong positive spatial autocorrelation is present everywhere on the map, it is reasonable to assume that more neighbours lead to more local smoothing. This is because when the neighbours are shown to be similar, more neighbours means that more data can be appropriately borrowed. However, when there is weak or no spatial autocorrelation, or when the (strong) spatial autocorrelation is only present in some part of the study region but not elsewhere, such a "global" effect of m_i on local shrinkage may not be appropriate. The latter situations lead to the discussion of adaptive smoothing, a topic that we shall return to in Section 8.4.

Finally, the variance parameter v controls the amount of local smoothing for the entire study region. Specifically, when v is large, the conditional variances for all the parameters in S are also large. The amount of local smoothing will, as a result, reduce for all areas, giving a less smooth-looking map of the posterior means of the parameters in S. This reduction of smoothing applies to every area regardless of its neighbourhood structure. This variance parameter v is the only unknown parameter in the ICAR model, and we will provide an example of its prior distribution in the analysis of the income data.

It is important to point out that local information sharing is reciprocated. This means that when the estimation of S_i borrows information from area j, a neighbouring area of i, then at the same time, the estimation of S_j borrows information from area i. This reciprocity is important and explains why the neighbourhood smoothing process is *not* limited to just those neighbours defined by the spatial weights matrix $\mathbf{W}$. When estimating S_i, information is borrowed not only from the neighbours of area i as defined in $\mathbf{W}$ but also from its neighbours' neighbours, and so on. However, it is also the case that the *amount* of information borrowed from areas further away is less than from areas that are nearby (see also Section 4.9.1).

8.2.1.1 The WinBUGS Implementation of the ICAR Model

In WinBUGS, the ICAR model is implemented via the `car.normal` function, and the syntax is given below:

```
S[1:N] ~ car.normal(adj[], weights[], num[], precision)
```

where N denotes the number of areas in the study region. The first three arguments of the `car.normal` function, `adj[]`, `weights[]` and `num[]`, together define the chosen spatial weights matrix $\mathbf{W}$. Specifically, `num[]` is an array of size N with the ith entry showing the number of neighbours of area i. The IDs of the neighbours and their associated weights are stored in `adj[]` and `weights[]`, respectively. See the Appendix 4.13.2 in Chapter 4 for the steps to create these three data arrays in R. We will provide an example of the three arrays derived from the Newcastle MSOA map later in Figure 8.1.

The last term in the `car.normal` function, `precision`, is the precision parameter, the inverse of the variance parameter v in Eq. 8.1 (i.e. *precision* $= 1/v$). This is a hyperparameter

```
 1   #  The WinBUGS code for specifying the model with the ICAR prior
 2   model {
 3     #  a for-loop to go through all the household level income data
 4     for (j in 1:nhhs) {
 5       #  defining the normal likelihood for each household
 6       y[j] ~ dnorm(theta[msoa[j]],prec.y)
 7     }
 8     for (i in 1:nmsoas) {
 9       #  each theta is the sum of the Newcastle average and the
10       #  corresponding S[i]
11       theta[i] <- alpha + S[i]
12     }
13     #  modelling all the parameters in S using the ICAR model
14     S[1:nmsoas] ~ car.normal(adj[],weights[],num[],prec.S)
15
16     #  the improper uniform prior on the whole real line for the
17     #  intercept
18     alpha ~ dflat()
19
20     #  a vague prior for the hyperparameter in the ICAR model
21     sigma.S ~ dunif(0.0001,1000)
22
23     #  a vague prior for the sampling standard deviation
24     sigma.y ~ dunif(0.0001,1000)
25
26     #  calculate the two precisions
27     prec.S <- pow(sigma.S,-2)
28     prec.y <- pow(sigma.y,-2)
29
30     #  calculate the household-level variance for the VPC calculation
31     var.y <- pow(sigma.y,2)
32     #  calculate the MSOA-level variance (the unconditional variance
33     #  of the S parameters) for the VPC calculation
34     var.S.un <- sd(S[1:nmsoas]) * sd(S[1:nmsoas])
35     #  calculate VPC defined in Eq. 7.8 in Chapter 7
36     vpc <- var.S.un / (var.S.un + var.y) * 100
37   }
38
39   #  household-level income data from survey with a spatial
40   #  neighbourhood structure
41   list(nhhs=760,
42   nmsoas=109,
43   y=c(501, 616, 472, 816, 637, 500, 506, 560, 542, 447, 644, 522,
44      ,487, 275,...),
45   msoa=c(1,1,1,1,1,2,2,4,4,4,4,4,5,5,...),
46   #  elements of the W matrix (defined via rook's move spatial
47   #  contiguity)
48   num=c(1,6,4,...),
49   adj=c(2,
50        1,3,4,11,23,24,
51        2,4,20,22,
52        ...),
53   weights=c(1,1,1,1,1,1,1,1,1,1,1,...)
54   )
55
56   #  initial values for chain 1
57   list(alpha=200,sigma.y=50,sigma.S=20,S=c(1,-1,0,0,0,...))
58
59   #  initial values for chain 2
60   list(alpha=700,sigma.y=80,sigma.S=50,S=c(-1,1,0,0,0,...))
```

FIGURE 8.1

The WinBUGS code, the data list and the two sets of initial values for fitting the local smoothing model defined in Eq. 8.2 with the ICAR model using rook's move contiguity. The data list and the sets of initial values are shown only in part for the purpose of illustration.

for which a hyperprior is required. The `car.normal` function specifies a joint distribution for the entire set of area-specific parameters, which is denoted as `S[1:N]` (on the left-hand side of the expression).

The `car.normal` function in WinBUGS imposes the so-called sum-to-zero constraint on S. That is, $\sum_{i=1}^{N} S_i = 0$. If we define the following line in WinBUGS and monitor the node `sum.S`, the value of this node is always 0.

$$\texttt{sum.S <- sum(S[1:N])}$$

This sum-to-zero constraint is required because the joint distribution for all the parameters in S (which can be derived from the conditional distributions given in Eq. 8.1) is an improper probability distribution, meaning that the joint distribution does not integrate to 1. We will defer discussion of the impropriety of the joint distribution to Section 8.2.2.2. Because of the sum-to-zero constraint, we always need to include a separate intercept term, say α. The prior distribution for the intercept must be the improper uniform prior defined on the whole real line, i.e. $\alpha \sim Uniform(-\infty, +\infty)$ (see Appendix 1 – under the Intrinsic CAR model section – of the GeoBUGS manual and also Section 8.2.2.2). This improper uniform distribution is coded as the `dflat()` function in WinBUGS.

We now turn to the Newcastle income example to illustrate the implementation of a Bayesian hierarchical model with the ICAR model using a binary W matrix derived from spatial contiguity.

8.2.1.2 Applying the ICAR Model Using Spatial Contiguity to the Newcastle Income Data

Recall y_{ij} is the income value for household j in MSOA i ($i=1, \ldots, 109$). In Eq. 8.2, the (unknown) average income in MSOA i, θ_i, is modelled as the sum of α, the Newcastle average, and S_i. Because of the sum-to-zero constraint, each S_i is either above or below (or equal to) 0, and thus θ_i will be either above or below (or the same as) the Newcastle average accordingly. In other words, the term S_i measures the deviation of the income level in area i from the global average – the set $S = (S_1, \ldots, S_{109})$ is also called a set of random effects because they vary from one area to another and they follow a common probability distribution with unknown hyperparameters. We can in fact rewrite the exchangeable model in Eq. 7.6 (Chapter 7) in the same format so that the MSOA average income is a sum of the global average and a local deviation. We will explore that formulation further when we discuss the BYM model in Section 8.5.

$$y_{ij} \sim N(\theta_i, \sigma_y^2)$$

$$\theta_i = \alpha + S_i$$

$$S_{1:109} \sim ICAR\left(W, \sigma_S^2\right)$$

$$\alpha \sim Uniform(-\infty, +\infty)$$

$$\sigma_y \sim Uniform(0.0001, 1000)$$

$$\sigma_S \sim Uniform(0.0001, 1000) \tag{8.2}$$

In this model, the prior distribution for $S = (S_1, \ldots, S_{109})$ is the ICAR model with the spatial structure defined by the spatial weights matrix W. In this example, W is defined using rook's move contiguity and the elements of the W matrix are entered into WinBUGS as data. σ_S^2 is the variance parameter v in Eq. 8.1. A vague uniform prior, Uniform $(0.0001, 1000)$, is assigned independently to the (conditional) standard deviation of the random effects, σ_S, and the household-level standard deviation, σ_y. The intercept α has the improper uniform prior, Uniform $(-\infty, +\infty)$.

Figure 8.1 summarises the WinBUGS code for the model, the data list and the initial values. For the model specification, Line 14 uses the `car.normal` function to assign the ICAR model to the vector of parameters $S = (S_1, \ldots, S_{109})$. Line 36 calculates the VPC to compare the partition of the variance in the data between the household-level and the MSOA-level. Note that the MSOA-level variance, measuring the variability of average income across MSOAs, is calculated using the unconditional variance of the parameters S (see Line 34; the function `sd` calculates the standard deviation of a vector of values). We cannot use σ_S^2 for the VPC calculation because it measures the variability of the *conditional distribution* for S (see Eq. 8.1), rather than the *unconditional* variability of S.

We now turn our attention to the three arrays, `adj[]`, `weights[]` and `num[]`, in the data list. The three arrays are derived from the off-diagonal elements in the W matrix. Using rook's move contiguity, $w_{ij} = 1$ if MSOAs i and j share a common border, and otherwise $w_{ij} = 0$ (for $i \neq j$). WinBUGS automatically assigns 0 to all the diagonal elements in W. Using the Newcastle map in Figure 7.1 (Chapter 7), we can see that MSOA 1, the polygon at the southern tip of Newcastle, has only one neighbour, which is MSOA 2. Hence, the first entry of `num[]` is 1, meaning one neighbour for MSOA 1, and the first entry of `adj[]` is 2, the label (or ID) for that single neighbour. MSOA 2 has six neighbours, namely, MSOAs 3, 4, 24, 11, 23 and 1. Thus the second entry in `num[]` is 6 and the second to the seventh entries of `adj[]` are the labels of its neighbours. The above procedure is then used to derive the elements in `num[]` and `adj[]` for all the remaining MSOAs.

The elements in the array `weights[]` define the weights assigned to the neighbours of each MSOA. With a binary weights matrix, the weights of all neighbours are set to 1. This assumes all the neighbours of MSOA i contribute equally towards the conditional mean, and hence the estimation of S_i. For example, the conditional mean for MSOA 1 is $\dfrac{S_2}{1}$, and the conditional mean for MSOA 2 is $\dfrac{S_1 + S_3 + S_4 + S_{11} + S_{23} + S_{24}}{6}$. When using a binary W matrix, the weights of the neighbours are fixed to 1. However, in some situations, we might want to assign different weights to the neighbours based on, for example, the length of the common border or some socio-economic factors so that the extent of information sharing depends on not only the neighbourhood structure but also the similarity between neighbours based on some chosen set of local characteristics that are considered to be relevant to the analysis. In Section 8.2.2, we will explore a more general specification of the ICAR model using a more general form of weighting.

The initial values for the set of parameters S also need to be subjected to the sum-to-zero constraint. A simple way to achieve this, as shown in Figure 8.1, is to set the first element of S to 1, the second element to −1 and then set the rest to 0 (Line 57). For a second set of initial values, the same trick is used, but this time the positions of the values 1 and −1 are swapped (i.e., −1 and 1 for the first and the second elements, respectively) so that the initial values of S in the second set are different from those in the first set.[5]

---

[5] For a set of parameters modelled using the ICAR model, WinBUGS does not require the user to supply a set of initial values that sum to 0. However, OpenBUGS does. Here, we recommend specifying the initial values to sum to 0 so that the same set of codes can also be fitted in OpenBUGS.

8.2.1.3 Results

Figure 8.2 compares the posterior means from the Bayesian hierarchical model with the ICAR prior (Eq. 8.2) to those from the non-hierarchical model with independent parameters (Eq. 7.4 and Eq. 7.5 in Chapter 7). The first thing to remark is that under the ICAR model, each point estimate is smoothed towards the mean of its neighbouring areas, instead of towards the global mean. Initially this can be spotted by noticing that the lines, going from left to right, do not necessarily converge towards the Newcastle average, which is indicated by the horizontal dashed line. For example, for the lines lying above the Newcastle mean, a close inspection of Figure 8.2 shows that some lines are tilted upward while some are tilted down. Using a selection of MSOAs, Figure 8.3 further illustrates the local smoothing nature of the ICAR model. For each MSOA, the posterior mean of the average income (the triangle at the bottom of the plot) always lies between the sample mean (the cross) and the mean of its neighbours (the inverted triangle). In addition to the point values, in Figure 8.3, we have also shown the posterior distributions of the average income estimated for MSOA *i* and the mean income of that MSOA's neighbours to emphasise that each of these two quantities is associated with a probability distribution.

Figure 8.4 demonstrates the effect of the number of neighbours on the amount of local shrinkage. The latter, as defined along the vertical axis of Figure 8.4, is measured by $r_i = d_{i1}/d_{i2}$, where d_{i1} is the absolute difference between the posterior mean of θ_i and the sample mean and d_{i2} is the absolute difference between the posterior mean of θ_i and the local mean, which is taken as the average of the posterior means of the contiguous neighbours of *i*. The calculation of d_{i1} and d_{i2} can be seen in each panel of Figure 8.3 as follows: d_{i1} measures how far the triangle is from the cross and d_{i2} measures how far the triangle is

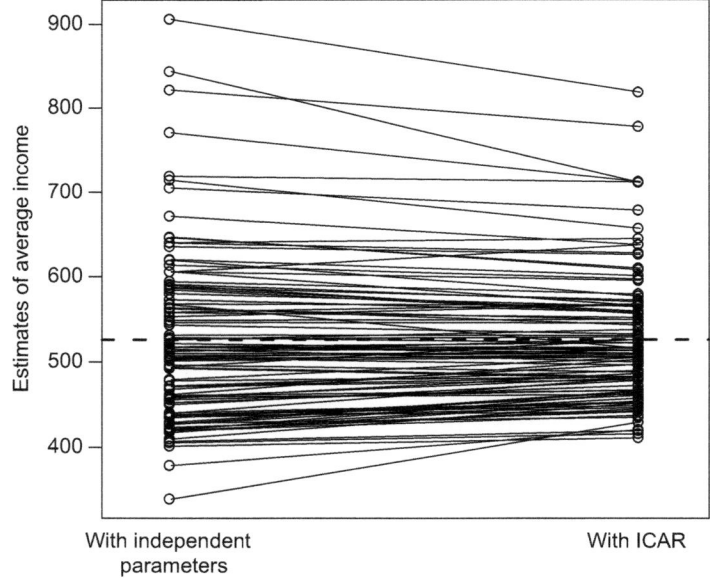

FIGURE 8.2
A shrinkage plot comparing the point estimates from the independent-parameters model (see Section 7.3.2) with those from the local smoothing model using the ICAR model and with the *W* matrix based on areas sharing a common border. This plot only shows the 99 MSOAs with survey data.

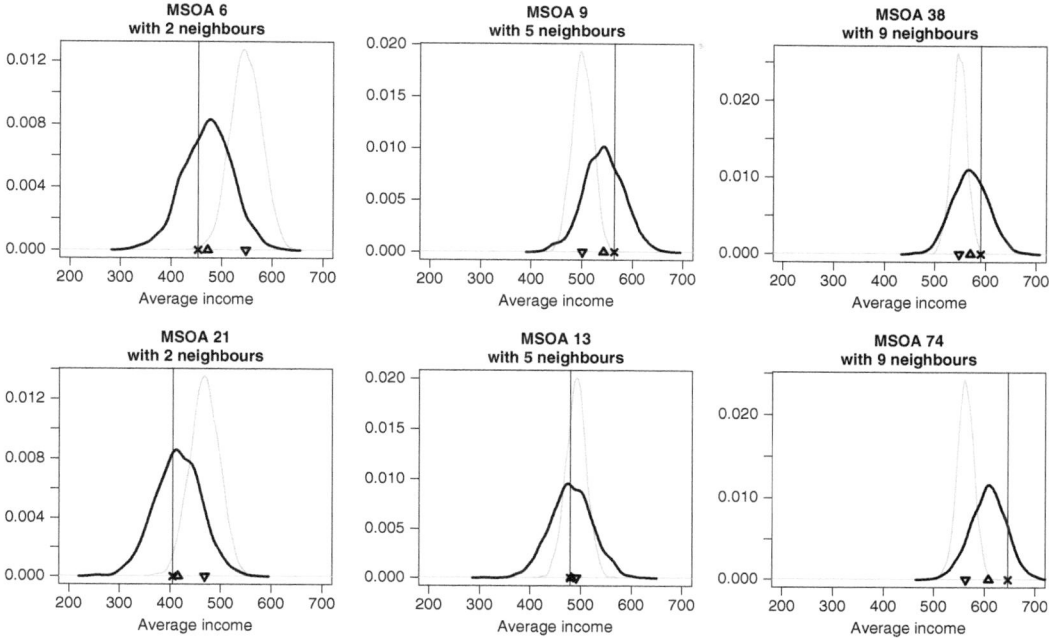

FIGURE 8.3

An illustration of the local smoothing nature under the ICAR model with six selected MSOAs. In each panel, the black curve denotes the posterior distribution of the average income θ_i in MSOA i, and it is constructed using the MCMC iterations for θ_i from WinBUGS. The grey curve is the corresponding conditional distribution of θ_i as defined in Eq. 8.1. That conditional distribution is calculated based on the posterior distributions of the θ_js from the neighbouring MSOAs ($j \in \Delta i$). That is, for each MCMC iteration, the mean of the sampled values of the θ_js is calculated, and the grey distribution is the distribution of those mean values. The vertical line indicates the sample mean from the survey data for MSOA i. At the bottom of each plot, the sample mean, the posterior mean and the conditional mean (given the neighbours) are denoted by a cross, a triangle and an inverted triangle, respectively. Because of local smoothing, the posterior mean (the triangle) always sits between the sample mean (the cross) and the mean of the neighbours (the inverted triangle).

from the inverted triangle. An r_i that is less than 1 means the posterior mean of θ_i is closer to the sample mean, and hence there is less local shrinkage, whereas r_i greater than 1 suggests more local shrinkage. It is evident from Figure 8.4 that the point estimate of an MSOA with more neighbours tends to be closer to the mean of its neighbours, thus receiving more local smoothing. On the other hand, an MSOA with fewer neighbours tends to receive less local shrinkage, so its posterior mean is closer to its sample mean.

Figure 8.4 also shows a considerable amount of variability in r_i even with the number of neighbours fixed. For example, with six neighbours, the values of r_i range from 0.28 to 2.30. As illustrated in Figure 8.5, such variability is primarily due to the varying sample size. With the number of neighbours fixed at 4 and 6 respectively in Figure 8.5(a) and Figure 8.5(b), the more data that an MSOA has, the smaller the value of r_i tends to be and the less information is borrowed from the neighbouring MSOAs. However, Figure 8.5 shows that there is still some variability left in the amount of local shrinkage even after accounting for the difference in sample size and in the number of neighbouring areas. For example, in Figure 8.5(a), MSOA 40 has a larger r_i than MSOA 48, although both have the

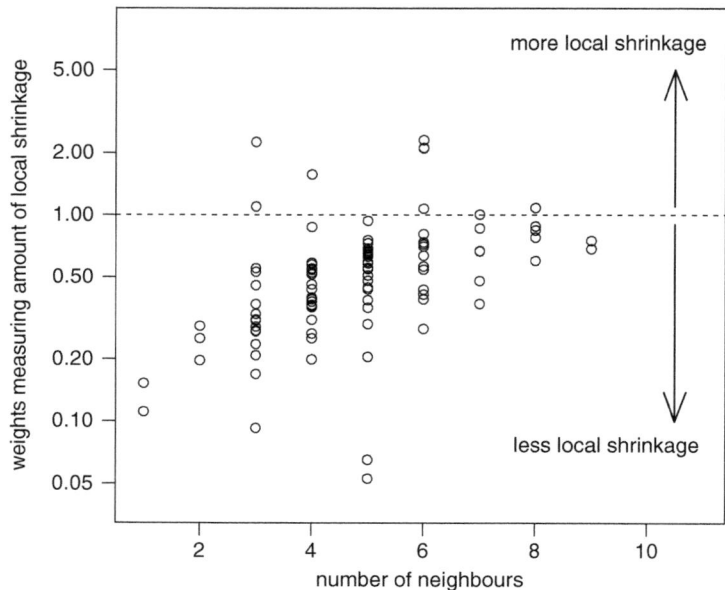

FIGURE 8.4
The effect of the number of neighbours on the amount of local shrinkage. The more neighbours that an area has, the more local shrinkage its parameter will receive.

same number of data values and the same number of neighbours. A closer inspection of their respective neighbours reveals that the income levels of the four neighbours of MSOA 40 are less variable than the income levels of the neighbours of MSOA 48 – for MSOA 40, the standard deviation (SD) of the posterior means of its neighbours is 5, while the SD for the neighbours of MSOA 48 is 26. This observation suggests that the consistency of information from the neighbouring areas also plays a role in the amount of local smoothing. For example, if the information provided from the neighbours is quite variable (e.g. some neighbours have high income levels while others have low), then the ICAR model borrows less from its neighbours. The two MSOAs highlighted in Figure 8.5(b) provide another example where MSOA 7 has more smoothing because its six neighbours show somewhat similar levels of income (the SD of the posterior means of the neighbours is 33). There is less local smoothing for MSOA 79 due to the more variable income levels of its neighbours (the SD of the neighbours is 53).

Figure 8.6 shows the summary of the posterior distributions for the average income across the 109 MSOAs. Because smoothing is local, the 10 MSOAs without data now have different posterior distributions, each of which depends on the mean of the parameters in their neighbouring areas. The map of the posterior means of the average income is given in Figure 8.7(a) while, for comparison, the map of the posterior means from the exchangeable model (Eq. 7.6 in Chapter 7) is given in Figure 8.7(b). Compared to the global smoothing model, the income estimates from the local smoothing model tend to be more similar in MSOAs that are spatially close. As a result, the spatial pattern in Figure 8.7(a) becomes more apparent. The MSOAs in the northwest and the northeast of Newcastle tend to have higher income, while those along the River Tyne show relatively lower income levels.

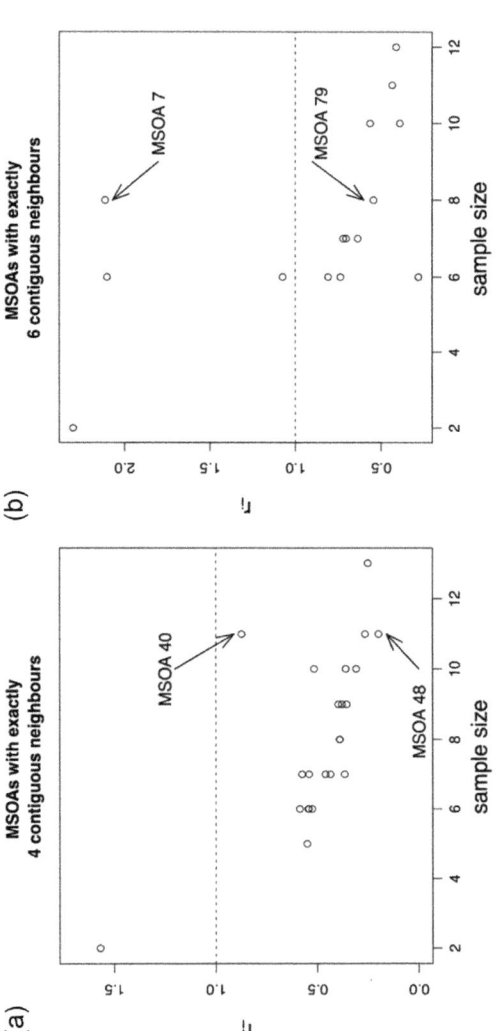

FIGURE 8.5
With the number of neighbours fixed at four in Panel (a) and at six in Panel (b), both plots show that the amount of local shrinkage tends to reduce with increasing sample size. The MSOAs highlighted in each panel demonstrate the additional variability in their weights (see the discussion in the main text).

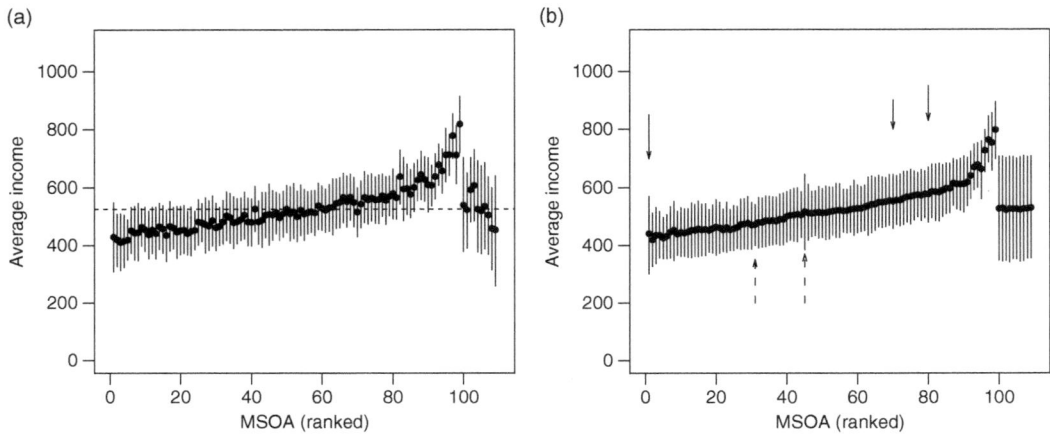

FIGURE 8.6
(a) A summary of the posterior distributions of the estimated average income for the 109 MSOAs obtained from Eq. 8.2, and for comparison, (b) posterior estimates for the MSOA-level average income based on the model with global smoothing (Eq. 7.6; an exact copy of Fig. 7.8(a) in Chapter 7 and see there for the explanation of the arrows). The solid dots are the posterior means and the vertical bars denote the 95% credible intervals.

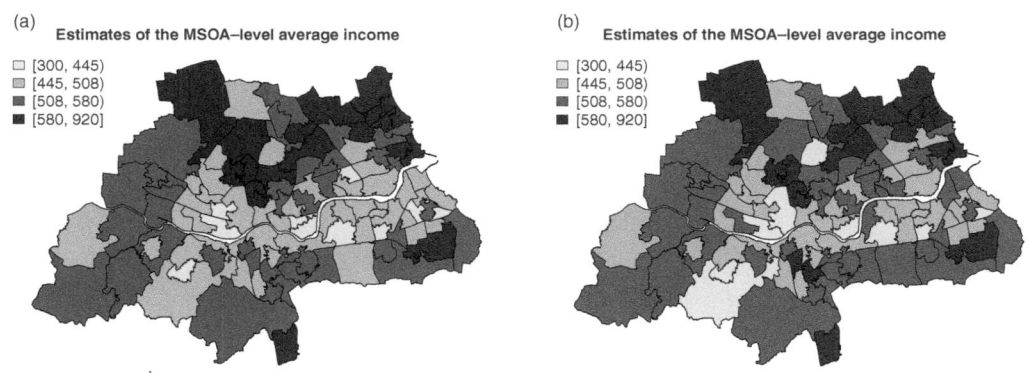

FIGURE 8.7
The maps of the posterior means of the average income (θ_is) from: (a) the local smoothing model using the ICAR model (Eq. 8.2) and (b) the global smoothing model based on exchangeability (Eq. 7.6 in Chapter 7).

8.2.1.4 A Summary of the Properties of the ICAR Model Using a Binary Spatial Weights Matrix

- The ICAR model assumes positive spatial autocorrelation and imposes a spatial neighbourhood structure on the area-specific parameters via the chosen spatial weights matrix.

- Through positive spatial autocorrelation, the ICAR model shares information locally (i.e. across a neighbourhood of nearby areas) and, as a result, produces estimates that are locally smoothed.

- The amount of local smoothing depends on the sample size in an area and the variability or consistency of the information from its neighbours.

- When using a spatial weights matrix with binary 0/1 entries, the number of neighbours of an area also affects the extent of local smoothing – the more neighbours an area has, the more local shrinkage its parameter will receive.

8.2.2 The ICAR Model with a General Weights Matrix

In Section 8.2.1, we introduced a version of the ICAR model where the spatial dependence structure is imposed via a binary spatial weights matrix with the weight associated with each neighbour fixed at 1. Using the equal weights of 1 has a direct implication on how information is shared locally: all neighbours of an area contribute the same amount of information towards the estimation of the parameter of this area. However, in addition to the spatial configuration of the areas, in some situations, other factors may also influence how similar an area is to each of its neighbours. For example, in estimating the income level, perhaps more information should be shared between two neighbouring areas if they are also similar in terms of their demographic structure. To do this, Eq. 8.3 presents the ICAR model that allows for a more general weights matrix:

$$S_i \mid S_{\{-i\}}, v, W \sim N\left(\frac{\sum_{j=1}^{N} w_{ij}S_j}{w_{i+}}, \frac{v}{w_{i+}}\right) \tag{8.3}$$

In Eq. 8.3, w_{ij} is the element of the ith row and the jth column in a spatial weights matrix W, where the diagonal elements of W are all 0. $w_{i+} = \sum_{j=1}^{N} w_{ij}$ is the sum of the elements in the ith row. Eq. 8.3 reduces to Eq. 8.1 when the weights of the neighbours are set to 1 and the weights of the non-neighbouring areas are set to 0. This specification of the ICAR model can accommodate a more general form of W while possessing all the features of the ICAR model that we have discussed in the previous section. However, as we shall now explain, the ICAR model can only permit a *symmetric* spatial weights matrix. The weights matrix derived from contiguity (as used in the previous section), as well as other definitions introduced in Chapter 4, are symmetric, but those defined via the k-nearest neighbours in Section 4.3 and through interactions/flows in Section 4.6 are not. We will also explain the reason behind the sum-to-zero constraint imposed on S that we alluded to in the previous section. The exposition of these two topics requires defining the ICAR model via its joint distribution.

8.2.2.1 Expressing the ICAR Model as a Joint Distribution and the Implied Restriction on W

Both Eq. 8.1 and Eq. 8.3 define the ICAR model via a set of N full *conditional* distributions (often referred to as the full conditionals), and each of these N full conditionals defines a probability distribution for each parameter given all the other parameters. Using the full conditionals in Eq. 8.3, one can derive *the joint probability distribution* for all the area-specific parameters. That is,

$$\Pr(S_1, \ldots, S_N) = \Pr(S)$$

$$\propto \exp\left\{-\frac{1}{2v}S'(D_w - W)S\right\} \tag{8.4}$$

In Eq. 8.4, the notation $\propto$ is the proportionality symbol, meaning that the expression on the second line of Eq. 8.4 ignores all the multiplicative constants (i.e. the terms that do not involve any elements of S) in the joint distribution. D_w is an $N \times N$ diagonal matrix with the diagonal elements equal to the row sums of the W matrix. That is, $(D_w)_{ii}$, the ith diagonal element of D_w, is equal to $w_{i+} = \sum_{j=1}^{N} w_{ij}$ for all $i = 1,\ldots,N$. The derivation from Eq. 8.3 to Eq. 8.4 is beyond the scope of this book, but readers are referred to Cressie (1991, p.410–419) for more details. It is, however, easier to go from Eq. 8.4 to Eq. 8.3 (see Exercise 8.1), which can be used to verify the derivation of Eq. 8.4.

Eq. 8.4 resembles the density of a multivariate normal distribution with a mean vector of 0 and a covariance matrix Σ specified on a set of N random variables $X = (X_1,\ldots,X_N)$, as given in Eq. 8.5:

$$\Pr(X) \propto \exp\left\{-\frac{1}{2}X'\Sigma^{-1}X\right\}$$

(8.5)

For the multivariate normal distribution, the covariance matrix Σ (or, equivalently, its inverse Σ^{-1}) is required to be symmetric. This is because the covariance between two random variables is symmetric, namely, $cov(X_i, X_j) = cov(X_j, X_i)$. For a joint probability distribution to exist under the ICAR model, this same requirement is applied to the matrix $(D_w - W)$ in Eq. 8.4. Since D_w is a diagonal matrix, the spatial weights matrix W must be symmetric, i.e. $w_{ij} = w_{ji}$ for all i and j, with $i \neq j$.

This symmetric restriction on W implies that the ICAR model assumes a spatial model in which the influence of area i on its neighbour j is the same as the influence of area j on i. However, if such an assumption is not appropriate (e.g. i has a larger influence on j than j has on i), then the ICAR model is not appropriate. For example, when modelling small area spatial variation in an economic variable, it is often the case that the economic relationships are asymmetric (see Chapter 2). In such cases, the simultaneous autoregressive model (see Chapter 10) or a spatial-temporal model (see Cressie, 1991, p.410) is a more suitable choice.

8.2.2.2 The Sum-to-Zero Constraint

As discussed in Section 8.2.1.1, the ICAR model places a sum-to-zero constraint on the area-specific parameters S, namely, $\sum_{i=1}^{N} S_i = 0$. This constraint is required because the joint distribution in Eq. 8.4 is improper, meaning that this distribution does not integrate to 1. Here, we will first explain why the joint distribution is improper, then how the sum-to-zero constraint resolves the impropriety problem and the implication of the constraint for the interpretation of the area-specific parameters.

To see why the joint distribution is improper, rewrite the joint distribution in Eq. 8.4 as follows (see Exercise 8.2):

$$\Pr(S_1,\ldots,S_N) \propto \exp\left\{-\frac{1}{2v}\sum_{i=1}^{N-1}\sum_{j=i+1}^{N} w_{ij}(S_i - S_j)^2\right\}$$

(8.6)

We can now see that $S_1,\ldots,S_N$ enter the joint distribution as pairwise differences, and the joint distribution is not affected by the addition (or subtraction) of a constant to the vector S. For any constant c, $\Pr(S_1,\ldots,S_N)$ is exactly the same as $\Pr(S_1 + c,\ldots,S_N + c)$. Put simply, under the ICAR model, the mean of S is not defined. This is why the ICAR model is only used as a prior model for a set of area-specific parameters but not used as a likelihood function to directly model the data. It is unrealistic to assume that the data we observe would have come from a process in which the overall mean is undefined. Besag et al. (1991, p.8) interpret Eq. 8.6 as a stochastic version of linear interpolation. Eq. 8.6, and hence the ICAR model belongs to "the so-called intrinsic models of geostatistics (Matheron, 1973) where all possible spatial differences are modelled simultaneously" (Künsch, 1987, p.517).

To solve the problem of an undefined overall mean, the sum-to-zero constraint fixes the mean of S to 0. Due to this constraint, a separate intercept is then required in the regression to represent the overall mean (of the outcome). The area-specific parameters are effectively measuring the deviations of the local estimates from the global mean. For example, in the income example, the parameters in S measure the deviations of the MSOA-level income from the Newcastle average (see the model in Eq. 8.2). The prior for the intercept must be the improper uniform distribution defined on the whole real line so that the constrained parameters with a separate intercept are equivalent to the unconstrained parameters (see Appendix 1 – under the Intrinsic CAR model section – of the GeoBUGS manual).

8.2.2.3 *Applying the ICAR Model Using General Weights to the Newcastle Income Data*

We now outline a Bayesian hierarchical model with the ICAR prior for the income example. We derive a general spatial weights matrix based on some demographic data. We explain the underlying assumption and the interpretation of results but leave the model fitting to Exercise 8.3.

Eq. 8.7 defines a local smoothing model similar to Eq. 8.2. In both models, the ICAR model with rook's contiguity is used to carry out local smoothing. However, in Eq. 8.7, the non-zero entries in the spatial weights matrix are derived based on h_i, an MSOA-level variable from the 2011 UK census measuring the number of residents aged 16 to 74 working full time for at least 31 hours per week (hereafter, for simplicity, we call this variable "the number of full-time working residents"). We use the notation W_{hour} to emphasise this specification. The use of W_{hour} in the ICAR model assumes that the income levels of two areas are more alike if (a) they share a common border *and* (b) they have similar numbers of full-time working residents. The latter seems reasonable since, as shown in Figure 8.8, an MSOA with a higher h_i tends to have a higher average income, although there are three MSOAs that appear to have high average incomes with relatively small full-time working populations. The global Moran's I based on the number of full-time working residents is 0.33, with a p-value of 0.001 derived from 999 random permutations, suggesting strong positive spatial autocorrelation.

$$y_{ij} \sim N(\theta_i, \sigma_y^2)$$

$$\theta_i = \alpha + S_i$$

$$S_{1:109} \sim ICAR\left(W_{hour}, \sigma_S^2\right)$$

$$\alpha \sim Uniform(-\infty, +\infty)$$

$$\sigma_y \sim Uniform(0.0001, 1000)$$

$$\sigma_S \sim Uniform(0.0001, 1000) \tag{8.7}$$

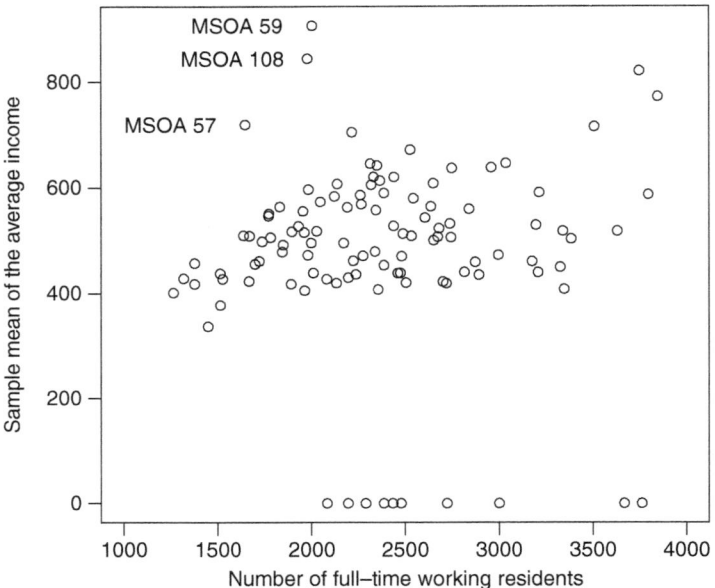

FIGURE 8.8
A scatter plot between the numbers of residents who worked full time with at least 31 hours and the sample means calculated from the survey data. The 10 data points at the bottom (with sample means equal to 0) are those MSOAs without income survey data. Also highlighted are the three MSOAs that appear to deviate from the overall positive relationship.

The diagonal elements in W_{hour} are 0, and the off-diagonal elements are set to be $w_{ij} = \dfrac{1}{|h_i - h_j|}$ if MSOAs i and j share a common border and $w_{ij} = 0$ otherwise.

8.2.2.4 Results

Figure 8.9 shows that for most of the MSOAs, the posterior means from the two specifications of the ICAR model are different but reasonably close, scattering around the diagonal line. The difference between the two sets of estimates is due to the different weights being assigned to the contiguous neighbours. Most of the estimates are similar between the two specifications because of (a) the strong positive spatial autocorrelation present in the sizes of the working populations and (b) the strong relationship between the working population characteristic and income. Point (a) implies that the weights of the neighbours tend to be similar while, even when large weights are assigned to some neighbours, point (b) suggests the resulting (weighted) mean of the neighbours tends not to be too different from the case where the weights of the neighbours are fixed at 1. If there is no apparent correlation between income and the size of the working population, the two ways of specifying weights can give quite different local conditional means.

There are, however, some exceptions. The two points above the diagonal line, MSOAs 28 and 33, in Figure 8.9 are two MSOAs with no income survey data. Compared to the estimates with weights fixed at 1, the posterior means of both MSOAs are higher when W_{hour} is used. This is because, according to W_{hour}, these two MSOAs are more similar to some of their respective neighbours that have high average income. For example, looking at the row standardised W_{hour}, two out of eight of the neighbours of MSOA 28 have weights of

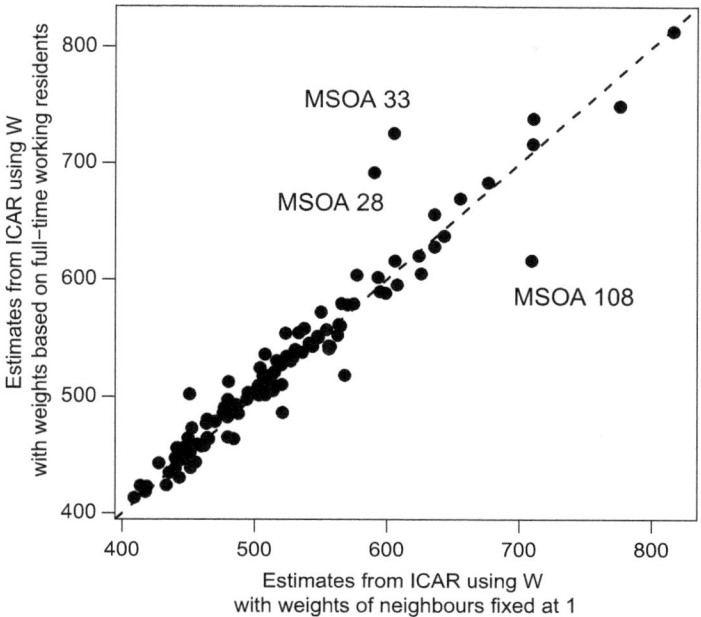

FIGURE 8.9
A scatter plot comparing the posterior means from the ICAR model with contiguity but where the weights between neighbours are either fixed at 1 (x-axis) or computed using data on the size of the working population (y-axis). The dashed line is the diagonal line.

0.31 and 0.26, and the posterior means of the average income of these two neighbours are quite large: 747 and 721. For MSOA 33, one neighbour (out of seven) has a weight of 0.79, and the posterior mean of this neighbour is 747. The non-equal weights in W_{hour} push the estimates of these MSOAs higher.

MSOA 108 is an MSOA that is located below and quite far away from the diagonal line in Figure 8.9. With a sample size of six, this MSOA has a particularly high sample mean of 844. It has five contiguous neighbours with relatively low incomes, and four of them have sample means less than 530. When using the fixed weight of 1 for the neighbours, the estimate of MSOA 108 is unsurprisingly shrunk towards the neighbours. When adding in the similarity measure based on the working population, the estimate of this MSOA is further smoothed towards the neighbours because the numbers of full-time working residents amongst MSOA 108 and its neighbours are similar and low.

8.3 The Proper CAR (pCAR) Model

As discussed in Section 8.2.2.2, the joint distribution of the ICAR model is an improper distribution because the overall mean is not defined. The ICAR model can be made proper by introducing a multiplicative parameter ρ to the spatial weights. This gives rise to the proper CAR model, which is defined via a set of N full conditionals as in Eq. 8.8 or, equivalently, via the joint multivariate distribution given in Eq. 8.9.

$$S_i \mid S_{\{-i\}}, v, W \sim N\left(\rho \frac{\sum_{j=1}^{N} w_{ij} S_j}{w_{i+}}, \frac{v}{w_{i+}}\right) \tag{8.8}$$

$$\Pr(S) \propto \exp\left\{-\frac{1}{2v} S'(D_w - \rho W) S\right\} \tag{8.9}$$

The notations in the above two distributions are the same as those in Eq. 8.3 and Eq. 8.4, the full conditionals and the joint distribution respectively under the ICAR model. Like the ICAR model, the spatial weights matrix W needs to be symmetric (see Section 8.2.2.1). Furthermore, for the joint distribution (Eq. 8.9) to be proper (i.e. integrating to 1), the precision matrix $Q = (D_w - \rho W)$ needs to be invertible so that the covariance matrix, $\Sigma = Q^{-1}$, exists. This requirement places a constraint on the multiplicative parameter ρ, under which ρ can only lie in the open interval $(1/\lambda_{min}, 1/\lambda_{max})$, where λ_{min} and λ_{max} respectively are the smallest and the largest eigenvalues of the row-standardised spatial weights matrix W^* (see Section 3.3.1 in Banerjee et al., 2004; Haining, 2003, p.300 and Appendix 1 – under the Conditional specification section – of the GeoBUGS manual).[6] Given ρ within the permissible range, the pCAR model specifies a multivariate normal distribution for S, namely,

$$S \sim MVN\left(0, v(D_w - \rho W)^{-1}\right) \tag{8.10}$$

The parameter ρ measures the strength of spatial autocorrelation. A value of ρ that is close to 0 signals very weak or no spatial autocorrelation. $\rho < 0$ implies negative spatial autocorrelation (i.e. things nearby in space tend to be dissimilar), while $\rho > 0$ implies positive spatial autocorrelation (i.e. things nearby in space tend to be similar).[7] Moreover, the largest eigenvalue of the row-standardised weights matrix is 1, while all other eigenvalues are between −1 and 1 (Exercise 8.4). Therefore, the bounds for the open interval within which ρ lies are typically simplified to −1 and 1.

The WinBUGS function to implement the proper CAR model is `car.proper` and the syntax is

```
S[1:N] ~ car.proper(a[],C[],adj[], num[], M[], tau, rho)
```

From the GeoBUGS manual, the definitions of the arguments of the `car.proper` function are given as follows:

---

[6] To derive the constraint on ρ, rewrite the precision matrix Q as $D_w(I - \rho W^*)$, where W^* denotes the row-standardised weights matrix and I is the identity matrix of size $N \times N$. A valid precision matrix is positive-definite, so its determinant, det(Q), is positive. So, det(Q) = det($D_w(I - \rho W^*)$) = det(D_w)det($I - \rho W^*$). Given that the row sums of W are all positive, det(D_W) > 0. Thus we require det($I - \rho W^*$) > 0. From matrix algebra, $\det(I - \rho W^*) = \prod_{k=1}^{N}(1 - \rho \lambda_k)$, where λ_1, ..., λ_N are the eigenvalues of W^* (Abadir and Magnus, 2005, Exercise 7.26). It is sufficient for det($I - \rho W^*$) > 0 if $\rho \in (1/\lambda_{min}, 1/\lambda_{max})$ where λ_{min} and λ_{max} are the smallest and largest eigenvalues of W^*. Ord (1975, Appendix C, p.125) shows that W^* and $D_w^{-1/2} W D_w^{-1/2}$ have identical eigenvalues – the latter is used in WinBUGS to derive the lower and the upper bounds for ρ. Haining (2003) uses the unstandardised spatial weights matrix, the largest eigenvalue of which is generally not equal to 1, but $\lambda_{min} < 0$ and $\lambda_{max} > 0$ are still true.

[7] It is worth noting that $\lambda_{min} < 0$ and $\lambda_{max} > 0$ with λ_{min} and λ_{max}, the smallest and the largest eigenvalues of $D_w^{-1/2} W D_w^{-1/2}$ or, equivalently, of W^*. This is because $tr(W^*) = 0$, i.e. the trace of W^* (the sum of the diagonal elements) is equal to 0 and, from Abadir and Magnus (2005, Exercise 7.27), $tr(W^*) = \sum_{k=1}^{N} \lambda_k$. Thus, $\sum_{k=1}^{N} \lambda_k = 0$, implying that some of the eigenvalues are negative and some are positive, so $\lambda_{min} < 0$ and $\lambda_{max} > 0$ (see also Section 3.3.1 in Banerjee et al., 2004).

- a[]: A vector of length N giving the mean for each area – each element in this vector can be (a) specified as a fixed value (thus, the mean is not estimated but a fixed constant), (b) assigned with a prior distribution then each mean is estimated or (c) specified deterministically within the model code, e.g. as a function of some covariates (see Section 8.3.2 below).

- C[]: A vector the same length as adj[] (defined below), storing the non-zero weights from the row-standardised weights matrix W^*.

- adj[]: A vector storing the IDs of the neighbouring areas (the same as in the car.normal function).

- num[]: A vector of length N storing the number of neighbours for each area (the same as in the car.normal function).

- M[]: A vector of length N specifying the inverse of the elements in the diagonal matrix D_w (i.e. the ith element in this vector is equal to $\frac{1}{w_{i+}}$, as defined in Eq. 8.4).

- tau: The unknown precision parameter controlling the overall local smoothing.

- rho: The unknown spatial autocorrelation parameter representing the overall degree of spatial dependence. See below for its prior specification.

8.3.1 Prior Choice for ρ

To ensure the constraint, a typical prior for ρ is a uniform distribution defined on the open interval $\left(1/\lambda_{min}, 1/\lambda_{max}\right)$, where λ_{min} and λ_{max} are the smallest and the largest eigenvalues of the row standardised weights matrix W^* or, equivalently, of the matrix $D_w^{-1/2}WD_w^{-1/2}$. The latter is the matrix used in WinBUGS to compute the two bounds via the functions min.bound (for $1/\lambda_{min}$) and max.bound (for $1/\lambda_{max}$). The syntax of the two functions is given as follows:

```
lower.bound <- min.bound(C[], adj[], num[], M[])
upper.bound <- max.bound(C[], adj[], num[], M[])
```

where the four arguments for both functions are as defined in the car.proper function. Then the uniform prior is

```
rho ~ dunif(lower.bound, upper.bound)
```

8.3.2 ICAR or pCAR?

Compared to the ICAR model, the pCAR model has three advantages. First, the joint distribution is a proper probability distribution, so the area-specific parameters under the pCAR model are not constrained to sum to zero. This allows the flexibility to model the mean of each S_i via the vector a[] in the car.proper function. Second, while the ICAR model assumes positive spatial autocorrelation, the pCAR model can accommodate both positive (i.e. when $\rho > 0$) and negative (i.e. when $\rho < 0$) spatial autocorrelation. Third, the pCAR model encompasses, to some extent, the global smoothing model with exchangeability introduced in Section 7.4. When ρ is estimated to be close to 0, the joint distribution in Eq. 8.10 effectively reduces to a multivariate normal distribution with a diagonal covariance matrix, i.e. $S \sim MVN\left(\mathbf{0}, vD_w^{-1}\right)$, which can be rewritten as $S_i \sim N\left(0, v \cdot w_{i+}^{-1}\right)$ for all i. Thus, conditioning on the variance v, the mean 0 and w_{i+}^{-1}, all areas are considered to be independent of each other. However, there is still a degree of "local smoothing" left in the pCAR model because each individual variance, $v \cdot w_{i+}^{-1}$, still depends on the spatial neighbourhood structure via the sum of the ith row in $\mathbf{W}$.

There are, however, a few limitations with the pCAR model. First is the interpretation of the spatial autocorrelation parameter ρ. Despite its permissible range lying between -1 and 1, ρ is not calibrated to have the same interpretation as a correlation coefficient. To illustrate, we follow the simulation outlined in Section 3.3.1 in Banerjee et al. (2004). For a range of values for ρ, we simulate 100 sets of values using the joint distribution in Eq. 8.10, where $\mathbf{W}$ defines the neighbourhood structure of the MSOAs in Newcastle based on rook's contiguity (Exercise 8.5). For each set, we compute the global Moran's I to measure the strength of spatial autocorrelation. Figure 8.10 shows the results of the simulation. The strength of spatial autocorrelation, as measured by Moran's I, increases as ρ increases, and a negative (or positive) ρ tends to result in a negative (or positive) value for Moran's I. Moran's I tends to be close to 0 when ρ is close to 0. However, even when ρ is 0.9, Moran's I merely ranges between 0.12 and 0.56 (see the right-hand plot in Figure 8.10). Only when $\rho = 0.9999$ does the average of Moran's I over the 100 sets of simulated data lie above 0.9. These simulations suggest that the interpretation of ρ remains qualitative rather than quantitative, at best suggesting how likely it is that positive spatial autocorrelation is present in the data rather than providing a measure of the strength of spatial autocorrelation. In a Bayesian analysis, this is equivalent to calculating the posterior probability $\Pr(\rho > 0 \mid \text{data})$. We will illustrate the interpretation of ρ in the income example.

The second limitation is the interpretation of the conditional mean in Eq. 8.8. Under the ICAR model, given the neighbours, S_i is similar to the average of the neighbours (Eq. 8.1 and Eq. 8.3), whereas in the pCAR model, S_i is similar to some proportion of the local average (Eq. 8.8), which, as pointed out by Banerjee et al. (2004, p.81), may not have any sensible interpretation.

A third limitation of the pCAR model is the need to estimate the parameter ρ, which, particularly in complex models, may introduce computational issues (e.g. slower convergence; maybe even non-convergence). We will return to this in Section 8.5.1 when the BYM model is discussed.

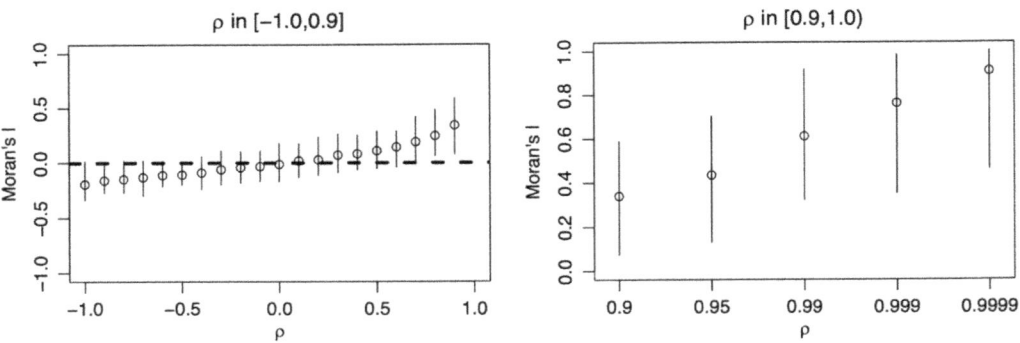

FIGURE 8.10
The strength of spatial autocorrelation of the values simulated from the proper CAR model using different values of ρ ranging from -1 to 1. With each ρ value, 100 sets of values are simulated based on the Newcastle MSOA map and the global Moran's I statistic is calculated for each set to measure the strength of spatial autocorrelation. Each open circle represents the mean value of Moran's I, and the vertical bar shows the 95% sampling variability (i.e. the values of the 2.5 and 97.5 percentiles). The left-hand panel shows $\rho \in [-1.0, 0.9]$, and the right-hand panel shows $\rho \in [0.9, 1.0)$.

8.3.3 Applying the pCAR Model to the Newcastle Income Data

We apply the pCAR model where the spatial neighbourhood structure is defined via rook's spatial contiguity and the weights of the neighbours are based on similarity in terms of the size of the full-time working population. That is, the weight of two neighbours i and j is set to be $w_{ij} = 1/|h_i - h_j|$, with h_i and h_j being the numbers of full-time working residents in the two areas. The model is given as follows:

$$y_{ij} \sim N\left(\theta_i, \sigma_y^2\right)$$

$$\theta_i = \alpha + S_i$$

$$S_{1:109} \sim pCAR\left(W_{hour}, \sigma_S^2, \rho\right)$$

$$\alpha \sim N\left(0, 100000000\right)$$

$$\sigma_y \sim Uniform\left(0.0001, 1000\right)$$

$$\sigma_S \sim Uniform\left(0.0001, 1000\right)$$

$$\rho \sim Uniform\left(a, b\right) \tag{8.11}$$

The lower and the upper bounds (denoted as a and b) of the uniform prior for ρ are computed using the built-in functions, as explained in Section 8.3.1. The WinBUGS implementation of this model is in Figure 8.11.

Lines 49 to 52 in Figure 8.11 illustrate the structure of the array C[], which contains the row standardised weights of the neighbours, and Line 53 shows the first few elements in the array M[]. To understand how the two arrays are constructed, take the example of the first MSOA. MSOA 2 is the only neighbour of MSOA 1, and their numbers of full-time working residents are $h_1 = 2650$ and $h_2 = 3385$, respectively, which gives the weight $w_{12} = 1/|2650 - 3385| = 0.00136$, and the corresponding row sum is $w_{1+} = 0.00136$. Therefore, the row-standardised weight becomes 1 (the first element in the C[] array) and the first element in M[] is $\dfrac{1}{w_{1+}} = \dfrac{1}{0.0014} = 735$. The same procedure applies to the calculation for the remaining MSOAs. Readers are encouraged to verify the values in the arrays C[] and M[] for MSOA 2 (3385), which has six neighbours – MSOAs 1 (2650), 3 (2289), 4 (2606), 11 (1723), 23 (2133) and 24 (2196); the numbers in brackets are their numbers of full-time working residents.

8.3.4 Results

Figure 8.12 compares the posterior means (Panel a) and the lengths of the 95% credible intervals (Panel b) of the MSOA-level average income from the pCAR model and from the ICAR model; the latter model was discussed in Section 8.2.2.3. The two models yield very similar posterior point and interval estimates. This is perhaps not surprising given that ρ is estimated to be close to 1. The posterior mean of ρ is 0.95 with a 95% credible interval of (0.85–1.00) and its posterior distribution in given in Figure 8.13.

At first sight it might seem that fitting the pCAR model where the spatial parameter is estimated should offer more flexibility in comparison to the ICAR model. However, as

```
1    #  The WinBUGS code for specifying the model with the pCAR prior
2    model {
3      #  a for-loop to go through all the household level income data
4      for (j in 1:nhhs) {
5        #  defining the normal likelihood for each household
6        y[j] ~ dnorm(theta[msoa[j]],prec.y)
7      }
8      for (i in 1:nmsoas) {
9        theta[i] <- alpha + S[i]
10       a[i] <- 0  #  set the mean of each MSOA to 0 in the pCAR model
11     }
12     #  modelling all the elements in S using the pCAR model
13     S[1:nmsoas] ~ car.proper(a[],C[],adj[], num[], M[], prec.S, rho)
14
15     #  a vague prior for the Newcastle average
16     alpha ~ dnorm(0,0.00000001)
17
18     #  a vague prior for the conditional SD in the pCAR model
19     sigma.S ~ dunif(0.0001,1000)
20
21     #  a vague prior for the sampling standard deviation
22     sigma.y ~ dunif(0.0001,1000)
23
24     #  calculate the two precisions
25     prec.S <- pow(sigma.S,-2)
26     prec.y <- pow(sigma.y,-2)
27
28     #  a uniform prior on rho and the calculation of the lower and
29     #  upper bounds
30     rho ~ dunif(lower.bound, upper.bound)
31     lower.bound <- min.bound(C[], adj[], num[], M[])
32     upper.bound <- max.bound(C[], adj[], num[], M[])
33   }
34
35   #  household-level income data from survey with a spatial
36   #  neighbourhood structure
37   list(nhhs=760
38       ,nmsoas=109
39       ,y=c(501,616,472,816,637,500,506,560,542,447,644,522
40       ,487,275,...)
41       ,msoa=c(1,1,1,1,1,2,2,4,4,4,4,4,5,5,...)
42   #  elements of the W matrix for pCAR (defined via rook's spatial
43   #  contiguity)
44       ,num=c(1,6,4,...)
45       ,adj=c(2
46             ,1,3,4,11,23,24
47             ,2,4,20,22
48             ,...)
49       ,C=c(1
50             ,0.235, 0.157, 0.221, 0.104, 0.138, 0.145
51             ,0.023, 0.078, 0.852, 0.048
52             ,...)
53       ,M=c(735.00,172.47,24.70,...)
54   )
55
56   #  initial values for chain 1
57   list(alpha=200,sigma.y=50,sigma.S=20,S=c(1,-1,0,0,0,...),rho=0.1)
58
59   #  initial values for chain 2
60   list(alpha=700,sigma.y=80,sigma.S=50,S=c(-1,1,0,0,0,...),rho=0.05)
```

FIGURE 8.11
The WinBUGS code for fitting the pCAR model (Eq. 8.11) to the Newcastle income data.

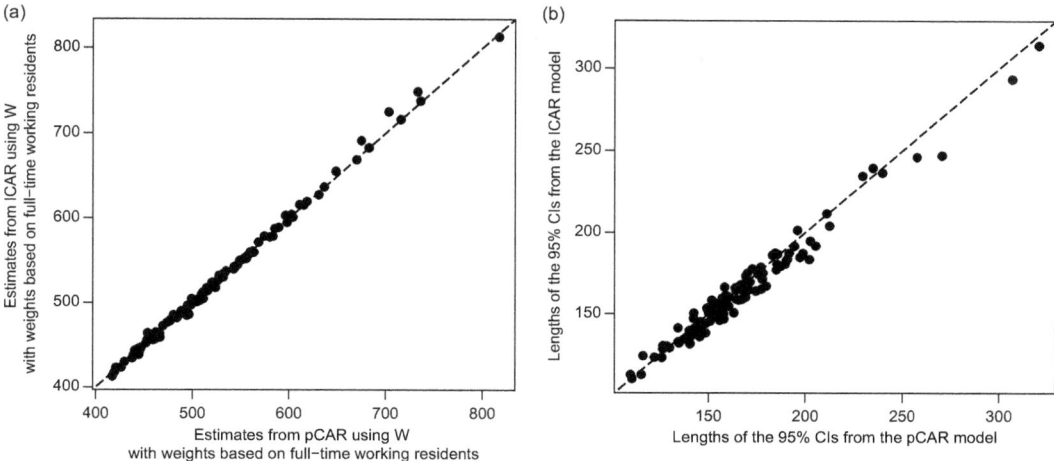

FIGURE 8.12
(a) Comparing the posterior means of the MSOA average income from the ICAR model and from the pCAR model. For both models, the spatial neighbourhood structure is defined via rook's contiguity and the weights of the neighbours are derived based on the similarity in the size of the full-time working population. Panel (b) compares the widths of the 95% credible intervals (the upper bound – the lower bound) from the two models. In both plots, the dashed line indicates the diagonal line.

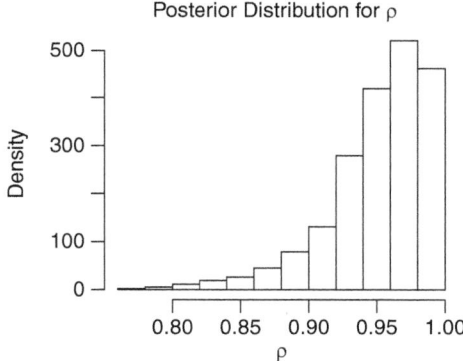

FIGURE 8.13
The posterior distribution of the parameter ρ from the pCAR model.

remarked in Gelfand and Vounatsou (2003, p.15), setting "$\rho=1$ is analogous to the non-stationary or random walk case in familiar autoregressive time series models and can be advantageous in accommodating *more irregular spatial behaviour*". Making a similar point, Banerjee et al. (2004, p.81) also remark that in the case of the pCAR model, "the breadth of spatial pattern may be too limited. In the case where a CAR model is applied to random effects, an improper choice may actually enable wider scope for posterior spatial pattern".

When positive spatial autocorrelation is found to be present, ρ is often estimated to be very close to 1 (as in, for example, the Newcastle income data). In many practical applications, this is another reason for adopting the ICAR model (with ρ fixed to 1) for imposing spatial structure on a set of area-specific parameters (Mollié, 1996).

However, if a localised pattern of spatial autocorrelation, a form of spatial heterogeneity, is suspected or its detection is of interest, one option is to employ locally adaptive models, a topic that we will discuss next.

8.4 Locally Adaptive Models

In some cases, the spatial data that we observe may exhibit more complex spatial depen-dence structures or even discontinuities. The evidence for the former may come from the use of local statistics such as the local Moran's I (Chapter 6). Such complexity is often seen in urban data when mapped at the scale of small spatial units such as census tracts. For example, some sections of an urban area may display quite a smooth-looking pattern of average household income values, while abrupt changes (or spatial discontinuities) may exist in other parts. In some cities, high income, affluent neighbourhoods are located geo-graphically close, even adjacent, to poorer neighbourhoods. Such abrupt changes may be evident by simply inspecting a map of the data. Wombling is a statistical method for determining the boundaries where spatial discontinuity or zones of rapid change occur (Womble, 1951; see also Section 4.10). In Section 8.4.2, we consider a method of wombling and we also refer readers to Gelfand and Banerjee (2015) and the references therein. For these types of situations, the CAR models that we have discussed so far may not be appro-priate because their strategy for borrowing information from the neighbours of an area is applied *uniformly* across the whole study region.

To demonstrate the issue of spatial discontinuity, consider again the Newcastle income data. Figure 8.14 shows an overall spatially-smooth map, where high-income (or low-income) MSOAs tend to be clustered together. There are, however, some sub-regions in Newcastle where spatial discontinuity is evident (see the two sets of cross-hatched areas). For example, for the pocket of MSOAs in the middle of Newcastle, there appears to be two clusters, where MSOAs 31, 32, 39 and 43 tend to have high-income levels, while the income levels of MSOAs 48, 50 and 55 can be seen to be low. While it is reasonable to share information amongst the MSOAs within each cluster, is it reasonable for MSOAs 32 and 50 to borrow information from each other or for 43 and 50 to do likewise? The same point applies to the subset of MSOAs in the northeast corner.

To address this problem, a number of locally adaptive spatial models have been pro-posed. The rationale behind these models is to recognise where it is appropriate to borrow information from one's neighbours and where it is not. Because the spatial neighbourhood structure is encoded in the W matrix, for most of these models adaptive spatial smoothing is achieved by modelling the non-zero elements in W. The basic modelling idea is to start with the W matrix defined by spatial contiguity with the weights of neighbours set to 1. However, instead of treating W as fixed, each $w_{ij} = 1$ in the matrix is potentially subject to modification so that two spatially-contiguous areas can be considered as neighbours (by confirming the weight as 1) or not neighbours (by re-setting the weight to 0). To operation-alise this idea, there are three main approaches: (1) choosing an optimal W matrix from all possible specifications; (2) modelling the elements in the W matrix; (3) grouping areas via mixture models. We will explore the first two approaches in this section and illustrate the mixture modelling approach through an application in Chapter 9.

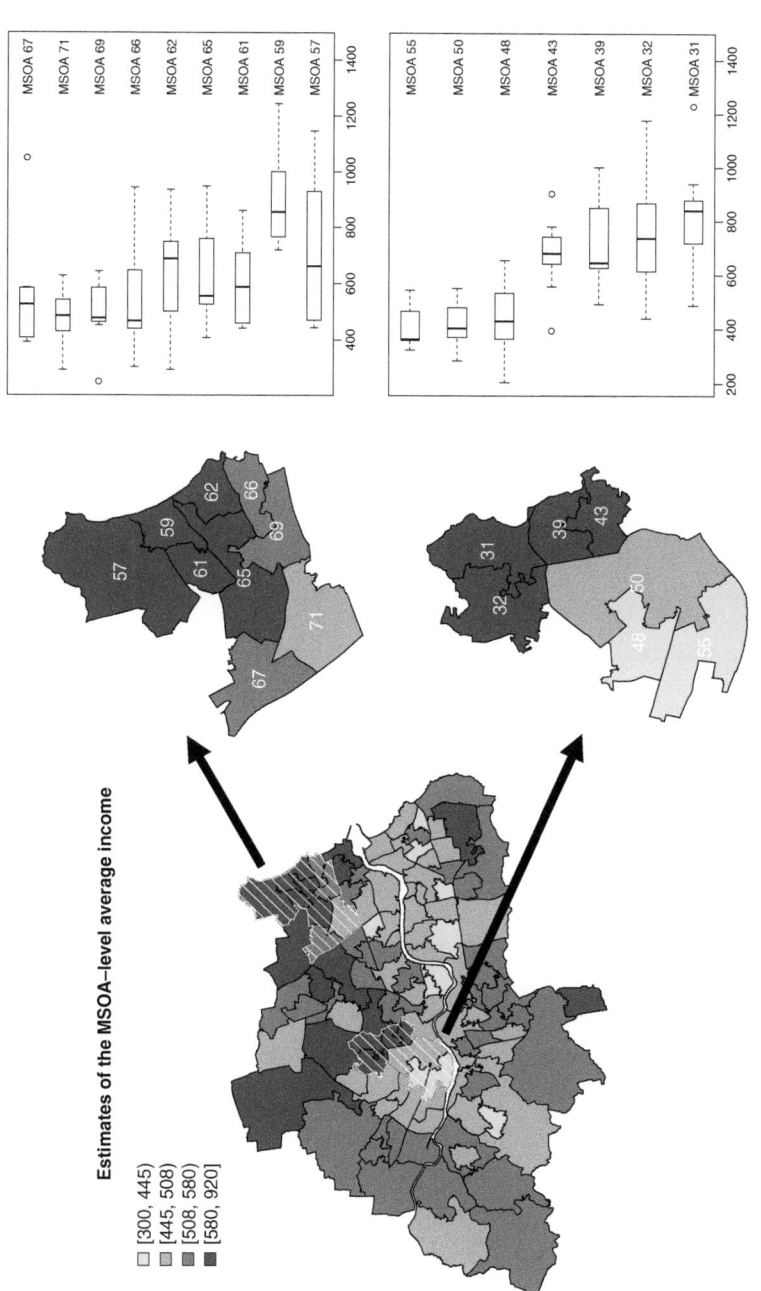

FIGURE 8.14
Examples of spatial discontinuity in the Newcastle income data at the MSOA scale. The left-hand map shows the posterior means of the MSOA-level income from the ICAR model in Eq. 8.2. The two cross-hatched sub regions are examples where spatial discontinuity is evident. The two boxplots on the right-hand side show the survey data for each MSOA. The labelling of the MSOAs is the same as that used in Figure 7.1 in Chapter 7.

8.4.1 Choosing an Optimal *W* Matrix from All Possible Specifications

The first approach is to consider all possible modifications to the *W* matrix derived from spatial contiguity. Different versions of the *W* matrix give rise to different models. Model choice is then based on, for example, the Bayesian information criterion (Li et al., 2011) or an automatic model selection algorithm using the reversible jump MCMC coupled with simulated annealing (Seya et al., 2013). However, an immediate challenge to this approach is the large number of *W* matrices, each requiring the fitting of a spatial model to the data. Since the weight of each pair of spatially-contiguous areas can take either 1 or 0, the number of possible *W* matrices and hence models is $2^{\mathbf{1W1'}/2}$, where *W* is derived from spatial contiguity with binary 0/1 weights, $\mathbf{1}$ denotes a vector of 1s of size $1 \times N$ and $\mathbf{1'}$ is its transpose. For the MSOA Newcastle map, $\mathbf{1W1'}/2$ gives 263 unique pairs of spatially-contiguous MSOAs, and hence the possible modifications to *W* is $2^{263} = 1.48 \times 10^{79}$ (!).

To circumvent this challenge, Anderson et al. (2014, 2016) proposed to reduce the set of *W* matrices by eliciting possible configurations of *W* using data observed from *q* time periods, say five years, immediately before the current study period. Specifically, a hierarchical agglomerative clustering algorithm is used to group areas together based on spatial contiguity and how similar they are in terms of the outcome values over the past *q* time periods. Over *N* areas, because of the hierarchical agglomerative nature of the clustering method, *N* cluster configurations are formed by, effectively, scanning from the bottom of the resulting dendrogram (which gives rise to a cluster configuration where each area is assigned to its own singleton cluster) to the top (which forms another cluster configuration where all areas are grouped into a single cluster). When modelling the data at the current period, the *W* matrices to consider are reduced vastly to the *N* cluster configurations derived from the past data. While reducing the computational burden, this method relies on the availability of past data. Deriving the potential *W* matrices from past data for the current time period also assumes that the clustering of areas is temporally stable. This assumption is perhaps justifiable for modelling chronic diseases whose risk factors are stable over time "but would be unsuitable for epidemic diseases such as influenza, where the spatial pattern in disease risk in the years prior to an outbreak would be vastly different to the pattern during an outbreak" (Anderson et al., 2016, p.12). This method is implemented in R; see Anderson et al. (2016) for more detail.

8.4.2 Modelling the Elements in the *W* Matrix

The second approach is to model directly the elements of the spatially-contiguous pairs in *W*. Recall that using rook's contiguity, if areas *i* and *j* share a common border, then $w_{ij} = 1$, they are considered to be neighbours and information is shared between them. However, if the two areas are shown to be sufficiently different based on some criteria, for example, one is a high average income area while the other is a low average income area, then there is little evidence to support the sharing of information between them. In this case, we might wish to set w_{ij} to 0, despite the fact that they are spatially contiguous. In other words, we are introducing a "hard boundary" or spatial discontinuity between the two areas. The methods discussed below aim to construct such "hard boundaries" and are referred to as boundary detection methods.[8]

Based on the above idea, Lu et al. (2007) model the weight of a pair of spatially-contiguous areas using a logistic regression with a single covariate, while the weight of two areas

---

[8] The introduction of impenetrable boundaries into models of geographical diffusion processes has a history (see for example Hagerstand, 1967).

that are not spatially-contiguous remains as 0. Under their model, w_{ij}, the weight of two *spatially-contiguous* areas, i and j, is modelled as

$$w_{ij} \sim \text{Bernoulli}\left(p_{ij}\right)$$

$$logit\left(p_{ij}\right) = \gamma_0 + \gamma_1 d_{ij}, \tag{8.12}$$

where γ_0 and γ_1 are regression coefficients to be estimated using the data and d_{ij} is a similarity measure defined as the difference between the two areas with respect to the chosen covariate. A small value of d_{ij} means that the two areas have similar values of the chosen covariate (e.g. average income). When Lu et al. (2007) applied this method to their Minnesota breast cancer late detection data (where the outcome values are the numbers of late detection cases across the 87 counties), several specifications of d_{ij} were considered. Using their labelling of the models (see Section 4 in Lu et al., 2007), these specifications are:

- Model 1: d_{ij} is the distance between the centroids of areas i and j scaled (divided) by the maximum distance on the map.
- Model 2: d_{ij} is the absolute difference in cancer mortality rate between the two areas, i.e. $d_{ij} = \left|\text{mortality}_i - \text{mortality}_j\right|$. The county-level mortality rates were available to the study as a set of covariate values.

Similar to Model 2, Models 3 and 4 were two additional specifications where the mortality rates in Model 2 were replaced by mammogram, a county-level covariate derived from a separate survey measuring the percent of patients aged 40 and over who reported having had a mammogram within the last two years, or by incidence, the county-level cancer incidence rate.[9]

In addition to d_{ij}, another important aspect of this modelling is the choice of priors for γ_0 and γ_1 in Eq. 8.12. In Lu et al. (2007), two moderately informative priors were chosen: $\gamma_0 \sim N(2, 0.5)$ and $\gamma_1 \sim N(-2, 0.5)$. Why are these two priors chosen? And given the priors on γ_0 and γ_1 and given a permissible value of d_{ij}, what is the equivalent prior on p_{ij} which determines the probability that two spatially-contiguous areas will be treated as neighbours ahead of seeing the data? To answer these two questions, we need to understand the roles of the two coefficients. First, γ_0 is the logit transform of p_{ij} when $d_{ij} = 0$ so it determines the probability that i and j are neighbours ($w_{ij} = 1$; hereafter referred to as the "neighbouring probability") when the two areas have the same covariate values (e.g. having the same mortality rates). Before seeing the data, when $d_{ij} = 0$, this neighbouring probability is expected to be high, which is reflected in the chosen prior for γ_0: the prior neighbouring probability has a mean of 0.86 and there is a 95% chance that this prior neighbouring

---

[9] In Lu et al. (2007), spatial contiguity was also considered for defining d_{ij}, where $d_{ij} = 0$ if areas i and j share a common boundary and $d_{ij} = 1$ otherwise. However, Lu et al. (2007, p. 445–446) considered the specification (their Model 0) where $d_{ij} = 1$ if i and j share a common boundary and 0 otherwise. When combined with their prior for $\gamma_1 \sim N(-2, 0.5)$, this latter specification of d_{ij} suggests that two spatially-contiguous areas are *less likely* to be considered as neighbours *a priori*. Such a (prior) assumption is not reasonable. See Figure 8.15 and the discussion of how the priors on γ_0 and γ_1 are translated to the prior on w_{ij}. Defining d_{ij} based on contiguity introduces some "general randomness or uncertainty" in deciding whether two contiguous areas will be allowed to remain as neighbours, albeit a strong prior belief (through the choice of prior by the authors) towards the two areas staying as neighbours. Such randomness/uncertainty may be difficult to quantify in the absence of covariate(s). This option would not be recommended as a candidate model in practice.

probability varies between 0.65 and 0.97 (see Figure 8.15 for the calculation where the results from Lines 18 and 20 give the values above). However, γ_1 quantifies the effect of d_{ij} on the neighbouring probability. Now, the smaller d_{ij} is, the more alike two areas are in terms of their attributes, in the case of Models 2–4, and thus the more likely it is that the two areas will be treated as neighbours, *a priori*. In the case of Model 1, the closer d_{ij} is to 0 the more likely it is the two areas will be treated as neighbours. To reflect these expectations, the prior on γ_1 is centred at a negative value so that a smaller d_{ij} gives a larger p_{ij}, and hence w_{ij} is more likely to take the value 1 (see Eq. 8.12). In the case of Model 1 where $d_{ij} = 1$ (i.e. for two areas the maximum distance apart), the prior mean of the neighbouring probability is around 0.50 and there is a 95% chance that the neighbouring probability varies between 0.12 and 0.88. The calculation of these values is in Figure 8.15 and the results from Lines 26 and 28 give the values above. The same calculation can be used to assess the prior information on p_{ij} with a given value of the absolute mortality difference (simply change the value multiplied with gamma1 in Line 24, Figure 8.15).

The exposition above gives an example of how we can investigate the behaviour of the model based on the chosen priors. When dealing with complex models such as the one above, this type of investigation is particularly important and should be carried out before

```
1    #   define the function expit, the inverse function of logit
2    #   (where logit(x) = log(x/(1-x)))
3    expit <- function(a) exp(a)/(1+exp(a))
4
5    #   define number of draws from the prior distributions
6    ndraws <- 100000
7    #   obtain a set of random draws (of size ndraws) from the prior
8    #   for gamma0
9    gamma0 <- rnorm(ndraws, 2, sqrt(0.5))
10   #   obtain a set of random draws (of size ndraws) from the prior
11   #   for gamma1
12   gamma1 <- rnorm(ndraws, -2, sqrt(0.5))
13
14   #   calculate the neighbouring probability when the similarity
15   #   measure is equal to 0
16   p0 <- expit(gamma0)
17   #   the prior mean of the neighbouring probability
18   mean(p0)
19   #   the 95% prior interval of the neighbouring probability
20   quantile(p0, c(0.025,0.975))
21
22   #   calculate the neighbouring probability when the similarity
23   #   measure is equal to 1
24   p1 <- expit(gamma0 + gamma1*1)
25   #   the prior mean of the neighbouring probability
26   mean(p1)
27   #   the 95% prior interval of the neighbouring probability
28   quantile(p1, c(0.025,0.975))
```

FIGURE 8.15
The R code to calculate the prior probability of treating two spatially-contiguous areas as neighbours given the value of the similarity measure, d_{ij}, and the moderately informative priors on γ_0 and γ_1. The basic idea is to carry out a Monte Carlo simulation, in which values are randomly drawn for γ_0 and for γ_1 from their respective prior distributions. For a given value of d_{ij} we can then calculate p_{ij} the (prior) neighbouring probability (of i and j being neighbours), and investigate its properties. Modify Lines 9 and 12 if different prior distributions are used. For example, change Line 9 to gamma0 <- runif(ndraws, -10, 10) if $\gamma_0 \sim$ Uniform$(-10,10)$.

fitting the model to the data to ensure that the model with the chosen priors behaves as expected. Readers are encouraged to explore the impact on the prior neighbouring probability if (much) less informative priors, say Uniform(–10,10) or $N(0,10)$, are chosen for both γ_0 and γ_1 (Exercise 8.7). However, when vaguely/weakly informative priors are used, the data may not have sufficient information to estimate the two parameters.

Lee and Mitchell (2012) extended the model by Lu et al. (2007) to incorporate several covariates in the logistic regression for the weights. Instead of treating w_{ij} as a Bernoulli random variable, Lee and Mitchell (2012) modelled it deterministically:

$$w_{ij} = \begin{cases} 1 & \text{if } \exp\left(-\sum_{k=1}^{K} d_{ijk}\gamma_k\right) \geq 0.5 \text{ and } i \text{ and } j \text{ share a common boundary} \\ 0 & \text{otherwise} \end{cases}$$

(8.13)

The above model measures the similarity between two spatially-contiguous areas using K covariates, as represented by d_{ijk}, with $k = 1,\ldots,K$. As in Lu et al. (2007), $d_{ijk} = 0$ if two areas take the same value for covariate k. The regression coefficients $\gamma_1,\ldots,\gamma_K$ quantify the importance of each covariate on the neighbouring weights. In their application of modelling the lung cancer risk across the 271 administrative units in Glasgow, three covariates were included – smoking prevalence; ethnicity, which is measured by the percentage of school children from ethnic minorities; and the natural log of the median house price (Lee and Mitchell, 2012). These three covariates were also used in the Poisson regression to explain the risks of lung cancer.

The prior for the regression coefficient γ_k ($k = 1,\ldots,K$) takes the form $\gamma_k \sim \text{Uniform}(0, M_k)$. This prior choice restricts the regression coefficients to be positive so that a large value of d_{ijk} corresponds to a small probability of w_{ij} being 1. In other words, before seeing the data, it is expected that if two spatially-contiguous areas are very different in terms of the chosen covariates, they have a small chance of being treated as neighbours. The upper bound of the uniform prior is chosen "so that at most 50% of borders in the study region can be classified as boundaries" (Lee and Mitchell, 2012, p.419). This corresponds to a restriction that at most 50% of all pairs of spatially-contiguous areas are considered to be non-neighbours. The fitting of their model is done in R.

Both the models by Lu et al. (2007) and by Lee and Mitchell (2012) rely on covariates to inform where the "hard boundaries" are, or equivalently, where spatial discontinuity occurs. However, such covariates may not be available in some cases due to either the availability of the covariate data or the available covariates being uninformative in determining the boundaries. In the absence of such covariate information, Lee and Mitchell (2013) developed another locally adaptive spatial model which sets the weights of the spatially contiguous areas based on the posterior estimates of the area-specific parameters. Their model was constructed in the context of disease mapping where the outcome values are the numbers of disease cases (Y_i with $i = 1,\ldots,N$) across N spatial units:

$$Y_i \sim \text{Poisson}(\mu_i)$$

$$\log(\mu_i) = \log(E_i) + \alpha + \sum_{k=1}^{K} \beta_k x_{ik} + S_i$$

(8.14)

where E_i is the expected number of cases and x_{ik} is the kth covariate value for area i with β_k being the associated coefficient. S_i is the area-specific parameter that represents the residual risk that is not explained by the covariates included in the model. The set of $S = (S_1,\ldots,S_N)$

is then modelled hierarchically using a spatial model in which the W matrix induces the spatial dependence structure. To allow adaptive smoothing, the modelling idea of Lee and Mitchell (2013) is to set two spatially-contiguous areas i and j as neighbours if the estimated S_i and S_j are similar (or, in other words, they have similar residual risks). If that is the case, it is then reasonable to borrow information between the two areas in estimating S_i and S_j. If, however, the estimated S_i and S_j are not similar, there is little support for borrowing information between them. The similarity between two estimated parameters is measured by whether their 95% credible intervals overlap or not. The weight of the two spatially-contiguous areas i and j is set as

- $w_{ij} = 1$ if the 95% credible intervals for S_i and S_j overlap.
- $w_{ij} = 0$ if the 95% credible intervals for S_i and S_j do not overlap.

This adaptive model is fitted iteratively. To start, the model in Eq. 8.14 with an exchangeable model for S (e.g. Eq. 7.6 in Chapter 7) is fitted to the data. The W matrix is then modified according to the rules given above and the model in Eq. 8.14 is refitted using the resulting W matrix via a chosen spatial model (e.g. the ICAR model or the Leroux model[10], and the latter model is used in Lee and Mitchell (2013)). This iterative process is carried out until either there is no change to the W matrix between two consecutive iterations or the current W matrix has already appeared in a previous iteration. To achieve computational efficiency, Lee and Mitchell (2013) used INLA for the iterative fitting of the spatial model. In the example below, we will apply this method to a subset of the Newcastle income data where WinBUGS is used.

8.4.3 Applying Some of the Locally Adaptive Spatial Models to a Subset of the Newcastle Income Data

To illustrate some of the locally adaptive smoothing models, we will use the income data for the seven MSOAs located in the middle of Newcastle (see Figure 8.14). For ease of discussion, the MSOAs are relabelled as MSOAs 1–7 (see Figure 8.16(a)) and the W matrix

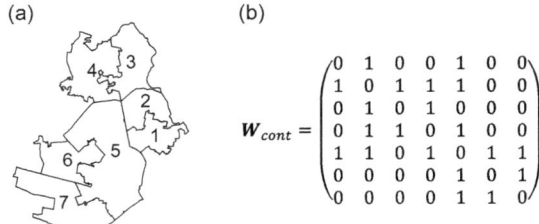

FIGURE 8.16

(a) Map of the selected seven MSOAs. For ease of discussion, the MSOAs are relabelled as MSOA 1 to MSOA 7; and (b) the W matrix is derived using rook's spatial contiguity. Note that in (a), due to imprecision in the drawing, MSOAs 2, 3, 4 and 5 may look as if they all join at a single point, but they do not (MSOAs 2 and 4 share a common border but MSOAs 3 and 5 do not).

---

[10] The Leroux model was proposed by Leroux et al. (2000) and is a prior model that combines both global and local smoothing for the area-specific parameters. The model structure is similar to the BYM model, which will be introduced in Section 8.5.

Some possible modifications	The implied spatial configuration	Notes
A global smoothing model: Eq. 8.15 but with $S_i \sim N(0, \sigma_S^2)$ for $i = 1, \ldots, 7$ implying the following neighbourhood structure $$W_a = \begin{pmatrix} 0 & 0 & 0 & 0 & 0 & 0 & 0 \\ 0 & 0 & 0 & 0 & 0 & 0 & 0 \\ 0 & 0 & 0 & 0 & 0 & 0 & 0 \\ 0 & 0 & 0 & 0 & 0 & 0 & 0 \\ 0 & 0 & 0 & 0 & 0 & 0 & 0 \\ 0 & 0 & 0 & 0 & 0 & 0 & 0 \\ 0 & 0 & 0 & 0 & 0 & 0 & 0 \end{pmatrix}$$		None of the spatially-contiguous MSOAs are treated as neighbours to each other but information is still shared globally through the unknown overall mean and variance.
A local smoothing model (Eq. 8.15) with $$W_b = \begin{pmatrix} 0 & 1 & 0 & 0 & 1 & 0 & 0 \\ 1 & 0 & 0 & 0 & 1 & 0 & 0 \\ 0 & 0 & 0 & 1 & 0 & 0 & 0 \\ 0 & 0 & 1 & 0 & 0 & 0 & 0 \\ 1 & 1 & 0 & 0 & 0 & 1 & 1 \\ 0 & 0 & 0 & 0 & 1 & 0 & 1 \\ 0 & 0 & 0 & 0 & 1 & 1 & 0 \end{pmatrix}$$		MSOAs 3 and 4 are neighbours but neither is a neighbour to other spatially-contiguous MSOAs (e.g. MSOAs 2 and 3 are not treated as neighbours).
A local smoothing model (Eq. 8.15) with $$W_c = \begin{pmatrix} 0 & 1 & 0 & 0 & 0 & 0 & 0 \\ 1 & 0 & 1 & 1 & 0 & 0 & 0 \\ 0 & 1 & 0 & 1 & 0 & 0 & 0 \\ 0 & 1 & 1 & 0 & 0 & 0 & 0 \\ 0 & 0 & 0 & 0 & 0 & 1 & 1 \\ 0 & 0 & 0 & 0 & 1 & 0 & 1 \\ 0 & 0 & 0 & 0 & 1 & 1 & 0 \end{pmatrix}$$		To reflect their differences in income level (see the bottom-panel boxplot in Figure 8.14), MSOA 5 is not considered as a neighbour to MSOAs 1, 2 and 4 but it is still a neighbour to MSOAs 6 and 7.

FIGURE 8.17
Three examples of possible modification to W_{cont}. The maps in the middle column illustrate the spatial configurations implied by the corresponding W matrices. Thicker black lines denote the boundaries where two spatially-contiguous MSOAs are not treated as neighbours.

derived from rook's contiguity (denoted as W_{cont}) is given in Figure 8.16(b). With 10 unique pairs of spatially-contiguous MSOAs, Figure 8.17 presents a selection of the 1024 possible modifications to W_{cont} that, together with W_{cont}, we will investigate. Readers may construct other versions and compare the fits with those presented here (Exercise 8.8). Eq. 8.15 is the Bayesian hierarchical model with the ICAR prior fitted to the data – this model is the same as that defined in Eq. 8.2 but only for the seven MSOAs.

$$y_{ij} \sim N\left(\theta_i, \sigma_y^2\right)$$

$$\theta_i = \alpha + S_i$$

$$S_{1:7} \sim ICAR\left(W, \sigma_S^2\right)$$

$$\alpha \sim Uniform(-\infty, +\infty)$$

$$\sigma_y \sim Uniform(0.0001, 1000)$$

$$\sigma_S \sim Uniform(0.0001, 1000) \tag{8.15}$$

To find an optimal spatial structure, we fit the model in Eq. 8.15 using each version of the W matrix in Figure 8.17, apart from W_a, for which the ICAR prior on S is replaced by $S_i \sim N(0, \sigma_S^2)$ for $i = 1, \ldots, 7$. Table 8.1 compares these models using the Deviance Information Criterion (DIC). Amongst the four models, the ICAR model with W_c gives the smallest DIC value of 974, suggesting it is the most parsimonious model for this set of data. With a difference in DIC value of only 1, the global smoothing (exchangeable) model performs equally well. However, the ICAR model with either W_{cont} or W_b yields a much poorer fit. The results therefore suggest that this dataset is modelled well by imposing either a global smoothing structure on the MSOA average income or a two-clusters neighbouring structure, as implied by W_c (see the spatial configuration in Figure 8.17). Of course, there are other configurations that should be considered. Instead of fitting them all here, readers are encouraged to fit other versions of the W matrix.

As an example of the second approach to locally adaptive smoothing, we take the method proposed by Lee and Mitchell (2013). The starting model is an exchangeable model for S_i with $S_i \sim N(0, \sigma_S^2)$ for $i = 1, \ldots, 7$, namely, the global smoothing model as presented in Figure 8.17.

For each of the 10 pairs of spatially-contiguous MSOAs, Figure 8.18(a) compares the 95% credible intervals (CIs) of the estimated S_is from the exchangeable model. For each pair, the corresponding element in W is set to 1 if the two CIs overlap and is set to 0 otherwise. The weight for a pair of MSOAs that do not share a common boundary is always fixed at 0. After Iteration 0, the weights of all contiguous pairs are set to 1, thus resulting in W_{cont}, as presented in Figure 8.16(b). Using W_{cont}, the next iteration fits the local smoothing model in Eq. 8.15 to the data. The 10 pairs of 95% CIs are shown in Figure 8.18(b) and the modification gives rise to W_c, as in Figure 8.17. Iteration 2 uses W_c in Eq. 8.15, and Figure 8.18(c) compares the 95% CIs, which results in the same weights matrix, namely W_c, as in Iteration 1. Thus, the iterative algorithm stops and W_c is the optimal spatial structure for these seven MSOAs, the same spatial configuration as determined from the first approach.

TABLE 8.1

Comparison of the Four Bayesian Hierarchical Models Fitted to a Subset of the Newcastle Income Data with Seven MSOAs

W used	Smoothing Type	$\bar{D}$	pD	DIC
W_{cont}	Local	972	13	985
W_a	Global	967	8	975
W_b	Local	972	13	985
W_c	Local	967	7	974

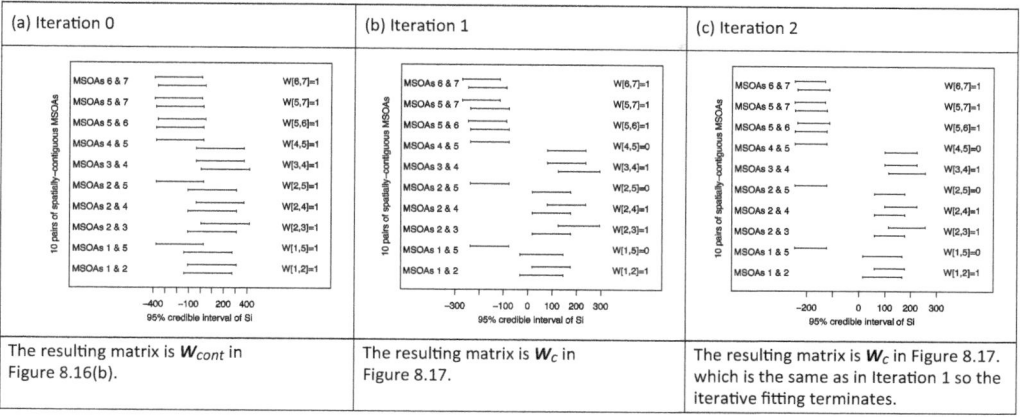

(a) Iteration 0	(b) Iteration 1	(c) Iteration 2
The resulting matrix is W_{cont} in Figure 8.16(b).	The resulting matrix is W_c in Figure 8.17.	The resulting matrix is W_c in Figure 8.17. which is the same as in Iteration 1 so the iterative fitting terminates.

FIGURE 8.18

Comparison of the 95% credible intervals (CIs) of the area-specific parameters S_is from two spatially-contiguous MSOAs. If the two 95% CIs overlap, the weight is set to 1 and is set to 0 otherwise. Note that $w_{ij} = w_{ji}$.

Finally, Figure 8.19 compares the posterior estimates of the MSOA-level average income across three models, the exchangeable model and the two versions of the ICAR model where the spatial structure is imposed either by W_{cont} or by W_c. Between the exchangeable model and the ICAR model with W_{cont}, there is little difference in the posterior means for MSOAs 1–4, 6 and 7 (Figure 8.19(a)), but the posterior mean for MSOA 5 under the ICAR model with W_{cont} is pulled upwards due to the influence from the three high-income neighbours, MSOAs 1, 2 and 4. However, when W_c is used, those three high-income MSOAs are not treated as neighbours with MSOA 5 and, as a result, the posterior mean of MSOA 5 is less affected by those high-income areas and as a result remains low. The posterior means

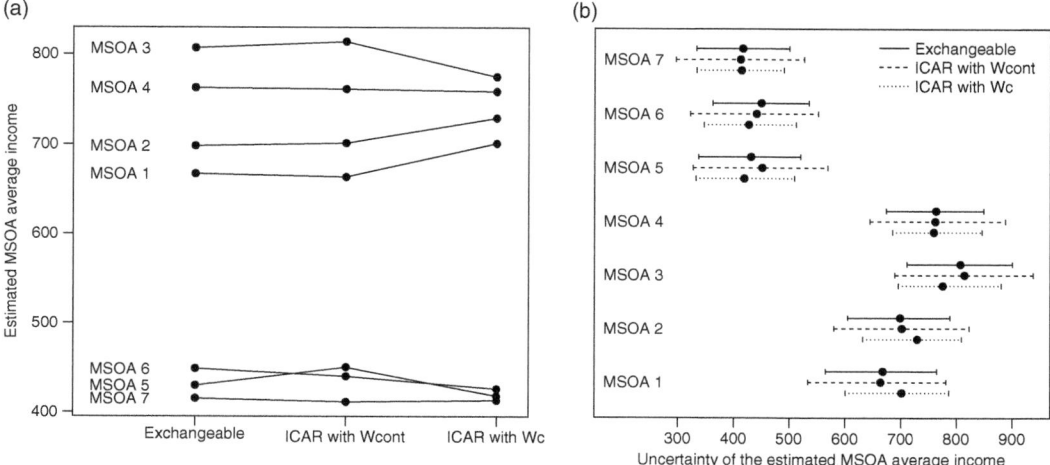

FIGURE 8.19

Comparison of (a) the posterior means and (b) the 95% credible intervals (with the dots denoting the posterior means) of the average income for the seven MSOAs from three models: the exchangeable model, the ICAR model with W_{cont} and the ICAR model with W_c. See Figure 8.16 and Figure 8.17 for the specifications of W_{cont} and W_c.

from the ICAR model with W_c (Figure 8.19(a)) better reflects the clustering feature, as evident in the exploratory plot in Figure 8.14 – the average income levels of MSOAs 1–4 tend to be similarly high, and those of MSOAs 5–7 tend to be similarly low, thus providing a better fit to the data.

As shown in Figure 8.19(b), the uncertainty intervals from the ICAR model with W_{cont} are wider compared to those from the other two models. The larger uncertainty may be due to the fact that this weights matrix imposes a spatial structure that is somewhat incompatible with the data. For the other two models, the uncertainty intervals are more comparable, although those from the ICAR model with W_c are slightly narrower, which may be due to the more appropriate sharing of local information through W_c.

Some practical implications follow from this example. Albeit simple with only seven areas, the above example highlights the fact that the spatial neighbourhood structure that we impose on the area-specific parameters is an *assumption* so that different definitions of the neighbourhood structure can result in different estimates of the parameters. From a Bayesian viewpoint, the specification of W can be thought of as comparable to a prior – a prior which is assigned to the neighbourhood structure of the area-specific parameter. In Section 5.6.2, we called this type of information *spatial knowledge*. As with any Bayesian analysis therefore, assessing the sensitivity of modelling results to different prior specifications is an important, indeed integral, part of any Bayesian spatial analysis. The locally adaptive spatial models discussed in this section allow us to systematically (and automatically) explore different specifications of the spatial structure in order to assess their potential impacts on the estimates as well as on the conclusions we draw from the analysis.

Although used in the demonstration here, WinBUGS may not be suitable for fitting these adaptive models for two reasons. First, the weights derived from the W matrix can only enter the `car.normal` function (via the `weights[]` argument) or the `car.proper` function (via the `C[]` and the `M[]` arguments) as data. The elements in these arguments cannot depend on unknown parameters. As a result, WinBUGS cannot fit an adaptive model where the weights are modelled as a function of covariates (e.g. those proposed by Lu et al. (2007) and Lee and Mitchell (2012)). Second, when dealing with a study region with a large number of spatial units, WinBUGS may need a long time in order to fit a spatial model. This makes WinBUGS inefficient for exploring different specifications of the W matrix. Other programmes for fitting Bayesian models such as INLA (e.g. in Lee and Mitchell (2013)) can be used or the fitting algorithm can be specially written in R (e.g. in Lee et al. (2014) and Anderson et al. (2016)) or in C (e.g. in Li et al. (2011)).

8.5 The Besag, York and Mollié (BYM) Model

When displayed on a map, a set of outcome values often exhibits positive, global (or whole-map) spatial autocorrelation. The spatial distribution of these values is said to be *spatially-structured*. For example, high-income (or low-income) areas tend to cluster in space because of the spatial structure of the urban housing market. However, there may also exist some pattern-elements in the data that are not spatially-structured and which are said to be *spatially-unstructured*. One source for such a pattern element might be associated with individual areas whose income levels differ from their neighbours due, for example, to localised patterns of gentrification or housing in-fill. Another source is where there is some element of randomness in the degree of similarity between otherwise quite similar contiguous areas. For example, some degree of income divergence might be expected in well-established housing areas as some streets or neighbourhoods improve and others decline, due to the effects of tenure history.

Such spatially-structured and spatially-unstructured patterns can be accounted for if the relevant covariates are available. Unfortunately, this may not be the case in practice, yet if our model is to provide a good basis for inference, it still needs to accommodate both the spatially-structured and the spatially-unstructured variation that may be present in the outcome data. In this section we introduce the BYM model, which combines two model components, one capturing spatially-structured variability and the other spatially-unstructured variability. This model is named after its authors, Besag, York and Mollié, who developed the model in their 1991 paper.

We formulate the BYM model in the context of the Newcastle income example. However, as we will illustrate subsequently in the book, this model can be applied to both continuous-valued and discrete-valued outcome data. As before y_{ij}, the income level from the survey for household j in MSOA i is modelled by $y_{ij} \sim N(\theta_i, \sigma_y^2)$. The aim of the analysis is to estimate the θ_is, the average income levels for the 109 MSOAs. Under the BYM model, the area-specific parameter θ_i is modelled as follows:

$$\theta_i = \alpha + S_i + U_i, \tag{8.16}$$

where S_i is a parameter from the spatially-structured component $S = (S_1, \ldots, S_N)$ and U_i is a parameter from the spatially-unstructured component $U = (U_1, \ldots, U_N)$. In the absence of all the necessary covariates in the model, S and U accommodate, respectively, the spatially-structured and the spatially-unstructured patterns present in the data. The parameters in each component are area-specific, and they are modelled hierarchically. For S, the ICAR model is used so that the parameters in S are structured spatially according to the chosen spatial weights matrix. The parameters in U are modelled through an exchangeable model, for which a common choice is a normal distribution centred at 0 with an unknown variance, i.e. $U_i \sim N(0, \sigma_u^2)$, independently for all $i = 1, \ldots, N$. The parameters in U are therefore spatially-unstructured.

Because of the sum-to-zero constraint on S and the prior mean of 0 for the elements in U, α, the intercept, is included to represent the overall level. As discussed in Section 8.2.1.2, $(S_i + U_i)$ can be interpreted as the deviation of the average income level in area i from the overall level α. Such local deviations can result from a combination of area-level characteristics that tend to be more alike in neighbouring areas and area-level characteristics that are specific to certain areas. In the absence of such covariate information, Besag et al. (1991, p.7) interpret the two components "as surrogates for unknown or unobserved covariates; the u_is represent variables that, if observed, would display substantial spatial structure in that the values for a pair of contiguous zones would be generally much more alike than for two arbitrary zones, whereas the v_is represent unstructured variables" – u_i and v_i in the quoted text correspond to S_i and U_i, respectively, in our notation.

When area-level covariates are available, Eq. 8.16 can be extended to

$$\theta_i = \alpha + \sum_{k=1}^{K} \beta_k x_{ik} + S_i + U_i \tag{8.17}$$

Then the two components measure the spatially-structured and the spatially-unstructured patterns in the *residuals* after the K covariate effects have been accounted for.

While the two components are included to represent the unmeasured/unobserved covariates, in terms of the estimation of the area-specific parameters, the hierarchical structures assigned to the two components imply that information is shared both locally and globally. Specifically, information is shared locally through the ICAR model on S while the exchangeable model on U allows information to be shared globally. In practice, for a given dataset, the relative importance of the two components is not known but can be quantified using the spatial fraction, a version of the variance partition coefficient that compares the variability in S to the variability of S and U combined. This relative importance has an implication on the estimates of the θ_is. As Besag et al. (1991) explain in the context of estimating small area disease risks, "If u [dominates], then the estimated risks will display spatial structure; if v, then the effect will be to shrink the estimated risks towards the overall mean." We will illustrate such a measure of importance in the example below.

8.5.1 Two Remarks on Applying the BYM Model in Practice

The first remark concerns the implementation of the BYM model. There are two equivalent formulations of the BYM model, the hierarchically-centred version and the non-hierarchically-centred version. The non-hierarchically-centred version is given in Eq. 8.16 (or Eq. 8.17 with covariates included), while the hierarchically-centred version with covariates is given below

$$\theta_i \sim N\left(\mu_i, \sigma_U^2\right)$$

$$\mu_i = \alpha + \sum_{k=1}^{K} \beta_k x_{ik} + S_i \tag{8.18}$$

The hierarchically-centred version is a re-parameterisation of the non-hierarchically-centred version so that the parameter estimates from both versions are virtually identical. Note that when fitted in WinBUGS, the posterior estimates from both versions will be very close but not identical because of the simulation nature of the MCMC method (see MC error in Section 5.3.5.3). From the hierarchically-centred version, we can recover the spatially-unstructured parameter U_i by calculating $\theta_i - \mu_i$ (see the income example below). When fitting a BYM model in WinBUGS, the hierarchically-centred version often yields faster convergence and better mixing.[11]

The second note is on the potential issue of non-identifiability. The BYM model partitions a single map of outcome values into two components. This partitioning causes a non-identifiability issue because each area has only one data value available to estimate two unknown parameters, S_i and U_i.[12] To overcome this problem, the two components are assumed to have different spatial structures *a priori*. The strong spatial positive autocorrelation from an ICAR model on S contrasts with the non-spatial structure from the exchangeable model on U so that the two components can be estimated separately. Although other

---

[11] First proposed in Gelfand et al. (1995), hierarchical centring is a method that re-parameterises a hierarchical model in order to achieve faster convergence when fitting the model using MCMC methods. Browne (2004) and Browne et al. (2009) further show the efficiency of the technique when fitting hierarchical models.

[12] In the context of count data, the single data value observed in each area refers to the number of crime or disease cases that occurred in the area (within a given period). In the case of the income survey example, although, for most of the areas, we have multiple data points, these data points are at the household-level. At the area-level, however, we only have one "observed value", which is the mean of the sample data in each area.

spatial models can be specified on S (e.g. a finite mixture model; see Chapter 9), the spatial dependence structure needs to be markedly different from the non-spatial structure of the exchangeable model. The proper CAR (pCAR) model for S, on the other hand, is generally not recommended to be used in the BYM model. This is to avoid the potential non-identifiability between the two components when spatial autocorrelation in the data is weak – when ρ is close to 0, both the pCAR model and an exchangeable model with a common normal distribution imply similar structures on the area-specific parameters (see Section 8.3.2). Non-identifiability can lead to non-convergence of the MCMC chains for the three hyperparameters: σ_u^2, the variance of the non-spatial component; ρ, the spatial autocorrelation parameter; σ_S^2, the conditional variance in the pCAR model; and/or poor mixing of the MCMC chains for the above parameters (i.e. the iterations within each MCMC chain are highly autocorrelated).[13]

8.5.2 Applying the BYM Model to the Newcastle Income Data

Eq. 8.19 specifies a Bayesian hierarchical model with the BYM structure on the average income levels across the MSOAs. The spatial neighbourhood structure is defined using rook's contiguity, and the neighbouring weights are set to 1. The last line of Eq. 8.19 defines the spatial fraction, the ratio of $\tilde{\sigma}_S^2$, the unconditional variance of the spatially-structured random effect terms (which is not equal to the conditional variance σ_S^2), to the sum of the two variances, $\tilde{\sigma}_S^2$ and $\tilde{\sigma}_U^2$, where the latter is the variance of all the spatially-unstructured random effect terms. The WinBUGS implementation is given in Figure 8.20, and Lines 35–38 calculate the spatial fraction.

$$y_{ij} \sim N\left(\theta_i, \sigma_y^2\right)$$

$$\theta_i \sim N\left(\mu_i, \sigma_u^2\right)$$

$$\mu_i = \alpha + S_i$$

$$S_{1:109} \sim ICAR\left(\mathbf{W}, \sigma_S^2\right)$$

$$a \sim Uniform\left(-\infty, +\infty\right)$$

$$\sigma_y \sim Uniform\left(0.0001, 1000\right)$$

$$\sigma_S \sim Uniform\left(0.0001, 1000\right)$$

$$\sigma_U \sim Uniform\left(0.0001, 1000\right)$$

$$spatial\ fraction = \frac{\tilde{\sigma}_S^2}{\tilde{\sigma}_U^2 + \tilde{\sigma}_S^2} \tag{8.19}$$

---

[13] Other modelling forms have been proposed to estimate the strength of spatial autocorrelation while simultaneously incorporating the spatially-unstructured component (see, for example, the Cressie model by Cressie (1992) and Stern and Cressie (2000) and the Leroux model first proposed by Leroux et al. (2000) then further explored in MacNab (2003)). Readers are referred to the aforementioned papers regarding the specifications of these models. MacNab (2003) and Lee (2011) both provide detailed comparisons of these models (including the BYM model) in simulation settings as well as in analysing real datasets.

```
1     #  The WinBUGS code for specifying the model with the BYM model
2     model {
3       #  a for-loop to go through all the household level income data
4       for (j in 1:nhhs) {
5         #  defining the normal likelihood for each household
6         y[j] ~ dnorm(theta[msoa[j]],prec.y)
7       }
8       #  the BYM model with hierarchical centring
9       for (i in 1:nmsoas) {
10        mu.theta[i] <- alpha + S[i]
11        theta[i] ~ dnorm(mu.theta[i],prec.U)
12        #  recovering the spatially-unstructured terms
13        U[i] <- theta[i] - mu.theta[i]
14      }
15      #  modelling all the elements in S using the ICAR model
16      S[1:nmsoas] ~ car.normal(adj[], weights[],num[], prec.S)
17
18      #  the improper uniform prior for the Newcastle average
19      alpha ~ dflat()
20
21      #  a vague prior for the sampling standard deviation
22      sigma.y ~ dunif(0.0001,1000)
23      #  a vague prior for the conditional SD in the ICAR model
24      sigma.S ~ dunif(0.0001,1000)
25      #  a vague prior for the SD in the exchangeable model
26      sigma.U ~ dunif(0.0001,1000)
27
28      #  convert the SD to the precision
29      prec.y <- pow(sigma.y,-2)
30      prec.S <- pow(sigma.S,-2)
31      prec.U <- pow(sigma.U,-2)
32
33      #  calculate the spatial fraction to measure the relative
34      #  importance of S and U
35      var.S.un <- sd(S[])*sd(S[]) #  the unconditional variance of S
36      var.U <- sd(U[])*sd(U[])    #  the variance of the spatially-
37                                  #  unstructured parameters
38      spatial.frac <- var.S.un / (var.S.un + var.U)
39
40      #  calculate the posterior probability that the average income
41      #  of MSOA i is above the Newcastle average
42      for (i in 1:nmsoas) {
43        diff[i] <- theta[i] - alpha
44        prob.gt.average[i] <- step(diff[i])
45      }
46    }
47
48    #  the data list is the same as that in Figure 8.1 for fitting
49    #  the model with the ICAR prior
50
51    #  initial values for chain 1
52    list(alpha=200,sigma.y=50,sigma.S=20,sigma.U=10
53         ,S=c(1,-1,0,0,0,…),theta=c(300,300,300,...))
54
55    #  initial values for chain 2
56    list(alpha=700,sigma.y=80,sigma.S=50,sigma.U=20
57         ,S=c(-1,1,0,0,0,…),theta=c(600,600,600,...))
```

FIGURE 8.20
The WinBUGS code for fitting the BYM model (Eq. 8.19) to the income data. The hierarchically centred version of the BYM model is used for the implementation.

We also calculate the posterior probability that the average income level of each MSOA exceeds the Newcastle average, i.e. $pp_i = \Pr(\theta_i > \alpha \mid \text{data})$. This posterior probability is calculated in Lines 42–45 using the `step` function. At each MCMC iteration, the `step` function returns 1 if its argument is non-negative (i.e. when θ_i is greater than α) and returns 0 otherwise. Thus, for each MSOA, the posterior mean of `prob.gt.average[i]` gives the proportion of iterations exceeding the Newcastle average and therefore the required posterior probability.

Figure 8.21 illustrates the structure of the BYM model. All maps in Figure 8.21 show the departures of the MSOA-level average income from the Newcastle average. For all

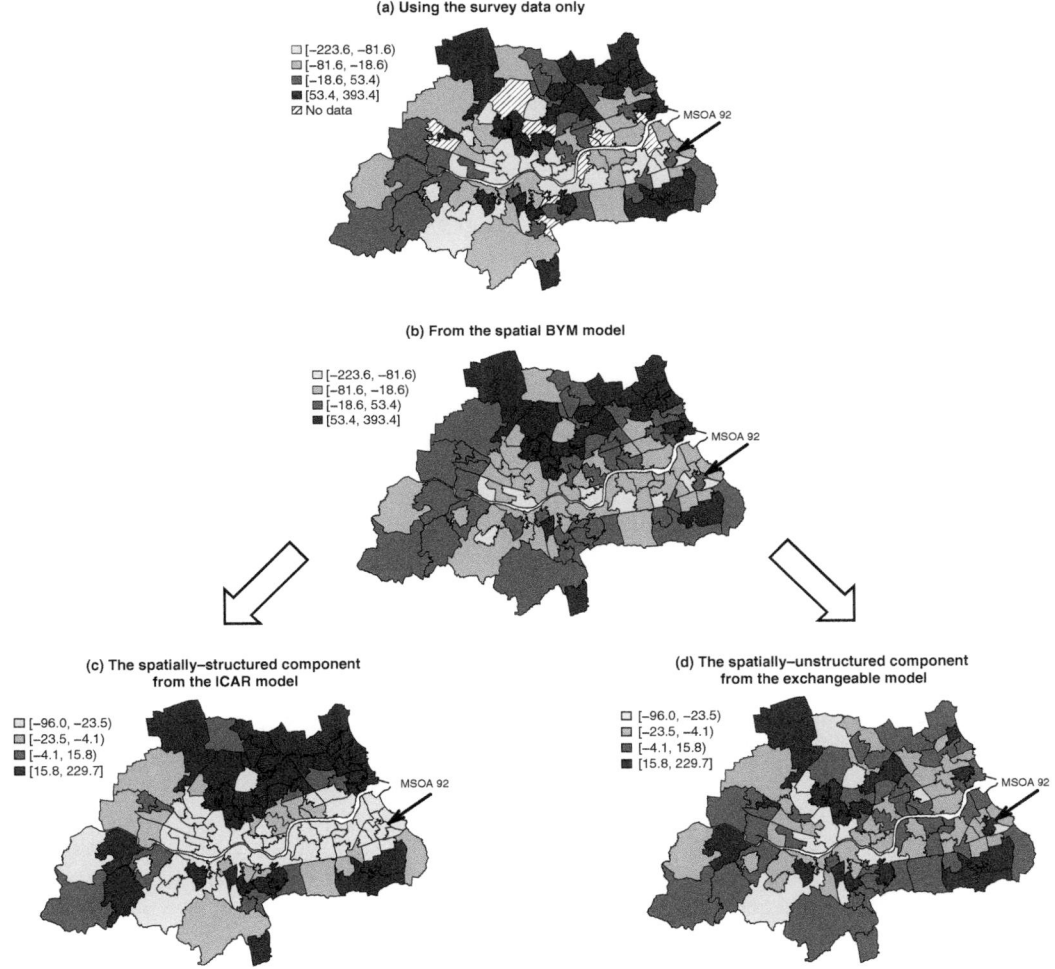

FIGURE 8.21
The maps of the local departures in average income from the Newcastle average (a) using only the survey data (no values assigned to the 10 MSOAs with no survey data) and (b) estimated based on the BYM model. The values in map (a) are calculated by subtracting the grand mean of the survey data, which is very close to the posterior mean of α from the BYM model, from the sample mean of each MSOA. The values in map (b) are the posterior means of $(S_i + U_i)$, for $i = 1, \dots, 109$. The posterior means of the spatially-structured and the spatially-unstructured components of the BYM model are shown in (c) and (d) respectively.

maps, an MSOA with a darker (or a lighter) colour is estimated to have an average income level above (or below) the Newcastle average. Based on the survey data, the BYM model estimates the average income levels across the MSOAs by combining the two model components, S and U. Therefore, adding the maps in panels (c) and (d) of Figure 8.21 gives the map in panel (b), which, together with the posterior estimate of the overall intercept α, gives the estimates of the average income levels across MSOAs. Compared to the raw data map in panel (a), the posterior means from the BYM model in map (b) are smoother spatially in the sense that the income levels amongst the MSOAs that are geographically close are more alike compared to those from MSOAs that are far apart. The locally smooth feature of map (b) is a direct result of the spatially-structured parameters, S, whose point estimates, as mapped in panel (c), show a spatial pattern that is broadly comparable to that in map (b). But there are clearly some differences between the two maps due to the spatially-unstructured component, U, whose job is to capture the local variability that does not conform to the spatially-smooth pattern from the ICAR model.

To provide an example, MSOA 92 is highlighted in each map. With a sample size of six, this MSOA has a sample mean of 563.79, which is slightly above the Newcastle average of 525.59 (the grand mean of the survey data) and is higher than those of its contiguous neighbours, which tend to have a light grey colour in map (a) of Figure 8.21. Because of the local smoothing nature, the posterior mean of this MSOA under the spatially-structured component is −23.9, meaning that its average income is lower than the Newcastle average. It has clearly been overly smoothed. But this over-smoothed estimate is counterbalanced by the spatially-unstructured component, under which the posterior mean is 22.4. Therefore, the sum of the two point estimates is −1.5, suggesting that this MSOA may have an average income level that is close to the Newcastle average, in line with the observed data. This illustrates the joint effort of the two components in the BYM model in describing the patterns in the observed data. Exercise 8.9 asks the reader to investigate the BYM model further in order to provide some general insight into the data properties that determine the global and local smoothing of this model.

Note also that the above calculation is based on posterior means, only one of the many ways to summarise a posterior distribution. An advantage of Bayesian inference is that, using the posterior distribution, we can compute the posterior probability of how likely it is that the average income level of an MSOA is above (or below) a certain threshold (see Exercise 7.4 in Chapter 7 as well as the calculation of the posterior probability, $pp_i = \Pr(\theta_i > \alpha \mid \text{data})$, as defined above).

As has been noted, the spatial fraction is a way to gauge the relative importance of the two spatial components. For this dataset, the posterior median of the spatial fraction is 76.8% with a 95% CI of (23.3%, 99.9%), suggesting that compared to the spatially-unstructured component, the spatially-structured component accounts for a large proportion (about 77%) of the between-MSOA variability in average income. In other words, S dominates U. This explains why the combined map (b) in Figure 8.21 is closer in pattern to the map of the posterior means of S than to that of U. Since the spatial fraction is estimated to lie between 0 and 1, modelling the spatial distribution of the MSOA-level income data requires both the spatially-structured component and the spatially-unstructured component, although the former is more important. Note also that we use the posterior median, as opposed to the posterior mean, as a point estimate for the spatial fraction due to its skewed posterior distribution (see Figure 8.22).

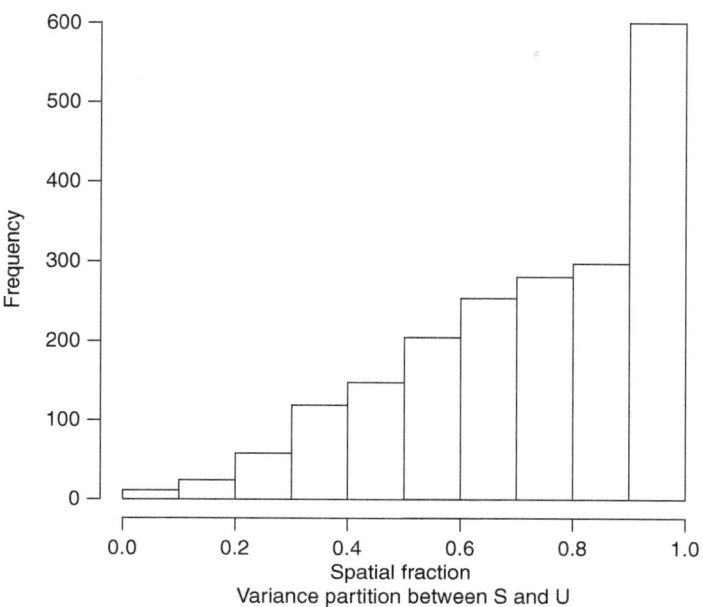

FIGURE 8.22
The posterior distribution of the spatial fraction.

Finally, Figure 8.23 shows the history plots for $\tilde{\sigma}_u^2$ and $\tilde{\sigma}_s^2$, the variance of the spatially-unstructured component and the unconditional variance of the spatially-structured component, respectively, and the spatial fraction. While the two chains have converged for all three quantities, the mixing is somewhat poor and the iterations within each chain are highly autocorrelated. Furthermore, a close inspection shows a negative correlation between the values for the two variances. For example, higher values of $\tilde{\sigma}_s^2$ tend to be associated with lower values of $\tilde{\sigma}_u^2$ (see, for example, the right-hand end of the black chains for $\tilde{\sigma}_s^2$ and $\tilde{\sigma}_u^2$). This is due to the identifiability issue between the two components discussed earlier. When the parameters in S vary more, then they explain more of the variability in the observed data, and hence the parameters in U are forced to vary less and vice versa. The two variances, $\tilde{\sigma}_s^2$ and $\tilde{\sigma}_u^2$, are thus intrinsically negatively correlated, and their correlation also results in the poor mixing of the MCMC chains. This poor mixing issue becomes less pronounced when the two components contribute to the combined map equally (or, equivalently, the spatial fraction is estimated to be well away from either 0 or 1). Figure 8.23 serves as an example of a typical mixing when one component dominates the other. While the autocorrelation of the MCMC iterations can be reduced via thinning whereby only every, say, 10th iteration is retained, we can alleviate the issue of poor mixing by simply running the MCMC chains longer. We can proceed with the posterior summary as soon as the requirements on convergence and efficiency are met (see Sections 5.3.4.2 and 5.3.4.3).

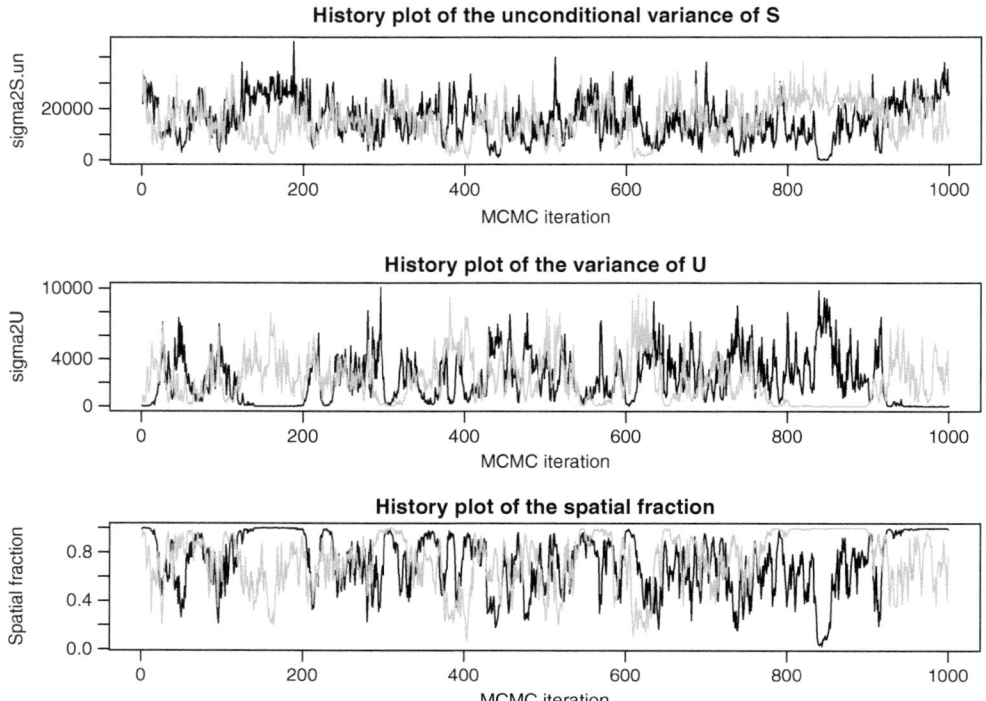

FIGURE 8.23
History plots of the unconditional variance of the spatially-structured component (top panel), the variance of the spatially-unstructured component (middle panel) and the spatial fraction (bottom panel). Each history plot is produced based on two MCMC chains run over 20000 iterations, with the first 10000 iterations discarded as burn-in. For the second half of each chain, every 10th iteration was stored, and each plot shows the resulting 1000 iterations from each chain. The two MCMC chains for each parameter have converged but show poor mixing with a high degree of autocorrelation within each chain.

8.6 Comparing the Fits of Different Bayesian Spatial Models

As we saw in Section 5.5, the Deviance Information Criterion (DIC), together with various other accuracy checks, can be used to compare Bayesian spatial models. We apply this approach now to the results we have obtained from fitting various models to the Newcastle income data.

8.6.1 DIC Comparison

Table 8.2 tabulates the values of $\bar{D}$, pD and DIC for five Bayesian models. Compared to the non-hierarchical model with identical parameters, the four Bayesian hierarchical models all have much lower DIC values and thus are more in line with the observed data. Amongst the four hierarchical models, the three models that include a form of local smoothing perform equally well and are all better than the model with exchangeable parameters, indicating the benefit of borrowing information locally in this application. In addition to the evidence from the raw data (see for example Figure 8.21(a)), this message has been consistently revealed by various models, for example, through the estimated ρ in the pCAR model (Figure 8.13) and the spatial fraction from the BYM model (Figure 8.22).

Inspecting the values more closely, the BYM model has a slightly lower $\bar{D}$ value (by 2) than the ICAR model, suggesting the addition of the spatially-unstructured component leads to a slightly better fit to the data. The BYM model, however, is more complex than the ICAR model, so the two models are not different in terms of DIC. The similarity between these two models is also evident from the fit of the BYM model, where the spatially-structured component dominates the spatially-unstructured component – including U matters, but it is less important than S. For the pCAR model, the spatial autocorrelation parameter ρ is estimated to be close to 1, thus the pCAR model and the ICAR model (where ρ is fixed at 1) have similar $\bar{D}$ and pD and the same DIC value.

With larger numbers of effective parameters (pD), the four hierarchical models are more complex than the non-hierarchical model, which has only two parameters, the overall average and the sampling variance – for a non-hierarchical model, pD is close to the actual number of parameters. However, for a hierarchical model, because of information sharing, pD is different from the actual number of parameters in the model. For example, for the exchangeable parameters model, pD is estimated to be 75, while there are 112 actual parameters (α, σ_θ^2, σ_y^2 and $\theta_1,\ldots,\theta_{109}$; see Eq. 7.6). See also the discussion in Section 5.5.

The DIC value of the non-hierarchical Bayesian model with independent parameters is 9569 (pD and $\bar{D}$ are 99 and 9469 respectively), much lower than those in Table 8.2. This is because the independent parameters model has a distinct feature of allowing the sampling variance to vary across areas (see Section 7.3.2 in Chapter 7, where the variance in the normal likelihood, σ_i^2, is indexed by i, the area indicator), while all the models in Table 8.2 assume a constant sampling variance (thus, there is only one variance parameter, σ_y^2, in, for example, Eq. 7.6 (Chapter 7) of the exchangeable model and in Eq. 8.2 of the ICAR model). The independent parameters model is a heteroscedastic model (also referred to as a heterogeneous variance model) and those in Table 8.2 are homoscedastic models. The lower DIC value from the independent parameters model suggests the benefit of allowing the within-area variance to be area-specific. Intuitively, the "varying-variance" assumption is more realistic since it seems very likely that the within-area variation in household income will differ from one area to another. The models in Table 8.2 can be extended to allow for area-specific within-area variances. One can plug in the sample variances as formulated in the independent parameters model, but this may give rise to problems

TABLE 8.2

Summary of the Posterior Mean Deviance ($\bar{D}$), the Effective Number of Parameters (pD) and the DIC Value for Each of the Five Bayesian Models Applied to the Newcastle Household-Level Income Data

Bayesian Models	Hierarchical?	$\bar{D}$	pD	DIC
With identical parameters (Eq. 7.1)	No	9910	2	9912
With exchangeable parameters (Eq. 7.6)	Yes	9647	75	9722
With an ICAR model (Eq. 8.2)	Yes	9647	69	9717
With a pCAR model (Eq. 8.11)	Yes	9646	71	9717
With a BYM model (Eq. 8.19)	Yes	9645	72	9717

The ICAR, pCAR and the BYM models all use the same spatial weights matrix, which is defined by rook's spatial contiguity with the weights of the neighbours set to 1.

when the data are sparse within each area. An alternative model-based approach is to model the (log) area-specific variances hierarchically, an idea that is similar to modelling the unknown area-level means. While such heterogeneous-variance models can be implemented in WinBUGS, they are beyond the scope of this book. Readers are referred to, for example, Leckie et al. (2014) for an exposition of such models. In addition, the reader is encouraged to study the paper by Best et al. (2005) where the authors compare a number of Bayesian spatial models in the context of disease mapping.

8.6.2 Model Comparison Based on the Quality of the MSOA-Level Average Income Estimates

Recall that the household-level income data were simulated using the known MSOA-level average (see the Appendix to Chapter 7). For each model, we can compare the posterior estimates of the MSOA-level average to the true values. The quality of the estimates is measured by three metrics, two of which are: (a) the average bias (avg. bias), the mean of the differences between the posterior means and the true values, and (b) the root mean square error (RMSE), the square root of the sum of the squared differences. Metrics (a) and (b) assess the quality of the point estimates so that a model with a smaller average bias or a smaller RMSE overall gives better estimates. Metric (c), the 95% coverage, measures how well a model represents parameter uncertainty. To calculate the 95% coverage, we assign the value 1 to an area if its 95% credible interval for the average income contains the true value and 0 if it does not. The mean of the assigned values of 0/1 across the areas gives the 95% coverage. If a model represents the uncertainty well, its 95% coverage should cover 95% of the true values. A 95% coverage that is higher (or lower) than 95% would suggest the estimates of this model are overly uncertain (or overly confident). These three metrics were calculated separately for the MSOAs with and without survey data, and the results are tabulated in Table 8.3.

For the 99 MSOAs with survey data, apart from the identical parameters model, all the other five models have similar summaries in terms of average bias, RMSE and 95% coverage. Due to the unrealistic assumption that all MSOAs have the same average income level, the identical parameters model performs poorly, with a high RMSE value and a very low 95% coverage. All the other five models allow the average income levels to vary across MSOAs, a feature that gives rise to posterior estimates of reasonably high quality. Across these five models, the point estimates are relatively accurate with small biases, and the

TABLE 8.3

Comparing the Quality of the Estimates for the MSOA-Level Average Income from Six Different Bayesian Models

	For the 99 MSOAs with Survey Data			For the 10 MSOAs with No Survey Data		
	Avg. bias	RMSE	95% Coverage	Avg. Bias	RMSE	95% Coverage
Independent	5.6	45.5	92.9%	–	–	–
Identical	5.8	95.4	9.1%	30.5	117.1	0.0%
Exchangeable	4.9	43.4	92.9%	29.9	117.4	80.0%
ICAR	5.4	41.6	92.9%	30.3	89.2	90.0%
pCAR	5.2	41.3	91.9%	29.2	95.6	80.0%
BYM	4.8	41.0	93.9%	30.1	92.8	90.0%

The comparison is carried out for the MSOAs with and without survey data separately.

95% CIs represent the uncertainty well since the 95% coverages are reasonably close to the nominal 95% level.

For the 10 MSOAs without survey data, there are no estimates from the independent parameters model. The identical parameters model has a poor 95% coverage of 0%. Although not yet hitting the 95% nominal level, the coverages from the four hierarchical models are much better, indicating a better representation of the uncertainty from these Bayesian hierarchical models. While these four hierarchical models do not differ much in terms of average bias, the three models that incorporate a form of local smoothing, i.e., the ICAR, the pCAR and the BYM models, yield similar values of RMSE and all are lower than the RMSE value from the exchangeable model. This highlights the benefit of borrowing information locally in this application, the same conclusion that we drew from the DIC comparison.

In a full simulation study, such model comparisons should be performed over a number of simulated datasets to obtain a more representative view on model performance (Exercise 8.10). The interested reader is referred to Gómez-Rubio et al. (2008a and 2008b) for a full investigation of using different Bayesian hierarchical models for small area estimation.

8.7 Concluding Remarks

At this point let us stand back from the detail of this chapter, and the previous one, to make clear where we have got to. Throughout these two chapters, our aim has been to provide estimates of a set of parameters, $\theta_1,\dots,\theta_N$, for N small geographical areas using "information borrowing". In our running example, these N parameters represent the average household income levels across the MSOAs in Newcastle, England. The data available to us for estimating these parameters consist of income levels for sample households in the N areas. However, sample sizes vary between areas and in some cases are small whilst some areas have no sample data at all – the challenge of data sparsity. As an important part of this endeavour, we have shown, for each spatial prior model and sample size, *the mechanism* under which information is borrowed across spatial units and the *consequences* in terms of their effect on parameter estimates.

We adopted the Bayesian hierarchical modelling approach to the estimation of these parameters. To this end we specified a *data model* for the sample data, $y_{ij} \sim N(\theta_i, \sigma_y^2)$, where y_{ij} is the income value for household j in area i, and θ_i, the parameter of interest, is the average household income for area i. These data values are assumed to be independent given the data generating process. So, in addition to the data model, we also specified a *process model* for the set of parameters, $\theta_1,\dots,\theta_N$. Typically, a process model includes an intercept term, a set of observable covariates that are thought to be associated with the outcome of interest and a set of area-specific *random effects*. It is the specification of the last of these three components that Chapters 7 and 8 have focused on. In Chapter 7, Strategy 3 specifies a *hierarchical* structure for these random effects in which the random effects are considered to be exchangeable. This exchangeable hierarchical structure allows *partial information borrowing* – a compromise between two non-hierarchical modelling structures, namely, complete information pooling (Strategy 1) and no information pooling (Strategy 2) – thereby addressing two of the challenges associated with modelling spatial data: spatial heterogeneity and data sparsity.

In this chapter, we have gone a stage further by modelling these N random effects according to the spatial dependence structure (amongst other properties) that we have reason to believe is present in the parameters. Of course, that spatial dependence is, by definition, unknown so that *how* it is specified is a modelling assumption. A natural starting point for modelling these random effects is to invoke the idea that parameters in areas that are near in geographical space tend to be more alike compared to those for areas that are far apart. We may arrive at such an assumption drawing on relevant research studies that have suggested the presence of this property in the system that we wish to study. In the absence of such relevant studies, such an assumption may come from general assertions, such as Tobler's First Law of Geography or the First Law of Geostatistics (Chapter 3), or it might come from a quite basic intuition we hold to, that in a world of continuous space the above property is more likely to arise than not and hence may be considered as a reasonable assumption to make. In a *Bayesian* hierarchical model, we represent this spatial assumption by assigning *a spatial prior probability distribution* (or, equivalently, a *spatial prior model*) to the random effects. In this chapter we have looked at various prior models to formalise such spatial assumptions: the ICAR, the pCAR and the BYM models. We have also considered some local adaptive models where spatial discontinuity is suspected.

The various spatial priors (ICAR, pCAR, BYM) on the random effects may be considered to be "somewhat" *informative* because they represent our prior assumptions (or beliefs) on the spatial structure of these parameters. However, depending on what we invoke as the source of our prior knowledge, the quality of this prior information is likely to be variable. Very often, when specifying these spatial priors, we are not drawing on the accumulated evidence of many previous experiments; often we are simply invoking some general ideas we have about how "parameters vary spatially" (spatial knowledge as we termed in Section 5.6.2). For this reason, as part of our analysis we should consider the sensitivity of our results (parameter estimates) to our modelling assumptions – both the type of spatial model and the choice of **W**. In doing so we should also recognise that these particular spatial models capture only a limited range of possibilities as to how attributes can vary spatially. We will pick up on this point in Chapter 17.

All the models considered in Chapters 7 and 8 are referred to as unit-level models because they deal with household-level outcome data (see Chapter 4 in Rao and Molina, 2015). However, our interest is in obtaining parameter estimates at the area level. Complications arise if we want to model outcome at the unit level (e.g. the household level, or in some cases the individual level) using covariates *and* derive area level estimates too. This is not the subject of this book, although we will discuss this in Chapter 17.

In the next chapter, Chapter 9, through a series of case studies we discuss several applications of Bayesian hierarchical modelling which will engage with other challenges in addition to the challenges and opportunities presented by spatial dependence. The applications we visit all involve the inclusion of covariates in the process model. Covariates are added to the process model in order to account for the variation in the observed outcome values so that we can make inferences about the underlying association between each covariate and the outcome of interest (the fixed effects). Such inference allows us to test theories or make predictions. Including spatial random effects in these circumstances means that information borrowing takes place across areas *after controlling for the effects of the included covariates*. In addition, including random effects enables us to acknowledge the presence of omitted covariates, which can display spatially-smooth and/or spatially-non-smooth patterns. Whilst our hope is that by including spatial random effects we shall obtain better estimates of the fixed effects, which usually are our principal interest, there is no guarantee that this will be so. As a warning, there remains the possibility of spatial confounding. See the discussion of this topic in Section 4.9.3.

8.8 Exercises

Exercise 8.1. Taking the joint probability distribution in Eq. 8.4 as the starting point, derive the conditional distributions under the ICAR model (Eq. 8.3) using the W matrix given in Figure 8.16(b).

Exercise 8.2. Show that the joint probability distribution for the ICAR model (Eq. 8.4) can be written in the form of Eq. 8.6, thereby illustrating the feature that only the pairwise differences of the random variables enter the ICAR model.

Exercise 8.3. Explore the relationship between income and the number of full-time working residents at the MSOA-level. Then fit the model presented in Eq. 8.7 to the Newcastle income data in WinBUGS, paying particular attention to the derivation of the spatial weights matrix W_{hour} whose non-zero elements are defined using the number of full-time working residents in each MSOA. (Hint: First derive W using rook's contiguity (see Appendix 4.13.2 in Chapter 4) then modify the non-zero weights accordingly; also set the difference to 10 if two MSOAs have the same number of full-time working residents.)

Exercise 8.4. Show that the largest eigenvalue of the row standardised weights matrix is 1 while all other eigenvalues are between −1 and 1.

Exercise 8.5. Carry out the simulation as described in Section 8.3.2 to investigate using ρ, the spatial autocorrelation parameter in the pCAR model, as a way to measure the strength of spatial autocorrelation. In the simulation, set v in Eq. 8.10 to 1.

Exercise 8.6. Fit the model in Eq. 8.11, using the pCAR prior for the spatially-structured random effects, to the Newcastle income data.

Exercise 8.7. The R code given in Figure 8.15 carries out a Monte Carlo simulation to investigate the impact of the choice of prior on the neighbouring probability in the locally adaptive spatial smoothing model (8.4.2) of Lu et al. (2007). Modify the R code in Figure 8.15 to explore the impact of the choice of prior on the neighbouring probability if the following two vague priors are used for both γ_0 and γ_1: (a) Uniform(−10,10) or (b) $N(0,10)$ assigned to both parameters.

Exercise 8.8. Follow the description in Section 8.4.3 to carry out the fitting of the model in Eq. 8.15 with several plausible specifications of the W matrix other than those given in Figure 8.17. Compare your model fits to those considered in Section 8.4.3 via DIC as well as through the posterior estimates of the MSOA-level average income (i.e. in the form of Figure 8.19).

Exercise 8.9. At various points in Chapters 7 and 8 we have provided insight into what is pulling estimates in different directions. For example, in the case of the exchangeable model (Section 7.4) we have commented on what determines the extent to which properties of the observed data pull the local estimates away from the sample mean and towards the global mean. In Section 8.2.1.4, we have indicated what influences the pull towards the local neighbours. In the case of the BYM model, provide some general insight into the data properties that determine the relative pull towards the sample mean, the global mean and the mean of the local neighbours. (Hint: A careful inspection of the results from fitting the model in Eq. 8.19 to the Newcastle income data may shed some light.)

Exercise 8.10. Carry out the model evaluation procedure as discussed in Section 8.6.2 on multiple (say 50) datasets, each simulated following the description given in the Appendix of Chapter 7.

Exercise 8.11. The zip file, `PHIA_with_covariates.zip`, on the book's website contains the shapefile of the Census Output Areas (COAs) in Sheffield with the binary 0/1 outcome values that indicate whether a COA was considered as a high intensity crime area (HIA) by the police (see Section 5.2.2). Four COA-level covariates are also included in the shapefile (see Exercise 5.11 in Chapter 5). Fit a logistic regression model to the binary outcome data with the four covariates and a set of area-specific random effects. Use the exchangeable model, the ICAR model, the pCAR model and the BYM model for the random effects. Compare the covariate effects estimated from the different models. Give reasons if you encounter non-convergence and/or poor mixing when fitting some of these models. Calculate residuals and test the correlation structure of the residuals of each model.

Exercise 8.12. Repeat the same analysis outlined in Exercise 8.11 for the data in the zip file, `EHIA_with_covariates.zip`, which contains the binary outcome data for the empirically-defined HIAs (see Exercise 5.12 in Chapter 5). Are the conclusions on the covariate effects different from those for the PHIAs? We will return to the analysis of the PHIA and EHIA data in Chapter 9.

9

Bayesian Hierarchical Models for Spatial Data: Applications

9.1 Introduction

In this chapter we shall overview a number of applications of spatial modelling that tackle a diverse range of substantive problems in the social and public health sciences. In the process, we show how, within the Bayesian approach to inference, certain statistical challenges can be addressed. All four applications involve the use of Bayesian hierarchical models. In the first application, an aim is to identify the covariates that explain why some areas of a city are classified as high intensity crime areas (HIAs) whilst others are not. In the second, the aim is to assess the relationship between exposure to nitrogen oxide and stroke mortality at the small area level. The third application is an analysis of small area counts of new cases of malaria in a small region of India. The fourth aims to model the spatial variation, at the small area scale, in the incidence of violent sexual assault in Stockholm. Each of these case studies presents certain statistical challenges, which is the reason for their inclusion. These challenges include: handling missing data, dealing with incompatible spatial units, handling overdispersion and zero inflation when modelling count data, dealing with spatially autocorrelated missing covariates, allowing for spatial heterogeneity in model parameters and providing reliable small area estimates.

As was remarked in an earlier chapter (Sections 1.4.2 and 1.4.3 in Chapter 1), Bayesian hierarchical models provide a flexible framework to analyse spatially dependent discrete outcome values in the form of either a binary indicator (e.g. a spatial unit is considered as an HIA or not) or a count (e.g. the number of stroke deaths or malaria cases in each spatial unit). Spatial dependence is dealt with in the process model using area-specific parameters, on which various spatial models (e.g. the exchangeable model, the proper CAR and ICAR models and the BYM model) that we discussed in Chapters 7 and 8 can be used to capture the dependence structure in the data.

All models considered in this chapter deal with outcome values at the area level (thus these models are often referred to as area-level models) and, in that regard, they are different from the unit-level models considered in the previous two chapters, where those models analyse outcome values at the individual (household) level. However, similar to the models in Chapters 7 and 8, area-specific random effects are included, and they are modelled using various spatial models so that information can be shared either globally, locally or both. The properties of information sharing from these spatial models still apply. For example, an exchangeable parameters model allows information to be shared globally, whilst the CAR-type of models (the ICAR or the pCAR model) share information locally. The amount of global or local smoothing (i.e. how much an area's estimate is influenced by the information from other areas) depends on the amount of information that this area has. For example, if an area has a large (small) at-risk population and a large (small) number of burglary cases, then its burglary risk estimate is less (more) likely to be pulled towards the estimates of its neighbouring areas in the case of local information sharing or towards the estimated overall risk of the study region in the case of global information sharing. As we shall see in application 2 (Section 9.3), information sharing both globally and locally helps address the problem of estimating small area smoking prevalence using survey data. The challenge in that application is similar to the challenge encountered in the income application that we discussed in Chapters 7 and 8. In both cases, data collected through a national or city-wide survey are disaggregated in order to provide estimates across a set of small areas within the country or within the city. Whilst the majority of areas at least have limited amounts of data, some areas have no data at all.

Apart from information sharing, another purpose of including area-specific random effects is to control for the effects of covariates that are not included in the model. When modelling the binary HIA outcome in application 1 (Section 9.2), random effects are included to acknowledge such missing covariates and to capture the correlation between two sets of related outcomes. The focus is less on information sharing. In applications 3 (Section 9.4) and 4 (Section 9.5), we also illustrate how one can use Bayesian hierarchical models to tackle the issues of overdispersion and zero-inflation (Section 6.4 in Chapter 6) when modelling spatial count data.

Through these four applications, we aim to demonstrate the thinking behind the choice of each model. This thinking is operationalised by a set of questions: what are the feature(s) of the data that lead us to consider a particular form of model? What are the key assumptions of the model, and are these assumptions justified by the data and/or suitable in order to achieve the goal of the study? Also: does the model describe the observed data well, and if not, which parts of the data do not agree with the model? How can the model be improved, or do we need to consider alternative modelling strategies for the data? These questions apply to every model that we consider. Exploratory data analysis, model evaluation and model comparison are crucial to answering these questions. Readers interested in following up the wider implications of each study are encouraged to refer to specific papers which are cited in the text at the appropriate points.

9.2 Application 1: Modelling the Distribution of High Intensity Crime Areas in a City

9.2.1 Background

As a prelude to possible inclusion in the grant allocation to police forces in England and Wales, the UK's Home Office defined high intensity crime areas (HIAs) as areas in the two countries' large metropolitan cities where there were high levels of violent crime, often involving the use of weapons, and where the resident population were either unwilling or too afraid to co-operate with police often because the perpetrators were also resident in the area (Home Office, 1997). The argument was put that these circumstances placed a special burden on the police forces of large metropolitan cities that called for additional funding. Such areas were more than just crime hotspots, but rather they represented an "extreme manifestation of the co-existence in an area of serious crimes, the victims of those crimes and the particularly dangerous individuals and groups responsible for committing those crimes" (Craglia et al., 2001, p.1925).

Craglia et al. (2005) report a study, using data from Sheffield, comparing the geographical distribution of HIAs identified by two different methodologies: HIAs based on police perceptions (hereafter referred to as PHIAs) and HIAs based on recorded crime data (hereafter referred to as EHIAs – E standing for "empirical"). PHIAs represented the views of senior police officers in South Yorkshire Police, whilst EHIAs were constructed using South Yorkshire's recorded crime database.[1] Both forms of HIA were defined in terms of the Census Output Areas (COAs) that partition the city, and each COA contains around 100 households. Each COA was classified according to both methods. The two sources of information can be considered, in a sense, complementary. The recorded crime database offers a more detailed spatial-temporal picture of serious crime than could be recalled by any group of senior police officers but tends to suffer from differential reporting rates across the city. Senior police officers, on the other hand, bring experience accumulated over many years of operational service, but their view may be influenced by particular past experiences, their own attitudes as well as what is and is not remembered.

In a later study using the same data, Haining and Law (2007) use Bayesian hierarchical models with the aim of developing a formal framework for combining "different types of knowledge in the analysis of crime and disorder maps" (Haining and Law, 2007, p.1019). In this application, we explore several methods to achieve this goal. First, we will revisit the two methods discussed in the Haining and Law study. We will then introduce the multivariate conditional autoregressive (MVCAR) model for a joint analysis of two correlated sets of spatial data. In the course of the exposition, we will comment on the strengths and limitations of these methods in the context of the HIA modelling. Towards the end of the section, we will describe the so-called shared component model, another method for joint modelling of data with multiple correlated outcomes.

9.2.2 Data and Exploratory Analysis

There are two sets of binary outcome values in this study that cover 337 COAs in Sheffield.[2] For each COA i, $y_i^{PHIA} = 1$ if it is a PHIA and 0 otherwise. Similarly, if this COA is considered to be an HIA based on recorded crime data, then $y_i^{EHIA} = 1$ and 0 otherwise. For each type of HIA, 47 of the 337 COAs were so defined (given the value 1). In this application, we have also included four COA-level covariates that were defined in Haining and Law (2007). These covariates are:

1. $x_{i,ethnic}$: index of ethnic heterogeneity that varies between 0 and 1, and the larger the value, the greater the ethnic mix in the COA

2. $x_{i,carvan}$: percentage of households in a COA without a car or van

3. $x_{i,lonepar}$: percentage of households in a COA classified as a single-parent family

4. $x_{i,turnover}$: percentage of population living in a COA who in the previous year were living at another address either in the same COA or elsewhere

---

[1] The interested reader is referred to Craglia et al. (2005) for details on how these two different types of HIA were identified. The number of COAs defined as EHIAs was constructed to be the same as the number of COAs defined as PHIAs.

[2] Only one of the Basic Command Units (BCUs) in Sheffield, at the time of the study, was identified as containing HIAs. All 337 COAs lay within this BCU. A BCU is the largest areal unit used for dividing a police force area. Sheffield is contained within the South Yorkshire Police force area.

The neighbourhood structure is defined using first order rook's contiguity, where two COAs are neighbours if they share a common boundary (see Chapter 4). Each covariate is mean-centred to achieve faster convergence and improve mixing (see Section 5.4.2 in Chapter 5).

The two sets of binary outcomes were analysed separately in Exercises 5.11 and 5.12 in Chapter 5, where a non-hierarchical logistic regression model with the four covariates but no random effects was used. Subsequently in Exercises 8.11 and 8.12 in Chapter 8, various Bayesian hierarchical models with different structures on the random effects were investigated. The findings of the univariate analyses are (a) all four covariates were strong predictors of EHIA, while only the ethnic heterogeneity index was associated with PHIA, the other three covariates having no association; (b) for EHIA, having accounted for the four covariates, a set of exchangeable random effects were sufficient to account for the residual variability, but for PHIA, both the spatially-structured and the spatially-unstructured random effects (i.e. the BYM model) were required.

To explore the correlation between the two outcomes, Figure 9.1 shows the spatial distributions of the PHIAs and EHIAs. While there is evidence of a broad similarity in the PHIA and EHIA distributions, 11 COAs (3.3% of the 337 COAs in total) are identified both as PHIAs and EHIAs. These 11 COAs are shown as the polygons in dark grey in both maps in Figure 9.1. The Phi coefficient (Yule, 1912), a statistic measuring the strength of correlation between two sets of binary (dichotomous) values, is 0.11 with the chi-square test of association giving a p-value of 0.07. The result suggests a weakly significant positive correlation. We further calculated the Clifford-Richardson modified correlation coefficient (Clifford and Richardson, 1985, and discussed in Section 6.3.2.1 in Chapter 6), which takes into account the spatial autocorrelation present in each outcome. The correlation coefficient is again 0.11 but now with a p-value of 0.13, giving a non-significant result even at the 10% significance level. The presence of spatial autocorrelation results in the "effective"

FIGURE 9.1
The spatial distributions of PHIAs and EHIAs. In each map, the polygons in light grey are those considered to be either a PHIA or an EHIA but not both, while those in dark grey are the COAs considered to be both PHIA and EHIA. The polygons in white are the COAs that are neither PHIAs nor EHIAs.

sample size (of around 190 from the output of the test; see the interpretation of the test results in Section 6.3.2.1) being much smaller than the "actual" sample size (of 337) so that the modified test shows that the correlation between the two outcomes is not significant. These exploratory findings, together with the theoretical arguments presented, to some extent, point towards a joint modelling of the two complementary maps with the need to account for spatial autocorrelation.

9.2.3 Methods Discussed in Haining and Law (2007) to Combine the PHIA and EHIA Maps

Haining and Law (2007) discuss two ways to combine the PHIA and EHIA maps. The first method is a two-stage process where the information in the PHIA data is considered to be the prior for the analysis of the EHIA data. Specifically, the first stage estimates the posterior distributions of the covariate effects by fitting a logistic regression model to the PHIA data. The resulting posterior distributions are then used as the prior distributions of the regression coefficients in the analysis of the EHIA data. This approach takes advantage of the Bayesian paradigm, where information from one study (here, the PHIA data) can be used in the analysis of data from another related study (the EHIA data) in the form of prior distributions. However, Haining and Law (2007) found that the results from this approach were not much different from those obtained when vague priors were used in the modelling of the EHIA data. This is mainly due to the dominance of the data so that the posterior inference is not much affected by the prior choice, informative or otherwise. See Exercise 9.1 for the implementation of this approach.

The second method considered by Haining and Law (2007) is to treat the PHIA and EHIA outcome values as two *independent* sets of observations (or so-called "repeated measurements"), both arising from the same underlying data generating process. Expressing this differently, this approach assumes that the two sets of outcome values are fully determined by the *same* underlying data generating process, and any difference between the two outcomes is due purely to chance and not to any factors that are specific to each outcome. Essentially, there is no substantive difference in a COA being a PHIA or an EHIA or both under this approach. For this method, we use HIA to refer to a COA being considered as an HIA by one or both definitions.

Under the repeated measurements model, the two outcome values of a COA are modelled by the same Bernoulli random variable. Thus, for $i = 1, \ldots, N$, $y_i^{PHIA} \sim Bern(\pi_i)$ and $y_i^{EHIA} \sim Bern(\pi_i)$, where π_i is the underlying unknown probability of COA i being identified as an HIA. Eq. 9.1 presents the full model with vague priors assigned to the model parameters. This model is an extension to the logistic regression model that we considered in Section 5.2.2, where we only modelled the PHIA outcomes with one covariate. In Eq. 9.1, both the PHIA and EHIA outcomes are modelled together through the same underlying process. This process depends on the four COA-level covariates (as opposed to one in Section 5.2.2) with the addition of a set of spatially-structured random effects. The inclusion of these random effects helps capture the variability in the outcome values that is not accounted for using the four covariates in the model. In Eq. 9.1, the random effects follow the ICAR prior, but Exercise 9.2 considers other specifications for the random effects. Note that the repeated measurements model fitted in Haining and Law (2007) includes the four covariates but no random effects. Here, the half normal distribution, $N_{+\infty}(0,10)$, is assigned to the random effect standard deviation σ_S as a vague prior distribution. This normal prior is bounded below by 0 (thus the subscript $+\infty$) to ensure σ_S is strictly positive. The reader is encouraged to consider other plausible alternative distributions as a vague prior distribution for σ_S (see Section 5.6.1).

$$y_i^{PHIA} \sim Bern(\pi_i)$$

$$y_i^{EHIA} \sim Bern(\pi_i)$$

$$logit(\pi_i) = \alpha + \beta_1 x_{i,ethnic} + \beta_2 x_{i,carvan} + \beta_3 x_{i,lonepar} + \beta_4 x_{i,turnover} + S_i$$

$$S_{1:N} \sim ICAR\left(W, \sigma_S^2\right) \tag{9.1}$$

$$\alpha \sim Uniform(-\infty, +\infty)$$

$$\beta_k \sim N(0, 1000000) \quad \text{for} \quad k = 1, \dots, 4$$

$$\sigma_S \sim N_{+\infty}(0, 10)$$

The WinBUGS code for fitting the repeated measurements model in Eq. 9.1 is given in Figure 9.2. Lines 31–34 calculate the odds ratios for the four risk factors. Since they are nonlinear transformations of the regression coefficients (beta[1], ..., beta[4]), these odds ratios need to be calculated within WinBUGS. See Section 5.2.2 for their interpretation. To evaluate the goodness of fit to the data from the model, we calculate the four quantities summarised in Table 9.1. The calculation is done separately for the observed PHIA data and the observed EHIA data. More specifically, n_{00} is the number of non-HIA COAs in the observed data that are correctly predicted to be non-HIAs by the model. Similarly, n_{11} is the number of HIA COAs in the observed data that are correctly predicted to be HIAs by the model. The other two quantities, n_{01} and n_{10}, represent respectively the number of COAs that are actually HIAs but are wrongly predicted to be non-HIAs by the model and the number of non-HIA COAs that are wrongly predicted to be HIAs. If the model fits the data well, the two quantities, n_{01} and n_{10}, will be small, ideally close to 0. Lines 38–54 in Figure 9.2 calculate these four quantities for PHIA, and Lines 56–66 carry out the same calculation for EHIA. The procedure will also be used to assess the goodness of fits from a number of other models, the results of which will be summarised in Section 9.2.5.

Figure 9.3(a) shows the combined HIA map using the posterior means of the HIA probabilities, $\pi_1, \dots, \pi_N$, in Eq. 9.1. Following Haining and Law (2007), the 47 COAs that have the highest posterior means of π_i are shown as the grey polygons in Figure 9.3(b). While the combined map (Figure 9.3(b)) shows a core area of HIAs across the central part of the study region, it misses some HIAs from each of the two individual maps of PHIAs and EHIAs (Figure 9.1), for example, the little cluster of EHIAs in the south, suggesting a somewhat poor fit to each set of observed data. This issue is partly due to the lack of an outcome-specific component in the model. Beyond the common underlying data generating process, perhaps there do exist other factors that significantly influence each type of outcome, as briefly suggested in Section 9.2.1. We now turn to investigating that.

9.2.4 A Joint Analysis of the PHIA and EHIA Data Using the MVCAR Model

In contrast to the repeated measurements approach, the MVCAR model assumes the two sets of spatial outcomes arise from two *different but possibly correlated* spatial processes. In the context of this application, the two forms of HIA can be affected by a set of common factors but, as Haining and Law (2007) argued, they may also contain different but

```
1    #  The WinBUGS code to implement the repeated measurement model
2    #  in Eq. 9.1
3    model {
4      for (i in 1:N) {         #  a for-loop to go through the 337 COAs
5        phia[i] ~ dbern(pi[i])  #  the Bernoulli likelihood for
6                                #  police-defined HIA
7        ehia[i] ~ dbern(pi[i])  #  the Bernoulli likelihood for
8                                #  empirical-defined HIA
9      #  modelling the logit-transformed probability using covariates
10     #  and spatially-structured random effects
11       logit(pi[i]) <- alpha
12                     + beta[1]*ethnic[i]
13                     + beta[2]*carvan[i]
14                     + beta[3]*turnover[i]
15                     + beta[4]*lonepar[i]
16                     + S[i]
17     }
18     #  priors
19     alpha ~ dflat()  #  the improper uniform prior defined on the whole
20                      #  real line for the intercept
21     #  ICAR prior on the random effects
22     S[1:N] ~ car.normal(adj[],weights[],num[],prec.S)
23     #  prior on the conditional standard deviation for the ICAR prior
24     prec.S <- pow(sigma.S,-2)
25     sigma.S ~ dnorm(0,0.1)I(0,)
26     for (k in 1:4) {
27       beta[k] ~ dnorm(0,0.000001)  #  a vague normal prior for each
28                                    #  regression coefficient
29     }
30     #  quantities of interest
31     for (i in 1:4) {
32       OR[i] <- exp(beta[i])  #  calculate the odds ratio for each
33                              #  covariate (see Section 5.2.2)
34     }
35
36     #  model evaluations
37     #     for PHIA
38     for (i in 1:N) {
39       phia.pred[i] ~ dbern(pi[i])  #  predicting the PHIA status for
40                                    #  COA   from the fitted model
41       #  the equals function (below) returns 1 if the two arguments
42       #  are the same otherwise returns 0;
43       #  phia.m1o1[i] = 1 if both the model and the observed data both
44       #  say that this COA is a PHIA otherwise phia.m1o1[i] = 0 (the same
45       #  calculation applies to phia.m0o1, phia.m1o0 and phia.m0o0)
46       phia.m1o1[i] <- equals(phia.pred[i],1)*equals(phia[i],1)
47       phia.m0o1[i] <- equals(phia.pred[i],0)*equals(phia[i],1)
48       phia.m1o0[i] <- equals(phia.pred[i],1)*equals(phia[i],0)
49       phia.m0o0[i] <- equals(phia.pred[i],0)*equals(phia[i],0)
50     }
51     phia.tab[1,1] <- sum(phia.m1o1[1:N])  #  calculating n11 in Table 9.1
52     phia.tab[2,2] <- sum(phia.m0o0[1:N])  #  calculating n00 in Table 9.1
53     phia.tab[1,2] <- sum(phia.m1o0[1:N])  #  calculating n10 in Table 9.1
54     phia.tab[2,1] <- sum(phia.m0o1[1:N])  #  calculating n01 in Table 9.1
55     #    for EHIA (the same calculation as for PHIA)
56     for (i in 1:N) {
57       ehia.pred[i] ~ dbern(pi[i])
58       ehia.m1o1[i] <- equals(ehia.pred[i],1)*equals(ehia[i],1)
59       ehia.m0o1[i] <- equals(ehia.pred[i],0)*equals(ehia[i],1)
60       ehia.m1o0[i] <- equals(ehia.pred[i],1)*equals(ehia[i],0)
61       ehia.m0o0[i] <- equals(ehia.pred[i],0)*equals(ehia[i],0)
62     }
63     ehia.tab[1,1] <- sum(ehia.m1o1[1:N])
64     ehia.tab[2,2] <- sum(ehia.m0o0[1:N])
65     ehia.tab[1,2] <- sum(ehia.m1o0[1:N])
66     ehia.tab[2,1] <- sum(ehia.m0o1[1:N])
67   }
```

FIGURE 9.2
The WinBUGS code for fitting the repeated measurements model for the PHIA and EHIA data.

TABLE 9.1

A Tabulation of the Four Quantities for Evaluating the Goodness of Fit to the Observed Data from the Model

	obs = 1 (actual HIA)	*obs* = 0 (actual non-HIA)
pred = 1 (predicted HIA)	n_{11}	n_{10}
pred = 0 (predicted non-HIA)	n_{01}	n_{00}

See the main text for interpretation.

FIGURE 9.3
(a) the map of the posterior means of the HIA probability estimated from the repeated measurements model; (b) the map of the 47 COAs that have the highest posterior probabilities.

complementary information. Therefore, the assumption of the MVCAR model appears to be closer to our understanding of how the two maps have been generated than the previous model. Using the MVCAR model here is primarily to (a) evaluate the strength of correlation between the two outcomes and (b) examine the effects of the observable risk factors. The method, however, places less emphasis on producing a combined map.

In terms of modelling, each set of outcome values is associated with a set of outcome-specific random effects. Modelled using the MVCAR model, the two sets of random effects are allowed to be correlated, and the strength of the correlation is estimated from the data. The version of the MVCAR model implemented in WinBUGS (and used here) is a multivariate extension of the intrinsic CAR model (see Chapter 8) and hence is an improper distribution, where the spatial autocorrelation parameter is fixed at 1. As a result, each set of outcome-specific random effects are spatially-structured *a priori*. We will return to this in Section 9.2.7. Gelfand and Vounatsou (2003) proposed a proper version of the MVCAR model where the spatial autocorrelation parameter is allowed to be estimated from the data.

To assess the correlation between the PHIA and EHIA maps, we first apply the MVCAR model without covariates. We then add the covariates back into the model to examine their effects. Using the same notations as in Eq. 9.1, the MVCAR model without covariates is given by

$$y_i^{PHIA} \sim Bern(\pi_{i1})$$

$$y_i^{EHIA} \sim Bern(\pi_{i2})$$

$$logit(\pi_{i1}) = \alpha + S_{i1}$$

$$logit(\pi_{i2}) = \alpha + S_{i2} \qquad (9.2)$$

$$S_{1:N,1:2} \sim MVCAR(W, \Sigma)$$

$$\alpha \sim Uniform(-\infty, +\infty)$$

For each COA i, its probability of being a PHIA (π_{i1}) and its probability of being an EHIA (π_{i2}) are not the same, as they are modelled respectively by the random effect terms, S_{i1} and S_{i2}, in addition to the common intercept α. The two sets of random effects, $\mathbf{S}_1 = (S_{11}, \ldots, S_{N1})$ and $\mathbf{S}_2 = (S_{12}, \ldots, S_{N2})$, are then modelled jointly through the MVCAR model to allow $\mathbf{S}_1$ and $\mathbf{S}_2$ to be correlated, and each set of random effects are spatially-structured. The spatial neighbourhood structure is defined through the spatial weights matrix $\mathbf{W}$. The cross-outcome correlation is estimated via the covariance matrix $\mathbf{\Sigma}$. For two outcomes, $\mathbf{\Sigma}$ takes the form $\begin{pmatrix} \sigma_1^2 & \lambda\sigma_1\sigma_2 \\ \lambda\sigma_1\sigma_2 & \sigma_2^2 \end{pmatrix}$. The parameter λ measures the strength of the correlation between the two outcomes, and σ_1^2 and σ_2^2 are the conditional variances of $\mathbf{S}_1$ and $\mathbf{S}_2$, respectively.

For priors, the improper uniform prior on the whole real line is assigned for the intercept α. For the 2 × 2 covariance matrix $\mathbf{\Sigma}$, the Wishart distribution, denoted as $Wishart_2(\mathbf{R}, d)$, is often used as a prior distribution on the precision matrix, $\mathbf{\Sigma}^{-1}$. The Wishart distribution is used to ensure that the covariance matrix is positive-(semi)definite and symmetric (see the footnote).[3] Here we choose a moderately informative prior for the precision matrix with $d = 10$ and $\mathbf{R} = \begin{pmatrix} 20 & 0 \\ 0 & 20 \end{pmatrix}$. This prior choice implies *a priori* that (a) there is 0.95 probability that the correlation between the two outcomes (i.e. λ) ranges from −0.6 to 0.6; and (b) there is 0.95 probability that the conditional standard derivation of each set of spatially-structured random effects lies between 1.1 and 7.3. Exercise 9.3 uses WinBUGS to show how to arrive at the above two probability statements.

---

[3] The Wishart distribution is a multivariate version of the Gamma distribution (recall in Chapter 5, the Gamma distribution is one of the prior choices for a precision parameter). If $X \sim Wishart_p(\mathbf{R}, d)$, then X is a symmetric, positive-definite matrix of size $p \times p$, satisfying the requirements of a covariance matrix. In addition, the Wishart distribution is the so-called semi-conjugate prior (conditioning on the mean vector) of the inverse covariance matrix of a multivariate normal distribution. The two parameters in the Wishart distribution are defined as follows. $\mathbf{R}$ is a positive definite and symmetric matrix of size $p \times p$, and d denotes the degrees of freedom, a scalar with a constraint that $d > p - 1$. When using the Wishart distribution as a prior distribution, we typically specify values for $\mathbf{R}$ and d. The following two properties are useful. First, the variances in X are inversely related to d, and hence a small d corresponds to a high variability (or uncertainty) about the random variables that have X^{-1} as their covariance matrix. Since d has to be greater than $p - 1$, setting $d = p$ represents a vague prior. Second, the mean of $X \sim Wishart_p(\mathbf{R}, d)$ is $d\mathbf{R}^{-1}$, so $\mathbf{R}$ is often set to be a prior guess at the unknown covariance matrix. See Lunn et al. (2012, p.226) for more detail on the Wishart prior.

The MVCAR model in Eq. 9.2 can be extended easily to incorporate covariates:

$$y_i^{PHIA} \sim Bern(\pi_{i1})$$

$$y_i^{EHIA} \sim Bern(\pi_{i2})$$

$$logit(\pi_{i1}) = \alpha + \beta_1 x_{i,ethnic} + \beta_2 x_{i,carvan} + \beta_3 x_{i,lonepar} + \beta_4 x_{i,turnover} + S_{i1} \qquad (9.3)$$

$$logit(\pi_{i2}) = \alpha + \beta_1 x_{i,ethnic} + \beta_2 x_{i,carvan} + \beta_3 x_{i,lonepar} + \beta_4 x_{i,turnover} + S_{i2}$$

$$S_{1:N,1:2} \sim MVCAR(\mathbf{W}, \Sigma)$$

The WinBUGS code for fitting the MVCAR model with covariates is given in Figure 9.4. Lines 36–66 in Figure 9.2, with slight modifications, can be added into the model code in Figure 9.4 (between Lines 39 and 40) to carry out model evaluation.

9.2.5 Results

From the MVCAR model without covariates, the parameter λ is estimated to be 0.33 with a 95% credible interval (denoted as 95% CI hereafter) of (−0.07, 0.67), suggesting little evidence of a correlation between the two types of HIA, the same conclusion as found when using the modified correlation test by Clifford et al. (1989). It is worth noting that similar to the Clifford-Richardson modified correlation test, the inference on the correlation using the MVCAR model also takes the spatial autocorrelation present in each map into account.

Using Deviance Information Criterion (DIC), Table 9.2 compares the two versions of the MVCAR model with the repeated measurements model. Clearly, the MVCAR model, with or without the covariates, is a better approach in terms of describing the two sets of outcomes. The inclusion of the four covariates in the MVCAR model helps better explain the outcomes (with a lower $\bar{D}$ value) and results in a simpler model (with a smaller pD value). The model is simpler because in the presence of the covariates the random effects become "less important" in explaining the observed variability.

Table 9.3 assesses how well the repeated measurements model and the MVCAR model with covariates predict the EHIA and PHIA status of each COA. For an additional comparison, Table 9.3 also shows the predicted results from the univariate analysis in the form of the modelling considered in Chapter 5 (Exercise 9.4). Compared to the repeated measurements model and the univariate analysis, the MVCAR model with covariates shows a better fit with fewer misclassifications (n_{01} and n_{10}, as in Table 9.1) and a greater number of correct classifications across the two outcomes. However, the number of wrongly classified COAs is still quite large, suggesting that further improvement (e.g. incorporating additional covariates) is needed.

The posterior estimates of the covariate effects are broadly similar between the repeated measurements model and the MVCAR model (Table 9.4). Both models suggest that a COA has a greater chance of being classified as either a PHIA or an EHIA or both if it has a larger population turnover and a greater ethnic mix. A high proportion of single-parent households and a large number of households without car or van may also increase the risk of a COA being considered as an HIA. After controlling for the risk factors, the posterior mean of λ is −0.434, with a 95% CI of (−0.795, 0.0870), showing little evidence of correlation between the two maps. It should be noted that λ has a different interpretation here compared to when no covariates are included. When no covariates are included, λ

```
1    model {
2      for (i in 1:N) {
3        phia[i] ~ dbern(pi1[i])
4        ehia[i] ~ dbern(pi2[i])
5        logit(pi1[i]) <- mu[i] + S[1,i]
6        logit(pi2[i]) <- mu[i] + S[2,i]
7        mu[i] <- alpha
8                + beta[1]*ethnic[i]
9                + beta[2]*carvan[i]
10               + beta[3]*turnover[i]
11               + beta[4]*lonepar[i]
12     }
13     #  the MVCAR model on the two sets of random effects
14     S[1:2,1:N] ~ mv.car(adj[],weights[],num[],omega[1:2,1:2])
15     #  the Wishart prior on the precision matrix of
16     #  the MVCAR model
17     omega[1:2,1:2] ~dwish(R[,],10)      #  the first argument is
18                                         #  the R matrix and the
19                                         #  second is the scalar d=10
20     R[1,1] <- 20
21     R[2,2] <- 20
22     R[1,2] <- 0
23     R[2,1] <- 0
24     #  calculate the covariance matrix (1/precision matrix);
25     #  the inverse function performs matrix inversion
26     Sigma[1:2,1:2] <- inverse(omega[1:2,1:2])
27     #  calculate the correlation of the two sets of random
28     #  effects (see the main text for its interpretation)
29     b <- pow(Sigma[1,1],0.5)*pow(Sigma[2,2],0.5)
30     lambda <- Sigma[1,2]/b
31
32     #  the improper prior for the intercept
33     alpha ~ dflat()
34     for (i in 1:4) {
35       #  vague prior on each regression coefficient
36       beta[i] ~ dnorm(0,0.0001)
37       #  calculate the odds ratio for each covariate
38       OR[i] <- exp(beta[i])
39     }
40   }
41
42   #  initial values for MCMC chain 1
43   list(alpha=-6, beta=c(0.1,0.1,0.1,0.1)
44       ,omega=structure(.Data=c(1,0
45                                ,0,1),.Dim=c(2,2)))
46
47   #  initial values for MCMC chain 2
48   list(alpha=-7, beta=c(0.05,0.05,0.05,0.05)
49       ,omega=structure(.Data=c(0.5,0
50                                ,0,0.5),.Dim=c(2,2)))
```

FIGURE 9.4
The WinBUGS code and two initial value lists for fitting the MVCAR model with covariates to the PHIA and EHIA data. Note that it is recommended to assign initial values to the precision matrix for each chain (see Lines 42–50).

TABLE 9.2

The DIC Comparison of the Repeated Measurements Model and the Two Versions of the MVCAR Models

Model	$\bar{D}$	pD	DIC
The repeated measurements model	372	22	394
MVCAR without covariates	302	83	385
MVCAR with covariates	257	65	322

TABLE 9.3

Assessing the Goodness of Fit to the EHIA and PHIA Data

	EHIA		PHIA	
Univariate analysis				
	obs = 1 (47 in total)	obs = 0 (290 in total)	obs = 1 (47 in total)	obs = 0 (290 in total)
pred = 1	14 (8, 20)	33 (21, 47)	27 (20, 34)	20 (13, 28)
pred = 0	33 (27, 39)	257 (243, 269)	20 (13, 27)	270 (262, 277)
The repeated measurements model				
	obs = 1 (47 in total)	obs = 0 (290 in total)	obs = 1 (47 in total)	obs = 0 (290 in total)
pred = 1	15 (9, 21)	33 (22, 44)	23 (16, 30)	24 (14, 35)
pred = 0	32 (26, 38)	257 (246, 268)	24 (17, 31)	266 (255, 276)
MVCAR with covariates				
	obs = 1 (47 in total)	obs = 0 (290 in total)	obs = 1 (47 in total)	obs = 0 (290 in total)
pred = 1	20 (12, 29)	22 (13, 33)	35 (27, 42)	17 (9, 26)
pred = 0	27 (18, 35)	269 (257, 277)	12 (5, 20)	273 (264, 281)

Each cell in the table reports the posterior mean and the 95% CI of each of the four quantities in Table 9.1.

TABLE 9.4

A Summary of the Posterior Estimates of the Covariate Effects

Posterior Estimate of the Odds Ratio Related to	The Repeated Measurements Model	The MVCAR Model with Covariates
Ethnicity	1.060 (1.007, 1.120)	1.069 (1.003, 1.141)
% households with no car or van	1.026 (1.001, 1.052)	1.033 (1.003, 1.064)
% single-parent households	1.082 (1.027, 1.140)	1.097 (1.029, 1.170)
Population turnover	1.058 (1.042, 1.077)	1.083 (1.056, 1.123)
λ	NA	−0.434 (−0.795, 0.087)

For each effect estimate, the posterior mean and the 95% CI are reported.

measures the correlation between the two sets of *outcome values*, whereas when covariates are included, λ measures the correlation between the two sets of *residuals*. Given the weak correlation, Exercise 9.5 considers a model that replaces the MVCAR model in Eq. 9.3 by two independent ICAR models, one for each outcome. Readers are encouraged to compare and contrast the results to those presented here.

9.2.6 Another Specification of the MVCAR Model and a Limitation of the MVCAR Approach

The results from the univariate analysis suggest that the effects of the risk factors may be different between the PHIA and EHIA maps (see Exercises 8.11 and 8.12 in Chapter 8). To account for this, the regression coefficients can be made to be specific to each outcome. This extension is given in Eq. 9.4.

$$logit(\pi_{i1}) = \alpha + \beta_{11}x_{i,ethnic} + \beta_{21}x_{i,carvan} + \beta_{31}x_{i,lonepar} + \beta_{41}x_{i,turnover} + S_{i1}$$

$$logit(\pi_{i2}) = \alpha + \beta_{12}x_{i,ethnic} + \beta_{22}x_{i,carvan} + \beta_{32}x_{i,lonepar} + \beta_{42}x_{i,turnover} + S_{i2}$$

$$(9.4)$$

However, the findings from Exercises 8.11 and 8.12 in Chapter 8 also show that after accounting for the four covariates, the residuals for the EHIA map are no longer spatially autocorrelated, thus making the use of the MVCAR model inappropriate (although as we shall discuss in Section 9.2.7, the shared component modelling framework can accommodate this).

The MVCAR model, as discussed earlier, assumes the outcomes represented by the PHIA and EHIA maps arise from two different but possibly correlated processes, and the use of two sets of random effects in the modelling reflects this assumption. A consequence is that the MVCAR model does not lend itself to combining the PHIA and the EHIA maps. In that sense, therefore, the repeated measurements model is better suited, although it does not appear to fit the data well. Overall, based on the results from the MVCAR model, the observed data may not support the idea of combining the two maps, as the correlation between the two outcomes appears to be weak. We now discuss two more approaches to the joint modelling of multiple outcomes. One of these, the shared component model, possesses the features of both the repeated measurements model and the MVCAR model.

9.2.7 Conclusion and Discussion

In this application, we have discussed three different methods for unifying different sources of evidence regarding the spatial distribution of serious crime neighbourhoods in Sheffield. Different types of knowledge, while each giving a partial picture, when appropriately combined, provide a better understanding of crime and disorder in a metropolitan city. In research into geographical crime patterns, this type of modelling, as pointed out by Haining and Law (2007, p.1030), is "important for tactical (short-term response) and strategic (longer-term) policing, as well as for academic research". It is important, not least, because the delivery of police services in England and Wales is territorial.

The methods discussed show different strengths and weaknesses. Both of the two approaches employed by Haining and Law (2007) place the emphasis on combining the two maps into one but ignore the individuality of each type of HIA. The MVCAR model, on the other hand, relaxes such a restriction by modelling the two types of HIA as two different but correlated processes. While achieving a better fit to the data, the MVCAR model does not lend itself to producing a unified map. Nevertheless, when modelling a set of spatial data with multiple, possibly correlated, outcomes, both of these methods are worth considering.

The shared component model (SCM) offers yet another way to model multiple outcomes jointly. The SCM model was first proposed by Knorr-Held and Best (2001) in the context of modelling two related disease outcomes and was subsequently extended by Held et al. (2005) to deal with four different diseases. Under the SCM model, each type of outcome is modelled as a combination of a shared component that is common (or shared) across different outcomes and an outcome-specific component, which is specific to that particular outcome. The rationale behind the modelling structure is that there exists a set of factors that are common to different types of outcome. In addition to the common factors, each type of outcome also possesses its own set of factors. Knorr-Held and Best (2001) applied the SCM model to jointly analyse the spatial distributions of oral cavity and oesophageal cancer mortality. They argued that the two cancer types may display a shared geographical pattern as they are both found to be associated with tobacco smoking and alcohol abuse. Beyond such a common/shared pattern, each cancer type may be associated with factors that only affect one but not the other – "the shared component model proposed here allows us to exploit the aetiological similarities between the two diseases, yet still to explore any differences in their respective geographical patterns of risk" (Knorr-Held and Best, 2001, p.79). Typically, the shared component and the outcome-specific components are modelled as random effects, representing respectively the common and the outcome-specific unobserved risk factors. Applying this model to the PHIA and EHIA data, the SCM model is given as follows.

$$y_{i1} \sim Bern\left(\pi_{i1}\right)$$

$$y_{i2} \sim Bern\left(\pi_{i2}\right)$$

$$logit\left(\pi_{i1}\right) = \alpha + \delta_1 S_i + M_{i1} \qquad (9.5)$$

$$logit\left(\pi_{i2}\right) = \alpha + \delta_2 S_i + M_{i2}$$

In Eq. 9.5, the set of random effects $\mathbf{S}_1 = (S_1,...,S_N)$ represents the shared component, while $\mathbf{M}_1 = (M_{11},...,M_{N1})$ and $\mathbf{M}_2 = (M_{12},...,M_{N2})$ are random effects specific to each of the two outcomes. The multiplicative scalars δ_1 and δ_2 allow the shared component to have different effects on the two outcomes but, for identifiability, they are constrained so that $\delta_1 = 1 / \delta_2$. The interested reader should see Knorr-Held and Best (2001) for more detail on this constraint as well as the prior specification on δ_1. Different models discussed in Chapter 8 (e.g. the exchangeable model, the ICAR or the BYM model) can be used as the prior model for each of the three sets of random effects, S, $\mathbf{M}_1$ and $\mathbf{M}_2$. In the context of the HIA data, the SCM model would be a reasonable alternative to the MVCAR model by the argument that both outcomes may share some common factors related to, for example, the socio- and demographic characteristics of the COAs, and observable covariates can be incorporated to investigate this. Using the small area count data reported in Greater London on four different crime types (burglary, robbery, vehicle crime and violent crime), Quick, M., et al. (2018) employed the shared component model to investigate the shared and the type-specific patterns. Covariates obtained from the 2011 UK census are included in the modelling so that, for each crime type, the variability in the observed data is decomposed into the two shared components (one component shared across all crime types and the other one shared amongst burglary, robbery and vehicle crime but not with violent crime), the type-specific component as well as the covariates. The variance partition coefficients (Section 8.5.2) are used to quantify the relative contributions of the above components in explaining the observed variability of each crime type. For the HIA data, however, with only a few overlapping COAs, there is perhaps insufficient information to estimate the three sets of random effects in the SCM model, resulting in non-convergence of the MCMC chains.

MacNab (2010) has highlighted some similarities as well as differences between the MVCAR and the SCM models. Both approaches offer a formal framework for studying the correlation structure amongst two or more sets of related spatial outcomes. Furthermore, placed within the Bayesian hierarchical modelling framework, a common advantage of both models is to strengthen the small area estimates through sharing information (or borrowing strength) not only across areas but also amongst the correlated outcomes. There are, however, some important distinctions between the two methods. As pointed out by MacNab (2010), SCM and MVCAR place different prior assumptions on the correlation structure of outcomes. A core assumption of the SCM model is that different outcomes share a common set of risk factors, and thus they are correlated *a priori*. As a result, it is important to assess, through exploratory analysis, the strength of pairwise correlations amongst different types of outcome: "…inappropriate assumption of pair-wise disease risks correlation and the subsequent formulation of shared- and disease-specific-risk components may lead to misspecification of u_k's [the random effects], lack of model identifiability, and failure of MCMC convergence" (MacNab, 2010, p.1242). The MVCAR model, however, makes no prior assumptions about the cross-outcome correlations, which, instead, are estimated using the observed data. In this regard, MVCAR is a more general approach in which we can make inference about the nature of the correlation between any two types of outcome.

Another difference is that the SCM model is more flexible than the MVCAR model. By construction, the random effects for each outcome are spatially-structured under the MVCAR model. The SCM model, on the other hand, allows consideration of various specifications of the shared and the outcome-specific random effects (e.g. the exchangeable model, the ICAR or the BYM models). Various specifications can be compared using, for example, DIC. Of course, how complex we can specify a model is dependent on the data available. In addition, in the SCM model, any two sets of random effects are typically assumed to be independent, implying that any interaction between (unobserved/omitted) covariates is not allowed (Held et al., 2005). Finally, in a similar spirit to the SCM model, Tzala and Best (2008) developed a Bayesian latent variable model for modelling risks associated with multiple cancer types over both space and time. Quick et al. (2018) extended the shared component model to jointly analyse a spatial-temporal dataset that contains the annual counts of cases from four different crime types from 2011 to 2015 at the census dissemination area (DA) scale, of which there are 655 DAs in the Regional Municipality of Waterloo, Canada.

All the above models are constructed using the Bernoulli likelihood as the data model (likelihood) for the two sets of binary data. Another approach is to combine the two sets of binary data into a single set of categorical outcomes. That is, each COA takes an outcome value between 1 and 4 such that

- 1 for a COA that is neither PHIA nor EHIA
- 2 for a COA that is PHIA but not EHIA
- 3 for a COA that is EHIA but not PHIA
- 4 for a COA that is both PHIA and EHIA

This single set of outcomes can then be modelled using the multinomial logistic regression in which the likelihood for the data is taken to be the multinomial distribution.[4] See,

---

[4] The multinomial distribution is a generalization of the binomial distribution when each trial can take one of K possible outcomes (for a binomial distribution, $K = 2$). The multinomial distribution has the parameter n denoting the number of trials and a vector of probabilities $\pi = (\pi_1,...,\pi_K)$, with π_k denoting the probability of each trial taking the kth possible outcome. For modelling a categorical outcome with K categories, n is fixed to 1 and we model the probability vector π.

for example, McCullagh and Nelder (1989) and Agresti (2013) for a detailed description of the non-spatial multinomial logistic regression. The interest then lies in the modelling of π_{1i}, π_{2i}, π_{3i} and π_{4i}, the probabilities of each COA being classified into each of the four outcome categories satisfying the constraint that $\sum_{k=1}^{4} \pi_{ki} = 1$ for every COA i. This combination of the two binary outcomes leads to the modelling of a set of categorical spatial data, the modelling of which can be done using, for example, a multinomial logistic model with spatially-structured random effects as proposed in Cao et al. (2011). Related to this approach, Wall and Liu (2009) developed a spatial latent class model for modelling multiple binary spatial outcomes. A latent class model that allocates the entire set of binary outcome values at each spatial location to one of the K latent classes is used to deal with the dependence structure amongst different types of outcome. To account for the spatial relationships between locations, the class allocation is spatially-structured so that nearby locations tend to have similar probabilities of being assigned to the same latent class.[5] The proposed method was implemented in WinBUGS, so the reader is encouraged to consult their paper for more information.

9.3 Application 2: Modelling the Association Between Air Pollution and Stroke Mortality

9.3.1 Background and Data

Ecological modelling in spatial (or geographical) epidemiology analyses the association between the presence of a disease in a population partitioned into small geographical areas and the exposure to some environmental risk factor that varies across that set of small areas. Such modelling typically focuses on either disease incidence, prevalence or mortality. The first aim of such modelling is to obtain reliable estimates of the impacts on disease outcomes arising from the environmental exposure *at the small area level*. Controlling (or adjusting) for confounders, covariates that impact on the outcome variable in ways similar to the exposure variable, is an essential part of any study design. Ecological modelling encounters all the challenges discussed in Chapter 1. Bayesian hierarchical modelling provides a methodology for meeting those challenges in order to obtain reliable estimates of the impact of the specified exposure on disease outcomes.

However, a second aim of such modelling may be to make inferences about the impact of the exposure variable *at the individual level*. This is known as ecological inference, an attempt to learn about the impact of exposure on disease outcome at the individual level using area level data. Uncritical use of risk estimates from an ecological model fitted to area level data to infer risk at the individual level commits the ecological fallacy (see Section 3.3.3). The difference between a risk factor effect estimated from an ecological model and the effect of the same risk factor but estimated from an individual level model is referred to as ecological bias. There are a number of different sources of ecological bias (see Greenland and Robins, 1994). An additional aim of ecological modelling, therefore, may be to try to

---

[5] Noted in Wall and Liu (2009, p.3060), "this type of model is often referred to as the linear model of co-regionalization (LMC) in geostatistics (Grzebyk and Wackernagel, 1994; Wackernagel, 2003, Chapter 26). For lattice or areal data, the multivariate conditional autoregressive (MCAR) model (Gelfand and Vounatsou, 2003) can similarly be described as a linear combination of independent univariate CAR models."

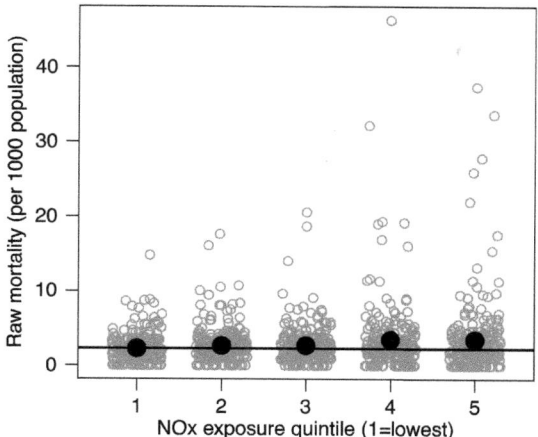

FIGURE 9.5

ED-level stroke mortality by NO_x exposure quintile. Each grey open circle represents the stroke mortality (number of observed stroke deaths divided by the population multiplied by 1000) of each ED between 1994 and 1998. The solid dots are the means of the stroke mortality over the exposure quintiles and the horizontal line indicates the mean mortality at the lowest exposure level.

reduce ecological bias in order that data and model can also be used for individual level inference. In this application we present an example of ecological modelling and discuss ways of reducing the ecological bias. Ecological inference is encountered in many areas of the social sciences (see King et al., 2004) as well as spatial epidemiology (Richardson, 1992; Richardson and Montfort, 2000).

In this application we describe an analysis of stroke mortality data at the enumeration district (ED) scale in Sheffield, England during the period 1994–98 carried out by Maheswaran et al. (2006).[6] Sheffield, an industrial city in the north of England, experienced considerable city-wide geographical variation in nitrogen oxide (NO_x) pollutants originating from factory emissions as well as housing estates (heating emissions) and road traffic. However, because EDs are generally small (especially in the densely populated parts of the city) intra-ED scale variation in NO_x exposure is not large, and this, together with the physical geography of the city and the spatial segregation of different income groups, makes Sheffield an interesting setting for this type of work.[7] There were just under 3000 stroke deaths during the five-year study period out of a population of just under 1000000 people aged 45 or more. This gives an average annual mortality rate of about 3.00 per 1000 population in Sheffield between 1994 and 1998. Evidence of a link between NO_x exposure and stroke mortality in the Sheffield data is given in Figure 9.5, which shows a plot of ED level mortality rates classified by NO_x quintile. Within quintiles 4 and 5 (the highest levels of exposure), there are some EDs with much higher stroke mortality rates, which raise their averages compared with quintiles 1–3. However, models are needed to assess this variability properly in relation to stroke mortality.

---

[6] International Classification of Diseases, 9th revision (ICD9), codes 430–438. Prior to the Census Output Area (COA) framework which was introduced in the 2001 UK census for reporting purposes, enumeration districts (EDs) were the smallest spatial unit comprising approximately 200 households (over 300 in Scotland).

[7] The city is located on hilly ground just to the east of the Peak District National Park, although much of the city's industrial and commercial activity is concentrated on lower ground in the Don Valley. Although an oversimplification, lower income groups tend to live centrally on the lower ground, whilst the more affluent groups tend to live on the higher ground on the western side of the city.

In discussing this project, we look in particular at how two statistical challenges with important implications for ecological modelling, namely the presence of missing data (in a covariate) and spatially misaligned data, were addressed. In an extension to the original analysis by Maheswaran et al. (2006), we also look at a methodology for reducing ecological bias, in the estimates of relative risk obtained from the ecological model, by accounting for within-area variation in NO_x exposure. We also briefly describe other approaches to reducing ecological bias.

Modelled NO_x data were obtained from a grid model in Indic Airviro (SMHI Inc) at a 200m grid square resolution. These modelled NO_x values were derived using a number of inputs including data on the point (factories), line (roads) and area (housing estate) sources of NO_x within Sheffield, photochemical processes as well as various meteorological variables (temperature, radiation, wind speed and direction). The modelled NO_x values were then validated by visual inspection and by comparing with monitored NO_x values obtained from the small number of street monitors in place at that time. To merge with all other ED-level variables (e.g. counts of stroke deaths) in the study, the grid square NO_x values were first assigned to all postcode centroids (or postpoints) within each grid square (see Figure 9.6). ED level NO_x values were then calculated by taking the average of the NO_x values across all postcode centroids within the ED, weighted by the number of domestic household delivery points for each postcode. Figure 9.7 shows the interpolated NO_x exposure in the enumeration district (ED) framework. For more details, the reader is referred to Brindley et al. (2004, 2005). Note that interpreting the NO_x values quantitatively on an absolute scale was thought to be too unreliable, both because the values were based on a model rather than directly observed and because of the need to transfer values from grid squares to EDs. So the ED-level NO_x values were converted into quintiles. Whilst this had the advantage at the modelling stage of avoiding assuming a linear association between NO_x exposure and stroke mortality risk, the modelling treated these NO_x values as "data"

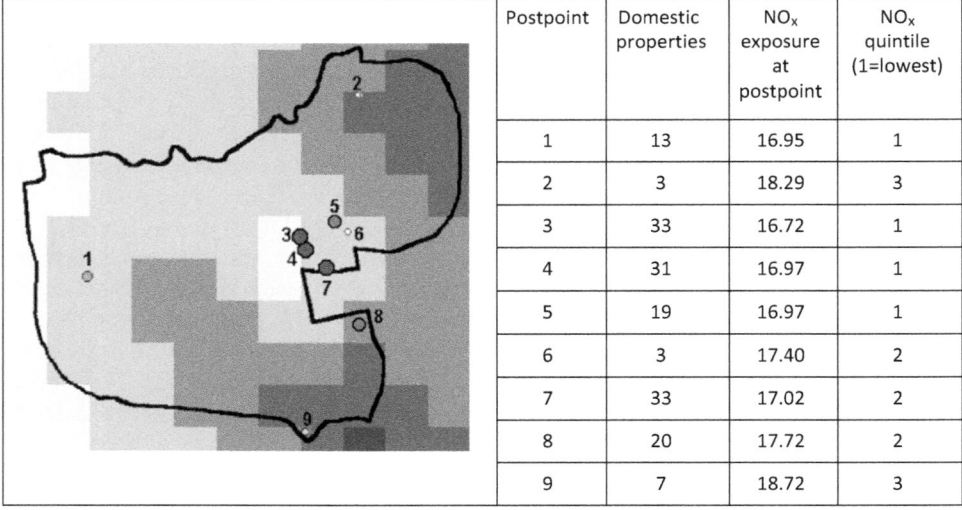

Postpoint	Domestic properties	NO$_x$ exposure at postpoint	NO$_x$ quintile (1=lowest)
1	13	16.95	1
2	3	18.29	3
3	33	16.72	1
4	31	16.97	1
5	19	16.97	1
6	3	17.40	2
7	33	17.02	2
8	20	17.72	2
9	7	18.72	3

FIGURE 9.6
An illustration of the within-ED distribution of population represented by postcode centroids (or postpoints) and NO_x pollution by grid cell. Each dot represents a postpoint with a known number of households with larger, darker dots indicating more households. The darker the cell shading, the higher the level of NO_x (after Brindley et al., 2004).

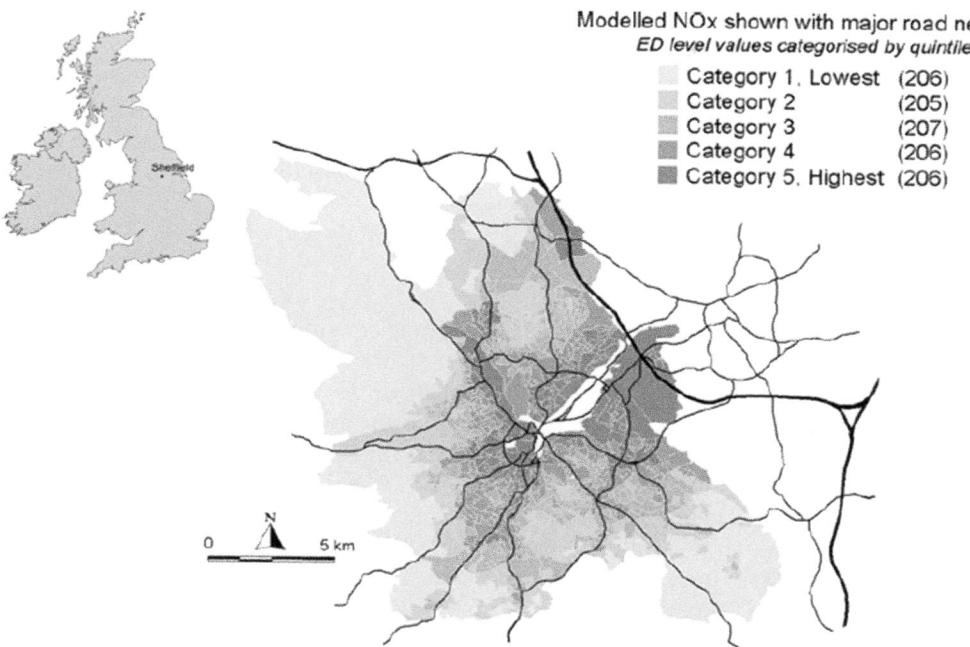

Modelled NOx shown with major road network
ED level values categorised by quintile

  Category 1, Lowest (206)
  Category 2   (205)
  Category 3   (207)
  Category 4   (206)
  Category 5, Highest (206)

FIGURE 9.7
Distribution of modelled ED-level outdoor NO_x categories by quintile in Sheffield (from Maheswaran et al., 2006).

and no allowance was made for the uncertainty associated with these values other than through this process of degrading the data to the ordinal scale.

Population data were obtained from the 1991 UK census by ED, age band and sex and corrected for under-enumeration. The 1991 census-based Townsend index was used as the measure of socio-economic deprivation at the ED level.[8] Expected counts of stroke deaths were obtained by standardizing for age (five-year bands), sex and Townsend deprivation quintile. Standardizing in this way allows adjustment for the confounding effect of deprivation, taking into account effect modification by age.[9] See Exercise 9.6 for the calculation of the expected counts.

Smoking prevalence is another potentially important confounder. Data on smoking prevalence came from a stratified random sample of the adult population resident in Sheffield. Out of the 9517 respondents, 2469 (about 26%) were smokers. The ED-level standardised smoking ratios, the number of smokers in each ED divided by the expected counts of smokers standardized for age and sex, were calculated. Across the EDs, the first and the third quartiles of the standardised smoking ratio are 0.50 and 1.49, respectively, showing some degree of between-ED variation in smoking prevalence. Whilst there were just under 10000 respondents in the survey, stratification was done by electoral ward (of

---

[8] Townsend et al. (1988). This index is a standardized combination of four census variables – the % of households with no car; where the head of household is unemployed; with more than one person per room; not owner occupied.

[9] Stroke mortality increases with increasing levels of deprivation at the area level, although the deprivation effect is modified by increasing age and may even reverse in the very elderly population (Maheswaran et al., 1997).

which there were 29 in Sheffield averaging about 5500 people per ward). This method of stratification resulted in a missing data problem since 14 EDs had no smoking prevalence data (no samples happened to have been taken in those EDs), an issue similar to that arising from the income survey data in Newcastle (see Chapter 7). Beyond the issue of missing data, the standardised smoking ratios are estimated based on a sample of the population, and so are inherently uncertain. We will now describe the two modelling strategies used in Maheswaran et al. (2006) to deal with these issues.

9.3.2 Modelling

Eq. 9.6 presents one of the Bayesian hierarchical models considered in Maheswaran et al. (2006). In this model, O_i is the observed number of stroke deaths in ED i. The mortality counts are assumed Poisson distributed with a mean of $\theta_i \cdot E_i$, where E_i is the expected number of stroke deaths in ED i standardized for age, sex and deprivation and θ_i denotes the underlying ED-specific relative risk of stroke mortality. The relative risk, θ_i, above (or below) 1 suggests that the stroke deaths observed in ED i exceeds (or is lower than) what might be expected given the age-sex-deprivation characteristics of this area. The regression relationship attempts to explain the spatial variability of the relative risks using two ED-level covariates, NO_x exposure level defined by quintile (x_{1i}) and smoking prevalence (x_{2i}), together with spatially-structured (S) and spatially-unstructured (U) random effects.

$$O_i \sim Poisson\left(\theta_i \cdot E_i\right)$$

$$log\left(\theta_i\right) = b_0 + b_{1,x_{1i}} + b_2 x_{2i} + S_i + U_i$$

$$S_{1:N} \sim ICAR\left(W, \sigma_S^2\right)$$

$$U_i \sim N\left(0, \sigma_U^2\right)$$

$$b_0 \sim Uniform(-\infty, +\infty)$$

$$b_{1,1} = 0 \text{ (reference category at the lowest } NOx)$$

$$b_{1,k} \sim N(0, 1000000) \text{ for } k = 2, \ldots, 5$$

$$b_2 \sim N(0, 1000000)$$

$$1/\sigma_S^2 \sim Gamma(0.5, 0.0005)$$

$$1/\sigma_U^2 \sim Gamma(0.5, 0.0005)$$

(9.6)

NO_x exposure is a categorical covariate with five levels (Figure 9.5), and it enters the model as a subscript of a regression coefficient, i.e. $b_{1,x_{1i}}$. As a result, b_1 represents five parameters, $b_{1,1}, \ldots, b_{1,5}$, one for each exposure level. This model fixes the lowest exposure level as the reference category ($b_{1,1} = 0$) but estimates the other four. Each $b_{1,k}$ ($k = 2, \ldots, 5$) compares the risk of stroke at the kth exposure category to the reference. See Section 5.4.1 for how to incorporate a categorical covariate in a regression model and its interpretation.

The set of spatially-structured random effects, S, is modelled using the ICAR prior, with W specified on the basis of ED contiguity. For the spatially-unstructured random effects, each U_i is assumed to follow a common normal distribution with mean 0 and variance σ_U^2.

The inclusion of S and U accommodates overdispersion, which should be anticipated because of intra-area population heterogeneity and the effects of omitted important covariates. While S represents the effects of any omitted important covariates that are spatially autocorrelated, U captures the effects of the omitted covariates that are uncorrelated spatially. Wakefield (2003) has shown that area level random effects are a natural way to model unmeasured confounders, although the resulting adjustment to estimated relative risk can be sensitive to the specification of the random effects, such as whether spatially-structured or not as well as the form of the spatial dependence structure imposed. Here spatial confounding is a potential issue, since NO_x levels are also spatially correlated (see Section 4.9.3).

Defining x_{2i}, the covariate measuring smoking prevalence, presents a challenge because this covariate is based on small numbers of smokers at the ED level, and some EDs had no data on smoking at all. Maheswaran et al. (2006) considered two versions of the model in Eq. 9.6, each dealing with x_{2i} differently. Version 1 expresses the regression relationship in Eq. 9.6 as

$$log\left(\theta_i\right) = b_0 + b_{1,x_{1i}} + b_2 x_{2i}^{AVG} + S_i + U_i, \tag{9.7}$$

which uses a spatially averaged standardized smoking ratio, x_{2i}^{AVE}, to capture smoking prevalence, x_{2i}. The spatially averaged ratio, x_{2i}^{AVE}, is calculated as the sum of the observed numbers of smokers in ED i *and its adjacent neighbours* divided by the sum of the expected numbers of smokers in ED i *and its adjacent neighbours* (see Section 6.2.2). In this version, the smoking prevalence values are included as data, that is as fixed values, a strategy that ignores the inherent uncertainty in these values as estimates of the unknown underlying smoking prevalence.

Version 2 treats the smoking prevalence, x_{2i}, as an unknown parameter, which is estimated using a Bayesian hierarchical model, referred to as the smoking model. Using the ED-level observed (O_i^{smoke}) and expected (E_i^{smoke}) counts of smokers, the smoking model estimates the full posterior distribution of the smoking prevalence covariate for each ED, and the uncertainty associated with the estimated smoking prevalence can then be fully accounted for when estimating the stroke-NO_x relationship. For this version, the (log) stroke relative risk is modelled by

$$log\left(\theta_i\right) = b_0 + b_{1,x_{1i}} + b_2 x_{2i}^{EST} + S_i + U_i, \tag{9.8}$$

where x_{2i}^{EST} denotes the estimated (log) smoking prevalence of ED i using the smoking model presented in Eq. 9.9.

$$O_i^{smoke} \sim Poisson\left(prev_i^{smoke} \cdot E_i^{smoke}\right) \ for \ i = 1,\ldots,N_1$$

$$log\left(prev_i^{smoke}\right) = a_0 + S_i^{smoke} + U_i^{smoke}$$

$$S_{1:N}^{smoke} \sim ICAR\left(\mathbf{W}, \sigma_{S.smoke}^2\right)$$

$$U_i^{smoke} \sim N\left(0, \sigma_{U.smoke}^2\right) \tag{9.9}$$

$$a_0 \sim Uniform\left(-\infty, +\infty\right)$$

$$1/\sigma_{S.smoke}^2 \sim Gamma\left(0.5, 0.0005\right)$$

$$1/\sigma_{U.smoke}^2 \sim Gamma\left(0.5, 0.0005\right)$$

The smoking model expresses the underlying (log) smoking prevalence, $prev_i^{smoke}$, as a sum of the overall intercept a_0 and the spatially-structured and the spatially-unstructured random effects, S_i^{smoke} and U_i^{smoke}. Similar to the model for stroke risk, the two sets of random effects account for random measurement error in the observed smoking counts, as well as the effects of unmeasured/unobserved covariates. But more importantly, in this model, the inclusion of S^{smoke} and U^{smoke} also enables information to be shared both locally and globally, strengthening the estimation of the ED-level smoking prevalence (see Chapter 8 for more detail on information sharing). For each of the 1030 EDs, $x_{2i}^{EST} = \log\left(prev_i^{smoke}\right)$.

Figure 9.8 shows the WinBUGS code for fitting the two models. There are two features of this implementation that are worth commenting on. First is the use of the cut function on Line 62. Both the smoking model and the stroke risk model are jointly fitted so that the entire posterior distribution of the smoking prevalence for each ED is used in the estimation of the stroke-NO_x relationship. In other words, the uncertainty in the smoking prevalence estimates is allowed to enter into the stroke model so that the estimates of the NO_x effect on stroke account for this source of uncertainty. However, we wish to estimate the smoking prevalence using only the smoking data but not affected by the data on stroke deaths. So the cut function in WinBUGS allows information (or uncertainty) to go "one way", from the smoking model to the stroke model but not the reverse. More specifically, at each MCMC iteration, values for the ED-level smoking prevalence are drawn from the posterior distribution obtained from the smoking model. Through the cut function, these values are then "fed" into the stroke model as a set of covariate values. Thus, over many MCMC iterations, the entire posterior distribution, and more importantly the uncertainty associated with the smoking prevalence estimates, is "fed" into the stroke model, achieving uncertainty propagation. This is a distinctive feature that the Bayesian approach to inference provides. We will see the use of the cut function again in other applications of temporal (Section 12.3.5) and spatial-temporal modelling (Sections 16.2 and 16.4).

Second, for the smoking model, the data likelihood is specified only for the $N_1 = 1016$ EDs with smoking data (Line 43, Figure 9.8; see also the first line in Eq. 9.9), but both sets of random effects are defined for *all* the $N = 1030$ EDs (Line 58 and Lines 64–73) so that the smoking model also produces estimates of the smoking prevalence for the 14 EDs without smoking data. This is the same setting as in the Newcastle MSOA income example throughout Chapters 7 and 8. More specifically, for each of the 1016 EDs with smoking data, Lines 64–67 calculate the (log) smoking prevalence as a sum of the three terms, the intercept, the spatially-structured random effect and the spatially-unstructured random effect. All the three terms are estimated using the smoking data through the smoking model. By contrast, for the 14 EDs with no smoking data (Lines 68–73), their spatially-structured random effects are predicted from the ICAR model, utilising information from their respective neighbours (Line 58; see also Section 8.2.1.3). As discussed in Section 7.4, the spatially-unstructured random effect for each of these 14 EDs takes the common normal distribution, $N\left(0, \sigma_{U.smoke}^2\right)$ (Eq. 9.9 and also Line 72 in Figure 9.8), as its posterior distribution, with $\sigma_{U.smoke}^2$ estimated from the smoking data from the other 1016 EDs.

9.3.3 Interpreting the Statistical Results

Table 9.5 reports the results from fitting the two versions of the model in Eq. 9.6.[10] For both versions, an association is observed between high levels of NO_x exposure (quintiles 4 and 5)

---

[10] Maheswaran et al. (2006, p.506–11) reports the fitting of several versions of this model.

```
1    #   The WinBUGS code to fit both the stroke-NOx model and the
2    #   smoking model together
3    #     the stroke-NOx model
4    model {
5      for (i in 1:N) {        #   loop through the (N=)1030 EDs
6        O[i] ~ dpois(mu[i])   #   the Poisson likelihood for the
7                              #   ED-level stroke death count
8        mu[i] <- theta[i]*E[i]
9        theta[i] <- exp(log.theta[i])
10       #   the hierarchically-centred version of the BYM model
11       #   (Section 8.5.1)
12       log.theta[i] ~ dnorm(mu.log.theta[i],prec.U)
13       mu.log.theta[i] <- b0 + b1[nox[i]] + b2*smoke[i] + S[i]
14       #   recovering the spatially-unstructured random effect
15       U[i] <- log.theta[i] - mu.log.theta[i]
16     }
17     ####    priors
18     #   the ICAR prior
19     S[1:N] ~ car.normal(adj[],weights[],num[],prec.S)
20     prec.U ~ dgamma(0.5,0.0005)
21     prec.S ~ dgamma(0.5,0.0005)
22     b0 ~ dflat()
23     b1[1] <- 0  #   setting the lowest NOx exposure as the reference
24     for (j in 2:5) {
25       b1[j] ~ dnorm(0,0.000001)    #   a vague prior for parameters
26                                    #   associated with higher levels
27                                    #   exposure
28     }
29     b2 ~ dnorm(0,0.000001) #   vague prior for the smoking effect
30     sd.S <- sd(S[1:N])     #   unconditional SD of the spatially-
31                            #   structured random effects
32     var.S <- pow(sd.S,2)   #   unconditional variance of the
33                            #   spatially-structured random effects
34     sd.U <- sd(U[1:N])     #   SD of the spatially-unstructured
35                            #   random effects
36     var.U <- pow(sd.U,2)   #   variance of the spatially-structured
37                            #   random effects
38     #   variance partition coefficient/spatial fraction
39     #   (Section 8.5)
40     vpc <- var.S/(var.S + var.U)
41     #   the smoking model to estimate the ED-level smoking
42     #   prevalence
43     for (i in 1:N1) {  #   a for-loop to go through the (N1=)1016
44                        #   EDs with data from the survey
45       smoke.O[i] ~ dpois(mu.smoke[i])
46       mu.smoke[i] <- prev.smoke[i] * smoke.E[i]
47       prev.smoke[i] <- exp(log.prev.smoke[i])
48       #   same as in the stroke model, the hierarchically-centred
49       #   version of the BYM model (Section 8.5.1)
50       log.prev.smoke[i] ~ dnorm(mu.log.prev.smoke[i],prec.U.smoke)
51       mu.log.prev.smoke[i] <- a0 + S.smoke[i]
52       #   recovering the spatially-unstructured term
53       U.smoke[i] <- log.prev.smoke[i] - mu.log.prev.smoke[i]
54     }
```

FIGURE 9.8
The WinBUGS code to fit jointly the stroke risk model (Eq. 9.6 with the regression relationship replaced by Eq. 9.8) and the smoking model (Eq. 9.9).

```
55   ####   priors
56   #  Note that the ICAR prior is assigned to all the 1030 EDs
57   #  (not just the 1016 with data; see main text)
58   S.smoke[1:N] ~ car.normal(adj[],weights[],num[],prec.S.smoke)
59   #  obtain the smoking prevalence estimates for all the 1030 EDs
60   for (i in 1:N) {
61      #  the use of the cut function (see main text)
62      smoke[i] <- cut(log.prev[i])
63   }
64   for (i in 1:N1) {   #  calculate the log smoking prevalence
65                       #   for those EDs with data
66      log.prev[i] <- a0 + S.smoke[i] + U.smoke[i]
67   }
68   for (i in (N1+1):N) {
69      #   calculate the log smoking prevalence for the EDs
70      #   without data (see main text)
71      mu.log.prev[i] <- a0 + S.smoke[i] + U.smoke[i]
72      U.smoke[i] ~ dnorm(0,prec.U.smoke)
73   }
74   #  priors for other parameters (intercept and variances)
75   a0 ~ dflat()
76   prec.U.smoke ~ dgamma(0.5,0.0005)
77   prec.S.smoke ~ dgamma(0.5,0.0005)
78   sd.U.smoke <- sd(U.smoke[1:N1])
79   var.U.smoke <- pow(sd.U.smoke,2)
80   sd.S.smoke <- sd(S.smoke[1:N1])
81   var.S.smoke <- pow(sd.S.smoke,2)
82   #  spatial fraction under the smoke model
83   vpc.smoke <- var.S.smoke/(var.U.smoke + var.S.smoke)
84   }
```

FIGURE 9.8

(Continued)

TABLE 9.5

Results from Fitting the Two Versions of the Model Presented in Eq. 9.6 Using Winbugs

Parameter	Version 1 Using spatially averaged smoking, x_{2i}^{AVG}	Version 2 Using the smoking model to estimate x_{2i}^{EST}
	Relative risk (95% credible interval)	Relative risk (95% credible interval)
NO_x category 5 (Highest exposure)	1.31 (1.09–1.56)	1.30 (1.08–1.55)
NO_x category 4	1.21 (1.02–1.45)	1.21 (1.01–1.44)
NO_x category 3	1.10 (0.86–1.22)	1.03 (0.87–1.22)
NO_x category 2	1.13 (0.89–1.28)	1.07 (0.90–1.27)
NO_x category 1 (Lowest exposure)	1	1
Smoking	1.03 (0.88, 1.20)	1.15 (0.90–1.49)
Spatial fraction for stroke death	0.02 (0.00, 0.06)	0.02 (0.00, 0.08)
Spatial fraction for smoking prevalence	NA	0.98 (0.92, 1.00)
DIC for stroke death	3925	3921

and stroke mortality. Version 2 suggests that compared to that with the lowest level of exposure, an ED that is in the highest exposure category is likely to have 30% (95% CI: 8%–55%) higher stroke mortality. The stroke mortality for an ED in the second highest category tends to be 21% (95% CI: 1%–44%) higher than an ED with the lowest exposure. Between the two versions, the effect estimates are largely similar, suggesting that the findings regarding the NO_x effect on stroke are not much affected by how smoking prevalence enters the model.

The random effects for explaining stroke risks were mainly confined to the spatially-unstructured component (the posterior mean of the spatial fraction = 0.02 from both versions), suggesting that the confounding effects of any unmeasured covariates did not have any spatial structure. Note that no association was found between smoking prevalence and stroke mortality in this set of data even though smoking is a known risk factor at the individual level. Note also that the effects of the unmeasured/unobserved covariates on smoking prevalence, unlike the stroke model, exhibit spatial structure, as the spatially-structured random effects dominate over the spatially unstructured random effects (spatial fraction of 0.98 with 95% CI of 0.92–1.00 from Version 2).

9.3.4 Conclusion and Discussion

Small area ecological studies have certain advantages over cohort studies when examining the health impacts of environmental exposure, particularly if the effect of the risk factor is expected to be small. Such ecological studies permit the study of large *populations* at relatively low cost. They may be less affected by measurement error when trying to quantify exposure than studies of individuals, since at the population level some of the individual level error might be expected to average out. Small areas may be relatively homogeneous in terms of population characteristics, and the census output areas used in the UK since 2001 have been designed with homogeneity as a criteria (see Section 3.3.1). Whilst the small number problem is often encountered in small area studies, Bayesian hierarchical models, through borrowing strength, offer a way forward, with additional benefits for handling data from different and incompatible spatial units and allowing for the uncertainty associated with merging such datasets. For more discussion around handling incompatible spatial units, see Gotway and Young (2002) and Young et al. (2009).[11] That said, the risk of ecological bias (or aggregation bias) underlies all ecological studies, which is why such studies are viewed with suspicion as a basis for individual-level inference. As we stated at the beginning of this application, uncritical use of estimates from an ecological model to measure risk at the *individual level* represents an example of the ecological fallacy (see Section 3.3.3).

In the Sheffield study, Haining et al. (2010) used information on within ED variation in NO_x exposure with the aim of reducing one source of ecological bias: within-area variation in exposure. A careful consideration of the within ED distribution of population and NO_x pollution by grid cells as illustrated in Figure 9.6 allows us not only to calculate a population weighted exposure but also to approximate the proportion of residents exposed to each of the quintiles of NO_x pollution in each ED.[12] For the ED in Figure 9.6, the proportions of domestic properties exposed to the five levels are 59% (at the lowest

---

[11] See also Banerjee et al. (2004) and Gelfand et al. (2001) for links to the change of support problem in geostatistics. Boulieri et al. (2017) link line and area data in a spatial-temporal analysis of UK road traffic accidents. Taylor et al. (2018) deal with the incompatible spatial unit problem using a point process modelling approach.

[12] On the assumption that residents are only exposed to the NO_x level of where they live and that this level remains fairly uniform over time.

exposure), 35%, 6%, 0% and 0% (at the highest exposure). It is such information on within-ED exposure that Haining et al. (2010) sought to incorporate with the stroke death data where the latter, of course, is only available at the ED level. But, how can we incorporate this information?

One way to achieve that is to build the underlying individual-level model, then aggregate up to deal with the area-level counts. At the individual level, the outcome is binary. That is, y_{ij}, the outcome on stroke death for person j living in ED i, is 1 if this person dies from stroke and 0 otherwise. Given the value of NO_x exposure, x_{ij}, of this person, the outcome is modelled as

$$y_{ij} \sim Bernoulli\left(\tilde{q}_{ij}\right)$$
$$logit\left(\tilde{q}_{ij}\right) = \alpha + \beta \cdot x_{ij} + h_i \tag{9.10}$$

In Eq. 9.10, the logit transform of the individual stroke risk, $\tilde{q}_{ij}$, is modelled as a linear combination of the overall intercept (α), the effect due to NO_x exposure ($\beta \cdot x_{ij}$) and the effect due to other ED-level covariates such as deprivation (h_i). Eq. 9.10 captures the essence of the model considered in Haining et al. (2010), in which they also dealt with age and sex effects as well as including separate terms to represent effects from observable covariates and those from omitted covariates. Since the NO_x exposure is classified into one of the quintiles, $\beta \cdot x_{ij}$ is replaced by β_k, the latter representing the effect of the kth exposure category. The stroke risk for *any* person in ED i assigned to the kth level of exposure becomes

$$q_{ik} = \text{expit}\left(\alpha + \beta_k + h_i\right), \tag{9.11}$$

where $\text{expit}(a) = \dfrac{\exp(a)}{1+\exp(a)}$, the inverse of the logit transform. Eq. 9.11 is the link for combining within-ED exposure with the ED-level counts.

Since stroke is not contagious, the independent Bernoulli random variables within an ED aggregate to a binomial random variable. Therefore, the observed number of stroke deaths, O_i, is modelled as $O_i \sim Bin(p_i, m_i)$ where m_i denotes the population at risk in ED i and p_i is the area-level risk, which is given by averaging over the individual-level risks with respect to the distribution of within-ED exposure. In the case of the Sheffield study with five exposure categories, it follows that p_i is given by

$$p_i = \sum_{k=1}^{5} q_{ik}\phi_{ik} \tag{9.12}$$

Essentially, p_i is the weighted average of the risks associated with each of the five exposure categories, (q_{ik}), with the weights given by the proportions of residents exposed to each category, (ϕ_{ik}). For example, the area level risk in the case of an ED where all of its residents are exposed to the highest level of NO_x is $p_i = q_{i5}$. On the other hand, if 40%, 30%, 20% and 10% of its residents were exposed to categories 2, 3, 4 and 5 respectively (and none in category 1), then $p_i = 0.4 \times q_{i2} + 0.3 \times q_{i3} + 0.2 \times q_{i4} + 0.1 \times q_{i5}$ for that ED. When the within-area variability in exposure is small (e.g. the first of the two examples), there will be little difference in risk estimate with/without the incorporation of the information on within-area exposure. Putting all together, Eq. 9.13 presents the complete model specification (see Exercise 9.7 for the fitting of this model).

$$O_i \sim Bin(p_i, m_i)$$

$$p_i = \sum_{k=1}^{5} q_{ik} \phi_{ik}$$

$$q_{ik} = \text{expit}(\alpha + \beta_k + h_i)$$

$$\beta_1 = 0$$

$$\beta_k \sim N(0, 1000000) \quad \text{for} \quad k = 2, \ldots, 5 \qquad (9.13)$$

$$\alpha \sim N(0, 1000000)$$

$$h_i \sim N(0, \sigma_h^2)$$

$$1/\sigma_h^2 \sim Gamma(0.5, 0.0005)$$

Based on a more elaborate version of the model in Eq. 9.13, Haining et al. (2010) found that in the case of the Sheffield data, the incorporation of within-area exposure variability did not change the risk estimates substantially compared to the estimates obtained when such information was not taken into account. This is mainly due to the small variability of within-ED NO_x exposure – 72% of EDs have more than 70% of their populations exposed to only one of the five NO_x categories. In such circumstances, as they concluded, the original ecological model that ignored within-ED exposure variation did not introduce significant ecological bias into the estimates. However, in circumstances where within-area exposure variability is more pronounced, this modification should be implemented. However, in dealing with this source of ecological bias, it is important to remember that there can be other sources. These include bias due to: non-linearity in the exposure-response relationship; between-area variation in the disease rate amongst those *not* exposed to the risk factor; effect modification, where the same exposure level has a different effect in different areas; confounders, which are unmeasured at the area level or which vary between individuals; the exposure effect at the individual level varying because of contextual or area-specific (group) effects (see Morganstern, 1995). The analyst needs to consider the extent to which these other sources of bias might be present.

Finally, if some sample individual level data are available,[13] another option to strengthen an ecological study for making individual level inference is to combine such individual level data with the small area data. The individual data should include both exposure and outcome information for each individual, but it is not necessary that such data are available for all small areas. Small samples lack power, but Bayesian hierarchical models enable individual level data to borrow strength from the small area data. The interested reader is referred to Haneuse and Wakefield (2007, 2008) and Jackson et al. (2006, 2008) for more information on such methodology.

---

[13] For confidentiality reasons, of course, such data may be hard to come by.

9.4 Application 3: Modelling the Village-Level Incidence of Malaria in a Small Region of India

9.4.1 Background

The World Health Organization (WHO) estimates that India accounts for 75% of all malaria cases in Southeast Asia, with some 95% of the Indian population residing in areas where malaria is endemic. Whilst the disease in India is particularly entrenched in the east and northeast of the country, the disease is also present in central and more arid parts, including the state of Karnataka, from where the data for this study has been taken. Karnataka is one of the country's high transmission zones. Whilst the introduction of vector (mosquito) control measures has done much to reduce the number of malaria cases in the state since the recorded peak in 1976 (around 630000 cases), from time to time increases in the number of cases stimulate renewed concern and renewed efforts to achieve sustainable reduction in the case number with the ultimate goal of elimination.

There are many risk factors associated with the incidence of malaria with rainfall level one of the key factors because it affects the survival of the mosquito and its ability to reproduce. Other factors found to be relevant in explaining spatial and temporal variation in the number of new cases of malaria include vegetation cover (e.g. amount of forest cover, wetlands, ditches and ponds), urbanization (e.g. the presence of construction sites where stagnant water can gather), population characteristics (e.g. population density, housing type) and the effectiveness of intervention programmes (see Shekhar et al., 2017).

The development of effective, targeted control measures depends on obtaining good estimates of the underlying spatial variation in the risk of malaria. However, because of the way malaria spreads, statistically estimating this risk surface raises a number of methodological challenges. First, the risk map is likely to display marked spatial heterogeneity due to varying levels of the risk factors (e.g. rainfall level) across the study region.[14] Second, because malaria is an infectious disease, cases are likely to cluster, with concentrations of cases associated with those areas where there was early infection. At the same time, there may be many areas with no cases of the disease, giving rise to what is termed zero inflation (see Section 6.4 in Chapter 6). Finally, and linked to both the previous points, the CAR-type of models for random effects may not be appropriate because these models may oversmooth the risk map. In this application we consider the problem of estimating the malaria risk at a specific time window, paying particular attention to these three methodological challenges.

9.4.2 Data and Exploratory Analysis

This application analyses the number of new malaria cases occurring in each of the 139 villages (the spatial units of this analysis) in Kalaburagi taluk, South India, in June and July 2012. These two months mark the beginning of the monsoon season. Kalaburagi City, the largest city/village in Kalaburagi taluk, has a population of over half a million, while, for other villages, the average population is about 2000. The aims of the analysis are to model the spatial distribution of the village-level incidence rates of malaria and to assess

---

[14] Another important type of spatial heterogeneity in some applications is where the outcome-covariate relationship of one (or more) risk factor varies spatially (see Chapter 1 and Section 9.5).

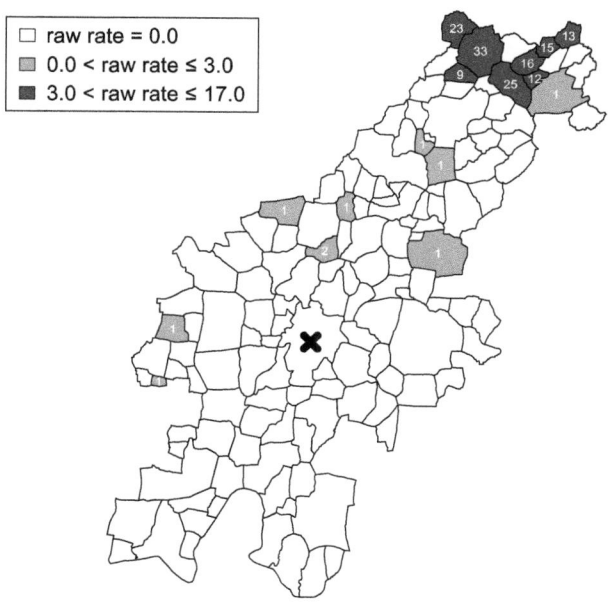

FIGURE 9.9
A map of the raw incidence rate (per 1000 people) in each village between June and July 2012. Kalaburagi City, the largest city in the study region, is located in the polygon with a cross. The numbers inside the village polygons are the (non-zero) counts of malaria cases. The polygons in white represent the villages with no cases reported in these two months. For scale purposes the NE to SW axis is approximately 80km.

the association between rainfall level and malaria incidence. The village population data are taken from the 2011 Indian census, and these data are taken to represent the at-risk population of each village. The rainfall covariate is the average rainfall level in June and July 2012 for each village.

Figure 9.9 shows the spatial distribution of reported new cases and raw incidence rates (per 1000 people). The latter are calculated as the reported case count in each village divided by its population then multiplied by 1000. As is evident from the map, most of the cases were concentrated in just a few villages in the north of the taluk. Only a small number of cases were reported in a scattering of other villages across the rest of the region. The global Moran's I applied to the raw incidence rates is 0.42, with a p-value of 0.001 (based on 999 random permutations of the original data). This confirms the presence of strong positive spatial autocorrelation in the incidence rates, largely due to the high concentration of cases in the north of the region. The positive relationship between rainfall and malaria risk is evident in Figure 9.10. A village with a high rainfall tends to have a high malaria incidence rate. The relationship is somewhat nonlinear, so the rainfall values are converted into quartiles and enter the model as a categorical covariate with four levels.

Another important feature of the data is that a large proportion of the villages (122 out of 139) had no reported malaria cases during these two months. In Section 6.4.2, we performed a test of zero-inflation based on a Poisson regression model where the village-level population is set as an offset and controls for the rainfall effect. The test statistic is 339.34, with a p-value that is very close to 0, suggesting that the fitted model cannot accommodate the number of zero observations in the dataset.

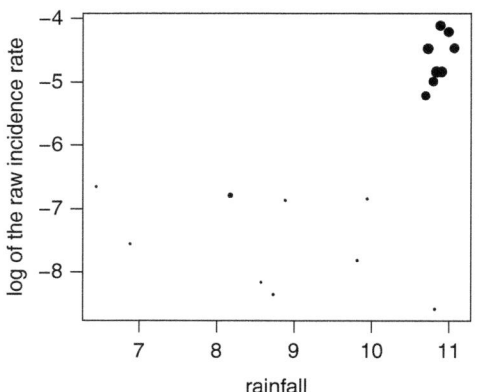

FIGURE 9.10
A positive relationship between malaria risk and rainfall level. A data point with a higher uncertainty (measured as the variance of the raw incidence rate) is represented by a smaller dot so that it asserts less influence visually. This figure is in the form that is presented in Figure 6.15(b).

9.4.3 Model I: A Poisson Regression Model with Random Effects

The first model is a Poisson regression that attempts to explain the observed cases of malaria using the covariate on village-level rainfall and a set of random effects. The latter serve as surrogate measures for the effects from the unmeasured/unobserved risk factors. This first model is given in Eq. 9.14:

$$y_i \sim Poisson\left(\theta_i \cdot pop_i\right)$$
$$\log\left(\theta_i\right) = \alpha + \beta_{x_{i,rain}} + U_i$$
$$\alpha \sim N(0,1000000)$$
$$\beta_1 = 0\left(\text{reference category at the lowest rain fall level}\right) \qquad (9.14)$$
$$\beta_k \sim N\left(0,1000000\right) \text{ for } k = 2,\ldots,4$$
$$U_i \sim N\left(0,\sigma^2\right) \text{ for } i = 1,\ldots,N$$
$$\sigma \sim N_{+\infty}\left(0,10\right)$$

where y_i and pop_i are, respectively, the number of malaria cases and the at-risk population in village i. On the log scale, θ_i, the underlying malaria risk in village i is modelled as a function of the intercept α, the rainfall effect $\beta_{x_{i,rain}}$ and U_i, where the latter is a spatially-unstructured random effect. We did not include a set of spatially-structured random effects in this model, as spatial discontinuity is evident from Figure 9.9, showing that the villages in the north with high raw incidence rates share common boundaries with some villages with zero, or very low, incidence rates. The categorical covariate on rainfall, $x_{i,rain}$, gives rise to four parameters, $\beta_1,\ldots,\beta_4$, where category 1, with the lowest rainfall level, is set as the reference category. Thus, each of the other three parameters $(\beta_2,\ldots,\beta_4)$ compares the malaria incidence of a higher rainfall level to that of the reference level. Vague priors are assigned to the unknown parameters α and $\beta_2,\ldots,\beta_4$, as well as to the random effect standard deviation, σ.

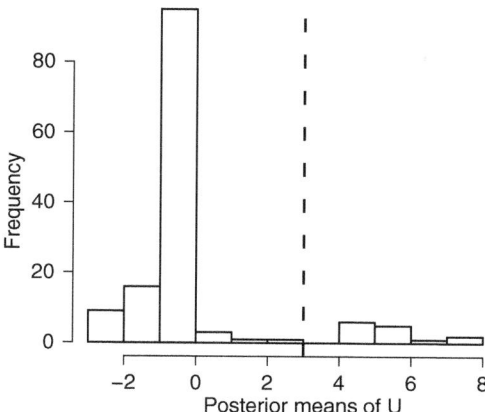

FIGURE 9.11
A histogram of the posterior means of the spatially-unstructured random effects estimated from the Poisson regression in Eq. 9.14. A dotted vertical line at 3.0 is added for reference.

Figure 9.11 shows the distribution of the posterior means of the spatially-unstructured random effects. The distribution is bimodal, showing two rather distinct groups of villages. One group of villages have posterior means of their random effects lying above 3.0, while the other group have posterior means of their random effects lying below 3.0. Although the random effects included in Eq. 9.14 are flexible enough to capture the type of variability observed in Figure 9.9, given the evidence from Figure 9.11, the assumption that all the random effect terms follow a single common distribution seems to be questionable.

9.4.4 Model II: A Two-Component Poisson Mixture Model

An alternative modelling strategy is the mixture approach. Mixture models have been applied extensively in modelling area-level spatial data, largely in the context of disease mapping (see, for example, Knorr-Held and Rasser, 2000; Green and Richardson, 2002; and Lawson and Clark, 2002) and, more recently, in the modelling of school exam results (Neelon et al., 2014). The mixture approach is typically employed to deal with multimodality in the distribution of the outcome values (or of the estimated random effects) and/or to capture spatial discontinuity (where abrupt changes are present on the map of the outcome values or the estimated random effects). The features of the malaria data, that is, the bimodal distribution of the estimated random effects and the evidence of spatial discontinuity, prompt us to consider the mixture approach.

Under a mixture model, each village is assigned to a cluster (or equivalently a risk group), and the villages in the same cluster are assumed to have the same risk. A mixture model makes no assumptions about the shape of the risk distribution nor on whether the underlying risk surface is smooth everywhere or whether abrupt changes exist. However, the mixture model we shall describe does make three key assumptions. First, the number of clusters is fixed at two – this is justified given our findings from EDA (Section 9.4.2) and from fitting the first model (Section 9.4.3). Second, having accounted for the rainfall effect, the allocation of each village to a risk cluster is independent of every other village. In other words, the residual risks (after accounting for the rainfall effect) in adjacent villages are not necessarily similar to each other (contrasting, for example, with the ICAR model, which assumes risks are similar locally). Third, every cluster has its own risk parameter, and this risk parameter is assigned to all the villages within the same cluster. These assumptions can be relaxed, but we will defer the discussion till the end of this application.

Based on the three assumptions, a two-component mixture model is given by

$$y_i \sim Poisson\left(\theta_i \cdot pop_i\right)$$
$$\log\left(\theta_i\right) = \alpha_{z_i} + \beta_{x_{i,rain}} \tag{9.15}$$

As in the first model, each outcome value y_i is modelled by a Poisson distribution. In addition to the rainfall effect ($\beta_{x_{i,rain}}$), the log malaria risk of village i, $\log(\theta_i)$, is modelled either by α_1 or α_2, and the selection between the two is carried out through the cluster indicator, z_i. Specifically, with two clusters, $z_i = 1$ or 2 for each village. Village i is assigned to cluster 1 if $z_i = 1$ then its log malaria risk is modelled as $\alpha_1 + \beta_{x_{i,rain}}$. On the other hand, if village i is assigned to cluster 2, then $z_i = 2$ and $\log\left(\mu_i\right) = \alpha_2 + \beta_{x_{i,rain}}$. The two parameters, α_1 and α_2 each represent the cluster-specific (residual) risks after accounting for the effect of rainfall. We impose the constraint $\alpha_1 \leq \alpha_2$ so that villages assigned to cluster 1 have a lower residual risk compared to those in cluster 2. When interpreting the two clusters, one should bear in mind that, because of the inclusion of the rainfall covariate, allocating a village to cluster 1 does not imply that this village has a low risk of malaria. Rather, it means that having accounted for the rainfall effect, the *residual risk* of this village is low, compared to those that are assigned to cluster 2. Similarly, a village assigned to cluster 2 means that having accounted for the rainfall effect, its residual risk is high relative to those in cluster 1.

To fully specify the model, we need to assign prior distributions to α_1, α_2, β_k (k = 1,...,4) and the set of village-specific cluster indicators z_i. For α_1 and α_2, a vague prior is assigned to α_1, but we set $\alpha_2 = \alpha_1 + \delta$ where the chosen prior for δ is $Uniform(0,100)$. This uniform (vague) prior ensures δ is non-negative-valued so that the constraint $\alpha_1 \leq \alpha_2$ is met.[15] Each of the cluster indicators z_i (for i = 1,...,N) is modelled using a common discrete uniform distribution so that $\Pr\left(z_i = 1\right) = p$ and $\Pr\left(z_i = 2\right) = 1 - p$. This is equivalent to allocating each village to one of the two clusters by flipping a coin, where one side of the coin shows the number 1 (thus allocating this village to cluster 1) and the other side shows 2 (allocating this village to cluster 2), and p is the probability of getting 1. We treat p as an unknown parameter so its posterior estimate will tell us the proportion of villages allocated to cluster 1. A vague prior $Beta(1,1)$ (equivalent to the $Uniform(0,1)$; see Section 5.2.1) is assigned to p. Finally, the specification for $\beta_1,...,\beta_4$ is the same as that for the model presented in Eq. 9.14.

The WinBUGS implementation of this model is given in Figure 9.12. Note that the WinBUGS function dcat on Line 13 specifies a discrete uniform distribution whose support is a set of h positive integers between 1 and h. The argument of the dcat function is an array of length h containing the probabilities of returning those h integer values, and the sum of the h probabilities equals one. So, Line 13 in Figure 9.12 models each cluster z_i as a random variable with a discrete uniform distribution so that the probability of z[i] = 1 is given by p[1] (this is the parameter p in the model specification) and the probability of z[i] = 2 is p[2] (this is $1 - p$). Also, the sum of p[1] and p[2] is always 1, so we only need to estimate one of the two probabilities (see Lines 16 and 18 in Figure 9.12).

---

[15] This trick can be used to impose an order constraint on a set of parameters. For example, to ensure $a_1 \leq a_2 \leq a_3$, we can assign a prior to a_1 (e.g. $N(0,1000000)$) and set $a_2 = a_1 + \delta_1$ and $a_3 = a_2 + \delta_2$, then assign a positive-valued distribution (e.g. $Uniform(0,100)$ or $N_{+\infty}(0,10)$) as the prior for δ_1 and δ_2.

```
1    #  the WinBUGS code to fit the two-component mixture model to
2    #  the malaria incidence data
3    model {
4      for (i in 1:N) {  #  looping through the (N=)139 villages
5        y[i] ~ dpois(mu[i])
6        mu[i] <- theta[i] * pop[i]      #  define the Poisson mean
7        log(theta[i]) <- beta[rain[i]]  #  the effect of rainfall
8                                        #  (over four categories)
9                      + alpha[z[i]]
10       #  the indicator z[i] assigns the village to either
11       #  cluster 1 (with low residual risk) or
12       #  cluster 2 (with high residual risk)
13       z[i] ~ dcat(p[1:2])    #  a discrete uniform distribution
14                              #  on the indicator z[i]
15     }
16     p[1] ~ dbeta(1,1)    #  estimate the proportion of villages
17                          #  in cluster 1
18     p[2] <- 1 - p[1]     #  the proportion of villages in cluster
19                          #  2 is 1 - the proportion of villages
20                          #  in cluster 1
21     alpha[1] ~ dnorm(0,0.000001)  #  estimate the residual risk
22                                   #  level of cluster 1
23     alpha[2] <- alpha[1] + delta  #  this specification of
24                                   #  alpha[2] ensures that
25                                   #  alpha[1]  alpha[2]
26     delta ~ dunif(0,100)  #  a vague non-negative-valued prior
27                           #  on delta
28     beta[1] <- 0   #  setting the reference category at
29                    #  the lowest rainfall level
30     for (k in 2:4) {
31       beta[k] ~ dnorm(0,0.000001)  #  the risk comparison between
32                                    #  category k and the reference
33     }
34   }
```

FIGURE 9.12
The WinBUGS implementation of the mixture model with two clusters.

9.4.5 Model III: A Two-Component Poisson Mixture Model with Zero-Inflation

We extend the two-component mixture model in order to handle zero-inflation, the dispro-portionately many zeros observed in the data (Section 9.4.2). Relevant to the modelling of the malaria data, Diggle and Giorgi (2016) discuss how an excess of zeros might arise in the context of disease mapping. They distinguish two possible (zero-generating) processes – "a zero (prevalence) estimate in a particular community can be either a chance finding, or a necessary consequence of the community being disease/infection-free" (Diggle and Giorgi 2016, p.1103). They provide an example for the latter: "For diseases that are environ-mentally driven", an excess of zeros can be because "some areas are fundamentally unsuit-able for disease transmission" (Diggle and Giorgi, 2016, p.1103). In the case of the present set of malaria data, it is not possible to say with any confidence the extent to which the excess of zero cases is due to chance on the one hand or environmental factors on the other. We are dealing with a very short interval of time and an infectious disease where initial infections "due to chance" can have a significant impact on spatial patterns of incidence, thereby lending some credence to the first process. It seems unlikely that areas within the taluk are "fundamentally unsuitable for disease transmission" for purely environmental reasons, but earlier intervention programmes, more effective in some parts of the taluk than in others, might have, even if only temporally, created conditions inimical to the emergence of the disease in those areas.

To deal with the excessive zeros, the third model we consider embeds the two-component mixture model in the zero-inflated Poisson (ZIP) model. The ZIP model is a mixture of a standard Poisson regression and a deterministic distribution at zero, a distribution that takes the value of zero with probability one (Lambert, 1992 and Neelon et al., 2010; see also Section 6.4.2 in Chapter 6). The ZIP model allows us to distinguish between two types of zero observations. The first type comes from a chance finding, where the underlying risk for a village is not exactly zero, but it so happens that no case was observed in the study period. This first type is modelled through the standard Poisson regression. The second type of zero comes from the "disease-free" villages with the underlying risk being exactly zero and is modelled by the deterministic distribution at zero. More generally, these two types of zero observations are also known as "sampling zeros" and "structural zeros" (see Fisher et al., 2017, and references therein).[16]

Following Lunn et al. (2012, p.285–286), the ZIP model is expressed as a mixture model. Each observed number of malaria cases y_i is assumed to come from a Poisson distribution with mean $\mu_i = g_i \cdot (\theta_i \cdot pop_i)$, where g_i is a binary 0/1 indicator. When $g_i = 0$, y_i comes from a Poisson distribution with mean $\mu_i = 0$, and this corresponds to the deterministic distribution at zero. In other words, village i is assigned to the "disease-free" group. When $g_i = 1$, the number of malaria cases y_i is modelled by a standard Poisson regression where the Poisson mean is $\mu_i = \theta_i \cdot pop_i$, the same as the Poisson mean expression in the two-component mixture model given by Eq. 9.15. Each indicator g_i (for $i = 1,...,N$) is modelled by a common Bernoulli distribution, i.e. $g_i \sim Bernoulli(\varphi)$, where $1 - \varphi$ is the unknown proportion of villages assigned to the "disease-free" group. The proportion parameter φ has a vague $Beta(1,1)$ prior. Figure 9.13 provides part of the WinBUGS code for fitting this model. The code in Figure 9.13 also features an evaluation of the model fit to the data (Lines 23–36) and a classification of the villages based on their risk levels (Lines 42–56). Both features will be discussed in detail in Section 9.4.6.

9.4.6 Results

Table 9.6 compares the three models, the Poisson regression with random effects, the two-component mixture model with and without zero-inflation, using DIC. The DIC values are fairly close and all the pairwise DIC differences are smaller than five, suggesting these three models are indistinguishable based on DIC. However, a careful inspection of $\bar{D}$ and pD, the constituent parts of DIC, sheds some useful insight into these models. The Poisson regression with spatially-unstructured random effects has the smallest value of $\bar{D}$ and thus fits the data best. This is due to the flexibility of the village-specific random effects. However, the large number of random effects also makes this model the most complex,

---

[16] An alternative method to handle the excess of zero observations is the hurdle model (Mullahy, 1986). Under a hurdle model, *all* the zero values are assumed to come from the deterministic distribution at zero, while all non-zero observations are modelled through a zero-truncated Poisson distribution (or a zero-truncated binomial distribution if one models probability). Diggle and Giorgi (2016, p.1104) argue that the hurdle model is unsuitable for disease mapping mainly because it "does not distinguish between observing no cases among sampled individuals as a chance finding or as a necessary consequence of the entire community being disease-free". In general, the difference between structural and sampling zeros, and therefore the choice between zero-inflated and hurdle models, may only be small (see, for example, Hu et al., 2011). Rose et al. (2006) emphasize that the choice of the two models depends on the nature of the experimental design and the outcome data.

```
1    #  part of the WinBUGS code to fit the two-component mixture
2    #  model with zero-inflation to the malaria incidence data
3    model {
4      for (i in 1:N) {    #  looping through the (N=)139 villages
5        y[i] ~ dpois(m[i])
6        #  implementing the zero-inflated Poisson part
7        m[i] <- g[i] * mu[i] * pop[i] #  define the Poisson mean
8        g[i] ~ dbern(phi)            #  the binary 0/1 indicator in
9                                     #  the ZIP component
10     }
11     phi ~ dbeta(1,1)  #  a vague prior on the proportion of
12                       #  villages not in the "disease-free" group
13
14     #  insert Lines 6 to 33 in Figure 9.12 after Line 17 (below)
15     #  to specify a two-component mixture with rainfall effect
16     #  for mu[i]
17     for (i in 1:N) {
18     #  remove this line then insert Lines 6 to 33 in Figure 9.12 here
19
20     ##  posterior predictive checks for this model
21     ##  (see Figure 9.15 and the associated discussion in the
22     ##  main text)
23     for (i in 1:N) {
24       y.pred[i] ~ dpois(m[i])      #  predict the number of cases
25                                    #  from the model for each village
26       #  check predicted number of cases against observed
27       ppp[i] <- step(y.pred[i] - y[i])  #  Pr(number of predicted
28                                    #  cases greater than or
29                                    #  equal to observed)
30       #  predicted proportion of villages with 0 cases vs observed
31       #  predict.zero[i]=1 if 0 cases predicted
32       #  otherwise predict.zero[i]=0
33       predict.zero[i] <- equals(y.pred[i],0)
34     }
35     #  proportion of villages with 0 predicted
36     pred.proportion.zero <- sum(predict.zero[1:N])/N
37
38     ##  classification of villages (see Table 9.7 and the
39     ##  associated text for definition);
40     ##  the posterior median of h[i] is used to produce
41     ##  Figure 9.14(a)
42     for (i in 1:N) {
43       h[i] <- 1*equals(g[i],0)  #  allocated to the
44                                 #  disease-free group
45          + 2*equals(g[i],1)*equals(z[i],1) #  allocated to the low
46                                 #  residual risk group
47          + 3*equals(g[i],1)*equals(z[i],2) #  allocated to the
48                                 #  high residual risk
49                                 #  group
50       for (k in 1:3) {
51       #  the posterior mean of prob.h (below) is the posterior
52       #  prob. of each village assigned to each of the three groups
53       #  (see Figure 9.14(b) and (c))
54         prob.h[i,k] <- equals(h[i],k)
55       }
56     }
57   }
```

FIGURE 9.13
Part of the WinBUGS code for fitting the two-component mixture model with zero-inflation.

TABLE 9.6

Comparing the Three Models Using DIC

Model	$\bar{D}$	pD	DIC^*
A Poisson regression with spatially-unstructured random effects	95	25	120
A two-component mixture model without zero-inflation	108	14	122
A two-component mixture model with zero-inflation	104	14	118

* WinBUGS does not return DIC for a mixture model, as there are different ways to calculate pD (see suggestions in Celeux et al., 2006, who also discuss some related issues). Here we use the posterior medians of z_i and g_i to calculate pD and DIC. Exercise 9.8 provides details of the calculation.

with the largest pD value. Compared to the Poisson regression model, both versions of the two-component mixture with and without the zero-inflation part are simpler in structure (they both have smaller pD values), but their larger $\bar{D}$ values suggest an inferior fit to the data. Between the two mixture models, the addition of the zero-inflated part provides a better data fit and yields a smaller DIC value. It is worth noting that the mixture model with zero inflation has more *actual* parameters (due to the additional zero-inflated part) than that without zero inflation, but the pD values of these two models are the same. This is because pD measures the *effective* number of parameters in the model, not the *actual* number of parameters in the model (see Section 5.5). Thus the two mixture models are deemed to have the same level of complexity.

Overall, although all three models are similar according to DIC, we prefer the model with two-component mixture and zero-inflation, as it explicitly models the two features of the data – the bimodal risk distribution and the excessive number of zeros in the outcome values. In addition, this model allows a classification of villages based on the risk level, an output that is relevant for disease surveillance and intervention – we will discuss this classification next.

In the model with a two-component mixture and zero-inflation, each village undergoes two different assignments. Under the ZIP part, a village is assigned either to the Poisson distribution or to the deterministic distribution at zero. This assignment is done via g_i. If the village is modelled by the Poisson regression, it is subsequently allocated, through z_i, to one of the two mixture components. Using g_i and z_i, each village is effectively assigned to one of the three groups as summarised in Table 9.7. Note that once a village is assigned to the "disease-free" group, there is no further cluster assignment via z_i. Lines 43–49 in Figure 9.13 implement this classification. For each village, a new indicator h_i is introduced so that at each MCMC iteration, $h_i = 1$ if $g_i = 0$ (i.e. allocating village i to the "disease-free"

TABLE 9.7

A Classification of Each Village into One of the Three Categories Based on the Two-Component Mixture Model Allowing for Zero-Inflation

	$g_i = 0$	$g_i = 1$
$z_i = 1$	"Disease-free" ($h_i = 1$)	Low (residual) risk ($h_i = 2$)
$z_i = 2$		High (residual) risk ($h_i = 3$)

The indicator h_i (that combines both z_i and g_i) allocates each village to one of the three groups. See the main text for its definition and see Lines 43–49 in Figure 9.13 for its implementation.

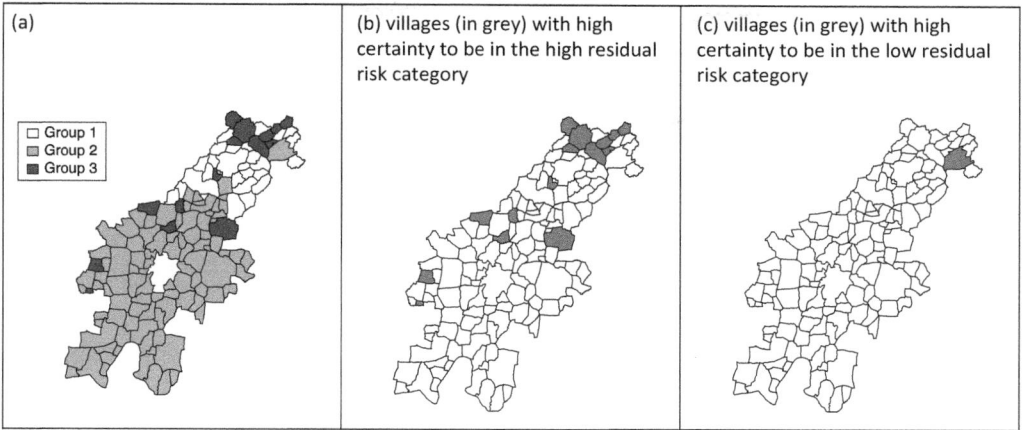

FIGURE 9.14

(a) Mapping the posterior medians of the indicator h_i across the villages from the two-component mixture model with zero-inflation. The polygons in grey in (b) and (c) are the villages with a *high* certainty of being allocated to the high residual risk cluster $\left(\Pr\left(h_i = 3 \mid data\right) > 0.8\right)$ and the low residual risk cluster $\left(\Pr\left(h_i = 2 \mid data\right) > 0.8\right)$, respectively.

group); $h_i = 2$ if $g_i = 1$ and $z_i = 1$ (i.e. allocating village i to cluster 1 with a low residual risk); and $h_i = 3$ if $g_i = 1$ and $z_i = 2$ (i.e. allocating village i to cluster 2 with a high residual risk). Figure 9.14(a) shows the allocation of the villages into the three categories. The villages allocated to the high residual risk category are those that warrant further investigation because they are shown to have much higher risks of malaria even after accounting for the rainfall effect.

It is important to recognise and present the uncertainty associated with the classification, indeed any classification. To quantify the uncertainty, we calculate the posterior probability of assigning each village to one of the three categories, namely, $\Pr\left(h_i = k \mid data\right)$ with k = 1, 2 and 3 (see Lines 50–55 in Figure 9.13). Figure 9.14(b) shows the villages which have a posterior probability of being in the high residual risk category above 0.8, i.e. $\Pr\left(h_i = 3 \mid data\right) > 0.8$. The assignment of all the 15 villages to this category is of high certainty. In other words, the model is certain about grouping these villages together into the high residual risk category. However, for the majority of the remaining villages, the assignment of each village into either the "disease-free" category or the low residual risk category is highly uncertain. There is only one village where $\Pr\left(h_i = 2 \mid data\right) > 0.8$ (see Figure 9.14(c)) and no village with a posterior probability of assignment to the "disease-free" group greater than 0.8.

Table 9.8 summarises the posterior estimates of the rainfall effect. There is little difference in malaria risk amongst the first three rainfall categories, as the 95% CIs of β_2 and β_3 all overlap with 0, the value indicating no risk difference. However, from the two-component mixture model with zero-inflation, the posterior mean of β_4 is 5.38 with a 95% CI of 2.96–8.39, implying a substantial increase in risk of malaria for the villages with the highest rainfall levels – this is also seen in the exploratory plot in Figure 9.10. Similar estimates are obtained from the other two models. While the estimates are consistent across the three models, it is worth noting that the uncertainty intervals from the two mixture models are narrower than those from the Poisson model with random effects. The lower

TABLE 9.8

The Posterior Estimates of the Rainfall Effect Across the Four Categories – the Villages in Category 4 had the Highest Rainfall Levels

Rainfall Category k	Poisson Regression with Random Effects	Two-Component Mixture without Zero-Inflation	Two-Component Mixture with Zero-Inflation
2	−0.73 (−6.51, 5.06)	−0.39 (−4.33, 3.24)	−0.50 (−4.39, 3.10)
3	3.15 (−1.12, 8.61)	2.01 (−0.50, 5.24)	1.85 (−0.72, 5.17)
4 (highest)	6.71 (2.95, 12.39)	5.56 (3.17, 8.64)	5.38 (2.96, 8.39)

Reported are the posterior means and the 95% CI of β_k ($k = 1,...,4$) from the three models. The lowest rainfall level is set as the reference, hence $\beta_1 = 0$ for all models.

uncertainty from the former two models may be due to the uncaptured within-cluster variability (see the posterior predictive checks below).

To assess how well the model fits the data, we compare the posterior predictive distribution to the observed number of malaria cases for each of the 17 villages where there was at least one case of malaria. As introduced in Section 5.4.2, the posterior predictive distribution is the posterior distribution of the number of malaria cases predicted by the model. If the model fits the data well, the observed number of cases should be around the middle of the corresponding posterior predictive distribution, as opposed to falling into one of the extreme tails of the distribution. For each village, we also compute the posterior predictive p-value as defined in the first line of Eq. 5.20, namely, $ppp_i = \Pr\left(y_i^{\text{pred}} \geq y_i^{\text{obs}} \mid \text{data}\right)$. A ppp_i either below 0.05 (the prediction is much lower than the observed) or above 0.95 (the prediction is much higher than the observed) suggests a poor fit. Otherwise, the model fits that observed count adequately. Figure 9.15 shows the comparison and reports the ppp_i for each village. Fifteen of the 17 villages have their posterior predictive probabilities between 0.05 and 0.95, suggesting an adequate fit to these observed counts. However, villages 9 and 12 are exceptions. In the case of village 12, the model predicts many more cases than observed, whereas far fewer cases are predicted from the model for village 9, which has the highest incidence rate in the raw data (16 cases from a population of 966, giving 16.5 cases per 1000 people). Both of the villages are assigned to the cluster with high residual risk. Their lack of fit suggests the within-cluster variability in risk has not been fully captured by the model. One way to account for such unexplained within-cluster variability is to include random effects (see Section 9.4.7). However, we do not pursue this option here because the lack of fit concerns only two villages. The addition of 139 random effects would result in an overparameterised model. Exercise 9.9 carries out a posterior predictive check to see how well the two-component mixture ZIP model captures the zero observed values in the data.

9.4.7 Conclusion and Model Extensions

In this application, we have discussed two modelling strategies to address several features present in the village-level malaria incidence data: the excess of zero observations, the bimodal risk distribution and spatial discontinuity. The Poisson regression with random effects appears to provide an adequate fit to the data, but the assumption that all random effect terms follow the same common distribution may not be appropriate when the risk distribution consists of distinct components. On the other hand, the two-component mixture model with zero-inflation captures the two distinctive components in the

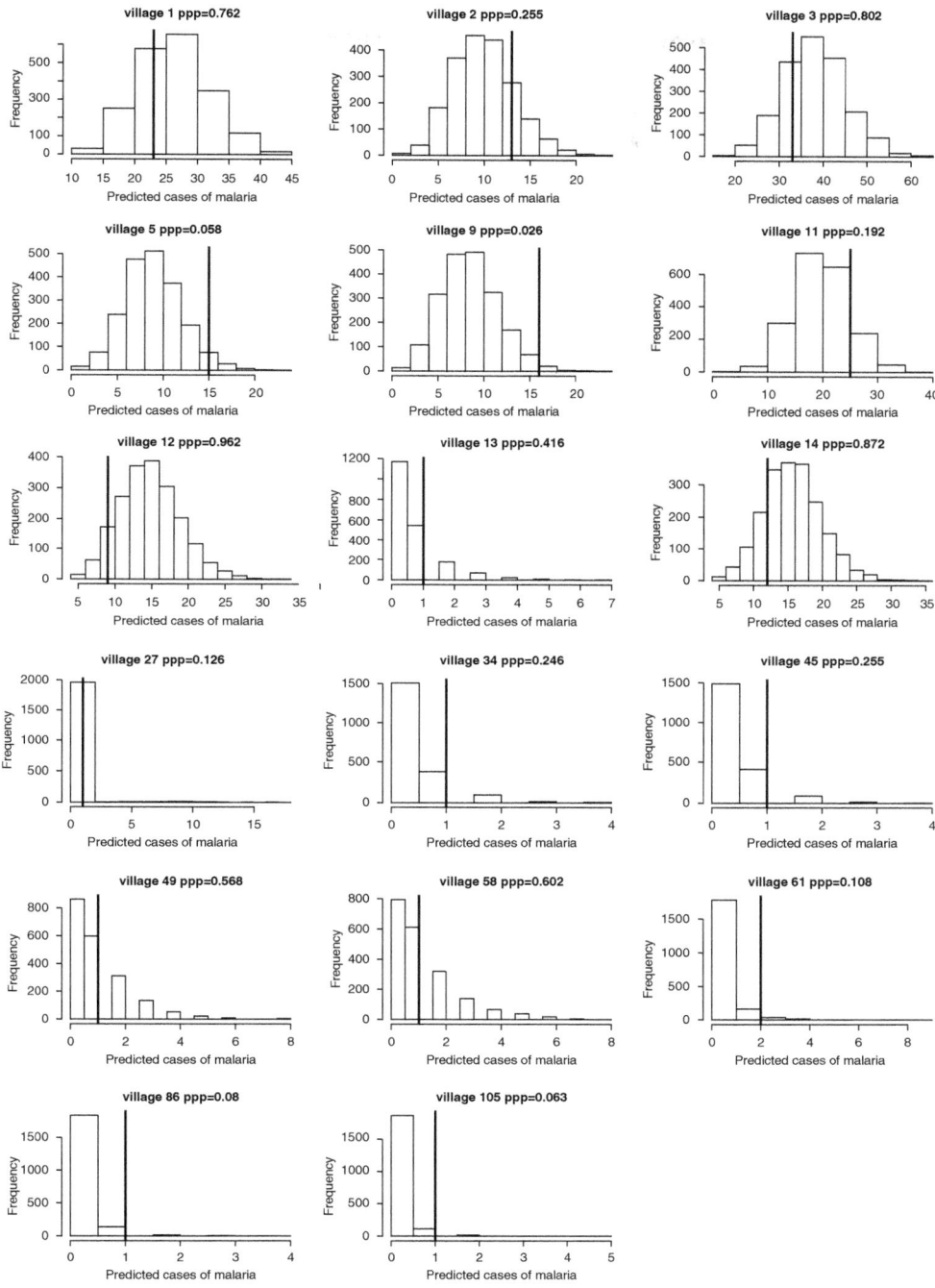

FIGURE 9.15

Posterior predictive checks comparing the observed (the vertical line) and predicted (the histogram) cases of malaria for each of the 17 villages with one or more malaria cases reported during the study period based on the two-component mixture model with zero-inflation. The posterior predictive p-value (the quantity ppp in Figure 9.13) of $\Pr(y_i^{pred} \geq y_i^{obs} \mid data)$ is reported in the title of each panel.

risk distribution through a two-component mixture *and* accommodates the excess of zero observations via the ZIP formulation. The cluster allocation does not assume the risk surface to be spatially continuous, reflecting the abrupt changes evident in the data. In addition to being a simpler approach (based on the DIC comparison), the mixture model with zero-inflation yields a classification map that can be used to inform policies for disease control and prevention (e.g. to help identify villages for targeted treatment). However, no model is perfect. In this application, we have demonstrated the use of posterior predictive checks to highlight where a model does not work well and where modifications could be made to improve the model.

There are a few extensions that can be made to the two-component mixture with zero-inflation. For example, to examine the characteristics of the disease-free villages, the analyst can incorporate covariates into the zero-inflated part through a logistic regression:

$$g_i \sim Bernoulli(\pi_i)$$
$$logit(\pi_i) = a + b_{x_{i,rain}}$$

$$(9.16)$$

As an example, the above model examines whether rainfall level has an impact on the probability of allocating village i. The regression coefficients $b_2,...,b_4$ measure such effects. A positive point estimate of b_4, for example, would suggest that a village in the category with highest rainfall is associated with a lower probability of being assigned to the disease-free category (recall that $1 - \pi_i$ is the probability of being allocated to the disease-free group).

The mixture models considered in this application assume that allocating villages does not depend on spatial contiguity. In other words, assigning area i to say cluster 1 does not affect the cluster allocation of the neighbouring villages of i. Alternatively, one may assume the cluster allocation to be spatially-dependent so that two spatially contiguous villages are more likely to be assigned to the same cluster. This assumption reflects the possibility that some risk factors affect not only area i but also its nearby areas. For example, villages that are close in space may have similar malaria risks due to, for example, transmission of the disease through movement of people between villages, or spatially-contiguous areas may have similar environmental conditions in the form of waterlogged or forested areas. Fernandez and Green (2002) induced spatial dependence in cluster allocation through the prior distribution of the weight parameters (i.e., $\pi_1,...,\pi_N$ in Eq. 9.16). Bayesian nonparametric methods provide another way to allow spatially-structured cluster allocation (see Gelfand et al., 2005; Kottas et al., 2008; and the references cited in Gershman and Blei (2012) on spatial modelling). We recommend interested readers consult the references above regarding the details of such modelling strategies.

We have fixed the cluster number to two in this analysis. In general, within the Bayesian framework, one can treat the cluster number as a random variable and estimate it from data using reversible jump MCMC (Green, 1995; Knorr-Held and Rasser, 2000; and Green and Richardson, 2002) or Bayesian nonparametric methods (Muller and Mitra, 2013; Hjort et al., 2010; Molitor et al., 2010; plus the references above). These Bayesian methods directly address and account for the uncertainty in the cluster number, which is typically fixed in a frequentist mixture model.

Finally, another extension to the mixture models is to incorporate random effects to better capture within-cluster variability. For example, in a non-spatial setting, Morgan et al. (2014) described a finite mixture model that incorporates zero-inflation and allows for heterogeneity by including random effects. In Chapter 16, we will extend the spatial two-component mixture model with zero-inflation to include random effects for analysing the malaria incidence data over both space and time.

9.5 Application 4: Modelling the Small Area Count of Cases of Rape in Stockholm, Sweden

9.5.1 Background and Data

In this application, largely drawn from Ceccato et al. (2019), we model variation in the number of cases of rape or attempted rape of females aged 15 or older in the Stockholm municipality, Sweden between 2008 and 2009. As in Ceccato et al. (2019), only rapes (or attempted rapes) committed outdoors (for example in streets, parks or transportation settings) are included in the modelling, of which there were 237 during the period.[17] Domestic rapes are excluded. Case counts are by *basområde*, the smallest spatial unit of analysis giving access to demographic, social and economic statistics in Sweden. The average population of a *basområde*, of which there are 407 in the Stockholm municipality, is approximately 2200.

The application is of interest here for two reasons. First, with only 237 cases distributed across 407 small areas, it is likely that many *basområdes* will have small counts, and in fact 63% of *basområdes* have no reported cases. This raises a challenge for statistical modelling similar to that encountered in the malaria application (Section 9.4). See Exercise 9.11 for an exploratory analysis of the small area case counts. Second, and more importantly, this application illustrates an approach to modelling spatial heterogeneity associated with model parameters where the same covariate may have a different effect on the outcome (the number of cases of outdoor rape in a *basområde*) in different parts of Stockholm (see Section 3.3.2.2).

In common with studies of other types of crime, including household burglary (Bernasco and Luykx, 2003), the justification for ecological modelling derives from the two-stage rational choice model of offender behaviour. At the first stage an offender identifies accessible areas in a city, drawn from their personal activity field or awareness space, where they believe the overall "risk-reward" trade-off is sufficiently in their favour. Then at the second stage they identify specific victims in specific "micro-environments" (a poorly lit street, an alleyway, a concealed entrance, proximity to bars) that provide sufficient opportunity and anonymity for them to carry out the criminal act (see Ceccato et al., 2019). At the data exploratory stage, geographically weighted regression can be used to explore the possibility of a spatially-heterogeneous covariate-outcome relationship. If the regression coefficients, associated with one (or more) of the covariates, are shown to vary spatially, a localised model, in addition to a global model, should be fitted.

Ceccato (2014), in a detailed study of individual cases of rape, observed that whilst some micro-environments in some ecological settings reported cases of rape, similar micro-environments in different ecological settings reported no cases of rape. The presence of bars in some neighbourhoods was associated with an increased risk of rape, in other neighbourhoods not. We aim to assess the extent to which Stockholm can be viewed as

---

[17] The Stockholm police recorded data codes are 0644, 0648, 0656 and 0660. Rape cases in Ceccato et al. (2019) refer to all police recorded cases of complete rape (sexual intercourse) and incomplete rape (attempted sexual intercourse and other types of sexual behaviour as defined by law) that happen outdoors (such as, for example, in streets, parks or transportation settings). For further details see BRÅ (2012).

comprising many local areas with their own "risk factor profiles", a list of the observed risk factor values for each area.

The modelling presented here seeks to test three hypotheses at the small area (*basområde*) scale:

i. High counts of rape are associated with areas offering *opportunity* to the offender.
ii. High counts of rape are associated with areas with poor surveillance and/or poor social control where residents have a high fear of crime (*meso-scale anonymity*). Neighbourhoods with buildings and street layouts providing areas for conceal-ment, perhaps because of inadequate lighting or quiet alleyways, may provide a number of suitable "micro-spaces" for the offence to be committed (*micro-scale anonymity*).
iii. High counts of rape are associated with areas that have good *accessibility*, because they bring potential targets into an area and offer a quick and easy escape route from the crime scene for the offender.

The interested reader is referred to Ceccato et al. (2019) for further discussion of these hypotheses and the covariates used to test them. The list of covariates are summarised in Table 9.9.

9.5.2 Modelling

To assess how risk factors *individually as well as jointly* affect the risk of outdoor rape we report two sets of results, one from a "whole" map analysis and the other from a "local-ised" analysis. The former uses a Poisson Bayesian hierarchical model to estimate the over-all effects of the risk factors on the number of cases of outdoor rape. We describe this first.

9.5.2.1 A "Whole-Map" Analysis Using Poisson Regression

Let O_i denote the number of outdoor rape cases in *basområde i* ($i = 1,...,407$). The data model specifies $O_i \sim Poisson(\mu_i)$. In the process model (Eq. 9.17), the Poisson mean μ_i is expressed as a function of area-level risk factors plus two random effect terms.

TABLE 9.9

Area- (*Basområde-*) Level Covariates Included to Test the Three Hypotheses

Covariate	Hypothesis	Type of Covariate
Number of street robberies in area	Micro-scale anonymity	Count/categorical
Population turnover	Meso-scale anonymity	Continuous
Average annual income (in 10000 Kronor)	Meso-scale anonymity	Continuous
Proportion who avoid going out in the neighbourhood (fear of crime)	Meso-scale anonymity	Continuous
Presence of forested area	Meso-scale anonymity	Binary
Presence of industrial area	Meso-scale anonymity	Binary
City centre	Accessibility	Binary
Presence of subway station in area	Accessibility	Binary
Presence of school(s)	Opportunity	Binary
Number of alcohol outlets in an area	Opportunity	Count/categorical
Young female population	Opportunity	Continuous

$$log\left(\mu_i\right) = \alpha + \sum_{k=1}^{11} \beta_k x_{ki} + U_i + S_i \tag{9.17}$$

The Poisson mean, μ_i, interpreted as the expected number of rape cases in *basområde i*, is log transformed to ensure that it is non-negative. The parameter α is the (log) overall number of cases in Stockholm municipality; $\sum_{k=1}^{11} \beta_k x_{ki}$ represents the effects arising from the risk factors in Table 9.9. The regression coefficient β_k ($k = 1,...,11$) quantifies the impact of the risk factor x_{ki} on rape occurrence. U_i and S_i are the random effect terms. The two sets of random effects take on the BYM structure (Section 8.5), where $S = (S_1,...,S_{407})$ is modelled through the ICAR model as a set of spatially-structured random effects, and $U = (U_1,...,U_{407})$ is modelled as a set of spatially-unstructured random effects, each of which follows a common normal distribution $N(0,\sigma_U^2)$.

The two sets of random effects are included to account for variability that is not explained by the covariates, and part of that variability may be due to the excessive number of zeros in the observed count data. The issue of (too) many zeros can be addressed using either zero-inflated models (as in Section 9.4) or the inclusion of random effects (as here). The former methodology focuses explicitly on accommodating the excessive number of zeros, whilst the latter is more general because it aims to capture unexplained variability in the model. Of course, random effects can be incorporated within a zero-inflated model in order to deal with more complex situations, such as where there are many zeros in the observed count data *and* correlation structures are also found to be present. See for example Min and Agresti (2005) in the case of modelling longitudinal count data and the comment made at the end of Section 9.4.6. The exploratory methods discussed in Chapter 6 will be useful in revealing such data features and in turn informing what type of model is called for. One interesting extension of the modelling presented in Eq. 9.17 is to incorporate covariates to understand the characteristics of places where there are *no* rape cases. This can be done by including a zero-inflated component, then modelling the zero-inflated part in the form of Eq. 9.16. We shall leave this extension for the reader to investigate.

Note that an offset term is not included in the Poisson regression model. First, it is difficult to define the "at-risk" population for outdoor rape. Second, using female residential population as an offset implies a smaller female population is associated with a lower risk of rape. For some *basområde*, as results will show, this assumption does not hold. Instead, the size of the female population is included as a covariate, whose effect on risk is then estimated from the data.

Finally, in the parameter model, we use vague priors to reflect the small (limited) amount of prior evidence we have from which to start the current modelling, although knowledge from existing studies informs the construction of the three hypotheses, which in turn inform the construction of the regression model (Section 5.6.2). For both the intercept and the regression coefficients, a normal prior $N(0,1000000)$ was used. To σ_s and σ_u, the standard deviations of the two sets of random effects, a truncated normal prior $N_{+\infty}(0,10)$, defined on the positive real line, is assigned (Gelman, 2006). Exercise 9.12 asks the reader to implement this regression model in WinBUGS.

9.5.2.2 A "Localised" Analysis Using Bayesian Profile Regression

For a local analysis we fit a Bayesian profile regression (BPR) model (Molitor et al., 2010). This method clusters or groups *basområdes* together based not only on their similarity in terms of number of rape cases but also on the similarity of their *risk factor profiles*, although spatial contiguity is not a criterion for grouping areas together. The risk factor profile of an area is simply a list of the observed risk factor values for that area. This modelling offers a number of advantages. First, through the use of risk factor profiles, BPR considers all risk factors *jointly* when assessing similarity. Two areas are considered to have similar risk factor profiles if their values for each of the covariates are similar. Clustering risk factor profiles places no restriction on the correlation structure amongst the covariates, some of which can be highly correlated, thereby resolving the multicollinearity problem that is often encountered when fitting a standard regression model.

Second, by using risk factor profiles, BPR does not require the specification of an explicit covariates-outcome relationship (as required when fitting a standard regression model), hence allowing for potential interactions *amongst* the covariates.

Third, BPR clusters areas based on their risk factor profiles *and* their risk levels, allowing us to differentiate, for example, two groups of high-risk areas, one group having a set of distinctly different *local characteristics* (as defined by their risk factor profiles) to the other. This "local-focus" of BPR enables us to examine spatially heterogeneous covariate effects. For example, two areas with similar risk factor profiles may be assigned to two different clusters due to their different risk levels.

BPR allows the inclusion of random effects, which deals with the overdispersion problem associated here with missing/unmeasured covariates and the large number of zero cases. Standard clustering methods do not include random effects and thus fail to account for this potential source of uncertainty in ecological modelling. Lastly, BPR allows the number of clusters to be estimated from the data, and this uncertainty is fully accounted for in estimating other parameters in the model. Non-Bayesian clustering methods fix the cluster number so that such uncertainty is ignored. The interested reader is referred to Molitor et al. (2010) for the mathematical description of BPR.

In this study, the risk factors are both continuous- and discrete-valued (binary and categorical). Within a cluster, the continuous-valued risk factors are modelled using a multivariate normal distribution, and the binary and categorical risk factors are modelled respectively using a vector of independent binary and categorical random variables. More specifically, each continuous-valued risk factor was transformed to be approximately normally distributed. The two discrete-valued covariates, numbers of alcohol outlets and robbery cases (Table 9.9 and also Table 9.10 below), were converted into categorical covariates because no suitable normality transformation was found for their highly positively-skewed distributions. Independent of the cluster allocation, we also specify two sets of random effects, as described in Eq. 9.17. Liverani et al. (2015) develops the R package PReMiuM, which fits the BPR model using efficient MCMC algorithms. We provide more detail about the fitting of the BPR model in Exercise 9.13.

BPR performs cluster allocation probabilistically, so that the *number* of clusters and the *composition* of each cluster (which areas are included in any one of the clusters) may differ from one MCMC iteration to the next. This probabilistic way of clustering incorporates uncertainty associated with the number of clusters and their composition into the modelling. Since both the number of clusters and their composition are changing, this poses a challenge when summarising the MCMC outputs. To post-process these MCMC outputs, Molitor et al. (2010) suggest three steps which are briefly explained here. The first step constructs the posterior similarity matrix M that summarises different partitions of areas over all MCMC iterations. Each off-diagonal entry in M, m_{ij} (for $i \neq j$), quantifies how likely it is for two *basområdes* i and j to be grouped together over all MCMC iterations. The second step finds a representative partition of the areas such that this partition is most consistent (according to a

TABLE 9.10

Posterior Estimates (Posterior Mean and 95% CI) of the Risk Factor Effects from the Poisson Regression Model (Models 1 and 2)

Hypotheses		Risk Factors		Posterior Estimates of Risk	
				Model 1	Model 2
Accessibility		Whether subway station is present or not		1.57 (1.09, 2.27)	1.43 (1.00, 2.14)
	Meso-scale	Whether area is close to the city centre or not		1.90 (1.22, 3.04)	1.90 (1.21, 3.00)
Anonymity		Average income (in 10,000 Kronor)		0.98 (0.97, 1.00)	0.98 (0.96, 1.00)
		Percentage of the population who fear crime and who avoid going out in the neighbourhood		1.00 (0.99, 1.02)	1.01 (0.98, 1.03)
		Population turnover		2.06 (1.16, 3.53)	1.86 (1.02, 3.22)
		Forest present or not		1.07 (0.45, 2.18)	1.08 (0.45, 2.19)
		Whether area is an industrial area or not		0.65 (0.35, 1.19)	0.69 (0.37, 1.19)
	Micro-scale	Counts of street robbery	0	Reference category	1.05 (1.02, 1.07)
			1–2	1.14 (0.69, 1.93)	
			3–4	1.63 (0.92, 2.89)	
			≥ 5	2.67 (1.55, 4.60)	
	Opportunity	Number of females aged 13–65 (in 1,000s)		1.16 (0.90, 1.52)	1.34 (1.05, 1.73)
		Whether educational institution(s) is(are) present or not		1.29 (0.92, 1.82)	1.34 (0.94, 1.80)
		Whether alcohol outlet(s) is(are) present or not	0	Reference category	1.03 (1.00, 1.05)
			1–2	0.97 (0.58, 1.60)	
			3–6	1.17 (0.69, 2.01)	
			≥7	1.55 (0.88, 2.78)	

criterion based on the so-called silhouette width; see Liverani et al., 2015, p.15) with the posterior similarity matrix M. Using all MCMC iterations, the third step then produces the estimates for the risk level and risk factor profile associated with each of the clusters within the representative partition. The results presented below are from the representative partition.

9.5.3 Results

9.5.3.1 "Whole Map" Associations for the Risk Factors

Table 9.10 reports the estimated effects of the 11 risk factors across Stockholm municipality from the Poisson regression model (Eq. 9.17). Two versions of that model were fitted. Both versions use the untransformed continuous-valued risk factors, but Model 1 incorporates number of alcohol outlets and number of street robbery cases in categorical form, whilst Model 2 uses those two covariates in the original count form. Here, $RR_k = \exp(\beta_k)$ represents the relative risk, which indicates how much the expected number of rape cases would change (increase/decrease) due to a one-unit increase in the kth risk factor (or a higher category compared to the reference category for binary/categorical risk factors) while holding all other risk factors fixed. This quantity $\exp(\beta_k)$ is not an odds-ratio (as in a logistic regression; see for example Section 9.2) since the Poisson regression model here deals directly with event occurrence, not probability.

The presence of a subway station in a *basområde* tends to increase the risk of outdoor rape in an area by over 50% (with 95% CI: 9%–127%). The risk of rape is nearly double in the "city centre" compared to *basområdes* outside the city centre. Ceccato et al. (2019) argued that together these findings provided strong support for the *accessibility* hypothesis. They also found some evidence to support the importance of *meso-scale anonymity*. As can be seen in Table 9.10, two covariates linked to this hypothesis, areas with low average incomes and those with high population turnover, contribute to explaining variation in the number of rape cases. However, the other three covariates under this hypothesis (presence/absence of forest, whether area is an industrial area or not and fear of crime proportion) are not found to be associated with the risk of rape. The covariate on street robbery counts, used as a surrogate measure of *micro-scale anonymity*, is positively associated with the risk of rape. Under Model 2, an increase in robbery count is found to be associated with an increase in the risk of rape (posterior mean of 1.05 with 95% CI: 1.02–1.07), indicating that one additional robbery case leads to a 5% (with 95% CI: 2%–7%) increase in the risk of rape. Such a linear trend is also seen when the covariate is included as a categorical variable in Model 1, although the effect is only significant when the number of street robberies is five or more, suggesting that the relationship is strongest where micro-scale anonymity is greatest. These results, taken together, indicate that both forms of anonymity contribute to our understanding of small area variation in the number of cases of rape.

Finally, Ceccato et al. (2019) concluded that they were unable to provide clear evidence to confirm the importance of the three risk factors associated directly with the *opportunity hypothesis*. None of the covariates was significant in Model 1, although the size of the female population and the number of alcohol outlets are significant in Model 2. There is an increasing trend between number of alcohol outlets in an area and the risk of outdoor rape, suggesting that risk in an area does increase when a large number of alcohol outlets are present.

9.5.3.2 "Local" Associations for the Risk Factors

The Bayesian profile regression model was applied to reveal the localised characteristics of places where rape took place. In Ceccato et al. (2019), where 200000 MCMC iterations were run to fit the model, about 99.7% of those MCMC iterations consisted of seven clusters, with 0.3% having eight clusters, showing very little variability (or uncertainty) in determining the cluster number. To summarise different clustering options of the areas, the representative partition of the *basområdes* was produced following the steps outlined in Section 9.5.2.2.

Figure 9.16 shows the seven clusters in the representative partition. There are three low-risk clusters (Clusters 1–3, all with the 95% CI for relative risk below and excluding 1). There are two medium-risk clusters (Clusters 4 and 5, with relative risks of 0.71 (95% CI: 0.40–1.27) and 0.98 (95% CI: 0.78–1.22) respectively. Finally, there are two high-risk clusters (Clusters 6 and 7, both with the 95% CI for relative risk above and excluding 1).

Figure 9.17 describes the estimated cluster-specific risk factor profiles. For the two high-risk clusters, Cluster 6 represents areas mainly in the city centre with high population turnover, small female residential populations and small numbers of schools. There are also large numbers of alcohol outlets (a high proportion of *basområdes* are in category 4) and large numbers of street robbery cases in this cluster. The characteristics of this cluster illustrate that it is inappropriate to use the female population as an offset. Cluster 7, on the other hand, represents relatively deprived suburban areas with large numbers of subway

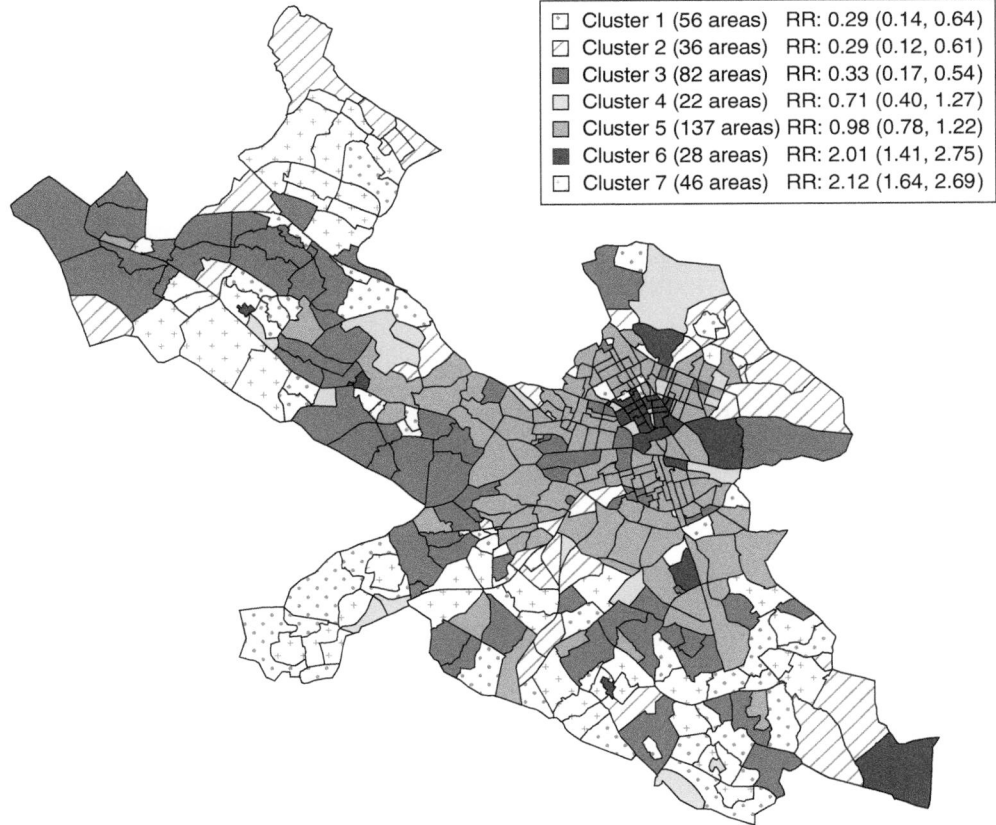

FIGURE 9.16
(See colour insert.) Membership of the *basområdes* under the representative partition using the Bayesian profile regression model. For each cluster or grouping of areas, the legend reports the number of *basområdes*, the posterior mean of the cluster-specific relative risk, together with the 95% CI.

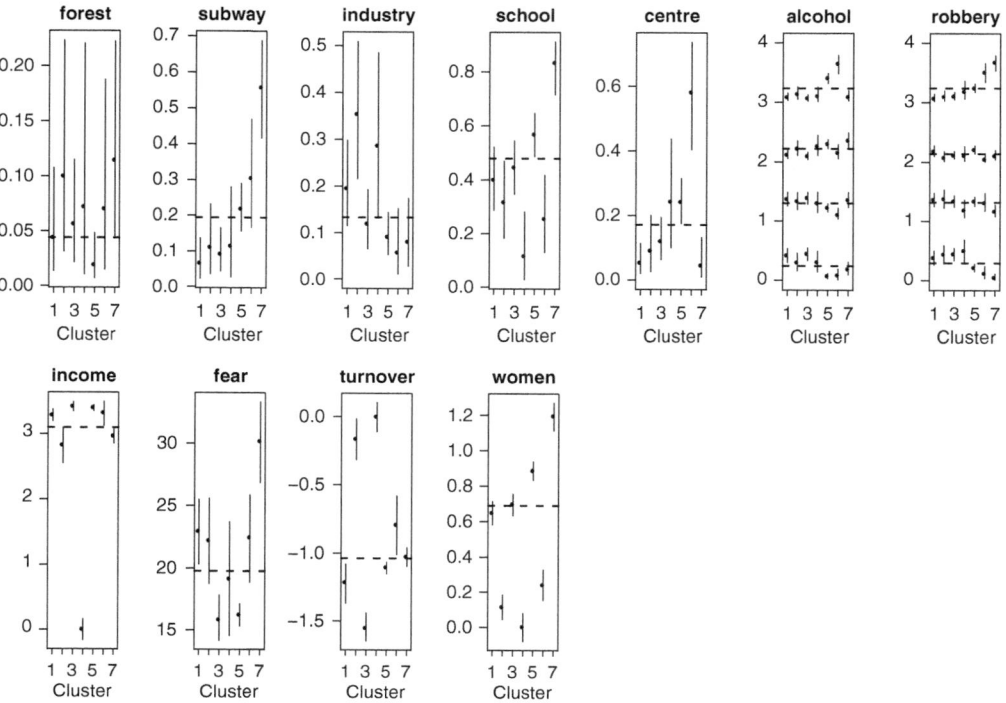

FIGURE 9.17
The estimated risk factor profile for each of the seven clusters in the representative partition. The dots correspond to the posterior means with the vertical bars representing the 95% credible intervals. The horizontal dashed line shows the Stockholm average for each risk factor. For the two categorical covariates, alcohol and robbery, the results show the estimated proportions of areas in each of the four categories across the seven clusters, and the horizontal dashed line shows the Stockholm average for each category.

stations and schools, relatively low income, high levels of fear of crime. The relatively large female residential populations in these areas present the would-be offenders with a large number of potential targets. All factors combined contribute towards the high risks of rape in these areas. Similar to Cluster 6, relatively large numbers of alcohol outlets and street robbery cases are found in these areas. The findings for these two groups of high-risk areas, each with distinctly different local characteristics, provide a better description of the clusters, illustrating a benefit of the Bayesian profile regression approach.

There are two medium-risk clusters, Clusters 4 and 5. Cluster 5 consists of areas mainly on the outskirts of the city centre (Figure 9.16). These areas tend to have a relatively large female residential population, with more schools and a generally low fear of crime amongst residents and good subway connections. There are large numbers of alcohol outlets present in these areas, but numbers of robbery cases are relatively low, compared to the levels in the two high-risk clusters. Cluster 4, the other medium-risk cluster, consists of 22 *basområdes*. These *basområdes* contain airports and shopping malls that have zero-values for income, population turnover and size of the female residential population. While the above three risk factors are all zero, areas within this cluster are mainly industrial areas with schools present in only a few.

Clusters 1 to 3 are the three low-risk clusters. Cluster 2 represents primarily industrial areas, whilst Clusters 1 and 3 are the affluent suburban areas with few alcohol outlets and few robbery cases, suggesting few micro-sites offering anonymity for the criminal act.

Findings from the local analysis in this study reveal two types of high-risk area within Stockholm. The first of these are the parts of the city centre where there are large numbers of alcohol outlets, high residential population turnover and high counts of robbery. The second type of high-risk area comprises poor suburban areas with schools, where subway stops are located, and where a large proportion of the population express fears about crime sufficient to make them avoid going out in the neighbourhood. So, in a local context, fear of crime in the population does present itself as a risk factor.

9.5.4 Conclusions

This application was included to present Bayesian profile regression that can be used to examine spatial heterogeneity in covariate effects. In concluding this section, we highlight a number of key points.

Bayesian profile regression reveals the localised characteristics of places where rapes occur and in particular the role of risk factor interactions in determining the risks in local areas. Risk factors that do not emerge as important in a whole map analysis may emerge in this form of regression modelling because of local context. In this application, two types of areas stood out where risks were high. Cluster 6 represents areas in the city centre with high population turnover, large numbers of alcohol outlets and large numbers of cases of street robbery. Cluster 7 represents relatively deprived suburban areas with large numbers of subway stations and schools, relatively low income levels, high levels of fear of crime and relatively large female residential populations. Relatively large numbers of robbery cases occur in these areas where there are also many alcohol outlets. In the case of the areas in Cluster 7, two covariates emerge as important risk factors locally but not in the "whole map" analysis. These are areas with high levels of fear of crime and areas where many schools are located. This observation highlights the importance of understanding the effect of a risk factor within the local context and jointly within the setting provided by other risk factors. This type of modelling will provide policymakers and police forces with information that can help them to develop a geographically tailored approach to preventing the crime of rape.

Other regression models have also been proposed to allow for heterogeneous covariate effects (Lloyd, 2011, Chapter 5). One such model is the geographically weighted regression model (Fotheringham et al., 2002) where a separate model is fitted to each locally defined subset of data. In Chapter 6, we considered geographically weighted regression (GWR) as an exploratory tool for investigating an outcome-covariate relationship that may vary spatially. However, as a basis for inference, GWR suffers some important limitations. Wheeler and Tiefelsdorf (2005) have shown that multicollinearity (correlations between the covariates in a regression model) gives rise to difficulties when interpreting regression coefficient estimates, and geographical weighting used within GWR makes the problem worse. Another model for consideration is the Bayesian spatially varying coefficient model (Gelfand et al., 2003) – a random coefficient model with a prior spatial structure. The regression coefficients associated with a covariate are allowed to vary spatially, and they are modelled as a set of random variables. Through the use of different spatial prior models (such as the ICAR or the BYM model) and/or different specifications of the spatial weights matrix W, spatial dependence structures can be imposed. However, in addition to the multicollinearity issue, the modeller would need to satisfy herself that spatially smooth variation in a regression coefficient was appropriate for their data. For example, it is not clear, on the evidence presented here (see Figure 9.16 noting in particular the mosaic-like quality of the map), that a spatially varying coefficients model would have been appropriate for the Stockholm rape data.

9.6 Exercises

Exercise 9.1. For the HIA application, fit a logistic regression model with the four covariates but no random effects to the PHIA data. Vague priors should be used for that model. Fit the same logistic regression model to the EHIA data but replace the vague priors for the regression coefficients by a set of (more informative) priors based on the posterior estimates from the PHIA fit. When fitting the EHIA data, the reader should pay particular attention to the following two issues: (a) how to derive a prior distribution of a parameter in the EHIA model using the posterior estimates for that parameter from the PHIA model; and (b) for each covariate, whether the EHIA data support the use of the PHIA estimate as a prior. The latter is referred to as checking for prior-data conflict.

Exercise 9.2. Fit the repeated measurements model exactly as given in Eq. 9.1 to the EHIA-PHIA data. Then consider other prior models for the random effects. Compare and comment on the performance from various model fits. Can you provide reason(s) if problems of slow (or even non-) convergence and/or poor mixing arise (hint: consider both the complexity of the model and the available data)?

Exercise 9.3. Fit a model in WinBUGS with only Lines 1, 17–30 and 40 in Figure 9.4 included. You do not need to use the observed data for running this model, so skip the "load data" step (i.e. ignore Step 4 in Section 5.3.4), but you need to load the two sets of initial values, i.e. Lines 43–50 in Figure 9.4 but removing `alpha` and `beta` from both lists. Monitor the appropriate quantities in order to verify the two probability statements on λ and the conditional standard derivation of each set of random effects in the MVCAR model based only on the prior specified. This is a Monte Carlo procedure (also referred to as forward simulation or forward sampling) to simulate values from the prior distribution, a procedure that is useful in order to understand the information on a parameter from its prior.

Exercise 9.4. As in Exercise 9.1, fit a logistic regression model with the four covariates but no random effects separately to the EHIA and the PHIA data. For both model fits, vague priors should be used. For each model fit, carry out model evaluation using Lines 36–66 in Figure 9.2.

Exercise 9.5. Fit the model as given in Eq. 9.3 but replace the MVCAR prior model by two independent ICAR models, one for each set of outcome values.

Exercise 9.6. Use the ED-level stroke death counts and the corresponding population stratified by age, sex and deprivation quintile to first calculate the mortality rate associated with each age-sex-deprivation stratum, then calculate the expected death count in each ED adjusting for the effects of those three factors. The data for this exercise are available on the book's website.

Exercise 9.7. Fit the model given by Eq. 9.13 using the ED-level data on counts of stroke deaths, population as well as the within-area variability of exposure. Explore the within-area exposure data in order to see how much exposure levels vary within these EDs, thus verifying the comments made towards the end of Section 9.3.4.

Exercise 9.8. Fit each of the three models presented in Table 9.6 to the malaria incidence data using the corresponding R script available online. Prior to running these R scripts, make sure you understand the fitting procedure, in particular, making sure you know what parameters are monitored. Save the results of each model fit. Then run the R script `malaria_spatial_mixture_DIC.R` to calculate the DIC values reported in Table 9.6. Go through the above R script to understand how each component of the DIC (in particular, the deviance) is calculated.

Exercise 9.9. Modify the R script used in Exercise 9.8 to fit the mixture model with zero-inflation to carry out a posterior predictive check on how well the model captures the zeros in the observed data (hint: see Lines 30–33 and Lines 35–36 in Figure 9.13). Do not forget to add the required quantity to the list of parameters to be monitored.

Exercise 9.10. When data for other time intervals are available, one possibility would be to replicate the spatial analysis presented in Section 9.4 to build up a picture of how the risk of malaria varied in space and time, which could further help to support intervention activities to limit the disease. So, perform the spatial analysis on each (or some) of the cross-sectional data from other two-month time windows. The data are available online in the csv file, `malaria_sptm.csv`.

Exercise 9.11. Perform an exploratory analysis on the small area rape case counts (suggestions: produce appropriate maps, summary tables of the data as well as carrying out some of the tests discussed in Chapter 6 to reveal features of the data).

Exercise 9.12. Implement the two versions of the Poisson regression model in Eq. 9.17 to the rape case count data, thus verifying the results summarised in Table 9.10. See also the model fitting script in Exercise 9.13 below on the transformations applied to the continuous-valued covariates (to satisfy normality) and the categorisation of the two discrete-valued covariates (counts of street robbery and alcohol outlets; see Table 9.10).

Exercise 9.13. The R script, `fit_BPR_stockholm.R`, available online fits the BPR model discussed in Section 9.5.2.2 to the small area counts of rape cases in Stockholm. Prior to running the script, make sure all the steps are clearly understood. The R script, `summarise_BPR_stockholm.R`, summarises the results from the fitting script to produce the cluster map in Figure 9.16 as well as the plot of the cluster-specific risk factor profiles in Figure 9.17.

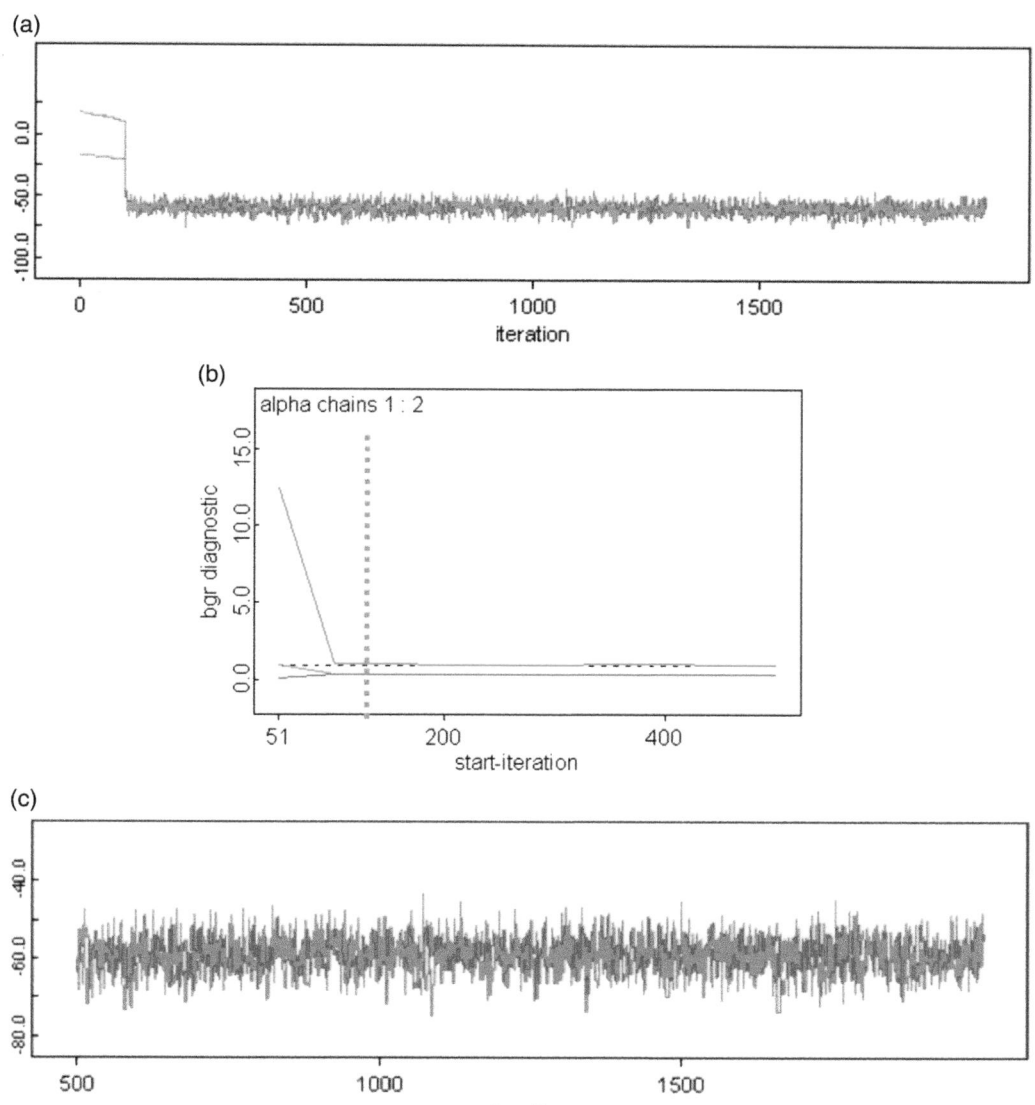

FIGURE 5.12
(a) A history plot of the 2000 iterations from two MCMC chains where the beginning of the two chains is clearly not from the target posterior distribution; (b) the BGR plot from WinBUGS showing that after the 150th iteration all three lines are stable and the red line is close to 1 (the grey vertical line is superimposed for ease of interpretation and is not part of the BGR plot); (c) after discarding the first 500 iterations the resulting history plot has the required form (more iterations than suggested by the BGR plot have been discarded to be on the safe side).

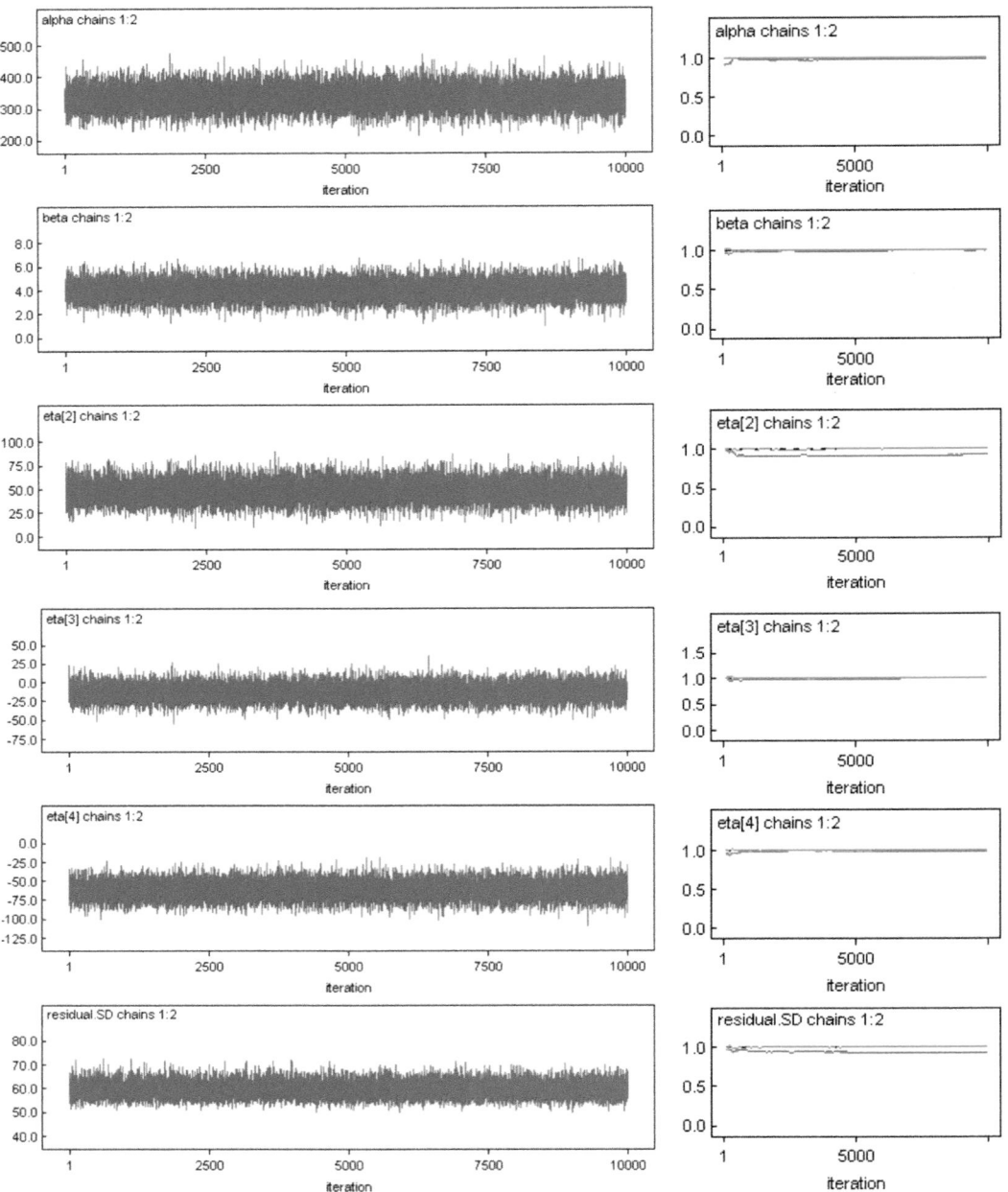

FIGURE 5.17
Checking convergence: history plots and the BGR diagnostic plots.

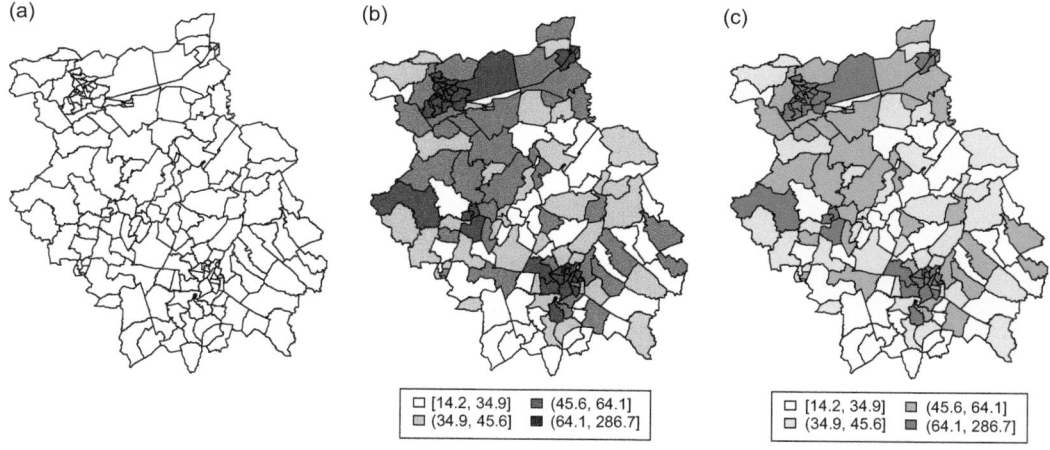

(a)

(b)

(c)

| | [14.2, 34.9] | | (45.6, 64.1] |
| | (34.9, 45.6] | | (64.1, 286.7] |

| | [14.2, 34.9] | | (45.6, 64.1] |
| | (34.9, 45.6] | | (64.1, 286.7] |

FIGURE 6.1

(a) A map of the 157 wards in Cambridgeshire, England; (b) and (c) are the choropleth maps of the burglary rates per 1000 houses in greyscale (b) and in shades of blue (c).

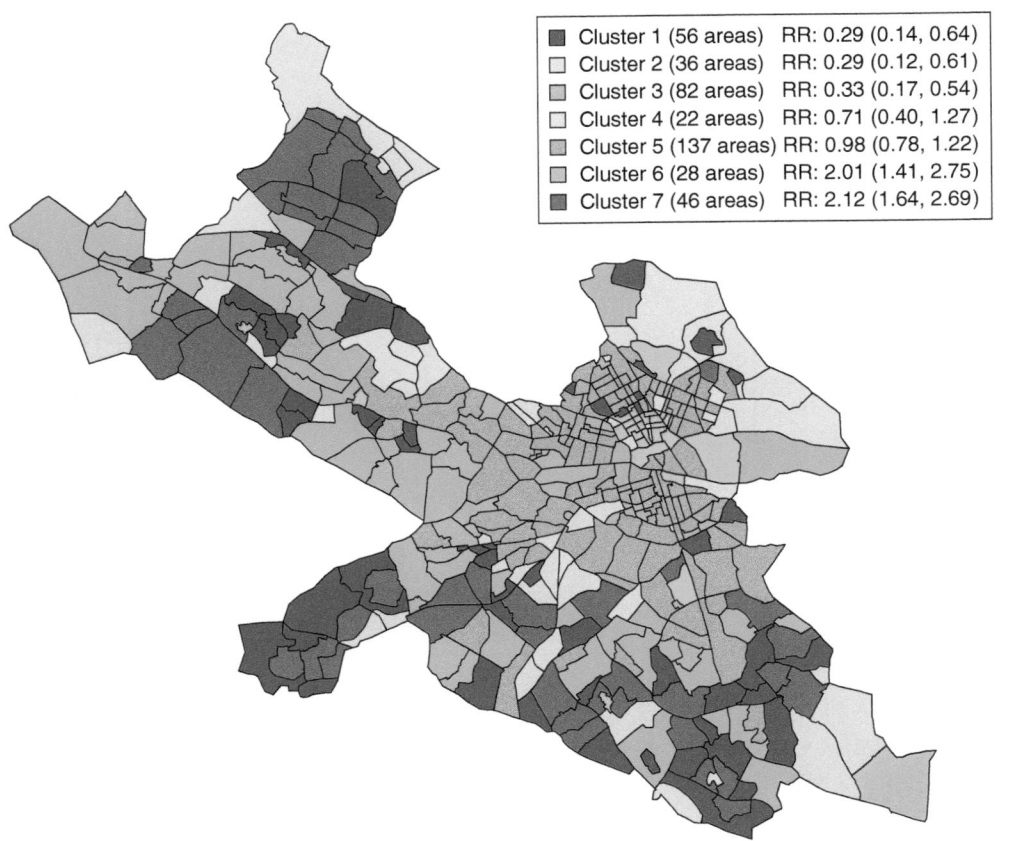

FIGURE 9.16
Membership of the *basområdes* under the representative partition using the Bayesian profile regression model. For each cluster or grouping of areas, the legend reports the number of *basområdes*, the posterior mean of the cluster-specific relative risk, together with the 95% CI.

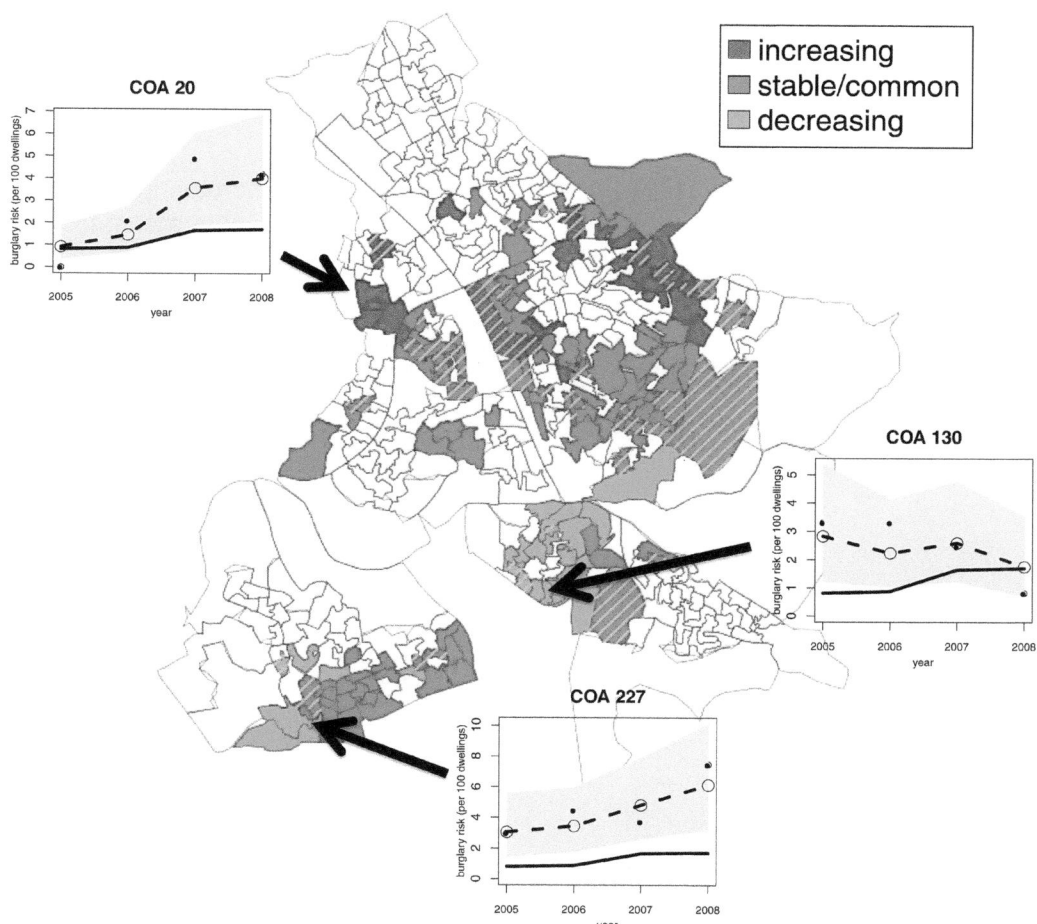

FIGURE 16.9
Persistently high-risk COAs (hotspots) in Peterborough, 2005–08. High-risk areas are further classified into those with an increasing or decreasing pattern relative to the common trend and those with a trend not differing from the common trend. For three COAs (one for each trend category), the inserted figures show the observed raw burglary rates (per 100 houses; black solid dots), the estimated burglary rates (open circles joined by dashed lines) with the 95% credible intervals (grey regions) and the estimated common trend (black solid line) over time. The crosshatched polygons are the COAs that cease to be hotspots after accounting for the effects of the included COA-level covariates. This figure is a revised version of Figure 2 in Li et al. (2014) where the inserted plot for COA 130 should be associated with the polygon indicated in this figure.

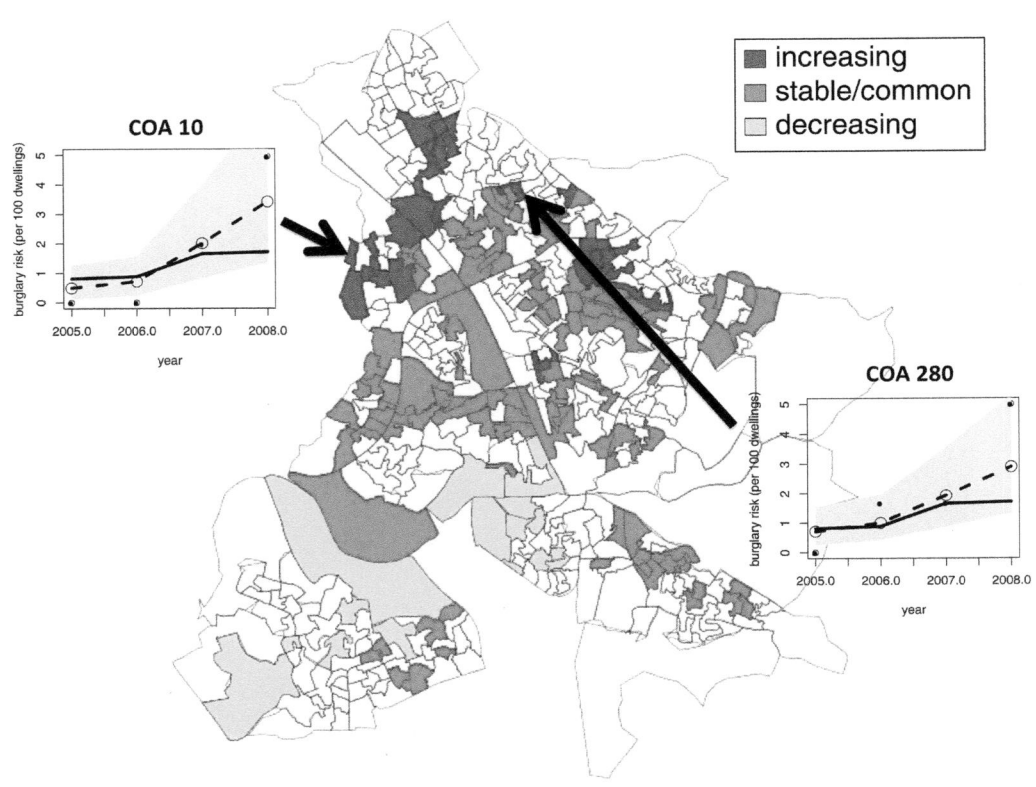

FIGURE 16.10
COAs in Peterborough with neither persistently high (hotspot) nor low (coldspot) risk of burglary. Two COAs are identified that are showing a tendency to becoming hotspots. (See Figure 16.9 for the explanation of the inserted figures.)

10

Spatial Econometric Models

10.1 Introduction

In this chapter we examine a family of models known as spatial econometric models. These models have been widely applied not only in regional and urban economics and regional science but also in the social sciences generally.[1] One of the central concerns of spatial econometric modelling is to adapt the standard normal linear regression model in order to address two of the fundamental challenges associated with spatial data, namely spatial dependence and spatial heterogeneity, and their implications for model specification, parameter estimation and hypothesis testing (Anselin, 1988, 2010; LeSage and Pace, 2009). Paelinck and Klaassen (1979, p.5–11) specified five important principles to guide the formulation of spatial econometric models, of which two are of particular relevance here: the role of spatial interdependence; the importance of explanatory factors "located in other spaces". These two principles refer to the challenge of assessing the extent to which outcomes observed in one area are a consequence of events or circumstances occurring in other areas, such as changes in outcomes and/or covariates in other areas. These are often referred to as spatial spillover effects, and the identification of spillover effects and associated feedbacks (in which spillovers from one space to another impact back on the originating space) represent the other central concern of spatial econometric modelling.

The plan of this chapter is as follows. In Section 10.2 we discuss the different forms spatial spillovers can take. We introduce four spatial econometric models for normally distributed data: the spatial lag model (SLM), the spatially-lagged covariates model (SLX), the spatial error model (SEM) and the spatial Durbin model (SDM). These models describe different types of spatial spillovers, and through an application, we explain the form(s) of spatial spillovers that each model captures and highlight issues associated with the interpretation of the covariate effects on outcomes. Such interpretation issues are the consequences of *effect propagation* and *spatial feedbacks* that are induced by the simultaneous autoregressive (SAR) modelling structure on the set of outcome variables. Essentially, the SAR structure creates a network connecting all areas in the study region. This connectivity allows a change that happens in one area to *propagate* to other areas as well as *return* to the area where the change originated (as well as to other areas), thereby forming feedback loops. We will explore the concept of spatial feedback in Section 10.3 and discuss its implications for how we interpret and measure the *direct* and the *indirect* (or *exogenous*) *spatial*

---

[1] The field of regional science makes quite extensive use of spatial econometric modelling. Regional science was a term coined by Isard (1960) to bring together theory and applications in spatial/regional economics and some elements of economic geography. Since that initial conception, the field has evolved to address a broader range of questions. Journals include the *Journal of Regional Science, Papers and Proceedings of the Regional Science Association* and the *Journal of Regional Science and Urban Economics*.

spillover effects of a covariate on the outcome. Section 10.4 deals with the computational issues that arise when fitting some of the spatial econometric models to observed data. Finally, in Section 10.5 we will provide some concluding remarks to help place this class of models in relation to the hierarchical models discussed in Chapters 7 and 8.

10.2 Spatial Econometric Models

10.2.1 Three Forms of Spatial Spillover

Conceptually, the spatial autocorrelation structure present in the observed outcome values may arise from the influence of a covariate (or set of covariates) whose values are spatially-autocorrelated. However, if this covariate is omitted from a regression model (for example, the relevance of the covariate is not recognized, or it is recognized but no data are available to the analyst on that variable), the residuals will be spatially autocorrelated. However, a well-defined model still needs to account for the autocorrelation structure in the residuals because it is an assumption of regression that errors must be uncorrelated (see Chapter 1).

Spatial autocorrelation in outcome values can also come from some process of spatial interaction where the outcome in any given area may influence the outcomes in neighbouring areas or where changes to observable covariates (explanatory variables) in one area have an impact not only on the outcome in the same area but also on the outcomes in other neighbouring areas. In spatial econometrics, such processes of spatial interaction are referred to as spatial spillovers (see, for example, LeSage and Pace, 2009, and Vega and Elhorst, 2015). In this chapter, we focus on the analysis of cross-sectional (i.e. purely spatial) data, where the outcome values are geo-referenced and are recorded at the same time or over the same time period. We will explore the modelling of panel (i.e. spatial-temporal) data in Part III.

For the purpose of modelling, spatial spillovers can be expressed in three different forms:

1. The outcome in an area (or at a location) depends on outcomes in neighbouring areas (or at neighbouring locations) – this form of spatial spillover is referred to as *endogenous interaction*.

2. The outcome in an area (or at a location) depends on the values of the observable covariates in neighbouring areas (or at neighbouring locations) – this is referred to as *exogenous interaction*.

3. The (regression) error for an area (or at a location) depends on the errors at neighbouring areas (or at neighbouring locations). As we alluded to earlier, this form of spatial spillover represents the spatial interactions associated with omitted covariates.

Spatial econometric models are constructed to accommodate certain form(s) of spatial spillover so that we can make inference about their presence (or absence). In subsequent sections, we will see four different spatial econometric models. Amongst them, the spatial lag model (SLM) is formulated to capture endogenous spillovers, while the spatially-lagged covariates model (SLX) focuses on the effects of exogenous spillovers.[2] The spatial

[2] We use SLX following Gibbons and Overman (2012, p.183).

error model (SEM) allows the error terms in a standard regression model to be spatially autocorrelated to acknowledge the autocorrelated nature of omitted covariates. These three spatial econometric models can be used as the basic building blocks for constructing more elaborate, complex models to reflect the fact that any combination of the three forms of spatial spillover listed above can coexist in a given spatial dataset. We will later introduce the spatial Durbin model (SDM), which features both the SLM and SLX models. We now describe these four models in turn.

10.2.2 The Spatial Lag Model (SLM)

10.2.2.1 Formulating the Model

The SLM model deals with spatial spillovers arising from the spatial interactions of the outcomes across different areas (or locations). For example, in the context of spatial price competition, the price for a good at a retail outlet may reflect not only the costs faced by that retailer but also the potential effect of competition through the prices for the same good set by the retail outlets that are nearby. In modelling the spatial spread of infectious diseases, the number of infected individuals in one village in a rural area may depend not only on the socio-demographic and environmental characteristics of that village but also on the numbers of infected individuals in nearby villages with whom the people interact socially or economically. The commonality in the two examples is that an area's outcome may influence the outcomes of other (possibly neighbouring) areas. The SLM model accommodates this feature by modelling the outcome variable Y_i of area i (for $i = 1,...,N$) as a function of the outcome variables in some or all the other areas in the study region. Eq. 10.1 gives the formulation of the SLM model specified over N areas within a study region:

$$Y_i = \alpha + \delta \sum_{j=1}^{N} w_{ij}^* Y_j + X_i \beta + e_i \tag{10.1}$$

In Eq. 10.1, α is the intercept, $X_i = (x_{i1},...,x_{iK})$ denotes a $1 \times K$ vector of covariate values and β is a $K \times 1$ vector of regression coefficients. The errors, $e_1,...,e_N$, are assumed to be independent, and each follows the same normal distribution, $N(0,\sigma^2)$.

The dependency of Y_i on the outcome variables of its neighbours is specified through the term $\delta \sum_{j=1}^{N} w_{ij}^* Y_j$. Specifically, w_{ij}^* is the entry on the ith row and the jth column in the row-standardised spatial weights matrix W^* (see Section 4.7). Using spatial contiguity to define the neighbourhood structure, $w_{ij}^* = 1 / m_i$ if areas i and j share a common boundary and $w_{ij}^* = 0$ otherwise, where m_i denotes the number of neighbours of i. Then $\delta \sum_{j=1}^{N} w_{ij}^* Y_j$ becomes $\delta (\sum_{j \in \Delta i} Y_j) / m_i$, a scaled version of the average of the outcome variables from the spatial neighbours of i (with Δi denoting area i's neighbours), which the outcome in area i depends upon. LeSage and Pace (2009, p.8) refer to $\sum_{j=1}^{N} w_{ij}^* Y_j$ as a spatial lag term. The use of W^* in Eq. 10.1 imposes a spatial dependence structure on the set of outcome variables, while δ, the spatial autocorrelation parameter, controls the strength of that dependence. Note that $w_{ii}^* = 0$ by definition so Y_i does not appear on both sides of Eq. 10.1. In other words, the outcome in an area cannot affect itself (at least not directly).

Applying Eq. 10.1 to all the N areas in the study region gives rise to a series of N simultaneous equations, as given in Eq. 10.2:

$$Y_1 = \alpha + \delta \sum_{j=1}^{N} w_{1j}^* Y_j + X_1 \beta + e_1$$

$$Y_2 = \alpha + \delta \sum_{j=1}^{N} w_{2j}^* Y_j + X_2 \beta + e_2 \qquad (10.2)$$

$$\vdots$$

$$Y_N = \alpha + \delta \sum_{j=1}^{N} w_{Nj}^* Y_j + X_N \beta + e_N$$

If two areas, say areas 1 and N, are neighbours of each other, the weights w_{1N}^* and w_{N1}^* are non-zero, then Y_1 will appear on the right-hand side of the equation for Y_N and vice versa. This highlights the *simultaneous nature* of these N equations. Because of their similar structure to the temporal autoregressive model of the form $X_t = c + \sum_{i=1}^{p} \phi_i X_{t-i} + e_t$, the set of equations defined by Eq. 10.2 specifies the so-called *Simultaneous AutoRegressive* (SAR) structure on the outcome variables. The SLM model imposes the SAR structure on the outcome variables to incorporate endogenous spillovers in the modelling where an area's outcome is expressed as a linear combination of the outcomes from its neighbours. Moreover, as we shall discuss later, the SLM model also allows us to measure the spillover effect due to changes to the covariates in area i on the outcomes of its neighbours. Another feature of the SLM model is that the interconnectivity of the areas creates *spatial feedbacks*, a phenomenon whereby a change occurring in an area's outcome and/or one (or some) of the observable covariates propagates to other areas via the spatial network *and* travels back to the area where the change originated. The presence of such spatial feedbacks requires us to think carefully how we measure the effect of a covariate change, a topic that we will come back to in Section 10.3.

10.2.2.2 An Example of the SLM

From the SLM model, Eq. 10.3 specifies the three simultaneous equations based on the map of three areas given in Figure 10.1.

$$Y_1 = \alpha + \qquad \delta Y_2 \quad + x_{11}\beta_1 + x_{12}\beta_2 + e_1$$

$$Y_2 = \alpha + \delta \cdot \frac{(Y_1 + Y_3)}{2} + x_{21}\beta_1 + x_{22}\beta_2 + e_2 \qquad (10.3)$$

$$Y_3 = \alpha + \qquad \delta Y_2 \quad + x_{31}\beta_1 + x_{32}\beta_2 + e_3$$

In the above equations, (x_{11}, x_{21}, x_{31}) and (x_{12}, x_{22}, x_{32}) are the values of two covariates. The spatial dependence structure of the outcome becomes more evident in Eq. 10.3. For example, in addition to the combined effect of the two covariates (i.e. $x_{21}\beta_1 + x_{22}\beta_2$), the outcome of area 2 also depends on $\delta \cdot \frac{(Y_1 + Y_3)}{2}$, the average of the outcomes from its two neighbours, areas 1 and 3, scaled by δ. Similarly, the outcome of area 1 depends on the outcome of area 2, the only neighbour that area 1 has. Note that the use of the row-standardised

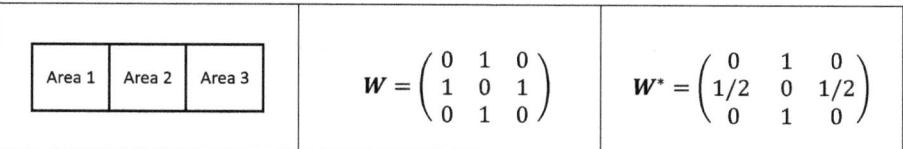

FIGURE 10.1
A map of three areas (left) with the associated spatial weights matrix W (middle), and its row-standardised version W^* (right), based on contiguity.

weights matrix W^* gives a natural interpretation of "an average of the spatial neighbours" as opposed to "a sum of the spatial neighbours" if the non-standardised weights matrix is used. Using W^* also accounts for the unequal numbers of spatial neighbours that each has. The nature (either positive or negative) and the strength of the spatial dependence is determined by δ, a parameter that is estimated from data.

10.2.2.3 The Reduced Form of the SLM and the Constraint on δ

To understand the properties of the SLM model, it is useful to write the system of N simultaneous equations in Eq. 10.2 in matrix notation. Starting with the simple case of Eq. 10.3, we can first express the three equations using matrices, then write the three equations into a single-line equation. The detail is shown in Eq. 10.4.

$$\begin{pmatrix} Y_1 \\ Y_2 \\ Y_3 \end{pmatrix} = \alpha \begin{pmatrix} 1 \\ 1 \\ 1 \end{pmatrix} + \delta \begin{pmatrix} 0 & 1 & 0 \\ 1/2 & 0 & 1/2 \\ 0 & 1 & 0 \end{pmatrix} \begin{pmatrix} Y_1 \\ Y_2 \\ Y_3 \end{pmatrix} + \begin{pmatrix} x_{11} & x_{12} \\ x_{21} & x_{22} \\ x_{31} & x_{23} \end{pmatrix} \begin{pmatrix} \beta_1 \\ \beta_2 \end{pmatrix} + \begin{pmatrix} e_1 \\ e_2 \\ e_3 \end{pmatrix}$$

$$Y = \alpha \mathbf{1}_3 + \delta W^* Y + X\beta + e \tag{10.4}$$

The terms in the second line of Eq. 10.4 are

- Y and e are vectors of size 3×1 containing, respectively, the three outcome variables, Y_1, Y_2 and Y_3, and the three error terms, e_1, e_2 and e_3; the specification of the independent and identically distributed (or IID) errors can be expressed as $e \sim MVN(0, \sigma^2 I_3)$, with I_3 denoting an identity matrix of size 3×3.
- $\mathbf{1}_3$ is a 3×1 vector of the value 1 (the subscript denotes the number of 1s in the vector).
- β is a vector of size 2×1 containing the two regression coefficients, β_1 and β_2.
- X represents the 3×2 matrix of covariate values.
- W^* is the 3×3 row-standardised weights matrix.

Using its matrix form, i.e. the second line of Eq. 10.4, we can derive the reduced form, where the vector of outcome variables, Y, only appears on the left-hand side of the equation, while all the other terms are on the right-hand side:

$$Y = \alpha \mathbf{1}_3 + \delta\, W^* Y + X\beta + e$$

$$Y - \delta\, W^* Y = \alpha \mathbf{1}_3 + X\beta + e$$

$$Y\left(I_3 - \delta\, W^*\right) = \alpha \mathbf{1}_3 + X\beta + e \tag{10.5}$$

$$Y = \left(I_3 - \delta\, W^*\right)^{-1}\left(\alpha \mathbf{1}_3 + X\beta\right) + \left(I_3 - \delta\, W^*\right)^{-1} e$$

For the reduced form to exist, the matrix $(I_3 - \delta W^*)$ needs to be invertible so that we can divide both sides of the third line in Eq. 10.5 by $(I_3 - \delta W^*)$ to obtain the reduced form. As discussed in Section 8.3 (Chapter 8) for the pCAR model, this requirement leads to a constraint on δ under which δ is bounded between -1 and 1, i.e. $|\delta| < 1$. A natural choice of the prior distribution for δ would be a uniform distribution between -1 and 1, and δ is estimated from the observed data. If δ is estimated to be large and positive, there is strong evidence of the presence of (positive) endogenous spatial spillovers – an increase in the outcome value for an area leads to an increase in outcome values in the neighbouring areas *and* in those areas beyond. An estimated value of δ that is close to 0 suggests weak, or an absence of, endogenous spillover.

The derivation in Eq. 10.5 can be generalised to N areas, and the reduced form of the SLM model is given by

$$Y = \left(I_N - \delta\, W^*\right)^{-1}\left(\alpha \mathbf{1}_N + X\beta\right) + \left(I_N - \delta\, W^*\right)^{-1} e, \tag{10.6}$$

where the dimensions of the matrix notations are modified to N accordingly (e.g. Y and e are vectors of size $N \times 1$, I_N is an $N \times N$ identity matrix and W^* is of size $N \times N$ and so on). The constraint on δ remains the same.

10.2.2.4 Specification of the Spatial Weights Matrix

From the reduced form in Eq. 10.6, it can be seen that the vector of outcome variables follows a multivariate normal distribution as given in Eq. 10.7 (Exercise 10.1):

$$Y \sim MVN\left(\mu_Y, \Sigma\right), \tag{10.7}$$

where the mean vector μ_y and the covariance matrix Σ are given by

$$\mu_Y = \left(I_N - \delta\, W^*\right)^{-1}\left(\alpha \mathbf{1}_N + X\beta\right)$$

$$\Sigma = \sigma^2\left(I_N - \delta\, W^*\right)^{-1}\left[\left(I_N - \delta\, W^*\right)^{-1}\right]^T \tag{10.8}$$

The point to note here is that the covariance matrix takes the form AA^T so that the resultant matrix is always symmetric. Therefore, as opposed to the CAR-type of models discussed in Chapter 8, the SAR structure places no restriction on the spatial weights matrix. Under the SAR structure, W can be either symmetric (e.g. based on contiguity or distance) or asymmetric (e.g. based on the k-nearest neighbours or flows/interactions).

10.2.2.5 *Issues with Model Fitting and Interpreting Coefficients*

When $\delta = 0$, the SLM model reduces to a standard linear regression with independent errors since $(I_N - \delta W^*)^{-1}e$ becomes e, a vector of IID errors. However, when $\delta \neq 0$, $(I_N - \delta W^*)^{-1}$ is a dense matrix, meaning that most of its elements are non-zero. The dense nature of $(I_N - \delta W^*)^{-1}$ has two implications. First, although the elements in e are independent, the elements in $(I_N - \delta W^*)^{-1}e$ are not. As a result, the SLM model has *dependent, correlated* errors. Because of these correlated errors, the SLM model cannot be fitted using the method of least squares, a conventional method of fitting a standard regression model. Also, from the reduced form of the SLM model, one can show that the errors, e, are not independent of the explanatory variable, W^*Y, a property of the model that is referred to as endogeneity (Arbia, 2014, p.66).[3] Due to the presence of endogeneity, the method of least squares should not be used. In Section 10.4, we will describe techniques for fitting the SLM model as well as other spatial econometric models in the Bayesian framework. The second implication is that under the SLM model, the covariate effects depend on not only β but also the matrix multiplier $(I_N - \delta W^*)^{-1}$. It is this multiplicative matrix that embeds the spatial feedbacks into the measurement of the covariate effects. In Section 10.3, we will describe a method proposed by LeSage and Pace (2009) where covariate effects are measured through partial derivatives.

10.2.3 The Spatially-Lagged Covariates Model (SLX)

10.2.3.1 *Formulating the Model*

The spatially-lagged covariates model (SLX) allows one or more covariates to have a *spatially-lagged effect* on an outcome. In the case of a single covariate, the outcome of area i depends not only on its own covariate value but also the average of the covariate values from the neighbouring areas of i – incorporating the second form of spatial spillover (exogenous spillover) into the modelling. Examples of such covariate spillovers include crime rates such as household burglary or car thefts in a city neighbourhood that may be explained by social, demographic and economic conditions both in that neighbourhood and other neighbourhoods that are geographically nearby (for example, the number of known burglary offenders in the nearby neighbourhoods and/or the number of young males living in problem families in neighbouring areas). In the case of modelling small area disease risks, respiratory disease rates in each area of a city may be affected by the air quality in each individual area as well as the air quality in other spatially contiguous areas due to daily or other movements of the resident populations. The underlying idea is that people move around, they have action fields, so are exposed to the risks present in areas other than where they live. When explaining cases of rape in public spaces, Ceccato et al. (2019) argue that the spatial impact of a risk factor need not be limited to where the risk factor is located, as in the case of city bars or subway stations. Geographical areas, because they are not "sealed", may interact with each other in many different ways. Explanatory factors may be "located in other spaces", as remarked by Paelinck and Klaassen (1979).

Under the SLX model, the outcome variable of area i (with $i = 1,...,N$) is modelled as

$$Y_i = \alpha + X_i\beta + \left(W^*X\right)_i \gamma + e_i, \tag{10.9}$$

---

[3] Using the reduced form of the SLM model in Eq. 10.6, $E[(W^*Y)e^T] = E[\{W^*(I_N - \delta W^*)^{-1}(\alpha 1_N + X\beta) + W^*(I_N - \delta W^*)^{-1}e\}e^T] = W^*(I_N - \delta W^*)^{-1}(\alpha 1_N + X\beta)E[e^T] + W^*(I_N - \delta W^*)^{-1}E[ee^T] = W^*(I_N - \delta W^*)^{-1}\sigma^2 \neq 0$, where E denotes expectation. $E[(W^*Y)e^T] \neq 0$ means that the errors, e, are not independent of the explanatory variable, W^*Y, which violates one of the assumptions of least squares estimation.

where $(W^*X)_i$ is a $1 \times K$ vector of the *average* covariate values from the neighbouring areas of i and $\gamma = (\gamma_1, \ldots, \gamma_K)$ is a vector of the associated coefficients. All other terms in Eq. 10.9 are the same as those defined in Eq. 10.1. As evident from Eq. 10.9, the spatially-lagged covariates, W^*X, are included into the SLX model simply as another set of covariates. As a result, the SLX model can be fitted as a standard regression model. However, if the values of a covariate are spatially autocorrelated then the covariate values (X) and their spatially-lagged values (W^*X) may well be correlated, creating a multicollinearity problem.

10.2.3.2 An Example of the SLX Model

By way of illustration of the SLX model, consider again the three areas given in Figure 10.1. Eq. 10.10 specifies the SLX model in the form of three equations, which are subsequently expressed in matrix form in Eq. 10.11.

$$Y_1 = \alpha + \left(x_{11}\beta_1 + x_{12}\beta_2\right) + \qquad x_{21}\gamma_1 + \qquad x_{22}\gamma_2 + e_1$$

$$Y_2 = \alpha + \left(x_{21}\beta_1 + x_{22}\beta_2\right) + \frac{x_{11}+x_{31}}{2}\gamma_1 + \frac{x_{12}+x_{32}}{2}\gamma_2 + e_2 \qquad (10.10)$$

$$Y_3 = \alpha + \left(x_{31}\beta_1 + x_{32}\beta_2\right) + \qquad x_{21}\gamma_1 + \qquad x_{22}\gamma_2 + e_3$$

$$\begin{pmatrix} Y_1 \\ Y_2 \\ Y_3 \end{pmatrix} = \alpha \begin{pmatrix} 1 \\ 1 \\ 1 \end{pmatrix} + \begin{pmatrix} x_{11} & x_{12} \\ x_{21} & x_{22} \\ x_{31} & x_{23} \end{pmatrix} \begin{pmatrix} \beta_1 \\ \beta_2 \end{pmatrix} + \begin{pmatrix} 0 & 1 & 0 \\ 1/2 & 0 & 1/2 \\ 0 & 1 & 0 \end{pmatrix} \begin{pmatrix} x_{11} & x_{12} \\ x_{21} & x_{22} \\ x_{31} & x_{23} \end{pmatrix} \begin{pmatrix} \gamma_1 \\ \gamma_2 \end{pmatrix} + \begin{pmatrix} e_1 \\ e_2 \\ e_3 \end{pmatrix}$$

$$Y = \alpha \mathbf{1}_3 + X\beta + W^*X\gamma + e \qquad (10.11)$$

Eq. 10.10 demonstrates how the spatially-lagged covariate values enter the model. For example, $\dfrac{x_{11}+x_{31}}{2}$ and $\dfrac{x_{12}+x_{32}}{2}$ are the average values of the two covariates from the two spatially-contiguous neighbours of area 2. Area 1 only has one neighbour, namely area 2, so its spatially-lagged covariate values are x_{21} and x_{22} for the two covariates. The same treatment applies to area 3. The second line of Eq. 10.11 gives the reduced form of the SLX model.

Vega and Elhorst (2015) have recently recommended using the SLX model as the starting point (in the absence of well-founded theory) for estimating spatial spillover effects in order to address the identifiability issues associated with spatial econometric models. We shall discuss the issue of identifiability in spatial econometric modelling in more detail in Section 10.5.1.

10.2.4 The Spatial Error Model (SEM)

The spatial error model (SEM) deals with the third form of spatial spillover, where the error terms are spatially autocorrelated due to the spatial dependence structure present in the omitted covariates. To deal with this spatial autocorrelation, the SEM model specifies the SAR structure on the error terms:

$$Y_i = \alpha + X_i \beta + u_i$$

$$u_i = \delta \sum_{j=1}^{N} w_{ij}^* u_j + e_i \qquad (10.12)$$

$$e_i \sim N\left(0, \sigma^2\right)$$

We can see that each error term u_i has a similar structure to that assigned to the outcome variable Y_i in the SLM model (Eq. 10.1). Therefore, the vector of errors $u = (u_1, \ldots, u_N)$ takes the following (reduced) form:

$$u = \left(I_N - \delta\, W^*\right)^{-1} e,$$

from which we can derive the reduced form of the SEM model as

$$Y = \alpha \mathbf{1}_N + X\beta + \left(I_N - \delta\, W^*\right)^{-1} e \qquad (10.13)$$

Eq. 10.13 suggests that, when $\delta = 0$, SEM reduces to a standard linear regression model with IID errors. Similar to the SLM model, the SEM model cannot be fitted via the least squares method because the errors are not independent, thus violating one of the assumptions of the least squares method of model fitting. However, as opposed to the SLM model, the covariate matrix is only multiplied by β without the additional multiplicative term $(I_N - \delta W^*)^{-1}$, thus avoiding the complication in interpreting covariate effects. Under the SEM model, each regression coefficient in β has the usual interpretation, namely, measuring the effect of the corresponding covariate on the outcome.

10.2.5 The Spatial Durbin Model (SDM)

10.2.5.1 Formulating the Model

The spatial Durbin model combines the features of the SLM and SLX models to account for both endogenous spillover, where the outcome in area i depends on the outcomes from the neighbouring areas, and exogenous spillover, in which the outcome of area i may be affected by the covariate values of neighbouring areas. Eq. 10.14 shows the SDM model in matrix form:

$$Y = \alpha \mathbf{1}_N + X\beta + \delta W^* Y + W^* X\gamma + e \qquad (10.14)$$

The reduced form can be derived as follows

$$Y = \alpha \mathbf{1}_N + X\beta + \delta W^* Y + W^* X\gamma + e$$

$$Y - \delta W^* Y = \alpha \mathbf{1}_N + X\beta + W^* X\gamma + e$$

$$Y\left(I_N - \delta W^*\right) = \alpha \mathbf{1}_N + X\beta + W^* X\gamma + e \qquad (10.15)$$

$$Y = \left(I_N - \delta W^*\right)^{-1}\left(\alpha \mathbf{1}_N + X\beta + W^* X\gamma\right) + \left(I_N - \delta W^*\right)^{-1} e$$

In this model, the parameters to be estimated are δ, α, β, γ and σ^2, where the latter is the variance in $e \sim MVN(0, \sigma^2 I_N)$.

10.2.5.2 Relating the SDM Model to the Other Three Spatial Econometric Models

It is useful to recognise that the SDM model encompasses the other three models as special cases. Firstly, the SDM model reduces to the SLM model by setting γ to 0. Thus, the SDM model inherits the features (such as spatial feedbacks) as well as the issues (of model fitting and effect interpretation) associated with the SLM model that we discussed in Section 10.2.2. Secondly, if we set $\delta=0$, the SDM model becomes the SLX model – and it reduces further to a standard regression model if γ is also set to 0. Thirdly, the SEM model is a restricted version of the SDM model, where the former model is obtained by restricting γ in the SDM model to take the form $-\delta\beta$. To prove the last point, recall the reduced form of the SEM model given in Eq. 10.13

$$Y = \alpha \mathbf{1}_N + X\beta + \left(I_N - \delta W^*\right)^{-1} e$$

Multiplying both sides by $(I_N - \delta W^*)$, we have

$$\left(I_N - \delta W^*\right)Y = \left(I_N - \delta W^*\right)\left(\alpha \mathbf{1}_N + X\beta\right) + e$$

$$Y - \delta W^* Y = \alpha \left(I_N \mathbf{1}_N\right) - \delta\alpha \left(W^* \mathbf{1}_N\right) + X\beta - \delta W^* X\beta + e$$

$$Y - \delta W^* Y = \alpha \mathbf{1}_N - \delta\alpha \mathbf{1}_N + X\beta - \delta W^* X\beta + e \qquad (10.16)$$

$$Y = \alpha \left(1 - \delta\right)\mathbf{1}_N + \delta W^* Y + X\beta - \delta W^* X\beta + e$$

If we write $-\delta\beta = \gamma$, the last line matches with the specification of the SDM model in Eq. 10.14. Hence, the SEM model is a restricted version of the SDM model where γ is forced to take the form $-\delta\beta$ (see also LeSage and Pace, 2009, p.51–52).

10.2.6 Prior Specifications

To complete the model specification under the Bayesian framework, we need to assign prior distributions to all the unknown parameters in the model. We specify priors to the parameters in the SDM model because it encompasses all the other three models (SLM, SLX and SEM). In the absence of substantive prior information about the parameters, vague (or weakly informative) priors are used in order to "let the data speak for themselves" (see Section 5.6.1). Specifically, we use a normal distribution centred at 0 with a large variance of 1000000 as a vague prior for each of the regression coefficients β and γ, i.e. $\beta_k \sim N(0,1000000)$ and $\gamma_k \sim N(0,1000000)$ for all $k=1,\ldots,K$. For the common variance of e, we assign a half normal with mean of 0 and variance of 10 but bounded below by 0 as a vague prior on the standard deviation, that is, $\sigma \sim N_{+\infty}(0,10)$. Finally, to ensure the constraint, a uniform distribution defined between -1 and 1 is assigned to δ as a vague prior distribution. These prior specifications are also used for the LM (standard regression), SLM, SLX and SEM models accordingly.

10.2.7 An Example: Modelling Cigarette Sales in 46 US States

In this section, we analyse a set of cigarette sales data across 46 US states in 1992.[4] The aims of this analysis are twofold. The first aim is to demonstrate the reasoning as to why a particular spatial econometric model is fitted. The second is to highlight some issues related to the interpretation of covariate effects. Here, we only focus on the discussion of the modelling results but defer the implementation of the models to Section 10.4.

The use of the spatial econometric approach is motivated by the so-called bootlegging effect, a phenomenon whereby consumers in one state are likely to purchase cigarettes from other neighbouring states if their prices are lower. If such effects exist, a change to the cigarette price in one state may have an impact on the sales of cigarettes in other nearby states. Thus, spatial econometric models are applied to assess whether the bootlegging effect is present (or absent) in the given data. Amongst the models introduced, the SLX model is perhaps of particular relevance, as the bootlegging effect can be seen as a result of exogenous interactions. The other three models, SLM, SEM and SDM, are also under consideration, as suggested by some features of the outcome data.

10.2.7.1 Data Description, Exploratory Analysis and Model Specifications

In this dataset, the outcome value of each state, denoted as y_i (for $i=1,...,46$), is the sales of cigarettes per capita, measured in packs per person aged 14 years and older. Each state is also associated with a covariate value, x_i, which is the average retail price of a pack of cigarettes in 1992. Both the outcome and the covariate values are log transformed. The log-transformed outcome values follow a normal distribution approximately (see Figure 10.2(a)). The row-standardised spatial weights matrix W^* (available on the book's website) defines the neighbourhood structure of the 46 states. The global Moran's I statistic calculated for the log cigarette sales is 0.34, with a p-value of 0.001 derived from 999 random permutations of

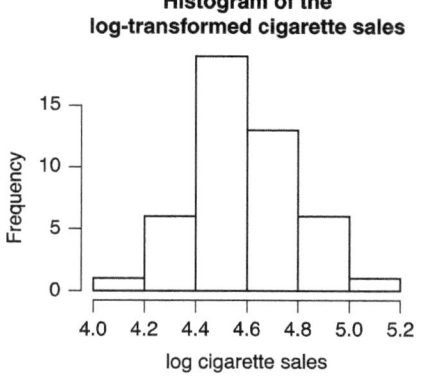
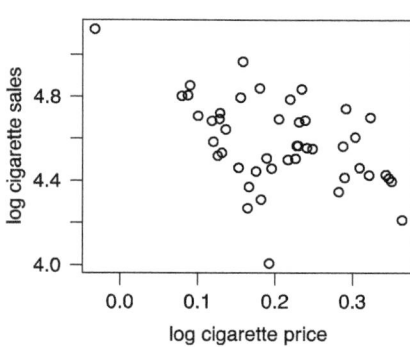

FIGURE 10.2
(a) the histogram of the log-transformed values on cigarette sales (on the left panel) and (b) a scatterplot of cigarette sales against cigarette price across the 46 US states (on the right panel).

---

[4] The data analysed here are part of a larger dataset that spans over 30 years from 1963 to 1992 for the 46 contiguous US states. The original panel data are available at http://spatial-panels.com/wp-content/uploads/2017/06/Files-SLX-paper.zip.

the data, providing strong evidence of positive spatial autocorrelation. In other words, the value of the log cigarette sales of a state tends to be similar to those from the neighbouring states, a feature of the data that needs to be accounted for in the modelling. The fit of a standard linear regression model $y_i = \alpha + \beta x_i + e_i$ in R gives a negative estimate for β with a point estimate of -1.176 and p-value of 0.000571. This negative sales-price association can also be seen in Figure 10.2(b) where a state with a higher cigarette price generally has lower cigarette sales. However, after controlling for the price effect, residual values are still strongly positively autocorrelated: the global Moran's I on the residual values is 0.317 with a p-value of 0.001 from 999 random permutations of the original data. Thus, the variability in price amongst the states is not enough to explain the spatial structure in cigarette sales.

The findings of the exploratory analysis, together with the aim of examining the bootlegging effect, lead us to the following modelling options. The first option is the use of the SLX model to model explicitly the spatial spillovers due to price changes in the set of neighbouring states. This is achieved by including an additional covariate measuring the average price in the neighbours of each state. The second option involves the application of either the SLM model or the SEM model to account for the strong spatial autocorrelation present in cigarette sales. These two models impose different assumptions on the mechanism that could give rise to the observed spatial autocorrelation. The SLM model assumes that endogenous interactions, where a state's cigarette sales may influence the sales in other neighbouring states (and vice versa), give rise to the spatial autocorrelation amongst cigarette sales across the states. The SEM model, on the other hand, assumes that spatial autocorrelation in the outcome values results from omitted covariates which themselves are spatially autocorrelated. In the absence of substantive support, both models seem to be appropriate to deal with the observed property of spatially autocorrelated outcome values. The third and final option is to consider the more general model, the SDM, that combines the features of both the SLX model (for capturing exogenous spillovers) and the SLM model (for dealing with correlated outcome values). The specifications in matrix form of the four spatial econometric models as well as the standard linear regression are summarised in Table 10.1. Each model is fitted in WinBUGS, and for each parameter we will

TABLE 10.1

Specifications of Various Models Fitted to the Cigarette Sales Data

	Specification by Construction	**The Residuals**
LM	$Y = \alpha \mathbf{1}_N + X\beta + e$	$y - \left[\tilde{\alpha} \mathbf{1}_N + X\tilde{\beta} \right]$
SLM	$Y = \alpha \mathbf{1}_N + \delta W^* Y + X\beta + e$	$y - \left[\tilde{\alpha} \mathbf{1}_N + \tilde{\delta} W^* y + X\tilde{\beta} \right]$
SLX	$Y = \alpha \mathbf{1}_N + X\beta + \gamma \left(W^* X \right) + e$	$y - \left[\tilde{\alpha} \mathbf{1}_N + X\tilde{\beta} + \tilde{\gamma} \left(W^* X \right) \right]$
SEM[a]	$Y = \alpha \mathbf{1}_N + X\beta + u$ and $u = \delta W^* u + e$	$y - \left[\tilde{\alpha} \left(1 - \tilde{\delta} \right) \mathbf{1}_N + \tilde{\delta} W^* y + X\tilde{\beta} - \tilde{\delta} \left(W^* X \right) \tilde{\beta} \right]$
SDM	$Y = \alpha \mathbf{1}_N + \delta W^* Y + X\beta + \gamma \left(W^* X \right) + e$	$y - \left[\tilde{\alpha} \mathbf{1}_N + \tilde{\delta} W^* y + X\tilde{\beta} + \tilde{\gamma} \left(W^* X \right) \right]$

[a] The form of the residuals for the SEM model comes from the observation that the SEM model is a special case of the SDM model (see Section 10.2.5.2).

In the definitions of the residuals, the tilde sign is added to each model parameter to emphasise that the residuals are calculated based on the posterior distributions of the parameters. Each residual, as a result, is associated with a posterior distribution. LM denotes the standard regression model. The notation y in the residuals column denotes the observed outcome values, i.e. the log-transformed cigarette sales. The expression of the residuals, as it turns out, comes out nicely from the derivation of the likelihood function, meaning that it is straightforward to calculate in WinBUGS (see Section 10.4.1 for more detail).

report its posterior mean and the 95% credible interval (denoted as 95% CI hereafter). We will return to the fitting algorithm in Section 10.4.

10.2.7.2 Results

We first assess how well each model deals with the spatial autocorrelation in the outcome values. To do this, we calculate the global Moran's I statistic for the residuals. Note that, under the Bayesian approach, each model parameter has a posterior distribution, so each residual, calculated as a function of the model parameters, has a posterior distribution as well. Here, Moran's I statistic is calculated using the posterior means of the residuals, and the corresponding p-value is calculated based on 999 random permutations of the posterior means.

Table 10.2 shows that the residuals from both the LM and the SLX models are highly spatially autocorrelated, indicating that the autocorrelation structure in the observed outcome values cannot be fully accounted for using the cigarette price data across states and their spatially-lagged values. There remain some underlying processes not captured by these two models. For the SLM, SEM and SDM models, on the other hand, the residuals are near to being independent as the values of Moran's I statistic are much smaller and close to 0. Furthermore, across these three models the spatial autocorrelation parameter δ is consistently estimated to be positive, with the 95% CI excluding 0. These two findings suggest that the spatial dependence structure in cigarette sales is adequately dealt with through the spatial-lagged structure either on the outcome variables as in SLM and SDM or through the error term as in SEM. All three models suggest some form of spatial spillover is present in the sales data.

Turning now to the price effect. Table 10.3 compares the posterior estimates of the regression coefficients across the five models. All five models yield a consistent estimate of β where the posterior means are all negative and the 95% CIs do not cover 0, indicative of a strong negative sales-price association as indicated by the exploratory analysis. However, the SLX and SDM also include the term $\gamma(W^*X)$ to represent the effect from the spatially-lagged cigarette prices. Between these two models, while both 95% CIs of γ include 0, the posterior means are quite different. Such apparent inconsistency is not unique to this analysis. As LeSage and Pace (2009) point out, that there *appears* to be inconsistency is a common misconception in the interpretation of covariate effects from different spatial econometric models. They remark: "We note[d] that invalid comparisons of point estimates from different spatial regression model specifications has led practitioners to conclude that changing the model specifications will lead to very different inferences" (LeSage and Pace 2009, p.74). LeSage and Pace develop a (more) valid basis for interpreting and comparing the effect of a covariate on the outcome across different spatial econometric models. Summarised briefly,

TABLE 10.2

A Summary of the Moran's I Statistic of the Residuals from Various Models and the Estimated Spatial Autocorrelation Parameter δ from the SLM, SEM and SDM Models

	LM	SLM	SLX	SEM	SDM
Moran's I on residuals	0.32	0.01	0.32	−0.04	0.03
	(0.002)	(0.387)	(0.001)	(0.546)	(0.319)
δ		0.41		0.48	0.38
		(0.07, 0.69)		(0.16, 0.76)	(0.08, 0.74)

The Moran's I statistic is calculated based on the posterior means of the residuals, and the value in the brackets is the p-value from 999 random permutations. For δ, the posterior mean and the 95% CI are reported. Under both the LM and the SLX models, δ is not part of the model formulation.

TABLE 10.3

The Posterior Means and the 95% CIs for the Regression Coefficients β and γ in the Standard Linear Regression Model and the Four Spatial Econometric Models Applied to the Cigarette Sales Data

	LM	SLM	SLX	SEM	SDM
β	−1.18	−1.01	−1.07	−1.07	−0.98
	(−1.76, −0.52)	(−1.65, −0.45)	(−1.80, −0.35)	(−1.70, −0.39)	(−1.66, −0.29)
γ			−0.42		−0.04
			(−1.73, 0.87)		(−1.46, 1.63)

The parameter γ is not part of the formulation in LM, SLM and SEM, and hence there is no estimate in the case of these models.

their theory uses "…the correct partial derivative interpretation of the parameters from various models [which] results in less divergence in the inferences from different model specifications. This result is related to the partial derivative interpretation of the impact from changes to the variables from different model specifications which represent a valid basis for these comparisons" (LeSage and Pace, 2009, p.74). This is the topic that we now turn to.

10.3 Interpreting Covariate Effects

10.3.1 Definitions of the Direct, Indirect and Total Effects of a Covariate

An objective of spatial econometric modelling is to make inference regarding the *effects* of a covariate on the outcome. The reason for putting "effect" in the plural form is that the spatial econometric approach allows us to measure two types of covariate effects, the *direct* effect and the *indirect* (or *spillover*) effect. A direct effect measures how the outcome in an area is affected due to a change to one (or more) of the covariates in the same area. Effectively, we want to measure the change to Y_i, the outcome of area i, with respect to a change in x_{ik}, the value of the kth covariate in this area (in the case of a model with K covariates). Mathematically, this is achieved by calculating the (partial) derivative of Y_i with respect to x_{ik} denoted as $\dfrac{\partial Y_i}{\partial x_{ik}}$. Using partial derivatives to quantify the covariate effects is the key idea underlying the theory proposed by LeSage and Pace (2009).

An indirect effect describes exogenous spillover, whereby a change to the covariate value in area i may influence not only the outcome in area i but also outcomes in the areas that are neighbours to i. Again, linking this to the idea of using partial derivatives, we measure the change to Y_j with respect to a change in x_{ik} using $\dfrac{\partial Y_j}{\partial x_{ik}}$. Following LeSage and Pace (2009, p.37), the total effect of a covariate on an outcome is defined as the sum of the direct and the indirect effects.

As alluded to in Sections 10.2.2 and 10.2.5, for both the SLM and SDM models, the SAR structure on the outcome variables induces spatial feedbacks. This feedback mechanism leads to some complications in the calculation of the direct and indirect effects of a covariate, and the estimated regression coefficients cannot be interpreted as in a standard regression model. This complication is evident from the reduced forms of the SLM (Eq. 10.6) and SDM (Eq. 10.15) models where the covariate matrix X in both models is multiplied not only by the regression coefficients, β, but also by a (dense) matrix, $(I_N - \delta W)^{-1}$. Similarly, the matrix of the spatially-lagged covariate values, W^*X, under the SDM model is multiplied by γ and $(I_N - \delta W^*)^{-1}$. By contrast, under the LM, SLX and SEM models, spatial feedbacks do not exist amongst the outcome variables, as all three models assume that the dependence structure in the outcomes arises from the covariates (observable or not), not amongst the outcomes themselves. While this assumption simplifies the calculation of a covariate's effects, it does place certain restrictions on how a covariate can affect outcomes in different areas. In what follows, we will first describe the calculation of the direct and indirect effects of the LM, SLX and SEM models using the partial derivatives method. Building upon that, we will then derive the more complex formulations for the SLM and SDM models.

10.3.2 Measuring Direct and Indirect Effects without the SAR Structure on the Outcome Variables

10.3.2.1 For the LM and SEM Models

To introduce the notations, first consider a standard regression model with K covariates of the form

$$Y_i = \alpha + \sum_{k=1}^{K} \beta_k x_{ik} + e_i \tag{10.17}$$

The partial derivative $\dfrac{\partial Y_i}{\partial x_{ik}}$ measures the direct effect of the kth covariate on the outcome in area i, i.e. the change in Y_i due to a change in x_{ik}. From Eq. 10.17, we can see that $\dfrac{\partial Y_i}{\partial x_{ik}} = \beta_k$ for all $i=1,\ldots,N$. This means that under a standard regression model, (a) the regression coefficient β_k measures the direct effect of covariate k on the set of outcomes; and (b) β_k represents a "whole map" effect in the sense that a unit change to the kth covariate in an area is always associated with a change to the area's outcome value by the same amount, β_k, regardless of where this area is in the study region. The latter is to be contrasted with the area-varying direct effects from the SLM and SDM models where the outcome variables are modelled using the SAR structure.

We use the partial derivative $\dfrac{\partial Y_i}{\partial x_{jk}}$ to measure the spillover effect of a change in x_{jk} on Y_i, i.e. how the outcome in area i is affected if there is a change in the kth covariate in a neighbouring area j. Since x_{jk} does not appear on the right-hand side of Eq. 10.17, $\dfrac{\partial Y_i}{\partial x_{jk}} = 0$. In fact, $\dfrac{\partial Y_i}{\partial x_{jk}} = 0$ is true for all $i \neq j$ and for all $k=1,\ldots,K$, meaning that the indirect effect of any covariate is always 0 under a standard regression model.

It turns out that the indirect effect of a covariate under the SEM model is also 0, and the direct effect of the kth covariate is given by the regression coefficient β_k. The SEM model is typically fitted to handle residual spatial autocorrelation arising after fitting Eq. 10.17 where residual spatial autocorrelation is thought to be due to the effects of omitted covariates. As a result, the SEM model provides no information about endogenous or exogenous spillovers and hence does not distinguish between direct and indirect effects.

10.3.2.2 For the SLX Model

To measure the direct and indirect effects under the SLX model, we apply the two partial derivatives, $\dfrac{\partial Y_i}{\partial x_{ik}}$ and $\dfrac{\partial Y_i}{\partial x_{jk}}$, to its model specification:

$$Y_i = \alpha + \sum_{k=1}^{K} \beta_k x_{ik} + \sum_{k=1}^{K} \gamma_k \left(\sum_{j=1}^{N} w_{ij}^* x_{jk} \right) + e_i \tag{10.18}$$

where w_{ij}^* denotes an off-diagonal element from the row-standardised W matrix. For all $i=1,\ldots,N$, $\dfrac{\partial Y_i}{\partial x_{ik}} = \beta_k$, so the direct effect of covariate k is still β_k and the same direct effect applies to all areas.

The calculation of indirect effects is a little more involved since $\dfrac{\partial Y_i}{\partial x_{jk}} = \gamma_k w_{ij}^*$, implying that the indirect effect of a covariate is not constant across the study region but instead is dependent upon the neighbourhood structure through W. To see this, consider the example of three areas as modelled in Eq. 10.10. Focus just on the first covariate (i.e. $k=1$), since the same reasoning also applies to any other covariates. To measure the indirect effects, we need to compute $\dfrac{\partial Y_i}{\partial x_{j1}}$ for all $i=1,\ldots,3$ and $j=1,\ldots,3$ but $i \neq j$, namely, $\dfrac{\partial Y_1}{\partial x_{21}}, \dfrac{\partial Y_1}{\partial x_{31}}, \dfrac{\partial Y_2}{\partial x_{11}}, \dfrac{\partial Y_2}{\partial x_{31}}, \dfrac{\partial Y_3}{\partial x_{11}}$ and $\dfrac{\partial Y_3}{\partial x_{21}}$. From vector-by-vector differentiation, these six terms are the off-diagonal elements in the resulting matrix, as shown in Eq. 10.19 when we differentiate the outcome vector $Y = (Y_1, Y_2, Y_3)$ by the covariate vector $X_1 = (x_{11}, x_{21}, x_{31})$ of the first covariate:

$$\frac{\partial Y}{\partial X_1} = \begin{pmatrix} \dfrac{\partial Y_1}{\partial x_{11}} & \dfrac{\partial Y_1}{\partial x_{21}} & \dfrac{\partial Y_1}{\partial x_{31}} \\[3mm] \dfrac{\partial Y_2}{\partial x_{11}} & \dfrac{\partial Y_2}{\partial x_{21}} & \dfrac{\partial Y_2}{\partial x_{31}} \\[3mm] \dfrac{\partial Y_3}{\partial x_{11}} & \dfrac{\partial Y_3}{\partial x_{21}} & \dfrac{\partial Y_3}{\partial x_{31}} \end{pmatrix} \tag{10.19}$$

From 10.19, we can also see that the diagonal elements in $\dfrac{\partial Y}{\partial X_1}$ correspond to the direct effects and under the SLX model $\dfrac{\partial Y_1}{\partial x_{11}} = \dfrac{\partial Y_2}{\partial x_{21}} = \dfrac{\partial Y_3}{\partial x_{31}} = \beta_1$. Using Eq. 10.10, we can calculate the partial derivatives in Eq. 10.19:

$$\frac{\partial Y}{\partial X_1} = \begin{pmatrix} \beta_1 & \gamma_1 & 0 \\[2mm] \dfrac{\gamma_1}{2} & \beta_1 & \dfrac{\gamma_1}{2} \\[2mm] 0 & \gamma_1 & \beta_1 \end{pmatrix} = \beta_1 I_3 + W^* \gamma_1 \tag{10.20}$$

The last equality emphasises the dependence of the indirect effects on W. Following the notation used in LeSage and Pace (2009), we denote $\dfrac{\partial Y}{\partial X_1}$ as $S_1(W^*)$, where the subscript 1 indicates the partial derivatives are with respect to the 1st covariate, and the dependence of the resulting matrix on W^* is emphasised through the argument. Based on the matrix in Eq. 10.20, we can draw two general points about the SLX model. First, the SLX model depicts "an indirect effect with a hard boundary" in the sense that the indirect effects only exist between area i, where the change to the covariate originates, and its neighbours but do not extend to areas that are not neighbours of i. For example, $\dfrac{\partial Y_1}{\partial x_{31}} = 0$ and $\dfrac{\partial Y_2}{\partial x_{31}} = \dfrac{\gamma_1}{2}$

mean that a change in the first covariate in area 3 only affects the outcome of its neighbour, area 2, but not the outcome of its non-neighbour, area 1. This observation emphasizes the importance of the choice of W^* on any inferences we make about covariate effects (see Section 4.9.2 in Chapter 4). This issue is further discussed in Vega and Elhorst (2015), who have proposed several formulations to model W^*. We will return to this in Section 10.3.3.3.

The second point is that in general the off-diagonal entries of $S_k(W^*)$ are not all the same, so it poses a challenge to present the estimated indirect effect of a covariate. To provide a single-value summary, LeSage and Pace (2009, p.36–37 and 39) define the *average direct effect* (or impact; impact and effect are used interchangeably) of covariate k as the average of the diagonal elements in $S_k(W^*)$. This quantity is denoted as $\bar{M}(k)_{\text{direct}}$. The *average total effect* of covariate k, denoted as $\bar{M}(k)_{\text{total}}$, is defined to be the average of the row sums of $S_k(W^*)$. LeSage and Pace (2009) interpret the sum of the ith row in $S_k(W^*)$ as the total effect to the ith area, resulting from changing the kth covariate by the same amount across all areas.[5] Thus, $\bar{M}(k)_{\text{total}}$ is the total effect averaged over all the total effects to each of the areas. Finally, the *average indirect effect* of covariate k is the difference between the average total effect and the average direct effect, i.e. $\bar{M}(k)_{\text{indirect}} = \bar{M}(k)_{\text{total}} - \bar{M}(k)_{\text{direct}}$. Eq.s 10.21–10.23 provide the formulations for computing the average direct effect, the average total effect and the average indirect effect respectively:

$$\bar{M}(k)_{\text{direct}} = N^{-1}\text{tr}\left(S_k\left(W^*\right)\right) \tag{10.21}$$

$$\bar{M}(k)_{\text{total}} = N^{-1}\mathbf{1}_N^T\left(S_k\left(W^*\right)\right)\mathbf{1}_N \tag{10.22}$$

$$\bar{M}(k)_{\text{indirect}} = \bar{M}(k)_{\text{total}} - \bar{M}(k)_{\text{direct}} \tag{10.23}$$

where $\text{tr}(S_k(W^*))$ denotes the trace of $S_k(W^*)$ (i.e. taking the sum of the diagonal elements) and $\mathbf{1}_N$ is an $N\times 1$ vector of ones. $\mathbf{1}_N^T$ is the transpose of $\mathbf{1}_N$. Using the matrix $S_1(W^*)$ in Eq. 10.20, for example, we have $\bar{M}(1)_{\text{direct}} = \beta_1$, $\bar{M}(1)_{\text{total}} = \beta_1 + \gamma_1$ and $\bar{M}(1)_{\text{indirect}} = \gamma_1$.

To provide an example, recall the estimates in Table 10.3. From the SLX model, the posterior mean of β is -1.07, implying that a one-unit increase in cigarette price in a state would lead to a reduction of its sales by 1.07 units; the same direct effect applies to every one of the states. Since the 95% CI of β does not include 0, the average direct effect is significant at the 5% level. The posterior mean of -0.42 for γ suggests that the cigarette sales in an area would decrease by 0.42 units if its price remains the same but all the other states raise their prices by one unit. However, such covariate spillovers are not found to be important, as the 95% CI of γ contains 0, the value indicative of no effect.

To summarise, the SLX model provides a simple interpretation: β_k and γ_k are, respectively, interpreted as the average direct effect and the average indirect effect of covariate k. As stated in Vega and Elhorst (2015, p.342), an advantage of the SLX model is that the spillover effects of a covariate are "more straightforward, both in terms of estimation and interpretation … than those from the commonly used SEM, SAR and SAC models" where their SAR model is referred to as the SLM model in our notation and the SAC model is one of the model extensions that we will return to in Section 10.5.1.

---

[5] The sum of the jth column of $S_k(W^*)$ represents the total effect to the outcomes of all areas resulting from a change to the kth covariate in area j, while the values of the same covariate in other areas remain the same. The average total effects calculated through either the row sums or the column sums have interpretations that are subtly different, although both give the same numerical value (see the explanation in LeSage and Pace, 2009, p.37, and references therein).

10.3.3 Measuring Direct and Indirect Effects When the Outcome Variables are Modelled by the SAR Structure

For both the SLM and SDM models, the idea of using partial derivatives to measure the direct and the indirect effects of a covariate remains the same. However, the calculation becomes more complicated because of the SAR structure on the outcome variables. More importantly, we need to think more carefully about what is meant by direct and indirect effects. This is because the SAR structure on the outcome variables induces *spatial feedbacks*, whereby a change to the outcome in an area may influence the outcomes of its spatial neighbours, which in turn may have an influence "pointing back to" the outcome of the area where the change originated.

10.3.3.1 Understanding Direct and Indirect Effects in the Presence of Spatial Feedback

To further understand this feedback mechanism and its implications for the interpretation of a covariate's effects, consider Eqs. 10.24 and 10.25 from the SLM model[6] for modelling the outcome variables of two neighbouring areas i and j. For simplicity, we only focus on two neighbouring areas while ignoring their dependency on other neighbours. We also place the discussion in the context of modelling the cigarette sales data so that the single covariate refers to cigarette prices across states.

$$Y_i = \alpha + \delta w_{ij}^* Y_j + \beta x_i + e_i \tag{10.24}$$

$$Y_j = \alpha + \delta w_{ji}^* Y_i + \beta x_j + e_j \tag{10.25}$$

Eq. 10.24 tells us that a change to the cigarette price in state i has an immediate (direct) effect on cigarette sales in the same state, and β quantifies this immediate effect. Furthermore, the coupling nature of the two equations implies that a change to Y_i may trigger a change to Y_j since Y_i appears on the right-hand side of Eq. 10.25. Following the same reasoning, because Y_j appears on the right-hand side of Eq. 10.24, the change that originated in state i may return to state i itself, creating a feedback loop. This feedback loop gives rise to an additional impact on the outcome Y_i, an impact that *goes beyond* what is measured by β. Thus, stating the implication generally, in the presence of spatial feedbacks, the direct effect of a covariate is the sum of two parts: (a) the immediate direct effect on Y_i due to the change to x_i and (b) the additional effect arising from the feedback mechanism where the change to a state's covariate propagates to other states via the neighbourhood structure but also travels back to the originating state, i itself. This convoluted process is the reason why, under the SLM and SDM models, a covariate's direct effect is not just equal to the regression coefficient β itself.

How about the indirect effect? Again, return to the two equations given by 10.24 and 10.25. From Eq. 10.24, we can see that a change to x_i influences Y_i, which in turn affects Y_j because of the SAR structure. Putting the two equations together, we can see that a change to x_i is allowed to affect the outcome of a neighbouring state indirectly. Although not modelled explicitly, the SLM model implicitly allows for the presence of exogenous spillovers (i.e. due to changes in covariate values). Furthermore, not only will the neighbouring states

---

[6] The same reasoning also applies to the SDM model since it contains the SLM model as a special case (see Section 10.2.5.2).

of the first order experience exogeneous spillovers, the neighbours' neighbours will too, and so on. Thus, such effects can propagate through the spatial network and return to state i, then go on to state j, again forming a feedback loop. Therefore, similar to the direct effect, the indirect effect of a covariate also consists of two parts, one part that accounts for the immediate effect on Y_j due to a change in x_i and the other part that results from the propagation of the impact through the spatial network and back to Y_j.

In the next section, we will describe how to use partial derivatives to quantify the direct and indirect effects of a covariate. By so doing, we can reveal some of the properties associated with the effects.

10.3.3.2 Calculating the Direct and Indirect Effects in the Presence of Spatial Feedback

To derive the direct and indirect effects, the reduced forms of SLM and SDM provide a good starting point. Recall the reduced form of SLM in Eq. 10.6:

$$Y = \left(I_N - \delta W^*\right)^{-1}\left(\alpha 1_N + X\beta\right) + \left(I_N - \delta W^*\right)^{-1}e$$

Differentiating each element in $Y = (Y_1, \ldots, Y_N)$ with respect to each element in the vector $X_k = (x_{1k}, \ldots, x_{Nk})$ associated with the kth covariate gives the following $N \times N$ matrix $S_k(W^*)$:

$$
S_k\left(W^*\right) = \frac{\partial Y}{\partial X_k} =
\begin{pmatrix}
\dfrac{\partial Y_1}{\partial x_{1k}} & \cdots & \dfrac{\partial Y_1}{\partial x_{Nk}} \\
\vdots & \ddots & \vdots \\
\dfrac{\partial Y_N}{\partial x_{1k}} & \cdots & \dfrac{\partial Y_N}{\partial x_{Nk}}
\end{pmatrix}
\tag{10.26}
$$

$$= \left(I_N - \delta W^*\right)^{-1}\beta_k$$

The second line of Eq. 10.26 is obtained by noting that (a) neither of the two terms $(I_N - \delta W^*)^{-1}\alpha 1_N$ and $(I_N - \delta W^*)^{-1}e$ in the reduced form involve X_k, so they both disappear after differentiation and (b) a term-by-term differentiation of $(I_N - \delta W^*)^{-1}X\beta$ in the reduced form gives $(I_N - \delta W^*)^{-1}\beta_k$. The latter is rather long-winded, so instead of working it out in its general form, Exercise 10.2 asks the reader to carry out the calculation using the simple map of three areas presented in Figure 10.1.

For the SDM model, the form of the matrix $S_k(W^*)$ is

$$S_k\left(W^*\right) = \frac{\partial Y}{\partial X_k} = \left(I_N - \delta W^*\right)^{-1}\left(I_N\beta_k + W^*\gamma_k\right) \tag{10.27}$$

Note that when $\gamma_k = 0$ for all $k = 1, \ldots, K$, the SDM model reduces to the SLM model and Eq. 10.27 becomes Eq. 10.26. It might also be useful to recognise that the matrix in Eq. 10.20 from the SLX model is a special case of $S_k(W^*)$ given in Eq. 10.27 when the SDM model is applied to the map in Figure 10.1 and δ is set to zero.

10.3.3.3 Some Properties of Direct and Indirect Effects

The diagonal elements of $S_k(W^*)$ represent the direct impacts of the kth covariate and the off-diagonal elements, $\dfrac{\partial Y_i}{\partial x_{jk}}$ (for all $i = 1, \ldots, N$, $j = 1, \ldots, N$ and $i \neq j$), give the indirect impacts.

Both Eqs. 10.26 and 10.27 show that the behaviours of the direct and the indirect effects depend on the nature of the matrix multiplier $(I_N - \delta W^*)^{-1}$. To explore this multiplier further, consider the following. Let W^* take the following form from the map in Figure 10.1:

$$W^* = \begin{pmatrix} 0 & 1 & 0 \\ 1/2 & 0 & 1/2 \\ 0 & 1 & 0 \end{pmatrix} \tag{10.28}$$

and let $\delta = 0.3$, then

$$\left(I_N - \delta W^*\right)^{-1} = \begin{pmatrix} 1.05 & 0.33 & 0.05 \\ 0.16 & 1.10 & 0.16 \\ 0.05 & 0.33 & 1.05 \end{pmatrix} \tag{10.29}$$

The matrix $(I_N - \delta W^*)^{-1}$ is dense, with no zero elements. Specifically, the off-diagonal elements are non-zero, implying the presence of exogenous spillovers, even though the SLM model does not explicitly involve any spatially-lagged covariates. This supports the reasoning given in Section 10.3.3.1.

The diagonal elements of $(I_N - \delta W^*)^{-1}$, and in turn those of $S_k(W^*)$, are not all the same, suggesting that the direct effect of a covariate may vary from one area to another. What is the reason behind this? To investigate, we follow LeSage and Pace (2009, p.14) to expand $(I_N - \delta W^*)^{-1}$ as the sum of an infinite series (see also Section 4.9.1 in Chapter 4):

$$\left(I_N - \delta W^*\right)^{-1} = I_N + \delta W^* + \delta^2 \left(W^*\right)^2 + \delta^3 \left(W^*\right)^3 + \delta^4 \left(W^*\right)^4 + \cdots, \tag{10.30}$$

where $(W^*)^2 = W^* \times W^*$ and $(W^*)^3 = (W^*)^2 \times W^* = W^* \times W^* \times W^*$ and so on. This expansion requires the absolute value of δ to be less than 1, i.e. $|\delta| < 1$. This requirement is met because of the constraint placed on δ (see Section 10.2.2.3). Applying this expansion to $S_k(W^*)$ in Eq. 10.26 from the SLM model, we have

$$S_k\left(W^*\right) = \left[I_N + \delta W^* + \delta^2 \left(W^*\right)^2 + \delta^3 \left(W^*\right)^3 + \delta^4 \left(W^*\right)^4 + \ldots\right]\beta_k \tag{10.31}$$

As discussed in Section 4.8 in Chapter 4, taking powers of the row-standardised weights matrix W^* derived using first order spatial neighbours represents neighbourhood structures at higher orders. Eq. 10.31 makes clear that both the direct and indirect effects depend on not only the first order neighbours but also the second order neighbours, the third order neighbours and so on. However, as the order of a neighbourhood structure increases, its influence on the direct and indirect effects diminishes. This is because $(W^*)^r$ is multiplied by δ^r, which converges to zero as the order r tends to infinity, providing $|\delta| < 1$. For example, from the analysis of the cigarette sales data, the posterior mean of δ from the SLM model is 0.37, so $\delta^3 = 0.05$, meaning that the third order neighbourhood structure has little influence on $S_k(W^*)$.

Now, we apply the expansion of $S_k(W^*)$ in Eq. 10.31 to the situation where the W^* matrix takes the form in Eq. 10.28 and $\delta = 0.3$. Eq. 10.32 gives an approximation of $S_k(W^*)$ up to and including δ^2:

$$S_k\left(W^*\right) \approx I_3\beta_k + \delta W^*\beta_k + \delta^2\left(W^*\right)^2\beta_k$$

$$= \begin{pmatrix} 1 & 0 & 0 \\ 0 & 1 & 0 \\ 0 & 0 & 1 \end{pmatrix}\beta_k + \delta\begin{pmatrix} 0 & 1 & 0 \\ 1/2 & 0 & 1/2 \\ 0 & 1 & 0 \end{pmatrix}\beta_k + \delta^2\begin{pmatrix} 1/2 & 0 & 1/2 \\ 0 & 1 & 0 \\ 1/2 & 0 & 1/2 \end{pmatrix}\beta_k \quad (10.32)$$

We can see now that each diagonal term in $S_k(W^*)$ is expressed in terms of the sum of the corresponding diagonal terms in the matrices $I_3\beta_k$, $\delta W^*\beta_k$ and $\delta^2(W^*)^2\beta_k$. Each diagonal term in $I_3\beta_k$ represents the "immediate" direct effect on the outcome for an area due to a change to its covariate value. For each of the other matrices, the diagonal terms essentially represent *additional* direct effects arising from the feedback loop(s) formed by the neighbourhood structure at the given order. For example, the diagonal terms in $\delta W^*\beta_k$ are all 0, meaning that there are no additional effects arising from the first order neighbours. This is because no feedback loop can be formed at the first order – you cannot go to a neighbour of area i and back in just one step.

However, the diagonal terms of $\delta^2(W^*)^2\beta_k$ are non-zero, meaning that spatial feedbacks exist at the second order neighbourhood structure, thus giving rise to some non-zero additional direct effects. For example, based on the spatial configuration in Figure 10.1, area 1 is a second order neighbour to itself via area 2, forming a feedback loop $(1 \to 2 \to 1)$. For area 2, on the other hand, two feedback loops can be formed at the second order, namely, $(2 \to 1 \to 2)$ and $(2 \to 3 \to 2)$. As a result, the value in the second diagonal term in $\delta^2(W^*)^2\beta_k$ is double that in the first diagonal term, reflecting the different number of feedback loops that each area has. Intuitively, the more feedback loops an area has, the stronger the additional effect becomes, and this is what is reflected in $\delta^2(W^*)^2\beta_k$. Since both areas 1 and 3 only have one feedback loop at the second order, the strength of the additional effect due to spatial feedbacks at that order is the same for the two areas. This answers the question that we originally posted: different areas may have different feedback loops and different numbers of feedback loops and, as a result, the direct effects may be different from one area to another.

It is also worth noting that in general, when δ is estimated to be positive, the spatial feedbacks help strengthen the effect of a covariate change, and thus the direct effects can be larger than the corresponding regression coefficient. We can see that, for example, the diagonal elements in Eq. 10.29 are all greater than 1, suggesting that the direct effects are all greater than the value of β_k. We will illustrate this in the cigarette sales example in Section 10.3.4.

The series expansion of $(I_N - \delta W^*)^{-1}$ helps us to understand the propagation of an indirect effect throughout the region, meaning that a change to the covariate in an area not only affects the outcomes of its neighbours but also affects those in other areas. This is in contrast with the "hard boundary" feature from the SLX model, in which the neighbourhood structure of W places a "hard boundary" as to how far the indirect effect can travel (see Section 10.3.2.2). To illustrate, consider $\partial Y_1 / \partial x_3$, the spillover effect on Y_1 resulting from a change to the covariate in area 3, which is not a first order neighbour to area 1. Recall Eq. 10.20: under the SLX model, this spillover effect is 0. Under the SAR structure on outcomes, a change to x_3 can have a spillover effect on the outcome in area 2, which then propagates to area 1. Eq. 10.32 illustrates this propagation in action. $\partial Y_1 / \partial x_3$ is approximately the sum of the elements in the first row and third column in each of the matrices $I_3\beta_k$, $\delta W^*\beta_k$ and $\delta^2(W^*)^2\beta_k$, so that, $\partial Y_1/\partial x_3 \approx 0+0+\delta^2\beta_k / 2$. The first two terms in the sum are 0 because (a) the change does not occur in area 1, so the element obtained from $I_3\beta_k$ is 0, and (b) the impact of a change initiated in area 3 cannot travel to area 1 in one step, so the element obtained from $\delta W^*\beta_k$ is also 0. However, $\delta^2(W^*)^2\beta_k$ makes a non-zero contribution to the sum, as the impact travels from area 3 to area 1 through the second order neighbourhood structure (i.e. via 2). Therefore, for both the SLM and SDM models, a change to an area's covariate can influence all areas in the study region, but different areas will experience different strengths of indirect effects. The closer an area is to where a change originates, the stronger the indirect effect will be. This is evident in Eq. 10.32 where $\partial Y_1 / \partial x_3 < \partial Y_1 / \partial x_2$ indicating that if a change occurred in area 2, as opposed to in area 3, the indirect effect experienced by area 1 will be stronger.

10.3.3.4 A Property (Limitation) of the Average Direct and Average Indirect Effects Under the SLM Model

Through a setting of a map with three areas, Elhorst (2010, p.22) illustrates that in the case of the SLM model, the ratio of the average direct effect to the average indirect effect of *every* covariate depends only on the spatial autocorrelation parameter δ and the chosen spatial weights matrix W^*, a property that is deemed to be unrealistic in practice (Elhorst, 2010, p.22). This limitation extends to a general situation with N areas.[7] Another implication of this property is that when the point estimate of δ is positive, for each of the covariates, the point estimates of both the average direct effect and the average indirect effect will have the same sign. Similarly, the point estimates of the two effects will have opposite signs if the point estimate of δ is negative. Such restrictions may suggest that the SLM model is only suitable for assessing endogenous spillovers, the basis on which the SLM model is formulated, but not appropriate for examining exogenous spillovers.

10.3.3.5 Summary

To summarise, for models where the simultaneous autoregressive (SAR) structure applies to the outcome variable (i.e. the SLM and SDM models):

- The SAR structure on the outcome induces propagation effects and spatial feedbacks so that a change to the outcome in an area, as a result of, for example, a change to one of its covariate values, influences not only the outcomes of its neighbours but also the neighbours' neighbours and so on, including returning to impact on the area where the change originated.

- In the presence of spatial feedbacks, the regression coefficients cannot be interpreted directly as a covariate's direct and indirect effects, but instead the method based on partial derivatives provides a more appropriate interpretation.

---

[7] As we shall see in Section 10.4.5, Eq. 10.46 and Eq. 10.49 present the general formulae for calculating, respectively, the average direct effect, $\bar{M}(k)_{\text{direct}}$, and the average total effect, $\bar{M}(k)_{\text{total}}$, of the kth covariate under SDM. By setting $\gamma_k = 0$ (thus reducing to SLM) in both equations above, β_k becomes a common factor in both $\bar{M}(k)_{\text{direct}}$ and $\bar{M}(k)_{\text{indirect}} = \bar{M}(k)_{\text{total}} - \bar{M}(k)_{\text{direct}}$ and thus is cancelled out when taking the ratio. As a result, $\dfrac{\bar{M}(k)_{\text{direct}}}{\bar{M}(k)_{\text{indirect}}}$ depends only on δ and $\mathbf{W^*}$.

- Different areas have different pathways over which spatial feedbacks can take place (as determined by the specification of the spatial weights matrix), and such differences lead to direct effects as well as indirect effects varying across areas.
- When a change (either to the outcome or to a covariate) has occurred in an area, this change will be experienced everywhere else in the study region, but the influence is strongest in areas that are geographically closest to the area where the change originates. By "geographically closest" we mean, "as defined by the construction of the **W** matrix".

For completeness, below are two points regarding models where the outcome variable is treated independently after accounting for the spatial structure through the observable covariates (i.e. the LM and SLX models) and the omitted covariates (i.e. the SEM model):

- There are no spatial feedbacks amongst the outcomes of the areas and, as a result, a change in an area does not propagate throughout the entire study region.
- The regression coefficients can be directly interpreted as a covariate's direct effect and, in the case of the SLX model, its indirect effect.

Table 10.4 summarises the formulas for calculating the direct and indirect effects of different models. For each model, the average direct effect, the average indirect effect and the average total effect of a covariate on the outcome can be calculated using the definitions in Eqs. 10.21–10.23.

10.3.4 The Estimated Effects from the Cigarette Sales Data

The posterior estimates of the average direct effect and the average indirect effect of cigarette price on sales from different models are summarised in Table 10.5, in which the posterior estimates of the regression coefficients β and γ are also presented for ease of comparison. All five models estimate that the cigarette sales in a state are strongly affected by its average price. A price increase would reduce the sales in the same state considerably – a strong and negative average direct impact. A close inspection reveals that, for both the SLM and SDM models, the posterior estimates of the average direct effects are slightly "more negative" compared to the corresponding estimates of the β parameter. This is the result of strengthening, arising from the additional effects due to spatial feedbacks. More

TABLE 10.4

The Direct and Indirect Effects of the kth Covariate (Assuming that there are K Covariates in the Study) from Different Regression Models

Model	Direct Effects	Indirect Effects
LM	β_k	0
SLX	β_k	The off-diagonal elements of $\mathbf{W}^*\gamma_k$
SEM	β_k	0
SLM	The diagonal elements of S_k, where $$S_k = \left(I_N - \delta \mathbf{W}^*\right)^{-1}\beta_k$$	The off-diagonal elements of S_k, where $$S_k = \left(I_N - \delta \mathbf{W}^*\right)^{-1}\beta_k$$
SDM	The diagonal elements of S_k, where $$S_k = \left(I_N - \delta \mathbf{W}^*\right)^{-1}\left(I_N\beta_k + \mathbf{W}^*\gamma_k\right)$$	The off-diagonal elements of S_k, where $$S_k = \left(I_N - \delta \mathbf{W}^*\right)^{-1}\left(I_N\beta_k + \mathbf{W}^*\gamma_k\right)$$

TABLE 10.5

The Posterior Means and the 95% CIs of the Average Direct Effect and the Average Indirect Effect of Cigarette Price on Cigarette Sales Estimated from Various Regression Models

	LM	SLM	SLX	SEM	SDM
Average direct effect	−1.18	−1.07	−1.07	−1.07	−1.01
	(−1.76, −0.52)	(−1.76, −0.47)	(−1.80, −0.31)	(−1.70, −0.39)	(−1.67, −0.32)
Average indirect effect	0	−0.76	−0.42	0	−0.54
		(−2.07, −0.07)	(−1.73, 0.87)		(−2.31, 2.16)
β	−1.18	−1.01	−1.07	−1.07	−0.98
	(−1.76, −0.52)	(−1.65, −0.45)	(−1.80, −0.35)	(−1.70, −0.39)	(−1.66, −0.29)
γ			−0.42		−0.04
			(−1.73, 0.87)		(−1.46, 1.63)
δ		0.41		0.48	0.38
		(0.07, 0.69)		(0.16, 0.76)	(0.08, 0.74)

For comparison, the estimated regression coefficients are also reported.

specifically, the negative estimate of β tells us that a state's cigarette sales reduce if the price of cigarettes is increased. As both models estimate δ to be positive and away from 0, a reduction in cigarette sales in one state would be associated with reductions in cigarette sales in the neighbouring states, which would in turn lead to a further reduction in sales in the originating state, resulting in a yet stronger direct effect. The modelling structures of LM, SLX and SEM do not allow for spatial feedbacks in the outcome variable, so the estimated direct effect is the same as the estimated β.

The estimated average indirect effects from the three models, SLM, SLX and SDM, are reasonably consistent. The posterior means are all negative, suggesting that the cigarette sales in a state would increase as a result of a reduction in cigarette prices in other states. Under both SLX and SDM, such an indirect effect does not appear to have a considerable impact, as the 95% CIs for these two models do include 0. For the SLM model, the upper bound of the 95% CI is on the borderline of 0, suggesting that there is some weak evidence of the indirect effect. However, these findings regarding covariate spillovers do not agree with any bootlegging hypothesis, the presence of which would yield a positive estimate of the indirect effect. A similar finding was reported by Vega and Elhorst (2015) when using W^* derived from spatial contiguity. When using some distance-based measures for defining W^*, the bootlegging effect becomes more evident. We refer readers to their paper for a more detailed discussion of the modelling.

10.4 Model Fitting in WinBUGS

This section deals with the implementation of the three spatial econometric models, SLM, SEM and SDM, in WinBUGS. The fitting of the SLX model is straightforward, as it is simply a standard regression model with the spatially-lagged covariates added as an additional set of covariates. For each covariate, the spatially-lagged values can be calculated in R, then imported into WinBUGS as part of the data. We refer readers to Section 5.4.2 in Chapter 5 for the WinBUGS code that can be adapted to fit the SLX model.

Fitting the SLM, SEM and SDM models in WinBUGS is more complicated due to the following three challenges. First, under these three models, the observed outcome values are not independent. Consequently, the likelihood for all the outcome values cannot be simplified to a product of the likelihood contributions from the individual outcome values. A solution to this is to recognise that the terms in the vector *e* are independent and they allow us to derive the expression for the likelihood contribution from each observed outcome. This solution comes largely from Section 6.2.3 in Anselin (1988). The second challenge is that the expression of the resulting likelihood is not in the form of a standard probability distribution (such as a normal distribution or a Poisson distribution). To implement this non-standard likelihood, we use the so-called zeros-trick in WinBUGS. The third challenge is that, as we shall see, the likelihood calculation involves manipulating potentially large matrices. To speed up the computation, a few simplifications can be made to the likelihood. Finally, we will attend to the calculation of the average direct effect, the average indirect effect and the average total effect of a covariate in WinBUGS. Putting all the procedures together, we will illustrate the fitting of the SDM model to the cigarette sales data in WinBUGS. The WinBUGS code provided can be easily adapted to fit the SLM and SEM models, as they are contained within the SDM model as special cases.

10.4.1 Derivation of the Likelihood Function

Let $Y = (Y_1,...,Y_N)$ be a vector of N outcome variables, each following the same probability distribution and $y = (y_1,...,y_N)$ be a set of observed outcome values of Y. If the outcome variables are assumed to be independent, then the joint probability distribution is given by

$$\Pr(y|\theta) = \prod_{i=1}^{N} \Pr(y_i|\theta) \qquad (10.33)$$

where θ denotes a set of parameters and $\Pr(y|\theta)$ is equivalent to $L(\theta|y)$, which is called the likelihood function – a function of the parameters given the observed data. Thus, $\Pr(y_i|\theta) \equiv L(\theta|y_i)$, the likelihood contribution from the outcome value y_i.

In the frequentist approach to parameter inference, those (point) estimates of the parameters θ are sought which maximise the likelihood function. In the Bayesian approach, the likelihood function is combined with the prior distributions of θ to form the posterior distributions, from which posterior summaries such as posterior means and 95% credible intervals are obtained – see also Section 1.4.2 in Chapter 1. Regardless which inferential framework the analyst chooses, a distinctive feature of Eq. 10.33 is that the likelihood function of y can be written as a product of the likelihood contributions from individual observed values. This can be done because the outcome variables are assumed to be independent. However, the outcome variables are dependent under the SLM, SEM and SDM models, and the likelihood function cannot be simplified into the product form given by Eq. 10.33.[8] However, the terms in the vector *e* are independent, and they offer the starting point for deriving the expression of the individual likelihood contributions.

---

[8] Note that an alternative treatment is to write down the likelihood function of y according to a multivariate normal (MVN) distribution, a result from Eq. 10.7 which states that the joint probability distribution of Y is a MVN distribution. However, from our experience, this alternative approach using the MVN likelihood runs very slowly in WinBUGS, even with the cigarette sales data that only have 46 spatial units. The lack of efficient algorithms to manipulate matrices in WinBUGS is part of the reason. This joint approach also involves inverting matrices of size $N \times N$. This may pose a challenge when N, the number of spatial units, is large (e.g. dealing with data at the US county level with over 3000 spatial units or at the MSOA level in England and Wales with over 7000 units). Some of the simplifications in Section 10.4.2 help bypass matrix inversion, but they require the expression of the likelihood contribution from each outcome value.

To start, given that $e \sim MVN(0, \sigma^2 I_N)$, the joint probability distribution for e is

$$\Pr(e \mid \sigma^2) = \left(2\pi\sigma^2\right)^{-N/2} \exp\left\{-\frac{1}{2\sigma^2} e^T e\right\} \qquad (10.34)$$

We rewrite the SDM model in Eq. 10.14 to provide the link between Y and e

$$Y - \left(\alpha \mathbf{1}_N + X\beta + \delta W^* Y + W^* X \gamma\right) = e \qquad (10.35)$$

Using the above two results, we are now in a position to write down the likelihood of y. There are two steps to achieve this.

First, we replace e in Eq. 10.34 by the expression $Y - (\alpha \mathbf{1}_N + X\beta + \delta W^* Y + W^* X \gamma)$. We denote this expression as $g(Y; W^*, X, \theta)$, where $\theta = (\delta, \alpha, \beta, \gamma)$. Here we make two remarks on $g(Y; W^*, X, \theta)$. The first is interpretational. Suppose δ, the spatial autocorrelation parameter, is known, then $g(Y; W^*, X, \theta)$ can be seen as a set of functions with W^*, X and θ as the arguments that transform the outcome variables Y into a set of independent residuals e. This interpretation is in the same spirit as for a standard regression model (where $g(Y; X, \theta) = Y - (\alpha \mathbf{1}_N + X\beta) = e$). In other words, with δ fixed, the SDM model is equivalent to a standard regression model where a set of spatially-filtered outcome variables, $Y - \delta W^* Y$, are regressed on a set of observable covariates, X, and their spatially-lagged values, $W^* X$. Following this reasoning, one can derive the least squares estimates for β and γ by minimising the sum of the squared residuals e (see, for example, Anselin, 2001, p.320). However, such least squares estimators are derived by holding δ, the spatial auto-correlation parameter, fixed, which is a rather unsatisfactory feature of this (non-Bayesian) approach. Instead of estimating β and γ at only one fixed value of δ, more recently, Bivand et al. (2014) proposed to estimate β and γ repeatedly across a range of values of δ. The different sets of estimates are then combined via Bayesian model averaging. The Bayesian procedure described here *jointly* estimates all the unknown parameters, namely, δ, α, β, γ and σ^2, under the SDM model.

The second remark is operational. $g(Y; W^*, X, \theta)$ is a collection of N transformation functions (thus the letter g is in boldface), one for each outcome variable. Each function $g_i(Y_i; W^*, X, \theta)$ takes the following form,

$$g_i\left(Y_i; W^*, X, \theta\right) = Y_i - \left(\alpha + \delta \sum_{j=1}^{N} w_{ij}^* Y_j + X_i \beta + \left(W^* X\right)_i \gamma\right) \qquad (10.36)$$

where X_i and $(W^* X)_i$ denote respectively the covariate values and the spatially-lagged covariate values of area i. We will see the use of this result later on. In addition, replacing the outcome variable Y_i by the corresponding observed value, Eq. 10.36, gives the residual (see Table 10.1). Therefore, the calculation of the residuals is simply the calculation of the $g(Y; W^*, X, \theta)$ functions, and we will see how that is done in WinBUGS in Figure 10.3 below.

Simply replacing e by $g(Y; W^*, X, \theta)$ in Eq. 10.34 does not yield the likelihood function of y since Eq. 10.34 is the joint probability distribution of e, not of Y. However, from the first remark above, the two sets of random variables e and Y are related through the

```
1    #  fitting the SDM model to the cigarette sales data using
2    #  the zeros-trick
3    model {
4      const <- 100    #  a large constant to ensure phi[i]>0
5      pi <- 3.14159   #  a mathematical constant
6      #  a for-loop going through the 46 states (where N=46)
7      for (i in 1:N) {
8        #  the zeros-trick
9        z[i] <- 0           #  the pseudo-observation
10       z[i] ~ dpois(phi[i])  #  the Poisson model for the pseudo-observation
11       #  Lines 12-13 calculates const - log likelihood (see Eq. 10.40)
12       phi[i] <- const - log.det.J/N + 0.5*log(2*pi*sigma2)
13                 + 0.5 * pow(g[i]/sigma,2)
14       #  the transformation function for each outcome value
15       #  (see Eq. 10.36) with Wy[i]=average cigarette sales of
16       #  the neighbours of i
17       g[i] <- y[i] - delta * Wy[i] - covariates.part[i]
18       #  define the regression relationship where
19       #  X[i]=cigarette price in state i and
20       #  WX[i]=average price of the spatial neighbours of i
21       covariates.part[i] <- alpha + X[i]*beta + WX[i]*gamma
22     }
23
24     #  compute the log determinant of the Jacobian matrix using
25     #  eigenvalues of W* (Eq.10.43)
26     for (i in 1:N) {
27       ei[i] <- log(1-delta*eigenvalues.of.W[i])
28     }
29     log.det.J <- sum(ei[1:N])
30     #  vague priors for regression coefficients
31     alpha ~ dnorm(0,0.000001)  #  intercept
32     beta ~ dnorm(0,0.000001)   #  coefficients of X, the cigarette price
33     gamma ~ dnorm(0,0.000001)  #  coefficients of WX, the local average
34                                #  cigarette price
35
36   ##  to fit SEM,
37   ##   (a) replace Line 33 by gamma <- -beta*delta;
38   ##   (b) remove gamma from the initial value lists; and
39   ##   (c) change alpha on Line 21 to alpha*(1-delta)
40   ##  to fit SLM,
41   ##   (a) replace Line 33 by gamma <- 0; and
42   ##   (b) remove gamma from the initial value lists
43
44     #  prior for spatial autocorrelation parameter
45     delta ~ dunif(delta.lim[1],delta.lim[2])
46     #  prior for error standard deviation
47     sigma ~ dnorm(0,0.1)I(0,)
48     sigma2 <- pow(sigma,2)
49
50     #  residuals (the g functions defined in Section 10.4.1
51     #  give the residuals)
52     for (i in 1:N) {
53       residual[i] <- g[i]
54     }
55   }
56
57   #  elements in the data list
58   list(N = 46  #  number of states
59       #  the array y contains the log cigarette sales across 46 states
60       ,y = c(4.692265, 4.369448, ..., 4.707727)
61       #  the array Wy contains the average log cigarette sales of
62       #  the neighbours of each state
63       ,Wy = c( 4.673927, 4.29706, ..., 4.390303)
64       #  the lower and upper bounds for delta (see Section 10.2.2.3)
65       ,delta.lim = c( -1, 1)
66       #  eigenvalues of W*
67       ,eigenvalues.of.W = c( 1, 0.9701024, ..., -0.02077245)
68       #  log cigarette prices across the 46 states
69       ,X = c( 0.2048736, 0.1663959, ..., 0.1002871)
70       #  average log cigarette price of the neighbours of each state
71       ,WX = c( 0.1768596, 0.2610551, ..., 0.1683004)
72   )
73
74   #  initial values for chain 1
75   list(alpha=4.1, beta=-0.5,gamma=0.1,sigma=1)
76   #  initial values for chain 2
77   list(alpha=3.7, beta=0.5,gamma=-0.1,sigma=0.2)
```

FIGURE 10.3
The WinBUGS code, the content of the data list and two sets of initial values for fitting the SDM model to the cigarette sales data via the zeros-trick. This set of codes can be adapted easily to fit the SEM and the SLM models, and the modifications required are given on Lines 36–39 and Lines 40–42 respectively for the two models.

transformations, *g*. Using a result from standard probability theory (see footnote[9]), to obtain the correct joint probability distribution of *Y* (and thus the likelihood function of *y*), we need to multiply the probability distribution of *e* by the determinant of the Jacobian matrix of the form given in Eq. 10.37:

$$
J = \begin{pmatrix} \dfrac{\partial g_1}{\partial Y_1} & \cdots & \dfrac{\partial g_1}{\partial Y_N} \\ \vdots & \ddots & \vdots \\ \dfrac{\partial g_N}{\partial Y_1} & \cdots & \dfrac{\partial g_N}{\partial Y_N} \end{pmatrix} \tag{10.37}
$$

Applying the definition of the Jacobian matrix to the form of g_i in Eq. 10.36 gives

$$
J = \left(I_N - \delta W^* \right)
$$

Thus, putting everything together, the likelihood function of *y* is

$$
L\left(\theta \mid y, W^*, X \right) \equiv \Pr\left(y \mid W^*, X, \theta \right)
$$

$$
= \left| I_N - \delta W^* \right| \cdot \left(2\pi\sigma^2 \right)^{-N/2} \exp\left\{ -\frac{1}{2\sigma^2} g(y)^T g(y) \right\} \tag{10.38}
$$

For clarity, we have suppressed the arguments *W**, *X* and θ that *g* depend on. Also note that the collection of model parameters, θ, now expands to include σ^2.

Taking logarithm to Eq. 10.38 yields the log likelihood function

$$
l\left(\theta \mid y, W^*, X \right) = \ln\left| I_N - \delta W^* \right| - \frac{N}{2}\ln\left(2\pi\sigma^2 \right) - \frac{g(y)^T g(y)}{2\sigma^2} \tag{10.39}
$$

(c.f. Eq. 3.6 on p.47 in LeSage and Pace (2009)).

Thus, the contribution from each outcome value y_i to the log likelihood is

$$
l\left(\theta \mid y_i, W^*, X \right) = \frac{1}{N}\ln\left| I_N - \delta W^* \right| - \frac{1}{2}\ln\left(2\pi\sigma^2 \right) - \frac{1}{2}\left[\frac{g_i(y_i)}{\sigma} \right]^2 \tag{10.40}
$$

where $g_i(y_i)$ is given in Eq. 10.36. The expression in Eq. 10.40 is what is required to fit the SDM model in WinBUGS through the zeros-trick.

---

[9] Consider two vectors of random variables, $A = (A_1,...,A_N)$ and $B = (B_1,...,B_N)$. Assume the two sets of random variables are related through a transformation, such that $A_i = u_i(B_i)$ for $i = 1,...,N$. Let $f_A(a)$ denote the joint distribution function of **A**, then the joint distribution function of **B** is given by (see for example Bishop, 2006, p.18):

$$
f_B(b) = |J| \cdot f_A\left(u(b) \right),
$$

where $|J|$ is the determinant of the Jacobian matrix, where the form of the Jacobian matrix is given in Eq. 10.37.

For both the SEM and SLM models, the contribution of y_i to the log likelihood takes the same form as in Eq. 10.40, but the transformation function $g_i(y_i)$ becomes

$$g_i\left(y_i; \boldsymbol{W}^*, \boldsymbol{X}, \boldsymbol{\theta}\right) = y_i - \left(\alpha\left(1-\delta\right) + \delta\sum_{j=1}^{N} w_{ij}^* y_j + \boldsymbol{X}_i\boldsymbol{\beta} - \delta\left(\boldsymbol{W}^*\boldsymbol{X}\right)_i\boldsymbol{\beta}\right) \qquad (10.41)$$

for the SEM model (by setting $\boldsymbol{\gamma}$ to $-\delta\boldsymbol{\beta}$ and multiplying α by $(1-\delta)$ in Eq. 10.36) and

$$g_i\left(y_i; \boldsymbol{W}^*, \boldsymbol{X}, \boldsymbol{\theta}\right) = y_i - \left(\alpha + \delta\sum_{j=1}^{N} w_{ij}^* y_j + \boldsymbol{X}_i\boldsymbol{\beta}\right) \qquad (10.42)$$

for the SLM model (by setting $\boldsymbol{\gamma}=\boldsymbol{0}$ in Eq. 10.36). The reader is referred to Section 10.2.5.2 for the connections amongst the SEM, SLM and SDM models.

10.4.2 Simplifications to the Likelihood Computation

Before developing the fitting code in WinBUGS, close inspections of Eq. 10.36 and Eq. 10.40 reveal some simplifications.[10] First, it is inefficient, especially for large spatial datasets, to calculate $\ln|\boldsymbol{I}_N - \delta\boldsymbol{W}^*|$ directly at each MCMC iteration. Instead, as originally proposed by Ord in 1975 (p.121), this log determinant can be computed as

$$\ln\left|\boldsymbol{I}_N - \delta\boldsymbol{W}^*\right| = \sum_{j=1}^{N}\ln\left(1-\delta\lambda_j\right) \qquad (10.43)$$

where $\lambda_1,\ldots,\lambda_N$ are the eigenvalues of $\boldsymbol{W}^*$. These eigenvalues can be precomputed outside WinBUGS (say, using the function `eigen` in R) and input into the fitting code as part of the data. At each iteration, we only need to compute the simple sum given by Eq. 10.43, as opposed to tackling the determinant directly.

The second simplification is that we can compute both the spatially-lagged outcome values $\boldsymbol{W}^*\boldsymbol{y}$ and the spatially-lagged covariates $\boldsymbol{W}^*\boldsymbol{X}$, appearing in the transformation function as $\sum_{j=1}^{N} w_{ij}^* y_j$ and $(\boldsymbol{W}^*\boldsymbol{X})_i$ respectively, outside WinBUGS. Taking both points together, the row-standardised spatial weights matrix $\boldsymbol{W}^*$ itself does not need to enter the WinBUGS fitting; instead only its eigenvalues and the two sets of spatially-lagged quantities are required.

10.4.3 The Zeros-Trick in WinBUGS

Using the log likelihood derived above, we can now employ the zeros-trick to fit the SDM model in WinBUGS. The zeros-trick is a feature of the BUGS language that allows users to specify a likelihood beyond those already included in WinBUGS. The description of the zeros-trick given below follows from Section 9.5.1 in Lunn et al. (2012, p.204).

---

[10] Due to the lack of efficient matrix operations, in general, it is advisable to avoid matrix manipulations when implementing a model in WinBUGS. We can speed up the running if some of the quantities can be precomputed and entered into WinBUGS as data.

The trick starts by introducing a set of independent "pseudo-observations" z_i (for $i = 1, ..., N$), and they all take the value 0. Each z_i is then modelled as an observed value of a Poisson distribution with mean ϕ_i (the Greek letter phi) so that each z_i has a likelihood contribution of $e^{-\phi_i}$. Thus the joint probability distribution of these pseudo-observations is given by

$$\Pr(z_1, ..., z_N) = \prod_{i=1}^{N} e^{-\phi_i} \tag{10.44}$$

Setting $\phi_i = -\ln L(\theta | y_i)$, Eq. 10.44 becomes

$$\prod_{i=1}^{N} e^{-\phi_i} = \prod_{i=1}^{N} e^{\ln L(\theta | y_i)} = \prod_{i=1}^{N} L(\theta | y_i) \tag{10.45}$$

Taking the logarithm of $\prod_{i=1}^{N} L(\theta | y_i)$, we have $\sum_{i=1}^{N} l(\theta | y_i)$, where the expression of the log likelihood of y_i is given in Eq. 10.40. Note that a large value, say 100, is added to $-\ln L(\theta | y_i)$ to ensure that the Poisson mean ϕ_i is positive.

Figure 10.3 illustrates the WinBUGS fitting of the SDM model to the cigarette sales data. Note that Lines 52–54 define the model residuals, which basically are the outputs of the transformation functions g. To fit the SEM model, replace Line 33 by gamma <- -delta*beta and multiply alpha on Line 21 by (1-delta). To fit the SLM model, replace Line 33 by gamma <- 0. For both SEM and SLM, remove gamma from the two initial value lists. The estimates from different models are shown and discussed in Section 10.2.7.2.

Note that when the zeros-trick is used, the DIC value reported from WinBUGS (denoted here as DIC_{pseudo}) is based on the pseudo-observations, not the actual observed outcome values (DIC_{obs}). The two DIC values only differ by a constant so that $\mathrm{DIC}_{obs} = \mathrm{DIC}_{pseudo} - 2 \times N \times 100$, where N is the number of spatial units in the study region (e.g. $N = 46$ in the cigarette sales data) and 100 is the large constant added to the -log likelihood (e.g. Line 4 in Figure 10.3). For more detail, see Lunn et al. (2012, p.284–285). Model comparison is then carried out using DIC_{obs}.

10.4.4 Calculating the Covariate Effects in WinBUGS

Eqs. 10.21–10.23 in Section 10.3.2.2 provide the formulations to compute the average direct effect, the average indirect effect and the average total effect of a covariate on the outcome. However, the three formulae involve matrix manipulations that cannot be dealt with efficiently in WinBUGS, and thus we need to make some necessary simplifications.

We start with the average direct impact, $\bar{M}(k)_{direct} = N^{-1} \mathrm{tr}(S_k(W^*))$. Replacing $S_k(W)$ by the expression under the SDM model in Table 10.4, we can rewrite the average direct impact as

$$\bar{M}(k)_{direct} = N^{-1} \left[\mathrm{tr}\left(\left(I_N - \delta W^* \right)^{-1} \right) \beta_k + \mathrm{tr}\left(\left(I_N - \delta W^* \right)^{-1} W^* \right) \gamma_k \right] \tag{10.46}$$

Using the series expansion in Eq. 10.30, namely, $(I_N - \delta W^*)^{-1} = I_N + \delta W^* + \delta^2 (W^*)^2 + \cdots$, the two traces in Eq. 10.46 can be approximated by the first $m + 1$ terms up to and including δ^m in the expansion (see also p.113 of LeSage and Pace, 2009, and Eqs. 29–30 in Bivand et al., 2014):

$$tr\left(\left(\mathbf{I}_N - \delta\mathbf{W}^*\right)^{-1}\right) \approx tr\left(\mathbf{I}_N + \delta\mathbf{W}^* + \delta^2\left(\mathbf{W}^*\right)^2 + \cdots + \delta^m\left(\mathbf{W}^*\right)^m\right)$$

$$= tr\left(\mathbf{I}_N\right) + \delta \cdot tr\left(\mathbf{W}^*\right) + \delta^2 \cdot tr\left(\left(\mathbf{W}^*\right)^2\right) + \cdots + \delta^m \cdot tr\left(\left(\mathbf{W}^*\right)^m\right)$$

$$= N + \sum_{l=1}^{m}\delta^l \cdot tr\left(\left(\mathbf{W}^*\right)^l\right)$$

and

$$tr\left(\left(\mathbf{I}_N - \delta\mathbf{W}^*\right)^{-1}\mathbf{W}^*\right) \approx tr\left(\mathbf{I}_N\mathbf{W}^* + \delta\left(\mathbf{W}^*\right)^2 + \delta^2\left(\mathbf{W}^*\right)^3 + \cdots + \delta^m\left(\mathbf{W}^*\right)^{m+1}\right)$$

$$= tr\left(\mathbf{I}_N\mathbf{W}^*\right) + \delta \cdot tr\left(\left(\mathbf{W}^*\right)^2\right) + \delta^2 \cdot tr\left(\left(\mathbf{W}^*\right)^3\right) + \cdots + \delta^m \cdot tr\left(\left(\mathbf{W}^*\right)^{m+1}\right)$$

$$= \sum_{l=1}^{m}\delta^l \cdot tr\left(\left(\mathbf{W}^*\right)^{l+1}\right)$$

Note that $tr(\mathbf{I}_N) = N$ and $tr(\mathbf{I}_N\mathbf{W}^*) = tr(\mathbf{W}^*) = 0$.
Hence

$$\bar{M}(k)_{\text{direct}} = \beta_k + N^{-1}\left[\beta_k\sum_{l=1}^{m}\delta^l \cdot tr\left(\left(\mathbf{W}^*\right)^l\right) + \gamma_k\sum_{l=1}^{m}\delta^l \cdot tr\left(\left(\mathbf{W}^*\right)^{l+1}\right)\right] \qquad (10.47)$$

The traces of different powers of $\mathbf{W}^*$ can be precomputed outside WinBUGS. Across all applications in this book, we choose $m = 10$. This value gives a reasonably good approximation when $|\delta| < 0.6$, as all higher order terms in δ are virtually 0. When δ is outside that range, one needs to set m larger.

For the average total impact, $\bar{M}(k)_{\text{total}} = N^{-1}\mathbf{1}_N^T(\mathbf{S}_k(\mathbf{W}^*))\mathbf{1}_N$, under the SDM model we rewrite it as

$$\bar{M}(k)_{\text{total}} = N^{-1}\mathbf{1}_N^T\left(\left(\mathbf{I}_N - \delta\mathbf{W}^*\right)^{-1}\left(\mathbf{I}_N\beta_k + \mathbf{W}^*\gamma_k\right)\right)\mathbf{1}_N$$

$$= N^{-1}\mathbf{1}_N^T\left(\left(\mathbf{I}_N - \delta\mathbf{W}^*\right)^{-1}\beta_k + \left(\mathbf{I}_N - \delta\mathbf{W}^*\right)^{-1}\mathbf{W}^*\gamma_k\right)\mathbf{1}_N \qquad (10.48)$$

$$= N^{-1}\left\{\begin{array}{l}\beta_k\left[\mathbf{1}_N^T\left(\mathbf{I}_N - \delta\mathbf{W}^*\right)^{-1}\mathbf{1}_N\right] \\[2ex] + \gamma_k\left[\mathbf{1}_N^T\left(\mathbf{I}_N - \delta\mathbf{W}^*\right)^{-1}\mathbf{W}^*\mathbf{1}_N\right]\end{array}\right\}$$

The last line in Eq. 10.48 can be simplified further. First consider the term $\mathbf{1}_N^T(\mathbf{I}_N - \delta \mathbf{W}^*)^{-1}\mathbf{1}_N$. Again, using the series expansion of $(\mathbf{I}_N - \delta \mathbf{W}^*)^{-1}$, we have

$$\mathbf{1}_N^T \left(\mathbf{I}_N - \delta \mathbf{W}^*\right)^{-1} \mathbf{1}_N = \mathbf{1}_N^T \left(\mathbf{I}_N + \delta \mathbf{W}^* + \delta^2 \left(\mathbf{W}^*\right)^2 + \cdots\right)\mathbf{1}_N$$

$$= N + \delta \mathbf{1}_N^T \mathbf{W}^* \mathbf{1}_N + \delta^2 \mathbf{1}_N^T \left(\mathbf{W}^*\right)^2 \mathbf{1}_N + \cdots$$

As shown in LeSage and Pace (2009, p.14), $\mathbf{W}^*\mathbf{1}_N = \mathbf{1}_N$ and $(\mathbf{W}^*)^2\mathbf{1}_N = \mathbf{W}^*(\mathbf{W}^*\mathbf{1}_N) = \mathbf{W}^*\mathbf{1}_N = \mathbf{1}_N$. More generally, for any power m of $\mathbf{W}^*$, $(\mathbf{W}^*)^m\mathbf{1}_N = \mathbf{1}_N$. Therefore, for $|\delta| < 1$,

$$\mathbf{1}_N^T \left(\mathbf{I}_N - \delta \mathbf{W}^*\right)^{-1} \mathbf{1}_N = N\left(1 + \delta + \delta^2 + \cdots\right)$$

$$= \frac{N}{1-\delta}$$

and

$$\beta_k \left[\mathbf{1}_N^T \left(\mathbf{I}_N - \delta \mathbf{W}^*\right)^{-1} \mathbf{1}_N\right] = \frac{N\beta_k}{1-\delta}$$

Using the same procedure,

$$\gamma_k \left[\mathbf{1}_N^T \left(\mathbf{I}_N - \delta \mathbf{W}^*\right)^{-1} \mathbf{W}^* \mathbf{1}_N\right] = \frac{N\gamma_k}{1-\delta}$$

Therefore,

$$\bar{M}(k)_{\text{total}} = \frac{\beta_k + \gamma_k}{1-\delta} \tag{10.49}$$

Warning: the above simplification to the average total impact only holds when the *row-standardised spatial weights matrix* is used. If this form of the matrix is not used, the average total impact needs to be calculated using $\bar{M}(k)_{\text{total}} = \mathbf{1}_k^T (S_k(\mathbf{W}^*))\mathbf{1}_N$.

Eq. 10.47 and Eq. 10.49 provide an efficient way to compute the average direct effect and the average total effect. The average indirect effect is simply $\bar{M}(k)_{\text{indirect}} = \bar{M}(k)_{\text{total}} - \bar{M}(k)_{\text{direct}}$. Table 10.6 summarises the calculations of $\bar{M}(k)_{\text{direct}}$, $\bar{M}(k)_{\text{indirect}}$ and $\bar{M}(k)_{\text{total}}$ for SLM, SEM

TABLE 10.6

The Average Direct Effect, the Average Indirect Effect and the Total Effect of the kth Covariate (Assuming that there are K Covariates in the Study) from Different Regression Models

Model	The Average Direct Effect $\bar{M}(k)_{\text{direct}}$	The Average Indirect Effect $\bar{M}(k)_{\text{indirect}} = \bar{M}(k)_{\text{total}} - \bar{M}(k)_{\text{direct}}$	The Average Total Effect $\bar{M}(k)_{\text{total}}$
LM	β_k	0	β_k
SLX	β_k	γ_k	$\beta_k + \gamma_k$
SEM	β_k	0	β_k
SLM	See Eq. 10.47 (with $\gamma_k=0$)	$\bar{M}(k)_{\text{total}} - \bar{M}(k)_{\text{direct}}$	$\dfrac{\beta_k}{1-\delta}$
SDM	See Eq. 10.47	$\bar{M}(k)_{\text{total}} - \bar{M}(k)_{\text{direct}}$	$\dfrac{\beta_k + \gamma_k}{1-\delta}$

```
1    #   computing the two traces for the average direct impact
2    #   note:
3    #   trace.of.powers.of.W[1] = trace of W* to the power of 1
4    #   trace.of.powers.of.W[2] = trace of W* to the power of 2 etc
5        for (i in 1:m) {
6           #  the calculation under the 1st summation sign in Eq. 10.47
7           direct.tr1[i] <- pow(delta,i) * trace.of.powers.of.W[i]
8           # the calculation under the 2nd summation sign in Eq. 10.47
9           direct.tr2[i] <- pow(delta,i) * trace.of.powers.of.W[i+1]
10       }
11       sum.direct.tr1 <- sum(direct.tr1[1:m])
12       sum.direct.tr2 <- sum(direct.tr2[1:m])
13   #   compute the average direct, total and indirect impacts
14   #   for a covariate
15       #  average direct impact from Eq. 10.47
16       direct <- beta
17              + (sum.direct.tr1 * beta + sum.direct.tr2 * gamma)/N
18       #  average total impact from Eq. 10.49
19       #   Warning: W must be a row-standardised weights matrix!
20       total <- (beta + gamma)/(1-delta)
21       #  average indirect impact
22       indirect <- total - direct
```

FIGURE 10.4
The WinBUGS code to calculate the average direct effect, the average indirect effect and the average total effect of a covariate on the outcome. The code in this figure is to be added to the main fitting code in Figure 10.3 (insert it between Lines 54 and 55 in Figure 10.3). Monitor the three nodes, `direct`, `total` and `indirect`, to obtain the posterior summaries of the effects. The traces of different powers of *W* are calculated in R, then entered into WinBUGS as data.

and SDM, also for LM and SLX for completeness. As an illustration, Figure 10.4 shows the WinBUGS code to calculate these three effects for the SDM. Note that the code in Figure 10.4 is generic to the SLM, SEM and SDM models as soon as the main fitting code in Figure 10.3 is modified as instructed. Readers are encouraged to use the code in Figure 10.3 and Figure 10.4 to reproduce the results from the cigarette sales analysis in Table 10.5 (see Exercise 10.3).

10.5 Concluding Remarks

10.5.1 Other Spatial Econometric Models and the Two Problems of Identifiability

Using the SLM, SLX and SEM models as basic building blocks, other spatial econometric models have been suggested. LeSage and Pace (2009) have proposed the "general spatial model" (also referred to as the spatial *a*utoregressive *c*ombined model or the SAC model) that combines the SLM and SEM models. The spatial weights matrix for the SLM part of the model and the spatial weights matrix for the SEM part may be set to be equal but are recommended to be different. It is not clear, however, that making the two weights matrices different will make a significant improvement to model identification. Referring to Manski's "reflection problem", Gibbons and Overman (2012, p.178) remark: "…only the *overall* effect of neighbours' characteristics is identified, not whether they work through exogenous or endogenous neighbourhood effects. These issues are intuitive: how can one distinguish between something unobserved and spatially correlated driving spatial correlation in *y*, from the situation where *y* is spatially autocorrelated because of direct interaction between outcomes?" This is one of the identifiability issues with spatial econometric modelling that we will return to later.

Vega and Elhorst (2015, p.344) specify the "general *nesting* spatial model [italics added for emphasis]" (or the GNS model) that adds the spatially-lagged values of the covariates (SLX) to the SAC model. The GNS model has the following form:

$$Y = \delta W_1^* Y + \alpha \mathbf{1}_N + X\beta + W_1^* X\gamma + u$$
$$u = \lambda W_2^* u + e \tag{10.50}$$

Again, W_1^* is recommended to be different from W_2^*, but the same remarks apply as above. Figure 1 in Vega and Elhorst (2015) succinctly summarises various models in the spatial econometric family and their relations.

Confronted by these different models, LeSage and Pace (2009) recommend taking a general model, such as the SDM model, as the point of departure, then deciding which (simpler) model is more suitable for the data (see Section 6.1 in Chapter 6 of LeSage and Pace (2009) for more detail). However, spatial econometric modelling suffers an identifiability issue whereby, as Gibbons and Overman (2012, p.177) have argued, "different specifications are generally impossible to distinguish without assuming prior knowledge about the true data-generating process that we often do not possess in practice." This identifiability issue arises because, as Gibbons and Overman (2012) have shown, the reduced forms for all these models (SLM, SEM, SLX and SDM) are essentially of the same form:

$$Y = X\beta + W^* X\pi_1 + \left(W^*\right)^2 X\pi_2 + \left(W^*\right)^3 X\pi_3 + \cdots + v \tag{10.51}$$

Eq. 10.51 gives rise to different spatial econometric models depending on: (a) how many spatial lags of X are included; (b) the constraints imposed on the parameters (of each spatial econometric model) that specify the composite parameters (π_1, π_2, ...); and (c) whether the error terms, v, are spatially autocorrelated or not. "Distinguishing which of these models generates the data ... is going to be difficult as the specification of W is often arbitrary, and because the spatial lags of X are just neighbor averages that are almost always very highly mutually correlated. ... In short contrasting motivations lead to models that cannot usually be distinguished" (Gibbons and Overman, 2012, p.177). For those reasons, we recommend developing models based on theoretical ground by stating explicitly the assumption(s) that each model places on the underlying data-generating mechanism and evaluate how consistent the estimates are across various model assumptions. Different models should be compared based on theory in addition to how well they fit the data (via for example the use of DIC).

Gibbons and Overman (2012) conclude that if the aim of data analysis is to identify causal associations in order to better understand economic processes or to make policy recommendations, then fitting spatial econometric models is of "limited value ... [and they] urge those considering embarking down this route to think again" (p.187).[11] Alternatively, spatial econometrics may be applied for the purpose of descriptive analyses to give clues

---

[11] By "causal", Gibbons and Overman (2012) mean questions of the type: "if we change x, what do we expect to happen to y?" In the context of the running example in this chapter, we mean a question of the type: "if cigarettes increase in price in one state (perhaps due to some state tax increase), what do we expect to happen to cigarette sales in the same state and in other states?" It is not clear, however, that causal analyses are possible using spatial data available at only one time period.

about associations (in the same way that spatial epidemiological analyses are employed to give clues about possible links between environmental characteristics and disease occurrence). However, even when used in that way, the identifiability issue described above is clearly a source of concern (just as is the need to ensure proper consideration of confounders in any spatial epidemiological analysis). Their solution is to recommend what they term the "experimentalist paradigm which puts issues of identification and causality at centre stage" (p.188), which includes, for example, exploiting natural experiments (see Chapter 3 of this book). The interested reader is encouraged to consult their paper.

Another identifiability issue in spatial econometric modelling is the specification of the spatial weights matrix, W. In fact, we have commented at length, in different parts of the book, on the issues that the W matrix raises for *any* form of spatial modelling. But for spatial econometric modelling specifically, the challenges associated with specifying the W matrix impact directly on the estimation of spillover and feedback effects – one of the main reasons for fitting these models. Many researchers have drawn attention to the critical role the W matrix plays in the determination of such impacts, and yet its structure is usually assumed by the modeler and is sometimes based purely on geometry. If the W matrix was estimated (say from other data identifying real interactions between places or sites) or grounded in strong theory, then economists and others might have more confidence in what these impacts are telling us in terms of "real" spillovers and their feedback effects (Corrado and Fingelton, 2012). For example, Vega and Elhorst (2015) considered different ways to model the elements in the W matrix in order to assess the bootlegging effect in the cigarette sales example. Within the Bayesian framework, Bayesian model averaging has been proposed to incorporate uncertainty into the many ways of specifying the W matrix (LeSage and Fischer, 2008; Chapter 6 in LeSage and Pace, 2009; Seya et al., 2013). The reader might also wish to revisit Section 4.9.2, where we discussed the implications of the W matrix for spatial econometric modelling. Bayesian model averaging, as proposed in the above mentioned references, also helps address another form of uncertainty associated with deciding what covariates should be included in the model.

Despite the issues discussed above, unique to the spatial econometric approach is the explicit modelling of spatial interactions on the outcome. This modelling feature enables us to investigate the mechanisms of spatial spillover and spatial feedbacks. Such modelling is important because we can explore how the outcome in one area could be a consequence of the outcomes and/or other circumstances in other spatially-close areas – two of the important principles that we referenced at the outset of this chapter from Paelinck and Klaassen (1979). The investigation of the bootlegging effect, as in the cigarette sales example, illustrates that idea. This kind of question cannot be answered using the hierarchical modelling approach. We therefore conclude this chapter by summarizing the key points of difference and similarity between the hierarchical and the spatial econometric approaches to modelling spatial data.

10.5.2 Comparing the Hierarchical Modelling Approach and the Spatial Econometric Approach: A Summary

Some of the models in the spatial econometric approach have their "equivalents" in the hierarchical modelling approach. For example, the form of the SLX model, i.e. $Y_i = \alpha + X_i\beta + (W^*X)_i\gamma + e_i$, from Eq. 10.9, is essentially a standard regression model with two sets of covariates, namely X and their spatially-lagged values, W^*X, together with a set of independent error terms. For the SEM model, if we rewrite its reduced form in Eq. 10.13, $Y = \alpha\mathbf{1}_N + X\beta + (I_N - \delta W^*)^{-1}e$, as $Y = \alpha\mathbf{1}_N + X\beta + S$ by letting $(I_N - \delta W^*)^{-1}e = S$, where S

denote a set of spatially-structured random effects, then the similarity between the SEM model and a hierarchical regression model with a set of spatially-structured random effects can be seen. However, when fitting the latter hierarchical model, we can avoid the complications arising when fitting the SEM model because the spatial dependence structure of the outcome values is not modelled through the data model (i.e. the likelihood) but instead through the process model. There are no "equivalents" to the SLM nor the SDM models within the hierarchical modelling approach since, under the latter approach, each outcome value is modelled independently given the process model.

The two approaches differ fundamentally in how spatial dependence in outcome values is modelled. This has an implication for model specification. To discuss this point further, consider first the case of using a normal distribution as the likelihood function for the outcome values. Both approaches seek to explain the variability in outcome values by expressing the mean of the normal distribution as a function of observable covariates. However, hierarchical modelling separates the data model (the likelihood) and the data-generating process (the process model) into two conditional probability models. Spatial dependence in outcome values is modelled in the process model (through the inclusion of covariates and spatially structured random effects) so that the outcome values (thus the outcome variables) can be assumed to be independent *given* the process model. This simplifies the model specification because each outcome value can be modelled independently using a *univariate* normal distribution. A (potentially) complex likelihood function jointly defined for all the outcome values is thus simplified into a product of a set of independent univariate probability distributions, one for each outcome value (see also Section 10.4.1).

By contrast, in spatial econometric modelling, there is no separation of the process model from the data model, rather the process model is embedded in the data model. To incorporate spatial dependence, all the outcome values must be modelled jointly through a *multivariate* normal distribution, and the covariance matrix of the multivariate normal distribution captures the spatial dependence structure of the outcome data. Modelling spatial dependence in the data model gives rise to an awkward normalizing term in the resulting likelihood (Besag, 1974, p.194–199). This normalizing term, taking the form of a determinant, complicates maximum likelihood estimation even in the case of normally distributed outcomes (see Besag, 1974, p.194–199; Cressie, 1991, p.427–433).[12] The fitting of some spatial econometric models also gives rise to complications within the Bayesian framework (see Section 10.4).

The challenges become more acute when extending spatial econometric models to the case of discrete-valued outcome data. Whilst there are joint likelihood functions for count data, "auto" models as they are called, in the case of the auto-Poisson and the auto-negative binomial, only negative spatial dependence is allowed (Besag, 1974, p.202; Cressie, 1991, p.427–428). This property severely limits the usefulness of this approach for modelling small area *count* data. Such a problem does not arise with hierarchical modelling since, given the process model, the likelihood consists of a product of N independent univariate probability distributions. When modelling binary outcome data, LeSage et al. (2011) describe a latent variable approach.

How we handle spatial dependence in the outcome data also impacts on how we utilize information when estimating model parameters. Placing dependence in the likelihood has the effect, in the case of positive spatial dependence, of reducing the amount of information

---

[12] One exception arises when the weights matrix, W, is upper or lower triangular. This occurs when the spatial relationships between locations are directional – if location i is a neighbour of location j, j is not a neighbour of i. For an example, see Section 4.6.

we have in our dataset for estimating model parameters. The term "effective sample size" has been used to describe the amount of information available for parameter estimation (Clifford et al., 1989). Denoting N as the actual number of spatial observations and N' as the equivalent number of *independent* observations, in the case of positive spatial dependence, $N' < N$, and how much less depends on the strength of the spatial correlation (see Section 6.3.2.1 and also Exercise 10.4). The effective sample size is smaller because of information duplication (or information overlap) – what we observe in one area may tell us something about other nearby areas (Section 1.2.1). So, for the spatial econometric approach, when N is small with strong positive spatial dependence present, loss of information may cause problems in parameter estimation.

By contrast, hierarchical modelling incorporates the spatial dependence structure of data in the process model so that it is the parameters that are spatially autocorrelated, not the data values. Thus, no such loss of information applies. In fact, when estimating the area-specific parameters, the reverse is the case. The spatial dependence structure that we impose on the parameters enables us to share information (in data) across spatial units. As we have seen in Chapters 7 and 8, the idea of information sharing helps address the data sparsity issue.

The two modelling approaches also appear to "take sides" in terms of using either a conditional autoregressive (CAR) model or a simultaneous autoregressive (SAR) model. In the case of hierarchical modelling, we use spatially-structured random effects to model unexplained spatial dependence, and these are specified in terms of different CAR models (ICAR, pCAR and BYM models). In the case of spatial econometrics, it is the simultaneous autoregressive (SAR) model that is most frequently used to define spatially structured errors to model unexplained spatial dependence (although other permissible spatial correlation models could be used). The SAR model is also used when modelling one form of explained spatial dependence – where outcomes in different spatial units interact with one another. In that case, covariate effects need to be interpreted correctly through the procedures discussed in Section 10.3.

Whether using a CAR or SAR model, both types of spatial model depend on the specification of the weights matrix (W). Unlike the SAR model, the W matrix must be symmetric in the case of the CAR model.[13] We have discussed the issue regarding the specification of W in the spatial econometric approach in Section 10.5.1. The specification of the W matrix is a significant issue for information borrowing in hierarchical modelling too. The form of the W matrix affects *how* information is borrowed between spatial units. Attention has been paid to looking at ways of defining the W matrix that goes beyond pure geometry, including developing adaptive methods that analyse data properties in order to reduce "inappropriate borrowing" (see Section 8.4). Because outcome values are modelled independently in the case of hierarchical modelling, it is not possible to estimate certain types of spillover and feedback effects so that the problems raised in the case of spatial econometric modelling do not apply. With that said, hierarchical models can be used to estimate covariate effects from other spaces on outcomes through including a set of spatially-lagged covariates, WX (the reader is referred back to the stroke mortality example in Sections 1.3.2.1 and 9.3), so the specification of W is an issue for hierarchical modelling of these types of spillover effect.

---

[13] The interested reader is referred to Besag (1974), where it is shown that whilst the first order, single parameter CAR model is first order Markovian, the first order, single parameter SAR model is third order Markovian. On this basis, the CAR model might be considered the simplest departure from spatial independence. For other treatments of this issue, see Cressie (1991, p.402–410, and Haining, 2003, p.297–302).

10.6 Exercises

Exercise 10.1. Express the reduced form of the SLM model (Eq. 10.6) as a multivariate normal distribution (Eq. 10.7 and Eq. 10.8).

Exercise 10.2. Write down the reduced form of the SLM model using the map of three areas in Figure 10.1, then carry out the term-by-term differentiation of $(I_N - \delta W^*)^{-1}X\beta$ as given in Eq. 10.26 to show that $\partial Y / \partial X_k = (I_N - \delta W^*)^{-1}\beta_k$.

Exercise 10.3. Fit the four spatial econometric models to the US states cigarette sales data. The WinBUGS implementation of the SLM, SEM and SDM models consists of two parts. The first part fits the model using the zeros-trick in WinBUGS (see Figure 10.3), and the second part calculates the average direct, the average indirect and the average total effects via partial derivatives (see Figure 10.4). The fitting of the SLX model, on the other hand, does not involve the above complications.

Exercise 10.4. As discussed in Section 10.5.2, handling positive spatial dependence in data via the likelihood (the data model) means we have *less information*, relative to the case of independent observations, when estimating model parameters. In exercises 4a and 4b we investigate some of the properties and consequences of such information loss if proper allowance is not made for spatial dependence in data values. In exercise 4a we consider testing the association between two spatial processes using the correlation coefficient as discussed in Section 6.3.2.1. In exercise 4b we consider testing for covariate effects in a regression model when errors are not independent (see also Section 1.2.1).

4a. We are interested in examining the relationship between two spatial processes represented by two sets of continuous valued random variables, $X = (X_1,\ldots,X_N)$ and $Y = (Y_1,\ldots,Y_N)$. Both sets of random variables are defined on the N areas within a study region where X might refer to N measures of some environmental exposure whilst Y might be a measure of some health outcome. Using the SLM model, spatial dependence on X is imposed via the joint multivariate normal distribution, i.e. $X \sim MVN(\mu_X,\Sigma_X)$, where $\mu_X = \alpha_X(I_N - \delta_X W^*)^{-1}$ and $\Sigma_X = \sigma_X^2(I_N - \delta_X W^*)^{-1}[(I_N - \delta_X W^*)^{-1}]^T$, as specified in Eq. 10.7 and Eq. 10.8, respectively. The other set of random variables, Y, is defined as $Y = bX + H$, where $H \sim MVN(\mu_H,\Sigma_H)$, with $\mu_H = \alpha_H(I_N - \delta_H W^*)^{-1}$ and $\Sigma_H = \sigma_H^2(I_N - \delta_H W^*)^{-1}[(I_N - \delta_H W^*)^{-1}]^T$. We assume that X and H are independent of each other. However, when the scalar parameter b differs from 0, X and Y are correlated. The interest of this exercise lies in testing the strength of correlation between X and Y under different assumptions about the strength of spatial autocorrelation in X and in Y, that is by assuming different values of the spatial autocorrelation parameters, δ_X and δ_H. With this aim in mind, this exercise performs a simulation study in which 100 datasets are to be simulated according to the following settings:

1) Set $\alpha_X = \alpha_H = 0$ (and hence $\mu_X = \mu_H = 0$, a vector of N zeros).

2) Set $\sigma_X^2 = \sigma_H^2 = 1$.

3) Define the row-standardised weights matrix, W^*, using the Cambridgeshire ward shapefile used in Section 6.2.1 in Chapter 6 with the neighbourhood structure defined using rook's move contiguity.

4) Set (δ_X, δ_H) to one of the following nine settings: (0.0, 0.0), (0.0, 0.4), (0.0, 0.8), (0.4, 0.0), (0.4, 0.4), (0.4, 0.8), (0.8, 0.0), (0.8, 0.4) or (0.8, 0.8).

5) Set b to 0.0 or 0.5 (see below).

6) Use the R function `mvnorm` in the `mvtnorm` package to simulate the values for X and the values for H independently, then the simulated values for Y are given by $bX + H$ (repeat this step 100 times to generate the 100 simulated datasets).

For each of the 100 simulated datasets, carry out (1) the modified t-test as described in Section 6.3.2.1 and (2) the unmodified t-test for bivariate correlation, which takes no account of the spatial dependence structures in the data on X and Y. For each of the two tests, record (a) whether the null hypothesis of no correlation (i.e. $b = 0.0$) is rejected or retained at the 5% significance level; and (b) N', the effective sample size.

If we now select different values of the spatial autocorrelation parameters, δ_X and δ_H, how are the outcomes of the two tests (the modified and the unmodified) affected, and how does N' change? Regarding the test outcome recorded through (a), when the data are simulated with $b = 0.0$, we are examining the risk of a type I error (rejecting the null hypothesis at the given significance level when the null hypothesis is true) when allowance is made (or not made) for spatial dependence when carrying out the test. When setting $b = 0.5$ in the simulation, if we set the null hypothesis to be $b = 0.0$, we are comparing the power of rejecting the null hypothesis at the given significance level between the modified and the unmodified tests, i.e. examining the risk of making a type II error (failing to reject a false null hypothesis at a given significance level). This is because the probability of committing a Type II error is one minus the power. This exercise is derived from the simulation setting outlined in Clifford and Richardson (1985) and Haining (1990, p.313–323). The interested reader is referred to the above references for more information, including some results.

4b. This part of the exercise considers the effect of unexplained spatial autocorrelation present in outcome values on estimating the regression coefficient associated with an observable covariate. Similar to the setting in 4a, we simulate values for X, the observable covariate, and values for Y, the outcome of interest, but in this case, we only simulate the values for X once, then use this set of simulated covariate values to generate multiple sets of outcome values for Y, with $Y = a + bX + H$. This setting considers the covariate values to be fixed quantities with no variability. In certain situations (for example, when modelling the exposure-outcome relationship in spatial epidemiology), one may consider the observed covariate values to be realizations of a set of random variables, X, in order to account for errors in the exposure measurements. We are not covering this topic in this book, but the interested reader is referred to Carroll et al. (2006) for detail.

Follow Steps 1 and 6 in exercise 4a and setting $\delta_X = 0$ to simulate the covariate values for X. Then use the settings below to simulate 100 sets of outcome values for Y where $Y = a + bX + H$:

1) Set $a = b = 0$.

2) $\sigma_H^2 = 1$.

3) Simulate a set of values for H by setting δ_H to one of the following values: 0.00, 0.40, 0.80 and 0.95 (see Steps 2, 3 and 6 in exercise 4a).

4) Calculate a set of values for $Y = a + bx + h$, where x and h are the simulated values for H and for the observable covariate (X) respectively.

Repeat Step 4 100 times for each value of δ_H. For each of the simulated datasets, fit a standard linear regression with a normal likelihood of the form $y_i = \alpha + \beta x_i + e_i$ with IID errors. Comment on how the risk of Type I error (i.e. wrongly rejecting the null hypothesis that $\beta = 0$ at the 5% level) changes as the spatial autocorrelation parameter δ_H increases. Also, for comparison, fit the SEM model to the simulated datasets and comment on your findings regarding the risk of a Type I error. Inflation of Type I errors due to not handling positive spatial dependence in model residuals properly is one of the main concerns of spatial econometric modelling.

11

Spatial Econometric Modelling: Applications

In this chapter, we present and discuss two applications of spatial econometric modelling. The first is concerned with testing for spillover effects in voting outcomes aggregated to the local authority district (LAD) level. The second tests for price competition effects between individual petrol retail outlets in a large city. In both examples, unlike the four in Chapter 9, interest focuses on estimating interaction effects between places – LADs and retail sites respectively. For this reason, both examples lend themselves to the spatial econometric approach in which models are expressed as a series of N simultaneous equations, one equation for each spatial unit. The spatial econometric approach also divides variables into endogenous (typically the outcome variable) and exogenous (typically the covariates) sets. However, as described in Chapter 10, there are spatial econometric models where the outcome variable in one area can depend on the outcome variables in neighbouring areas (e.g. the SLM/SDM model), giving rise to what is termed the problem of endogeneity.

In both applications, the outcome variables are assumed to be normally distributed, a distributional assumption commonly made in spatial econometrics on the observed outcome values. However, the outcome data in both applications show features that may not satisfy the normality assumption, and we will discuss various other alternative approaches. For each application, we first describe the background to the problem, then discuss the data, the modelling issues and present an exploratory analysis of the data. We present the results of the modelling followed by a summary of some of the key statistical findings.

11.1 Application 1: Modelling the Voting Outcomes at the Local Authority District Level in England from the 2016 EU Referendum

11.1.1 Introduction

In this application, we use the spatial econometric approach to model the spatial variation in voting outcomes from the 2016 referendum on the UK's membership of the European Union (EU). In this application we attach particular interest to identifying the presence of local spatial spillover effects in voting patterns, a topic that has attracted some interest in political science in the context of other UK elections.

In the referendum, voters were asked whether they wished to remain in, or leave, the EU, and the data analysed below is a record, at the local authority district (LAD) scale, of the percentage of the voters who voted to leave. In Parliamentary elections in the UK, which involve the election of Members of Parliament (MPs), voting is organized on a constituency basis with the various candidates competing for the votes of the electorate in their constituency. The candidate receiving the most votes in that constituency is elected – a system referred to as "first past the post". In such a system, there is the potential, in some constituencies, for some people to not bother voting ("it won't make any difference if I do") or to vote tactically.

The outcome of the EU referendum, by contrast, was based on simply counting up the total number of votes cast for each of the two outcomes across the whole of the UK. In a sense, therefore, unlike Parliamentary voting, every vote counted regardless of whether the voter lived in an LAD that was predominantly made up of remain voters or leave voters. Every vote would count towards the final, UK-wide, tally. It was the final tally that mattered as opposed to the number of constituencies (LADs) voting one way or the other. Such a simple binary outcome with no reason to engage in tactical voting, or not voting at all, provides an interesting case study for UK election studies and for testing for spillover effects.[1]

The compositional characteristics of the electorate in a constituency often help us to better understand the geography of election results. Goodwin and Heath (2016) carried out a LAD-level analysis of the EU referendum result and showed that LADs with more older people and more people with a lower level of educational attainment were more likely to vote to leave the EU.[2] Immigration was also shown to play a part in voting outcomes at the LAD scale. These factors should therefore be taken into account when modelling LAD-level variation. Research into UK voting patterns has also indicated the presence of certain types of spillover effects between neighbouring voting areas. There are two, possibly linked, sources for such spillover effects: (i) political sources, such as campaign spending (Cutts and Webber, 2010) and incumbency effects (Jensen et al., 2013); and (ii) sources linked to social interaction, such as network effects amongst groups of people where the networks overlap voting area boundaries (Cutts et al., 2014). Because of the nature of the referendum, it is possible that social network effects might have played a role in influencing voting decisions, as networked individuals shared opinions.[3] If these processes are at work, a spatial spillover effect may be present between voting areas with the consequence that voting outcomes between adjacent LADs are correlated. By the same reasoning, of course, such interactions between people are also likely to create intra-LAD clustering of voting preferences, and we shall return to this point in Section 11.1.6.

In exploring these ideas, we construct several spatial econometric models to analyse the proportion of leave voters that allow us to examine spatial spillover effects and the effects of LAD-level compositional characteristics on the decision to vote leave.

11.1.2 Data

The outcome variable is the proportion of votes for leaving the EU for each of the 326 LADs in England. Drawing on the findings of Goodwin and Heath (2016), three district-level covariates are included to represent the demographic and immigration characteristics of each district i (with $i = 1,...,N$ with $N = 326$):

1. Proportion of residents aged over 65 (denoted as $x_{i,65+}$)
2. Proportion of residents with no qualifications (denoted as $x_{i,noQu}$)
3. Proportion of residents during July 2015 to June 2016 who were aged 16–64 and were born in another state of the European Union (denoted as $x_{i,EU}$)

---

[1] Although the paper by Manley et al. (2017) found that the predictive ability of their model was improved by including data on voter turnout. As it turned out, about a third of the electorate did not vote.
[2] Beware the ecological fallacy! See Chapter 3.
[3] This is also, sometimes, referred to as "group learning".

The first two covariates were obtained from the 2011 UK census while the third is from the published estimates produced by the Office for National Statistics.[4] The spatial weights matrix is constructed by connecting each LAD to those neighbouring LADs with which it shares a common border. One issue of interest is what to do with two LADs, the Isle of Wight (an island off the south coast of England) and the Isles of Scilly (a group of small islands off the southwest tip of England), that are isolated from all other LADs in England. Here, we have joined each of these isolated LADs to its nearest LAD so that the Isle of Wight is connected to the Southampton LAD and the Isles of Scilly to the Cornwall LAD. By this device, all LADs are fully connected. Another option is to leave the two LADs as isolated polygons. This option corresponds to setting all entries in the weights matrix for the Isle of Wight and the Isles of Scilly to 0. Such a treatment, however, implies that the voting outcomes of these two LADs depend only on their corresponding covariates, but not on the outcomes from their respective neighbouring areas. In other words, spatial spillover (if present in the country as a whole) does not apply to either of these two LADs. Isolating these two districts from the rest of the country does not seem appropriate. Social interaction, the mechanism that may give rise to spatial spillover effects, is unlikely in our view to be inhibited by such geographical boundaries.[5]

11.1.3 Exploratory Data Analysis

Figure 11.1(a) shows the spatial distribution of the leave vote proportion. However, this may be a rather poor visualisation, as it does not take into account population size. For example, the districts in Greater London have large populations, yet their geographical sizes are small. The map in Figure 11.1(a) may play down the importance of these districts. A cartogram version of the same map is shown in Figure 11.2, where the polygons are "distorted" to reflect the population size of each district. The histogram of the leave vote proportions is reasonably close to a normal distribution, although the histogram has a slightly heavy left-hand tail (Figure 11.1(b)). The global Moran's I statistic on the leave vote proportions is 0.57, and the p-value from 999 random permutations is 0.001, showing strong evidence of the presence of positive spatial autocorrelation in the outcome values.

11.1.4 Modelling Using Spatial Econometric Models

The SLM and the SDM models are both considered in this analysis for the purpose of explicitly modelling spatial interactions in voting outcomes across LADs. Specifically, both models assign the simultaneous autoregressive (SAR) structure to the outcome variable so

---

[4] https://www.ons.gov.uk/peoplepopulationandcommunity/populationandmigration/internationalmigration/datasets/populationoftheunitedkingdombycountryofbirthandnationality

[5] However, treating the polygons as isolated may be more realistic in other circumstances. Consider the case where the outcome of interest is the number of cases of household burglary and the underlying mechanism that is hypothesized to give rise to spatial spillover is the movement behaviour of offenders. Offenders who tend to operate in, say, the Isle of Wight, might be thought to be reluctant to travel to Southampton to commit a crime, and vice versa. Conceptually, therefore, there may be little to support the presence of a spatial spillover of risk between these two districts. In terms of analysis, when a polygon is unconnected to any other polygon on the map, all the entries in the W matrix relating to this polygon are set to 0. In R, this is done by setting the argument zero.policy in the function nb2mat (or nb2listw) to TRUE. In the resulting W matrix, the corresponding row contains only 0 values. To calculate the global Moran's I, the argument zero.policy in moran.mc (or moran.test) also needs to be set to TRUE. For preparing the spatial data for WinBUGS, there is no change to the use of the nb2WB function (see Appendix 4.13.2 in Chapter 4).

(a)   **(b)**

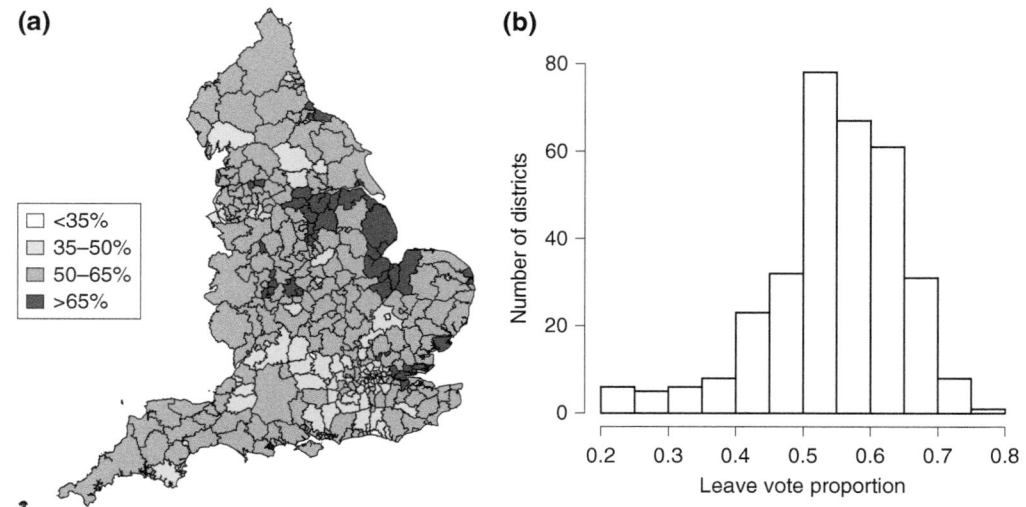

FIGURE 11.1
Exploratory plots of the district-level leave vote proportions from the EU referendum in England.

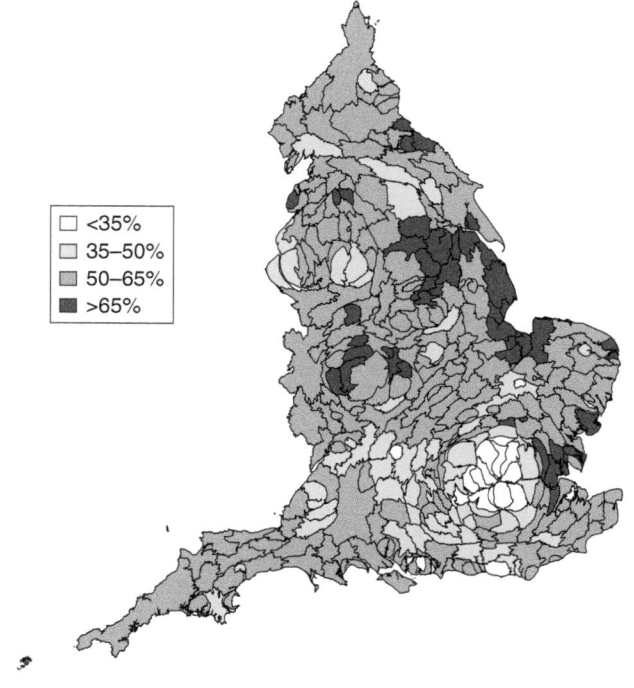

FIGURE 11.2
A cartogram showing the district-level leave vote proportions from the EU referendum in England with the polygons (districts) resized according to the population sizes from the 2011 UK census.

that the voting outcome of a district is dependent upon the voting outcomes of its neighbouring districts. By fitting these models to the observed leave vote proportions, we seek to make inference about the strength and the nature of spatial interaction via the spatial autocorrelation parameter δ. Both models also include the three covariates chosen to represent the LAD-level demographic and immigration structure, while the SDM model includes an additional set of spatially-lagged covariates. The forms of the two models are given in Table 11.1.

Another way to model the autocorrelation structure in the voting outcomes is through the SEM model. The SEM model, in contrast to the SLM and SDM models, assumes there are omitted covariates that are spatially autocorrelated, giving rise to correlated residuals, which in turn induce the spatial autocorrelation structure in the outcome data. In modelling voting behaviours, this is likely to be the case. For example, such omitted covariates can take the form of campaign spending and/or the social networking of people, both of which have been shown to be associated with spatial spillovers. In the absence of this information, the SEM model is also included in our analysis. The SLX model is not considered here, as exogenous spillovers (i.e. the effects of a covariate change in one area on outcomes in other neighbouring areas) are not thought to be important. However, we will later discuss a model extension to examine the possibility that a district's voting tendency to leave the EU may be affected by the presence of large EU populations in neighbouring districts.

With no substantive prior information, vague priors are used throughout the analysis. Specifically, a vague normal prior N (0,1000000) is assigned to the intercept α and every regression coefficient, i.e. β_{age65}, β_{noQu} and β_{EU}, across the three models and γ_{age65}, γ_{noQu} and γ_{EU} for the SDM model. For the spatial autocorrelation parameter δ, a vague uniform prior, *Uniform* (−1,1), is assigned. Finally, for the error standard deviation, σ, a half normal prior

TABLE 11.1

Descriptions of the Three Spatial Econometric Models Applied to the EU Referendum Data

Model	Specification
SLM	$Y_i = \alpha + \delta \sum_{j=1}^{N} w_{ij}^* Y_j + x_{i,age65} \beta_{age65} + x_{i,noQu} \beta_{noQu} + x_{i,EU} \beta_{EU} + e_i$
SEM	$Y_i = \alpha + x_{i,age65} \beta_{age65} + x_{i,noQu} \beta_{noQu} + x_{i,EU} \beta_{EU} + u_i$ and $u_i = \delta \sum_{j=1}^{N} w_{ij}^* u_j + e_i$
SDM	$Y_i = \alpha + \delta \sum_{j=1}^{N} w_{ij}^* Y_j + x_{i,age65} \beta_{age65} + x_{i,noQu} \beta_{noQu} + x_{i,EU} \beta_{EU}$ $+ \left(W^* x_{age65} \right)_i \gamma_{age65} + \left(W^* x_{noQu} \right)_i \gamma_{noQu} + \left(W^* x_{EU} \right)_i \gamma_{EU} + e_i$

In all models, Y_i denotes the outcome variable associated with the leave vote percentage of district i. The district-level errors $e = (e_1,...,e_N)$ are assumed to be independent, and each e_i follows the same distribution $N\left(0,\sigma^2\right)$. The term w_{ij}^* comes from the row-standardised version of W, the spatial weights matrix defined using rook's move contiguity. For the SDM model, $(W^* x)_i$ represents the spatially-lagged value of the corresponding covariate for district i.

with mean 0 and variance 10 is assigned – this half normal prior is truncated so that it has positive support that ensures σ is non-negative (see Section 5.6). The WinBUGS code for fitting the SDM model is given in Figure 11.3. There are two parts to the code. The first part fits the SDM model using the zeros-trick and the second part estimates the average direct effect, the average indirect effect and the average total effect for each covariate. See Sections 10.3 and 10.4 for more details on effect calculations and model fitting.

11.1.5 Results

Table 11.2 compares the three models using the Deviance Information Criterion (DIC). Both the SEM and SDM models perform much better than the SLM model. Between SEM and SDM, the latter model has a lower $\bar{D}$ value (or, equivalently, a higher log likelihood value), meaning it provides a better fit to the data. But, with a larger pD value, the SDM model is more complex and has three additional regression coefficients associated with the spatially-lagged covariates. Taking goodness of fit and model complexity together, the SEM and the SDM models are indistinguishable, as their DIC values differ only by 1. Thus, we present the results from both models.

From the SDM model, the spatial autocorrelation parameter δ is estimated to be positive with a posterior mean of 0.70 with (0.60, 0.79) as the 95% credible interval (denoted as 95% CI hereafter). The SEM model gives a very close estimate where the posterior mean of δ is 0.71 and the 95% CI is (0.62, 0.80). Although the two models have different assumptions regarding the form of spatial interaction, they both consistently suggest that, after accounting for the demographic and immigration characteristics, spatial interactions are present in the voting behaviour across LADs. The positive estimate of δ implies that a LAD tends to have a high leave vote proportion if the adjacent LADs also show high proportions of leave votes.

As discussed in Chapter 10, the regression coefficients from both the SEM model and the SDM model are not directly comparable. Moreover, for the SDM model, it is not appropriate to interpret the β coefficients as the direct effects and the γ coefficients as the indirect effects. The more appropriate method is, as proposed by LeSage and Pace (2009), to compute the average direct effects, the average indirect effects and the average total effects using partial derivatives (see more detail in Section 10.3). Table 11.3 summarises the estimated effects of the three covariates.

Both models reveal a strong age effect on the tendency to vote to leave the EU. For both models, the direct effect of the proportion of residents aged over 65 is estimated to be positive, with the 95% CIs excluding 0, suggesting that a district with more people aged 65 or over tends to have a higher proportion of leave votes. However, under the SDM model, the 95% CI of the indirect effect overlaps with 0, showing little evidence that the voting behaviour of a district is influenced by the age structure of the neighbouring districts. Note that the SEM does not permit an indirect effect of a covariate, hence the value of which is zero. Adding together the direct and the indirect effects, the average total effect of age is positive and important.

Education is also shown to have a considerable impact on the leave vote. For both models, the estimates of the direct effect from the proportion of people without formal qualifications are positive and significant, in the sense that the 95% CIs exclude 0. The implication is that a district with more people with no formal qualifications tends to have a higher percentage of leave votes. Under the SDM model, there is no indirect effect of education, suggesting that the voting outcome of a district is only affected by its own population's education characteristics but not by those of its neighbours.

```
 1    model {
 2    ###################################################################
 3    ## part 1: fitting the SDM model to the Brexit data
 4    ###################################################################
 5      const <- 100 # a large constant to ensure phi[i]>0
 6      pi <- 3.14159 # a mathematical constant
 7      # a for-loop going through the 326 districts (where N=326)
 8      for (i in 1:N) {
 9        # the zero-trick
10        z[i] <- 0            # the pseudo-observation
11        z[i] ~ dpois(phi[i]) # the Poisson model for the
12                             # pseudo-observation
13      # Lines 14-15 calculates const - log likelihood (see Eq. 10.40)
14      phi[i] <- const - log.det.J/N + 0.5*log(2*pi*sigma2)
15              + 0.5 * pow(g[i]/sigma,2)
16      # the transformation function for each outcome value
17      # (see Eq. 10.36) with Wy[i]=average leave votes of
18      # the neighbours of i
19      g[i] <- y[i] - delta * Wy[i] - covariates.part[i]
20
21      # define the regression relationship
22      covariates.part[i] <- alpha
23                          + X.age65[i]*beta[1]   # original covariates
24                          + X.noQu[i]*beta[2]
25                          + X.eu[i]*beta[3]
26                          + WX.age65[i]*gamma[1] # spatially lagged
27                          + WX.noQu[i]*gamma[2]  # covariates
28                          + WX.eu[i]*gamma[3]
29      }
30      # compute the log determinant of the Jacobian matrix using the
31      # eigenvalues of W* (Eq.10.43)
32      for (i in 1:N) {
33        ei[i] <- log(1-delta*eigenvalues.of.W[i])
34      }
35      log.det.J <- sum(ei[1:N])
36
37      # vague priors for regression coefficients
38      alpha ~ dnorm(0,0.000001) # intercept
39      for (i in 1:3) {
40        beta[i] ~ dnorm(0,0.000001)
41        gamma[i] ~ dnorm(0,0.000001)
42      }
43
44    ## to fit SEM,
45    ## (a) replace Line 41 by gamma[i] <- -beta[i]*delta;
46    ## (b) remove gamma from the initial value lists; and
47    ## (c) change alpha on Line 22 to alpha*(1-delta)
48    ## to fit SLM,
49    ## (a) replace Line 41 by gamma[i] <- 0 and
50    ## (b) remove gamma from the initial value lists
51
52      # prior for spatial autocorrelation parameter
53      delta ~ dunif(-1,1)
54      # prior for error standard deviation
55      sigma ~ dnorm(0,0.1)I(0,)
56      sigma2 <- pow(sigma,2)
57      # residuals
58      for (i in 1:N) {
59        residual[i] <- g[i]
60      }
61
```

FIGURE 11.3
The WinBUGS code to fit the SDM model. Follow the instructions on Lines 44–47 and Lines 48–50 to adapt the code for fitting the SEM and the SLM models, respectively.

```
62  ################################################################
63  ## part 2: calculate the average effects
64  ################################################################
65  # computing the two traces for the average direct impact
66  # note:
67  # trace.of.powers.of.W[1] = trace of W* to the power of 1
68  # trace.of.powers.of.W[2] = trace of W* to the power of 2 etc.
69    for (i in 1:m) {
70    #   the calculation under the 1st summation sign in Eq. 10.47
71    direct.tr1[i] <- pow(delta,i) * trace.of.powers.of.W[i]
72    #   the calculation under the 2nd summation sign in Eq. 10.47
73    direct.tr2[i] <- pow(delta,i) * trace.of.powers.of.W[i+1]
74    }
75    sum.direct.tr1 <- sum(direct.tr1[1:m])
76    sum.direct.tr2 <- sum(direct.tr2[1:m])
77    # compute the average direct, total and indirect impacts
78    # for each covariate
79    for (i in 1:3) {
80      # average direct impact from Eq. 10.47
81      direct[i] <- beta[i]
82        + (sum.direct.tr1 * beta[i] + sum.direct.tr2 * gamma[i])/N
83      # average total impact from Eq. 10.49
84      # Warning: W must be a row-standardised weights matrix!
85      total[i] <- (beta[i] + gamma[i])/(1-delta)
86      # average indirect impact
87      indirect[i] <- total[i] - direct[i]
88    }
89  }
```

FIGURE 11.3 *(Continued)*

TABLE 11.2

Comparing the Three Spatial Econometric
Models Using the DIC

Model	$\bar{D}$	pD	DIC
SLM	−1015	6	−1010
SEM	−1075	6	−1069
SDM	−1077	9	−1068

Note that the DIC value reported from WinBUGS (denoted as DIC_{Pseudo} at the end of Section 10.4.3) is based on the pseudo-observations. Reported here are the DIC values based on the observed leave vote percentage, and they are obtained by taking a constant factor away from DIC_{Pseudo}. That is, $DIC_{pseudo} - 2 \cdot N \cdot 100$ (where $N = 326$ LADs and 100 is the large constant added to the -log likelihood (see Lines 5 and 14 in Figure 11.3)). See Section 10.4.3 for more detail.

The proportion of EU residents in a district does not appear to play a role in either the leave vote proportion in the same district or those in the neighbouring districts. The same conclusion was found in Goodwin and Heath (2016). But their modelling showed that "those places which experienced an increase in EU migration over the last ten years tended to be somewhat more likely to vote Leave" (Goodwin and Heath, 2016, p.329). Also related to immigration, Harris and Charlton (2016) examined another interesting idea that the "anti-EU sentiment arises in places that themselves have relatively low levels of immigration

TABLE 11.3

A Summary of the Posterior Estimates (Posterior Mean with its 95% Credible Interval) of the Average Direct Effects, the Average Indirect Effects and the Average Total Effects of the Three Covariates

	Model	% Residents Aged ≥ 65	% Residents with No Formal Qualifications	% Residents Aged 16–64 Born in the EU
Average direct effects	SEM	0.0042 (0.0024, 0.0059)	0.0146 (0.0132, 0.0159)	−0.0006 (−0.0019, 0.0006)
	SDM	0.0045 (0.0028, 0.0062)	0.0142 (0.0129, 0.0156)	−0.0008 (−0.0021, 0.0005)
Average indirect effects	SEM	0	0	0
	SDM	0.0041 (−0.0021, 0.0106)	−0.0012 (−0.0060, 0.0037)	−0.0022 (−0.0086, 0.0037)
Average total effects	SEM	0.0042 (0.0024, 0.0059)	0.0146 (0.0132, 0.0159)	−0.0006 (−0.0019, 0.0006)
	SDM	0.0086 (0.0020, 0.0156)	0.0131 (0.0082, 0.0181)	−0.0029 (−0.0101, 0.0035)

The indirect effects under the SEM models are set to 0, and thus, the average direct effect of a covariate is the same as its average total effect, and both are equal to the estimate of the corresponding β coefficient.

but share a border with places where the levels are higher" (p. 2123). Exercise 11.1 is set out to explore these two possibilities by extending the SEM model to include two additional LAD-level covariates: one, the change in EU migrant populations (the idea of Goodwin and Heath, 2016) and the other, the difference between a district's EU migrant proportion and the average proportion of its spatial neighbours (the idea of Harris and Charlton, 2016).

11.1.6 Conclusion and Discussion

In the context of the 2016 EU referendum, the modelling results show the presence of spatial interactions in the voting pattern at the LAD level. Both the SEM and SDM models estimate the spatial autocorrelation parameter δ to be positive and away from 0, implying that the voting outcome of a district tends to be similar to those of neighbouring districts. The results also suggest that the leave vote tendency of the residents in a district is strongly affected by the age and education structure of the district but less influenced by that of the neighbouring districts. In addition, the proportion of EU residents in a district does not appear to much influence the voting outcome. These findings are largely in line with published literature.

However, in the case of the present dataset, the DIC comparison does not allow us to distinguish the SEM and the SDM models, which is regrettable since each model represents a somewhat different form of spatial interaction. One reason may be that the three covariates included in the analysis, while chosen based on existing research, are seriously under-representing the complexity of why LADs voted in the way they did. One may seek to include more observable covariates, particularly economic and political covariates, so that the explanatory power becomes less reliant on the correlated residuals in the SEM model, which would make the SDM model more attractive. Another interesting extension would be to apply the general nesting spatial (or GNS) model that incorporates the SEM model into the SDM model (see Section 10.5.1 and Vega and Elhorst, 2015), acknowledging the presence of omitted covariates that are spatially-autocorrelated. A clear strength of the spatial econometric approach is that the suite of models allows us to formulate and evaluate various forms of spatial spillover, providing useful tools to explore the complex patterns of spatial interaction in the voting outcome.

In conclusion, there are two issues that are worth noting. First, under SLM/SEM/SDM, it is difficult to produce the "conventional residual map" to visualise the variability left after accounting for covariate effects. The residuals as defined in Table 10.1 in Chapter 10 (apart from those for the SLX model) cannot be used for such a purpose because they represent the variability remaining after accounting for both the covariate effects *and* the spatial autocorrelation amongst the outcome values. While they are useful for checking whether a given model can adequately account for the spatial autocorrelation in outcome values, they cannot be used "as if" they are the area-specific random effects in a hierarchical model to, for example, highlight areas that have unusually high (or low) leave vote percentages.

The second issue stems from how the leave vote percentage is modelled in the present application. This issue applies when modelling binomial data generally. Here, the outcome values are taken to be the proportions of leave votes and are assumed to come from a multivariate normal distribution. This, however, ignores the binomial variability of the data due to the varying number of votes. For example, although both Stockton-on-Tees and West Lindsey had about 61% of votes to leave the EU, the two proportions are associated with different levels of variability: there were about 100000 valid votes in total in Stockton-on-Tees but only about 50000 votes in West Lindsey. One way of dealing with this source of variability would be to model the count of leave votes directly using a binomial distribution. Spatial structure could be imposed by specifying a spatially correlated binomial (or auto-binomial) distribution. This option, however, poses some significant challenges in estimation (Besag, 1974; Huffer and Wu, 1998; see also Section 10.5.2).

An alternative approach to address the above issue is to specify the error variance σ^2 as $\kappa \cdot v_i$. Thus, $e_i \sim N(0, \kappa \cdot v_i)$ in Table 11.1. The term v_i accounts for the binomial variability and is given by $v_i = p_i(1 - p_i) / n_i$, where n_i is the total number of votes in district i and p_i is the leave vote proportion calculated from the data. The term κ is a parameter to be estimated from the data, and it can be interpreted as a measure of dispersion, the amount of variability in the leave vote counts that exceeds binomial variability. $\kappa = 0$ suggests no overdispersion. $\kappa > 0$ (or $\kappa < 0$) is indicative of over (or under) dispersion. Overdispersion is more commonly encountered in practice. In the context of voting, overdispersion may result from, for example, intra-area correlation, where the votes cast by individual voters are correlated due to social network effects amongst groups of voters living in the same LAD. A heterogeneous voting tendency within an area, for example multiple groups of voters each having different views on the Brexit issue, would also lead to overdispersion. The two processes need not be unrelated. This model, that allows the error variance to vary across areas, belongs to so-called robust heteroscedastic spatial regression discussed in Section 5.6.1 of LeSage and Pace (2009). Dealing with non-normal outcome data is under active development in spatial econometrics (see, for example, LeSage et al. (2011) for modelling binary outcome data; Lambert et al. (2010) and Liesenfeld et al. (2017) for dealing with Poisson count data).

11.2 Application 2: Modelling Price Competition Between Petrol Retail Outlets in a Large City

11.2.1 Introduction

In this application we analyse petrol price data for the city of Sheffield, England. The aim of the analysis is to test for price competition effects between neighbouring retail outlets.

The period of this study coincided with the final stages of a price war where petrol prices had been falling steeply over a three-month period (January to March 1982), driven by falling oil prices on the Rotterdam spot market. This provides a particularly interesting context for studying competition effects because the circumstances possess some of the characteristics of a natural experiment (see Section 3.2). Petrol is a homogeneous good and in 1982 was sold from a large number of sites, some of which were owned by international brands (for whom market share was critical) whilst others were "cut price" outlets whose business model centred on selling cheaply, sometimes from less attractive sites (Haining, 1986). Whilst all petrol retailers were "treated" (in the language of the natural experiment), the latter group of retailers had a strong incentive to adjust their prices, as did some, but not all, of the branded sites that were critical to market share for the parent company. These outlets might act as market disruptors. Customers are somewhat price sensitive and highly mobile within a single urban area and because of price posting can compare prices at virtually no cost, often as part of a multi-purpose trip. The same price posting enables retailers to know, at little cost, what prices their competitors are offering. For these reasons it can be argued that what we are observing is a "network of interdependent, overlapping and possibly volatile market regions" within which there is considerable potential for price competition, which is the main basis for competition in this sector (Haining, 1983, p.517).[6]

11.2.2 Data

The outcome values of this study are the prices recorded at 63 petrol stations collected on a single day in March 1982. For this analysis, each petrol station i ($i = 1,...,N$ with $N = 63$) has two covariates:

1. Whether a station is on a main road (=1) or not (=0), denoted as $x_{i,main}$
2. The prices collected on a single day from the same 63 petrol stations about four weeks earlier, in February 1982, denoted as $x_{i,prev}$

The connectivity matrix that defines the neighbourhood structure across the petrol stations is derived from the map on p.377 in Haining 1990 (see Appendix to this chapter). The neighbours of site i are those other sites to which i is directly linked on the map. Two rules were used to construct the set of linkages: (a) proximity along the principal radial routeways that lead to Sheffield's city centre and (b) proximity at important road intersections. Certainly, other spatial connectivity matrices between the set of petrol stations could be considered.

   The 63 stations include overlapping clusters of stations that are close enough to each other so that it is plausible that they would be in direct competition for the same set of customers. Some of the sites off main roads in Sheffield were excluded. These were sites that combined petrol retailing with other functions such as car repairs and car sales so that it was likely that they would not be in competition with the large petrol-only stations located on the main roads.

11.2.3 Exploratory Data Analysis

Figure 11.4(a) shows a heavily positively skewed histogram of the prices in March. While the prices from most stations were set below £1.50 per gallon, station 25 shows a

---

[6] See also Haining (1984), and for a different theoretical perspective but again emphasizing market interdependence, see Sheppard et al. (1992).

considerably higher price at £1.628. The boxplot in Figure 11.4(b) suggests no difference in price between stations that were on a main road from those that were not. The prices set in March are correlated with the prices set in the previous month (Figure 11.4(c)), although the correlation is somewhat affected by the high price set at Station 25. The set of prices are spatially autocorrelated: global Moran's I, based on the above connectivity matrix, is 0.24, with a p-value = 0.001 derived from 999 random permutations. After controlling for the two covariates, the residuals of a standard linear regression model are still highly autocorrelated with a positive Moran's I statistic and a p-value that is lower than the usual 0.05 significance level. This suggests the need to account for the spatial interdependence in petrol prices across the 63 stations.

11.2.4 Spatial Econometric Modelling and Results

Guided by the above findings, we fit three spatial econometric models, SLM, SEM and SDM, in order to investigate the spatial structure in the pattern of petrol prices across the stations. Table 11.4 describes the three models. Note that the SDM model includes only $x_{i,main}$, the binary covariate on whether a station is on a main road or not, but not its spatial lag, as the two are highly correlated, leading to non-convergence in the WinBUGS fitting. In addition, to improve mixing, the covariate $x_{i,prev}$ is mean-centred, whereby each covariate value is subtracted from its mean (see Section 5.4.2). As in the Brexit application, vague priors are used on all model parameters so that the findings are minimally dependent on the chosen priors.

Table 11.5 shows that all three models have similar DIC values, indicating similar fits to the petrol price data. The posterior estimates of δ from both the SLM and SDM are broadly similar, and the 95% CIs from both models overlap with 0. On the other hand, the estimate from the SEM model shows some weak evidence of positive spatial interaction. The evidence is weak here because the lower bound of the 95% CI is on the borderline of 0. Across the three models, the distribution of the residuals reveals a large positive residual associated with Station 25 (Figure 11.5). As we noted earlier, the petrol price of this station was particularly high in March 1982. The presence of this large residual leads us to consider an alternative model which uses the Student's t distribution with 4 degrees of freedom (denoted as a t_4 distribution hereafter) as the likelihood.[7] Compared to the normal distri-

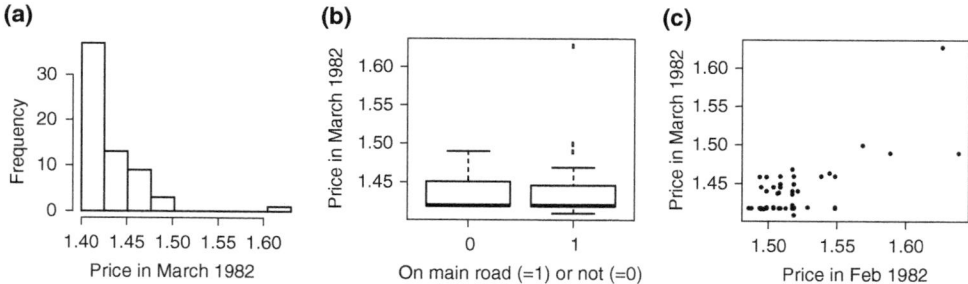

(a) **(b)** **(c)**

FIGURE 11.4
Some exploratory plots of the petrol price data.

---

[7] The higher the degrees of freedom, the closer the t distribution is to the normal distribution. A t_{30} distribution is virtually identical to the corresponding normal distribution. Here we fix the degrees of freedom to 4, but if more extreme values (i.e. outliers) are present, one can estimate the degrees of freedom from the observed data in the Bayesian approach (see for example Gelman and Hill, 2007, p.372).

TABLE 11.4

Descriptions of the Three Spatial Econometric Models Applied to the Sheffield Petrol Price Data

Model	Specification
SLM	$Y_i = \alpha + \delta \sum_{j=1}^{N} w_{ij}^* Y_j + x_{i,main} \beta_{main} + x_{i,prev} \beta_{prev} + e_i$
SEM	$Y_i = \alpha + x_{i,main} \beta_{main} + x_{i,prev} \beta_{prev} + u_i$ and $u_i = \delta \sum_{j=1}^{N} w_{ij}^* u_j + e_i$
SDM	$Y_i = \alpha + \delta \sum_{j=1}^{N} w_{ij}^* Y_j + x_{i,main} \beta_{main} + x_{i,prev} \beta_{prev} + \left(W^* x_{prev} \right)_i \gamma_{prev} + e_i$

In all models, Y_i denotes the outcome variable associated with the petrol price of station i in March 1982, and the vector of errors $e = (e_1, ..., e_N)$ are assumed to be independent; each e_i follows the same distribution $N(0, \sigma^2)$. W^* is the row-standardised spatial weights matrix based on the connectivity described in the map in Appendix. For the SDM model, $(W^* x_{prev})_i$ represents the average price of the neighbouring stations of i in February 1982.

TABLE 11.5

The DIC Comparison of the Three Spatial Econometric Models and the Posterior Estimates of the Spatial Autocorrelation Parameter, δ.

Model	$\bar{D}$	pD	DIC	Posterior Mean and 95% CI of δ
SLM	−302	5	−297	0.11 (−0.07, 0.27)
SEM	−303	5	−298	0.18 (−0.02, 0.38)
SDM	−302	6	−295	0.16 (−0.09, 0.36)

bution, the t_4 distribution has two heavier tails (Figure 11.6), implying that extreme values are more likely to occur under the t_4 distribution. Thus the t_4 distribution is more robust to outliers in the outcome values.

11.2.5 A Spatial Hierarchical Model with t_4 Likelihood

We implement this option as a spatial hierarchical model because currently there is no spatial econometric model that uses a t distribution as its likelihood. Eq. 11.1 describes this model, which is referred to as the t_4 model.

$$y_i \sim t_4 \left(\mu_i, \sigma^2 \right)$$
$$\mu_i = \alpha + x_{i,main} \beta_{main} + x_{i,prev} \beta_{prev} + S_i \tag{11.1}$$

This model assumes the petrol price of station i comes from a t_4 distribution with mean μ_i and variance σ^2. The regression relationship is specified so that the petrol price of station i depends on whether or not the station was on a main road ($x_{i,main} \beta_{prev}$), the price

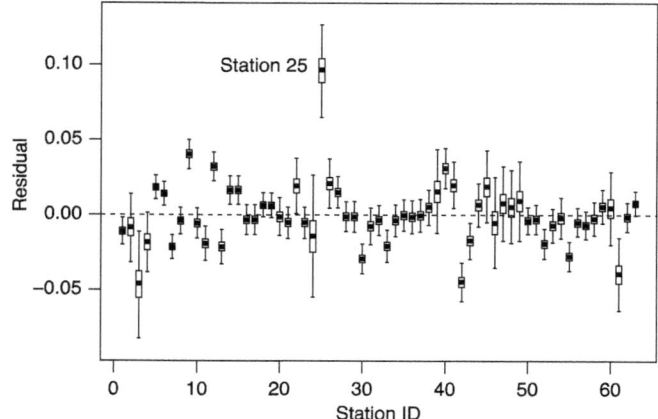

FIGURE 11.5
Boxplots representing the posterior distributions of the residuals from the SDM model. The residuals from the SLM and SEM models are similar to those presented here.

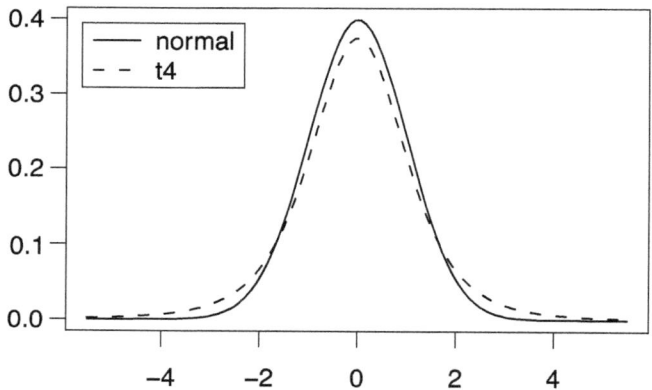

FIGURE 11.6
Comparison of the t_4 distribution with the standard normal distribution.

at the same station in the previous month ($x_{i,main}\beta_{prev}$) and a random effect term S_i. The set of station-specific random effect terms, $S = (S_1,...,S_N)$ accommodates between-station variability beyond that explained by the two covariates. This part of the variability may come from the effects of omitted covariates and/or price competition amongst the stations. However, unlike the SLM or the SDM model, this model does not allow us to say anything regarding whether a competition effect exists or not. The random effects S are modelled using the proper CAR model so that the spatial properties of these random effects can be assessed by estimating the spatial autocorrelation parameter ρ. Vague priors are chosen. Specifically, a normal distribution with mean 0 and a large variance of 1000000 is used as the prior distribution for the intercept α and the β coefficients. A normal distribution $N_{+\infty}(0,10)$ is assigned to σ – the subscript $+\infty$ means that this distribution is bounded below by 0 to ensure σ is positive. The prior to the spatial autocorrelation parameter ρ is *Uniform* (a,b), where a and b are calculated using the built-in functions in WinBUGS (see Section 8.3.1) based on the spatial weights matrix W as explained in Section 11.2.2. See Exercise 11.2 for the WinBUGS implementation of the t_4 model.

The DIC value of the t_4 model is -308 ($pD = 30$ and $\bar{D} = -338$), much smaller than the DIC values of the spatial econometric models (Table 11.5), suggesting that the t_4 model is the best amongst the models that we have considered. Although the t_4 model is a more complex model (with a larger pD value), it yields a much better fit to the data, as its $\bar{D}$ value is smaller than those of the spatial econometric models. Readers are encouraged to ponder the reasons for a better fit from the t_4 model.

Table 11.6 presents the posterior estimates of the regression coefficients. The results suggest that the petrol prices of the stations that are on main roads are no different from those of the stations that are not on main roads. The prices in the current month are shown to be correlated with the prices set in the previous month. The positive posterior mean of β_{prev} is indicative of price continuity across the two consecutive months – a station that set a high price in February tended to set the price high in March. This may reflect differences in the underlying fixed costs each petrol retailer faces. The above two findings are consistent with what was observed in the exploratory analysis of the data, but of course the modelling allows us to quantify the effects and assess how certain (or, equivalently, uncertain) we are regarding the effect of each covariate. For the spatial autocorrelation parameter in the proper CAR model, although the posterior mean is positive, the estimate is highly uncertain, with its 95% CI spanning between -0.435 and 0.786. This result suggests that the random effects are not spatially autocorrelated. However, we should interpret this finding with care because, as we stated earlier, this model does not explicitly model any price competition amongst the stations. Weak or no spatial autocorrelation in these random effects does not imply an absence of price competition.

Finally, we examine the standardised residuals from the t_4 model in order to assess model adequacy. The posterior means of the standardised residuals have a global Moran's I statistic of 0.133 and p-value of 0.07 (from 999 random permutations), suggesting some, albeit weak, evidence of spatial autocorrelation amongst the residuals. The inclusion of the two covariates with a set of random effects may not be sufficient to account for the spatial dependence structure in the observed data, a limitation of this model. Assigning the ICAR model on the random effects would better account for the spatial structure, but then we would already have assumed that the random effects had a spatial structure *a priori*. Other modifications may be worth considering, e.g. using a distance-based definition for the W matrix. The possibility of price competition happening *locally* rather than globally (the latter being assumed and investigated by all the models considered here) may also be of interest.

TABLE 11.6

Summary of the Posterior Estimates (Posterior Mean and 95% Credible Interval) from the t_4 Model Fitted to the Sheffield Petrol Price Data

β_{main}	0.002 (–0.016, 0.019)
β_{prev}	0.576 (0.362, 0.813)
ρ	0.287 (–0.435, 0.786)

11.2.6 Conclusion and Discussion

In this second application, we have presented an analysis to investigate the spatial dependence structure in petrol prices amongst retail outlets in Sheffield. We are unable to arrive at any firm conclusion about a price competition effect. The modelling presented here is by no means complete – quite the opposite. However, using this analysis, we have highlighted some issues that are specific to the spatial econometric approach and to the hierarchical modelling approach. Spatial econometric models are capable of modelling spatial dependence in the outcome. However, the use of a multivariate normal distribution as the likelihood makes these models more likely to be affected by outliers (the outlying high price from station 25). By contrast, the hierarchical modelling approach allows us to construct models that are more robust against the presence of outliers. The t_4 model gives a better fit to the data, although it may still not be able to capture the spatial dependence structure in the data well. Another issue is that, without the SAR structure on the outcome variables, the t_4 model does not allow for effect propagation and spatial feedbacks. Conceptually, the evidence supporting the existence of local price competition implies that price changes at any one site will ripple across the set of interconnected sites (Section 4.9). These are distinctive features of spatial econometric modelling but are lacking in hierarchical modelling (Section 10.5.2).

Haining (1983) collected data over a three-month period, from January to March in 1982. One possibility to strengthen the study is to carry out a space-time analysis of the price data. A more complex space-time dependence structure could be specified where the price at any station on a given day in a month may depend on a number of spatial and/or temporal lags such as: (i) the prices set at the neighbouring stations in the same month (a spatial lag); (ii) the price at the same station but during the previous month (or a number of previous months; a number of temporal lags); (iii) prices at the neighbouring stations but over the previous month(s) (a number of spatial-temporal lags).[8] Elhorst (2010, p.24–26) provides a concise summary of spatial-temporal modelling in spatial econometrics. With these issues in mind, in Chapter 16 we will describe a Bayesian hierarchical spatial-temporal model for disease data that includes time-lagged covariates that provide a basis for informing and supporting prevention strategies in a timely fashion.

11.3 Final Remarks on Spatial Econometric Modelling of Spatial Data

We conclude this chapter by drawing attention to some of the issues that arise when applying spatial econometric modelling. First, the SEM can be fitted as part of a "coping strategy" in the event that the residuals from some standard regression model are found to be spatially autocorrelated, and it is reasonable to assume that there are spatially autocorrelated missing covariates. The aim in fitting the SEM in this case is to

---

[8] However, in the case of these data, given the likely speed with which prices in a highly competitive market (such as petrol retailing) adjust to some price disturbance, time lags are likely to be measured in days, perhaps even hours, rather than weeks, so much would depend on being able to access price data at a much greater temporal frequency than considered here.

satisfy modelling assumptions in the hope of strengthening inference on those observable covariates that are in the model. But as we have noted in other contexts, there remains a risk of spatial confounding (see Section 4.9.3). Second, all spatial econometric models give clues about possible spatial interaction effects, but whatever models are fitted, they must have some grounding in substantive theory or substantive understanding of what is driving outcomes. The analyst should resist the temptation to engage in a statistical "fishing expedition" using the set of econometric models to find the model that fits best. Third, and linked to the above, because of non-identifiability concerns (see Gibbons and Overman, 2012 and also Section 10.5.1), the analyst should resist associating interaction effects with the mechanism implied by any particular spatial econometric model merely on the basis of the model providing the best statistical goodness of fit to the data. Fourth, the analyst should resist making claims that any apparent association is causal.

In both the applications considered here, aimed at evaluating spillover and feedback effects, we can make a case for setting aside the temporal aspects of the underlying processes driving outcomes. In the voting study we have data on the aggregation of a myriad of processes that have culminated in the decision, on the part of individual voters, to vote either to leave or to remain. There is an *end state* that is realised on the voting day. We can argue a similar case when analysing the petrol price data if we accept the idea that markets adapt very quickly to achieve a *competitive equilibrium* and we can control for the underlying costs each retailer faces.

In Part III of the book we look at how analysing spatial-temporal data, rather than purely spatial data, may help us to better understand interaction effects, thereby providing insights into how outcomes at one spatial location can effect outcomes at other locations.

11.4 Exercises

Exercise 11.1 For the Brexit application, extend the SEM model to include two additional LAD-level covariates: (a) the change in EU migrant populations and (b) the difference between a district's EU migrant proportion and the average proportion of its spatial neighbours. Comment on the posterior estimates of the two covariates. The data on these two covariates can be found on the book's website.

Exercise 11.2 Implement the t_4 model in Eq. 11.1 for the petrol price data and evaluate the model fit by calculating the standardised residuals. Why does the t_4 model provide a better fit to the data than the spatial econometric models considered in Section 11.2? (Hint: the WinBUGS syntax for the t distribution with 4 degrees of freedom is dt(mu,prec,4), where the first argument is the mean, the second argument relates to the variance and the third argument specifies the degrees of freedom. The variance is $k\sigma^2/(k-2)$, where k is the degree of freedom and σ^2 is the scaled variance (=1/prec, the inverse of the second argument in the dt function). Thus, with k = 4, the standardised residuals are defined as $r_i = (y_i - \mu_i)/(\sqrt{2}\sigma)$.)

Exercise 11.3 To what extent do the circumstances of the petrol price study meet the requirements of a natural experiment?

Appendix: Petrol Retail Price Data

Station ID	Price in March 1982 (in £)	On Main Road (=1) or Not (=0)	Price in February 1982	IDs of the Neighbouring Stations
1	1.419	1	1.546	2
2	1.419	1	1.546	1 3
3	1.490	1	1.640	2 4
4	1.419	1	1.559	3 5
5	1.450	1	1.699	4 6
6	1.446	1	1.546	5 7
7	1.419	1	1.560	6 8
8	1.437	1	1.560	7 9
9	1.459	1	1.545	8 10 11
10	1.420	1	1.550	9 11
11	1.419	1	1.550	9 10
12	1.469	1	1.596	13
13	1.423	1	1.578	12
14	1.460	1	1.590	15
15	1.460	1	1.590	14
16	1.420	1	1.550	17
17	1.420	1	1.550	16
18	1.439	1	1.569	19
19	1.438	1	1.548	18
20	1.419	1	1.546	49
21	1.420	1	1.550	22 23
22	1.500	1	1.570	21 23
23	1.420	1	1.550	21 22
24	1.464	1	1.578	25
25	1.628	1	1.664	24 26
26	1.460	1	1.587	25 27 38
27	1.441	1	1.560	26 28
28	1.419	1	1.569	27 29 32 33
29	1.418	1	1.560	28 32 33
30	1.410	1	1.570	31 35
31	1.419	1	1.559	30
32	1.419	1	1.580	28 29 33
33	1.419	1	1.580	28 29 32
34	1.420	1	1.580	35 37
35	1.419	1	1.560	30 34 36 37
36	1.419	1	1.560	35 37
37	1.419	1	1.560	34 35 36
38	1.450	1	1.600	26
39	1.460	0	1.630	61
40	1.459	1	1.569	41
41	1.446	1	1.546	40
42	1.419	1	1.578	43 44

(Continued)

Station ID	Price in March 1982 (in £)	On Main Road (=1) or Not (=0)	Price in February 1982	IDs of the Neighbouring Stations
43	1.419	1	1.550	42 44
44	1.441	1	1.591	42 43
45	1.446	1	1.569	46
46	1.490	0	1.640	45
47	1.419	0	1.550	48 61
48	1.420	0	1.550	47 61
49	1.419	0	1.539	20 59
50	1.418	1	1.544	51 60
51	1.419	1	1.545	50 52
52	1.420	1	1.560	51 53
53	1.419	1	1.555	52 54
54	1.460	1	1.578	53 55
55	1.420	1	1.580	54 56
56	1.420	1	1.550	55 57
57	1.420	1	1.555	56
58	1.441	1	1.555	59 62 63
59	1.419	1	1.546	49 58
60	1.441	0	1.596	50
61	1.420	0	1.570	39 47 48
62	1.419	1	1.546	58
63	1.439	1	1.555	58

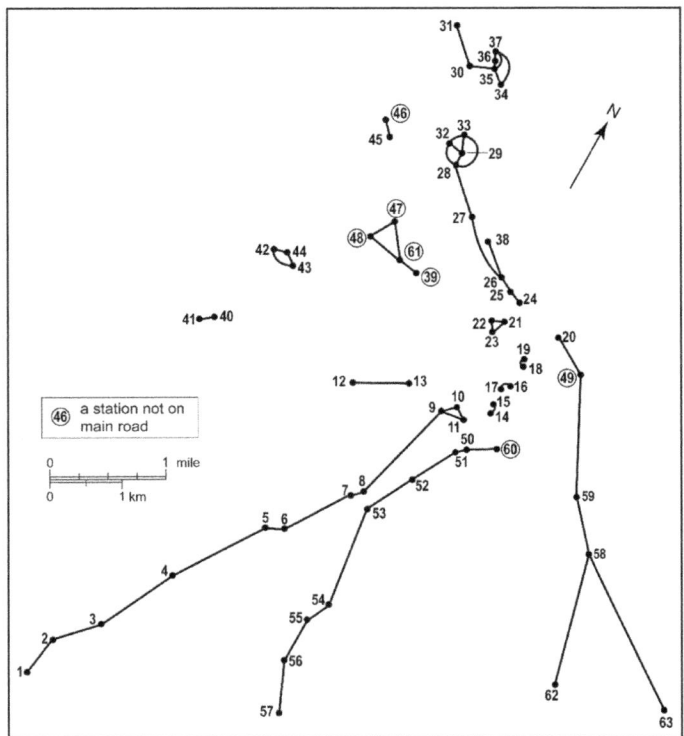

Connectivity map linking the 63 retail petrol retail sites in Sheffield, England in March 1982. The connecting lines indicate which sites are treated as neighbours ($w_{ij} > 0$). All sites are on main roads leading into/out of the city except those circled. For comparison with the "real" road map for Sheffield, see Figure 1.11.

Part III

Modelling Spatial-Temporal Data

12

Modelling Spatial-Temporal Data: An Introduction

12.1 Introduction

Modelling purely-spatial (cross-sectional) data, as we have seen in Part II, provides insights into the spatial *distribution* of some event (outcome) for a single time frame (a week, a month, a year). Our examples have included modelling the distribution of annual burglary rates across the 157 wards in Cambridgeshire (Chapter 6); the distribution of average annual income levels per household for the 109 MSOAs in Newcastle (Chapters 7 and 8). We can also model purely spatial data to examine the "here and now" *relationship* between the distribution of the outcome and a set of covariates. This sort of analysis can be useful if the underlying processes generating the outcome can be considered to be in temporal "equilibrium" (e.g. certain diseases that are linked to persistent environmental conditions; economic outcomes, such as prices or per capita income levels, which are in some form of market equilibrium). It can also be useful if the process has culminated in a final outcome, as in the case of the 2016 Brexit referendum in the UK or a Presidential election in the US – or indeed any election, or other event, where there is a "final state".[1] The geo-reference attached to each outcome and covariate value is important and, as we have seen, provides information which we can use in various ways to model the observed variation in the outcome.

In Part III, we turn our attention to the situation where data are available over both space and time for a set of areas or sites. Adding the time dimension to a spatial analysis gives us a dynamic view of the system. This dynamic view enables us to investigate the temporal evolution not only of the study region as a whole but also of each of the small areas which partition it. The availability of spatial-temporal data means we can ask questions such as: "how does the crime rate associated with a crime hotspot vary over time?"; "is the crime hotspot stable, both spatially and temporally, during the period of observation"; and "are areas that are not crime hotspots showing evidence of a rising level of crime (emerging hotspots)?" So, we can ask questions about the spatial-temporal *distribution* of an outcome over the set of small areas in order to assess, for example, how stable that distribution is over time and its *temporal trajectory*. This is important in several ways. It allows us to make inferences about the *effects* of a policy intervention, including geographically targeted policies (e.g. a crime reduction programme targeted at selected areas – has it been effective?). It also allows us to make inferences about possible changes taking place with a view to

---

[1] The earliest applications of spatial statistical modelling, using models that are of the spatial econometric type, placing spatial dependency in the data likelihood, were in analysing crop yields (see Whittle, 1954, Mead, 1967). At the geographical scale of counties, Haining (1978) used the same class of models to analyse crop yields for an area of the High Plains, US. Crop yields at the point of harvest constitute a "final state".

informing an intervention (e.g. early detection of rising numbers of cases of a disease in certain areas – when should we intervene and commit resources to tackling what might be an emerging problem?)

Other possibilities also open up. We are no longer constrained to argue that outcomes are in some "equilibrium" or "stable" state or have reached some "final state". We can study the *relationship* between outcomes observed over a set of space-time observations in which the outcome at location i, time t is modelled as a function of covariates, possibly including the outcome itself, at location i at time $t - 1$ (and further back in time if necessary), as well as covariate and outcome values at other locations at time $t - 1$. This becomes possible if observations over time are taken with a frequency that is sufficiently often in relation to the response times in the system that we are studying. Depending on the field of research, this might require spatial data to be collected very frequently. If modelling the number of new cases of malaria by small area as a function of rainfall levels is of interest, we might need to have rainfall data for quite short intervals of time (previous seven days? 14 days?) depending on our understanding of how quickly new or enlarged spaces where mosquitoes can breed translate into higher levels of risk. No longer modelling an outcome as a function of outcomes at other locations *at the same point in time*, as we were doing in Chapter 10 when testing for spillover and feedback effects, means we can avoid some of the difficulties we encountered when undertaking spatial econometric modelling (see Section 10.5) whilst also opening up the possibility of fitting hierarchical models (as well as econometric models) to tackle such substantive questions. The possibility of establishing a relationship as causal is enhanced, relative to an analysis of purely spatial data, because one of the preconditions for establishing a causal relationship is to be able to demonstrate temporal precedence (cause, A, comes before effect, B) – but there are other issues to consider (see Section 3.2).

At the same time, however, we are confronted by a number of additional challenges to those discussed in Chapter 1 when we move from space-only to spatial-temporal data modelling. The first challenge is the increase in data: from a vector of size N in a space-only dataset for a single set of outcome values, we now have to work with a matrix of outcome values of size $N \times T$ (T time periods) in the case of a space-time dataset. This results in a huge increase in the number of model parameters to be estimated. Just to place these dimensions into context, there are 6791 MSOAs in England (according to the 2011 UK census), and thus a space-time analysis estimating MSOA level average income over four years would contain over 27,000 space-time units. There has been a dramatic increase in the number of parameters, and estimating these parameters, particularly when confronting issues of data sparsity, requires specialised computer-intensive algorithms and software.

The second challenge lies in measuring population, as in measuring the size of the population at risk in the case of an epidemiological study. In a space-only setting, population is static, whereas in a space-time setting, population is dynamic. It is exactly because space-time data enables us to capture dynamics that makes their analysis exciting, but as a consequence, population dynamics need to be considered. For example, local populations change over time not only because of births and deaths but also because of population movements. For example, any subnational population forecasting model needs to include a migration component (Bijak, 2010). In practice, the challenge is made all the greater if data on migration and population turnover within a country are not available or are only available at very coarse spatial-temporal scales.

The third challenge relates to the issue of defining the space-time unit for analysis. In some applications, the unit of analysis is determined by the research question, for example:

"how did district level burglary rates change annually between 2010 and 2014?" or "what was average income per household at the MSOA level in 2014?" The challenge becomes a methodological one, namely, to decide how models can help us tackle these questions at the chosen space-time level. For pattern discovery, however, there is no pre-defined space-time framework. Furthermore, different granularities at the space-time level are likely to result in different space-time patterns being revealed. For example, the dynamics of burglary will depend on whether we analyse district-month data or street-annual data. This is the space-time extension of the modifiable areal unit problem described in Section 3.2.3. Ye and Rey (2013) propose a systematic framework for exploring space-time dynamics and illustrate it using income distribution data for the US and China. Different tasks can be investigated at different space-time data (dis)aggregation levels.

The fourth challenge is to capture the complexity and the inter-correlated nature of space-time observations via statistical models. In addition to spatial dependence and spatial heterogeneity (Chapter 1), we now have to consider temporal dependence and temporal heterogeneity and their spatial-temporal equivalents. Events may cluster in particular places (space-only analysis) or at particular times in a place (time-only analysis). But we also need to consider the existence of clusters of events in time and space: detecting clusters in particular places during particular time periods. One of the biggest challenges we shall need to consider is how to model space-time interaction in which during the study period patterns (such as the existence of clusters) shift in space *and* time. In Chapter 2 we described five spatial-temporal interaction processes (see Section 2.4.1) that may lie behind such space-time patterns. We shall develop the statistical theory in Chapters 14 and 15. As we shall see in the applications presented in Chapter 16, statistical modelling of space-time interaction plays an important role in policy evaluation and in surveillance methodology, where a common goal is to detect extreme (or "unusual") clusters of events occurring in particular places at particular times.

Last, but by no means least, we have the challenge of how to store and visualise results from a spatial-temporal analysis. We shall touch on this problem in Chapter 13 when we consider exploratory methods for spatial-temporal data and in later chapters when we present results.

To tackle any of these questions, a modelling framework is needed, and that is what we shall begin to describe in this chapter and develop further in Chapters 14 and 15. In this chapter, we describe the frameworks proposed by Knorr-Held (2000) for modelling spatial-temporal data. His space-time inseparable framework partitions the space-time variation in the observed outcome values into three components or models: a spatial model that captures an overall spatial distribution; a temporal model that captures an overall temporal pattern; and a space-time interaction model that accommodates space-time interaction. This separation into the three components allows us to build a complex spatial-temporal model in a "modular-fashion", thereby simplifying the modelling task, which may start by fitting a space-time separable model that comprises just the first two components.

As we shall see, this modelling framework is formulated via the Bayesian hierarchical modelling approach so that spatial, temporal and spatial-temporal dependence structures in the observed outcome values are dealt with in the process model. As a result of taking this approach, the spatial models that we have seen in Part II of the book (Chapter 8) can be directly used to model overall spatial variation. What is left is to specify temporal models for describing temporal patterns and models for space-time interaction. A fully-specified spatial-temporal model also requires us to define a data model (i.e. the likelihood for the observed outcome data) and a parameter model (i.e. the prior distributions for the hyperparameters, the regression coefficients and the parameters in the data model), but the reasoning behind the specification of these two components is the same as that employed in modelling a set of purely spatial data.

This chapter is structured as follows. In Section 12.2, we discuss the space-time modelling frameworks used in this part of the book and outline the three components described above. The discussion is placed within an application of modelling a set of small area annual burglary count data from Peterborough, England between 2005 and 2008. In Section 12.3, we describe a set of models for describing overall temporal variation, paying particular attention to dealing with temporal dependence and temporal heterogeneity and addressing the issues of data sparsity and uncertainty. As we shall see, the idea of information borrowing is central. Amongst the temporal models that we describe, some, for example the linear time trend model and the random walk model of order 1, are widely used in modelling time-series data. However, these (perhaps) familiar time-series models are now applied in the context of the hierarchical modelling of spatial-temporal data, which has implications not only for parameter estimation but also for model specification. We also describe the interrupted time series model (important in policy evaluation) and show how to fit these various temporal models in WinBUGS.

12.2 Modelling Annual Counts of Burglary Cases at the Small Area Level: A Motivating Example and Frameworks for Modelling Spatial-Temporal Data

Spatial-temporal data on offence counts derived from police recorded crime databases provide us with the opportunity to understand the spatial-temporal characteristics of the risk of experiencing a crime event of a particular type. Modelling such spatial-temporal crime data is important because not only does it help to reveal "unusual" patterns in the spatial-temporal distribution of crime risk (e.g. detection of crime hotspots and coldspots) with implications for policing, but also allows us to identify trends in crime patterns and factors that can help us to explain the spatial-temporal distribution of crime events as observed in the data.

Figure 12.1 shows a sequence of maps of the raw annual burglary rates (per 1000 houses) across 452 Census Output Areas (COAs) in Peterborough, England over the four-year period from 2005 to 2008. Table 12.1 summarises the burglary data. As suggested by both the map sequence and the summary table, burglary rates vary over both space *and* time. Within the same year, different COAs show different burglary rates. The burglary rates of a COA can vary from one year to another, whilst the temporal pattern of burglary rates can differ from one COA to another. Describing and modelling these spatial-temporal patterns may seem, at first sight, to present us with insurmountable challenges.

However, a careful inspection of the four maps in Figure 12.1 reveals *an overall spatial pattern* in which COAs in the southwest of Peterborough, those in the centre of Peterborough and small pockets of COAs in the west and northeast tend to show higher burglary rates than other COAs in Peterborough. The burglary rates in those COAs tend to be consistently higher than those in the rest of Peterborough across all four years. Now, if we scan through the four maps in Figure 12.1 over time, there appears to be more COAs in dark

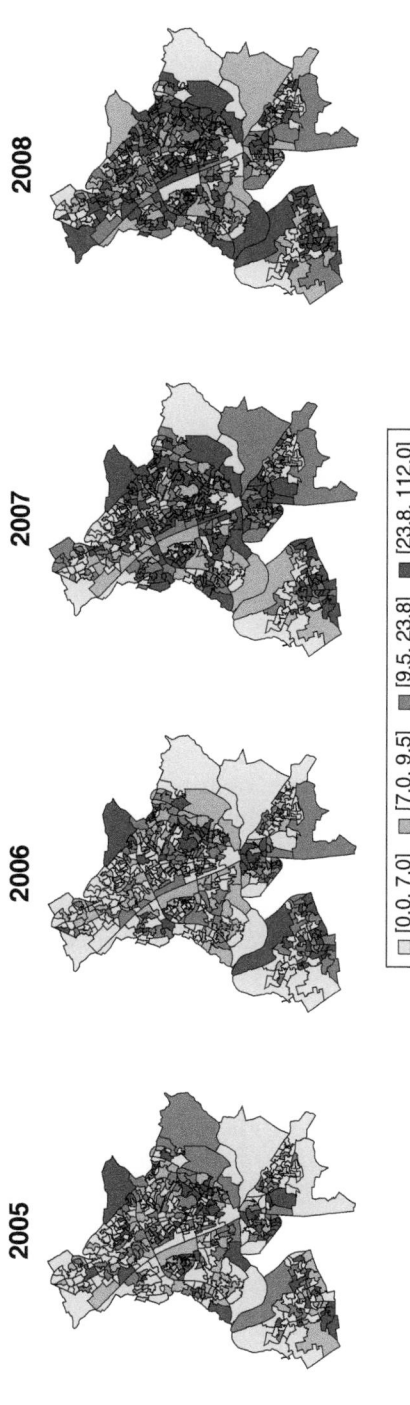

FIGURE 12.1
Maps of the raw annual burglary rates (per 1000 houses) across the 452 Census Output Areas (COAs) in Peterborough, England from 2005 to 2008. The raw burglary rate per 1000 houses for each COA in each year is calculated by multiplying the ratio of the observed number of burglary cases to the number of at-risk houses in that COA in that year by 1000. The legend below the maps applies to all years.

TABLE 12.1

Summary of the Burglary Data in Peterborough from 2005 to 2008

	2005	**2006**	**2007**	**2008**
Number of COA-level burglary cases observed	1.37 (2.00; 0.00; 8.00)	1.43 (2.00; 0.00; 9.00)	2.68 (3.00; 0.00; 12.00)	2.82 (3.00; 0.00; 16.00)
Number of COA-level at-risk houses		130 (16; 78; 239)		
Raw annual burglary rates (per 1000 houses) at the COA-level	10.5 (16.1; 0.0; 58.0)	11.0 (16.4; 0.0; 75.0)	20.6 (22.2; 0.0; 96.0)	21.8 (24.0; 0.0; 112.0)

Each cell reports the mean value, and the three numbers in brackets are the interquartile range, the minimum and the maximum, respectively.

grey in the 2008 map compared to the 2005 map. This implies *an overall increasing time trend* in the burglary rate for Peterborough as a whole. This overall increasing time trend is also evident in the mean values of the COA-level annual burglary rates in Table 12.1. If we now bring together the above two observations, we can say that the spatial-temporal variability in the small-area annual burglary rates can be expressed as the sum of an overall spatial pattern *and* an overall temporal pattern. This describes a so-called *space-time separable* structure, a modelling framework for spatial-temporal data introduced in Knorr-Held and Besag (1998) and Knorr-Held (2000).

However, variability in a set of spatial-temporal data often cannot be fully described by such a (overall spatial + overall temporal) structure – and further close inspection of our Peterborough crime dataset shows this to be the case. In the Peterborough data, we can see that not all the COAs follow the Peterborough *time* trend. As shown in Figure 12.2, some COAs display a steeper increasing trend over time compared to the Peterborough trend, whilst the trends of some COAs may either remain relatively stable (or steady) or even decrease. Similarly, the *spatial* distribution of burglary rates for a particular year may display some departures from the overall spatial pattern, i.e., the annual burglary rates of a COA whose average burglary rate per year is high (low) may not always stay high (low) across the four years. Figure 12.3 illustrates some of these COAs. What the above examples illustrate is a form of space-time variability termed *space-time interaction*, which, essentially, is the variability in the observed data that is *not* captured by the space-time separable structure. In terms of modelling, the presence of space-time interaction requires a *space-time inseparable* structure (Knorr-Held, 2000) to allow for *space-time inseparability*. Figure 12.4 illustrates the structures of the two modelling frameworks schematically. The key difference between the two is whether or not a component is included to allow for space-time interaction. Therefore, a space-time inseparable model can always be simplified to a space-time separable model if space-time interactions are found to be absent. Likewise, a space-time separable model can be extended to allow for space-time inseparability when space-time interactions are detected. To inform the choice between these two modelling frameworks, Chapter 13 will discuss various exploratory techniques for examining the presence (or absence) of space-time interaction. Throughout model building, model comparison and model evaluation are carried out continuously in order to help refine the specification of the components in the chosen framework. We will discuss this in detail in Chapters 14 and 15. Before we can construct a spatial-temporal model, we first need to consider models for describing the overall temporal pattern as well as for capturing the temporal variability in space-time interactions. We will discuss these temporal models next.

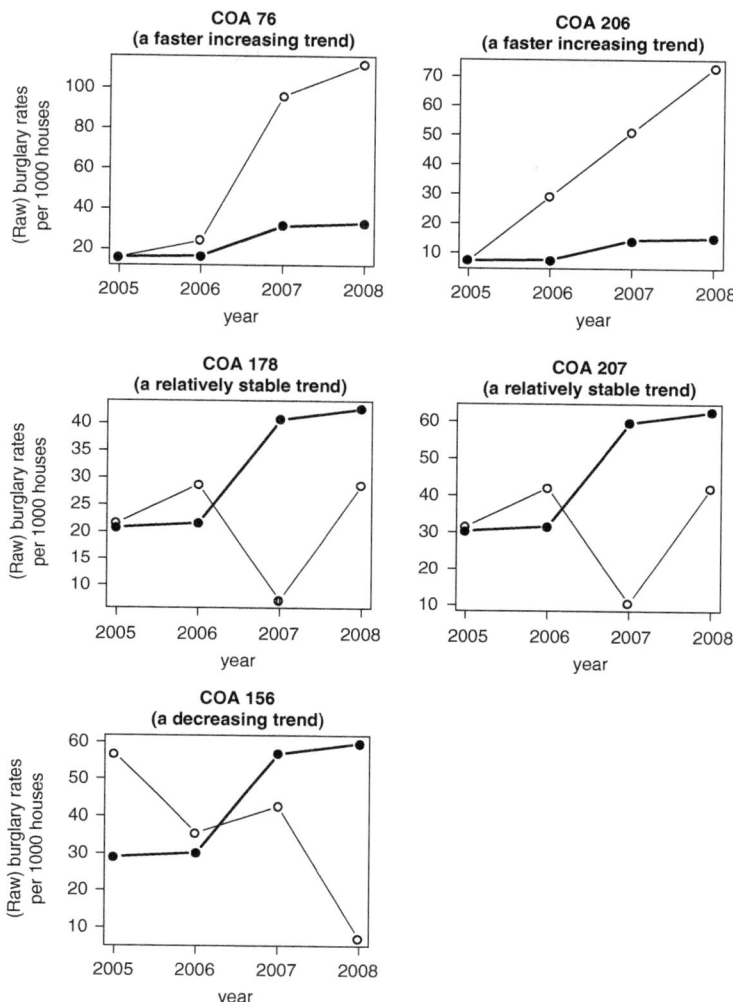

FIGURE 12.2
Examples of COA-level trends that differ from the Peterborough trend. In each panel, the thinner line with open circles shows the raw burglary rates per 1000 houses (number of burglary cases in each year divided by the number of at-risk houses and multiplied by 1000) of the selected COA. The thicker line with dots represents the burglary rates of that COA if it were to follow the overall Peterborough trend.

12.3 Modelling Small Area Temporal Data

This section focuses on modelling time patterns in spatial-temporal data. We describe three temporal models: linear trend models, random walk models and interrupted time series models. These three models have been used widely in modelling time-series data, and their specifications may be familiar to some readers. However, when using these models in the small-area context, there are issues that we need to pay particular attention to. We will first discuss these issues before turning to the specifications of each of the three temporal models.

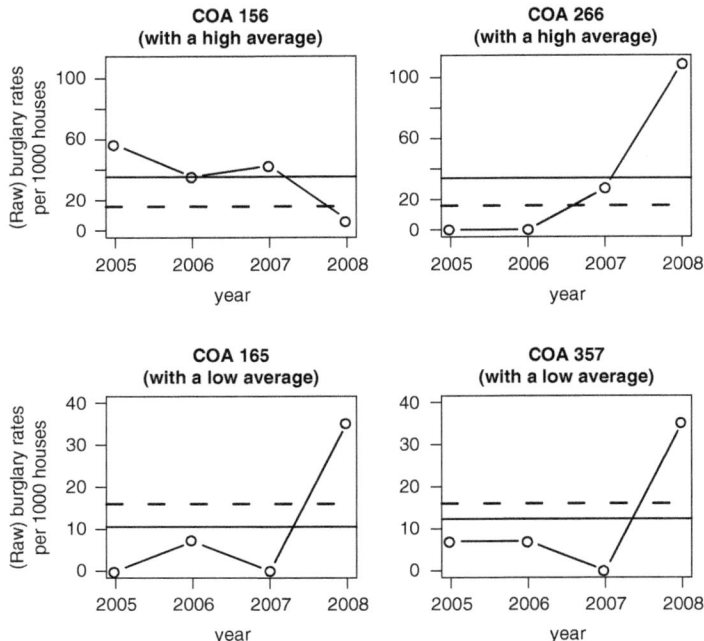

FIGURE 12.3
Annual burglary rates of a COA with a high (low) average may not stay high (low) across all four years. The average burglary rates (the solid horizontal lines) of both COAs 156 and 266 are higher than the Peterborough average (the dashed line; calculated via $\frac{O}{A\times4}\times1000$, where O is the total number of cases across all COAs over four years, A is the total number of houses in Peterborough), but the annual burglary rates of these two COAs clearly are not above the Peterborough average across all four years. Moreover, they display two quite different time trends. In the case of COAs 165 and 357, their average burglary rates per year are lower than the Peterborough average. However, there appears to be a sudden increase in 2008 for both COAs.

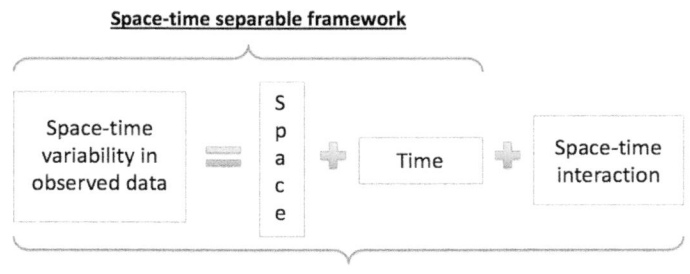

FIGURE 12.4
A schematic diagram presenting the structures of the space-time separable framework and the space-time inseparable framework proposed by Knorr-Held (2000).

12.3.1 Issues to Consider When Modelling Temporal Patterns in the Small Area Setting

12.3.1.1 *Issues Relating to Temporal Dependence*

As discussed in Chapter 3, dependency (or lack of independence) is a fundamental property of spatial and spatial-temporal data. Just as in the spatial setting, data values observed over time tend to be positively autocorrelated: values observed between consecutive points in time tend to be more alike compared to values observed at two time points that are further apart. Take the time series of burglary rates in Figure 12.2 and Figure 12.3 as an example. For a given COA, the burglary rates in any two consecutive years are generally closer together compared to the burglary rates that are, say, three years apart, e.g. between 2005 and 2008.[2] The presence of temporal dependency has two implications for modelling. First, any temporal model needs to incorporate temporal dependence. For a Bayesian hierarchical model, temporal dependence is dealt with in the *process model* by introducing a set of time-point-specific parameters. We have discussed some benefits of this approach in the purely spatial case (see Section 10.5.2), and these benefits carry over to the temporal case. Second, the presence of positive temporal autocorrelation in the observed values provides a basis for information sharing when estimating the parameters associated with each time point, which we now describe.

When estimating the levels of annual burglary risk for each COA, the number of burglary cases observed at time point t may give us some information about how many burglary cases occur, and hence the underlying burglary risks, at time point $t - 1$ or $t + 1$. In other words, analogous to the spatial setting, when estimating the risk of burglary at one time point we can borrow information from the data observed in its *temporal neighbours*. It is important to note that "temporal neighbours" of a time point t include time points that are both before *and* after t, given that both $t - 1$ and $t + 1$ are time points within the observation period – see Section 12.3.4 for a full discussion of specifying temporal neighbours.

In addition to sharing information over time (as discussed above) and over space (as discussed in Chapters 7 and 8), when modelling spatial-temporal data, information can be shared across both space *and* time. To illustrate, consider the following question: "does the occurrence of six and two burglary cases in two consecutive years, t and $t + 1$ respectively, in a COA of 100 houses indicate a real decrease in burglary risk in this area?" Given the data, a reduction of four cases from six to two does not imply a real change in burglary risk from t to $t + 1$ because such a difference in the observed case numbers can be accounted for solely by inherent Poisson variability. In other words, there is a high level of uncertainty associated with the estimated level of burglary risk in each year, and if we calculate the 95% credible intervals for both years, they overlap (see Exercise 12.1). Suppose longer time series data are available for this COA and the data show that the observed numbers of burglary cases a few years prior to year t ranged between five and ten, while the observed numbers of cases a few years after $t + 1$ ranged between zero and three. This temporal information reduces the uncertainty associated with the risk estimates before (and including) t and after (and including) $t + 1$. Thus, the claim of a reduction in the underlying annual burglary risks between t and $t + 1$ is strengthened as a result of borrowing information over time. When similar temporal patterns in the counts of burglary

---

[2] When the time sequence of data displays a "periodic" pattern, for example, burglary rates tend to be high during the summer months but low in the winter months, then data values observed over some regular intervals (say, every 12 months) tend to be similar. But values observed at time points that are close together still tend to be more alike.

cases are observed in neighbouring COAs (or in COAs with similar characteristics), borrowing information spatially from these COAs can further reduce the uncertainty about the underlying time trend in the risk of burglary in this COA, thus further strengthening inference. Again, the assumption of similarity is central to information sharing. However, care must be taken in the validation of this assumption. We will see more detail when we discuss the policy evaluation application in Section 16.2 in Chapter 16 where the process just described is placed within an interrupted time series (ITS) model, which then becomes crucial to the estimation of the policy's impacts at the small-area level.

12.3.1.2 Issues Relating to Temporal Heterogeneity and Spatial Heterogeneity in Modelling Small Area Temporal Patterns

As we have seen in the Peterborough data, the raw burglary rates for any COA vary from one year to another, showing temporal heterogeneity in the mean (i.e. the risk level). For data observed over a short period of time (say, over two to four years), such year-on-year differences in the burglary rate can often be adequately described by a simple linear function of time (a linear trend model). However, in the small-area context, different areas may display different temporal patterns (see, for example, Figure 12.2 and Figure 12.3). To accommodate such spatially heterogeneous time trend patterns, a linear trend model needs to be specified for each area. So, for N areas, there will be N linear trend models with a total of at least $2N$ unknown parameters (N intercepts plus N slopes) to estimate. With sparse data, the estimation of such a large number of parameters will suffer, as the estimation of each parameter relies on (very) limited data (see Chapter 1 and Chapter 7 for the case when modelling spatial data). When the time patterns that we wish to describe are nonlinear, this estimation issue becomes more acute, as nonlinear models (e.g. a model with a quadratic or cubic function of time or the random walk model) involve more parameters than the simple linear trend model. Therefore, information sharing becomes more important. As we have seen in Chapters 7 and 8 and will see again in Chapters 14 and 15, the Bayesian hierarchical modelling approach is a natural framework to achieve information sharing. When, for example, applying the linear trend model to the temporal data of each area, the set of area-specific intercepts and the set of area-specific slopes are each modelled hierarchically by using the spatial models that we have discussed in Chapters 7 and 8, allowing information to be shared globally, locally or a combination of both. The Bayesian paradigm enables us to consider different spatial models, allowing us to evaluate how different assumptions on information borrowing affect parameter estimation and modelling conclusions.

The need to estimate a large number of parameters in a spatial-temporal model also raises another question for time trend modelling: "how complex a temporal model do we want to consider?" – a question to which we now turn.

12.3.1.3 Issues Relating to Flexibility of a Temporal Model

There is an extensive suite of statistical models for analysing time-series data (see, for example, Box et al., 2015 and Prado and West, 2010). Here, we have chosen to focus on three temporal models: a linear time trend model, a random walk (RW) model and an interrupted time series (ITS) model. The reason for this focus is that these three temporal models, although few in number, are versatile enough to be applicable to a wide variety of situations including: the case of disease surveillance (Li et al., 2012) and crime surveillance (Li et al., 2014); joint modelling of spatial-temporal data on multiple crimes

(Quick et al., 2019); small-area mortality forecasting (Bennett et al., 2015 and Kontis et al., 2017) and for evaluating policing policy at the small area level (Li et al., 2013).

For a given application, the following two factors play a role in determining which temporal model to use: (a) the goal of the analysis and (b) whether or not the parameters of the chosen model can be estimated using the available data. For example, a RW model is a flexible model for describing small area time trend patterns that are nonlinear. In the context of policy evaluation (Li et al., 2013), the RW model of order 1 (denoted hereafter as the RW1 model) is used to describe the time trend pattern of burglary rates for the control areas (areas in which the policing policy under evaluation was not implemented) over a period of eight years. As shown in Figure 12.5, the time trend under study is clearly nonlinear. In Quick et al. (2019), the RW1 model is used to capture the temporal pattern that is common to four different crime types and the temporal patterns that are specific to each type. The estimated time trend pattern common to all crime types appears to deviate from linearity (see Figure 3 in Quick et al., 2019). An alternative flexible model that is related to the RW1 model is the autoregressive model of order 1 (denoted as the AR1 model hereafter). Compared to the RW1 model, an AR1 model has an additional unknown parameter that measures the strength of temporal (serial) autocorrelation (c.f. the spatial autocorrelation parameter ρ in the pCAR model in Section 8.3 in Chapter 8). This parameter is fixed at 1 in the RW1 model. If the aim is to estimate small-area time trends and the temporal autocorrelation parameter is not of interest, then the RW1 model, with one less parameter to estimate, will be preferable. However, this model choice depends on the assumption that the temporal parameters are positively autocorrelated (see Section 12.3.4). An AR1 model is more suitable if that assumption is not satisfied and/or if our interest lies in inferring the nature of the temporal autocorrelation structure (i.e. positive/negative autocorrelation) as well as the strength of autocorrelation.

For some applications, interpretability, rather than flexibility, is the more important feature of a temporal model. For example, for monitoring small area burglary risks, an inferential goal in Li et al. (2013) is to identify COAs that display an "unusual" local time trend that deviates from the overall trend. In that situation, a linear trend model

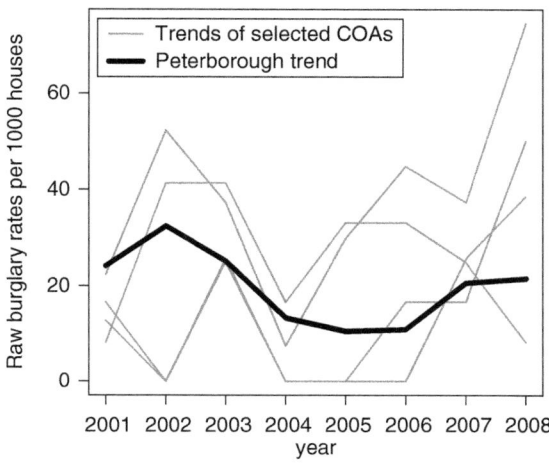

FIGURE 12.5
Temporal patterns of the raw burglary rates per 1000 houses for some COAs (the grey lines) and for the whole of Peterborough (the black line) from 2001 to 2008. Each raw burglary rate is calculated by dividing the number of burglary cases by the number of at-risk houses and multiplied by 1000.

is used as a temporal model because: (a) the study period is only four years (four time points), hence short, and the time trends under study are reasonably linear; (b) the comparison of a local trend to the overall/global trend can be easily translated into the comparison of the slope of a local trend to the slope of the global trend. Based on the slope comparison, an estimated local trend can then be interpreted as a "faster increasing trend", a "slower increasing/decreasing trend" or a trend that is similar to the overall trend, which, in the dataset considered in Li et al. (2013), is increasing. While flexible, the RW1 model, by contrast, does not yield the above straightforward interpretation. We will see detail about this modelling in Section 16.3 in Chapter 16. However, the linearity assumption may not be appropriate when modelling longer time series data with, say, 6 to 10 data points. In that case, the RW1 model is more appropriate, but the comparison of a local trend with the global trend becomes more complicated – as opposed to comparing two slope parameters, we need to compare two nonlinear trends. To do that, BaySTDetect, a space-time detection model proposed by Li et al. (2012), performs the trend comparison using a Bayesian model comparison approach. We defer the discussion of this method to Section 16.4 in Chapter 16. The method developed by Boulieri et al. (2018) deals with even longer small area time series data with 10 to 15 time points, and again we defer discussion of this method until Section 16.4 in Chapter 16. We will touch upon spline modelling for spatial-temporal data in Chapter 15 (see, for example, Section 15.7.1).

Interrupted time series (ITS) models (also known as segmented regression or regression discontinuity analysis) serve a different purpose from linear trend models and RW models (see for example O'Keeffe et al., 2014; Kontopantelis et al., 2015; Bernal et al., 2017). ITS models are designed to investigate abrupt changes in time series data due to one or more interventions (e.g. the introduction of new policies or new medical treatments) implemented in a quasi-experimental setting. In that setting, individuals are not allocated into the treatment group or the control group randomly (see Section 3.2.2 in Chapter 3). When assessing the effect of the intervention(s) on an outcome of interest (e.g. crime or disease risk), care needs to be taken. We will defer the discussion, as well as the specification of ITS models, till Section 12.3.5 and Section 16.2 in Chapter 16.

In the remainder of this chapter, we will look at the three temporal models, the linear trend models, the RW models and the ITS models, in more detail.

12.3.2 Modelling Small Area Temporal Patterns: Setting the Scene

The exploration of the three temporal models is placed in the context of the Peterborough burglary data. In this chapter, we only focus on the modelling of the temporal burglary counts either from some selected COAs or from Peterborough as a whole. Although the spatial configuration of the COAs does not enter the modelling here (but we shall return to it in subsequent chapters), we have selected these COAs in order to illustrate some of the issues that we have discussed in Section 12.3.1.

Under the Bayesian hierarchical modelling approach, as we have now seen many times, we need to specify three models: a *data model* for the observed outcome values, a *process model* to describe the underlying data generating mechanism and a *parameter model* that specifies a set of prior distributions for the unknown parameters in the data model and in the process model. We consider the case of a single time series, which could for example be burglary counts for the whole of Peterborough or some selected COA. In either case, the observed outcome values are $y_1,...,y_T$ the number of burglary cases reported in each of the T years. Since we are only modelling a time series, it is not necessary to index data values by location. We will deal with the full space-time dataset for all 452 COAs in subsequent chapters; then it will be necessary to index an outcome value by both t and i, namely, y_{it}. For modelling count data, a Poisson distribution is a suitable choice for the data model (the likelihood). Thus, each observed burglary count, y_t (for $t = 1,...,T$), is modelled as an observation from a Poisson random variable with mean μ_t, and we write:

$$y_t \sim Poisson(\mu_t)$$

$$\mu_t = n \cdot \theta_t$$

(12.1)

The Poisson mean μ_t is a product of n, the number of houses at risk in a COA, and θ_t, a parameter representing the burglary risk in year t. Here, the number of at-risk houses is obtained from the 2001 UK census and, because of data limitations, this number is assumed to be constant over the study period (see Section 12.1).

In this and subsequent chapters, we call θ_t the burglary rate parameter or simply the burglary rate at time t because under the Poisson model given in Eq. 12.1, θ_t measures the frequency with which a burglary event occurs in a defined population (either a COA or Peterborough as a whole) in a given year, t. Hereafter, we use risk and rate interchangeably. One way to obtain a set of point estimates for the burglary rate parameters, $\boldsymbol{\theta} = (\theta_1,...,\theta_T)$, is to calculate the *raw* burglary rates, the observed numbers of burglary cases divided by the size of the at-risk populations (i.e. $\frac{y_t}{n}$ for $t = 1,...,T$). We call them raw burglary rates because their calculation uses only data available at each time point with no information sharing involved. Whilst useful for exploring properties of the data (as we did in Section 12.2), at the COA-level the uncertainty associated with these raw rate estimates is likely to be very large (see for example Figure 1.6 and Section 12.3.1.1) because both the size of the at-risk population and the number of burglary cases observed in each year are small. Such sparse data also give rise to the small number problem that we discussed in Section 3.3.3. However, none of the above problems arise when analysing the aggregated data for the whole of Peterborough. The rest of this chapter therefore focuses on how we construct temporal models for these burglary rates so that we can invoke the idea of information sharing to improve parameter estimates.

To proceed, we consider a process model in the following form to model the burglary rates, $\boldsymbol{\theta}$:

$$\log(\theta_t) = \alpha + v_t$$

(12.2)

In Eq. 12.2, the log link function ensures θ_t, the burglary rate, is non-negative. The parameter α is the intercept term, and its interpretation depends on how $v = (v_1,...,v_T)$, a set of time-point-specific parameters, are specified. In what follows, we consider each of the three models, the linear trend model, the RW model and the ITS model, for v.

12.3.3 A Linear Time Trend Model

12.3.3.1 Model Formulations

A linear time trend model is the simplest of the three, and the resulting time trend estimate has a straightforward interpretation: an increasing trend, a decreasing trend or a trend that remains stable (or steady) over time. Eq. 12.3 specifies this temporal model. For $t = 1,...,T$,

$$v_t = d \cdot \left(t - t^* \right) \tag{12.3}$$

In Eq. 12.3, d is the slope parameter that measures the rate of change. If d is estimated to be positive (negative), the resulting time trend is increasing (decreasing), whilst if d is estimated to be close to 0, the resulting time trend is stable.

The time index t enters the model as $(t - t^*)$, where t^* is a fixed value set at the midpoint of the observation period, namely, $t^* = (T+1)/2$. The reason is that the set of time indices, $1,...,T$, can be considered as a covariate in a regression model. The formulation in Eq. 12.3 is then equivalent to mean-centring this covariate. Then, α is interpreted as the burglary rate on the log scale at the midpoint of the observation period. When fitting the model using MCMC algorithms, mean-centring a covariate may lead to faster convergence and/or help improve the mixing of the MCMC chains (see Section 5.4.2 in Chapter 5). In addition to the midpoint, t^* can also be set at other values between 1 and T (say, $t^* = 1$ or $t^* = T$). The slope estimate will be very similar (not exactly the same because of the simulation nature of the MCMC fitting) to that from using the midpoint, but the estimate of α will be different because α will be interpreted differently: α is interpreted as the burglary rate on the log scale at the first (last) time point when setting $t^* = 1$ ($t^* = T$).

Eq. 12.3 can be extended to include a set of "noise terms" to allow for departures from linearity:

$$v_t = d \left(t - t^* \right) + \varepsilon_t \tag{12.4}$$

The temporal noise terms, $\boldsymbol{\varepsilon} = (\varepsilon_1,...,\varepsilon_T)$, are IID, meaning that each ε_t independently follows an identical normal distribution, $N(0, \sigma^2)$, the same specification as for the "white noise" process in time series analysis.[3] Each ε_t measures the deviation of the estimate at time t from the linear pattern defined by the estimated intercept and slope. Therefore, if ε_t is estimated to be close to zero for all time points, then the resulting trend will be close to being linear. If, on the other hand, some of the noise terms are estimated to be away from zero, then the estimated time trend exhibits some non-linear pattern (e.g. a sudden change at certain time point(s) or a trend that shows some curvature). The unknown variance, σ^2, determines the overall level of departure from linearity – a larger (smaller) σ^2 results in a time trend that is moving further away (closer towards) linearity. When modelling data observed over a relatively short time period (over two to four time points), the noise terms are typically constrained to sum to zero, namely, $\sum_{t=1}^{T} \varepsilon_t = 0$, to avoid a potential non-identifiability issue between the intercept α (as in Eq. 12.2) and the overall mean of the noise terms. Consider the formulation $\log(\theta_t) = \alpha + d \cdot \left(t - t^* \right) + \varepsilon_t$; adding an arbitrary constant, say C, to α and subtracting the same constant from all the terms in $\boldsymbol{\varepsilon} = (\varepsilon_1,...,\varepsilon_T)$ yields the same model. Section 12.3.3.2.2 gives another reason for imposing the sum-to-zero constraint.

In addition to allowing for nonlinearity, $\boldsymbol{\varepsilon}$ can also be seen as a set of surrogate measures of the effects from omitted covariates. Because the noise terms are IID, these effects are time-independent, i.e. the effect size at one time point does not affect effect sizes at

---

[3] Other distributions can also be used to specify ε_t. For example, one can use the t_4 distribution to deal with the presence of outlier(s) (see Section 11.2.5 in Chapter 11). The Bayesian approach makes it easy to consider alternative (non-standard) specifications for the model components such as the noise terms and the likelihood function.

other time points. In other words, the time-specific parameters, $\varepsilon_1, ..., \varepsilon_T$, can be considered as *exchangeable* over time. This allows us to link the temporal noise component, ε, to the exchangeable structure that we have discussed in the context of modelling spatial data (Section 7.4 in Chapter 7). As a result, all the properties discussed there, in particular those relating to information borrowing and global smoothing, also apply to the modelling of ε. This observation becomes particularly relevant when we discuss RW models where, similar to the spatial ICAR model, information sharing is carried out *locally* (see Section 12.3.4).

To complete the model specification, a vague prior distribution $N(0, 1000000)$ is independently assigned for the intercept α and the slope d. For the variance σ^2, a vague prior *Uniform*$(0.00001, 10)$ or $N_{+\infty}(0, 10)$ is assigned to σ, the standard deviation. The reader is encouraged to justify, in the context of modelling the burglary data, why the above priors are vague for their respective parameters. Putting together the data, process and parameter models, Table 12.2 summarises a Bayesian temporal model with linear time trend with/without temporal noise.

Figure 12.6 shows the WinBUGS code for fitting the temporal model with a linear trend and temporal noise. The code can be modified to fit the model without temporal noise by simply removing all the terms, i.e. epsilon and g, related to the temporal noise. Note that in Line 10 in Figure 12.6, the term g[t] represents the corresponding noise term epsilon[t] but after imposing the sum-to-zero constraint. This constraint is carried out in Line 14 by subtracting each epsilon[t] from the mean of all the epsilon terms, which is calculated in Line 18. In addition, for each time point, Lines 23 to 31 calculate the posterior predictive p-value of the likelihood of obtaining the observed number of burglary cases, y_t, under the model: when $y_t = 0$, $ppp_t = \Pr\left(y_t^{pred} = y_t \mid data\right)$, while $y_t > 0$, $ppp_t = \Pr\left(y_t^{pred} \geq y_t \mid data\right)$ where y_t^{pred} is the predicted number of burglary cases at time point t from the fitted model. A value of ppp_t between 0.05 and 0.95 is indicative of an adequate fit to the observed case count at that time point. On the other hand, a ppp_t value below 0.05 (above 0.95) suggests that, compared to the observed count, the model predicts too few (many) cases for that time point, indicating that the model fits the data poorly. See

TABLE 12.2

Two Versions of a Bayesian Temporal Model with a Linear Time Trend with/without Temporal Noise

A Bayesian temporal model with a linear time trend without temporal noise (for $t = 1, ..., T$)	A Bayesian temporal model with a linear time trend with temporal noise (for $t = 1, ..., T$)
$y_t \sim Poisson\left(\mu_t\right)$	$y_t \sim Poisson\left(\mu_t\right)$
$\mu_t = n \cdot \theta_t$	$\mu_t = n \cdot \theta_t$
$\log\left(\theta_t\right) = \alpha + d \cdot \left(t - t^*\right)$	$\log\left(\theta_t\right) = \alpha + d \cdot \left(t - t^*\right) + \varepsilon_t$
$\alpha \sim N(0, 1000000)$	$\varepsilon_t \sim N\left(0, \sigma^2\right)$
$d \sim N(0, 1000000)$	$\alpha \sim N(0, 1000000)$
	$d \sim N(0, 1000000)$
	$\sigma \sim Uniform(0.00001, 10)$

Note that the temporal noise terms are constrained to sum to zero, i.e. $\displaystyle\sum_{t=1}^{T} \varepsilon_t = 0$.

```
 1    #  The WinBUGS code for fitting a Bayesian temporal
 2    #  model with linear trend and noise
 3    model {
 4      #  a for-loop to go through all the time points
 5      for (t in 1:T) {
 6        #  defining the Poisson likelihood for each
 7        #  count value
 8        y[t] ~ dpois(mu[t])
 9        mu[t] <- n*theta[t]
10        log(theta[t]) <- alpha + d*(t-t.star) + g[t]
11        #  temporal noise without the sum-to-zero constraint
12        epsilon[t] ~ dnorm(0,prec.epsilon)
13        #  impose the sum-to-zero constraint on the noise
14        g[t] <- epsilon[t] - mean.epsilon
15      }
16    #  calculate the mean of the unconstrained
17    #  temporal noise
18    mean.epsilon <- mean(epsilon[1:T])
19    #  calculate the posterior probability for
20    #  checking model fit
21    for (t in 1:T) {
22        #  predicted number of cases at time t
23        y.pred[t] ~ dpois(mu[t])
24        #  difference between predicted and the
25        #  observed case counts
26        diff[t] <- y.pred[t] - y[t]
27        #  calculate the posterior predictive p-values
28        #  ppp[t] = Pr(pred = obs | data) when y[t]=0
29        #  ppp[t] = Pr(pred >= obs | data) when y[t]>0
30        ppp[t] <- equals(y[t],0) * equals(y[t],y.pred[t])
31               + (1-equals(y[t],0)) * step(diff[t])
32    }
33    #  define the midpoint of the study period
34    t.star <- (T+1)/2
35
36    #  vague priors
37    alpha ~ dnorm(0,0.000001)
38    d ~ dnorm(0,0.000001)
39    sigma.epsilon ~ dunif(0.00001,10)
40    prec.epsilon <- pow(sigma.epsilon,-2)
41    #  calculate the average burglary rate per 1000
42    #  houses per year
43    overall <- mean(theta[1:T])*1000
44  }
45
```

FIGURE 12.6
The WinBUGS code, the data and the two sets of initial values for fitting the Bayesian temporal model with linear trend and noise (presented in the second column of Table 12.2) to the annual burglary count data from COA 3 in Table 12.3. The same code can be used to fit the aggregated burglary data for Peterborough as a whole (i.e. the data in the last row of Table 12.3).

```
46  #  the burglary data from a selected COA
47  #   (COA 3 in Table 12.3)
48  list(T=4              #  4 years
49      ,y=c(4,6,5,10) #  numbers of burglary cases
50                       #  across the four years
51      ,n=134           #  number of at-risk houses
52                       #  in this COA
53  )
54
55  #  initial values for chain 1
56  list(alpha=-4,d=0.1,sigma.epsilon=0.1
57      ,epsilon=c(0.001,0.001,0.001,0.001)
58      ,y.pred=c(0,0,0,0))
59  #  initial values for chain 2
60  list(alpha=-5,d=0.05,sigma.epsilon=0.2
61      ,epsilon=c(-0.001,-0.001,-0.001,-0.001)
62      ,y.pred=c(1,1,1,1))
```

FIGURE 12.6 (*Continued*)

Section 5.4.2 for more detail. Finally, Line 43 calculates the average burglary rate per year per 1000 houses over the four-year period. The reader is encouraged to make the distinction between the two calculations of the average burglary rate per year per 1000 houses: (a) using the calculation as on Line 43 or (b) via the calculation of $\exp(\alpha)\cdot 1000$ (hint: the two formulations give the same results when the models in Table 12.2 only contain the intercept, α, but they give different results when the log burglary rates are allowed to vary over time – why is that the case?).

12.3.3.2 Modelling Trends in the Peterborough Burglary Data

We apply the two models presented in Table 12.2 to the data in each of the four COAs as well as the aggregated Peterborough data shown in Table 12.3. In doing so, we ask the following two questions. First, does any COA have a higher average level of burglary risk (per 1000 houses per year) than the Peterborough average? Second, does any COA show a time trend that is increasing faster or slower than the Peterborough trend? In the course of this

TABLE 12.3

The Annual Burglary Counts Between 2005 and 2008 and the Number of Houses at Risk for Each of the Four Selected COAs as well as for Peterborough as a Whole

Unit of analysis	2005	2006	2007	2008	Number of at-risk houses
COA 1	4	4	3	1	121
COA 2	0	0	2	3	78
COA 3	4	6	5	10	134
COA 4	0	2	2	6	120
Peterborough	620	644	1213	1275	58567

investigation, we aim to study properties of the linear trend model and also to illustrate some of the issues that we have discussed in Section 12.3.1.

At first glance, the COA-level case numbers and population sizes are quite similar to those we discussed at the end of Section 12.3.1.1, implying that the uncertainty of the estimated burglary risks will be similarly high. As we will see later, this has an implication for parameter estimation of the more complex linear trend model with temporal noise.

12.3.3.2.1 Results from Fitting the Linear Trend Model without Temporal Noise

Figure 12.7 shows the posterior mean and the 95% credible interval (denoted as 95% CI hereafter) of the average burglary rate for each of the four COAs and those for Peterborough as a whole. The estimated average rates differ from one COA to another. However, the 95% CIs for COAs 1, 2 and 4 (represented by the horizontal lines in Figure 12.7) overlap with that for the Peterborough average, suggesting that the overall burglary risks of these three COAs are similar to that of Peterborough. Note that the horizontal bar representing the 95% CI for the Peterborough average (95% CI: 15.50–16.54) cannot be seen in Figure 12.7 because it is much too narrow compared to those for the COAs. This raises an interesting point when modelling count data. When both the number of cases and the population size are as large as in the case of the whole of Peterborough, it is perhaps not surprising to see a very narrow uncertainty interval. This is because, recall Section 3.3.3 in Chapter 3, the standard error (SE) for the average burglary rate estimate (per 1000 houses per year) can be calculated by $\left(\sum_{t=1}^{4} y_t\right)^{0.5} \times 1000/(n \times 4)$. For the whole of Peterborough, this equates to 0.261, which, because we use vague priors, is very close to the posterior standard deviation of 0.263 obtained from WinBUGS. However, such a discussion becomes more complicated at the COA level because, according to the above SE expression, the

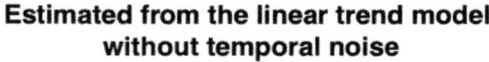

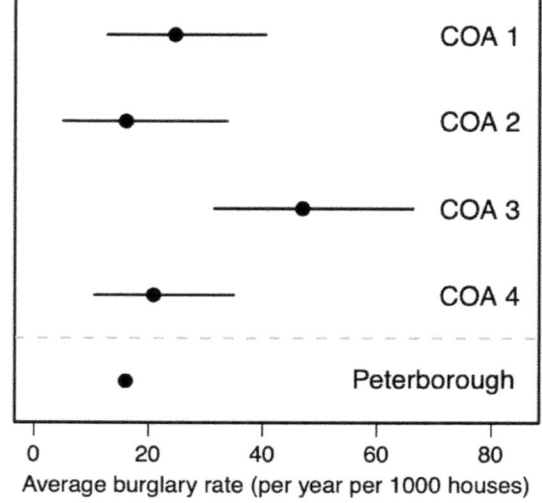

FIGURE 12.7
The posterior mean (the dot) and the 95% CI (the horizontal bar) of the average burglary rate per 1000 houses per year for each of the four COAs as well as for the whole of Peterborough estimated using the linear trend model without temporal noise.

uncertainty of the estimated average rate depends on both the number of cases *and* the population size. For each of these two quantities, the values across the COAs tend to be quite similar, and thus we have to consider both together in the uncertainty comparison. For example, compared to COA 1, COA 3 has more cases and more houses at risk, but its 95% CI is wider than that for COA 1. For these two COAs, the SEs calculated using the formula $\left(\sum_{t=1}^{4} y_t \right)^{0.5} \times 1000 / (n \times 4)$ are 7.15 for COA 1 and 9.33 for COA 3, and again, they are reasonably close to 7.05 and 9.15, the respective posterior standard deviations of the average rates estimated from WinBUGS.

Interestingly, this observation implies that the form of spatial smoothing discussed in Section 6.2.2 in Chapter 6, whereby a spatially-smoothed rate is obtained by adding the cases and the populations from multiple nearby spatial units, tends to reduce the uncertainty of the resulting spatially-smoothed rate. Circumstances can arise where this form of spatial smoothing can *increase the uncertainty*. However, the calculation in Exercise 12.2 shows that the latter effect rarely occurs in practice.

Turning our attention now to the estimated slopes, these are summarised in Figure 12.8. For the aggregated Peterborough data, from 2005 to 2008 the estimated burglary rates were increasing, as shown by the positive posterior mean of the slope parameter, and the associated 95% CI does not cover 0. Similar to the estimated average burglary rate, when aggregating all the data for the whole of Peterborough, the uncertainty of the slope estimate is much lower compared to the slope estimates for the individual COAs. For the individual COAs, both COAs 2 and 4 are estimated to have an increasing trend, and the positive slopes for both COAs are significant in the sense that their 95% CIs exclude 0 (Figure 12.8).

Estimated from the linear trend model without temporal noise

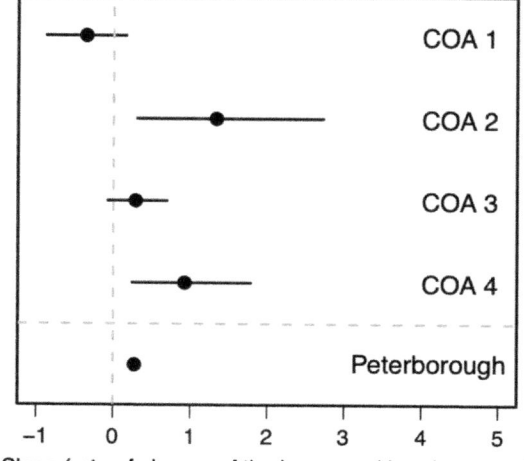

Slope (rate of change of the log annual burglary rates)

FIGURE 12.8
The posterior mean (the dot) and the 95% CI (the horizontal bar) for the slope for each of the four COAs as well as for the whole of Peterborough estimated from the linear trend model without temporal noise. The dashed vertical line indicates 0. A 95% CI that includes 0 suggests the estimated annual burglary rates are relatively constant over the four years.

On the other hand, the 95% CIs for COAs 1 and 3 do overlap with 0, suggesting their burglary trends remain relatively stable over the four years.

The slope estimates across these four COAs suggest that a slope estimate is more uncertain if that COA has fewer burglary cases or a smaller at-risk population or if both the case number *and* the at-risk population are small. For example, the slope estimate of COA 2 (95% CI: 0.30–2.72) is the most uncertain amongst the four COAs because both its population and the total number of burglary cases are the smallest (Table 12.3). For COA 3, by contrast, the uncertainty of its slope estimate is much less (95% CI: −0.08–0.69), as it has a larger population and more cases of burglary over the four years. The above observations can also be seen when extending the analysis to all the other 448 COAs in Peterborough (see Exercise 12.3).

Figure 12.9 examines the fit of the model to the observed number of cases. Even without temporal noise, the linear trend model fits the COA-level data reasonably well, as none of the posterior predictive p-values exceeds 0.95 or is below 0.05. The observed numbers of cases for each COA are within the 95% uncertainty region. Note that the dashed lines (the posterior means of the predicted numbers) in Figure 12.9 are not linear because of the exponential transformation (the burglary trend is linear on the log scale; see the specifications in Table 12.2).

In the case of the data for the whole of Peterborough, using only a linear trend clearly does not describe these (nonlinear) data well. As a result, the model predicts too many cases in 2006 with a posterior predictive probability of 1.00 and too few cases in 2007 where the posterior predictive probability is 0.00. This prompts us to add the temporal noise terms to the linear model. However, would this flexible, but more complex, model be supported by the COA-level data when each COA is taken on its own? We shall illustrate in the next section.

12.3.3.2.2 *Results from Fitting the Linear Trend Model with Temporal Noise*

Table 12.4 summarises the posterior estimates (posterior means and 95% CIs) for the overall burglary rates and the slopes for the four COAs as well as those for the whole of Peterborough. For ease of comparison, Table 12.4 also shows the corresponding posterior estimates from the linear trend model without temporal noise.

For the average burglary rates (the second and the third columns in Table 12.4), both the posterior means and the 95% CIs are comparable between the two models. The posterior means of the slopes are also similar between the two models, but the 95% CIs of the slope estimates are consistently wider when a set of temporal noise terms are added in. Such an increase in uncertainty is more pronounced for COA 2, where there are only a few cases of burglary and the population is small. So, why does the uncertainty in the slope estimate increase when a set of temporal noise terms are added?

The reason is that there are more parameters to estimate based on limited data at the small-area level. The linear trend model with temporal noise, in fact, seeks to partition the variability in the time series data into three parts: a linear time trend, a set of temporal noise terms and the Poisson variability from the data model. When the count data are sparse (i.e. small numbers of cases with a small at-risk population) and the data are only observed for a short period of time, it is difficult enough to distinguish the inherent Poisson variability from a step-change pattern, as discussed in Section 12.3.1.1, or a linear trend pattern, let alone the temporal noise part with more parameters to estimate. The lack of sufficient information at the COA-level leads to the increase in uncertainty in the slope estimates. In addition, the partitioning of the three parts also explains why the temporal noise terms are typically constrained to sum to zero (see Section 12.3.3.1). Essentially, this constraint restricts the flexibility of the temporal noise terms and, in turn, helps the

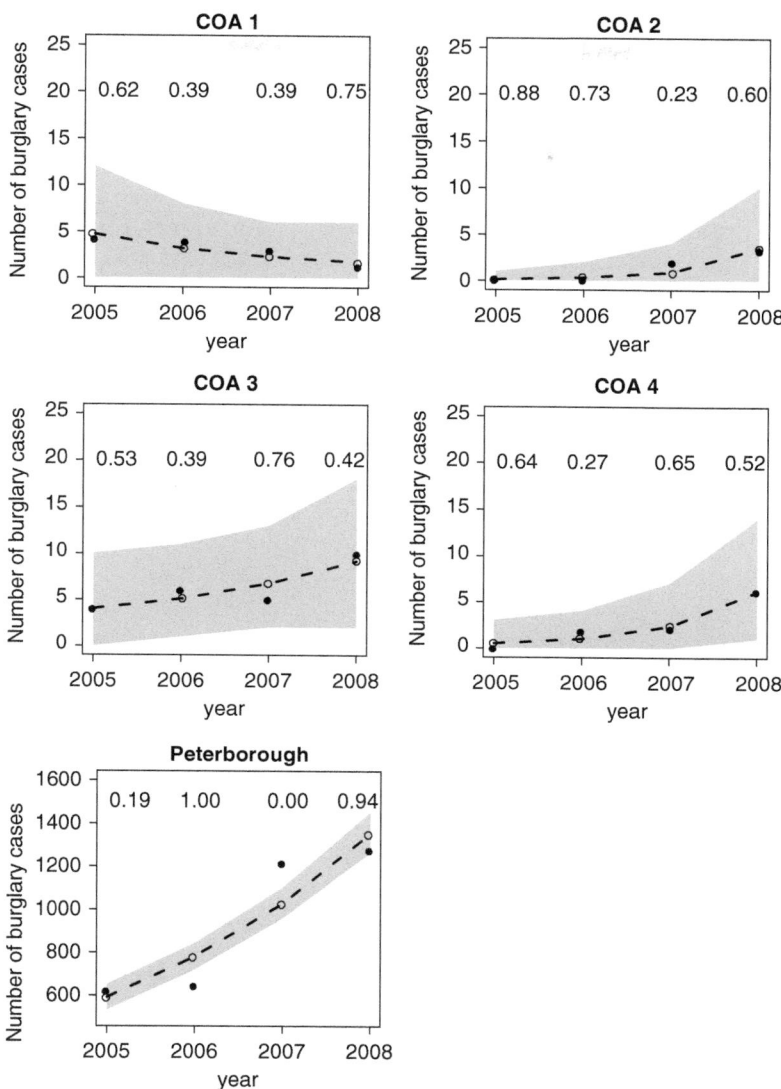

FIGURE 12.9

Comparisons of the observed numbers of burglary cases (the dots) with the predicted numbers from the linear time trend model without temporal noise. In each panel, the open circles with the dashed line represent the posterior means of the predicted numbers of cases, and the grey region shows the 95% CIs across the four years. The numbers in the upper part of each panel are the posterior predictive p-values. A ppp_t value that is between 0.05 and 0.95 indicates a good fit of the model to the data, otherwise the fit is poor.

identification/estimation of each of the three parts. However, when just using the data from each COA on its own, the linear trend model with temporal noise is not well supported, illustrating the data sparsity issue as well as the model complexity issue that we discussed in Sections 12.3.1.2 and 12.3.1.3.

The linear trend model with temporal noise, on the other hand, works nicely with the Peterborough data, as there are a lot more burglary cases and a much larger at-risk population. As shown in Figure 12.10, the nonlinear pattern in the observed burglary counts is

TABLE 12.4

A Summary of the Posterior Estimates (Posterior Means and 95% CIs) for the Average Burglary
Rates and the Slopes from the Linear Trend Model With and Without Temporal Noise

	Average burglary rate (per 1000 houses per year)		Slope (rate of change of the log annual burglary rates)	
	With noise	Without noise	With noise	Without noise
COA 1	24.81 (13.20–39.95)	24.61 (12.34–40.44)	−0.39 (−1.39–1.11)	−0.35 (−0.88–0.16)
COA 2	15.76 (5.14–31.94)	16.06 (4.96–33.77)	2.22 (0.33–5.68)	1.34 (0.30–2.72)
COA 3	46.64 (30.65–68.37)	46.97 (31.55–66.29)	0.29 (−0.19–0.88)	0.29 (−0.08–0.69)
COA 4	20.65 (9.85–35.27)	20.89 (10.61–34.93)	0.95 (0.11–2.03)	0.92 (0.24–1.79)
Peterborough	16.02 (15.49–16.54)	16.01 (15.50–16.54)	0.29 (0.00–0.72)	0.28 (0.25–0.30)

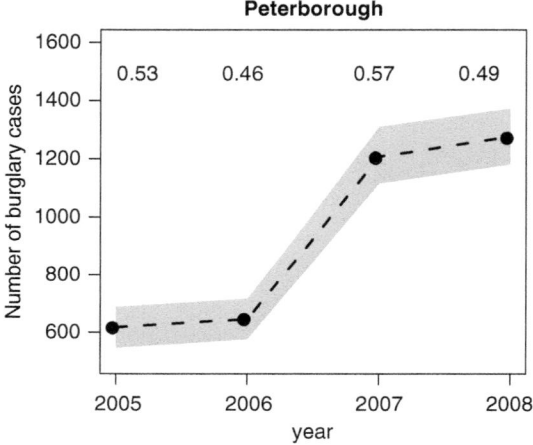

FIGURE 12.10
The aggregated Peterborough data are much better described using the linear time trend model with temporal
noise. See Figure 12.9 for the description of the labelling. The open circles (the posterior means of the burglary counts predicted from the model) do not appear in the figure because they overlap with the dots, i.e. the
observed case counts.

better modelled by the linear trend plus temporal noise structure. This is the modelling
structure used in the crime surveillance model for capturing the overall temporal burglary pattern in Peterborough (Section 16.3).

12.3.4 Random Walk Models

As illustrated in Figure 12.5, the temporal patterns of the raw burglary rates for some COAs
and for the whole of Peterborough over an eight-years period from 2001 to 2008 are clearly
not linear – a frequently encountered feature when temporal data are observed over five
or more time points. In addition, note that the raw burglary rates in each time sequence
vary quite smoothly over time, suggesting the presence of positive temporal/serial autocorrelation. The aggregated Peterborough trend varies even more smoothly than the COA
trends because Peterborough as a whole has a much larger number of burglary cases each
year and its at-risk population is also much larger. The COA trends shown in Figure 12.5
give us an idea of the scale of space-time variability that we need to be able to deal with

when modelling time series data at the small area level. The random walk model, to be discussed in this section, is important because it enables us to capture nonlinear temporal patterns whilst retaining the smoothly-varying feature that is evident in much observed time series data.

12.3.4.1 Model Formulations

Following the setting in Section 12.3.2, Eq. 12.5 specifies the random walk model of order 1 (the RW1 model) on the set of time-point-specific parameters, $v = (v_1,...,v_T)$, in the process model. For $t = 2,...,T$,

$$v_t = v_{t-1} + \varepsilon_t, \tag{12.5}$$

where $\varepsilon_t \sim N(0, \sigma^2)$ denotes the noise term and $\varepsilon_2,...,\varepsilon_T$ are independent. Following the Bayesian formulation of the RW1 model in Fahrmeir and Lang (2001), v_1, the parameter at the first time point, is assigned a vague/diffuse prior, for example $v_1 \sim Uniform(-100000,100000)$.

The RW1 model estimates the temporal pattern in the observed time series data through essentially *a temporal adaptive process*. To see its adaptive nature, according to Eq. 12.5, the RW1 model produces v_t by adding or subtracting an independent noise term to v_{t-1}, the parameter at the previous time point. Thus, if more (fewer) burglary cases are observed in t compared to $t-1$, then a positive value will be added to (subtracted from) v_{t-1} to produce v_t so that the burglary rate at t is higher (lower) than that at $t-1$. Following the same logic, if the numbers of burglary cases are the same between $t-1$ and t, v_t will be kept the same as v_{t-1}, i.e., the noise term ε_t is 0. The RW1 model is able to capture a nonlinear temporal pattern because of this temporally adaptive feature. Each noise term ε_t can be positive, negative or 0 according to the observed data.

At the same time, the adaptive feature is coupled with *temporal smoothing*. The amount of temporal smoothing depends partly on the noise variance σ^2 – there are also other factors that can affect the amount of temporal smoothing, but we will defer the discussion to Section 12.3.4.2. When σ^2 is large, v_t (for any $t \geq 2$) is allowed to have a large deviation from v_{t-1}. The resulting temporal pattern is less smooth but more flexible, since it is able to follow temporal data with large variation. On the other hand, when σ^2 is small, v_t can only have a small deviation from v_{t-1} (for any $t \geq 2$), producing a smooth temporal pattern which is less flexible because it is only able to follow temporal data with small variation. In the extreme where σ^2 is 0, equivalent to removing the noise terms all together, the corresponding time trend is constant over time. This noise variance, σ^2, is estimated from the observed data typically with a vague/noninformative prior assigned. In some situations, however, one may want to control the amount of temporal smoothing or, equivalently, the flexibility of the RW1 model by using a moderately informative prior (see Section 16.4).

Although the RW1 model in Eq. 12.5 is "seemingly defined in an asymmetric directed way", this prior model on the set of time-point specific parameters, v, "can always be rewritten in an undirected symmetric form" (Fahrmeir and Lang, 2001, p.206). As a result, the estimation of v_t borrows information not only from the past time points (e.g. $t-1$ and $t-2$) but also from the future time points (e.g. $t+1$ and $t+2$), providing that the past and the future time points are within the study period. In Section 12.3.1.1, we have argued that such a bi-directional (undirected symmetric) form is a more efficient way to share information over time. In the next section, we study this feature of the RW1 model by rewriting Eq. 12.5 into a set of conditional probability distributions. We will also see that these conditional distributions take the same form as for the conditional distributions under the ICAR model applied to a set of area-specific parameters. This means the temporal smoothing properties of the RW1 model are similar to the spatial smoothing properties under the ICAR model. In addition, the RW1 model can be fitted easily in WinBUGS using the `car.normal` function, just as for the spatial ICAR model.

12.3.4.2 *The RW1 Model: Its Formulation Via the Full Conditionals and Its Properties*

The aim here is to rewrite the RW1 model defined in Eq. 12.5 into a set of T conditional probability distributions, one for each v_t conditioning on all other parameters in v and the noise variance σ^2. That is, we want to write down the distribution of $v_t \mid v_{\{-t\}}, \sigma^2$, where $v_{\{-t\}}$ denotes all the time-point-specific parameters excluding v_t. Exercise 12.4 shows that, for $T \geq 3$, these T conditional probability distributions take the following form:

$$v_t \mid v_{\{-t\}}, \sigma^2 \sim \begin{cases} N\left(v_{t+1}, \sigma^2\right) & \text{for } t = 1 \\ N\left(\dfrac{v_{t-1} + v_{t+1}}{2}, \dfrac{\sigma^2}{2}\right) & \text{for } t = 2,\ldots,T-1 \\ N\left(v_{t-1}, \sigma^2\right) & \text{for } t = T \end{cases} \tag{12.6}$$

Furthermore, we can express Eq. 12.6 as Eq. 12.7, which now takes the form of Eq. 8.1 (Chapter 8) that defines a set of conditional probability distributions under the spatial ICAR model with a binary 0/1 spatial weights matrix.

$$v_t \mid v_{\{-t\}}, \sigma^2, W_{RW1} \sim N\left(\dfrac{\sum_{g \in \Delta t} v_g}{m_t}, \dfrac{\sigma^2}{m_t}\right) \tag{12.7}$$

where

- Δt denotes the set of temporal neighbours of t. When $t = 2,\ldots,T-1$, time point t has two neighbours, $t-1$ and $t+1$, and when $t = 1$ or $t = T$, time point t has only one neighbour, 2 or $T-1$, respectively.
- m_t is the number of neighbours that time point t has. When $t = 2,\ldots,T-1$, $m_t = 2$, and when $t = 1$ or $t = T$, $m_t = 1$.
- W_{RW1} represents the temporal weights matrix, from which the values of Δt and m_t are derived. The subscript $RW1$ emphasises that this weights matrix is defined according to the temporal neighbourhood structure imposed by the RW1 model. For example, with $T = 5$, W_{RW1} is given by

$$W_{RW1} = \begin{pmatrix} 0 & 1 & 0 & 0 & 0 \\ 1 & 0 & 1 & 0 & 0 \\ 0 & 1 & 0 & 1 & 0 \\ 0 & 0 & 1 & 0 & 1 \\ 0 & 0 & 0 & 1 & 0 \end{pmatrix} \tag{12.8}$$

The above derivation shows that the RW1 model can be seen as a one-dimensional version of the spatial ICAR model (Fahrmeir and Lang, 2001). In other words, for a map with T areas forming a "linear" structure, the conditional distributions for a set of area-specific parameters, $S = (S_1,...,S_T)$, under the ICAR model with a binary weights matrix based on first order spatial contiguity are the same as those defined for a set of time-point-specific parameters, $v = (v_1,...,v_T)$, under the RW1 model. Figure 12.11 gives an example with $T = 5$.

The above connection between the temporal RW1 model and the spatial ICAR model is important because, when used as a prior model for a set of time-point-specific parameters, the RW1 model possesses the same set of properties that we discussed for the spatial ICAR model (Section 8.2.1.4 in Chapter 8). We now express these properties in terms of the RW1 model.

First, the RW1 model assumes positive temporal autocorrelation (serial correlation) and imposes a temporal neighbourhood structure on the time-point-specific parameters via the chosen temporal weights matrix. The positive temporal autocorrelation assumption can be seen from the conditional distribution for each v_t (Eq. 12.6, Eq. 12.7 or in Figure 12.11) where the conditional distribution of v_t is centred at the average of the parameters of its temporal neighbours. In other words, v_t is assumed to be more similar to v_{t-1} and v_{t+1} than to the parameters at time points that are further away from t (i.e. $v_{t-2}, v_{t-3}, v_{t+2}, v_{t+3}$ and so on). This assumption needs to be checked at the exploratory data analysis stage by calculating, for example, the temporal autocorrelation coefficient (Section 13.4.3.2 in Chapter 13). As in the ICAR model, the extent to which information is borrowed from other time points

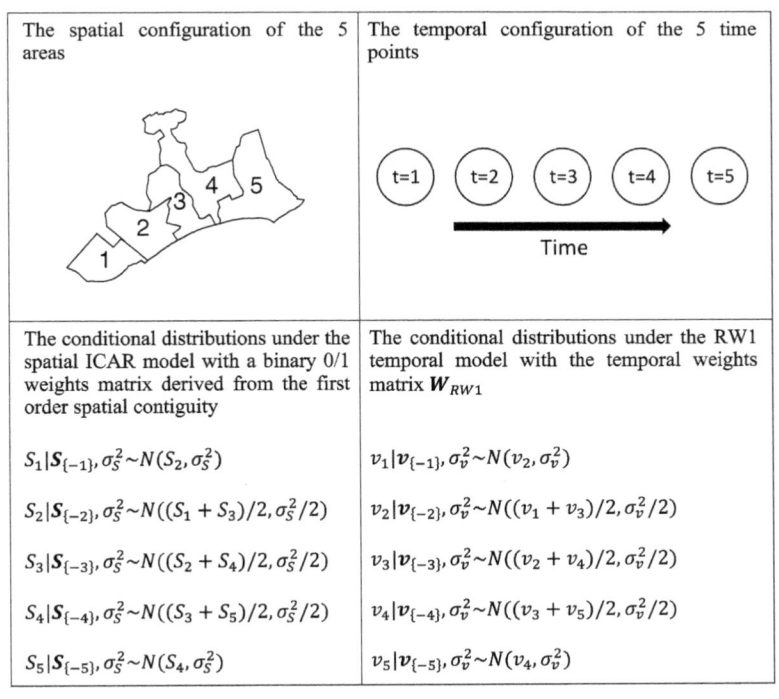

FIGURE 12.11
The RW1 model on a set of temporal parameters can be considered as a one-dimensional version of the spatial ICAR model on a set of spatial parameters where the spatial units form a "linear" structure. Here $T = 5$ (five time points/areas).

depends on the temporal neighbourhood structure defined in the temporal weights matrix W_{RW1}. Again, the choice of the temporal neighbourhood structure is a modelling assumption. Its impact on parameter estimation can be assessed by considering other specifications, for example using W_{RW2}, the temporal weights matrix based on the random walk model of order 2. The discussion of that model is deferred until Section 12.3.4.5.

Second, through positive temporal autocorrelation, the RW1 model shares information temporally across a set of neighbouring time points, thus producing a temporally smooth time pattern. We discussed temporal smoothing of the RW1 model in Section 12.3.4.1, but the conditional distributions in Eq. 12.6 highlight the bi-directional nature when sharing information over time. For each $t = 2, \ldots, T - 1$, the conditional mean of v_t depends on the parameters at both $t - 1$ *and* $t + 1$. However, the estimation of the parameters at the first and the last time points primarily borrows information from their respective individual temporal neighbours, namely time point 2 and time point $T - 1$ (compared to the boundary problem in spatial analysis where there may be more neighbours).

Third, the amount of temporal smoothing depends on the sample size and the variability or consistency of the information coming from the temporal neighbours. As for spatial smoothing, the more information contained in the data at a time point, the less the estimate of its parameter will be affected by the observations from its temporal neighbours. However, unlike the spatial case, population sizes (e.g. the number of at-risk houses in each COA) are often assumed to be time invariant.[4] In addition, phenomena of interest often change smoothly over time. Thus, in practice, compared to the effect of the noise variance σ^2, the impact of these two factors on temporal smoothing is small.

Lastly, when using a temporal weights matrix with binary 0/1 entries, the number of neighbours of an area also affects the extent of temporal smoothing – the more neighbours a time point has, the more temporal shrinkage its parameter will receive. This can be seen from Eq. 12.7 that the conditional variance for v_t is defined as σ^2 / m_t. So, all other things being equal, the parameters associated with the first and the last time points receive less temporal smoothing compared to those at time points between 2 and $T - 1$. We shall also see that, compared to the RW1 model, the RW2 model generally yields a smoother temporal pattern, since the RW2 model assigns more neighbours to each time point (see Section 12.3.4.5). However, compared to the spatial setting, the number of temporal neighbours does not vary greatly – for the RW1 model, most time points have two neighbours, while the first and the last time points only have one. As a result, holding everything else fixed, different numbers of neighbours across different time points does not much affect the amount of temporal smoothing.

In addition to the above properties, the RW1 model is also subject to the following two constraints, the same as for the spatial ICAR model. Following Section 8.2.2.1 in Chapter 8, the joint probability distribution for $v = (v_1, \ldots, v_T)$, which can be derived using the T conditional probability distributions in Eq. 12.6, is given by

$$\Pr\left(v_1, \ldots, v_T\right) = \Pr\left(v\right)$$

$$\propto \exp\left\{-\frac{1}{2\sigma^2} v^T \left(D_W - W_{RW1}\right) v\right\} \tag{12.9}$$

---

[4] Sometimes a quite reasonable assumption, because over the period 1 to T there have been no new houses put up or old ones demolished. But where the at-risk population is specified in terms of people, there may be no data on population shifts.

where D_W is a $T \times T$ diagonal matrix with the tth diagonal element, $(D_W)_{tt}$, being the number of temporal neighbours, i.e. $(D_W)_{tt} = 2$ for $t = 2,...,T-1$ and $(D_W)_{tt} = 1$ for $t = 1$ and $t = T$ under the RW1 model. For the joint distribution in Eq. 12.9 to exist, $(D_W - W_{RW1})$ needs to be symmetric, and thus W_{RW1}, the temporal weights matrix, needs to be symmetric, a constraint that is automatically satisfied by construction.

The joint distribution in Eq. 12.9 is improper because the set of parameters $v = (v_1,...,v_T)$ enter the distribution as pairwise differences. This can be seen by rewriting Eq. 12.9 into Eq. 12.10:

$$\Pr(v_1,...,v_T) \propto \exp\left\{-\frac{1}{2\sigma^2} \sum_{t=1}^{T-1} \sum_{g=t+1}^{T} (v_t - v_g)^2\right\} \tag{12.10}$$

As a result, adding an arbitrary constant c to all terms in v yields a joint distribution exactly the same as that in Eq. 12.10, i.e. $\Pr(v_1 + c,...,v_T + c) = \Pr(v_1,...,v_T)$. In other words, the overall mean of v is undefined. Similar to the ICAR model, the RW1 model is only used as a prior model on v but not for modelling the observed outcome values in the likelihood (as it is unrealistic to assume the overall mean of the underlying data generating process is undefined). To resolve the issue of the undefined mean, a sum-to-zero constraint is imposed on v, i.e. $\sum_{t=1}^{T} v_t = 0$. Because the RW1 model can be written in the form of the ICAR model, the RW1 model can be implemented using the car.normal function, which automatically imposes the sum-to-zero constraint on v.

As for the spatial ICAR model, an intercept term α must be included and the prior for α must be the improper uniform distribution defined on the whole real line, i.e. $\alpha \sim Uniform(-\infty,+\infty)$. We detail the WinBUGS fitting of the RW1 model next.

12.3.4.3 WinBUGS Implementation of the RW1 Model

In WinBUGS, the line below applies the RW1 smoothing model to $v = (v_1,...,v_T)$:

```
v[1:T] ~ car.normal(adj[],weights[],num[],prec)
```

The three arguments, adj[], weights[] and num[], together define the temporal neighbourhood structure, and the last argument is the precision parameter, which is the inverse of σ^2. Table 12.5 describes the three arguments and illustrates their specifications based on the temporal weights matrix in Eq. 12.8 derived from the RW1 model with $T = 5$. The R script RW_weights_matrix.R (available on the book's website) can be used to produce the temporal neighbourhood structure over T time points as required by WinBUGS.

12.3.4.4 Example: Modelling Burglary Trends Using the Peterborough Data

One of the purposes of modelling small area time trends is to identify "unusual" local time trends that differ markedly from the overall time trend of the study region. This type of detection analysis is often used in disease (or crime) surveillance, a topic that we will discuss in more detail in Section 16.4. The aim of this example is two-fold. First is to demonstrate the modelling of a small area time trend using the RW1 model. Second is to initiate the discussion of how to construct models for detecting unusual local time trends.

TABLE 12.5

Specification of the Temporal Neighbourhood Structure for WinBUGS Derived from the RW1 Model on a Set of Temporal Parameters, $v = (v_1,...,v_5)$, over $T = 5$ Time Points

	t = 1	t = 2	t = 3	t = 4	t = 5	Note
num[]	1,	2,	2,	2,	1	num[] is an array of length T where the tth entry denotes the number of temporal neighbours that time point t has.
adj[]	2,	1, 3,	2, 4,	3, 5,	4	adj[] defines the temporal neighbours. For example, the first entry adj[] defines time point 2 as the (only) temporal neighbour of 1. The following two entries in adj[] tell us that the two neighbours of time point 2 are 1 and 3 and so on.
weights[]	1,	1, 1,	1, 1,	1, 1,	1	weights[] specifies the weights assigned to each pair of neighbours. For the RW1 model, all neighbouring weights are set to 1.

The corresponding temporal weights matrix is in Eq. 12.8.

The data used in this example contain the annual burglary counts from three selected COAs as well as the aggregated annual burglary counts for the whole of Peterborough. Each of these four sets of time series count data is modelled separately using the model given as follows:

$$y_t \sim Poisson(\mu_t)$$

$$\mu_t = n \cdot \theta_t$$

$$log(\theta_t) = \alpha + v_t$$

$$v_{1:T} \sim ICAR\left(W_{RW1}, \sigma_v^2\right)$$

$$\sigma_v \sim N_{+\infty}(0,10)$$

$$\alpha \sim Uniform(-\infty,+\infty)$$

(12.11)

where y_t is the number of burglary cases reported in year t for the COA in question (or for Peterborough as a whole) and n is the corresponding number of houses at-risk. The model in Eq. 12.11 takes the same form as that in Table 12.2 but with a RW1 prior model on $v = (v_1,...,v_T)$ for modelling the time trend. For each of the three COAs and Peterborough as a whole, the quantities of interest in this model are (a) the average burglary rate per 1000 houses per year over the eight-year period and (b) the burglary rate per 1000 houses for each of the eight years. The WinBUGS implementation of this model is given in Figure 12.12.

The posterior estimates of the average level of burglary risk and the burglary trend at the COA level are much more uncertain (with much wider uncertainty intervals) compared to those estimated using the aggregated data for the whole of Peterborough. This should not come as a surprise because there are a lot more cases over a much larger at-risk population at the Peterborough level. The average burglary rate for COA 2 is estimated to be 30.63 (the posterior mean) cases per 1000 houses per year with a 95% CI of (20.89, 43.07), somewhat higher than that for Peterborough as a whole (19.79 with a 95% CI of

```
 1   #  The WinBUGS code for fitting the Poisson
 2   #  model in Eq. 12.11.
 3   model {
 4     #  a for-loop to go through all the time points
 5     for (t in 1:T) {
 6       #  defining the Poisson likelihood for
 7       #  each count value
 8       y[t] ~ dpois(mu[t])
 9       mu[t] <- n *theta[t]
10       log(theta[t]) <- alpha + v[t]
11     }
12     #  RW1 model on v[1:T]
13     v[1:T] ~ car.normal(adj[],weights[],num[],prec.v)
14     #  priors
15     alpha ~ dflat()
16     sd.v ~ dnorm(0,0.1)I(0,)
17     prec.v <- pow(sd.v,-2)
18     ##  quantities of interest
19     #  the average burglary rate per 1000
20     #  houses per year
21     overall <- mean(theta[1:T])*1000
22     #  the burglary rates (per 1000 houses) over time
23     for (t in 1:T){burglary.rates[t]<-theta[t]*1000}
24   }
25
26   #  the burglary data from a selected COA
27   list(T = 8          #  8 years
28       ,y = c(7,6,5,3,3,3,7,2)  #  burglary counts for COA 1
29       ,n = 145 #  number of at-risk houses for COA 1
30                #  (remains the same over the 8 years)
31     #  defining the temporal neighbourhood structure
32     #  based on the RW1 model
33       ,num = c(1,2,2,2,2,2,2,1) #  number of temporal
34                                 #  neighbours for each
35                                 #  time point
36     #  the IDs of the temporal neighbours
37       ,adj = c(2,1,3,2,4,3,5,4,6,5,7,6,8,7)
38     #  weights of the temporal neighbours
39       ,weights = c(1,1,1,1,1,1,1,1,1,1,1,1,1,1)
40   )
41
42   #  initial values for chain 1
43   list(alpha =-4,v=c(0.01,0,0,0,0,0,0,-0.01),sd.v=0.1)
44   #  initial values for chain 2
45   list(alpha =-5,v=c(-0.01,0,0,0,0,0,0,0.01),sd.v=0.2)
46
47   ##  data for COA 2
48   #  y = c(2,4,3,2,6,1,7,10)
49   #  n = 143
50   ##  data for COA 3
51   #  y = c(2,4,3,2,6,1,7,10)
52   #  n = 143
53   ##  aggregated data for the whole of Peterborough
54   #  y = c(1413,1897,1471,778,614,639,1204,1258)
55   #  n = 58567
```

FIGURE 12.12

The WinBUGS implementation of the model in Eq. 12.11 with a RW1 structure on the time trend. Replace the data on Lines 28 and 29 by the corresponding ones on Lines 47 to 55 to estimate the burglary trends for the other two COAs and for the whole of Peterborough.

(19.31, 20.27)). The average burglary rates for both COAs 1 and 3 are not different from that for Peterborough.

However, as shown in Figure 12.13, the estimated time trend patterns for these three COAs are quite different. The estimated burglary trend for COA 1 follows the estimated Peterborough trend closely, although the uncertainty associated with the COA trend is considerably greater compared to that of the Peterborough trend. The estimated trends for COAs 2 and 3, on the other hand, show deviations from the estimated Peterborough trend. For COA 2, an increasing trend is estimated, whereas the trend for COA 3 appears to be more variable with higher peaks in 2002 and perhaps again in 2007 compared to the Peterborough trend. Such visual inspection is useful to provide initial assessment as to whether a local trend is similar or different from the overall trend. However, we need to construct models in order to assess evidence statistically, i.e. are the departures large enough to declare a local trend unusual? Moreover, as evident in Figure 12.13, we need models to reduce the large uncertainty associated with the resulting local trend estimates when each area is treated separately, a challenge arising from data sparsity. Again, as in the case of spatial modelling (Chapter 8), we call upon the idea of information sharing. The modelling of small area time trends is placed within the hierarchical modelling approach so that various spatial-temporal prior models can be constructed for sharing information over space, time and/or both space and time. We will discuss the formulations and the properties of these spatial-temporal prior models in Chapter 15 and their applications in Chapter 16.

12.3.4.5 The Random Walk Model of Order 2

The RW1 model can be easily extended to higher orders. Eq. 12.12 specifies the random walk model of order 2 (the RW2 model) on a set of temporal parameters $v = (v_1,...,v_T)$ with $T \geq 3$. For $t = 3,...,T$,

$$v_t = 2v_{t-1} - v_{t-2} + \varepsilon_t \tag{12.12}$$

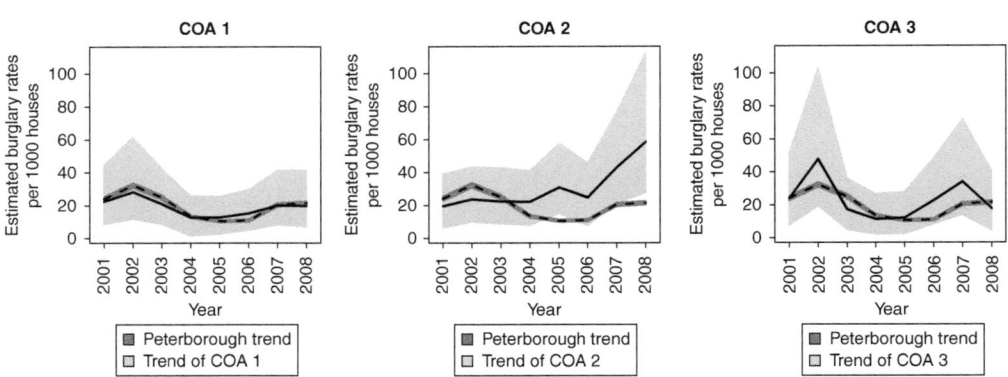

FIGURE 12.13
The estimated burglary rates per 1000 houses between 2001 and 2008 for each of the three selected COAs. In each plot, the solid line shows the posterior means (of $\theta_t \cdot 1000$ in Eq. 12.11) and the light grey band represents the 95% uncertainty region. Superimposed in each plot is the estimated Peterborough trend, where the posterior mean and the 95% uncertainty region are represented respectively by the dashed line and the dark grey band.

A vague uniform prior, say *Uniform*(–100000,100000), is assigned to v_1 and v_2 independently. As for the RW1 model, we can express the RW2 model in terms of a set of T conditional distributions as given below:

$$
v_t \mid v_{\{-t\}}, \sigma^2 \sim \begin{cases} N\left(2v_{t+1} - v_{t+2}, \sigma^2\right) & \text{for } t = 1 \\ N\left(\dfrac{2v_{t-1} + 4v_{t+1} - v_{t+2}}{5}, \dfrac{\sigma^2}{5}\right) & \text{for } t = 2 \\ N\left(\dfrac{-v_{t-2} + 4v_{t-1} + 4v_{t+1} - v_{t+2}}{6}, \dfrac{\sigma^2}{6}\right) & \text{for } t = 3,\dots,T-2 \\ N\left(\dfrac{-v_{t-2} + 4v_{t-1} + 2v_{t+1}}{5}, \dfrac{\sigma^2}{5}\right) & \text{for } t = T-2 \\ N\left(-v_{t-2} + 2v_{t-1}, \sigma^2\right) & \text{for } t = T \end{cases} \quad (12.13)
$$

We refer the reader to Exercise 12.4 for the derivation of Eq. 12.13. We can still show that the RW2 model is a one-dimensional version of the ICAR model, but the elements in the temporal weights matrix of the RW2 model are no longer binary. Instead, the conditional distributions in Eq. 12.13 take the form of the ICAR model with a general weights matrix (see Eq. 8.3 in Chapter 8):

$$
v_t \mid v_{\{-t\}}, \sigma^2, \mathbf{W}_{RW2} \sim N\left(\dfrac{\sum_{g=1}^{T} w_{tg} v_g}{w_{t+}}, \dfrac{\sigma^2}{w_{t+}}\right) \quad (12.14)
$$

where w_{tg} is the element in the tth row and the gth column in $\mathbf{W}_{RW2}$ and w_{t+} is the sum of the elements in the tth row. $\mathbf{W}_{RW2}$ is more complex than $\mathbf{W}_{RW1}$. For example, with $T = 5$,

$$
\mathbf{W}_{RW2} = \begin{pmatrix} 0 & 2 & -1 & 0 & 0 \\ 2 & 0 & 4 & -1 & 0 \\ -1 & 4 & 0 & 4 & -1 \\ 0 & -1 & 4 & 0 & 2 \\ 0 & 0 & -1 & 2 & 0 \end{pmatrix} \quad (12.15)
$$

The elements in $\mathbf{W}_{RW2}$ are derived from the conditional distributions in Eq. 12.13. For each row t in $\mathbf{W}_{RW2}$, the elements are basically the values multiplied by each of the T parameters $(v_1,\dots,v_T)$ in the numerator of the conditional mean for v_t. For example, for $t = 2$, the numerator of the conditional mean for v_2 is $2v_1 + 4v_3 - v_4$, so the elements in row 2 of $\mathbf{W}_{RW2}$ are 2, 0, 4, –1 and 0. The second and the fifth entries are 0 because v_2 and v_5 do not appear in $2v_1 + 4v_3 - v_4$.

We can use the `car.normal` function in WinBUGS to implement the RW2 model and use the weights matrix in Eq. 12.15 to specify the three arguments, `adj[]`, `weights[]`, and `num[]`. For example, with $T = 5$, the three arguments for the RW2 model are

adj=c(2,3,	num=c(2,	weights=c(2,-1,
1,3,4,	3,	2,4,-1,
1,2,4,5,	4,	-1,4,4,-1,
2,3,5,	3,	-1,4,2,
3,4)	2)	-1,2)

In practice, the key difference between RW1 and RW2 is that the estimated trend from RW2 tends to be smoother than that from RW1. Exercise 12.5 illustrates this by carrying out the same analysis as in Section 12.3.4.4 but using the RW2 model instead. In addition, when time points represent overlapping time intervals – for example, $t = 1$ represents the three-year interval between 1990 and 1992, $t = 2$ represents the interval between 1991 and 1993 and so on – then the RW2 model is typically used to reflect the overlapping nature of these time intervals (Clayton, 1996).

12.3.5 Interrupted Time Series (ITS) Models

12.3.5.1 Quasi-Experimental Designs and the Purpose of ITS Modelling

Interrupted time series (ITS) modelling is based on the time series quasi-experimental designs initially proposed by Campbell et al. (1963) that aim to assess the impact of an intervention on a system of interest. We focus on the so-called *multiple time series design* and Figure 12.14 illustrates the design schematically. As its name suggests, the multiple time series design requires time series observations, each denoted by O in Figure 12.14, from (at least) two groups, the treatment group, to which the treatment under study (denoted by X in Figure 12.14) is introduced at one or more time points, and the control group to which the treatment is not introduced. Amongst other quasi-experimental designs that they discuss, the multiple time series design is considered to be "[in general,] an excellent quasi-experimental design, perhaps the best of the more feasible designs" (Campbell et al., 1963, p.57). This is because, as they argue (p.55), "the experimental effect is in a sense twice demonstrated, once against the control and once against the pre-X values in its own series". Whilst under a quasi-experimental design the allocation of subjects into the treatment and the control groups is not random, Campbell et al. (1963, p.47–48) point out that "it should be recognized that the addition of even an un-matched or non-equivalent control group reduces greatly the equivocality of interpretation over what is obtained in Design 2, the One-Group Pretest-Posttest Design" – where Design 2 makes the pretest-posttest comparison using only the treatment group data – "The more similar the experimental and the control groups are in their recruitment, and the more this similarity is confirmed

The time series data from the treatment group: O O O OXO O O O

The time series data from the control group: O O O O O O O O

FIGURE 12.14
A schematic diagram of the multiple time series quasi-experimental design. For each of the two time series, each O denotes an observed outcome value, and for the time series of the treatment group, X denotes the intervention under study.

by the scores on the pretest, the more effective this control becomes." The remark in the last sentence emphasizes the importance of considering various ways of constructing the control group, as in Li et al. (2013) – an application of ITS that we will return to in Section 16.2 in Chapter 16.

To provide an example, Figure 12.15 shows the temporal sequences of the raw annual burglary rates per 1000 houses from two groups of COAs in Peterborough. The treatment group consists of 10 COAs in which the so-called "no cold calling" (NCC) scheme – the X in Figure 12.14 – was implemented. Note that not all the 10 COAs received the NCC scheme in the same year. Some were "treated" in 2005 and some in 2006. The control group comprises all the other 442 COAs in Peterborough in which the NCC scheme was not implemented. Each of two aggregated time trends in Figure 12.15 is obtained by aggregating the burglary cases to form a single time series of data, which are then divided by the total number of at-risk houses in the corresponding group to obtain the raw annual burglary rates. The comparison of the two temporal trends shows that from 2006 onward the two trends appear to gradually diverge, where the burglary rates of the treatment group remained stable but the burglary rates of the control group went up. This initial assessment of the data suggests a positive impact of the NCC scheme, but it also generates a series of questions, such as: "how big are differences in the post-NCC burglary rates between the two groups?"; "based on the observed data, how certain (or uncertain) are we that the post-NCC burglary rates in the NCC group are lower than those in the non-NCC group, thus inferring a positive impact of the policy?"; and "how sensitive are the findings to different constructions of the control group?" ITS modelling provides us with the analytical tool to tackle such questions.

ITS modelling can be considered as a modelling framework, as opposed to a model. This is because, as we shall see in the next section, an interrupted time series analysis requires the specification of at least two models, one for each group of time series data. Different temporal models are then considered depending on the nature of the time series data that we have observed (e.g. linear or nonlinear, observed over a short or a long period of time

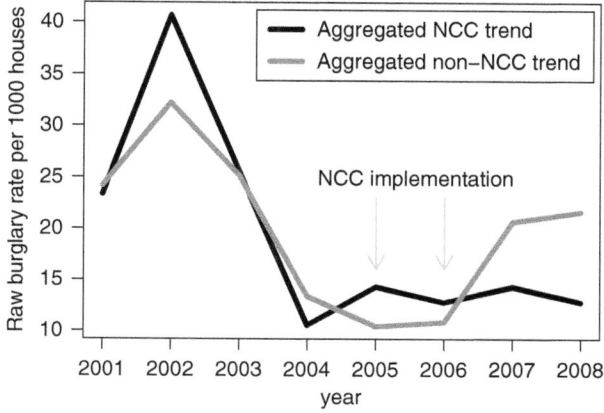

FIGURE 12.15

A comparison of the aggregated no cold calling (NCC) time trend (the black curve) to the aggregated non-NCC (non-treatment) time trend (the grey curve). The NCC scheme was implemented in some of the COAs in 2005 and some in 2006. The NCC trend is calculated by aggregating the burglary counts in the 10 NCC COAs to a single time series data and dividing the resulting burglary counts by the total number of houses in those 10 NCC COAs. The same calculation is applied to the non-NCC trend but for all other COAs in Peterborough where the NCC scheme was not implemented.

and so on). The next section focuses on the specification for an interrupted time series analysis and the subsequent section applies ITS to assess the NCC impact using the aggregated time series data presented in Figure 12.15.

12.3.5.2 Model Formulations

We describe ITS modelling in the context of evaluating the NCC scheme with the outcome of interest being burglary counts. To fix notation, denote $y^{treat} = \left(y_1^{treat}, \ldots, y_T^{treat}\right)$, the time sequence of burglary counts from the treatment group over a study period of T years. Similarly, denote $y^{control} = \left(y_1^{control}, \ldots, y_T^{control}\right)$, the time sequence of burglary counts from the control group over the same period of time. Denote also t_p, the time point at which the treatment/intervention is introduced. Here, for ease of exploration, we have made a few simplifications to the NCC evaluation. First, we only deal with the aggregated data. Second, we only deal with the situation where the intervention is introduced once at a single time point, although the NCC scheme was introduced over two years. Third, we only consider one plausible definition of the control group, namely, all the COAs in Peterborough without the NCC implementation. Section 16.2 in Chapter 16 describes the full NCC evaluation at the COA-level with a sensitivity analysis of the conclusions to different ways of constructing the control group. Although here the ITS analysis is set up for outcome data in the form of counts, it can also be adapted to model continuous-valued outcome data, as in Bloom (2003), for example, where ITS was used to analyse students' exam scores in order to measure the impacts of school reforms.

An ITS analysis comprises two models: the baseline model and the assessment model. The baseline model estimates the control trend pattern using the time series data from the control group. The assessment model analyses the treatment group data to perform the pre-post intervention comparison *after* controlling for the estimated control trend. For evaluating the NCC scheme, Table 12.6 presents the specifications of the two models. Specifically, both models describe the observed burglary counts using a Poisson likelihood, where $n^{control}$ and n^{treat} respectively denote the numbers of at-risk houses in the control and the treatment groups and both are assumed to be constant over the eight-year period. The two models differ in the modelling of the underlying burglary rates.

TABLE 12.6

The Specifications of the Baseline Model and the Assessment Model for an ITS Analysis to Assess the Impact of the NCC Scheme

The baseline model (for $t = 1,\ldots,T$)	The assessment model (for $t = 1,\ldots,T$)
$y_t^{control} \sim Poisson\left(\mu_t^{control}\right)$	$y_t^{treat} \sim Poisson\left(\mu_t^{treat}\right)$
$\mu_t^{control} = n^{control} \cdot \theta_t^{control}$	$\mu_t^{treat} = n^{treat} \cdot \theta_t^{treat}$
$\log\left(\theta_t^{control}\right) = \alpha^{control} + v_t$	$\log\left(\theta_t^{treat}\right) = \alpha^{treat} + v_t + I\left(t \geq t_p\right) \cdot f(t,b)$
$v_{1:T} \sim RW1\left(W_{RW1}, \sigma^2\right)$	$f(t,b) = b$
$\alpha^{control} \sim Uniform(-\infty, +\infty)$	$\alpha^{treat} \sim Uniform(-\infty, +\infty)$
$\sigma \sim U(0.00001, 10)$	$b \sim N(0, 1000)$

Under the baseline model, the burglary rates, $\theta_1^{control},\ldots,\theta_T^{control}$, are modelled, on the log scale, as a sum of an overall intercept, $\alpha^{control}$, and a set of time-point-specific parameters, $v = (v_1,\ldots,v_T)$. As suggested in Figure 12.15, the burglary trend in the non-NCC group appears to be nonlinear, so the RW1 model is a suitable candidate for v. Vague priors are chosen for the intercept, $\alpha^{control}$, and the conditional standard deviation of the RW1 model, σ (see Table 12.6).

Turning now to the assessment model, the log burglary rates are modelled by α^{treat}, an overall intercept specific to the treatment group, $v = (v_1,\ldots,v_T)$, the control trend pattern estimated from the baseline model, and $f(t,b)$, the impact function which, first introduced by Box and Tiao (1975), measures the deviations from the control trend. However, we want to measure such deviations *after* the implementation of the NCC scheme. To do that, the impact function is multiplied by the binary indicator function $I(t \geq t_p)$, which returns 1 when $t \geq t_p$, i.e. in and after the year of the NCC implementation (t_p), and returns 0 when $t < t_p$, i.e. during the pre-NCC period. Thus, this binary indicator ensures that the impact function comes into play only once the NCC scheme has been implemented. The impact function $f(t,b)$ in Table 12.6 takes the form of a step change where the scalar parameter b measures how big the change is. If b is estimated to be close to 0, then the post-NCC burglary rates between the two groups are shown to be similar, which then suggests no measurable impact of the NCC scheme from the observed data. However, if b is estimated to be negative (positive), then the post-NCC burglary rates in the NCC group are, on average, lower (higher) than the post-NCC burglary rates in the control group. In particular, a negative estimate for b would suggest the scheme has been successful. In the Bayesian framework, we can calculate the posterior probability of b being non-positive, i.e. $\Pr(b \leq 0 \mid data)$, to represent the probability of the policy's success.

In addition to quantifying the policy's impact, we can also consider different functional forms for the impact function in order to study the "pattern" of the impact. The Appendix shows three functional forms for $f(t,b)$ – the step change impact function, the linear impact function and the generalised impact function that encompasses the first two as special cases – and the details associated with each functional form. The next section employs the step change impact function to evaluate the NCC's impact, while Exercise 12.6 applies the other two functional forms. We can then compare model fits to select the most appropriate pattern of impact. The aggregated data do not raise any issues in estimating the parameter(s) in these forms of impact. However, when modelling the COA-level data, data sparsity poses challenges in parameter estimation (see Section 16.2 in Chapter 16). But first, we will apply the model setting presented in Table 12.6 to the aggregated data.

12.3.5.3 WinBUGS Implementation

Figure 12.16 presents the WinBUGS code for fitting the two models in Table 12.6 to the aggregated burglary count data from the NCC (treatment) and the non-NCC (control) groups. The code consists of two parts, one for the baseline model (Lines 9 to 22) and one for the assessment model (Lines 27 to 52). The code for the baseline model is similar to that for fitting the RW1 model to individual time series data in Figure 12.12. Some explanations are needed for the assessment model. Line 32 uses the `step` function to implement the binary indicator function $I(t \geq t_p)$, where t_p is set to 5, i.e. setting the NCC implementation year in 2005 (see Line 78 in the data list). The step change impact function is simply implemented by multiplying the parameter b by the binary indicator (Line 32). To implement the other two impact functions, modify Lines 31–32 according to the code given in the Appendix and specify the prior(s) on the associated parameter(s) accordingly.

```
 1     #  The WinBUGS code for fitting the ITS model to the
 2     #  aggregated data from the NCC (treatment) group and
 3     #  from the non-NCC (control) group
 4     model {
 5        #########################################################
 6        #  the baseline model for the aggregated non-NCC data
 7        #########################################################
 8        #  a loop to go through all the time points
 9        for (t in 1:T) {
10           #  defining the Poisson likelihood for each count
11           y.control[t] ~ dpois(mu.control[t])
12           mu.control[t] <- n.control*theta.control[t]
13           log(theta.control[t]) <- alpha.control + v[t]
14           #  predict burglary cases of the non-NCC group
15           y.pred.control[t] ~ dpois(mu.control[t])
16        }
17        #  RW1 model on v[1:T]
18        v[1:T] ~ car.normal(adj[],weights[],num[],prec.v)
19        #  priors
20        alpha.control ~ dflat()
21        sd.v ~ dunif(0.00001,10)
22        prec.v <- pow(sd.v,-2)
23        #########################################################
24        #  the assessment model for the aggregated NCC data
25        #########################################################
26        #  a loop to go through all the time points
27        for (t in 1:T) {
28           #  defining the Poisson likelihood for each count value
29           y.treat[t] ~ dpois(mu.treat[t])
30           mu.treat[t] <- n.treat*theta.treat[t]
31           log(theta.treat[t]) <- alpha.treat + v.cut[t]
32                             + step(t-ncc.year) * b
33           #  predict the burglary cases of the NCC group
34           #  with the impact function
35           pred.y.treat.with.impact[t] ~ dpois(mu.treat[t])
36           #  predict the burglary cases of the NCC group
37           #  without the impact function
38           mu.treat.no.impact[t] <- n.treat*theta.treat.no.impact[t]
39           log(theta.treat.no.impact[t]) <- alpha.treat + v.cut[t]
40           pred.y.treat.no.impact[t] ~ dpois(mu.treat.no.impact[t])
41        }
42        #  apply the cut function to each v[t] so that the
43        #  control trend is only estimated using the data
44        #  from the non-NCC group (without the cut function the
45        #  estimation of v[1:T] would be using data from
```

FIGURE 12.16
The WinBUGS code for fitting the two models under ITS modelling (Table 12.6) to the aggregated data on burglary counts from both the NCC and the non-NCC groups from 2001 to 2008. The initial values for two MCMC chains are also provided. For simplicity, the year of the NCC implementation is set at 2005.

```
46   #  both the control group and the treatment group)
47   for (t in 1:T) {
48     v.cut[t] <- cut(v[t])
49   }
50   #  priors
51   alpha.treat ~ dflat()
52   b ~ dnorm(0,0.0001)
53   #  calculate the posterior probability of Pr(b≤0|data)
54   prob.success <- step(-b)
55   #  ratio of the overall burglary rate of the NCC group
56   #  to that of the non-NCC group over the
57   #  post-policy period (i.e. 2005 to 2008)
58   RR <- exp(b)
59   #  the overall percentage change to the overall NCC rate
60   #  compared to the overall non-NCC rate over
61   #  the post-policy period (a negative value implies a
62   #  relative reduction in burglary rate in the NCC group)
63   percentage.change <- (RR-1)*100
64   }
65
66   #  the burglary data from a selected COA
67   list(T=8          #  8 years
68      #  burglary counts from the control group
69      ,y.control=c(1387,1849,1442,766,597,623,1187,1241)
70      ,n.control=57365  #  total number of at-risk houses
71                        #  in the control group
72      #  burglary counts from the NCC group
73      ,y.treat=c(31,54,34,14,19,17,19,17)
74      ,n.treat=1328     #  total number of at-risk houses
75                        #  in the NCC group
76      #  set the year in which NCC was implemented to the
77      #  fifth time point, i.e. 2005
78      ,ncc.year=5
79      #  defining the temporal neighbourhood structure based
80      #  on the RW1 model
81      ,num=c(1,2,2,2,2,2,2,1)  #  number of temporal neighbours
82                               #  for each time point
83      #   the IDs of the temporal neighbours
84      ,adj=c(2,1,3,2,4,3,5,4,6,5,7,6,8,7)
85      #  weights of the temporal neighbours
86      ,weights=c(1,1,1,1,1,1,1,1,1,1,1,1,1,1)
87   )
88
89   #  initial values for chain 1
90   list(alpha.control=-4,v=c(0.01,0,0,0,0,0,0,-0.01)
91       ,sd.v=0.1,alpha.treat=-5,b=0.1)
92   #  initial values for chain 2
93   list(alpha.control=-5,v=c(-0.01,0,0,0,0,0,0,0.01)
94       ,sd.v=0.2,alpha.treat=-4,b=-0.1)
```

FIGURE 12.16 *(Continued)*

It should also be highlighted that the estimated control trend, v, enters the assessment model (v.cut[t] on Line 31) after the cut function is applied to each parameter in v (Lines 47 to 49). The cut function ensures that the control trend is only estimated using the data from the control group but not influenced by the data from the NCC group. Specifically, at each MCMC iteration, the control trend v is estimated under the baseline model. This estimated trend is then "fed" into the assessment model while the cut function stops any information from the treatment data affecting the estimation of the control trend. More importantly, a benefit of placing the ITS analysis in the Bayesian approach is that the two models can be jointly fitted so that the *uncertainty* in the control trend estimate is fully accounted for when estimating the NCC's impact. By contrast, a non-Bayesian approach would typically take a two-stage approach in which the control trend is first estimated using the control group data, then *only* the point estimates of the control trend pattern are entered into the assessment model, effectively ignoring the uncertainty associated with the trend estimates. The cut function plays an important role here. For more information on the cut function, see "Tricks: Advanced Use of the BUGS Language" in the WinBUGS manual and also Lunn et al. (2012, p.201–203).

In addition to implementing the two models, we also calculate a number of quantities to help answer the questions at the end of Section 12.3.5.1. On Line 54, we use the step function to compute $\Pr(b \leq 0 \,|\, data)$, the posterior probability that the step change parameter b is less than or equal to 0. Line 58 exponentiates the parameter b so that the resulting quantity, RR, represents the ratio of the average burglary rate between the NCC group and the non-NCC group in the post-NCC period. Using RR, Line 63 calculates the percentage change to the overall burglary rate in the NCC group relative to that in the non-NCC group. We will come back to the interpretation of these quantities when we look at the results. We also predict the burglary counts for the non-NCC group (Line 15) and the burglary counts for the NCC group either using only the control trend without the impact function (Lines 38 to 40) or using both the control trend with the impact function (Line 35). These predictions will be compared to the observed count data from both groups to gain a better understanding of ITS modelling.

12.3.5.4 Results

The step change parameter, b, in the assessment model is estimated to be −0.23, with the 95% CI going from −0.51 to +0.05. The negative posterior mean suggests the average burglary rate for the NCC group over the post-NCC period (i.e. from 2005 to 2008) is lower than that for the non-NCC group, although the 95% CI covers 0. The posterior probability $\Pr(b \leq 0 \,|\, data)$ is estimated to be 0.94. This suggests there is a high probability (a "94% chance") that the NCC scheme, implemented in selected areas of Peterborough, is associated with a reduction in the overall burglary rate in those areas relative to the overall burglary rate in the non-NCC group. We cannot, of course, claim to know that the association between the introduction of the scheme and the relative reduction in the number of burglaries is causal (see Section 3.2).

To quantify the policy's impact, we have calculated the percentage change to the average post-NCC burglary rate in the NCC group relative to that in the non-NCC group, namely, $(RR - 1) \times 100$, where $RR = \exp(b)$ (see Lines 58 and 63 in Figure 12.16). The posterior mean of the percentage change is −19.5%, with a 95% CI going from −40.0% to +4.7%. This estimate suggests that after NCC implementation, the average burglary rate in the NCC group is 19.5% lower compared to that in the non-NCC group. We have 95% confidence that the average post-NCC burglary rate in the NCC group can be as low as 40.0% of that in the non-NCC group, but it can also be 4.7% higher than the corresponding burglary rate in the non-NCC group. There is a very small chance, about 6% (= 100%–94%, with 94% coming from $\Pr(b \leq 0 \,|\, data)$), of an increased burglary rate in the NCC group.

To better understand the process of evaluation, Figure 12.17 shows the estimates for the different components in the ITS analysis. As shown in Figure 12.17(a), the control trend fits the observed count data from the control group reasonably well. However, for the NCC group, using the estimated control trend alone overestimates the observed number of burglary cases in 2008 (Figure 12.17(b)). To account for this lack of fit, the posterior mean of b in the step change impact function is estimated to be negative so that, from 2005 onwards, the control trend is pulled down to provide a better fit to the observed numbers of cases in the NCC group, in particular to the number observed in 2008 (Figure 12.17(c)). A careful inspection of Figure 12.17(b) and (c) suggests the departure of the NCC trend from the non-NCC trend is gradual rather than abrupt. Thus, a linear impact function may be more appropriate – Exercise 12.6 investigates such an option.

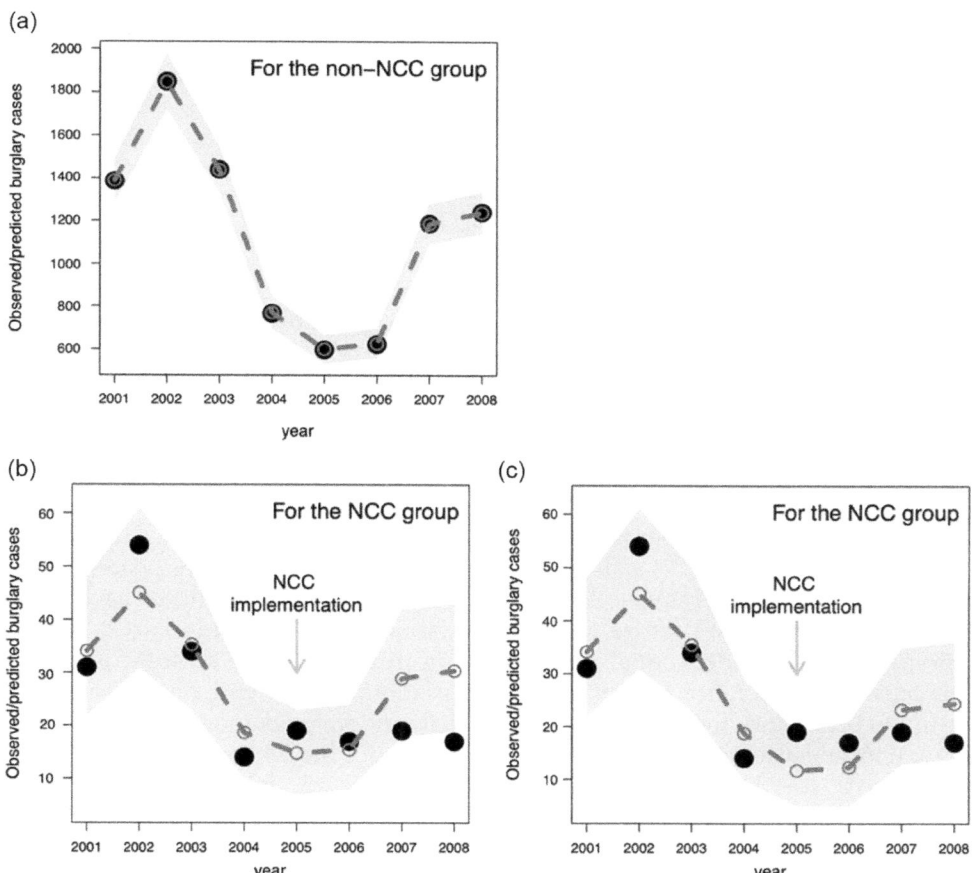

FIGURE 12.17
The estimated components from the baseline model and the assessment model. In each plot, the black dots represent the observed numbers of burglary cases while the circles with the dashed line show the posterior means of the predicted numbers from either the baseline model (in (a)) or the assessment model (in (b) and (c)). The grey region represents the 95% CIs across the eight years. In (b), the observed burglary case counts are fitted using only the control trend estimated from the control group data without the step change impact function. In (c), the step change impact function is added back in. The parameter b is estimated to be negative, thus lowering the control trend from 2005 onwards and giving a better fit to the observed numbers of burglary cases in the NCC group over the post-NCC period.

In conclusion, the analysis presented here suggests some evidence to support the overall success of the NCC scheme. It should be noted that the success of the scheme is shown to be stabilising the average burglary rate in the NCC group, whilst the burglary rates in the non-NCC group appear to be increasing. This highlights the importance of evaluating the policy's impact *in relation to* the control group. In a quasi-experimental design, whilst we know which subjects are in the treatment group, how to construct a control group is typically not known. Given its importance, it is central in an ITS analysis to consider different ways of constructing the control group so that we can examine their impacts on the policy's evaluation.

In addition to an overall assessment, assessing the policy's impact *locally* is equally important, as it helps us to address questions such as: "Do different COAs where the NCC scheme was implemented experience different levels of impact?"; "If so, what is the extent to which the policy's impact varies from one treated COA to another?"; "If the policy's impact does vary from one treated COA to another, what would be the factors that affect its local impact?" and equally important, "Is there any COA in which the NCC scheme did not show an impact in the sense that its post-NCC burglary rates are similar to those in the non-NCC group?" These questions help us to understand the *spatial variation* of the policy's impact, to examine factors that are associated with such spatial variation and ultimately to improve the local implementation of the policy. In this section, we have demonstrated the use of time series data in conjunction with the ITS modelling framework to evaluate the overall impact of a locally-implemented policing policy. In Section 16.2 in Chapter 16, we will extend this evaluation framework to the spatial-temporal setting in order to evaluate not only the policy's overall impact but also its impact on individual treated areas.

12.4 Concluding Remarks

This chapter has described the two frameworks proposed by Knorr-Held (2000) for modelling spatial-temporal data: the space-time separable framework and the space-time inseparable framework. Both frameworks include a model (or component) to capture spatial variation and a model to capture temporal variation. The material in Part II of the book has prepared us to model spatial variation, so in this chapter we have focused on models for capturing different forms of temporal variation: linear trend, non-linear trend and interrupted time series models. We have seen that random walk models capture smoothly varying non-linear temporal variation, the second order model being the smoother of the two because it uses more temporal neighbours. The RW1 model represents a one-dimensional analogue of the ICAR model (see Section 8.2) with binary weights and first order contiguity, whilst the RW2 model is a one-dimensional analogue of the ICAR model with general weights.

But we should not assume space and time are independent dimensions of variation. The term space-time interaction is used to refer to the existence in a dataset of clusters of events in particular places at particular times. It often arises that events that occur close together in time also occur close together in space, giving rise to space-time clusters or what is called *space-time interaction*. Maps of the incidence of infectious diseases in humans and animals often display this property. Cases of certain types of crime cluster in certain places at certain times, as do traffic accidents and earthquakes. The spread of a rumour (who knows and who does not) may display characteristics that are similar to the spread of an infectious disease. The reader is referred back to Section 2.4.1 for further discussion and some relevant literature on spatial-temporal interaction processes that may generate this space-time interaction property. To handle this property when it is present in a spatial-temporal dataset, we need to use the space-time inseparable modelling framework.

Clearly the inclusion or exclusion of the space-time interaction component marks a significant point of demarcation in modelling variation in a spatial-temporal dataset. For this reason, it will be useful to have exploratory techniques that will help to alert us to the possible presence of space-time interaction in our data. This will be the subject of the next chapter.

12.5 Exercises

Exercise 12.1. For an area with 100 houses, if six and two cases of burglary were reported over two consecutive years, construct a Bayesian model to produce (a) an estimate of the level of burglary risk in each year and (b) the ratio of the two estimated risk levels. Is there any evidence of a change in burglary risk between the two years?

Exercise 12.2. Consider the following situation. A spatially-smoothed rate is calculated by adding the cases and the populations from two areas, A and B, together, where area A has 10 cases with 100 houses and area B has one case with any number of houses between one and four. Show that the spatially-smoothed rate estimate for area A becomes more uncertain compared to producing a rate estimate using only the data in area A alone. Would such a situation be likely to occur in practice?

Exercise 12.3. Fit the linear trend model without temporal noise to the burglary count data over four years from each of the 452 COAs in Peterborough. Produce a map of the slope estimates and investigate the factors that may affect the uncertainty associated with the slope estimates.

Exercise 12.4. Starting with the formulation of the RW1 model in Eq. 12.5, show that, for $T \geq 3$, the conditional probability distribution for each v_t, given $v_1,...,v_{t-1},v_{t+1},...,v_T$, takes the form of Eq. 12.6. (Hint: the proof can start with a simple case with $T = 3$, then

$$\Pr(v_2 \mid v_1, v_3) = \frac{\Pr(v_1, v_2, v_3)}{\Pr(v_1, v_3)} = \frac{\Pr(v_3 \mid v_2)\Pr(v_2 \mid v_1)\Pr(v_1)}{\Pr(v_3 \mid v_1)\Pr(v_1)} = \frac{\Pr(v_3 \mid v_2)\Pr(v_2 \mid v_1)}{\Pr(v_3 \mid v_1)}. \text{ One}$$

can then use Eq. 12.5 to write down the normal probability density functions for the two terms in the numerator, i.e. $v_3 \mid v_2 \sim N(v_2, \sigma^2)$ and $v_2 \mid v_1 \sim N(v_1, \sigma^2)$. The normal probability density function in the denominator can be written down by realising $v_3 = v_2 + \varepsilon_3 = v_1 + \varepsilon_2 + \varepsilon_3$, and thus $v_3 \mid v_1 \sim N(v_1, 2\sigma^2)$). Follow the above derivation for the RW1 model; derive the conditional distributions in Eq. 12.13 for the RW2 model (hint: again, start with a simple case with $T = 5$).

Exercise 12.5. Produce the analysis of the COA-level trend in Peterborough as given in Section 12.3.4.4. Repeat the same analysis but replace the RW1 prior model on the set of temporal random variables by the RW2 model. Compare the estimated COA trends and the estimated Peterborough trend from the two models (one using an RW1 prior and the other using an RW2 prior). Comment on the smoothness of the trend estimates from the two models. Is there any noticeable difference in the Peterborough trend from the two models?

Exercise 12.6. Carry out the group level NCC evaluation as presented in Sections 12.3.5.2 and 12.3.5.3. Then repeat the same analysis with the step change impact function replaced by (a) the linear impact function and (b) the generalised impact function. Is the evaluation result robust against different forms for the impact function? Modify the WinBUGS code in order to assess how well the model with each impact function describes the post-policy data from the NCC group.

Appendix: Three Different Forms for Specifying the Impact Function f.

	A Step Change Impact Function	A Linear Impact Function	A Generalised Impact Function
The functional form for f	$f(t,b) = b$	$f(t,d) = d \cdot (t - t_p + 1)$	Let $f(t,a,s) = R_t$, where R_t is defined iteratively as follows: $$R_t = \begin{cases} a \cdot R_{t-1} + m & \text{when } t \geq t_p \\ 0 & \text{otherwise} \end{cases}$$ and $m = \alpha^{treat} - s$ where α^{treat} is the intercept term in the assessment model (see Table 12.6).
Description of the represented pattern of impact	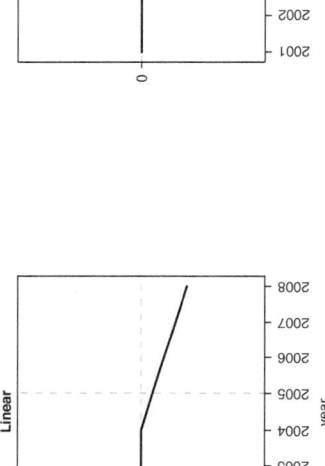 Abrupt and permanent	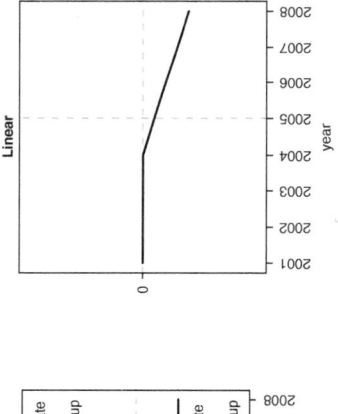 Gradual, permanent and linear (on the log scale)	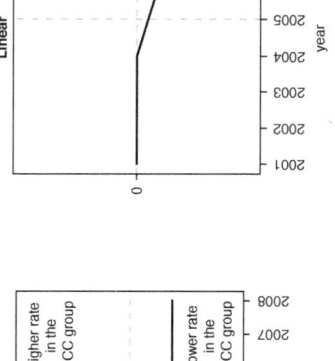 Gradual, permanent but nonlinear (on the log scale)
Parameter interpretation	b measures the size of the step change. In the context of the NCC evaluation, a negative (positive) estimate for b suggests a lower (higher) average burglary rate in the NCC group compared to the non-NCC group over the post-NCC period.	d measures the slope of a linear departure from the control trend on and after the implementation of the policy. A negative (positive) estimate for b would suggest a positive (negative) policy impact, i.e. a relative reduction (elevation) in the burglary rate.	The parameter $m = \alpha^{treat} - s$ estimates the departure of the NCC trend from the control trend, and the parameter a, constrained to be $0 < a < 1$, measures the degree of nonlinearity in the departure. In the limiting cases, when $a = 0$, the generalised function reduces to the step change function while when $a = 1$, we have the linear impact function. However, the parameters m and s do not directly correspond to b and d in the step change and the linear impact functions.

	b	*d*	*a*, *m*
Interpretation of transformations of parameters	• exp(*b*) is the ratio of the average post-NCC burglary rates between the two groups; • (exp(*b*) − 1)×100 represents the percentage change to the average burglary rate in the NCC group in relation to the non-NCC group over the post-NCC period.	• $\left(\exp(d \cdot t^*) - 1\right) \times 100$ is the percentage change to the burglary rate in the NCC group relative to the non-NCC group *t** years after the NCC implementation. For example, one year after the NCC implementation, the percentage change to the burglary rate in the NCC group relative to the controls is quantified by (exp(*d*) − 1)×100.	• $\left[\exp\left(\sum_{k=1}^{t} m \cdot a^k\right) - 1\right] \times 100$ is the percentage change to the burglary rate in the NCC group relative to the non-NCC group *t** years after the NCC implementation. For example, after *t** = 1 year of NCC implementation, the percentage change to the burglary rate in the NCC group relative to the controls is quantified by (exp(*ma*) − 1)×100.
An overall NCC success is indicated when	$b \leq 0$ or, equivalently, exp(*b*) ≤ 1 or (exp(*b*)−1)×100 ≤ 0.	$d \leq 0$ or, equivalently, $\left(\exp(d \cdot t^*) - 1\right) \times 100 \leq 0$	$ma \leq 0$ or, equivalently, $\left[\exp\left(\sum_{k=1}^{t^*} m \cdot a^k\right) - 1\right] \times 100 \leq 0.$
Choice of vague prior(s)	$b \sim N(0, 1000000)$	$d \sim N(0, 1000000)$	$a \sim Uniform(0.00001, 0.99999)$ $s \sim N(0, 1000000)$

WinBUGS implementation	See Figure 12.16	Replace Lines 31 and 32 by
		`log(theta.treat[t]) <- alpha.treat + step(t-ncc.year) * d * (t - ncc.year + 1)`
		
		and add the prior specification for *d*, e.g. `d ~ dnorm(0, 0.000001).`
		
		Remove Line 52 and modify Lines 54 to 58 and the two sets of initial values accordingly.
			Replace Lines 31 and 32 by
			`log(theta.treat[t]) <- alpha.treat + v.cut[t] + step(t-ncc.year) * R[t].`
			
			Then, add the following immediately after Line 41:
			<pre>for (t in 1:(ncc.year-1)) { # pre-NCC
			R[t] <- 0
			}
			for (t in ncc.year:T) { # post-NCC
			R[t] <- a*R[t-1] + m
			}
			m <- alpha.treat - s
			# vague prior for a and s
			s ~ dnorm(0, 0.000001)
			a ~ dunif(0.00001, 0.99999).</pre>
			Remove Line 52 and modify Lines 54 to 58 and the two sets of initial values accordingly.

13

Exploratory Analysis of Spatial-Temporal Data

13.1 Introduction

In this chapter we describe methods for analysing a new set of spatial-temporal data. In Chapter 6 we presented methods for exploratory spatial data analysis (ESDA) which support small area spatial data modelling. In this chapter we turn to methods for exploratory spatial-temporal data analysis (ESTDA) which support small area space-time data modelling. All our earlier comments regarding exploratory data analysis (EDA) carry over to the space-time context, and we refer the reader back to Section 6.1 if a reminder is needed about the principles that underlie the methods used in EDA and what its objectives are.

In Chapter 12 (in Section 12.1 in particular) we made several important points about working with space-time data and modelling space-time data variability, three of which have a direct bearing on the conduct of ESTDA. First, we often encounter a substantial increase in the amount of data. Whilst this is not necessarily a challenge for the methods we shall want to use, the amount of data and the fact that it is partitioned into the "third" dimension of time can present a challenge for how to visualise both the original data and the results of analysis including model fit effectively. The static page is not always the best medium for data visualisation, but it can be particularly inhibiting when it comes to engaging with space-time data. Whilst we will discuss space-time data visualization in this chapter, we will not pursue the broader question of how best to visualize space-time data. Instead we refer the reader to some specialist treatments (see for example: Andrienko and Andrienko, 2006; Ye and Rey, 2013; Cressie and Wikle, 2011, p.243–259).

Second, whilst dependency and heterogeneity are two fundamental properties of spatial data, we must now contend with these properties in time as well as dependency and heterogeneity in both space *and* time. The heterogeneity property tells us that a characteristic (say the crime risk or per capita income level) in a particular area at a particular time point can be different from the same characteristic in another area at the same time point and/or in the same area but at a different time point. But to further complicate the picture, things tend to vary in a "dependent" fashion, showing autocorrelation over space, time or over both space *and* time. It is the combination of these two properties that leads to interesting patterns in spatial-temporal data that we need to understand through the help of modelling. Examining model fit now includes evaluating whether there are any spatial-temporal heterogeneity and/or dependency patterns in model residuals.

Third, in Chapter 12 we drew on the modelling frameworks of Knorr-Held (2000) in which space-time variability in a dataset is partitioned into three components: the common (or overall) spatial component, the common (or overall) temporal component and the space-time interaction component. A space-time separable model consists of the first two components, whilst a space-time inseparable model comprises all three components (see

Figure 12.4). These two modelling frameworks inform the direction of this chapter. We are particularly interested in exploratory methods that provide evidence for the presence of space-time interaction in our data, implying the need for a space-time inseparable model rather than a space-time separable model.

In Section 13.2, we look at some space-time data patterns that can be seen to follow either a space-time separable structure or a space-time inseparable structure. These example patterns will help the reader to visualise several important concepts such as space-time separability versus space-time inseparability; presence versus absence of space-time interaction in a dataset; what we mean by purely spatial clustering, purely temporal clustering and localised space-time clusters; and two different forms of space-time interaction (one form referring to the existence of clusters of cases in particular places at particular times whilst the other form refers to the presence of different local time trends across different areas). A key purpose of ESTDA is to identify whether space-time interaction is present. This question can be addressed graphically as well as numerically using test statistics. In Section 13.3, we discuss ways to visualise a spatial-temporal dataset. In Section 13.4, we describe test statistics. In Section 13.4.1, we describe Knox's *global test* for space-time *clustering*. This test establishes whether there is a whole map tendency for cases to cluster in space-time (whether large (small) values are generally found near other large (small) values in space-time). Section 13.4.2 describes Kulldorff's space-time scan statistic for detecting *localized* space-time *clusters*. This test seeks to identify local space-time clusters, each of which contains large values (or small values) located closely in both space and time. Section 13.4.3 describes methods for assessing the presence of space-time interaction where different areas display different time trends but where these time trends may be spatially autocorrelated so that information borrowing can be employed at the modelling stage. We illustrate all these exploratory techniques with worked examples.

13.2 Patterns of Spatial-Temporal Data

In Section 6.2 we suggested that Tukey's (1977) "smooth"/"rough" distinction could be applied to spatial data, providing an organizing principle for describing ESDA techniques. We suggest the same applies in the case of spatial-temporal data:

$$\text{SPATIAL-TEMPORAL DATA} = \text{SPATIAL-TEMPORAL SMOOTH}$$
$$+\text{SPATIAL-TEMPORAL ROUGH} \tag{13.1}$$

The spatial-temporal smooth part describes the regular and predictable pattern in the observed data. Following the space-time separable structure discussed in Chapter 12 (Figure 12.4), this smooth part of the data can be further decomposed into two components, the overall spatial component and the overall temporal component. These two components are assumed to be independent (or separable) in the sense that all the spatial units are assumed to display the same time trend pattern as described by the overall temporal component and, equivalently, the sequence of maps over time are assumed to all show the same spatial pattern as depicted by the overall spatial component. If a spatial-temporal dataset can be described adequately by these two components, then the dataset is said to be space-time separable, and space-time interaction is not present in the data.

Figure 13.1 illustrates two examples of a space-time separable dataset, each of which can be well represented by an overall time trend and an overall spatial distribution. However, the two space-time patterns in Figure 13.1 are slightly different. In the case of Pattern 1, all areas follow the same increasing time trend, but there is evidence of *purely spatial clustering* because the outcome values of the three areas in the north are consistently higher than those in the other areas. Purely spatial clustering may arise for example in a situation where different areas have different income levels at the beginning of the observation period (due to differences in local circumstances) but all the areas in the region show the same increasing pattern in income levels over time. Since all areas only differ by level, the overall map is then sufficient to capture between-area differences.

Pattern 2 in Figure 13.1, on the other hand, shows that although overall levels across the areas are all very similar, it is the time trend pattern that changes. The outcome values for the last part of the observation period are all higher as compared to the earlier period. Every single area in the study region experiences exactly the same increase. This is referred to as *purely temporal clustering*. For example, suppose a new policy is implemented uniformly across all the areas and on the same date, and results in the same impact across all the treated areas. In that case, a purely temporal analysis of the overall time trend (through the use of, for example, interrupted time series modelling) is sufficient to make inference regarding the policy's impact. Finally, we can combine these two patterns to form a general space-time separable pattern in which different areas have different overall levels (as in Pattern 1) but the same temporal change happens in every single area (as in Pattern 2). This general pattern can still be explained just using the overall spatial and the overall temporal components, i.e. what we have called the spatial-temporal smooth part in Eq. 13.1.

Often, however, real spatial-temporal data also possess a "rough" part. This spatial-temporal rough part corresponds to the spatial-temporal interaction component in the

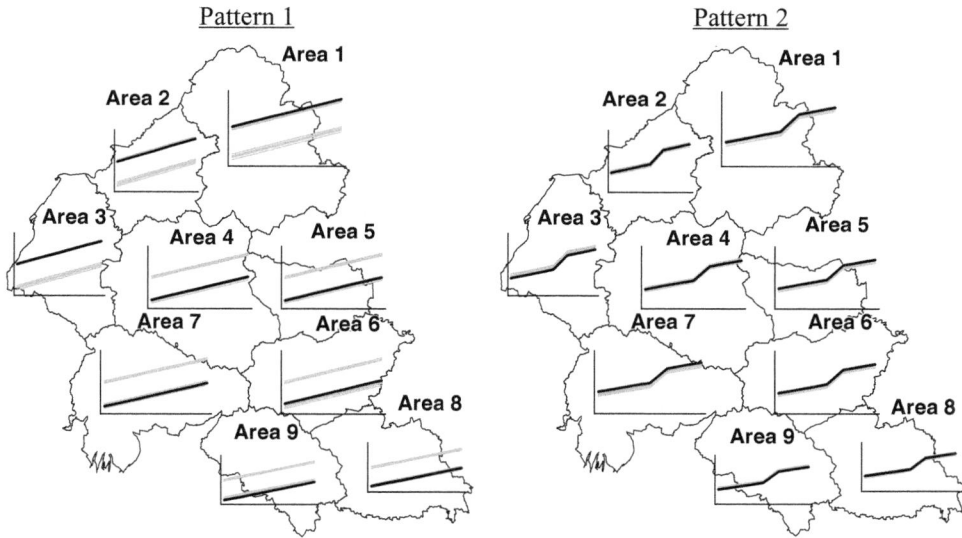

FIGURE 13.1
Two examples of a space-time separable pattern. The inserted plot for each area shows the trend of that area (the black line) and the trends of all other areas (the grey lines).

space-time inseparable modelling framework (Figure 12.4). When spatial-temporal inter-action is present in a dataset, we can no longer describe space-time variation in the data by drawing on only the overall spatial and the overall temporal components. A space-time interaction component needs to be added. That component deals with the part of the data variability that deviates from the variability which can be explained by the space-time separable structure. Different from purely spatial or purely temporal clus-tering, spatial-temporal interaction describes *localised spatial-temporal patterns* that only appear in a subset of areas and/or time points. Figure 13.2 illustrates two examples of a space-time inseparable structure. Pattern 3 in Figure 13.2 shows that whilst all areas display a linear time trend pattern, the slopes of these linear time trends vary spatially. In such a case, an overall time trend pattern can no longer describe the local time trends well, but instead we need to incorporate the space-time interaction component to allow each local trend to deviate from the overall time trend. In a more complex situation, areas can also differ by the overall level, in addition to displaying different time trend patterns.

A different example of space-time inseparability is shown in Pattern 4 in Figure 13.2. In that case, some areas in the study region with higher outcome values (say higher disease rates) emerge at a later stage in the observation period. Again, the spatial-temporal smooth part alone can no longer describe this space-time pattern well but needs to be modified locally, through the addition of the space-time interaction component, to capture the local increase in the two areas in the north.

The emergence of such localised patterns, in the form of either Pattern 3 or Pattern 4, may be due to changes to local characteristics, for example, the emergence of a new localised risk factor, changes to population demographics, a sudden outbreak of cases of a disease or the introduction of an intervention or early prevention policy. An exploratory analysis

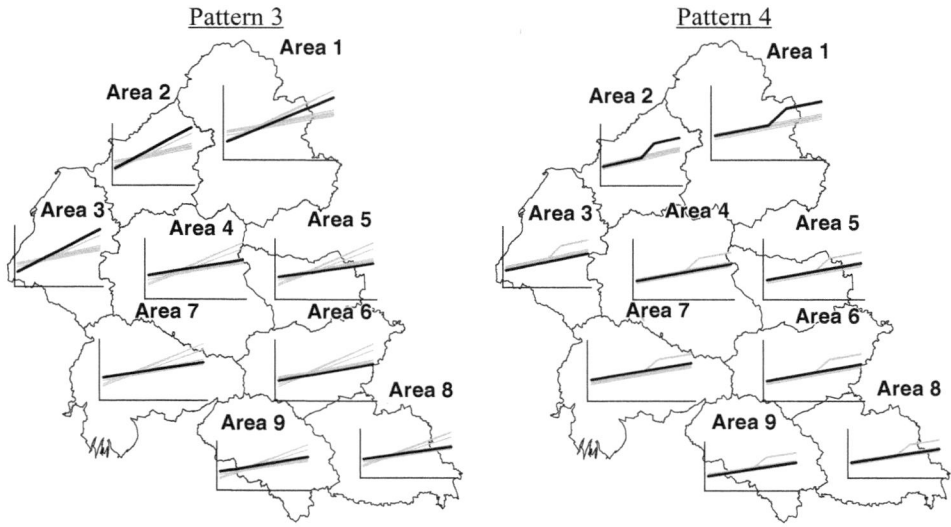

FIGURE 13.2
Two examples of a space-time inseparable pattern. The labelling of these two patterns follows on from Figure 13.1. The inserted plot for each area shows the trend of that area (the black line) and the trends of all other areas (the grey lines).

of a given space-time dataset will help us to decide what type of space-time structure we are dealing with.

Before leaving this section, it is worth pointing out that space-time interaction, as a concept, involves identifying space-time deviations from "some norm". In other words, if space-time interaction is found to be present in a dataset, then there are values in the dataset deviating markedly from a space-time pattern that we consider to be usual or expected. Typically, the definition of the "norm" is taken to be the space-time separable structure (thus space-time interaction is absent in the two patterns in Figure 13.1) but not always. In certain applications, we need to pay particular attention to the definition and the construction of the norm – see, for example, Section 16.2. Quite often, as we shall illustrate through various applications in Chapter 16, an inferential goal of modelling space-time data is to detect any departures of the data from the norm, then to learn about the nature of such space-time departures. Modelling and detection of such localised patterns is important in epidemiology (Lawson, 2013), disease surveillance (Corberan-Vallet and Lawson, 2016) and criminology (Li et al., 2014).

In practice we rarely encounter the sorts of "tidy" space-time patterns shown in Figure 13.1 and Figure 13.2 because data sparsity often masks the underlying spatial-temporal patterns in the data. Nonetheless, visualisation, through mapping and plotting the small area time series data, still has a potentially important role to play in the exploratory analysis of a spatial-temporal dataset. It may help us to get a better understanding of our data ("a picture is worth a thousand words") and may help us spot possibly interesting "signals or departures" in the data. With these remarks and qualifications in mind, we now describe how we can visualise a space-time dataset.

13.3 Visualising Spatial-Temporal Data

A simple, but effective, way to visualise a set of space-time data is to produce a sequence of maps in chronological order. Figure 12.1 in Chapter 12 provides an example showing a time sequence of maps of the annual COA-level raw burglary rates in Peterborough from 2005 to 2008. These maps can be easily produced following the five basic steps for producing a single choropleth map described in Section 6.2.1. The only modification is that we need to use the same set of cut points so that the sequence of maps is comparable. We now take the malaria incidence data introduced in Section 1.6 as an example to illustrate these five steps in a space-time setting. We want to visualise the raw malaria incidence rates (per 1000 people) across the 139 villages (small areas) in Kalaburagi taluk, South India, for each of the 12 two-month periods from February 2012 to January 2014 (Feb–Mar 2012, Apr–May 2012 and so on). The block of R code below first reads the shapefile of the malaria data into R, then calculates the raw space-time incidence rates.

```
1  # load the maptools package in for reading a shape file into R
2  library(maptools)
3  # read in the shapefile of Kalaburagi taluk containing the
4  # village-level malaria incidence cases across the 12 two-month
5  # periods
```

```
 6  malaria <- readShapePoly('malaria.shp')
 7  ####################################################################
 8  # Step 1: calculate the raw malaria incidence rates
 9  # per 1000 people
10  ####################################################################
11  cases <- malaria@data[c('O1','O2','O3','O4','O5','O6'
12                          ,'O7','O8','O9','O10','O11','O12')]
13  pop <- malaria@data$TotPop
14  # calculate the raw malaria incidence rates per 1000 people
15  rawRate <- cases/pop *1000
```

On Lines 11–12, the syntax `malaria@data` is used to access the dataset stored in the shapefile (see Appendix 4.13.2 in Chapter 4 for more detail), then the malaria cases in the 12 columns, O1, ..., O12 (one for each two-month time period), are extracted and stored in the object called `cases` – a data matrix of size 139 × 12. The array `pop` (Line 13) contains the numbers of at-risk people across the 139 villages, and population sizes are assumed to be constant over the entire study period.

Step 2 defines a set of cut points to assign each continuous-valued raw incidence rate into a category, which is then associated with a colour shade in Step 3 for mapping. But what cut points should we use? A feature of this dataset is the excessive number of zeroes in the count data. As summarised in Table 13.1, at least 86% of the 139 villages have no reported cases of malaria in each of the 12 time windows. The cut points for the risk categorisation are chosen to be 0.00, 0.09, 0.73, 4.01, 9.19 and 34.90, where the last four cut points are the minimum, the first, second and third quartiles and the maximum of *all* the non-zero raw incidence rates. Therefore, the first (lowest) risk category accommodates the zero values in

TABLE 13.1

Descriptive Summary of the Village-Level Case Counts Between February 2012 and January 2014

2012/13	Feb–Mar	Apr–May	Jun–Jul	Aug–Sept	Oct–Nov	Dec–Jan
Min	0	0	0	0	0	0
Mean	1.10	0.64	1.12	1.16	1.29	0.93
Max	30	30	33	40	70	38
% villages with 0 cases	92.8	92.1	87.8	91.4	93.5	89.9
2013/14	Feb–Mar	Apr–May	Jun–Jul	Aug–Sept	Oct–Nov	Dec–Jan
Min	0	0	0	0	0	0
Mean	0.73	0.81	1.55	0.86	0.96	0.60
Max	19	22	40	20	32	15
% villages with 0 cases	89.9	91.4	86.3	90.6	91.4	91.4

the observed case counts. The associated R code for summarising the count data and for Step 2 is given below.

```
1   ###################################################################
2   # summarising the case data as reported in Table 13.1
3   ###################################################################
4   summary(cases)
5   # Line 13 below uses the apply function to go through each column
6   # (i.e. time point) to obtain the proportion of villages with 0
7   # cases reported. Specifically, the 1st and 2nd arguments in the
8   # apply function specify the data matrix and the margin on which
9   # the function specified in the 3rd argument is applied over
10  # (1=over rows and 2= over columns), the 3rd argument specifies a
11  # function to calculate the proportion of villages with 0 cases at
12  # each time point.
13  apply(cases,2,function(x){length(which(x==0))/139})
14
15  ###################################################################
16  # Step 2: define a set of cut points to convert the
17  # continuous-valued raw incidence rates into categories
18  ###################################################################
19  # (a) obtain the minimum, the first, second and third quartiles
20  # and the maximum of the non-zero raw incidence rates
21  rawRate.array <- c(unlist(rawRate)) # make the matrix into
22                              # a one-dimensional array
23  # using the which function to select the non-zero values
24  non.zero.rawRate <- rawRate.array[which(rawRate.array>0)]
25  # Line 26 (below) returns 5 values: min, Q1, Q2, Q3 and max
26  q <- quantile(non.zero.rawRate)
27  cutpoints <- c(0,q) # obtain the cut points used next
28
29  # (b) for each two-month time period, assign each raw incidence
30  # rate to one of the five categories defined by the cut points
31  rate.level <- matrix(0,nrow=139, ncol=12) # construct a matrix of
32                              # size 139 ×12 to store
33                              # the resulting categories
34  for (tt in 1:12) {
35   rate.level[,tt] <- cut(rawRate[,tt]
36    ,cutpoints,labels=1:5,include.lowest=TRUE)
37  }
```

Step 3 defines five shadings on the greyscale, one for each category.

```
1   ####################################################################
2   # Step 3: define shadings
3   ####################################################################
4   shadings <- grey(c(1,0.8,0.6,0.4,0.2))
```

Step 4 then produces the map sequence for the 12 time periods. For ease of comparison, the 12 maps are arranged into two rows, one for each year. This is done using the mfrow option in the par function (see Line 4 below). A legend is added at Step 5. The R code under both Steps 4 and 5 is given below, and Figure 13.3 shows the resulting map sequence.

```
1    #####################################################################
2    # Step 4: create the choropleth maps, one for each time window
3    #####################################################################
4    par(mfrow=c(2,6)) # create a plot window with 2 rows and 6
5      # columns to display the 12 maps
6    # set title for each plot/time window
7    titles <- c('Feb-Mar 2012','Apr-May 2012','Jun-Jul 2012'
8                 ,'Aug-Sep 2012','Oct-Nov 2012','Dec-Jan 2012/3'
9                 ,'Feb-Mar 2013','Apr-May 2013','Jun-Jul 2013'
10                ,'Aug-Sep 2013','Oct-Nov 2013','Dec-Jan 2013/4')
11   # produce the plots
12
13   for (tt in 1:12) {
14     plot(malaria,col=shadings[rate.level[,tt]],main=titles[tt])
15   }
16   #####################################################################
17   # Step 5: add a legend
18   #####################################################################
19   intervals <- c('Zero risk','[0.09, 0.73)','[0.73, 4.01)'
20                   ,'[4.01, 9.19)','[9.19, 34.90]')
21   legend('topleft',legend=intervals, fill=shadings)
```

It is evident from Figure 13.3 that (1) most of the villages did not have any reported malaria cases during the two years; (2) the raw incidence rates of the group of villages concentrated in the north are consistently high and (3) there are some villages, scattered across the study region, showing low, but non-zero, incidence rates, and the occurrence of malaria cases in these villages appears to be sporadic. Overall, the mean number of cases remains relatively stable at around one case per village per two months (Table 13.1). Figure 13.4 visualises the space-time incidence rates from the "temporal" perspective. Figure 13.4(a) shows the overall stable time trend as well as the time trends of the villages where there is at least one case reported over the two-year period. In Figure 13.4(a), there is a clear separation of those villages into two distinct groups, one group consisting of villages with incidence

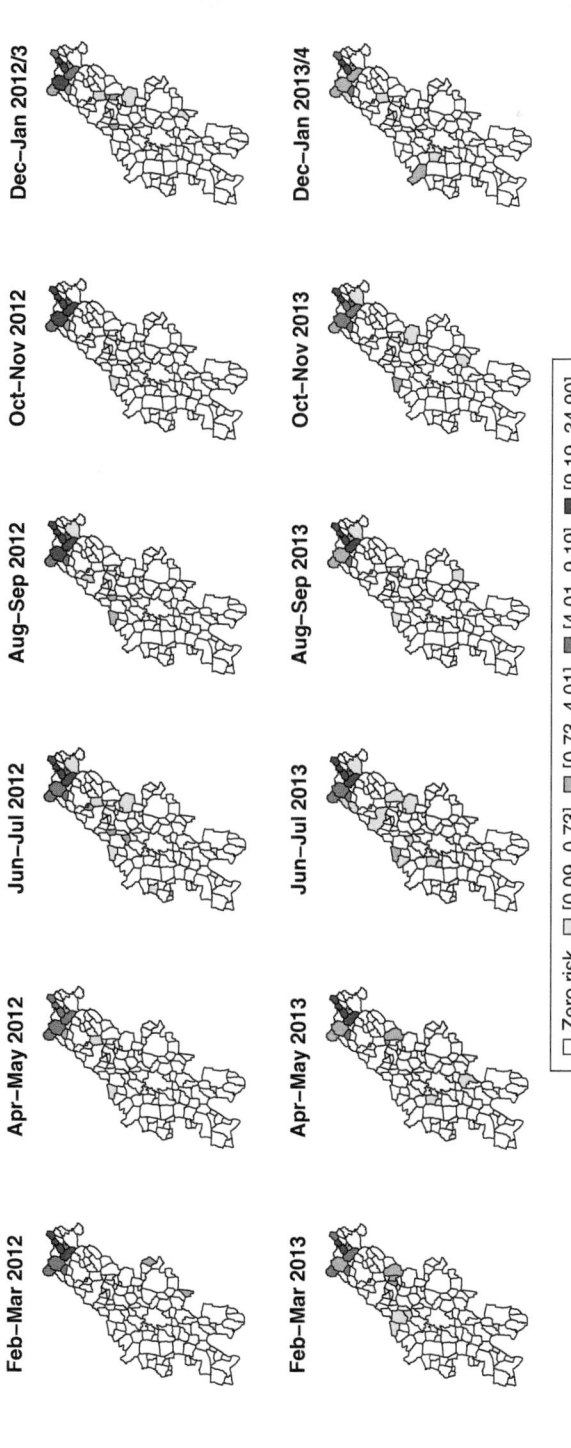

FIGURE 13.3

A map sequence of the raw malaria incidence rates (per 1000 people) across the 139 villages in Kalaburagi taluk, South India, between February 2012 and January 2014.

(a) 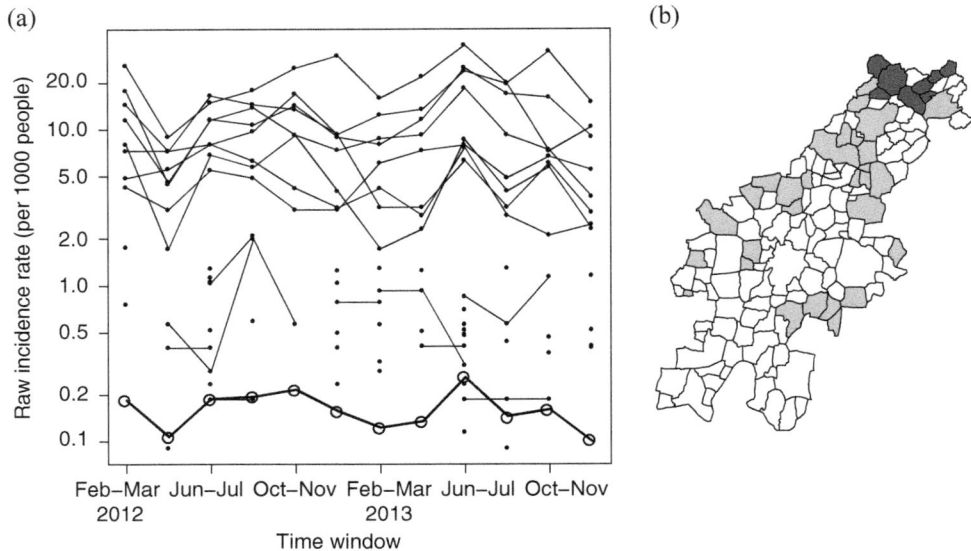 (b)

FIGURE 13.4

Panel (a) shows the time series of the raw incidence rates from villages with at least one case reported during the two years. Cases occurring in consecutive time windows are joined by a line, while an isolated black dot represents a "one-off" occurrence of at least one case in a village. The thick black curve with open circles is the time trend for the entire study region. In panel (b), the villages with non-zero incidence rates in all the 12 time windows are shown in dark grey. These are the eight villages with continuous time trends in the upper part of panel (a). Shown in light grey are those villages with either "discontinuous" time trends or isolated dots in panel (a). The polygons in white are villages with no cases reported throughout the two years.

rates consistently high across the two years while the other group contains villages with lower incidence rates. These two groups of villages are colour-coded in Figure 13.4(b), showing the concentration of the high-risk villages in the north of the taluk. See Exercise 13.1 for detail on producing the two plots in Figure 13.4.

At first glance, the time series plot in Figure 13.4(a) might suggest the malaria dataset possesses a space-time separable structure similar to Pattern 1 in Figure 13.1 – or does it? There is considerable variability in the observed data which casts doubt on any claim about the underlying dependence structure in the data, and in turn about the data generating process itself. Moreover, one may wonder, "what are the factors that might be associated with the rate differences between the two groups of villages and indeed amongst all the villages across the taluk?" To answer the second question, modelling is required. But before doing that, we need to tackle the first question: what is the structure of the data; is it space-time separable or space-time inseparable? The answer to that will help inform the framework that we use for the modelling. So in the next section, we look at some exploratory methods for examining whether space-time interaction exists or not.

13.4 Tests of Space-Time Interaction

One of the aims of an exploratory analysis of a spatial-temporal dataset is to test *globally* whether space-time interaction is present and, if so, to make an initial assessment of the *local* characteristics of any space-time departures (e.g. to identify where and when the departures from the space-time separable structure might have occurred as well as to indicate the patterns and the magnitudes of the departures). In the following sections, we will first look at a global test of space-time interaction, then describe a number of methods that aim to explore local behaviours.

13.4.1 The Knox Test

In a series of papers (Knox, 1963, 1964a and 1964b), a global test, known as the Knox test, was developed to assess the presence of space-time interaction. The Knox test decomposes a space-time dataset into three components, an overall distribution over space, an overall distribution over time and a space-time interaction component, following essentially the space-time inseparable modelling framework presented in Figure 12.4. The Knox test is designed to detect whether the number of case pairs, deemed close together in both space and time, exceeds the number to be expected under the space-time separable structure. As a result, the test will not declare the presence of space-time interaction under the two space-time separable patterns shown in Figure 13.1 but will declare that space-time interaction is present under the two patterns shown in Figure 13.2.

The Knox test is formulated as follows. For a set of n cases observed over space and time, given a definition of "closeness in space" and a definition of "closeness in time", each pair of cases is allocated to one of the four categories in the 2×2 contingency table in Table 13.2. In Table 13.2, N_S and N_T are, respectively, the numbers of case pairs that are close in space (irrespective of their distance in time) and close in time (irrespective of their distance in space), and the quantity X is the number of case pairs that are close in *both* space *and* time. X is the test statistic.

When space-time interaction is absent, X will be close to $\mu = (N_S \cdot N_T) / N$, which is the expected number of case pairs that are close in both space and time under the null hypothesis of space-time separability or independence (Knox, 1963). The quantity $N = n(n-1)/2$ is the total number of case pairs over a total of n cases. A large difference between X and μ, however, would be indicative of the presence of space-time interaction in the observed data. Thus, the Knox test looks at all possible case pairs to see "whether pairs of cases which are relatively close in time are also relatively close in space" (Knox, 1963, p.122). Before we can apply the Knox test to a dataset, two issues need to be considered: (a) how large a difference between X and μ needs to be so that the presence of space-time interaction can be declared and (b) how to define closeness in space and closeness in time. We will now discuss each of these points in turn.

To derive a p-value to assess the statistical evidence, a Monte Carlo procedure through random permutation of the data is commonly used. Proposed by David and Barton (1966), the Monte Carlo procedure involves calculating the test statistic X across a large number of simulated datasets (usually taken to be 999), each being generated by randomly permuting

TABLE 13.2

Summarising All the Case Pairs Into a 2×2 Contingency Table Depending on Whether Two Cases are Close in Space and/or in Time

	Close in Space	Not Close in Space	Row Total
Close in time	X	$N_T - X$	N_T
Not close in time	$N_S - X$	$N - N_T - N_S + X$	$N - N_T$
Column total	N_S	$N - N_S$	N

the time labels of the cases while holding the spatial labels fixed. This random permutation effectively breaks the space-time structure in the observed data. The value of X obtained from the original dataset is then compared to those calculated from the simulated datasets to derive a (empirical) p-value using Eq. 6.2 in Chapter 6, i.e.

$$\frac{k+1}{m+1} \tag{13.2}$$

In Eq. 13.2, m is the total number of simulated datasets and k is the number of these simulated datasets that produce a test statistic greater than *or equal to* the test statistic calculated using the original data. Note that, slightly different from that in Eq. 6.2, for the Knox test, since the test statistic X is discrete, the definition of k also includes those simulated datasets that result in a test statistic equal to the test statistic calculated using the original data (Davison and Hinkley, 1997, p.141).

A p-value can also be derived based on different approximations to the distribution of X under the null hypothesis of space-time separability (e.g. the Poisson approximation by Knox (1964a) and the normal approximation by David and Barton (1966)). However, as we shall illustrate in Section 13.4.1.1, there are issues associated with these approximations. In addition, extensions to the Monte Carlo procedure described above have been proposed to account for population shifts (in which case the size of the at-risk population is not time invariant, a factor that would give rise to space-time interaction) and other observable risk factors (see Kulldorff and Hjalmars (1999) and Schmertmann (2015)).

We now turn to the second issue of defining closeness in space and closeness in time. For a set of n cases, each observed at spatial location s_i and time point t_i, the test statistic X, the number of case pairs close in both space and time, is calculated as follows

$$X = \sum_{i=1}^{n-1} \sum_{j=i+1}^{n} a_S\left(s_i, s_j\right) a_T\left(t_i, t_j\right) \tag{13.3}$$

where a_S is an indicator function for spatial proximity and takes the value 1 if cases i and j are close together in space and 0 otherwise. Spatial proximity is defined by the user and can be based on distance (e.g. two cases are close if they are less than, say, 1km apart) or contiguity (e.g. two cases are close if they are observed in two different spatial units that share a common boundary). Similarly, a_T in Eq. 13.3 is an indicator function for defining temporal proximity. It takes the value 1 if cases i and j are considered to occur close together in time and 0 otherwise. Again, the user needs to define a temporal window within which case pairs are deemed temporally close. Since the space-time scale at which space-time interaction might be present is unknown, the common practice is to carry out the Knox test using various spatial and temporal distance thresholds (Gilman and Knox, 1995 and Grubesic and Mack, 2008). The same applies if proximity is defined through contiguity. The resulting table of p-values can then be combined using the procedure proposed by Kulldorff and Hjalmars (1999), a procedure that we will return to in Section 13.4.1.2. In addition, N_S and N_T in Table 13.2 can be obtained by summing the values of the spatial indicator a_S and the temporal indicator a_T respectively across all the case pairs. It is also worth noting that a number of space-time interaction tests (e.g. the Mantel test by Mantel (1967) and the space-time K function by Diggle et al. (1995)) can be expressed in the form of Eq. 13.3. For more detail, see Meyer et al. (2016) and Diggle (2013, p.214–216).

The Knox test is implemented in the R package `surveillance` (Meyer et al., 2017) via the `knox` function. The `knox` function is formulated to analyse individual-level data and cannot be applied directly to areal count data. We have written an R function `KnoxTestForCountData.R` (available on the book's website) that first converts a count dataset to the format required by the `knox` function then subsequently performs the test.

13.4.1.1 An Instructive Example of the Knox Test and Different Methods to Derive a p-Value

This example uses a simple space-time dataset with six cases observed in three areas over four weeks (Figure 13.5). Here, two cases are deemed to be close in space if they are observed either in the same area or in two areas that share a common boundary. For temporal proximity, two cases are deemed close if they occur either in the same week or in two consecutive weeks. Table 13.3 lists all the case pairs ($N = 15$), together with their proximity in space, in time and in both space and time. Thus, $N_S = 13$, $N_T = 7$, and $X = 7$. The value of the test statistic X is slightly greater than the expected value $\mu = \dfrac{N_S N_T}{N} = 6.1$. A Monte Carlo test with 999 random permutations of the original data gives an empirical p-value of 0.28, indicating the absence of space-time interaction. Table 13.4 tabulates the frequencies of the values of X calculated from the 999 simulated datasets. When dealing with count data, only the first three columns in Figure 13.5 are available. The R function `KnoxTestForCountData.R` creates the case labels, as in the fourth column in Figure 13.5, so that the required quantities for the Knox test can then be calculated.

Under the null hypothesis of space-time independence, the Poisson approximation (Knox, 1964a) assumes that X follows a Poisson distribution with mean $\mu = 6.1$. The resulting p-value is 0.41 (i.e. the probability of observing seven or more case pairs that are close in both space and time under the null hypothesis). Although the results from the Poisson approximation and the Monte Carlo test both point to the absence of space-time interaction, the Poisson approximation is not appropriate here because neither N_S nor N_T are small compared to N (Kulldorff and Hjalmars, 1999). We will return to this point later.

David and Barton (1966) derived a normal approximation where, under the null hypothesis of space-time independence, X follows a normal distribution with mean μ and variance v. The normal mean is defined as before, and the calculation of the variance v involves not only N_S and N_T but also N_{2S} and N_{2T}, which are, respectively, the number of pairs of case pairs close in space that share one case in common and the number of pairs of case pairs

Area	Time	Number of cases observed	Case labels
A1	Week 1	2	a, b
A2	Week 3	3	c, d, e
A3	Week 4	1	f

FIGURE 13.5
A simple space-time dataset for illustrating the Knox test.

TABLE 13.3

Calculation for Various Quantities Required to Carry Out the Knox Test

Case Pair	Spatial Locations of Two Cases		Spatial Proximity of the Case Pair (a_S)	Time of Occurrence of the Two Cases		Temporal Proximity of the Case Pair (a_T)	Spatial-Temporal Proximity $a_S \cdot a_T$
(a, b)	A1	A1	1	Week1	Week1	1	1
(a, c)	A1	A2	1	Week1	Week3	0	0
(a, d)	A1	A2	1	Week1	Week3	0	0
(a, e)	A1	A2	1	Week1	Week3	0	0
(a, f)	A1	A3	0	Week1	Week4	0	0
(b, c)	A1	A2	1	Week1	Week3	0	0
(b, d)	A1	A2	1	Week1	Week3	0	0
(b, e)	A1	A2	1	Week1	Week3	0	0
(b, f)	A1	A3	0	Week1	Week4	0	0
(c, d)	A2	A2	1	Week3	Week3	1	1
(c, e)	A2	A2	1	Week3	Week3	1	1
(c, f)	A2	A3	1	Week3	Week4	1	1
(d, e)	A2	A2	1	Week3	Week3	1	1
(d, f)	A2	A3	1	Week3	Week4	1	1
(e, f)	A2	A3	1	Week3	Week4	1	1

TABLE 13.4

Frequency of the X Values Calculated from the 999 Simulated Datasets

X	5	6	7
Frequency	206	516	277

close in time that share one case in common. From David and Barton (1966), the formula for v is

$$v = \frac{N_S N_T}{N} + \frac{4 N_{2S} N_{2T}}{n(n-1)(n-2)}$$

$$+ \frac{4 \left[N_S (N_S - 1) - N_{2S} \right] \left[N_T (N_T - 1) - N_{2T} \right]}{n(n-1)(n-2)(n-3)} - \left(\frac{N_S N_T}{N} \right)^2 \tag{13.4}$$

The calculation of N_{2S} can be done using the formula from David and Barton (1966):

$$\left(\frac{1}{2} \sum_{i=1}^{n} b_i^2 \right) - N_S \tag{13.5}$$

where b_i is the number of cases that are close to case i in space. The same formula can be used to calculate N_{2T}. In that situation, N_S is replaced by N_T and b_i counts the number of cases that are close to case i in time. So, for the data in Figure 13.5, to calculate v, the numbers of spatially close cases across the six cases are 4, 4, 5, 5, 5 and 3 (i.e. there are four cases that are close in space to case a, four cases close in space to case b, five cases close in space to case c and so on). Likewise, the numbers for temporally close cases are 1, 1, 3, 3, 3 and 3, respectively, for each of the six cases. Therefore, $N_{2S} = 45$, $N_{2T} = 12$ and $v = 24.3$. Then the p-value derived from the normal approximation is 0.43, a much larger value compared to that derived from the Monte Carlo test. The problem is that when n, the total number of cases in the dataset, is small, the distribution of X under the null hypothesis tends to be discrete (see Table 13.4), a feature that is not approximated well by a normal distribution. This calculation also shows that under the null hypothesis of space-time independence, the variance of X ($v = 24.3$) is much larger than the mean of X ($\mu = 6.1$), meaning that the Poisson approximation (which assumes the mean is equal to the variance) proposed by Knox (1964a) is invalid for this dataset.

Thus, in general, deriving a p-value via a Monte Carlo test is often preferred. It avoids making any distributional assumption about the test statistic. In addition, Johnson et al. (2007) suggest that the use of the permutation procedure could minimise the potential impact of edge effects.

13.4.1.2 *Applying the Knox Test to the Malaria Data*

Over the two years period, a total of (n =) 1633 cases of malaria were reported across the 139 villages in Kalaburagi taluk, giving a total of (N =) 1332528 case pairs. The median distance of these case pairs is about 6km, and 50% of these case pairs are within three time periods. Table 13.5 summarises the p-values across different thresholds used for defining spatial and temporal closeness. All the tests with a spatial proximity threshold set at 10km or less are statistically significant, with p-values below the 5% significance level. When the spatial threshold is set at 20km, the test results become non-significant. The above finding holds across the different temporal thresholds. These results suggest that (a) space-time interaction is shown to be present within 10km but not beyond and (b) space-time interaction is shown to be present over long time spans. Point (a) matches with what we observe in the map in Figure 13.4(b). The majority of the malaria cases were reported in the eight villages in the north (the darker coloured polygons in Figure 13.4(b)), and most of these villages are no further than 10km apart in terms of the distance between their centroids. The malaria risks in those eight villages appear to be high and relatively stable (Figure 13.4(a)), which echoes the presence of space-time interaction over quite a long time span.

TABLE 13.5

Testing Space-Time Interaction in the Village-Level Malaria Incidence Data Using the Knox Test Carried Out with Spatial and Temporal Proximity Set at Different Threshold Values

	Distance (km)				
Time (Months)	0	2	6	10	20
2	0.001	0.001	0.018	0.001	0.165
4	0.001	0.001	0.001	0.002	0.233
8	0.001	0.001	0.001	0.019	0.403
10	0.001	0.001	0.001	0.030	0.187

Tabulated are the p-values derived from 999 random permutations across these spatial/temporal thresholds. The spatial distance between two cases is measured as the centroid distance of the villages (polygons) that the two cases were reported in. Two cases reported in the same village are given a distance of 0km. Two cases reported in the same time window are considered to have a temporal distance of two months, the maximum possible temporal distance between them. Two cases reported in two consecutive time windows then have a temporal distance of four months.

However, care needs to be taken when interpreting the p-values in Table 13.5 because of the problem of multiple testing. To adjust for multiple testing, Kulldorff and Hjalmars (1999) proposed a procedure to combine the p-values across different thresholds of space and time closeness to form what they call a "combined" p-value. Their procedure is as follows:

1. For both the original dataset and each of the simulated datasets, calculate the test statistic X using each of the space-time proximity settings (e.g. setting the spatial and the temporal thresholds at 0km and two months, at 2km and two months and so on).

2. For each space-time proximity setting, normalise the resultant test statistic X using the following formula,

$$\tilde{X} = \frac{X - \mu}{v}$$

with $\mu = N_S N_T / N$ and calculating v as given in Eq. 13.4.

3. For the original dataset and each of the simulated datasets, find the maximum value of the set of normalised test statistics across the space-time proximity settings.

4. Calculate the empirical p-value using Eq. 13.2 based on the set of maximum values obtained from Step 3.

Using the space-time proximity settings in Table 13.5, the combined p-value is 0.001. Thus, after accounting for multiple testing and across various proximity settings, there is space-time interaction present in the village-level malaria data over the two-year period. Then the next question to ask would be: "to what extent can we explain the presence of space-time interaction through incorporating observable risk factors (such as rainfall level) and by modelling the infectious nature of malaria over space and/or time?" We will return to this question in Section 16.5, where we carry out a Bayesian space-time modelling of this dataset. Exercise 13.2 provides detail on the implementation of the analysis carried out in this section.

13.4.2 Kulldorff's Space-Time Scan Statistic

When space-time interaction is shown to be present, we would subsequently want to know when and where departures from the space-time separable structure might have occurred. Various *local* tests have been developed to detect specific space-time departures and to study the properties of such departures. Here, we focus on the space-time scan statistic proposed by Kulldorff (2001).

Kulldorff's space-time scan statistic is an extension of the spatial scan statistic discussed in Section 6.2.5.3. Instead of creating circular windows of varying size to scan a map, the space-time scan statistic creates a set of cylinders to scan through the three-dimensional space-time data cube. The circular base of each cylindrical window covers the spatial dimension (i.e. indicating the "where"), whilst the height of the cylinder, defined by the start and the end times of the scanning period, scans through time (i.e. indicating the "when"). These cylindrical windows vary in size. The circular base ranges from 0 to an upper bound that is typically set to cover no more than 50% of the total population.

Similarly, the height goes from 0 to an upper bound that covers no more than 50% of the observation period. For area-level data aggregated over discrete time points, a space-time cell falls within a cylindrical scanning window if the centroid of the corresponding area lies inside the circular base and the time point of that space-time cell falls within the scanning period.

For each of such cylindrical windows, the case rate *inside* the cylinder is compared to the case rate *outside* the cylinder using the likelihood ratio given in Eq. 6.7 in Chapter 6. A likelihood ratio value greater than 1 would then suggest a higher case rate inside the window compared to outside, indicative of the presence of local space-time interaction (or a space-time cluster). Across all the scanning windows, the one that yields the largest value of the likelihood ratio is considered to be the most likely space-time cluster. A permutation test is then carried out to assess how likely (or unlikely) the detected space-time cluster is a chance occurrence. Under the permutation test, the likelihood ratio of the most likely cluster is calculated over a large number of simulated datasets, each generated by randomly permuting the time labels of the cases but holding their spatial locations fixed – a procedure that is the same as that for the Knox test. The values of the likelihood ratio calculated from both the original dataset and the simulated datasets are then used to obtain the empirical p-value using Eq. 13.2. A p-value that is less than the usual 5% significance level provides evidence for the presence of space-time interaction within the detected cylindrical scanning window. In addition to the most likely cluster, secondary clusters are also reported. A cylindrical window is declared as a secondary space-time cluster if its p-value is less than 0.05 and the window does not overlap with that of the most likely cluster (Kulldorff, 2001).

Kulldorff's space-time scan statistic has been widely used, partly owing to SaTScan™ (www.satscan.org), an open-source software that facilitates the application of the methodology. But before illustrating the use of SaTScan, there are a few features of the space-time scan statistic that are worth mentioning. First, because the scanning windows are typically restricted to cover no more than 50% of the time points and no more than 50% of the population in space, the detection of the purely spatial cluster or the purely temporal cluster as shown in Figure 13.1 is not of interest. The restriction on the temporal span also implies that the space-time scan statistic would not be powerful to detect an area for which its entire time trend is different from all other areas (e.g. as in Pattern 3 in Figure 13.2). It is more suited to investigate situations in the form of Pattern 4 in Figure 13.2.

Second, the rate calculation as discussed above requires the availability of data on the population at risk. Kulldorff et al. (2005) developed a version of the space-time scan statistic for case-only data. In that circumstance, the likelihood ratio for a scanning window is calculated by comparing the observed numbers of cases to the corresponding *expected counts*. Specifically, denoting o_{it} to be the observed number of cases in area i at time point t in a dataset with N areas over T time points and C to be the total number of observed cases (i.e. $C = \sum_i \sum_t o_{it}$), Kulldorff et al. (2005) define μ_{it}, the expected count for that space-time cell as

$$\mu_{it} = \frac{1}{C}\left(\sum_{i=1}^{N} o_{it}\right)\left(\sum_{t=1}^{T} o_{it}\right)$$

In other words, μ_{it} is the expected number of cases that would have occurred in area i at time t under space-time independence. Therefore, any large deviation of the observed count from the corresponding expected count would be indicative of space-time interaction. When data are available, effects of observable risk factors can be also controlled for through the calculation of expected counts. Then the space-time detection is carried out after adjusting for the observable risk factors – see Klassen et al. (2005) for an example. In Section 16.3 we will present a modelling approach to the detection problem with control for covariate effects.

Finally, the space-time scan statistic can be performed in either prospective or retrospective mode. A prospective analysis aims to detect the so-called *alive* space-time clusters, where a detected cluster is defined to be alive if the end of the corresponding scanning window reaches the end of the study period. A retrospective analysis, on the other hand, searches for all potential space-time clusters in the dataset, whether alive or not. Prospective space-time detection tends to be performed sequentially in the situation where new data become available routinely over a regular time interval and has been used for detecting disease outbreaks (see, for example, Hughes and Gorton, 2013 and Greene et al., 2016). However, Correa et al. (2015) raises some concerns regarding the multiple testing issue when using the space-time scan statistic in a sequential/prospective fashion. Interested readers are also referred to Woodall et al. (2008), who discuss prospective disease surveillance in a more general context.

13.4.2.1 Application: The Simulated Small Area COPD Mortality Data

In this section, Kulldorff's space-time scan statistic is applied to a simulated mortality dataset on chronic obstructive pulmonary disease (COPD) across the 354 local authority districts (LADs) in England over an eight-year period from 1990 to 1997. The dataset is simulated based on a space-time inseparable pattern that is similar to Pattern 4 in Figure 13.2. Most of the LADs follow the same (common) time trend (the grey line in Figure 13.6), but 15 LADs are chosen to follow an unusual trend where the mortality is elevated for two years between 1992 and 1993 (the black line in Figure 13.6). Figure 13.7 shows the locations of the 15 unusual areas, which are scattered across the study region. The aim is to identify these 15 unusual areas. We also use this analysis to illustrate how the performance of a space-time detection method is assessed in a simulation study.

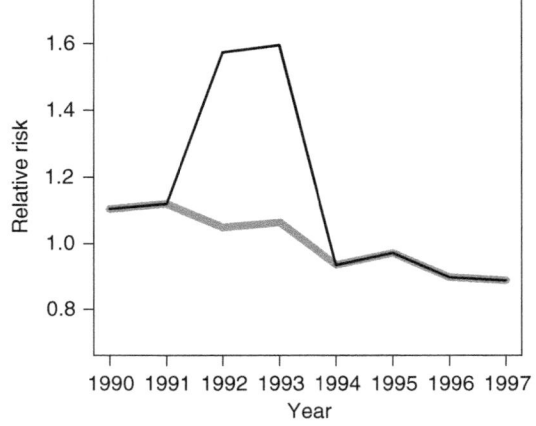

FIGURE 13.6
The common time trend (the grey line) and the unusual time trend (the black line) used for generating the annual mortality counts across the 354 local authority districts in England between 1990 and 1997.

FIGURE 13.7
Locations of the 15 local authority districts (the polygons in dark grey) that are selected to possess the unusual time trend (the black line shown in Figure 13.6).

There are two challenges. First is the issue of data sparsity. Figure 13.8 shows the local trends estimated using only the observed data – the time sequence of observed numbers of deaths in each LAD divided by the corresponding expected counts. The unusual trend pattern is obscured by the excessive amount of variation in the data. The second challenge is specific to Kulldorff's space-time scan statistic, for which the use of cylindrical scanning windows may be inefficient when needing to detect *isolated* unusual areas (i.e. risk elevated in a single area or in a cluster consisting of only a few areas).

SaTScan was used to carry out the analysis. Specifically, the maximum size of the circular spatial window was set at 50% of the at-risk population, and the maximum temporal window was restricted to cover no more than 50% of the eight time points (years). The empirical p-values were derived using 999 simulated datasets obtained through random permutation. Two clusters were identified, and they are shown in Figure 13.9.

Detected as the most likely cluster, the large cluster in the north of England contains 115 LADs and has a log likelihood ratio of 984.3. The p-value of 0.001 is well below the 0.05 level. A careful comparison between Figure 13.9 and Figure 13.7 reveals that only three districts in this detected cluster follow an unusual time trend, whilst all the other 112 detected areas are in fact false positives, i.e. areas that do not exhibit any departure from the common time trend. The time window of this detected cluster spans four years, from 1990 to 1993. Compared to the unusual trend pattern in Figure 13.6, two additional years are falsely identified as part of the space-time cluster. An elevated risk is estimated for this cluster, a result that matches with the unusual time trend pattern.

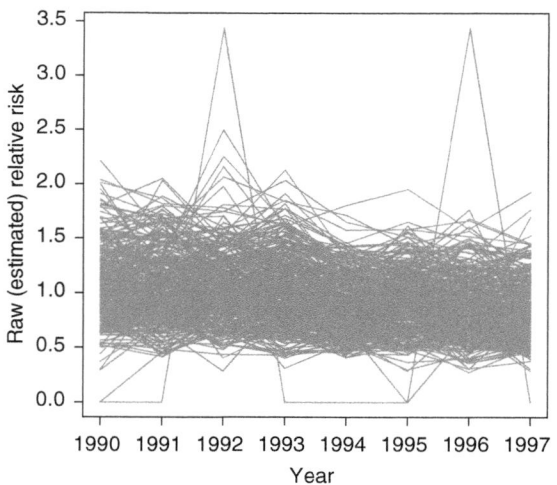

FIGURE 13.8
Time trends of the individual districts estimated using only the observed mortality data. The trend of each district is calculated by dividing the observed numbers of deaths (across the eight years) by the corresponding expected counts.

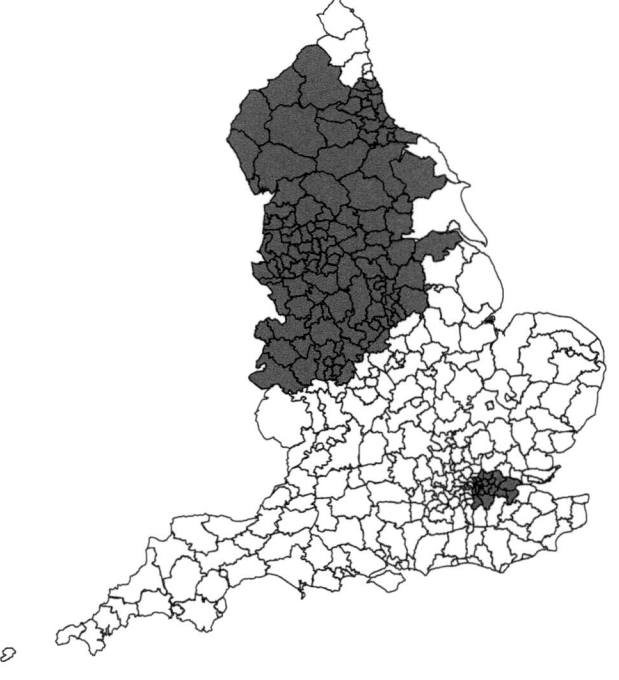

FIGURE 13.9
Two clusters of LADs identified by SaTScan.

A second, smaller cluster is detected in the south of England (Figure 13.9). This cluster contains 19 districts with an increased risk estimated between 1990 and 1993. Similar to the most likely cluster, a majority of the areas contained in this second cluster are false positives and only two are truly unusual areas.

In a simulation setting, detection performance can be assessed through sensitivity, the percentage of the truly unusual areas (i.e. areas that follow the unusual time trend) being detected, and the empirical false discovery rate (empirical FDR), the percentage of the detected areas that are in fact *not* unusual (i.e. areas that follow the common time trend). From the detection results, the power is 33.3% (= 5/15 × 100) and the empirical FDR is 96.3% (= 129/134 × 100), a rather poor performance from the space-time scan statistic under this simulated dataset. However, this may not be unexpected. As alluded to earlier, the low power is partly due to the isolated nature of the unusual areas. The use of a scanning window will be more successful if the truly unusual areas are close together in space. The lack of power also illustrates the conservative nature of the space-time scan statistic in detecting subsequent clusters beyond the most likely cluster (Haining, 2003, p.257). As a result, the test fails to report other truly unusual areas at the specified 5% significance level. Power can be increased by setting a larger significance level, say to 10% or even 20%. However, this is not recommended because the empirical FDR is already high at the 5% level. The empirical FDR at 96.3% implies a high level of uncertainty in the detection result – over 96% of the detected areas are in fact false positives, i.e. they do not follow the unusual time trend. Increasing the significance level would lead to an even higher FDR/uncertainty. See Exercise 13.5 for the implementation detail.

One should, however, be careful not to overinterpret the results from a single simulated dataset. The detection performance of a statistical method should be assessed through multiple simulated datasets (see Exercise 16.10 in Chapter 16). In addition, performance depends on a number of factors, such as different geographical distributions of the unusual areas (e.g. most of the unusual areas are isolated or they are mostly positioned within a contiguous cluster), different departure magnitudes from the common trend (i.e. larger or smaller departures of the unusual trend from the common trend in Figure 13.6), different types of departure patterns and whether the dataset is sparse (only a few case counts per space-time cell) or not (many case counts per space-time cell). Thus, a simulation study needs to be carefully thought-out to examine a space-time detection method under relevant, realistic scenarios. The analysis presented in this section is part of the study in Li et al. (2012). We will return to the topic of space-time detection in Section 16.4, where we will present a model-based space-time detection method developed in the Bayesian hierarchical modelling framework and apply it to the simulated dataset analysed here.

13.4.3 Assessing Space-Time Interaction in the Form of Varying Local Time Trend Patterns

In this section, we focus on assessing the presence of space-time interaction in the form of Pattern 3 in Figure 13.2, where different areas display different time trend patterns. Moreover, these heterogeneous local time trends may also be spatially autocorrelated in the sense that the time trends of two areas that are close in space may be more alike compared to the situation where the two areas are far apart. If that is the case, the estimation of the local time trend of an area can borrow information from other (nearby) areas, in addition to borrowing information temporally across time points. We look at some exploratory techniques that investigate the spatial distribution of local time trends.

One way to achieve that objective is first to summarise the temporal observations of each area by a scalar quantity (e.g. the slope parameter in the linear trend model or the temporal autocorrelation parameter in the AR1 model). Then various spatial exploratory analysis methods discussed in Chapter 6 can be applied to the resulting set of (spatial) values. In what follows, we illustrate this idea using two datasets, the annual burglary count data at the Census Output Area (COA) level in Peterborough from 2005 to 2008 and the (simulated) annual mortality data on chronic obstructive pulmonary disease (COPD) across the 354 local authority districts (LADs) in England between 1990 and 1997. The reason for selecting these two sets of data is that with four time points, the burglary data gives an example of a set of short time series data for which a linear trend model can typically be used to summarise the temporal observations from each area. The COPD mortality data, on the other hand, represent a longer time series with five or more time points. In that case, a more flexible AR1 model is used to summarise the local temporal data.

13.4.3.1 Exploratory Analysis of the Local Trends in the Peterborough Burglary Data

As we have seen in Section 12.2, whilst the raw burglary rates appear to be increasing from 2005 to 2008 for Peterborough as a whole, there is evidence of local variation in the temporal evolution of burglary risk at the COA level (see Figures 12.2 and 12.3). Compared to the Peterborough trend, the data of some COAs show a faster (or slower) increasing level of risk. Some COAs display a decreasing risk. For relatively short time series (no more than four time points), the time trend pattern of each area can be summarised using the slope of a linear trend model fitted to the temporal data for that area. Thus, for each of the 452 COAs, a Poisson regression is fitted to the observed burglary counts over four years, including the time index $t = 1,2,3$ and 4 as a single covariate. The spatial properties of the COA-level time trends can then be studied using the slope estimates. The R code in Figure 13.10 estimates the COA-level slopes, then produces a map of the estimates shown in Figure 13.11.

It is evident from Figure 13.11 that these local slopes and hence the local time trends vary over space. Not all the slope estimates are close to the slope of the Peterborough trend, which is estimated to be 0.13. Furthermore, the slope estimates appear to be spatially autocorrelated, where the slopes in COAs that are close in space tend to be similar. The global Moran's I calculated on the slopes confirms this: using the first order rook's spatial contiguity, the Moran's I is 0.059 and the p-value derived from the Monte Carlo test with 999 random permutations is 0.014. This finding, as we shall see in Section 16.3, supports the use of local spatial smoothing when estimating the COA-level slopes. As a result, the estimation of the burglary trend in one area can borrow information from other nearby areas, a feature that helps stabilise the small area trend estimates.

Figure 13.11 also illustrates a problem that can arise when estimating the slope for an area using only the data from that area. The estimates on the map vary considerably, ranging from −23.44 to 23.75. The large positive estimates come from a few COAs where there were no cases of burglary for the first three years, then a small number of cases reported in the last year. The large negative slope estimates a rise for a handful of COAs with the opposite pattern, i.e. a few cases reported in the first year followed by no cases in the subsequent three years. We shall see how spatial modelling helps tackle this problem.

13.4.3.2 Exploratory Analysis of the Local Time Trends in the England COPD Mortality Data

This example deals with longer time series data, which is data with five or more time points. When that is the case, nonlinear time series models are needed to summarise the

```
1    library(maptools)
2    #  read in the Peterborough shape file containing the
3    #  burglary data
4    peterborough <- readShapePoly('peterborough.shp')
5
6    ##  extract the COA-level annual burglary counts and the
7    ##  at-risk population (number of houses) from the shapefile
8    N <- 452   #  number of COAs
9    T <- 4     #  number of years
10   #  create a matrix of size NT to store the burglary counts
11   burglary.counts <- matrix(0,nrow=N,ncol=T)
12   #  extract the annual burglary counts and store them in
13   #  the matrix burglary.counts
14   burglary.counts[,1] <- peterborough@data[,'cases2005']
15   burglary.counts[,2] <- peterborough@data[,'cases2006']
16   burglary.counts[,3] <- peterborough@data[,'cases2007']
17   burglary.counts[,4] <- peterborough@data[,'cases2008']
18   #  number of at-risk houses in each COA
19   nhouses <- peterborough@data[,'houses']
20
21   ##  fit a Poisson regression to the time sequence of burglary
22   ##  counts from each COA
23   #  first create an array to store the estimated slopes
24   slopes <- rep(0,N)
25   years <- 1:T  #  define a single covariate of time index
26   for (i in 1:N) {
27     #  then using the glm function to fit a Poisson regression to
28     #  the temporal burglary counts for each COA the offset
29     #  argument specifies the log population at-risk for each year
30     #  (which remains constant over time) so that we are modelling
31     #  the burglary rates
32     m <- glm(burglary.counts[i,] ~ years
33             , family=poisson,offset=rep(log(nhouses[i]),T))
34     #  extracting the slope estimate from the results
35     slopes[i] <- m$coefficients[2]
36   }
37
38   ##  produce a choropleth map of the estimated slopes (see the
39   ##  five steps in Section 6.2.1)
40   cutpoints <- quantile(slopes)
41   slopes.level <- cut(slopes,cutpoints,labels=1:4
42                       ,include.lowest=TRUE)
43   shadings <- grey(c(1,0.7,0.4,0.2))
44   plot(peterborough,col=shadings[slopes.level])
45   intervals <- c('[-23.44,  0.00]','(0.00,  0.27]','(0.27,  0.56]'
46                  ,'(0.56,  23.75]')
47   legend('topright',legend=intervals,fill=shadings,bty='n')
```

FIGURE 13.10
The R code to estimate the COA-level slopes and produce a map of the slope estimates.

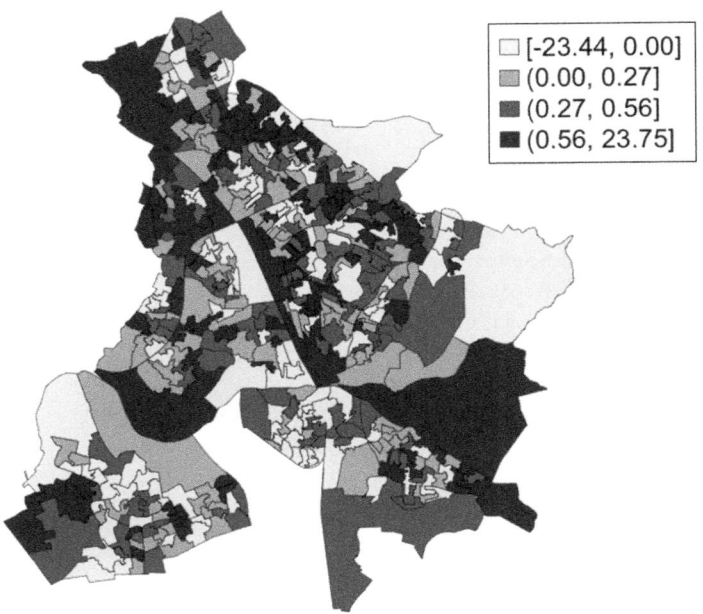

□ [-23.44, 0.00]
▨ (0.00, 0.27]
▰ (0.27, 0.56]
■ (0.56, 23.75]

FIGURE 13.11
A map of the estimated slopes using the annual burglary counts from 2005 to 2008 in Peterborough at the COA level. A positive (negative) slope estimate corresponds to an increase (decrease) in burglary rates over the four-year period.

local temporal data before exploring the spatial properties of the local time trend patterns. Here, we follow the exploratory procedure proposed by Shand et al. (2018) where, first, the AR1 model is fitted to the time series data attached to each of the areas. Spatial exploratory methods are then applied to the set of estimated temporal autocorrelation parameters.

In the case of the COPD data, the AR1 model is fitted separately to the raw estimates of the relative risks – the observed case counts divided by the expected numbers of cases – over the eight years for each of the 354 LADs in England. A map of these estimates is shown in Figure 13.12. It appears that districts that are close in space tend to have similar values of the temporal autocorrelation parameter, implying that if the relative risks of an area are shown to be positively (negatively) autocorrelated over time, then the relative risks of its spatial neighbours are also likely to be positively (negatively) autocorrelated. The global Moran's I calculated for the estimated temporal autocorrelation parameters gives a value of 0.056 with a p-value of 0.045 derived from 999 random permutations. This result confirms the observation made about the map in Figure 13.12.

Whilst the foregoing illustrates the use of this exploratory technique, care needs to be taken when interpreting results. Similar to the slope estimates in Figure 13.11, the estimated temporal autocorrelation parameters are also highly uncertain. Only 27 districts yield an estimated temporal autocorrelation parameter greater than twice its standard error (a criterion typically used to declare statistically significant temporal autocorrelation). This implies that (a) positive spatial autocorrelation may not be present if this uncertainty is taken into account and (b) results from a "unit-by-unit" modelling of the small area temporal data would simply be too uncertain to enable us to draw firm conclusions. In the subsequent chapters (Chapters 14 to 16), we will address this challenge, amongst others, through Bayesian hierarchical modelling.

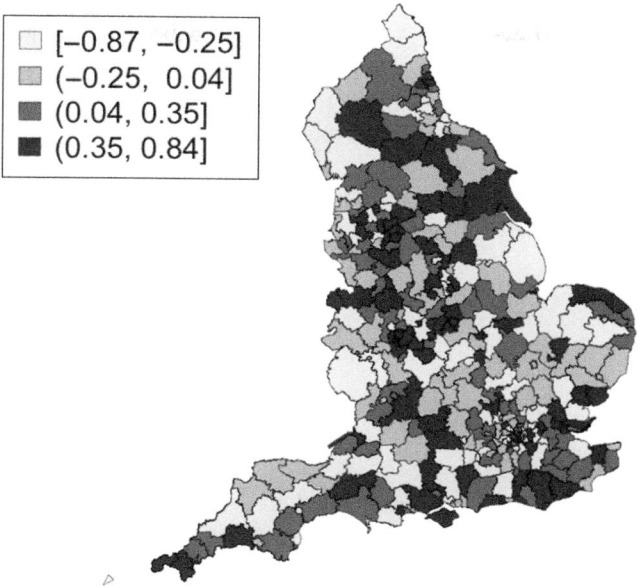

FIGURE 13.12

A map of the temporal autocorrelation parameters estimated by fitting an AR1 model (see Section 12.3.1.3) to the temporal relative risks (calculated by dividing the observed case counts over eight years by the corresponding expected case counts) from each of the 354 local authority districts (LADs) in England.

13.5 Concluding Remarks

ESTDA techniques applied to the space-time data in hand provide insights that can help to suggest the appropriate framework for modelling the data. These techniques can also be used in model checking. Model checking includes checking whether statistical assumptions have been met and visualising space-time residuals to identify lack of fit. We shall have more to say on these uses in later chapters in the context of specific models. The Knox test can be used to test for space-time independence globally, whilst the other local tests and exploratory techniques can be used to highlight local space-time dependence structures. Informative though these insights can be, it needs to be borne in mind that ESTDA techniques, whether applied to data or model residuals, ignore uncertainty. All the residuals from a Bayesian analysis have uncertainty attached via their posterior distributions. When uncertainty associated with parameter estimates is taken into account, some properties of the estimates may not be statistically significant, as noted in the example described in Section 13.4.3.2.

In the rest of Part III we shall be modelling spatial-temporal count data. The Poisson regression model is frequently used to model such data. This is a model where the mean and variance are assumed to be equal. However, this assumption is often violated when working with spatial-temporal data, just as we saw in the case with purely spatial data (Section 6.4). Typically, the variance in the observed outcome values exceeds the mean of those values. This situation is referred to as overdispersion. In the regression setting, overdispersion can arise due to the absence of some important risk factors (because of a lack of data or a lack of knowledge about their existence!) from the regression model. A Bayesian hierarchical model uses random effects that serve as a surrogate measure of the effect from these unobserved risk factors on the outcome. If the unobserved risk factors are assumed to be spatially and/or temporally autocorrelated, these random effect terms can be structured accordingly. We will explore these spatial-temporal hierarchical models in the subsequent chapters but refer the reader to Section 6.4 for a more detailed discussion on overdispersion.

13.6 Exercises

Exercise 13.1. Produce the time series plot of the raw malaria incidence rates across the 139 villages as shown in Figure 13.4(a). The time trend for the entire study region (the black curve in Figure 13.4(a)) is obtained by summing the malaria cases at each time point, then dividing the resulting value by the size of the at-risk population in the entire study region. Follow the steps outlined in Section 6.2.1 to draw the map in Figure 13.4(b) – some modifications are required.

Exercise 13.2. Write an R script to perform the Knox test on the village-level malaria incidence data. Pay particular attention to the calculation of the combined p-value set out in Section 13.4.1.2. (Hint: first carry out the Knox test based on each one of the definitions of spatial-temporal proximity in Table 13.5, save the resulting test statistic, then follow the procedure outlined in Section 13.4.1.2 to combine those test statistics to form the combined p-value. The variance of the test statistic, given in Eq. 13.4 and Eq. 13.5, also needs to be calculated.)

Exercise 13.3. Explain how to adjust for an observable risk factor when carrying out Kulldorff's space-time scan statistic (hint: see Klassen et al., 2005).

Exercise 13.4. Explain how the issue of multiple testing arises in a prospective surveillance scenario.

Exercise 13.5. The R script, `kulldorff_sptm_satscan.R`, runs Kulldorff's space-time scan statistic in SaTScan through R. The detection results are then fed into another R script, `summarise_sptm_satscan.R`, to calculate the power and the empirical FDR. Go through the two R scripts to understand the procedures involved, then run the analysis to obtain the detection results reported in Section 13.4.2.1. Both R scripts are available on the book's website.

Exercise 13.6. For the simulated COPD data, calculate the raw estimates of the relative risks, then, for each district, compute the temporal autocorrelation at lag 1 using the time sequence of risk values (hint: use the `acf` function in R). Then explore the properties of the resulting temporal autocorrelations: are they all the same and, if not, are they spatially autocorrelated? Produce a map of those values.

14

Bayesian Hierarchical Models for Spatial-Temporal Data I: Space-Time Separable Models

14.1 Introduction

A principal goal of spatial-temporal modelling is to adequately describe the variability in the observed space-time data so that we can address the questions in hand. Observed outcome values vary over space, across time and over both space *and* time, suggesting that the underlying data generating process can be spatially, temporally and quite often spatially-temporally varying. At first sight, the construction of a spatial-temporal model can appear a daunting task. However, the Bayesian hierarchical modelling framework allows us to build a model for such a complex data structure, in a modular fashion. In fact, we can create our first space-time models by using the models that we have discussed up to this point – the spatial models of Chapters 7 and 8 to describe the overall spatial pattern common to all time points and the temporal models of Chapter 12 to describe the overall temporal pattern common to all spatial units – treating the two components rather like "Lego blocks".

So, a sensible starting point in the model building process is to combine a spatial model with a temporal model and build a model that belongs to the class of space-time separable models. This is what we do here, and the chapter is organized as follows. In Section 14.2 we present the dataset that will be used to illustrate the modelling methodology and define the goal of the analysis, which is to estimate annual small area burglary rates over the period 2001 to 2008. In Section 14.3 we describe a number of space-time separable models that are candidates for meeting the goal of the analysis and fit one of the candidate models. In the course of this section we also consider alternative ways of combining the two models – additively or multiplicatively – and explain why the spatial and the temporal models are, in most practical applications, combined additively. Through the discussion of the modelling results, we reveal the importance of a space-time separable model to the process of modelling spatial-temporal data.

14.2 Estimating Small Area Burglary Rates Over Time: Setting the Scene

We discuss various space-time models using the Peterborough burglary data, where the outcome of interest is a set of annual burglary counts in each of the 452 Census Output Areas (COAs) over an eight-year period from 2001 to 2008 in Peterborough, England. Parts of this dataset have been analysed in previous chapters. In Chapter 12, we modelled the temporal patterns of individual COAs as well as the temporal pattern for Peterborough as a whole using a set of temporal models. In Chapter 13, we investigated some of the spatial-temporal characteristics of this dataset using a suite of exploratory methods. Now we apply a set of Bayesian hierarchical space-time models to analyse these small-area burglary counts jointly over both space *and* time. In so doing, not only can we describe the overall spatial and the overall temporal patterns in the data but, more importantly, we can identify space-time interaction.

Table 14.1 summarises the annual burglary counts, the numbers of at-risk houses and the raw annual burglary rates (per 1000 houses) across the 452 COAs. Each raw burglary rate is computed as the ratio of the reported burglary count to the number of at-risk houses. We call it a "raw" burglary rate to distinguish it from a "modelled" burglary rate, which is estimated using a statistical model involving information sharing. The last row of Table 14.1 suggests a general trend in average burglary rates, which start high during the first three years from 2001 to 2003, followed by a drop between 2004 and 2006 and then a fresh increase in 2007 and 2008. As part of an exploratory analysis, the researcher can construct a sequence of maps of the annual raw burglary rates in order to assess (a) the spatial pattern within each map; (b) whether there exists some commonality amongst these maps (e.g. some parts of the city tend to have relatively high (low) burglary rates across some or all of the eight years); and (c) whether there are some parts of the city where sudden changes (e.g. a sudden increase or decrease) in burglary rates are evident. We refer the reader to Exercise 14.1 for the construction and the exploration of these eight maps.

At this point, our inferential goal is to estimate the underlying annual burglary rates at the COA-level over the study period. To achieve this goal, we follow the Bayesian hierarchical modelling approach. To model each outcome value y_{it}, the burglary count observed in year t in COA i ($i=1,...,N$ and $t=1,...,T$), the *data model* takes a Poisson distribution as the likelihood:

$$y_{it} \sim Poisson(\mu_{it}) \qquad (14.1)$$

The Poisson mean, μ_{it}, is expressed as $\mu_{it}=n_i \cdot \theta_{it}$, a product of n_i, the number of at-risk houses in COA i, and θ_{it}, the underlying unknown burglary rate in COA i during year t. The number of at-risk houses, n_i, is obtained from the 2001 UK census and is assumed to be time-invariant. Interest lies in modelling the set of spatial-temporal burglary rates, $\boldsymbol{\theta}=(\theta_{11},...,\theta_{1T}...,\theta_{N1},...,\theta_{NT})$, in the *process model*.

TABLE 14.1

A Summary of the Peterborough Burglary Data at the COA Level from 2001 to 2008

	2001	2002	2003	2004	2005	2006	2007	2008
Number of burglary cases	3.13 (3.25; 0.00; 19.00)	4.20 (4.00; 0.00; 26.00)	3.25 (4.00; 0.00; 15.00)	1.72 (2.00; 0.00; 12.00)	1.36 (2.00; 0.00; 8.00)	1.41 (2.00; 0.00; 9.00)	2.66 (3.00; 0.00; 12.00)	2.78 (3.00; 0.00; 16.00)
Number of at-risk houses	130 (16; 78; 239)							
Raw annual burglary rate per 1000 houses	24.00 (26.85; 0.00; 165.22)	32.32 (29.83; 0.00; 201.68)	25.00 (26.51; 0.00; 119.05)	13.20 (19.66; 0.00; 57.97)	10.41 (16.13; 0.00; 57.97)	10.90 (16.29; 0.00; 75.00)	20.49 (22.23; 0.00; 96.00)	21.53 (23.57; 0.00; 112.00)

Each cell reports the mean value and the three numbers in brackets are the interquartile range, the minimum and the maximum values, respectively.

As evident from the map sequence in Figure 12.1 (see also Exercise 14.1), it is not appropriate to assume all the burglary rates are the same, which would reduce the $N \times T$ parameters to just a single scalar parameter, because the raw COA-level annual burglary rates (the data) vary both spatially and temporally. This observation suggests that the space-time distribution of burglary risk is heterogenous, varying both spatially and temporally. Neither is it appropriate to model these burglary rates independently because there is evidence suggesting the presence of (positive) spatial autocorrelation and (positive) serial (or temporal) autocorrelation. There are two additional issues when modelling burglary rates independently. First, each burglary rate is estimated only using the observed data in the corresponding "space-time cell". Given the small number of burglary cases and the small size of the population at-risk – both reflecting the issue of data sparsity – we must doubt the reliability of estimates obtained by this method (see also Section 7.5 in Chapter 7). Second, we do not gain any insights into the space-time distribution of burglary risk in Peterborough over the study period. Using the independence option, we simply cannot tackle questions such as "what are the overall spatial and temporal patterns in the burglary rates?" and "does any COA display a burglary trend that is different from the overall Peterborough trend?" For the above reasons, we need to engage in modelling these parameters at the "space-time cell" level. In other words, each parameter can be specific to each area at a specific time point, giving rise to a matrix of $N \times T$, potentially different and correlated, parameters for us to model. Eq. 14.2 shows the form of this parameter matrix.

$$\theta = \begin{pmatrix} \theta_{11} & \cdots & \theta_{1T} \\ \theta_{21} & \cdots & \theta_{2T} \\ \vdots & \ddots & \vdots \\ \theta_{N1} & \cdots & \theta_{NT} \end{pmatrix} \tag{14.2}$$

For the Peterborough burglary data, we model $\log(\theta_{it})$ to ensure the burglary rates in θ are non-negative. In the next section, we start the modelling process by considering the space-time separable framework.

14.3 The Space-Time Separable Modelling Framework

14.3.1 Model Formulations

The structure of a space-time separable model separates the observed space-time variability into two modelling components: the overall *spatial* component that describes the spatial pattern that remains relatively stable over time (e.g. the spatial pattern common to the time sequence of maps; see Exercise 14.1) and the overall *temporal* component that captures the general trend pattern (e.g. the general, overall, temporal pattern of burglary rates in Peterborough discussed in Section 14.2). Based on this space-time separable structure, Eq. 14.3 specifies the *process model* using a generic formulation for the log burglary rates:

$$\log(\theta_{it}) = \alpha + SP_i + TM_t \tag{14.3}$$

In the above formulation, the two sets of terms, $\boldsymbol{SP} = (SP_1, \ldots, SP_N)$ and $\boldsymbol{TM} = (TM_1, \ldots, TM_T)$, represent respectively the overall spatial component and the overall temporal component, whilst α is the intercept. SP is also referred to as the spatial main effects, the effects of space (location) that remain constant across the time points. Likewise, TM is referred to as the temporal main effects, the effects of time that remain the same across the areas. The formulation in Eq. 14.3 is generic because the modelling of the spatial and the temporal main effects can be done using various space-only and time-only models. Table 14.2 provides a summary of some candidate models for each of these two components. We refer the reader to the corresponding section for the detail and properties of each model in Table 14.2.

To study the implications of the space-time separable structure in Eq. 14.3 for the space-time parameters, we rewrite the parameter matrix in Eq. 14.2 as

$$
\log\begin{pmatrix} \theta_{11} & \cdots & \theta_{1T} \\ \theta_{21} & \cdots & \theta_{2T} \\ \vdots & \ddots & \vdots \\ \theta_{N1} & \cdots & \theta_{NT} \end{pmatrix} = \alpha \mathbf{1}_{N\times T} + \begin{pmatrix} SP_1 + TM_1 & \cdots & SP_1 + TM_T \\ SP_2 + TM_1 & \cdots & SP_2 + TM_T \\ \vdots & \ddots & \vdots \\ SP_N + TM_1 & \cdots & SP_N + TM_T \end{pmatrix}, \tag{14.4}
$$

where $\mathbf{1}_{N\times T}$ denotes a matrix of ones of size $N \times T$. Now we can see from the matrix on the right-hand side of Eq. 14.4 that the temporal trends of all COAs in Peterborough are identical to that described by the overall temporal component, $TM = (TM_1,\ldots,TM_T)$. Because SP_i, the term added to TM, is COA-specific, the COA trends can differ by level. So, essentially, the space-time separable model in Eq. 14.3 produces a temporal sequence of burglary rates for each COA by either lifting the entire overall trend up (if that COA has a high average burglary rate over time) or pushing it down (if it has a low average burglary rate). As a result, *all the COA trends on the log scale are parallel to each other, making the same pattern as the overall trend described by* TM.

Equivalently, the matrix on the right-hand side of Eq. 14.4 implies that the spatial pattern depicted by the spatial main effects, $SP = (SP_1,\ldots,SP_N)$, remains the same from one time point to another. In other words, *if we rank the COAs according to* SP, *this ranking does not change over time* – and this has to be the case because the COA trends do not intersect but are parallel to each other. This implication is equivalent to assuming that if a COA has the highest average burglary rate (averaged over years), its burglary rate is also the highest in

TABLE 14.2

Some Candidate Models for Specifying the Overall Spatial Component, $SP = (SP_1,\ldots,SP_N)$, and the Overall Temporal Component, $TM = (TM_1,\ldots,TM_T)$

For the Overall Spatial Component SP		For the Overall Temporal Component TM	
An exchangeable model (Section 7.4)	$SP_i = U_i$ with $U_i \sim N\left(0, \sigma_U^2\right)$	The linear trend model without temporal noise (Section 12.3.3.1)	$TM_t = v_t$ with $v_t = d \cdot \left(t - t^*\right)$
The ICAR model (Section 8.2)	$SP_i = S_i$ with $S_{1:N} \sim ICAR\left(W_{sp}, \sigma_S^2\right)$	The linear trend model with temporal noise (Section 12.3.3.1)	$TM_t = v_t$ with $v_t = d \cdot \left(t - t^*\right) + \varepsilon_t$
The pCAR model (Section 8.3)	$SP_i = S_i$ with $S_{1:N} \sim pCAR\left(W_{sp}, \sigma_S^2, \rho\right)$	The RW1 model (Section 12.3.4.1)	$TM_t = v_t$ with $v_{1:T} \sim ICAR\left(W_{RW1}, \sigma_v^2\right)$
The BYM model (Section 8.5)	$SP_i = S_i + U_i$ with $S_{1:N} \sim ICAR\left(W_{sp}, \sigma_S^2\right)$ and $U_i \sim N\left(0, \sigma_U^2\right)$	The RW2 model (Section 12.3.4.5)	$TM_t = v_t$ with $v_{1:T} \sim ICAR\left(W_{RW2}, \sigma_v^2\right)$

For the detail of each model, see the corresponding section.

every one of the eight years. This assumption applies to all COAs. In Section 14.3.2, we will visualise the patterns of the temporal burglary rates at the COA level.

Under the space-time separable structure, the estimation of the spatial and the temporal main effects generally does not raise any problems since the number of parameters to estimate is fewer than the number of data points available from the spatial-temporal dataset. For example, the space-time separable model that uses the BYM model for the overall spatial component and the RW1 model for the overall temporal component has $(2N + T + 4)$ *actual* parameters, which are estimated using $(N \times T)$ data points. Specifically, the area-specific parameters in the overall spatial component are estimated essentially by collapsing the temporal dimension of the space-time dataset to form a single set of "aggregated" spatial data. Similarly, the estimation of the time-point-specific parameters in the overall temporal component is based on a single set of time series data in which the data values at each time point are basically formed by combining the data values across the spatial units. The variances and the overall intercept are estimated using all the outcome values in the dataset. It is not until the space-time interaction component is introduced that parameter estimation will become a challenge in terms of the number of data points relative to the number of parameters that need to be estimated.

14.3.2 Do We Combine the Space and Time Components Additively or Multiplicatively?

The formulation in Eq. 14.3 combines the overall spatial and the overall temporal components additively on the log scale. Compared to the multiplicative structure, i.e. $SP_i \times TM_t$, the additive structure is more often used in practice for two reasons.

First, when modelling count data, it is easier to interpret the two components with the additive structure either on the log scale (when the Poisson likelihood is used) or on the logit scale (when the binomial likelihood is used). It should also be noted that adding the two components together on the log scale is equivalent to combining the two components multiplicatively on the *rate* scale (i.e. $\theta_{it} = \exp(\alpha) \times \exp(SP_i) \times \exp(TM_t)$). So, to clarify, "additive" here means that we add the two components together on the *log* (or *logit*) scale. When using the Poisson likelihood, $\exp(SP_i)$ and $\exp(TM_t)$ are interpreted as the rate (or risk) ratios so that, for example, $\exp(SP_i) > 1$ would suggest a higher rate (or risk) in area *i* compared to the grand mean. When using the binomial likelihood, we then model the probability of event occurrence, so $\exp(SP_i)$ and $\exp(TM_t)$ are interpreted as odds ratios. Thus, $\exp(SP_i) > 1$, for example, would suggest the probability of having an event of interest in area *i* is greater when compared to the corresponding probability for the whole study region.

Second, the implied structure for the space-time parameters does not depend on how the two overall components are specified. As discussed earlier, the additive structure implies that the resulting trends at the small area level are parallel to each other and, equivalently, the spatial pattern does not differ from one time point to another. The form of the resulting parameter matrix in Eq. 14.4 suggests that these implications are generic and do not depend on how we specify SP and TM. However, this is not the case when the multiplicative structure is used. Consider the situation of modelling a set of continuous-valued outcome data using a normal distribution as the likelihood, e.g. $y_{it} \sim N(\mu_{it}, \sigma^2)$, and consider the multiplicative formulation, $\mu_{it} = \alpha + (SP_i \times TM_t)$. The implication for the resulting small area time trends is rather complicated, as it depends on how SP_i and TM_t are specified. For example, if the temporal parameters in TM are constrained to sum to zero, i.e. $\sum_{t=1}^{T} TM_t = 0$

(e.g. under the RW1 or RW2 prior model), then, as shown in Figure 14.1(a), all areas have the same level (anchored at α) but the time trend of each area is obtained by either compressing (if $0 < SP_i < 1$) or stretching (if $SP_i > 1$) the overall trend pattern along the y-axis. If $SP_i = 1$, then that area follows exactly the overall trend, but if $SP_i < 0$, the resulting trend for that area goes opposite to the overall pattern, as illustrated in Figure 14.1(a). If, however, we set $TM_1 = 0$, then as Figure 14.1(b) shows, although starting at the same value, the time trends of different areas not only differ by pattern (a compressed or stretched overall trend) but also by level. In fact, the multiplicative formulation of $\mu_{it} = \alpha + (SP_i \times TM_t)$ can be seen as a regression model with interaction between space and time but without spatial and temporal main effects, a specification that is not recommended in regression modelling (Nelder, 1977, p.49–50).[1] For these reasons, we only consider the additive structure for combining the overall spatial and the overall temporal components (and later the space-time interaction component).[2]

14.3.3 Analysing the Peterborough Burglary Data Using a Space-Time Separable Model

We consider the space-time separable model in Eq. 14.5 in which the overall spatial component is modelled using the BYM model and the overall temporal component is assigned the RW1 model. The nonlinear overall time trend pattern as seen from Table 14.1 suggests the use of the RW1 model for time, whilst the reader is encouraged to justify, through examining the properties of the data, the use of the BYM model for space (see Exercise 14.2). Model

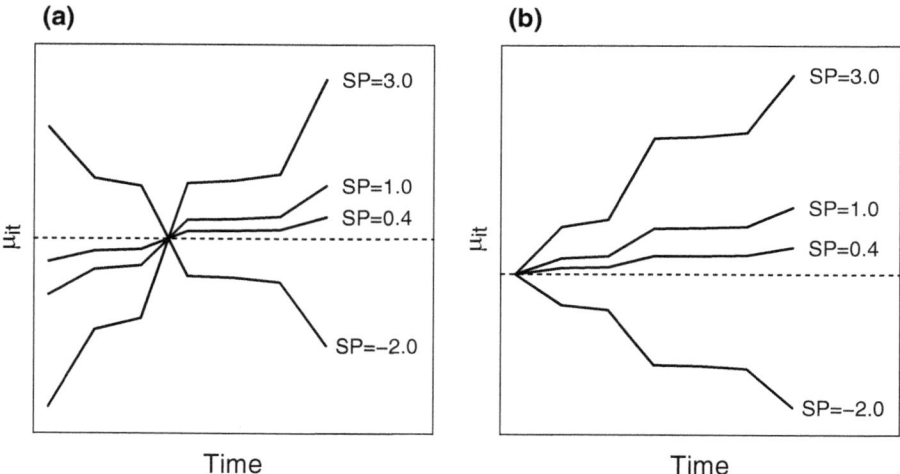

FIGURE 14.1
Illustration of the different implied structures on the space-time parameters μ from the multiplicative structure of combining the overall spatial and the overall temporal components in the space-time separable framework. (a) shows the implied structure when a sum-to-zero constraint is imposed on the overall temporal component, whereas (b) shows the implied structure when the first element in the temporal component, TM_1, is set to 0.

---

[1] Nelder (1977, p.50) has argued that "models where an interaction, $A \cdot B$ say, is postulated to exist whose marginal effects, A or B, are to be null... are of no practical interest".
[2] The interested reader is referred to Bennett et al. (2015) – the specification of the Lee-Carter model – for an example of the multiplicative structure (on the log scale).

comparison can also assist the specification of model components (Exercise 14.3). By now, the reader will be comfortable with the three models: *the data model, the process model* and *the parameter model*, in Bayesian hierarchical modelling and will also be familiar with the specification of each model in Eq. 14.5. The spatial weights matrix, W_{sp}, is derived using the first order rook's move contiguity, so $(W_{sp})_{ij} = 1$ if COAs i and j $(i \neq j)$ share a common border and $(W_{sp})_{ij} = 0$ otherwise. Each diagonal entry in W_{sp} is 0, i.e. $(W_{sp})_{ii} = 0$ for all i.

$$
\begin{array}{ll}
y_{it} \sim Poisson(\mu_{it}) & \\
\mu_{it} = n_i \cdot \theta_{it} & \text{The data model} \\
\hline
\log(\theta_{it}) = \alpha + (S_i + U_i) + v_t & \\
S_{1:452} \sim ICAR(W_{sp}, \sigma_S^2) & \\
U_i \sim N(0, \sigma_U^2) & \text{The process model} \qquad (14.5) \\
v_{1:8} \sim ICAR(W_{RW1}, \sigma_v^2) & \\
\hline
\sigma_S \sim Uniform(0.0001, 10) & \\
\sigma_U \sim Uniform(0.0001, 10) & \\
\sigma_v \sim Uniform(0.0001, 10) & \text{The parameter model} \\
\alpha \sim Uniform(-\infty, +\infty) &
\end{array}
$$

Figure 14.2 shows the WinBUGS code, the data structure and the two sets of initial values for fitting the model in Eq. 14.5. The WinBUGS specification for a space-time model becomes more complicated compared to that for a purely-spatial or a purely-temporal model. But we can start by modifying the WinBUGS code for the spatial model – in this case Figure 8.20 in Chapter 8 for the BYM model. In addition to going through all the areas (i.e. the for-loop on Line 5), Lines 8 to 18 in Figure 14.2 go through all the eight time points for each area. Line 17, in particular, specifies the space-time separable structure. The implementation of the spatial BYM model is given on Lines 22–26. Note that this is the hierarchically-centred version of the BYM model (see Section 8.5.1). Line 29 in Figure 14.2 adds in the specification of the RW1 model for the overall temporal component (see Section 12.3.4.4).

The structure of the outcome data now becomes an $N \times T$ matrix. This data matrix is specified using the WinBUGS function structure on Lines 61 to 65 in Figure 14.2. The structure function takes on two arguments. The argument .Data – be aware of the . in front of the capital letter D – is an array of length $N \times T$ storing all the space-time burglary counts so that the first T entries are the temporal burglary counts from the first COA and the subsequent T entries are the burglary counts from the second COA and so on. The argument .Dim – again, D must be capital with a . in front – defines the dimension of the matrix, which in this case has 452 rows and 8 columns. In the first set of initial values, the first two values in S, as well as in v, are set to be 0.1 and −0.1, whilst all other values are set to 0 so that the initial values sum to zero (Line 79 and Line 81, respectively). The same setting is applied for the second set of initial values, but the positions of 0.1 and −0.1 are swapped so that the two sets of initial values for S and for v are different (Line 85 and Line 87; see also Section 8.2.1.2).

Exercise 14.4 extends the code in Figure 14.2 to calculate some additional quantities of interest such as the variance partition coefficient, the predicted numbers of burglary cases and the posterior predictive p-values. In Exercise 14.4, we also calculate the deviation residuals, which help us to identify any spatial or temporal patterns in that part of the

```
1     # The WinBUGS code for fitting a space-time separable BHM
2     # with BYM (space) + RW1 (time)
3     model {
4       # an outer for-loop to go through all the (N=)452 COAs
5       for (i in 1:N) {
6         # an inner for-loop to go through all
7       #the (T=)8 time points
8         for (t in 1:T) {
9           # defining the Poisson likelihood for each count value
10          y[i,t] ~ dpois(mu[i,t]) # the Poisson likelihood for
11                                  # the burglary count in area
12                                  # i at time t
13          mu[i,t] <- n[i]*theta[i,t] # define the mean of the
14                                  #   Poisson distribution
15          # the additive space+time structure on the log
16          # burglary rate
17          log(theta[i,t]) <- SP[i] + v[t]
18       }   # close the for-loop over T
19     }     # close the for-loop over N
20     # BYM for the overall spatial component
21     # a spatial ICAR on S[1:N]
22     S[1:N] ~ car.normal(sp.adj[],sp.weights[],sp.num[],prec.S)
23     for (i in 1:N) {
24     mu.SP[i] <- alpha + S[i]
25     SP[i] ~ dnorm(mu.SP[i], prec.U) # adding U[i] in
26       }
27      # RW1 for the overall temporal component
28      # a temporal ICAR on v[1:T]
29      v[1:T] ~ car.normal(tm.adj[],tm.weights[],tm.num[],prec.v)
30
31      # specification of (vague) priors
32      alpha ~ dflat()
33      sigma.S ~ dunif(0.0001,10)
34      sigma.U ~ dunif(0.0001,10)
35      sigma.v ~ dunif(0.0001,10)
36      prec.S <- pow(sigma.S,-2)
37      prec.U <- pow(sigma.U,-2)
38      prec.v <- pow(sigma.v,-2)
39      # Some quantities of interest
40      # the average burglary rate per 1000 houses per year in
41      #   Peterborough
42      peterborough.average <- mean(theta[,])*1000
43      # rate ratio (RR) of the average burglary rate (over all
44      # COAs) in each year to the Peterborough average
45      for (t in 1:T) {
46        temporal.RR[t] <- exp(v[t])
47      }
```

FIGURE 14.2
The WinBUGS code, the data structure and two sets of initial values for fitting the space-time separable model in Eq. 14.5 to the Peterborough burglary data across 452 COAs over eight years.

```
48      # rate ratio: COA-average compared to Peterborough
49      # Note SP[i] includes alpha (see Line 24)
50      for (i in 1:N) {
51        spatial.RR[i] <- exp(SP[i] - alpha)
52      }
53    }
54
55    # the structure of the space-time burglary data
56    # in Peterborough
57    list(T=8 # 8 years
58        ,N=452 # 452 COAs
59      # the NxT matrix containing the space-time
60      # burglary counts
61      ,y=structure(.Data=c(15,3,9,0,0,4,7,5
62                          ,2,2,3,2,2,1,5,6
63                                ...
64                          ,5,7,6,0,1,0,1,1)
65                ,.Dim=c(452,8))
66      # the three arguments in the car.normal function to
67      # define the RW1 neighbourhood structure
68        ,tm.num=c(1,2,2,2,2,2,2,1)
69        ,tm.adj=c(2,1,3,2,4,3,5,4,6,5,7,6,8,7)
70        ,tm.weights=c(1,1,1,1,1,1,1,1,1,1,1,1,1,1)
71      # the three arguments in the car.normal function
72      # to define the spatial neighbourhood structure
73        ,sp.num=c(6,5,4,...)
74        ,sp.adj=c(6,7,9,11,16,17,...)
75        ,sp.weights=c(1,1,1,1,1,...)
76    )
77    # initial values for chain 1
78    list(alpha=-3
79        ,S=c(0.1,-0.1,0,0,...,0)
80        ,SP=c(0.01,0.01,...,0.01)
81        ,v=c(0.1,-0.1,0,0,0,0,0,0)
82        ,sigma.S=0.1,sigma.U=0.1,sigma.v=0.1)
83    # initial values for chain 2
84    list(alpha=-1
85        ,S=c(-0.1,0.1,0,0,...,0)
86        ,SP=c(0.02,0.02,...,0.02)
87        ,v=c(-0.1,0.1,0,0,0,0,0,0)
88        ,sigma.S=0.5,sigma.U=0.5,sigma.v=0.5)
```

FIGURE 14.2 (*Continued*)

observed variability *not* explained by the space-time separable model. For a Poisson likelihood, the deviance residual in area i at time point t is given by

$$
dr_{it} = \begin{cases} sign(y_{it} - \mu_{it})\sqrt{2\left[y_{it}\log\left(\dfrac{y_{it}}{\mu_{it}}\right) - (y_{it} - \mu_{it})\right]} & \text{if } y_{it} > 0 \\[2ex] -\sqrt{2\mu_{it}} & \text{if } y_{it} = 0 \end{cases}
\tag{14.6}
$$

where μ_{it} is the fitted value from the model and $sign(y_{it} - \mu_{it}) = 1$ if $y_{it} \geq \mu_{it}$; $sign(y_{it} - \mu_{it}) = -1$
otherwise. The deviance residuals, together with the posterior predictive p-values, will be
used for model assessment.

14.3.4 Results

We first look at whether the space-time separable model is a better model compared to
a model where each burglary rate, θ_{it}, is modelling independently with a vague Gamma
prior, $\theta_{it} \sim Gamma(0.01, 0.01)$. The latter model is basically a space-time version of the independent parameters model discussed in Section 7.3.2 in Chapter 7 (see Exercise 14.5), so
each burglary rate, θ_{it}, is estimated using only the data (y_{it} and n_i) in the corresponding
space-time cell.

The DIC value from the space-time separable model is 13387, which is lower than 14043,
the DIC value from the space-time independent parameters model, suggesting the former model is a more parsimonious model for the given dataset. The space-time separable
model is preferred mainly because it is a simpler model: the pD value from the space-time
separable model of 342 is lower than that from the independent parameters model, which
gives a pD of 3020. However, the space-time separable model yields an inferior fit to the
data, as its $\bar{D}$ value is 13045, higher than that from the independent-parameters model
where $\bar{D} = 11022$. We will return to the fits of the space-time separable model later.

We turn now to the parameter estimates from the space-time separable model. Between
2001 and 2008, the average annual burglary rate in Peterborough is estimated to be
19.7 (95% credible interval: 19.3–20.2) cases per 1000 houses per COA. Compared to this
Peterborough average, Figure 14.3(a) shows a map of the posterior means of the rate ratios,
$RR_i^{SP} = \exp(S_i + U_i)$, each comparing the annual average burglary rate of a COA to the
Peterborough average. The superscript in RR_i^{SP} emphasises that the rate ratio is computed
for each COA, so it is related to the spatial dimension. For each COA in Figure 14.3(a), the
darker its colour, the higher its average burglary rate across all eight years. While giving us an idea of where the COAs with high average burglary rates might be, the map in
Figure 14.3(a) does not take into account the uncertainty associated with the rate ratio estimates. One way to represent uncertainty is to calculate the posterior probability of the rate
ratio in each COA being above one, i.e. $pp_i = \Pr(\exp(S_i + U_i) > 1 \mid data)$ for all i. As shown in
Figure 14.3(b), a COA with a dark grey colour has a high certainty that its average burglary
rate is higher than the Peterborough average over the study period. A COA shown with
a light grey colour, on the other hand, is an area where the model is certain that its average annual burglary rate is lower than the Peterborough average. The COAs in white are
considered to have average burglary rates similar to the Peterborough average. The above
description is effectively using the posterior probability, pp_i, together with some chosen
cut-off values to classify each COA into one of the three categories. A COA is considered
to be a high-risk COA (coloured in dark grey in Figure 14.3(b)) if $pp_i > 0.8$; a low-risk COA

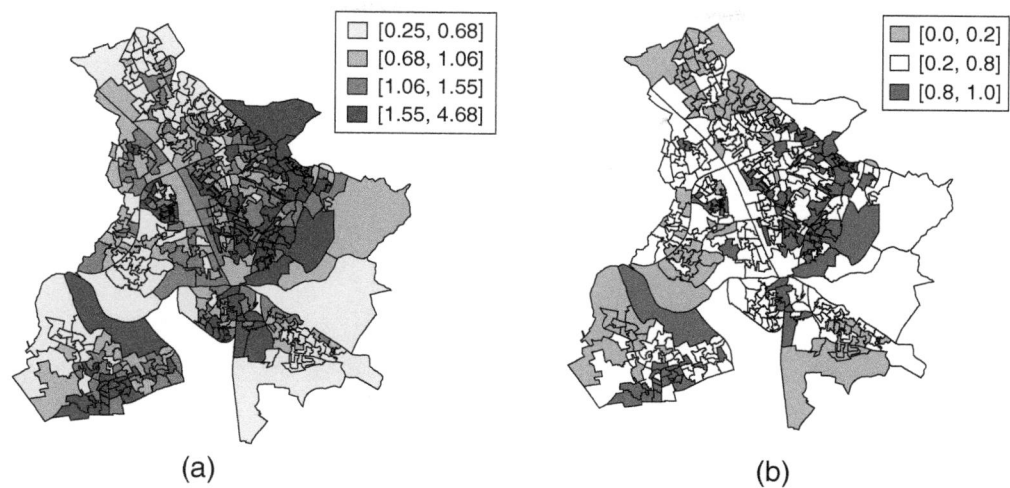

(a)   (b)

FIGURE 14.3
(a) a map of the posterior means of the rate ratios comparing the average burglary rate of each COA to the Peterborough average, $RR_i^{SP} = \exp(S_i + U_i)$. A COA shown in a dark colour is estimated to have a higher average burglary rate across the eight years and a COA with a rate ratio of 1 suggests its average burglary rate is similar to that of the whole of Peterborough. (b) A map of the posterior probabilities of each COA having an average burglary rate higher than the Peterborough average, $pp_i = \Pr\left(RR_i^{SP} > 1 \mid data\right)$. See the main text for interpretation of the three shadings in (b).

(coloured in light grey in Figure 14.3(b)) if $pp_i < 0.2$; or a COA with a risk level similar to the Peterborough average (coloured in white in Figure 14.3(b)) if $0.2 < pp_i < 0.8$. The cut-off values, 0.2 and 0.8, were proposed by Richardson et al. (2004). We will discuss this classification rule in more detail in Section 16.3. A close inspection of the two maps in Figure 14.3 reveals that some of the COAs in dark grey in (a), i.e. the COAs with posterior means of $RR_i^{SP} > 1$, are white in (b). These include some polygons in the northeast and some in the southwest. This suggests that while the posterior mean is high for each of those COAs, the posterior estimate is still considered to be uncertain according to the classification rule with over 20% (=(1−0.8) × 100%) of the posterior distribution lying below the value 1. In other words, based on this classification rule and this model, we are not confident enough to declare these COAs as high-risk areas.

Figure 14.4 presents the posterior means and the 95% uncertainty band of the rate ratios, $RR_t^{TM} = \exp(v_t)$, across the eight years. The estimated temporal pattern follows closely the description from the exploratory analysis in Section 14.2. The nonlinear pattern it reveals justifies the use of the RW1 model for the overall temporal component. Now, under this space-time separable model, what do the estimated temporal patterns for different COAs look like?

Figure 14.5(a) shows the posterior means of the modelled temporal burglary rates at the COA-level. As we discussed earlier, when plotted on the log scale, the temporal pattern of the estimated burglary rates for each COA (the thinner line) is exactly identical to the temporal pattern of the temporal burglary rates estimated for Peterborough as a whole (the thicker line). For some COAs, the time sequences of burglary rates are above (below) the Peterborough trend because their average burglary rates are higher (lower) than the Peterborough average. When compared to the raw burglary rates (calculated as $\frac{y_{it}}{n_i} \times 1000$)

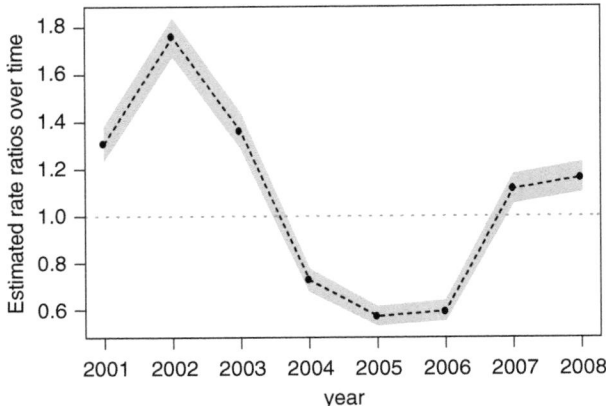

FIGURE 14.4
The posterior estimates of the rate ratios over time, $\exp(v_t)$, for $t = 1,\ldots,8$. The dots and the dashed line show the posterior means and the 95% credible intervals are shown by the grey band. The grey dotted horizontal line at 1 indicates no difference between the average burglary rate (averaged across all COAs) of a given year and the Peterborough average.

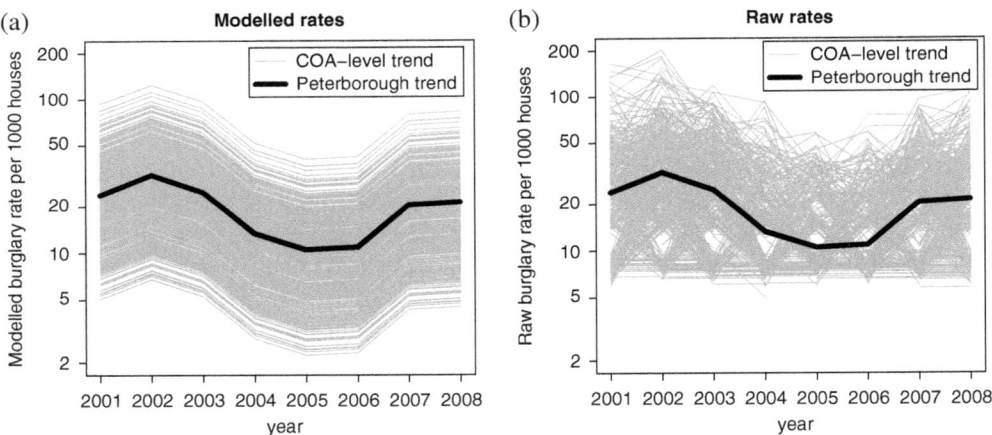

FIGURE 14.5
(a) the posterior means of the COA-level burglary rates from 2001 to 2008 and (b) the raw burglary rates obtained by dividing the observed burglary counts in each COA by its number of at-risk houses then multiplied by 1000. While the Peterborough trends in both plots, represented by the thicker line, are very similar, the modelled burglary rates at the COA level may oversmooth the observed variability evident in (b). Some lines (b) are "broken" near the bottom of the plot because the raw rate with a zero case count is not defined when plotted on the log scale.

shown in Figure 14.5(b), the COA-level estimates produced by this space-time separable model seem to be overly "structured". The space-time separable structure is unable to accommodate the part of the variability in the observed space-time data that does not follow what is presented in Figure 14.5(a). This results in an inferior fit to the data compared to the space-time version of the independent parameters model.

We inspect the fit (or the lack of fit) of the space-time separable model in more detail in order to understand where and why it does not fit well. Such information is useful in informing the next stage of the modelling process. Using the estimated posterior predictive

p-values, the space-time separable model adequately describes the observed annual burglary cases in 190 COAs (out of a total of 452), in which all of the posterior predictive
p-values are between 0.05 and 0.95. However, for 193 COAs, the model fits the observed
case count at one of the eight time points poorly, with a posterior predictive p-value either
above 0.95 or below 0.05. For 56 COAs, the lack of fit occurs at two time points, and in the
case of 13 COAs, lack of fit occurs at three or four time points. To inspect any temporal
pattern in the unexplained variability, Figure 14.6 and Figure 14.7 respectively show the
posterior means of the deviance residuals for a selection of the 11 COAs with poor model
fits at three time points and the two COAs with poor model fits at four time points. There
is no obvious temporal pattern presented for the COAs in those two figures, apart from
COAs 83 and 113 in Figure 14.6 and COA 107 in Figure 14.7, where the residuals for each of
the two COAs may suggest a decreasing pattern. Exercise 14.6 uses the `acf` function in R
to further examine the serial correlation at lag one for each of the COAs where poor model
fit occurs at least at one time point. The results suggest little temporal autocorrelation in
the residuals of each COA.

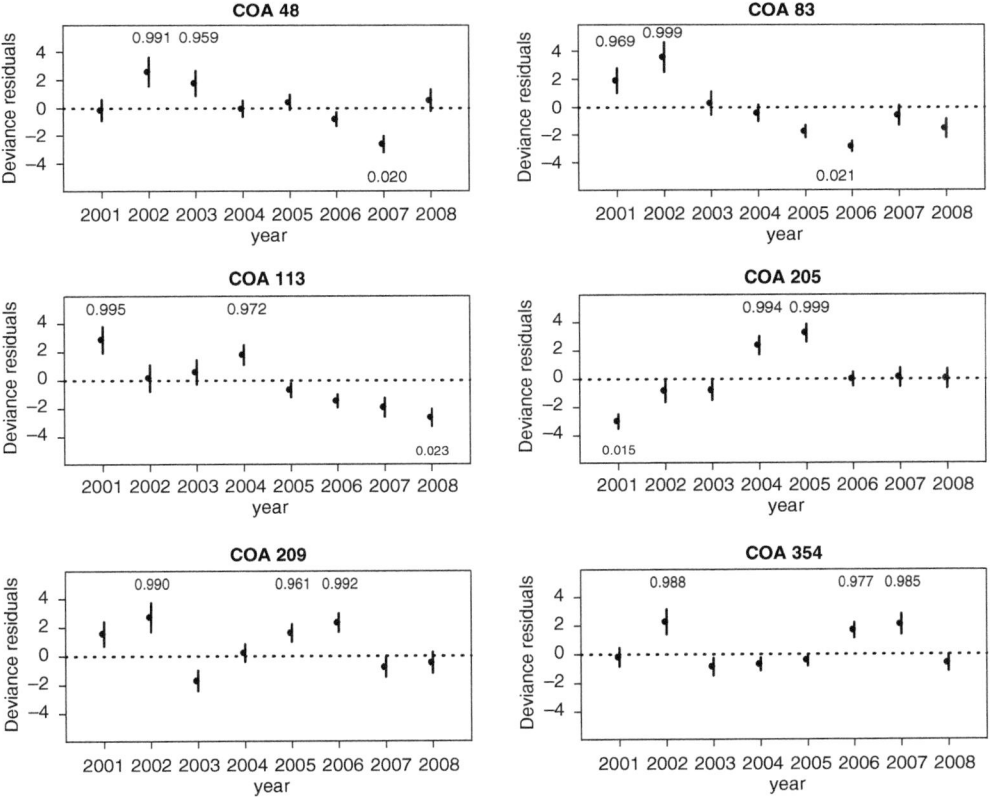

FIGURE 14.6
Assessing the temporal pattern of the deviance residuals of a selection of the 11 COAs in which the space-time
separable model fits poorly three of the eight observed burglary counts. In each plot, the dots represent the
posterior means while the vertical bars are the 95% credible intervals. For the time points when the model does
not fit well, the corresponding posterior predictive p-values are either shown on top of the plot (when greater
than 0.95) or at the bottom of the plot (when less than 0.05).

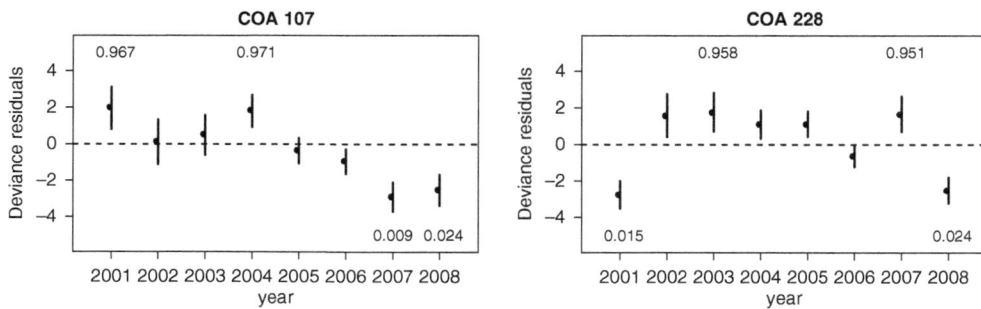

FIGURE 14.7
Assessing the temporal pattern of the deviance residuals of each of the two COAs in which the space-time separable model fits poorly four of the eight observed burglary counts. The setting of the plots here is the same as those in Figure 14.6.

Are the residuals across the COAs at each time point spatially autocorrelated? For each of the eight years, the number of COAs for which the space-time separable model does not provide a good fit (with posterior predictive p-values being either above 0.95 or below 0.05) to the observed burglary counts ranges from 34 (in 2002) to 56 (in 2008). To assess any spatial autocorrelation in the unexplained variability, we calculate the global Moran's I using the posterior means of the deviance residuals for each year. All Moran's I values are positive, and the associated p-values, each derived from 9999 random permutations, are all less than 0.05, suggesting the presence of positive spatial autocorrelation in the deviance residuals in each year.

As informed by the results from the above residual analysis, the next step in modelling needs to allow for the variability in the observed space-time outcome data that is not captured by the space-time separable structure. Such unexplained variability exhibits strong positive spatial autocorrelation within each year. There is, however, little temporal autocorrelation in the COA-level temporal deviance residuals.

14.4 Concluding Remarks

In general, the space-time separable framework provides a sensible starting point in the modelling of a spatial-temporal dataset. The application above shows that compared to an independent parameters model, a space-time separable model is to be preferred as it not only summarises the overall spatial and the overall temporal patterns in the observed data but also provides a modelling structure that allows us to consider various spatial models and temporal models. However, a space-time separable model may not yield a good fit to the temporal data at the small area level. Such poor model fits should not come as a surprise, as they are a consequence of the rather restrictive assumptions of space-time separability. All areas have the same time trend pattern or, equivalently, the spatial pattern is the same across all time points. Yet, assessing the fits (or the lack of fits) to the observed outcome data from a space-time separable model helps us to understand where the model fails. The type of residual analysis presented in Section 14.3.4 plays an important role in informing the next stage of the modelling process. In the case of modelling the Peterborough burglary data, having accounted for the spatial and the temporal main effects, the residual variability in each year shows strong positive spatial autocorrelation, but there is little temporal structure in the temporal deviance residuals at the COA level. We shall see in Chapter 15 how such findings from fitting a space-time separable model help not only inform the types of space-time inseparable models that we build for the Peterborough data but also understand the behaviours of these space-time inseparable models.

14.5 Exercises

Exercise 14.1. Produce a sequence of maps for the raw annual COA-level burglary rates (per 1000 houses) in Peterborough across the eight years from 2001 to 2008. Comment on the spatial patterns and the temporal variability across the eight maps (hint: see Section 13.3 for more detail on visualising space-time data).

Exercise 14.2. Obtain the average raw burglary rate over the eight years for a COA by summing over the annual burglary counts in each COA, then dividing the total count by the product of the number of at-risk houses and the value eight (years). Explore the properties of the resulting averages using methods discussed in Chapter 6. How does the finding of this exploratory analysis help specify the overall spatial component in Eq. 14.5?

Exercise 14.3. Construct various versions of the space-time separable model for the Peterborough data (Eq. 14.5) using other suitable choices for the overall spatial and the overall temporal components (e.g. some of those in Table 14.2). Compare these models together with the one in Eq. 14.5 using DIC.

Exercise 14.4. Extend the WinBUGS code in Figure 14.2 to calculate the following quantities of interest: (a) the posterior predictive p-values as defined in Section 5.4.2; and (b) the deviance residuals in Eq. 14.6. Also, how can we measure the relative importance of the overall spatial component and the overall temporal component in explaining the space-time variability in the observed data? Save the modelling results, as Exercises 14.6 and 14.7 (below) will examine the deviance residuals for some of the COAs.

Exercise 14.5. Write down an independent parameters model for analysing the Peterborough burglary count data so that the burglary rate in each space-time cell is estimated using the data available in that cell. What are the limitation(s) and potential strength(s) of this modelling strategy? Fit the model in WinBUGS and comment on the results.

Exercise 14.6. Examine the temporal autocorrelation of the deviance residuals for the COAs where the model gives a poor fit to at least one of the observed burglary counts. (Hint: the `acf` function in R calculates temporal autocorrelation of a time sequence of values.) Comment on the results. As we shall see in Chapter 15, the next step in the modelling process is to include a space-time interaction component to deal with departures from the space-time separable structure. This is done by introducing a set of space-time parameters, one for each space-time cell. How would the findings from this residual analysis inform the specification of the structure for these space-time parameters, e.g. should the temporal parameters in each COA be temporally correlated?

Exercise 14.7. Carry out the Moran's *I* calculation on the deviance residuals at each time point (as discussed towards the end of Section 14.3.4). As in Exercise 14.6, how does this finding inform the structure we should consider for modelling those space-time parameters in the space-time interaction component?

15

Bayesian Hierarchical Models for Spatial-Temporal Data II: Space-Time Inseparable Models

15.1 Introduction

In many applications, significant discrepancies are often found between the spatial-temporal data that we observe and the predictions from a space-time separable model. Such departures from space-time separability, indicative of the presence of space-time interaction, can arise for many different reasons. For example, because of a locally-targeted policy that aims to boost local economies, the income levels of some areas show a temporal trajectory different from the national trend. In the context of disease monitoring or surveillance, the emergence of a localised risk factor may lead to an increase in the disease risk in the affected areas while the disease trend remains relatively stable nationally.

As was discussed at the beginning of Chapter 12, in order to move from a space-time separable model to a space-time inseparable model, we need to add a space-time interaction component. Following our earlier "Lego block" analogy, we now need to add a new "block" that will accommodate space-time interaction, and as before, the specification of such a block will draw on the two fundamental properties of spatial-temporal data: dependency (the basis for information borrowing) and heterogeneity (the form of which informs the specification of the space-time interaction component).

Each strategy that we shall examine in this chapter is a modelling assumption that imposes a dependence structure on the parameters in the space-time interaction component. We will study the implications of each strategy (or assumed dependence structure) for the purposes of information sharing and parameter estimation. These strategies will be illustrated using the COA-level annual burglary count data in Peterborough from 2001 to 2008. We will also highlight the computational challenge that arises when fitting some of these space-time inseparable models. We will discuss applications of these space-time inseparable models in Chapter 16.

15.2 From Space-Time Separability to Space-Time Inseparability: The Big Picture

Knorr-Held (2000) proposes a space-time inseparable modelling framework in which a space-time interaction component is added to the space-time separable structure. Following the setup for modelling space-time burglary rates, θ, in Section 14.2, Eq. 15.1 shows the general structure under Knorr-Held's space-time inseparable framework:

$$\log(\theta_{it}) = \alpha + SP_i + TM_t + \delta_{it} \tag{15.1}$$

In Eq. 15.1, the three terms, α, SP_i and TM_t, are the same as those defined in Eq. 14.3, the generic structure under space-time separability. However, a new term, δ_{it}, is added, which, as opposed to the overall spatial and the overall temporal components, is indexed by *both* space (i) *and* time (t). As a result, δ_{it} is specific to area i at time point t (or what we label as "space-time cell (i,t)"). In other words, the parameters, $\delta = (\delta_{11},...,\delta_{1T},...,\delta_{N1},...,\delta_{NT})$, in the space-time interaction component vary jointly over *both* space *and* time. It is these space-time varying parameters that allow for departures from the $(SP_i + TM_t)$ space-time separable structure, enabling us to model, for example, temporal trajectories that are specific to just some of the areas or a dynamic spatial pattern that changes (smoothly) from one time point to another.

The introduction of these space-time parameters, however, raises an estimation challenge: there are now many more parameters to estimate. For example, Eq. 15.1 has at least $(1 + N + T + (N \times T))$ parameters, more than the $N \times T$ data points available in the space-time dataset. In the context of estimating small area disease risks over time, a similar challenge is pointed out by Knorr-Held (2000, p.2555): "The problem with such [spatial-temporal] data is that the number of cases and the corresponding population at risk in any single unit of space x time are too small to produce a reliable estimate of the underlying disease risk without 'borrowing strength' from neighbouring cells."

Once again, central to tackling this problem is information borrowing. But, how do we borrow information? The answer is to impose a dependence structure on the space-time parameters, an idea that we have discussed at length in Chapters 7 and 8 in the case of spatial modelling and subsequently in Chapter 12 when modelling a set of temporally-varying parameters. Now, recall that in the case of spatial modelling, we constructed a spatial dependence structure on a set of area-specific parameters based on an assumption about how we thought these area-specific parameters were related to one another. For example, parameters might be considered to be *globally similar* (e.g. under an exchangeable model), *locally similar* (e.g. under various CAR specifications) or *both globally and locally similar* (e.g. under the BYM model). The same line of thinking can be applied to construct a dependence structure, thus a prior model, for the space-time interaction parameters, δ. That is, we derive a dependence structure based on how we think the parameter associated with any space-time cell is related to (or dependent on) the parameters in all the other space-time cells. Then, a prior model for the space-time parameters can be constructed based on this space-time dependence structure. Unsurprisingly however, this is a rather challenging task.[1]

To operationalise this model building process, the aim of this chapter is to discuss four different space-time dependence structures for modelling the $N \times T$ parameters in the space-time interaction component, $\delta = (\delta_{11},...,\delta_{1T},...,\delta_{N1},...,\delta_{NT})$, and to show how each

---

[1] As Waller et al. (1997, p.610) put it, "In general, defining a space-time component is thorny, because it is not clear how to reconcile the different scales." However, there has been a large body of work developing various space-time models with the feature of space-time inseparability, some of which we will discuss in this chapter and the next.

structure can be used to develop a suite of prior models for δ. These four space-time dependence structures are as follows:

I. All parameters are similar regardless of where they are in "the space-time cube".

II. The temporal parameters in each area i, i.e. $\delta_{i1},...,\delta_{iT}$, are *temporally smooth*, but the time trend pattern represented by the temporal parameters in area i and the time trend pattern represented by the temporal parameters in area j (for $i \neq j$) are not assumed to be dependent upon (or similar to) each other, even in the case where i and j are spatially contiguous. So, this dependence structure allows for the situation where, for example, area i displays an increasing trend whilst area j displays a completely opposite, decreasing, trend, and the patterns of both local trends are different from the relatively constant/stable overall trend.

III. The area-specific parameters at each time point t, i.e. $\delta_{1t},...,\delta_{Nt}$, are *spatially smooth*, but the spatial pattern represented by the area-specific parameters at time point t and the spatial pattern represented by the area-specific parameters at time point g $(t \neq g)$ are not assumed to be dependent on (or similar to) each other, even in the case where t and g are close (e.g. $g = t + 1$). So, this dependence structure can be used when, for example, the spatial patterns differ from one time point to another and the spatial patterns at two consecutive time points are not similar to one another.

IV. The dependence structure for δ is no longer just over time (as in II) or just over space (as in III) but is fully space *and* time. So, for example, the time trends of two spatially contiguous areas can be different, but they are assumed to resemble each other and, likewise, the spatial patterns at two consecutive time points can be different but they are assumed to resemble each other.

These four space-time dependence structures are referred to as Types I to IV space-time interaction in Knorr-Held (2000) – thus the use of Roman numerals in the list above. Following Knorr-Held's convention, we refer to these four space-time dependence structures as Types I to IV space-time interaction and use "space-time dependence structure" and "space-time interaction" interchangeably.

Each of the four types of space-time dependence structure provides an instruction to construct a *prior* model for the parameters in the space-time interaction component. It is important to recognise that each type is *an assumption* on the interrelatedness amongst these parameters, $\delta_{11},...,\delta_{1T}...,\delta_{N1},...,\delta_{NT}$. Different dependence structures have different implications for parameter estimation. For example, as we shall see, a prior model following Type II space-time interaction tends to produce smoothly-varying time trends for each area, but the time trends of two spatially-contiguous areas are not necessarily similar (Exercise 15.1). By contrast, a prior model following Type III space-time interaction tends to produce a sequence of smooth maps, but the spatial patterns of two maps at two consecutive time points are not necessarily similar. Type IV space-time interaction assumes yet a different space-time dependence structure, which is likely to produce smooth time trends that are spatially autocorrelated. While the goal of the analysis often helps determine which of these space-time dependence structures to follow, it is, as we have emphasised throughout this book, important (a) to check modelling assumption(s) against the observed data (via exploratory data analysis and model evaluation); (b) to consider and compare various models (via the use of DIC); and (c) to assess how sensitive modelling results are if an alternative prior specification is selected. When defining the space-time interaction component, such prior specifications correspond to different space-time dependence structures imposed on δ.

In the following sections, we study each of these four types of space-time interaction in detail. For each type, we first discuss how to construct a prior model for δ under the given dependence structure. We then describe how to fit an example model to the Peterborough burglary data (as introduced in Section 14.2) in WinBUGS. We signpost the reader to a set of exercises in order to carry out the actual WinBUGS fitting. We will summarise and discuss the modelling results of Types I to III in Section 15.6. In Section 15.7, we will discuss some ideas about constructing models under the Type IV structure, but we defer applications of some of these models to Chapter 16.

15.3 Type I Space-Time Interaction

Type I space-time interaction is the simplest of the four. It assumes that the parameters in the space-time interaction component are all similar to each other or, expressed differently, the space-time configuration plays no role in determining the dependence structure amongst the space-time parameters. So, for example, despite the fact that space-time cell (i,t) is close to space-time cell $(i,t + 1)$ in time, or that two space-time cells (i,t) and (j,t) are close in space, because areas i and j share a common border, neither of these circumstances play any part in specifying the dependence structure. A space-time model with Type I space-time interaction essentially assumes that having accounted for the spatial and the temporal main effects, the space-time residuals do not vary smoothly, neither over space nor time. In other words, the effects of the unobserved/unmeasured covariates that the space-time interaction component represents lack any space-time structure. In the case of modelling small area burglary rates, if an area in a specific year experienced an unexpectedly high (or low) burglary rate, such an "anomaly", under Type I space-time interaction, would not be associated with any change in the burglary rate in the same area in subsequent years. Neither would that anomaly be associated with any change in burglary rates in neighbouring areas during the same year. As a result, Type I space-time interaction does not represent any of the space-time processes discussed in Section 2.4.1 in Chapter 2 that involve, for example, space-time diffusion or space-time dispersal. Rather, any departure from the space-time separable structure is assumed to be "one-off" in both space and time.

15.3.1 Example: A Space-Time Model with Type I Space-Time Interaction

For modelling the Peterborough space-time burglary count data, Eq. 15.2 extends the space-time separable model in Eq. 14.3 in Chapter 14 to incorporate a space-time interaction component with Type I space-time interaction. All the space-time interaction parameters in δ are independent of each other, but each δ_{it} follows the same normal distribution, $N\left(0,\sigma_\delta^2\right)$, where σ_δ^2 is an unknown (hyper)parameter. In the parameter model, a vague uniform prior distribution, *Uniform*(0.0001,10), is assigned to the standard deviation, σ_δ. This prior model on δ under Type I space-time interaction can be considered as a space-time extension of the spatial exchangeable model discussed in Section 7.4 in Chapter 7.

$$y_{it} \sim Poisson\left(\mu_{it}\right)$$

$$\mu_{it} = n_i \cdot \theta_{it}$$

The data model

$$\log\left(\theta_{it}\right) = \alpha + \left(S_i + U_i\right) + v_t + \delta_{it}$$

$$S_{1:452} \sim ICAR\left(\mathbf{W}_{sp}, \sigma_S^2\right)$$

$$U_i \sim N\left(0, \sigma_u^2\right)$$

The process model   (15.2)

$$v_{1:8} \sim ICAR\left(\mathbf{W}_{RW1}, \sigma_v^2\right)$$

$$\delta_{it} \sim N\left(0, \sigma_\delta^2\right)$$

$$\sigma_S \sim Uniform(0.0001, 10)$$

$$\sigma_U \sim Uniform(0.0001, 10)$$

$$\sigma_v \sim Uniform(0.0001, 10)$$

The parameter model

$$\sigma_\delta \sim Uniform(0.0001, 10)$$

$$\alpha \sim Uniform(-\infty, +\infty)$$

The space-time model in Eq. 15.2 is actually quite flexible in fitting the outcome values observed in space and time. Its flexibility comes from the fact that the Type I structure places no restrictions on where and when a space-time anomaly (in the form of a departure from the space-time separable structure) can occur. However, in practice we may not expect to have too many such anomalies, implying that not all the N X T space-time interaction parameters are needed. If that is so, the model in Eq. 15.2 may be over-parameterised, which implies that some parameters in δ are redundant.

Mixture modelling is one way of making this model more parsimonious. In the context of examining the stability of space-time patterns of disease outcomes, Abellan et al. (2008) proposed a mixture modelling approach where each space-time interaction parameter, δ_{it}, is allocated to one of two normal distributions, $N\left(0, \sigma_1^2\right)$ or $N\left(0, \sigma_2^2\right)$, as opposed to using just a single distribution, as in Eq. 15.2. The two variances, σ_1^2 and σ_2^2, are constrained so that $\sigma_1^2 < \sigma_2^2$ and, as a result, the variability of the second normal distribution is (much) larger than that of the first. Their rationale for this formulation is that, with the aim of assessing the stability of the local disease risks, if an area's observed data over time can be adequately described by the space-time separable structure with some small departures added, then this area is said to have a *stable* risk pattern. However, if some of those departures (from the space-time separable structure) are large, then the area's risk pattern is said to be *unstable*. Therefore, the first normal distribution (constrained to have a smaller variance) represents the stable component that accommodates small space-time departures, while the second normal distribution (with a larger variance) is the unstable component that captures the large space-time departures. When combined with data, the posterior probability of each space-time interaction parameter being allocated to either the stable (first) or the unstable (second) component can be used to examine the stability (or instability) of the space-time distribution of the disease risk. In their model, the allocation of δ_{it} to either the first or the second normal distribution is done via a binary 0/1 indicator variable z_{it}, such that $\delta_{it} \sim z_{it} \cdot N\left(0, \sigma_1^2\right) + \left(1 - z_{it}\right) \cdot N\left(0, \sigma_2^2\right)$. This specification is essentially the same as that used in the two-component Poisson mixture model discussed in Section 9.4 – the reader is encouraged to rewrite the above formulation into the form presented in Eq. 9.15. In Abellan et al. (2008), the proposed mixture model was illustrated through an investigation of the spatial-temporal risk patterns in nonchromosomal congenital anomalies in England from 1983 to 1998. We refer the reader to their paper for details regarding the full specification of the model, the application as well as the WinBUGS implementation. We shall return to the use of mixture modelling in Section 16.5.

15.3.2 WinBUGS Implementation

The WinBUGS implementation of the space-time inseparable model presented in Eq. 15.2 is shown in Figure 15.1. The WinBUGS code is very similar to that in Figure 14.2 in Chapter 14, where the corresponding space-time separable model, with δ_{it} in Eq. 15.2 removed, was fitted. The only modifications are: on Line 13 (in Figure 15.1), where delta[i,t] is added in; on Line 19, where a common normal distribution is assigned to every one of the parameters in δ; and on Lines 38 and 42, where a vague prior is assigned to σ_δ, which is then converted to the corresponding precision, $1/\sigma_\delta^2$. Also note that, for the two sets of initial values, the WinBUGS function structure is used to assign initial values to the matrix of parameters in delta (see Lines 66 to 71 and Lines 77 to 82 in Figure 15.1).

Prior to carrying out the fitting, Exercise 15.2 adds in the calculations of the predicted numbers of cases, y_{it}^{pred}, and the posterior predictive p-values, $\Pr\left(y_{it}^{obs} \geq y_{it}^{pred} \mid data\right)$, when $y_{it}^{obs} > 0$ or $\Pr\left(y_{it}^{obs} = y_{it}^{pred} \mid data\right)$ when $y_{it}^{obs} = 0$, to the WinBUGS code in Figure 15.1 for evaluating the model fit. We also calculate the variance partition coefficients (VPCs) to quantify the relative contribution of each of the three components, namely, the overall spatial component, the overall temporal component and the space-time interaction component, in explaining the observed variability. These three VPCs are defined as

$$VPC_{SP} = \frac{\tilde{\sigma}_{SP}^2}{\tilde{\sigma}_{SP}^2 + \tilde{\sigma}_{TM}^2 + \tilde{\sigma}_{SPTM}^2}$$

$$VPC_{TM} = \frac{\tilde{\sigma}_{TM}^2}{\tilde{\sigma}_{SP}^2 + \tilde{\sigma}_{TM}^2 + \tilde{\sigma}_{SPTM}^2} \quad (15.3)$$

$$VPC_{SPTM} = \frac{\tilde{\sigma}_{SPTM}^2}{\tilde{\sigma}_{SP}^2 + \tilde{\sigma}_{TM}^2 + \tilde{\sigma}_{SPTM}^2}$$

where $\tilde{\sigma}_{SP}^2$, $\tilde{\sigma}_{TM}^2$ and $\tilde{\sigma}_{SPTM}^2$ are the unconditional variances of the overall spatial, the overall temporal and the interaction components (see Exercise 15.2). Finally, we also monitor the DIC calculation, the results of which will be used for model comparison in Section 15.6. Exercise 15.2 examines the posterior distribution of each δ_{it} to indicate how likely it is that the observed outcome value of that space-time cell departs from the space-time separable structure.

15.4 Type II Space-Time Interaction

Under Type II space-time interaction, the temporal parameters in each area i, $\delta_{i,1:T} = (\delta_{i1},...,\delta_{iT})$, are assumed to be temporally smooth so that each set of temporal parameters can

```
1    # The WinBUGS code for fitting a space-time inseparable
2    # model with BYM (space) + RW1 (time) + Type I interaction
3    model {
4      # an outer for-loop to go through all the (N=)452 COAs
5      for (i in 1:N) {
6        # an inner for-loop to go through all the
7        # (T=)8 time points
8        for (t in 1:T) {
9          # defining the Poisson likelihood for each count value
10         y[i,t] ~ dpois(mu[i,t])
11         mu[i,t] <- n[i]*theta[i,t]
12         log(theta[i,t]) <- alpha + S[i] + U[i]
13                         + v[t] + delta[i,t]
14       # On Line 13 (above), note the addition of the
15       # space-time interaction term delta[i,t]
16       #
17       # assign Type I space-time interaction structure to
18       # each space-time interaction parameter, delta[i,t]
19       delta[i,t] ~ dnorm(0,prec.delta)
20        } # close the for-loop over T
21      } # close the for-loop over N
22
23      ## BYM for the overall spatial component
24      # a spatial ICAR on S[1:N]
25      S[1:N] ~ car.normal(sp.adj[],sp.weights[],sp.num[],prec.S)
26      # an exchangeable model on U[i]
27      for (i in 1:N) {
28        U[i] ~ dnorm(0,prec.U)
29      }
30      ## RW1 for the overall temporal component
31      # a temporal ICAR on v[1:T]
32      v[1:T] ~ car.normal(tm.adj[],tm.weights[],tm.num[],prec.v)
33      ## specification of other (vague) priors
34      alpha ~ dflat()
35      sigma.S ~ dunif(0.0001,10)
36      sigma.U ~ dunif(0.0001,10)
37      sigma.v ~ dunif(0.0001,10)
38      sigma.delta ~ dunif(0.0001,10)
39      prec.S <- pow(sigma.S,-2)
40      prec.U <- pow(sigma.U,-2)
41      prec.v <- pow(sigma.v,-2)
42      prec.delta <- pow(sigma.delta,-2)
43      ## Some quantities of interest
44      # the average burglary rate per 1000 houses per year
45      # in Peterborough
46      peterborough.average <- mean(theta[,])*1000
```

FIGURE 15.1

The WinBUGS code and two sets of initial values for fitting the Bayesian space-time inseparable model with Type I space-time interaction in Eq. 15.2 to the Peterborough burglary data across 452 COAs over eight years. The data structure for fitting this model is the same as that presented in Figure 14.2 in Chapter 14.

```
47    # rate ratio (RR) of the average burglary rate
48    # (over all COAs) in each year to the Peterborough
49    # average
50    for (t in 1:T) {
51      temporal.RR[t] <- exp(v[t])
52    }
53    # rate ratio (RR) of the average burglary rate
54    # (over the 8 years) in each COA to the
55    # Peterborough average
56    for (i in 1:N) {
57      spatial.RR[i] <- exp(S[i]+U[i])
58    }
59  }
60
61  # initial values for chain 1
62  list(alpha=-3
63    ,S=c(0.1,-0.1,0,0,...,0),U=c(0.01,0.01,...,0.01)
64    ,v=c(0.1,-0.1,0,0,0,0,0,0)
65      ,sigma.S=0.1,sigma.U=0.1,sigma.v=0.1
66      ,delta=structure(
67            .Data=c(0.01,0.01,0.01,0.01,0.01,0.01,0.01,0.01
68                  ,0.01,0.01,0.01,0.01,0.01,0.01,0.01,0.01
69                  ,...
70                  ,0.01,0.01,0.01,0.01,0.01,0.01,0.01,0.01)
71          ,.Dim=c(452,8))
72      ,sigma.delta=0.1)
73  # initial values for chain 2
74  list(alpha=-1,S=c(-0.1,0.1,0,0,...,0),U=c(0.02,0.02,...,0.02)
75    ,v=c(-0.1,0.1,0,0,0,0,0,0)
76    ,sigma.S=0.5,sigma.U=0.5,sigma.v=0.5
77    ,delta=structure(
78            .Data=c(-0.01,-0.01,-0.01,-0.01,-0.01,-0.01,-0.01,-0.01
79                  ,-0.01,-0.01,-0.01,-0.01,-0.01,-0.01,-0.01,-0.01
80                  ,...
81                  ,-0.01,-0.01,-0.01,-0.01,-0.01,-0.01,-0.01,-0.01)
82          ,.Dim=c(452,8))
83    ,sigma.delta=0.05)
```

FIGURE 15.1 (*Continued*)

be modelled using, for example, the linear trend model, the RW1 or RW2 model.[2] However, Type II space-time interaction also assumes that the temporal pattern, represented by the parameters $\delta_{i,1:T}$ in area i, is not dependent on the temporal pattern represented by the parameters $\delta_{j,1:T}$ in another area j regardless of whether i is close to j or not. In terms of setting up a prior model for δ, the N sets of temporal parameters – one for each area – are

---

[2] When using the linear trend model, we typically use the formulation presented in Eq. 12.3 in Chapter 12, i.e. $v_t = d \cdot (t - t^*)$, as opposed to the one in Eq. 12.4: $v_t = d \cdot (t - t^*) + \varepsilon_t$. The temporal noise terms, $\varepsilon_1,...,\varepsilon_T$, are excluded so that the resulting model is not overparameterized.

assumed to be independent of each other, but each set follows the same temporal model (e.g. the RW1 model). In the case of modelling small area burglary rates, the implication of the Type II structure is that the annual burglary rates in each area follow their own temporal smoothly-varying pattern, but the time trend patterns associated with spatially contiguous areas may not necessarily resemble each other. In other words, having accounted for the spatial and temporal main effects, the effects of the unobserved/unmeasured covariates that δ represents operate independently from one area to another, although their effects on the outcome of interest vary smoothly over time. This may be due to circumstances that are specific to each area. Under Type II interaction, effects are assumed to be "localised" in the sense that they would not be associated with any change in burglary rates in other areas. The Type II dependence structure would therefore be suitable for capturing unusual local time trends which result from the effects of some highly localised risk factors whose effects do not spread spatially. In the absence of information on such localised risk factors, the space-time interaction parameters then act as surrogate measures for these effects.

15.4.1 Example: Two Space-Time Models with Type II Space-Time Interaction

Following the structure of Type II space-time interaction, Eq. 15.4 and Eq. 15.5 present two different prior models for the space-time interaction parameters, δ. These two prior models assign the linear trend model in Eq. 15.4 or the RW1 in Eq. 15.5, respectively, to the temporal parameters of each area independently.

$$\delta_{it} = d_i \cdot \left(t - t^* \right)$$

$$d_i \sim N\left(0, 1000000 \right)$$

(15.4)

Specifically, in Eq. 15.4, t^* denotes the midpoint of the observation period. Each slope parameter, d_i, is area-specific. A vague prior distribution, $N(0,1000000)$, is assigned to each of the slopes independently so that each slope d_i is estimated using only data observed in that area, and no information sharing across areas is allowed. The reader is encouraged to compare and contrast Eq. 15.4 and Eq. 15.7 (later in Section 15.7.1) in order to understand the different assumptions underlying the two prior models on δ.

For modelling the Peterborough burglary data over eight years, however, the specification in Eq. 15.4 using a linear trend model may not be suitable. This is because, as shown in Figure 14.6 and Figure 14.7 in Chapter 14, the deviance residuals obtained after fitting the corresponding space-time separable model did not show any linear pattern in general. For this reason, we do not consider this specification in this chapter, but we will return to it in Section 16.3 in Chapter 16 when modelling the same set of burglary data in Peterborough but over a shorter time period (2005 to 2008).

$$\delta_{i,1:T} \sim ICAR\left(W_{RW1}, \sigma_{\delta}^2 \right)$$

$$\sigma_{\delta} \sim Uniform(0.0001, 10)$$

(15.5)

The prior model in Eq. 15.5 associates each $\delta_{i,1:T}$ with the RW1 model, which, compared to the linear trend specification, is more flexible in capturing nonlinear departures from the overall time trend pattern. Although the RW1 model is independently assigned to each set of temporal parameters, under the specification in Eq. 15.5, the N sets of temporal parameters are not estimated *truly* independently of each other. The reason is that the unknown conditional variance σ_{δ}^2 of each RW1 model is set to be the same for all areas. To explain, consider the situation where a majority of areas have the same temporal pattern and there is only a small number of areas showing temporal patterns that are different from the common temporal pattern. As we shall see in Section 16.4 in Chapter 16, this is akin to the space-time disease surveillance situation considered in Li et al. (2012). If that is the case, σ_δ^2 will be estimated to be small (and close to 0) because a majority of the space-time interaction parameters are not needed. Because this variance is set to be the same for all areas, any deviations from the common temporal pattern are likely to be pulled towards 0, potentially resulting in a set of trend estimates that look much the same. As a consequence, the prior model in Eq. 15.5 may fail to capture the unusual temporal patterns that occur in the handful of areas. The space-time detection model proposed by Li et al. (2012) allows the random walk variance to be area-specific, i.e. using $\sigma_{\delta,i}^2$ instead of σ_δ^2, so that each RW1 model has additional flexibility to capture an unusual local time trend pattern. These area-specific variances are then modelled hierarchically via a common probability distribution. We will return to that modelling in Section 16.4 in Chapter 16. The prior model on δ in Eq. 15.5 is more appropriate when there is a relatively large pool of areas displaying temporal patterns different from the overall trend pattern.

15.4.2 WinBUGS Implementation

To implement the prior model on δ in Eq. 15.5 in WinBUGS, we only need to remove Line 19 in Figure 15.1 and add the following lines immediately after Line 21:

```
# specify the Type II space-time interaction using the RW1 model
for (i in 1:N) {
  delta[i,1:T] ~ car.normal(tm.adj[],tm.weights[],tm.num[],prec.delta)
}
```

The above for-loop assigns the RW1 model to the temporal parameters delta[i,1:T] for each of the N areas. It should be noted that although the temporal neighbourhood structure (as defined through the three arguments tm.adj[], tm.weights[] and tm.num[]) is the same across the areas, the estimates of the temporal parameters for any area will be largely based on the observed outcome data from that area. We say "largely" because, in addition to using its own temporal data, the estimation of delta[i,1:T] for area i also depends on the RW1 variance, which is estimated using the data from all areas. Consequently, the posterior estimates of these parameters can be different from one area to another. Exercise 15.3 provides additional detail on the WinBUGS fitting of this model, and the results will be discussed in Section 15.6.

15.5 Type III Space-Time Interaction

Type III space-time interaction smooths the area-specific parameters, $\delta_{1:N,t} = (\delta_{1t},..,\delta_{Nt})$, spatially at each time point t, but spatial smoothing is carried out independently across the time points. As a result, after accounting for the spatial and the temporal main effects, each time point can exhibit a residual spatial pattern different from those at other time points. Under Type III interaction, these residual spatial patterns are not assumed to change smoothly over time. The implication under this dependence structure is that δ_{it} is assumed to be similar to δ_{jt} when areas i and j are geographically close but is not assumed to be similar to $\delta_{i,t+1}$ or to $\delta_{i,t-1}$ even though they are close in time. Therefore, the effects of the unmeasured/unobserved risk factors represented by the space-time interaction parameters δ display spatial patterns, but such effects are not assumed to transfer across time points. In the context of burglary rate modelling, for a given year, say t, any departures from the overall spatial main effects (such as an unexpected increase or decrease in the burglary rate) tend to appear in areas that are geographically close. However, under the Type III structure, these departures are not expected to appear in years close to t, e.g. $t-1$ and/or $t+1$. In other words, such departures are "spatially-structured but short-lived", and their presence is not expected to go beyond the time window (e.g. a year in the case of the Peterborough burglary data) defined for t.

Type III space-time interaction is therefore suitable to describe a space-time process in which the effect of a change initiated in an area at, say time point t, spreads to other neighbouring areas – for example, through a diffusion process and/or a between-area interaction (see Section 2.3.1). But such effects are not associated with the outcome of interest at the next time point $t+1$ (or, for that matter, time point $t-1$) when the observed outcome values display a spatial pattern that is different from the spatial pattern at t. In the Peterborough data where a time point corresponds to one year, such a space-time process, if it exists, is assumed to start and finish within a year. If, compared to the models based on other space-time interaction structures, a space-time model with Type III space-time interaction is better supported by the observed data, then it implies that the process discussed above (as well as other processes, of course) may be at work. If the investigation of such a process is of interest, one then needs to analyse a set of space-time data obtained at a finer temporal resolution (say, at the weekly or monthly level).

15.5.1 Example: A Space-Time Model with Type III Space-Time Interaction

Eq. 15.6 presents the prior model on δ under Type III interaction considered in Knorr-Held (2000).

$$\delta_{1:N,t} \sim ICAR\left(W_{sp}, \sigma_\delta^2\right)$$

$$\sigma_\delta \sim Uniform(0.0001, 10)$$

(15.6)

Specifically, the spatial parameters at one time point are independent of the spatial parameters at a different time point regardless how close or far apart the two time points are, but each set of spatial parameters follows a common ICAR model, with W_{sp} being the chosen spatial weights matrix (see Chapter 4). A vague prior distribution, $Uniform(0.0001, 10)$, is assigned to the standard deviation σ_δ.

Similar to the prior model in Eq. 15.5, using a common variance σ_δ^2 across all time points may limit the flexibility of this model in capturing large space-time residuals when they only appear infrequently. An extension to the prior model in Eq. 15.6 would be to allow the ICAR variance to be time-point specific. That is, $\delta_{1:N,t} \sim ICAR\left(W_{sp}, \sigma_{\delta,t}^2\right)$. We can then model these time-point-specific variances independently by assigning a vague prior to each, e.g. $\sigma_{\delta,t} \sim Uniform(0.0001, 10)$. For example, to model a set of small area lung cancer death counts in Ohio between 1968 and 1988, Xia and Carlin (1998, Eq. 8 on p.2034) modelled the space-time log mortality rates using a set of socio-demographic covariates (age, race, gender and smoking), the temporal main effects specified as a linear function of time and a space-time (interaction) component specified in the form of Eq. 15.6. But in their model, the ICAR variance is time-point specific (i.e. $\sigma^2_{\delta,t}$), and a gamma prior is assigned to each of the precisions independently. That is, $\lambda_t \sim Gamma(c,d)$, independently across all t where $\lambda_t = 1/\sigma^2_{\delta,t}$ and c and d are some chosen fixed constants (see Xia and Carlin, 1998 for more detail). Waller et al. (1997) used a similar approach to model the same set of lung cancer mortality data, but they modelled each set of area-specific parameters, $\delta_{1:N,t}$, using the BYM model. While the model formulation in Waller et al. (1997) accounts for gender and race effects together with the space-time interaction component, neither spatial main effects nor temporal main effects were included. We encourage the reader to consult both papers to understand the similarities and the differences between the two modelling strategies.

It is worth noting that when using the ICAR model, as in Eq. 15.6, the time-point-specific ICAR variances, $\sigma^2_{\delta,t}$, may not need to be modelled hierarchically. This is because each $\sigma^2_{\delta,t}$ is estimated using the observed data from N areas and N, the number of areas in the study, is typically large. By contrast, however, the area-specific RW1 variances often need to be modelled hierarchically as in, for example, Section 16.4. This is because the estimation of each RW1 variance, $\sigma^2_{\delta,i}$, relies on a much smaller set of data over T time points – T is usually (much) smaller than N. In that case, information sharing through hierarchical modelling is necessary.

15.5.2 WinBUGS Implementation

To implement the prior model on δ in Eq. 15.6 to the Peterborough burglary data, we need the following two modifications to the code in Figure 15.1. First add the following lines immediately after Line 21 in Figure 15.1:

```
# specify the Type III space-time interaction structure using
# the spatial ICAR model
for (t in 1:T) {
 delta.transposed[t,1:N] ~ car.normal(sp.adj[],sp.weights[],sp.num[],prec.delta)
}
```

Then replace Line 19 in Figure 15.1 by the following:

```
delta[i,t] <- delta.transposed[t,i]
```

Similar to the implementation of the Type II structure, the first block of code uses a for-loop to independently assign the ICAR model to the set of area-specific parameters at each time point. However, in WinBUGS, when assigning a multivariate distribution to a set of parameters that are stored in a column of a matrix – in this case, assigning the ICAR model to the N area-specific parameters stored in the tth column of the $N \times T$ matrix delta (δ) – we have to first transpose the matrix, then assign the multivariate distribution to the corresponding parameters which, after the transpose operation, are now stored in the tth row in the resulting matrix – a rather peculiar syntax restriction. Thus, the first block of code assigns the ICAR model to the area-specific parameters in each row of delta.transposed, a $T \times N$ matrix which is the transpose of the matrix delta. Then the second block of code, `delta[i,t] <- delta.transposed[t,i]`, performs the matrix transpose (i.e. switching the row and column indices).[3]

We also need to remove `delta` in each set of initial values, then add `delta.transposed` to the initial value list as given below

```
delta.transposed= structure(.Data=c(0.01,-0.01,0,0,...,0
                            ,0.01,-0.01,0,0,...,0
                            ,...
                            ,0.01,-0.01,0,0,...,0)
                ,.Dim=c(8,452))
```

Note that the first two columns of `delta.transposed` are set to the same value of 0.01 but of opposite signs. This is to ensure the values in each row sum to zero, the constraint imposed by the `car.normal` function. A second set of initial values for `delta.transposed` can be obtained by simply swapping the first two columns. With these modifications, the fitting is performed in Exercise 15.4.

15.6 Results from Analysing the Peterborough Burglary Data

In this section, we summarise and discuss the results from fitting the four space-time models presented in Table 15.1 to the Peterborough burglary data. The data and the process models are the same across these four space-time models, but they differ in terms of whether space-time interaction is allowed and, if so, how space-time interaction is modelled.

Table 15.2 provides an overall comparison across the four space-time models using the Deviance Information Criterion (DIC). The $\bar{D}$ values from the three space-time inseparable models are all smaller than that from the space-time separable model, suggesting that the added space-time interaction component improves the fit to the observed burglary data.

TABLE 15.1

A Summary of the Four Bayesian Space-Time Hierarchical Models Fitted to the Peterborough Burglary Count Data from 2001 to 2008

	Space-Time Separable	Type I	Type II (with RW1)	Type III
Data model			$y_{it} \sim Poisson(\mu_{it})$ $\mu_{it} = n_i \cdot \theta_{it}$	
Process model			$\log(\theta_{it}) = \alpha + (S_i + U_i) + v_t + \delta_{it}$	
The space-time interaction component	$\delta_{it} = 0$	$\delta_{it} \sim N(0, \sigma_\delta^2)$	$\delta_{i,1:T} \sim RW(W_{RW1}, \sigma_\delta^2)$	$\delta_{1:N,t} \sim ICAR(W_{SP}, \sigma_\delta^2)$
Full model specification	Eq. 14.5 in Chapter 14	Eq. 15.2	As in Eq. 15.2, but δ is specified in Eq. 15.5	As in Eq. 15.2, but δ is specified in Eq. 15.6

---

[3] If we directly assign the ICAR to each column of parameters in delta, i.e. `delta[1:N,t] ~ car.normal(sp.adj[], sp.weights[], sp.num[], prec.delta)`, WinBUGS will return the error message: "Vector valued relation delta must involve consecutive elements of variable" in the bottom left-hand corner.

TABLE 15.2

A Comparison of the Four Space-Time Models Using DIC

Model	Space-Time Inseparable?	$\bar{D}$	pD	DIC
Space-time separable	No	13045	342	13387
Type I	Yes	11969	1054	13023
Type II	Yes	12259	847	13016
Type III	Yes	12087	847	12934

This additional space-time interaction component, at the same time, increases the complexity of the three space-time inseparable models, whose *pD* values – a measure of model complexity – are larger than that of the space-time separable model. Bringing together both the goodness of fit and the measure of model complexity, the (much) smaller *DIC* values from the three space-time inseparable models suggest that these three models are better supported by the observed data, confirming the need to incorporate the space-time interaction component – a conclusion that we drew in Section 14.3.4 after analysing the data using the simpler space-time separable model.

We now study the three space-time inseparable models in more detail. The *DIC* values in Table 15.2 suggest that the space-time inseparable model with Type III space-time interaction is the most parsimonious, as it has the smallest *DIC* value. The comparisons of the $\bar{D}$ and *pD* values provide additional information about these three models. The space-time inseparable model with Type I interaction has the smallest $\bar{D}$ value, suggesting that it fits the data better than the other two models. As discussed in Section 15.3, the better data fit is due to the lack of structural restrictions, which enables the corresponding space-time model to better capture deviations from space-time separability. However, this lack of a dependence structure under Type I interaction also makes the model the most (and unnecessarily) complex of the three inseparable models, as its *pD* value is the largest, which consequently makes this model inferior to the other two for this dataset.

Both Type II and Type III space-time interaction impose some smoothing/dependence structure on the interaction parameters, thus making the resulting models less complex –the space-time model with either Type II or Type III interaction has a *pD* value smaller than the model with Type I interaction. Compared to that with Type II interaction, the space-time model with Type III interaction is ultimately the better model because it yields a better fit to the data. This observation is also supported by the results from Section 14.3.4 that showed, after fitting a space-time separable model, that the deviance residuals at each time point are positively spatially autocorrelated, thus suggesting the appropriateness of the Type III dependence structure. The Type II structure is less appropriate because the temporal deviance residuals at the COA level are not found to be temporally autocorrelated. Looking ahead, Type IV space-time interaction structures the space-time interaction parameters both spatially *and* temporally. The complexity of the corresponding model would be further reduced. The question then becomes how such a reduction in model complexity (i.e. making the model less flexible) would impact on the fit to the data. We will investigate that in using some space-time inseparable models with Type IV space-time interaction in Exercises 15.5, 15.6 and 15.9 after introducing Type IV interaction

Figure 15.2 compares the posterior distributions of the unconditional variances of δ, the set of parameters in the space-time interaction component, across the three space-time inseparable models. It should be noted that here we are not comparing the conditional

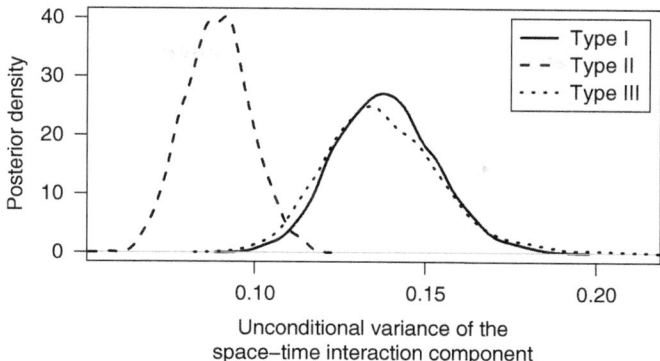

FIGURE 15.2

A comparison of the posterior distributions of the unconditional variances of the space-time interaction component under Type I, II and III space-time interaction. Compared to Type II, the models with either Type I or Type III space-time interaction resulted in a set of space-time interaction parameters that are more variable, thus allowing them to better capture the deviations from the space-time separable structure.

variances (i.e. σ_δ^2 in each of the models presented in Table 15.1) because they are conditioning on either the spatial neighbourhood structure or the temporal neighbourhood structure. These conditional variances do not represent the variability of the estimated parameters in δ (see also Section 8.5.2 in Chapter 8). Instead, the unconditional (or empirical) variances represent how variable these estimated space-time interaction parameters are, or, in other words, how effective they are at capturing deviations from the space-time separable structure. It is the evidence from Figure 15.2, showing that the unconditional variances from both the Type I and Type III structures are larger than those from the Type II structure, which tells us that the interaction parameters from the Type I and III space-time interaction structures are better at accommodating the space-time deviations (from the separable model) than the Type II structure.

Table 15.3 reports the posterior estimates of the variance partition coefficients (VPCs) across the three space-time inseparable models. Consistent across the three models, the overall spatial component makes the biggest contribution to explaining the spatial-temporal variability in the data, followed by the overall temporal component and finally the space-time interaction component. The VPC of the space-time interaction component under Type II space-time interaction is the lowest, while that of the overall spatial component is the highest of the three. This reflects the fact that, as we discussed above, the Type II structure is less flexible for capturing space-time deviations, compared to the other two dependence structures. So, the overall spatial component in the model with Type II interaction needs to "work harder" to explain the observed data. It should be noted that VPC is a measure of *relative* contributions of different components *within* a model (i.e. it indicates

TABLE 15.3

The Posterior Means and the Associated 95% Credible Intervals of the Variance Partition Coefficients Across the Three Space-Time Inseparable Models

Model	VPC_{SP}	VPC_{TM}	VPC_{SPTM}
Type I	55.6% (51.5%–59.4%)	24.7% (21.7%–27.8%)	19.7% (16.3%–23.4%)
Type II	62.1% (58.6%–65.5%)	24.6% (21.4%–27.8%)	16.3% (10.9%–15.7%)
Type III	55.7% (50.6%–60.5%)	24.7% (20.8%–28.4%)	19.6% (15.7%–23.8%)

which component is more important within a particular model). Cross-model comparison using VPC only makes sense if these models yield similar fits to the observed data (i.e. they have similar $\bar{D}$ values).

To investigate model fits further, Figure 15.3 compares the observed case counts with the predicted counts from each of the three space-time inseparable models for three selected COAs. These three COAs are selected from Figure 14.6 and Figure 14.7 in Section 14.3.4, and each COA was shown to have three (COAs 113 and 205) or four (COA 107) time points where the space-time separable model fitted the burglary counts poorly. For ease of comparison, the fits from the space-time separable model are also included in Figure 15.3. While the three space-time inseparable models all fit the observed burglary counts reasonably well and all better than the space-time separable model, these three models capture space-time interaction differently. Specifically, for each COA i, the Type I structure effectively modifies the overall time trend (the dotted lines in Figure 15.3) to adapt to the observations (the dots in Figure 15.3) through $\delta_{i1},\ldots,\delta_{iT}$, which can be considered as a set of independent (exchangeable) noise terms. So, for example, for the case of COA 107, the posterior means of both $\delta_{107,1}$ (for year 2001) and $\delta_{107,4}$ (for year 2004) are positive so that the corresponding burglary rates are lifted up (from the overall time trend) to better fit the large burglary counts observed in those two years. Similarly, to fit the small observed burglary count in 2007 and 2008, $\delta_{107,7}$ and $\delta_{107,8}$ (for years 2007 and 2008, respectively) have negative posterior means to pull the corresponding burglary rates down. In fact, such an adaptive process takes place "whenever and wherever required", not just for those time points where the space-time separable model fits poorly. As a result, the posterior means of the fitted trends (the solid lines with open circles in Figure 15.3) from the Type I structure follow the observed data reasonably well. The property that these space-time interaction parameters are independent also allows the model to capture the sudden increase in the number of burglary cases between 2004 and 2005 observed in COA 205. While the burglary counts on either side of the period 2004 and 2005 are fairly similar, the Type I structure is unaffected by such temporal smoothness and is able to lift the burglary rates up during those two years. By contrast, as we discuss below, Type II and Type III space-time interaction, because they impose either a temporal or a spatial dependence structure, fail to capture one (or both) of the two departures because of the dissimilarity of the outcome values to either their temporal or spatial neighbours.

The Type II dependence structure smooths the space-time interaction parameters temporally. The temporal smoothing is able to capture the relatively smoothly varying burglary counts in COA 107 and COA 113 reasonably well (Figure 15.3). For COA 205, however, the model with Type II interaction under-predicts the unexpectedly large burglary count in 2005 due to temporal smoothing, although it fits the slightly lower burglary count in 2004 well. It is generally the case that when a departure from the overall trend occurs around the middle of the observation period, the bi-directional information borrowing of the RW1 model (similarly, of the RW2 model) tends to oversmooth the departure. However, if the departure happens either at the beginning or at the end of the observation period (e.g. in 2001 across the three COAs in Figure 15.3), the RW1 model is still able to capture the departure. We will see this property of the RW1 model again when we discuss the space-time disease surveillance model in Section 16.4 in Chapter 16.

Type III space-time interaction spatially smooths the space-time interaction parameters independently across different time points. As a result, an unusually high (low) number of burglary cases in a COA can be well described if the burglary rates in the neighbouring COAs are similarly high (low). Take the example of COA 107 in 2001 in Figure 15.3.

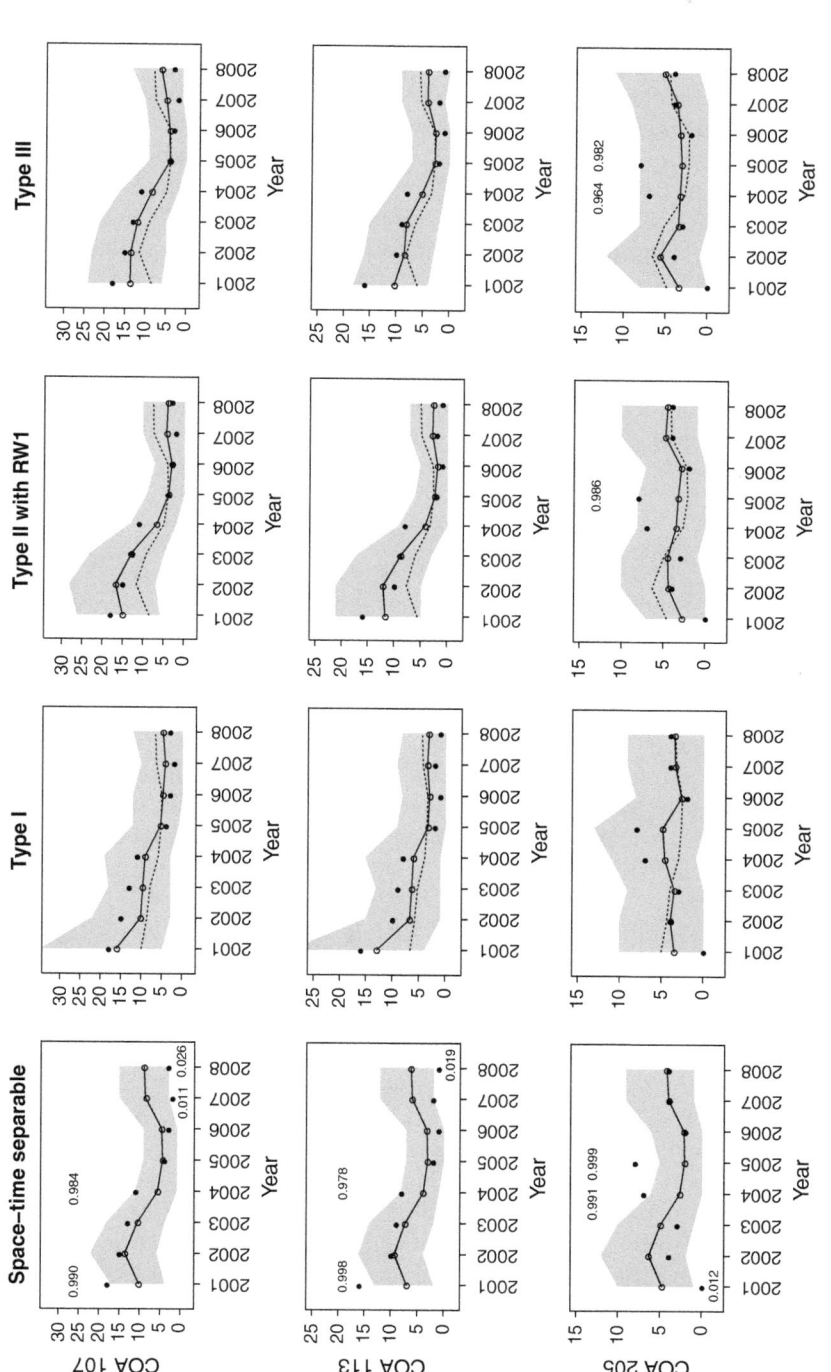

FIGURE 15.3

For a selection of three COAs, each with three (COAs 113 and 205) or four (COA 107) time points at which the space-time separable model does not fit the observed burglary counts well, the observed annual burglary case counts are compared to the predicted case counts from four space-time models. In each plot, the dots represent the observed burglary counts. The open circles connected with a solid line show the posterior means of the predicted counts, and the grey band shows the 95% credible intervals. The fitted curve from the overall temporal pattern is shown by the dotted line, which is formed using the posterior means of $\exp\left(a + (S_i + U_i) + v_t\right) \cdot n_t$. We report a posterior predictive p-value if it either exceeds 0.95 (appearing on the top of the plot) or below 0.05 (appearing at the bottom of the plot) to indicate an observed burglary count that is fitted poorly.

The Type III structure is able to lift the burglary rate up from the overall trend for that year because the (raw) burglary rates from the five neighbouring COAs in the same year are also high: the raw burglary rates of the five neighbours of COA 107 in 2001 are 21, 46, 46, 54 and 111 cases per 1000 houses, and the raw burglary rate of COA 107 in the same year is 149 cases per 1000 houses, all but one being higher than the Peterborough average in 2001, which is 24 cases per 1000 houses. For COA 205, however, the Type III structure does not capture the large case counts in both 2004 and 2005. This is because the raw burglary rates in the five spatial neighbours of COA 205 are no higher than 17 cases per 1000 houses in both years. In each of those two years, two of the neighbouring COAs of COA 205 have zero cases, and the other three neighbouring COAs have only one or two burglary cases. By contrast, there are seven and eight cases of burglary reported in COA 205 in 2004 and 2005, respectively. As a result, there is virtually no (useful) information that can be borrowed from its spatial neighbours in order to capture the large observed case counts in 2004 and 2005. Following the discussion above, we encourage the reader to carry out the same investigation of the fits from the four space-time models for other COAs presented in Figures 14.6 and 14.7 in Chapter 14.

15.7 Type IV Space-Time Interaction

Type IV space-time interaction describes a full spatial-temporal dependence structure so that the space-time interaction parameters, $\delta_{11},\ldots,\delta_{1T},\ldots,\delta_{N1},\ldots,\delta_{NT}$, are no longer structured into either N independent blocks of temporal parameters under the Type II structure or T independent blocks of area-specific parameters under the Type III structure. This form of space-time interaction is interesting "from a theoretical point of view" (Knorr-Held, 2000, p.2560). This is perhaps in part because Type IV space-time interaction gives rise to a rich set of flexible space-time models with a wide range of applications, some of which will be discussed in this section as well as in Chapter 16.

Compared to the previous three types, the construction of a prior model for δ under Type IV interaction is more challenging. The reason is that, as opposed to considering how parameters are related primarily either over space or over time, the Type IV dependence structure is fully spatial and temporal, meaning that each parameter in δ can depend not only on its spatial neighbours at the same time point but also on its temporal neighbours in the same area. And, as we shall soon see, such a dependence structure can extend to a space-time cell's higher order neighbours (e.g. the temporal neighbours' spatial neighbours).

To help develop a prior model for δ under Type IV interaction, we consider three strategies in the following sections. In particular, to achieve a full space-time dependence structure, Strategies 1 and 2 share the same rationale. The idea is to use either the Type II structure or the Type III structure as a starting point, then to link the corresponding independent blocks of parameters together so that the N blocks of temporal parameters are spatially structured (under Strategy 1) and the T blocks of spatial parameters are temporally structured (under Strategy 2). Strategy 3 follows a procedure that was originally suggested by Clayton (1996) and further developed by Knorr-Held (2000), who called it Clayton's rule. This strategy (or Clayton's rule) forms a prior model for δ under Type IV interaction by embedding the Kronecker product of two *structure matrices*, one defining the dependence structure of the spatial main effects and the other one defining the dependence structure of the temporal main effects, into a so-called Gaussian Markov random field (GMRF) prior.

The above description contains a number of new and important concepts, the discussion of which is deferred until Section 15.7.3.

Different from the previous sections in this chapter, we place less emphasis on the WinBUGS implementation of the models that we will discuss under Type IV interaction. This is for two reasons. First, we believe that by now the reader should be comfortable to implement some of these space-time models in WinBUGS. Second, when there are many areas and/or time points – the Peterborough burglary dataset is considered to be a case in point – WinBUGS takes a long time to fit some of these models. WinBUGS is great but it has its limitations! There have been a number of computational advances to address this issue, and we will discuss some of them in Section 15.7.4.

15.7.1 Strategy 1: Extending Type II to Type IV

Recall that Type II space-time interaction smooths the temporal parameters $\delta_{i,1:T} = (\delta_{i,1}, \ldots, \delta_{i,T})$ of each area i, and the temporal smoothing is carried out independently from one area to another. Now, suppose we impose a spatial dependence structure on these N blocks of temporal parameters so that if two areas, i and j, are geographically close (e.g. they share a common border), then the two time trends represented by $\delta_{i,1:T}$ and $\delta_{j,1:T}$ tend to be more alike. Then, the parameters in δ are smoothed not only temporally (according to the Type II structure) but also spatially (because of the spatial structure imposed on the blocks of temporal parameters), or, in other words, a full space-time dependence structure. But the question is "how can we impose a spatial dependence structure on a set of area-specific time trends?" For some time series models, such as polynomial functions of time and spline models, we can spatially structure the area-specific trends by modelling the parameters that control the time trend pattern hierarchically.[4]

To provide an example of this strategy, Eq. 15.7 extends the prior model in Eq. 15.4 under Type II space-time interaction so that the area-specific slopes are modelled hierarchically via the ICAR model with a chosen spatial weights matrix, $\mathbf{W}_{sp}$.

$$\delta_{it} = d_i \cdot \left(t - t^*\right)$$
$$d_{1:N} \sim ICAR\left(\mathbf{W}_{sp}, \sigma_d^2\right)$$

(15.7)

As opposed to assuming independence, as in Eq. 15.4, the area-specific slopes and hence the area-specific trends are spatially structured. As a result, the prior model in Eq. 15.7 assumes that the two linear trends, $\delta_{i,1:T}$ and $\delta_{j,1:T}$, are similar to each other if, for example, areas i and j are spatially contiguous. So, the implication is that if, for example, area i is estimated to have an increasing trend pattern, the trends of the areas that are neighbours to i are more likely to show an increasing pattern. Although we do not pursue this prior model further in this chapter, as we will return to it in more detail in Section 16.3 in Chapter 16, it demonstrates the idea that we can construct a prior model for δ with a full space-time dependence structure by (simply) imposing a spatial dependence structure on a set of area-specific parameters, each controlling the time trend pattern of that area.

When more flexible nonlinear time series models are required, the same idea applies. For example, Assunção et al. (2001) employed a quadratic formulation to model the annual

---

[4] Note that the random walk models (e.g. RW1 and RW2) and the autoregressive models (e.g AR1) cannot be structured spatially using Strategy 1 because the time trend pattern under these models is governed by the entire set of temporal parameters, not just a reduced set such as the slope in the linear time trend model. But we will return to these temporal models in Strategy 2 in Section 15.7.2 and Strategy 3 in Section 15.7.3.

human visceral Leishmaniasis (HVL) cases, a typically rural zoonotic disease, observed across 117 health zones in the Brazilian city of Belo Horizonte over a three-year period from 1994/5 to 1996/7. Using a Poisson likelihood for the observed case counts, they modelled the log disease rate, $\log(\theta_{it})$, in area i over year t as

$$\log\left(\theta_{it}\right) = \alpha_i + \beta_i\left(t-1\right) + \gamma_i\left(t-1\right)^2 \tag{15.8}$$

The intercept (α_i) and the two parameters (β_i and γ_i) which control the shape of the quadratic pattern are area-specific. Each of the three sets of parameters, namely, the α_i's, the β_i's and the γ_i's, is modelled independently using the spatial ICAR model, implying that areas that are geographically close tend to have similar temporal patterns. In their study, three models were applied to the observed HVL data: (1) the constant model with only the intercepts; (2) the linear model with γ_i's all set to 0; and (3) the quadratic model as presented in Eq. 15.8. Using a leave-one-out cross validation procedure,[5] the authors concluded that both the linear and the quadratic models were more appropriate for their observed data.

   More generally, when modelling disease or crime rates using case counts observed over space and time, MacNab and Dean (2001) proposed a general log-linear formulation that encompasses both the prior models above with the linear (Eq. 15.7) and the quadratic (Eq. 15.8) functions. Following their notation, the formulation is given by

$$y_{rt} \sim Poisson\left(\mu_{rt}\right)$$
$$\log\left(\mu_{rt}\right) = \log\left(n_{rt}\right) + \log\left(m\right) + a\left(t\right) + a_r + \beta_r\left(t\right) \tag{15.9}$$

where μ_{rt} is the mean of the Poisson likelihood for y_{rt}, the observed number of cases in area r at time t; n_{rt} is the population at risk; log (m) is "the mean rate over the time and region considered" (MacNab and Dean, 2001, p.950), which corresponds to α in Eq. 15.1; $a(t)$ and a_r represent respectively the temporal and the spatial main effects (i.e. TM_t and SP_i in Eq. 15.1); and β_r (t) accounts for space-time interaction. Expressing the space-time interaction as β_r (t) places the emphasis on the spatial modelling of the set of N time series, what Cressie and Wikle (2011, p.342) refer to as "a spatial process of time series". MacNab and Dean (2001, p.950) point out that "without assuming rigid forms for $a(t)$ and β_r (t)'s", the formulation in Eq. 15.9 "is extremely flexible for the modelling of spatiotemporal rates", opening up the possibility of using more flexible time series models in spatial-temporal modelling. Spline models, in particular, have gained in popularity due to their flexibility. A wide variety of spatial-temporal models using splines have been developed and applied in various studies (MacNab and Dean, 2001; Kneib and Fahrmeir, 2006; MacNab and Gustafson, 2007; MacNab, 2007; Ugarte et al., 2010; and Ugarte et al., 2017, just to name a few). Although the details about these models are outside the scope of our exploration, as illustrated through the simpler formulations in both Eq. 15.7 and 15.8, these spatial-temporal models all share a common modelling strategy. That is, the parameters governing the chosen temporal model (e.g. the set of area-specific slopes in the linear trend model or multiple sets of spline coefficients in a spline model) are made area-specific. These area-specific parameters are then modelled hierarchically through a spatial prior (e.g. the ICAR model or the BYM model) so that the resulting area-specific trends vary: as MacNab (2007, p.728) puts it, "in a spatially structured or spatially unstructured manner, the interplay between spatial and temporal smoothing may be explored".

---

[5] A leave-one-out cross validation is an iterative procedure where each time a model is fitted to the entire dataset but with one of the observed outcome values omitted. Then a prediction for the held-out value is obtained from the fitted model. A good model should yield a good prediction on average.

A prior model for $\boldsymbol{\delta}$ constructed following Strategy 1 can also be considered as a spatially varying coefficient model (Gelfand et al., 2003) where the coefficients are associated with, for example, the linear and/or quadratic (or even higher order) functions of time or the basis functions in spline modelling. Following this logic of model building, if the observed data display some seasonal patterns at the small area level, and these seasonal patterns tend to be more alike in neighbouring areas, then one can include a model component to capture the seasonal patterns and subsequently structure the associated parameters spatially so that the resulting seasonal patterns vary smoothly over space (see, for example, Benth et al. (2007)).

15.7.2 Strategy 2: Extending Type III to Type IV

Similar to Strategy 1, Strategy 2 uses the Type III dependence structure as a starting point to build a space-time dependence structure under Type IV interaction. The Type III dependence structure assumes that the area-specific parameters in each of the T blocks, one for each time point, are spatially autocorrelated, but these blocks of area-specific parameters are modelled independently over time. The idea of Strategy 2 is to impose a temporal dependence structure on these T blocks of spatial parameters such that the spatial patterns of any two consecutive time points can be different, but these two spatial patterns are assumed to be more alike compared to the case where the two time points are far apart. To turn this strategy into a prior model for $\boldsymbol{\delta}$, we need to temporally structure the T blocks of spatial parameters, $\boldsymbol{\delta}_{1:N,1}, \boldsymbol{\delta}_{1:N,2}, \ldots, \boldsymbol{\delta}_{1:N,T}$, where $\boldsymbol{\delta}_{1:N,t} = (\delta_{1t}, \ldots, \delta_{Nt})$. Now, if we label $\boldsymbol{\delta}_{1:N,t}$ as $\boldsymbol{\delta}_t$ (i.e. suppressing the sequence of area indices), then we are essentially back to the context of time series modelling. In the case here, however, the complication is that instead of univariate time series modelling whereby each time point is associated with only one single parameter (random variable), we have a vector of spatially-autocorrelated parameters (random variables) at each time point, leading to the modelling of a multivariate spatially-autocorrelated time series.

An intuitive yet flexible way to model $\boldsymbol{\delta}_1, \boldsymbol{\delta}_2, \ldots, \boldsymbol{\delta}_T$ is to use a vector autoregressive (VAR) model. A VAR model utilises an autoregressive structure to describe the temporal evolution across the vectors of parameters. Eq. 15.10 shows the structure of a VAR model of order 1, which we denote as a VAR1 model:

$$\boldsymbol{\delta}_t = \begin{cases} a \cdot \boldsymbol{\delta}_{t-1} + \boldsymbol{b}_t & \text{for } t = 2, \ldots, T \\ \boldsymbol{b}_1 & \text{for } t = 1 \end{cases} \qquad (15.10)[6]$$

If we ignore the boldface font type for the moment, Eq. 15.10 basically defines a standard AR1 process as in univariate time series modelling. So, a is a scalar parameter (known as the temporal autocorrelation parameter) that determines the nature (either positive or negative) and strength of the serial correlation. A constraint of $|a| < 1$ is typically placed so that the AR1 process is stationary (e.g. in the case of the univariate AR1 process, both the mean and the variance of each time-indexed variable are finite and constant over time). The AR1 structure in Eq. 15.10 implies that the vector of parameters at one time point is dependent upon the vector of parameters at the previous time point. If a is estimated to be positive and away from 0, then the vector of parameters at t tend to be similar to those at the previous time point. And of course, when a is estimated to be close to 0, then the strength of such a temporal dependence structure is shown to be weak (or absent).

---

[6] The VAR1 model in Eq. 15.10 is a special case of the model presented in Eq. 6.90 in Cressie and Wikle (2011, p.342), with their matrix M set to just a scalar parameter a. Cressie and Wikle (2011, Section 6.4.2) also presents a wider class of space-time autoregressive moving-average (STARMA) models, of which the VAR1 model in Eq. 15.10 is a special case.

Under the Type IV dependence structure, in addition to temporal dependence, the parameters in each δ_t are required to be spatially autocorrelated. To induce spatial autocorrelation is the job of the vector of noise terms (also known as the innovation terms in time series modelling), b_t, at each time point t. The procedure is as follows. If the noise terms in b_1 are spatially autocorrelated, then δ_1, the set of space-time interaction parameters at the first time point, are also spatially autocorrelated simply because $\delta_1 = b_1$. If we now add a set of spatially-autocorrelated noise terms, b_2, to $a \cdot \delta_1$, then $\delta_2 = a \cdot \delta_1 + b_2$, which are not only temporally autocorrelated with δ_1 (because of the AR1 structure) but the parameters in δ_2 are themselves spatially autocorrelated.[7] By continuing the above process over time, all the parameters in δ are connected over both space and time, which is the distinctive feature of Type IV space-time interaction. As we shall later see in Section 15.7.3, each space-time interaction parameter δ_{it} is dependent upon not only the parameters associated with its spatial neighbours at the same time point and its temporal neighbours in the same area but also the parameters of its second order neighbours in space and time – that is, its spatial neighbours' temporal neighbours.

Similar to a standard univariate AR1 model, the T vectors of noise terms $b_1, \ldots, b_T$ are independent, but each vector follows a common multivariate probability distribution. Thus, various spatial smoothing models, e.g. the ICAR model or the BYM model, can be used for b_t. We will see a number of examples below, whilst Exercise 15.5 uses the ICAR model for the vectors of noise terms in Eq. 15.10. It is worth noting that the prior model in Eq. 15.10 reduces to the prior model in Eq. 15.5 (under Type II interaction with RW1) when (a) each b_t is modelled by a multivariate normal distribution with a mean vector of 0 and a covariance matrix $\sigma^2 I_N$, where I_N is an identity matrix of size $N \times N$, i.e. $b_t \sim MVN\left(0, \sigma^2 I_N\right)$, independently across all t, and (b) the temporal autocorrelation parameter a is fixed at 1. In that case, the resulting prior model specifies the Type II dependence structure in which the N sequences of time series parameters are uncorrelated, but each sequence follows the RW1 temporal dependence structure. When $b_t \sim MVN\left(0, \sigma^2 I_N\right)$ and $a = 0$, we then have the prior model under Type I interaction given in Eq. 15.2.

15.7.2.1 Examples of Strategy 2

Some recently proposed spatial-temporal models can be considered to follow Strategy 2. For example, Rushworth et al. (2014) modelled the number of hospital admissions for respiratory disease over 624 electoral wards in Greater London from 2003 to 2009 as a function of three spatially-temporally varying covariates (one measuring the level of air pollution and two acting as proxy measures of deprivation and socio-economic status) and a set of space-time residuals. The latter, in their modelling, act as a set of surrogate measures of the risk factors that have not been included in the model. Since their inferential goal was to investigate the long-term effects of air pollution (which was included as a covariate) on hospital admissions for respiratory conditions, the space-time residuals were not specified to include an overall spatial and an overall temporal component (e.g. as in Eq. 15.1) but only as a space-time interaction component. Thus, the modelling of their space-time residuals is equivalent to the modelling of our space-time interaction parameters, δ. In their prior model for δ (or ϕ in their notation), the T sets of area-specific parameters, $\delta_1, \ldots, \delta_T$ (or

---

[7] Because b_1 and b_2 are independent and follow the same (multivariate) distribution, the covariance matrix of $\delta_2 = ab_1 + b_2$ is $2a^2\Sigma$, where Σ is the covariance matrix of the distribution for the noise vector. Thus, the parameters in δ_2 are also structured spatially under the VAR1 structure in Eq. 15.10.

$\phi, ..., \phi_T$ in their notation), follow the AR1 structure as given in Eq. 15.10, and each set of the area-specific noise terms, b_t, are structured spatially using the spatial model developed by Leroux et al. (2000). Similar to the BYM model, the Leroux model combines both the ICAR model and the Gaussian exchangeable model, where the latter was discussed in Section 7.4 in Chapter 7 (also see Rushworth et al. (2014) and Leroux et al. (2000) for more detail). For computational efficiency, the authors implemented their MCMC algorithm in C++ to fit the proposed Bayesian space-time model.

The VAR1 construct given in Eq. 15.10 is actually quite flexible in terms of developing a prior model for δ to accommodate some "non-standard" features of the observed data. This is done by (simply) modifying the spatial model on the noise terms, b_t. For example, Rushworth et al. (2017) extend the space-time model that they proposed in 2014 (as discussed above) to carry out adaptive spatial smoothing to deal with cases where the residual surfaces are not spatially smooth everywhere. When spatial discontinuity is found to be present, they argued that (Rushworth et al., 2017, p.146) "a pair of adjacent areas exhibiting substantially different levels of unexplained risk will have those risks wrongly smoothed towards each other, masking the step change to be identified." Their adaptive space-time model still follows the structure of their 2014 model, in which the space-time residuals follow the VAR1 structure, as in Eq. 15.10. However, they used the ICAR model instead of the Leroux model for each vector of noise terms, b_t. To achieve adaptive smoothing, each element in the spatial weights matrix W_{sp} that corresponds to the weight of two spatially contiguous areas was treated as unknown, as opposed to fixing it to 1. Each of these unknown weights was then modelled as a random variable that can take any value between 0 and 1, and these weights were estimated from the observed data. Therefore, two spatially contiguous areas are considered to be neighbours if the corresponding weight is estimated to be close to 1, but they are not considered to be neighbours if the estimated weight is close to 0 (see also Section 8.4 in Chapter 8). As they explained (Rushworth et al., 2017, p.147), the ICAR model was used instead of the Leroux model because estimating both the spatial weights matrix and the parameter ρ in the Leroux model, which weighs the relative importance of the ICAR and exchangeable models, "could result in high posterior correlation and multimodality, because the random effects are spatially independent if either $\rho = 0$ or all the adjacency elements of the spatial weights matrix, W, equal 0". In the application to modelling the circulatory and respiratory disease cases at the local and unitary authority level in England from 2001 and 2010, this adaptive spatial-temporal model led to the identification of some spatial discontinuities (step changes) in the residual spatial surfaces. The fitting of this adaptive space-time model was carried out using the R package CARBayesST. We recommend the interested reader consult this package (see Lee et al., 2018), as it offers the functionalities to fit a number of spatial-temporal models with conditional autoregressive (CAR) priors.

Our final example features the spatially varying autoregressive model proposed by Shand et al. (2018). The goal of their modelling is to predict the rates of newly diagnosed cases with the human immunodeficiency virus (HIV). As the name suggests, their model employs the familiar AR1 structure as presented in Eq. 15.10 to describe the temporal dynamics of the rates of HIV at the small area level, but there are a number of specific features to their model that are worth discussing here. First, the actual case counts were not analysed for confidentiality reasons. Instead, the observed outcome values that they model are the rates of newly diagnosed HIV cases – in the format of number of HIV cases per 100000 people. This has two modelling implications. First, the observed log-transformed and mean-centred HIV rate in area i in year t, denoted as Z_{it} in their model, is modelled via a normal likelihood. That is, $Z_{it} \sim N\left(\eta_{it}, \sigma^2 q_{it}\right)$, where σ^2 is an unknown variance common across all areas and years. Second, one needs to account for the variability associated with Z_{it} due to the effect of population size: the rate varies more (less) when it is associated with a smaller (larger) population (see also Section 3.3.3). While this variability is directly accounted for when modelling the observed case counts using a Poisson likelihood, it is not the case when using a normal likelihood on the rate. To do that, the authors included a diagonal matrix Q of size $(N \cdot T) \times (N \cdot T)$ in the normal likelihood. Each diagonal element in Q, q_{it} as appeared in the likelihood above, is a fixed constant proportional to $1/\left(n_{it} y_{it}\right)$, where n_{it} and y_{it} are respectively the population size and the observed case count. Thus, the rate Z_{it} associated with a large n_{it} and/or y_{it} will have a lower variance.

The second feature of their modelling lies in the specification of the process model for η_{it}, which takes the form

$$\eta_{it} = X_{i,t-1}^T \beta + \psi_{it} \rho_i \left(Z_{i,t-1} - X_{i,t-2}^T \beta\right) \tag{15.11}$$

The above formulation, albeit a rather complicated one, models each η_{it} as a linear combination of $X_{i,t-1}^T \beta$, the effects of the previous year's covariates in area i, and $\psi_{it} \rho_i \left(Z_{i,t-1} - X_{i,t-2}^T \beta\right)$, the residual at the previous year scaled by $\psi_{it} \rho_i$. The quantity, ρ_i, can be considered as the serial correlation parameter in the AR1 formulation in Eq. 15.10, and ψ_{it} is a constant related to the matrix Q in the normal likelihood discussed above. The reason to add in the constant ψ_{it}, as they explained, is to ensure ρ_i varies between –1 and 1 and hence can be interpreted as a serial correlation parameter (see p.1009 in Shand et al., 2018). Essentially, Eq. 15.11 imposes an AR1 structure on the temporal residuals of each area.

To induce spatial dependency, the area-specific serial correlation parameters, $\rho_1, \ldots, \rho_N$, are modelled hierarchically. The spatial modelling of these parameters gives rise to a rather challenging yet interesting problem: each ρ_i is restricted to vary between –1 and 1, so the typical CAR models that we have seen throughout this book are not applicable. A novelty of the work by Shand et al. (2018) is the use of a copula approach to construct a joint probability model for $\rho_1, \ldots, \rho_N$ so that one can impose a spatial dependence structure on these parameters and, at the same time, satisfy the constraint on each ρ_i. We do not pursue the discussion further as it goes beyond our scope. However, we recommend the interested reader consult their paper for the details of the copula construction, while Nelsen (2007) studies copulas in great detail.

Finally, the third feature of the model by Shand et al. (2018) is the use of temporally-lagged covariate values and temporally-lagged residuals. This modelling strategy reflects their inferential goal to produce a set of one-year-ahead rates of newly diagnosed HIV cases at the small area level. We will return to this modelling feature in Section 16.5 in the next chapter.

15.7.3 Strategy 3: Clayton's Rule

Suggested by Clayton (1996), Strategy 3 constructs a prior model for δ, the space-time interaction parameters, with Type IV interaction by "multiplying" the spatial main effects and the temporal main effects together. This strategy, essentially, follows the usual way to investigate the effect of interaction between two observable covariates, say X_{1i} and X_{2i}. In that case, we include a new covariate into the regression model, where this new covariate is formed by multiplying X_{1i} and X_{2i} together. The procedure here is more complicated because (a) both the spatial and the temporal main effects are non-observable, but they are modelled as unknown parameters (random effects) in the regression; and (b) these two sets of random effects are of different lengths – one with N parameters whilst the other has T parameters – thus creating a problem in multiplication. What Clayton (1996) proposed is to define a prior model for $\boldsymbol{\delta}$ (the "new covariate") following two steps. First, we take *the Kronecker product* of *the two structure matrices*, each defining the dependence structure of the spatial main effects or the temporal main effects, to form a structure matrix for $\boldsymbol{\delta}$. The resulting structure matrix is then embedded in the formulation of a so-called Gaussian Markov random field (GMRF) prior to give rise to a prior model for $\boldsymbol{\delta}$.

In what follows, we first discuss what a structure matrix is, what the formulation of a GMRF prior looks like and how we can bring the two elements together to achieve our goal, i.e. to form a prior model for $\boldsymbol{\delta}$. We then study the spatial-temporal dependence structure induced by the resulting prior model to ensure that the resulting structure is indeed of Type IV space-time interaction.

15.7.3.1 Structure Matrices and Gaussian Markov Random Fields

Strategy 3 starts with a space-time separable model. For exploration, we consider a process model with a space-time separable structure where the spatial main effects, $\boldsymbol{S} = (S_1,...,S_N)$, are modelled using the ICAR model with a spatial weights matrix $\boldsymbol{W}_{sp}$ and the temporal main effects, $\boldsymbol{v} = (v_1,...,v_T)$, are modelled through the RW1 model with a temporal weights matrix, $\boldsymbol{W}_{RW1}$. The structure matrices, $\boldsymbol{K}_{sp}$ for $\boldsymbol{S}$ and $\boldsymbol{K}_{RW1}$ for $\boldsymbol{v}$, are given respectively by

$$\boldsymbol{K}_{SP} = \boldsymbol{D}_{SP} - \boldsymbol{W}_{SP}$$

$$\boldsymbol{K}_{RW1} = \boldsymbol{D}_{RW1} - \boldsymbol{W}_{RW1}$$

(15.12)

where $\boldsymbol{D}_{SP}$ and $\boldsymbol{D}_{RW1}$ are both diagonal matrices with respective sizes of $N \times N$ and $T \times T$. For $\boldsymbol{D}_{SP}$, the diagonal element $(\boldsymbol{D}_{SP})_{ii}$ is the sum of the ith row in $\boldsymbol{W}_{SP}$. The same definition applies to $\boldsymbol{D}_{RW1}$. Now, if we look at the joint probability distribution of the ICAR model in Eq. 8.4 and that of the RW1 model in Eq. 12.9, the structure matrices are nothing new but part of the respective precision matrices. That is, $\boldsymbol{Q}_{SP} = \dfrac{1}{\sigma_S^2}\boldsymbol{K}_{SP}$ for the ICAR model and $\boldsymbol{Q}_{RW1} = \dfrac{1}{\sigma_v^2}\boldsymbol{K}_{RW1}$ for the RW1 model, where $\boldsymbol{Q}_{SP}$ and $\boldsymbol{Q}_{RW1}$ are the precision matrices of the two models.

The reason for calling $\boldsymbol{K}_{SP}$ (and likewise $\boldsymbol{K}_{RW1}$) a *structure* matrix is that this matrix *solely* specifies the dependence structure of the associated random effects. To see that more clearly, we express the joint probability distribution of the ICAR model in Eq. 8.4 in terms of its structure matrix:

$$\Pr(S_1,...,S_N) = \Pr(\boldsymbol{S})$$

$$\propto \exp\left\{-\frac{1}{2\sigma_S^2}\boldsymbol{S}^T\boldsymbol{K}_{SP}\boldsymbol{S}\right\}$$

(15.13)

As we discussed in Section 8.2.2.1 in Chapter 8, the joint probability distribution in Eq. 15.13 takes the form of the density function of a multivariate normal distribution.

Therefore, the structure matrix K_{SP} acts as the inverse covariance matrix that controls how the parameters in S are related to one another. More generally, Eq. 15.13 gives an example of a Gaussian Markov random field (GMRF). As defined in Rue and Held (2005, p.1), "a GMRF is really a simple construct: It is just a (finite-dimensional) random vector following a multivariate normal (or Gaussian) distribution." According to this definition, many of the spatial models – the exchangeable model with a normal (Gaussian) common distribution, the ICAR model, the pCAR model and the BYM model – and the temporal models – the RW1 and the RW2 models and the AR1 model – that we have discussed so far are GMRFs. The joint distributions of these prior models all take the following general form:

$$\Pr(a) \propto \exp\left\{-\frac{1}{2\sigma^2} a^T K a\right\} \tag{15.14}$$

In terms of constructing a prior model for δ, the space-time interaction parameters, under Type IV interaction, it is important to invoke the formulation of a GMRF. This is because, conceptually, suppose we can construct a structure matrix, say K_{SPTM}; we can then write down the joint probability distribution in the form of a GMRF, i.e.

$\Pr(\delta) \propto \exp\left\{-\frac{1}{2\sigma_\delta^2} \delta^T K_{SPTM} \delta\right\}$, thus giving us a prior model for δ. As in the cases of the spatial ICAR model and the temporal RW1 model, the resulting prior model imposes a space-time dependence structure on δ according to K_{SPTM}. Clayton's rule tells us to form K_{SPTM} by taking the Kronecker product of K_{SP} and K_{RW1}, and we will return to that in the next section. In fact, the above procedure is quite general in the sense that it can be used to construct a purely-spatial and a purely-temporal model. The derivation of a spatial (or a temporal) structure matrix is much more intuitive: see Chapter 4 for the spatial case and Section 12.3.4.2 in Chapter 12 for the temporal case. In both cases, the structure matrix is constructed via a weights matrix. Constructing a weights matrix over both space and time, however, is not easy, so we, as suggested by Clayton (1996), form the structure matrix instead.

15.7.3.2 Taking the Kronecker Product

The Kronecker product, denoted by $\otimes$, is a matrix operation that multiples two arbitrary matrices, A and B, together. For example, if A is a 2×2 matrix of the form $A = \begin{pmatrix} a_{11} & a_{12} \\ a_{21} & a_{22} \end{pmatrix}$ and B is a matrix of an arbitrary size, then the Kronecker product of A and B is

$$A \otimes B = \begin{pmatrix} a_{11}B & a_{12}B \\ a_{21}B & a_{22}B \end{pmatrix},$$

where $a_{11}B$, for example, corresponds to multiplying all the elements in B by the scalar quantity a_{11}. More generally, if A is an $n \times m$ matrix and B is a $p \times q$ matrix, then $A \otimes B$ is a matrix of size $(np) \times (mq)$.

Turning now to the modelling of δ, we form K_{SPTM}, a structure matrix for δ, by taking the Kronecker product of K_{SP} and K_{TM}, the structure matrices of the spatial and the temporal main effects. To illustrate the calculation, Figure 15.4 presents K_{SP} associated with a map of four areas and K_{TM} associated with the RW1 model over five time points (thus $K_{TM} = K_{RW1}$). Then K_{SPTM} is given by

$$K_{SPTM} = K_{SP} \otimes K_{RW1}$$

$$= \begin{pmatrix} 2 \cdot K_{RW1} & -K_{RW1} & -K_{RW1} & 0 \cdot K_{RW1} \\ -K_{RW1} & 2 \cdot K_{RW1} & -K_{RW1} & 0 \cdot K_{RW1} \\ -K_{RW1} & -K_{RW1} & 3 \cdot K_{RW1} & -K_{RW1} \\ 0 \cdot K_{RW1} & 0 \cdot K_{RW1} & -K_{RW1} & K_{RW1} \end{pmatrix}$$

$$= \left(\begin{array}{ccccc|ccccc|ccccc|ccccc}
2 & -2 & 0 & 0 & 0 & -1 & 1 & 0 & 0 & 0 & -1 & 1 & 0 & 0 & 0 & 0 & 0 & 0 & 0 & 0 \\
-2 & 4 & -2 & 0 & 0 & 1 & -2 & 1 & 0 & 0 & 1 & -2 & 1 & 0 & 0 & 0 & 0 & 0 & 0 & 0 \\
0 & -2 & 4 & -2 & 0 & 0 & 1 & -2 & 1 & 0 & 0 & 1 & -2 & 1 & 0 & 0 & 0 & 0 & 0 & 0 \\
0 & 0 & -2 & 4 & -2 & 0 & 0 & 1 & -2 & 1 & 0 & 0 & 1 & -2 & 1 & 0 & 0 & 0 & 0 & 0 \\
0 & 0 & 0 & -2 & 2 & 0 & 0 & 0 & 1 & -1 & 0 & 0 & 0 & 1 & -1 & 0 & 0 & 0 & 0 & 0 \\ \hline
-1 & 1 & 0 & 0 & 0 & 2 & -2 & 0 & 0 & 0 & -1 & 1 & 0 & 0 & 0 & 0 & 0 & 0 & 0 & 0 \\
1 & -2 & 1 & 0 & 0 & -2 & 4 & -2 & 0 & 0 & 1 & -2 & 1 & 0 & 0 & 0 & 0 & 0 & 0 & 0 \\
0 & 1 & -2 & 1 & 0 & 0 & -2 & 4 & -2 & 0 & 0 & 1 & -2 & 1 & 0 & 0 & 0 & 0 & 0 & 0 \\
0 & 0 & 1 & -2 & 1 & 0 & 0 & -2 & 4 & -2 & 0 & 0 & 1 & -2 & 1 & 0 & 0 & 0 & 0 & 0 \\
0 & 0 & 0 & 1 & -1 & 0 & 0 & 0 & -2 & 2 & 0 & 0 & 0 & 1 & -1 & 0 & 0 & 0 & 0 & 0 \\ \hline
-1 & 1 & 0 & 0 & 0 & -1 & 1 & 0 & 0 & 0 & 3 & -3 & 0 & 0 & 0 & -1 & 1 & 0 & 0 & 0 \\
1 & -2 & 1 & 0 & 0 & 1 & -2 & 1 & 0 & 0 & -3 & 6 & -3 & 0 & 0 & 1 & -2 & 1 & 0 & 0 \\
0 & 1 & -2 & 1 & 0 & 0 & 1 & -2 & 1 & 0 & 0 & -3 & 6 & -3 & 0 & 0 & 1 & -2 & 1 & 0 \\
0 & 0 & 1 & -2 & 1 & 0 & 0 & 1 & -2 & 1 & 0 & 0 & -3 & 6 & -3 & 0 & 0 & 1 & -2 & 1 \\
0 & 0 & 0 & 1 & -1 & 0 & 0 & 0 & 1 & -1 & 0 & 0 & 0 & -3 & 3 & 0 & 0 & 0 & 1 & -1 \\ \hline
0 & 0 & 0 & 0 & 0 & 0 & 0 & 0 & 0 & 0 & -1 & 1 & 0 & 0 & 0 & 1 & -1 & 0 & 0 & 0 \\
0 & 0 & 0 & 0 & 0 & 0 & 0 & 0 & 0 & 0 & 1 & -2 & 1 & 0 & 0 & -1 & 2 & -1 & 0 & 0 \\
0 & 0 & 0 & 0 & 0 & 0 & 0 & 0 & 0 & 0 & 0 & 1 & -2 & 1 & 0 & 0 & -1 & 2 & -1 & 0 \\
0 & 0 & 0 & 0 & 0 & 0 & 0 & 0 & 0 & 0 & 0 & 0 & 1 & -2 & 1 & 0 & 0 & -1 & 2 & -1 \\
0 & 0 & 0 & 0 & 0 & 0 & 0 & 0 & 0 & 0 & 0 & 0 & 0 & 1 & -1 & 0 & 0 & 0 & -1 & 1
\end{array} \right) \qquad (15.15)$$

The resulting structure matrix is of size 20×20, which matches with the 20 parameters that we have in δ. The resulting matrix, K_{SPTM}, is a permissible structure matrix in the sense that it is symmetric, and both the row sums and the column sums are all zero – the reader is encouraged to verify that. We have added the dotted lines in K_{SPTM} in order to make the "block" structure clear. Each block takes the form of a scalar entry in K_{SP}, multiplying the structure matrix K_{RW1}. According to Clayton's rule, a prior model for δ is then given by

$$\Pr(\delta) \propto \exp\left\{ -\frac{1}{2\sigma_\delta^2} \delta^T K_{SPTM} \delta \right\} \qquad (15.16)$$

Now the question is "how are the parameters in δ related to one another under this prior model?" This question is important because not only do we need to ensure that this prior model follows the requirement of Type IV space-time interaction, but also the dependence

	Configuration	The associated structure matrix
Spatial		$K_{SP} = \begin{pmatrix} 2 & -1 & -1 & 0 \\ -1 & 2 & -1 & 0 \\ -1 & -1 & 3 & -1 \\ 0 & 0 & -1 & 1 \end{pmatrix}$
Temporal	t=1　t=2　t=3　t=4　t=5　 Time	$K_{RW1} = \begin{pmatrix} 1 & -1 & 0 & 0 & 0 \\ -1 & 2 & -1 & 0 & 0 \\ 0 & -1 & 2 & -1 & 0 \\ 0 & 0 & -1 & 2 & -1 \\ 0 & 0 & 0 & -1 & -1 \end{pmatrix}$

FIGURE 15.4
The structure matrix K_{sp}, based on the first order rook's move of spatial contiguity over a map of four areas, and the structure matrix K_{RW1}, associated with the RW1 model over five time points.

structure has implications regarding how information is shared across these space-time parameters, ultimately affecting the parameter estimates. But before we investigate this, we bring the following to the reader's attention.

Warning: it is important to correctly match the parameters in $\boldsymbol{\delta}$ with the rows in the structure matrix. When taking the Kronecker product as $K_{SP} \otimes K_{RW1}$, as we did in Eq. 15.16, the first T rows (equivalently the first T columns) in K_{SPTM} correspond to the temporal parameters of the first area, $\delta_{11},\dots,\delta_{1T}$, then the subsequent T rows (T columns) correspond to the temporal parameters of the second area, $\delta_{21},\dots,\delta_{2T}$, and so on. However, if the Kronecker product is taken as $K_{RW1} \otimes K_{SP}$ (i.e. swapping the positions of the two structure matrices), then the labelling of the rows will be different, whereby the first N rows (equivalently the first N columns) in K_{SPTM} correspond to the area-specific parameters of the first time point, $\delta_{11},\dots,\delta_{N1}$, and the subsequent N rows (N columns) correspond to the area-specific parameters of the second time point, $\delta_{12},\dots,\delta_{N2}$, and so on. Whilst the induced space-time dependence structure remains the same regardless of whether we are taking $K_{SP} \otimes K_{RW1}$ or $K_{RW1} \otimes K_{SP}$, a mismatched labelling of the rows (columns) would lead to a wrong dependency structure.

15.7.3.3 Exploring the Induced Space-Time Dependence Structure via the Full Conditionals

To understand the dependence structure of the prior model we constructed in the previous section, we study the *full conditionals*, a process that we have followed to study the behaviours of various CAR models in Chapter 8 and the RW1 model in Section 12.3.4.2 in Chapter 12. These full conditionals, each defined as the probability distribution of each parameter given all the others, can be derived using the joint probability distribution, but, in general, the derivation is rather complicated. However, as we shall see, a useful property of a GMRF prior is that we can write down the full conditionals directly from the structure matrix without invoking any algebraic manipulation. In what follows, we first derive the full conditionals using the joint probability distributions of the ICAR model and of the RW1 model algebraically. We then go on to show that these full conditionals can be derived directly from the corresponding structure matrix without algebraic manipulation.

Consider a general setting in which we have a set of parameters (random effects), $\boldsymbol{a} = (a_1,\dots,a_M)$, of size M. Eq. 15.17 shows that the full conditional for each a_i ($i = 1,\dots,M$) is simply proportional to the joint probability distribution for $\boldsymbol{a}$.

$$\Pr\left(a_i \mid \boldsymbol{a}_{\{-i\}}\right) = \frac{\Pr\left(a_i, \boldsymbol{a}_{\{-i\}}\right)}{\Pr\left(\boldsymbol{a}_{\{-i\}}\right)} \tag{15.17}$$

$$\propto \Pr(\boldsymbol{a})$$

As usual, $\boldsymbol{a}_{\{-i\}}$ denotes all the parameters in $\boldsymbol{a}$ but excluding a_i. The second line of Eq. 15.17 is obtained by noticing that (a) in the numerator, putting a_i into $\boldsymbol{a}_{\{-i\}}$, we have the full set of $\boldsymbol{a}$ and (b) in the denominator, $\Pr\left(\boldsymbol{a}_{\{-i\}}\right)$ does not depend on a_i, so it is just a multiplicative constant that can be ignored on the right-hand side of the proportionality symbol $\propto$. We are now in position to derive the full conditionals for the ICAR model as well as for the RW1 model. Recall Eq. 8.6 and Eq. 12.10, under both models, the joint probability distribution can be expressed in the following form:

$$\Pr(\boldsymbol{a}) \propto \exp\left\{-\frac{1}{2\sigma^2}\sum_{i=1}^{M-1}\sum_{j=i+1}^{M} w_{ij}\left(a_i - a_j\right)^2\right\} \tag{15.18}$$

where w_{ij} is the weight attached to i and j. Then the full conditional for a_i, $\Pr\left(a_i \mid \boldsymbol{a}_{\{-i\}}\right)$, can be simplified further to include only terms that involve a_i, because all other terms that do not involve a_i are simply multiplicative constants that can be ignored. Thus, combining Eq. 15.17 with Eq. 15.18 and using binary weights (i.e. $w_{ij} = 1$ if i and j are considered to be neighbours and $w_{ij} = 0$ otherwise), we have

$$\Pr\left(a_i \mid \boldsymbol{a}_{\{-i\}}\right) \propto \Pr(\boldsymbol{a})$$

$$\propto \exp\left\{-\frac{1}{2\sigma^2}\sum_{j\in\Delta i}\left(a_i - a_j\right)^2\right\} \tag{15.19}$$

$$\propto \exp\left\{-\frac{m_i}{2\sigma^2}\left(a_i - \frac{\sum_{j\in\Delta i} a_j}{m_i}\right)^2\right\}$$

where Δi denotes the set of neighbours of i and m_i is the number of neighbours that i has. The third line of Eq. 15.19 is obtained by expanding the brackets in the second line, then arranging terms to form a perfect-square trinomial in a_i.[8] Therefore, given all other parameters in $\boldsymbol{a}$, the conditional distribution of a_i is a normal distribution with mean $\frac{\sum_{j\in\Delta i} a_j}{m_i}$ and variance $\frac{\sigma^2}{m_i}$, and this is exactly the form given in Eq. 8.1 for the ICAR model.[9] One can also use Eq. 15.19 to easily obtain the full conditionals given in Eq. 12.7 for the RW1 model (Exercise 15.8).

Applying the above derivation on the space-time interaction parameters $\boldsymbol{\delta}$ is a lot more difficult. However, for a GMRF prior, there is a direct correspondence between the structure matrix and the full conditionals. This correspondence is expressed in Eq. 15.20, which can be used to write down the full conditionals directly using the elements in the structure matrix.

$$a_i \mid \boldsymbol{a}_{\{-i\}} \sim N\left(\sum_{j\in\Delta i} -\frac{K_{ij}}{K_{ii}} a_j, \frac{\sigma^2}{K_{ii}}\right) \tag{15.20}$$

In Eq. 15.20, K_{ij} is the element on the ith row and the jth column of a structure matrix $\boldsymbol{K}$. Using the illustrations in Figure 15.4, the reader is encouraged to verify Eq. 15.20 such that the full conditionals for both the ICAR model and the RW1 derived through Eq. 15.19 can indeed be obtained using Eq. 15.20 via the corresponding structure matrix. More generally, Rue and Held (2005) show that the structure matrix encodes *conditional independence*, an important concept in the theory of GMRF. On p.2 in Rue and Held (2005), they state a

---

[8] A perfect-square trinomial in a is in the form of $a^2 \pm 2ab + b^2$, which gives $(a \pm b)^2$.

[9] If the density of a univariate random variable X is given by $\Pr(X) \propto \exp\left\{-\frac{1}{2\sigma^2}(X-\mu)^2\right\}$, then $X \sim N\left(\mu, \sigma^2\right)$.

general rule that holds for any GMRF: "If $Q_{ij} = 0$ for $i \neq j$, then x_i and x_j are conditionally independent given the other variables and vice versa", and Q_{ij} is the element in the precision matrix which is the structure matrix scaled by an unknown variance, i.e. $Q = \dfrac{1}{\sigma^2} K$.

Equipped with Eq. 15.20, we can now write down the full conditionals straightforwardly using the structure matrix. As an illustration, consider K_{SPTM} given in Eq. 15.15. The first row is associated with δ_{11} (the space-time interaction parameter associated with area 1 at time point 1), and using Eq. 15.20, its full conditional is given by

$$\delta_{11} \mid \delta_{\{-11\}} \sim N\left(\frac{2\delta_{12} + \delta_{21} - \delta_{22} + \delta_{31} - \delta_{32}}{2}, \frac{\sigma_\delta^2}{2} \right) \tag{15.21}$$

Note that the indices i and j in Eq. 15.20 correspond respectively to the ith row and the jth column in a structure matrix; they are not indices of area and time. So, for writing down the full conditionals, it helps to label the rows and the columns as explained at the end of Section 15.7.3.2. For K_{SPTM} in Eq.15.15, the row labels (which are the same as the column labels) are $\delta_{11}, \delta_{12}, \ldots, \delta_{15}, \delta_{21}, \ldots, \delta_{45}$ (recalling that the first subscript is for area and the second subscript is for time).

The mean of the full conditional in Eq.15.21 implies that, given all other parameters in δ, δ_{11} depends (only) on the parameters in five other space-time cells, two associated with its spatial neighbours at the same time point, i.e. δ_{21} and δ_{31}, one associated with its temporal neighbour of the same area, i.e. δ_{12}, and two associated with its spatial neighbours' temporal neighbours, i.e. δ_{22} and δ_{32}. In other words, the full conditional of δ_{11} shows a *full* space-time dependence structure in which the dependency is not only on both space and time but, more interestingly, on the spatial neighbours' temporal neighbours, which are referred to as the second order space-time neighbours in the space-time cube. The spatial neighbours at the same time point and the temporal neighbours in the same area are referred to as the first order space-time neighbours. This full space-time dependence structure holds for every parameter in δ. Take another example: δ_{22}, the space-time interaction parameter associated with area 2 at time point 2. Using the 7th row in K_{SPTM}, together with Eq. 15.20, the full conditional is given by

$$\delta_{22} \mid \delta_{\{-22\}} \sim N\left(\frac{-\delta_{11} + 2\delta_{12} - \delta_{13} + 2\delta_{21} + 2\delta_{23} - \delta_{31} + 2\delta_{32} - \delta_{33}}{4}, \frac{\sigma_\delta^2}{4} \right) \tag{15.22}$$

A careful inspection of the conditional mean reveals that given all other parameters, δ_{22} depends on the parameters of the first order space-time neighbours, namely δ_{12} and δ_{32} in its spatial neighbours and δ_{21} and δ_{23} in its temporal neighbours, and those of the second order space-time neighbours (i.e. the spatial neighbours' temporal neighbours or equivalently, the temporal neighbours' spatial neighbours), and they are $\delta_{11}, \delta_{13}, \delta_{31}$ and δ_{33}. Note also that the structure matrix K_{SPTM} retains the bi-directional nature of the temporal dependence under the RW1 model.

Knorr-Held (2000, p.2560) provides a general specification of the mean and the variance of the full conditional distribution for each δ_{it} over N areas and T time points. Again, the ICAR model and the RW1 model are used respectively for the modelling of the spatial and the temporal main effects. This general specification is given as follows:

$$\delta_{it} \mid \delta_{\{-it\}} \sim N\left(\mu_{it}, \sigma_{it}^2 \right), \tag{15.23}$$

where the conditional mean μ_{it} is

$$
\mu_{it} = \begin{cases} \delta_{i,t+1} + \dfrac{1}{m_i}\displaystyle\sum_{j\in\Delta i}\delta_{jt} - \dfrac{1}{m_i}\displaystyle\sum_{j\in\Delta i}\delta_{j,t+1} & \text{for } t=1 \\[2ex] \delta_{i,t-1} + \dfrac{1}{m_i}\displaystyle\sum_{j\in\Delta i}\delta_{jt} - \dfrac{1}{m_i}\displaystyle\sum_{j\in\Delta i}\delta_{j,t-1} & \text{for } t=T \\[2ex] \dfrac{1}{2}\left(\delta_{i,t-1}+\delta_{i,t+1}\right) + \dfrac{1}{m_i}\displaystyle\sum_{j\in\Delta i}\delta_{jt} - \dfrac{1}{2m_i}\displaystyle\sum_{j\in\Delta i}\left(\delta_{j,t-1}+\delta_{j,t+1}\right) & \text{for } t=2,\ldots,T-1 \end{cases}
\tag{15.24}
$$

and the conditional variance, σ_{it}^2, is

$$
\sigma_{it}^2 = \begin{cases} \dfrac{\sigma_\delta^2}{m_i} & \text{for } t=1 \text{ or } t=T \\[2ex] \dfrac{\sigma_\delta^2}{2m_i} & \text{for } t=2,\ldots,T-1 \end{cases}
\tag{15.25}
$$

As before, Δi and m_i denote the set of spatial neighbours and the number of spatial neighbours that area i has.

The formulation above given by Knorr-Held (2000) better summarises the dependence properties where, as he explained (p.2560), "not only (first-order) temporal ($\delta_{i,t-1}$ and/or $\delta_{i,t+1}$) and spatial (δ_{jt}, j ϵ Δi) neighbours enter in the full conditional for δ_{it}, but also second order neighbours ($\delta_{j,t-1}$ and/or $\delta_{j,t+1}$, j ϵ Δi), that is, spatial neighbours of temporal neighbours or, equivalently, temporal neighbours of spatial neighbours." Knorr-Held went on to point out the impact of this dependence structure on parameter estimation (p.2560): "this prior 'borrows strength' from spatial neighbours as it assumes that the temporal trend in county [area] i (in terms of first differences) is similar to the average trend in neighbouring counties." Effectively, the full conditionals in Eqs. 15.23–15.25 impose a prior assumption that a space-time surface in the space-time three-dimensional space varies smoothly with no sudden jumps nor any discontinuity. Such smoothly varying behaviour is also the reason that some of the off-diagonal elements in the structure matrix are negative and some are positive (but the diagonal elements are always positive and non-zero).

We encourage the reader to derive the full set of full conditionals using the structure matrix K_{SPTM} in Eq. 15.15 together with Eq. 15.20, then verify that these full conditionals can also be derived using Eq. 15.23 to Eq. 15.25, the (more compact) formulation given by Knorr-Held (2000).

Exercise 15.9 carries out the WinBUGS fitting of the space-time inseparable model in which the spatial and the temporal main effects are modelled using the ICAR model and the RW1 model, respectively, and the prior model for δ is derived using Clayton's rule. We do not consider this fitting in the main text here because WinBUGS takes too long to fit the model to the Peterborough data.[10] In general, when dealing with complex space-time models with large numbers of areas and/or time points, WinBUGS (and, to some extent, iterative MCMC algorithms in general) is shown to be inefficient, a limitation that has led to the development of various advanced computational techniques that we will mention in the next section.

---

[10] It took a week to run two MCMC chains, each with 5000 iterations on an iMac machine with a 4GHz Intel Core i7 processor and 8 GB of RAM.

15.7.4 Summary on Type IV Space-Time Interaction

In the last three sections, we have seen three different strategies to construct a prior model for the space-time interaction parameters, δ, under Type IV space-time interaction. To achieve a full space-time dependence structure, Strategy 1 models the N time series spatially through imposing a spatial dependence structure on the parameters that determine the shapes of the area-specific temporal trends. To some extent, Strategy 1 retains the *individuality* of the areas in the sense that individual areas still possess their own time trends. Thus, as we will illustrate in Sections 16.3 and 16.4, Strategy 1 (and Type II space-time interaction that it associates with) is better suited when the inferential goal is to detect *individual areas* that display unusual time trends.

By contrast, Strategy 2 connects the T spatial patterns together via a temporal dependence structure. As a result, the focus is more on the *temporal evolution* of the underlying space-time process, making Strategy 2 more appropriate to investigate the question of "how things evolve from one time point to the next". This emphasis on temporal evolution has also been exploited to develop space-time models for small-area (short-term) prediction (e.g. as in Shand et al., 2018 and the malaria application that we will discuss in Section 16.5 in Chapter 16).

Compared to the previous two strategies, Strategy 3 using Clayton's rule places a more "equal balance" between space and time. A prior model for δ constructed via Strategy 3 is theoretically interesting because, by studying the resulting model, albeit challenging at times, allows us not only to gain understanding of the fully spatial-temporal dependence structure induced but also to discuss important concepts such as structure matrices, conditional independence and Gaussian Markov random fields (GMRFs) that underlie some of the spatial and/or temporal smoothing models. However, most of the prior models for δ under Type IV interaction discussed here contain a large number of parameters (random effects), which raises issues in terms of computation as well as achieving model parsimony. As Bauer et al. (2016, p.1853) remarked, critically: "such a form contains a large number of random effects, $N \times T$ to be exact, which is not a parsimonious representation. In addition, the interaction term adds great complexity to the model and makes the implementation for Bayesian inference, usually via Markov chain Monte Carlo (MCMC), very difficult for data with many areas and/or time points." However, realising the limitations of existing methods is important, as these limitations (always) lead to new methodological developments.

On the computational side, a number of techniques have been developed to achieve efficient fitting of these space-time models. For example, Knorr-Held and Rue (2002) proposed to block update the correlated random effects, as opposed to updating each of them in turn, when running MCMC chains. Developed by Rue, Martino and Chopin (2009), approximate Bayesian inference through the integrated nested Laplace approximation (INLA) reduces the computational time massively, as posterior summaries are produced based on approximation instead of running iterative MCMC chains. More recently, Gómez-Rubio and Rue (2018) combine MCMC and INLA to widen the scope of models – including spatial econometric models – that can be fitted using INLA. There have been a number of books on INLA, such as Blangiardo and Cameletti (2015), Wang et al. (2018) and Krainski et al. (2019). This is also the reason that we do not pursue the WinBUGS fitting of most of the models presented under Type IV space-time interaction, for which INLA is a better and more efficient package to use. Table 15.4 summarises the implementation methods for fitting some of the space-time inseparable models that we have discussed under Type IV space-time interaction.

TABLE 15.4

The Methods for Fitting Some of the Space-Time Inseparable Models with Type IV Space-Time Interaction

Construction strategy	Space-time inseparable models under Type IV interaction	Can be fitted efficiently in WinBUGS? If not, what is the more efficient way?
Strategy 1	The space-time interaction component specified as a set of spatially-structured linear trends (Eq. 15.7)	Yes, see Section 16.3 in Chapter 16.
	The space-time interaction component specified as a set of spatially-structured quadratic trends (Eq. 15.8)	Yes, see Exercise 15.6.
Strategy 2	The space-time model proposed by Rushworth et al. (2014) with the VAR1 structure and the Leroux model	No. The fitting algorithm was implemented in C++.
	The adaptive space-time model proposed by Rushworth et al. (2017)	No. The fitting of the model is done through the R package CARBayesST.
	The spatially varying autoregressive model by Shand et al. (2018)	No. Purpose-built MCMC algorithms were developed to fit the model.
Strategy 3	The space-time interaction component specified as a GMRF prior with a space-time structure matrix constructed via the Kronecker product of the structure matrices of the spatial ICAR model and the temporal RW1 model	No. INLA is a more efficient package to use (see, for example p.244–246 in Blangiardo and Cameletti (2015)). See also Exercise 15.9.

To achieve model parsimony, Bauer et al. (2016, p.1853) proposed to "model the space-time interaction surface with a bivariate spline and place a GMRF prior on the coefficients associated with the basis functions". Alternatively, another way to construct a more parsimonious model for the space-time interaction component is using the mixture modelling approach that we briefly discussed in Section 15.3.1.

15.8 Concluding Remarks

In Chapter 14, we saw how *space-time separable* models provide a sensible starting point when modelling a spatial-temporal dataset. But a space-time separable model may not yield a good fit to data, not least because space-time separability makes some rather restrictive assumptions such as: all areas have the same time trend pattern, or the spatial pattern is the same across all time points. Assessing the lack of fit between observed outcome data and predictions from a space-time separable model should help us both to understand where the model fails and how the model might be improved. The direction of improvement we have described in this chapter is to fit one or more models from the class of *space-time inseparable* models. These, as we have seen, are models that add a space-time interaction component (or "module") to the space-time separable model. This component draws on two of the fundamental properties of spatial-temporal data: dependency and heterogeneity. The former is the basis for information borrowing, the latter informs the specification of the space-time interaction component.

Following Knorr-Held (2000), we have examined four space-time interaction structures (Types I to IV). Each of the four structures captures some different departure from space-time separability. Each of these structures is defined by a set of space-time random effects, consistently referred to throughout this chapter as the space-time interaction parameters, $\delta = (\delta_{11},\ldots,\delta_{1T}\ldots,\delta_{N1},\ldots,\delta_{NT})$, and it is how these space-time interaction parameters are specified that determines how we model the departure. As we have seen in earlier parts of the book, we include random effects in a hierarchical model in order to model the impact of missing covariates on the outcome. In Table 15.5 we provide a summary of the four different types of space-time interaction and what they imply about the space-time structure of such missing covariates.

We have now concluded our presentation of the theory behind Bayesian space-time hierarchical modelling, and we have seen how to fit some of these models using WinBUGS. In the next chapter we turn to some practical applications of the theory described in Chapters 14 and 15, covering areas that include space-time disease surveillance and policy evaluation. In two of the applications (application 2: modelling crime hotspots (Section 16.3) and application 4: investigating spatial, temporal and space-time spillover effects in the spread of malaria (Section 16.5)), we have covariates that can be included in the space-time hierarchical model. Including covariates will help us to *explain* the spatial, temporal and space-time patterns that we observe in the data. However, even in the circumstances where the goal includes explaining space-time variation, there is still a need for the random effects models we have described in Chapters 14 and 15 simply because it is unlikely that we will

TABLE 15.5

A Summary of the Four Types of Space-Time Interaction

Space-Time Interaction	Interpretation in Terms of Events*	Interpretation in Terms of Missing Covariates
Type I	An unexpectedly high or low number of events in area i at time t is not associated with higher or lower counts in area i in other time periods, nor in neighbouring areas. A *"one-off"* event.	Missing covariates lack space-time structure.
Type II	An unexpectedly high or low number of events in area i at time t is associated with higher or lower counts in area i in *other* time periods but not in the neighbouring areas. The variation in the number of events in an area i over time tends to be smooth. *"Temporally persistent but spatially localized".*	Missing covariates have smoothly varying structure through time but are highly localized in their effect on the outcome and so have no structure over space.
Type III	An unexpectedly high or low number of events in area i at time t is associated with higher or lower counts in neighbouring areas but not in other time periods. The variation in the number of events over space tends to be smooth. *"Spatially structured but short-lived".*	Missing covariates have smoothly varying structure across space but are short-lived in time (effects are restricted to the time period specified by t).
Type IV	An unexpectedly high or low number of events in area i at time t is associated with higher or lower counts in neighbouring areas, in other time periods and in other time periods of other neighbours. The variation in the number of events over space, time and space-time tends to be smooth. *"Spatially structured and temporally persistent".*	Missing covariates have smoothly varying structure across space, time and space-time.

*By the phrase "an unexpectedly high or low number of events", we mean after allowing for spatial and temporal main effects.

entirely account for the presence of all the spatial and temporal main effects and space-time interaction effects through the inclusion of covariates. The situation echoes the challenges we encountered in Part II of the book and which motivated the inclusion of random effects in (purely) spatial models.

15.9 Exercises

Exercise 15.1. Explore the COA-level burglary count data in Peterborough from 2001 to 2008 to identify spatially contiguous areas that (a) show similar time trend patterns in raw burglary rates and (b) show different time trend patterns in raw burglary rates.

Exercise 15.2. Modify the WinBUGS code presented in Figure 15.1 that fits the space-time model in Eq. 15.2 to the Peterborough burglary data in order to: (a) evaluate model fit via the posterior predictive p-values; (b) calculate the variance partition coefficients in Eq. 15.3; (c) examine how likely it is that the observed value in each space-time cell departs from the space-time separable structure; and (d) calculate the deviance residuals as in Eq. 14.6 in Chapter 14. (Hint: point (c) can be done by calculating the posterior probability of each parameter in the space-time interaction component lying above zero.) Also monitor DIC.

Exercise 15.3. Make the necessary modifications to the WinBUGS code in Figure 15.1 so as to fit a space-time model to the Peterborough burglary data with an area-specific RW1 model for space-time interaction (see Eq. 15.5). Modify the code to calculate: (a) the posterior predictive p-values, to show how well the model describes the space-time count data; and (b) the variance partition coefficients (VPCs) given by Eq.15.3. Monitor the posterior estimates of the space-time burglary risk, the posterior predictive p-values, the random effect variances, the VPCs, the deviance residuals as well as DIC.

Exercise 15.4. Carry out the fitting of the space-time model in Eq. 15.2 but with the space-time interaction component following the Type III prior specification in Eq. 15.6. Monitor the quantities of interest as listed in Exercise 15.3. When writing the WinBUGS code, pay particular attention to the points discussed in Section 15.5.2.

Exercise 15.5. Implement the space-time model in Eq. 15.2 but using the spatial vector autoregressive prior model (Eq. 15.10) for the space-time interaction component. In particular, model each vector of noise terms b_t (for $t = 1,...,T$) independently using the ICAR model and assign a uniform prior between -1 and 1 (excluding the two boundary values) to the scalar parameter a.

Exercise 15.6. Carry out the same analysis as in Exercise 15.3 but replacing the RW1 model by the quadratic model in Eq. 15.8 as the time trend model. Note that the model in this exercise has a Type IV space-time interaction structure whilst that in Exercise 15.3 has a Type II space-time interaction structure.

Exercise 15.7. Write down first the spatial weights matrix and the temporal weights matrix based on the map and the time sequence configuration in Figure 15.4. Transform those two weights matrices into the corresponding structure matrices (through Eq. 15.12), then calculate their Kronecker product. Write an R function to take a given spatial weights matrix and a given temporal weights matrix as input and return the resulting space-time weights matrix using Clayton's rule. One can subsequently format the resulting space-time weights matrix into the required form to be entered as data into WinBUGS.

Exercise 15.8. Derive the full conditional distributions for the RW1 model (Eq. 12.6 in Chapter 12) using Eq. 15.19.

Exercise 15.9. The WinBUGS code in `TypeIV_ICAR_RW1_Peterborough.txt` (available online) implements a space-time inseparable model in which the spatial and the temporal main effects are modelled using the ICAR model and the RW1 model respectively, and the prior model for δ is derived using Clayton's rule. The space-time weights matrix is derived using the R script `contruct_TypeIV_weights.R` (available online), or you can use your own function from Exercise 15.7. When running this model, you will notice how slow WinBUGS updates each MCMC iteration. This is the computation issue that we highlighted at the end of Section 15.7.4. The interested reader should consult alternative software such as INLA (see for example Chapter 7 in Blangiardo and Cameletti, 2015) and STAN (Carpenter et al., 2017).

16

Applications in Modelling Spatial-Temporal Data

16.1 Introduction

This chapter provides four applications of spatial-temporal data modelling. The first three applications are all based on the underlying space-time inseparable model structure, discussed in Chapter 15. These three applications all share a common inferential goal of detecting "unusual" areas with local time trends that differ markedly from the "common" time trend. However, these applications are not all the same. Placed in the context of policy evaluation, Application 1 assesses whether a geographically targeted crime prevention policy has had a measurable impact on the crime rates in the targeted areas. Under this setting, the locations of these "unusual" areas are known to us, but what we are trying to assess is whether the crime rates of these targeted areas, after the implementation of the policy, were different from the crime rates in other non-targeted areas. Both Applications 2 and 3 are topics within space-time surveillance, in which there is no subset of areas of interest prior to the analysis. The aim of the analysis is to identify *any* *areas* whose event rates differ considerably from the general time trend. Distinguishing the two applications is whether we are dealing with shorter (Application 2) or longer (Application 3) time series. Different time series models are then specified accordingly. As we shall see, underlying all three applications is a modelling strategy in which the common/general time trend is obtained through the space-time separable structure, and unusual local behaviours are captured through the space-time interaction component. In each application, Bayesian hierarchical models are constructed in order to obtain reliable small area estimates and account for various sources of uncertainty. A common problem in surveillance is the issue of multiple testing (or multiple comparisons), and this will be discussed in Applications 2 and 3.

Different from the first three, the fourth application describes a spatial-temporal model to investigate the presence (or absence) of spatial-temporal spillover effects on village-level malaria risk within a district in India. This extends the (static) spatial modelling described in Chapter 9 (Section 9.4) to a space-time model of the malaria incidence data.

Each section follows a similar pattern. We describe the background to the problem before discussing the data and the form of the analysis model. We then summarize some of the key statistical findings. Readers interested in following up the wider implications of the studies are encouraged to refer to specific papers which are cited in the text at the appropriate points.

16.2 Application 1: Evaluating a Targeted Crime Reduction Intervention

16.2.1 Background and Data

In small area policy evaluation, we are placed in a scenario where a policy or intervention has been implemented in a set of areas, and the aim is to evaluate whether it has had a measurable impact on event rates in the targeted areas. To illustrate, we report a study assessing the No Cold Calling (NCC) intervention introduced by Cambridgeshire Constabulary, in partnership with Trading Standards in the period 2005–06 (Li et al., 2013). The aim of the NCC intervention by the police was to reduce door-step cold calls by "rogue traders" or "bogus callers" – people coming to the door, uninvited, pretending to offer services that in fact provide a smoke screen for criminal activity including household burglary (either at the time of the cold call or at a later time) and fraud (charging high prices for often unnecessary and substandard work). Older householders were, and still are, particularly at risk from this type of scam. The police response involved a number of activities, including setting up street and household signage, to discourage would-be criminals, support services to householders, including advice on how to protect themselves from becoming victims, as well as advice on simple but effective household security devices. The NCC scheme is an example of situational crime prevention (Clarke, 1997) combined with re-assurance policing (Tuffin et al., 2006).

The assessment focused initially on Peterborough, a city then with a population of around 170,000 within Cambridgeshire in Eastern England. Household burglary data (Home Office Codes 28 and 29) were made available at the Census Output Area (COA) level for the period between 2001 and 2008. Unlike many other police forces in England and Wales, rather than offering the service to every householder interested in joining the scheme, Cambridgeshire's scheme was based on selecting particular streets and clusters of houses where householders were believed to be at most risk. Between 2005 and 2006, 11 streets in Peterborough were selected, and these streets were located in 10 COAs. Table 16.1 lists the 10 NCC-COAs and the percentage of households in each COA included in the NCC scheme (coverage). Figure 1.12(a) shows the locations of these 10 NCC-COAs.

TABLE 16.1

Summary of the No Cold Calling Scheme in Peterborough
Started in 2005 and 2006 at the Census Output Area (COA) Level

Start Date of Scheme	COA Code	Number of Dwellings (or Houses) in the Scheme	Number of Dwellings (or Houses) in the COA	Coverage (%)
2005	00JANC0016	42	122	34
	00JANE0006	48	150	32
	00JANE0010	28	151	19
	00JANG0013	12	131	9
	00JANQ0023	54	103	52
	00JANT0027	45	127	35
	00JANY0010	10	122	8
2006	00JANG0025	100	168	60
	00JAPB0010	36	128	28
	00JANH0003	36	126	29

Unsurprisingly, NCC areas did not match the COA boundaries, and some coverage rates are small. We comment on this spatial misalignment issue and the steps taken to explore its effect on the analysis.

The availability of time series recorded crime data enables us to measure any pre-post policy difference in crime rates. However, construction of appropriate control groups is essential in order to distinguish genuine policy impacts from other external factors, thereby strengthening internal validity. The evaluation is thus placed within the multiple time series (quasi-experimental) design, on which the interrupted time series (ITS) modelling is based (Campbell and Stanley, 1963, and Section 12.3.5). The group level analysis carried out in Section 12.3.5 using an ITS model provided some evidence for the overall success of the NCC scheme: the implementation of the scheme was shown to stabilise the burglary rates in the NCC group while the burglary rates elsewhere in Peterborough were going up. The modelling here goes beyond the group-level analysis to focus on: (a) evaluating the scheme's impact locally (see also the end of Section 12.3.5.4); (b) examining the robustness of the evaluation outcomes against various definitions of the control group since the police had not allocated NCC treatments at random to areas of Peterborough; and (c) assessing the external validity of findings (see Section 3.2.1) by extending the Peterborough study to all such schemes throughout Cambridgeshire. Whilst this example is specific to the evaluation of the NCC intervention, the methodology has wider relevance in policy evaluation at the small area level (Li et al., 2013).

When carrying out policy evaluation at the small area level, data sparsity is a particular concern. Burglary events in COAs can be too few in number to provide a reliable estimate of trends, so any policy effects could be masked by high levels of noise in the data. Figure 16.1 shows a plot of the individual COA level raw burglary rates that illustrates the

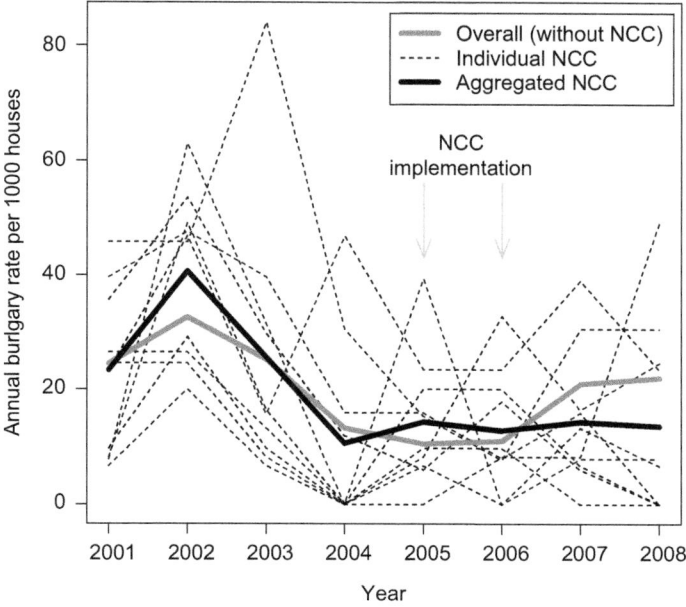

FIGURE 16.1
Raw data: temporal burglary profiles for the individual NCC targeted Census Output Areas (COAs) implemented in 2005/06. Superimposed is the overall burglary trend from all COAs in Peterborough excluding NCC areas and the trend from the group of all NCC-COAs.

noisiness of the data, although the deviation of two aggregated trends after 2006 gives a first indication of a possible NCC policy effect. The reader is referred back to Section 12.3.5, where we showed the beneficial effect associated with the NCC policy at the group level (i.e. aggregating all the NCC areas). As part of their exploratory data analysis, Li et al. (2013, p.2016) use a difference in differences (DID) regression model to carry out an initial assessment of the policy's impact.

Hierarchical models that combine information from multiple areas and time periods (i.e. across time and space) provide a natural framework to strengthen the estimation of impact. *Bayesian* hierarchical models allow parameter inference to be based on the full joint posterior distribution. In this application, there are three benefits that follow from the Bayesian approach: (a) the estimation of the policy's impacts takes into account the uncertainty associated with other model parameters, particularly the estimated control time trend; (b) parameters with direct evaluation relevance can be constructed from their posterior distributions; (c) assessments can be expressed in probability terms.

16.2.2 Constructing Different Control Groups

To account for external factors that may affect burglary rates, control areas are selected to match certain characteristics of the NCC areas. Five different control groups are shown in Table 16.2. Unlike a randomised controlled experiment, where the control group is formed before the experiment is carried out, there is no predefined control group under a quasi-experimental design. Hence, it is important to consider different plausible ways to form control groups in order to assess the robustness of findings regarding estimated policy impacts. Using an ITS model, the NCC treated areas can then be compared against each of those control groups.

One issue arises. When using a control criterion with more stringent matching (e.g. in RateMatch_MDI in Table 16.2, where matching is based on both pre-policy burglary rate

TABLE 16.2

Five Control Groups Based on Similarity of Local Characteristics

Group ID	Criterion	Description	Number of Selected COAs
1	All COAs	All non NCC-COAs and the NCC-COAs with the NCC-related data extracted.	449
2	NB	Spatial proximity. All other COAs in the same lower super output area (LSOA) in which the NCC-COAs were located.	46
3	MDI	COAs with similar multiple deprivation index (MDI) scores to those in the NCC-COAs in 2004: all non NCC-COAs with MDI between 5.98 and 51.70, the 10th and the 90th percentiles of the MDI values of the 10 NCC-COAs	416
4	RateMatch	Individual matching using the three-year average burglary rate prior to the implementation of the NCC intervention. Average burglary rates to be within ±10% of that for the NCC-COA.	18(min)–54 (max) over the 10 NCC-COAs
5	RateMatch_MDI	Matching both MDI and pre-policy burglary rates as described in RateMatch.	10–20

and multiple deprivation index), the number of control areas tends to be quite small. To strengthen the estimation of the control trend, the burglary data from the NCC areas *prior to* the NCC implementation are also used in the estimation of the control trend, in addition to using the data from the control areas. This applies to all the five control groups. Thus, for example, the 449 selected COAs under the criterion All COAs (Group ID 1) in Table 16.2 includes the 439 non-NCC-COAs and the 10 NCC-COAs, but the post-policy data of the 10 NCC-COAs are removed when estimating the control trend. We show in Section 16.2.4 how this is done.

16.2.3 Evaluation Using ITS

Following the formulation described in Section 12.3.5.2, the ITS model presented in Figure 16.2 models the COA-level time series counts of burglary cases in both the baseline model (with the addition of the COA indicator i) and the assessment model (with the addition of the COA indicator k). Similar to the group analysis in Section 12.3.5, the baseline model estimates the control trend pattern through a set of temporal random effects, $v_1, \ldots, v_T$. These random effects are modelled via a prior model with a first order random walk (RW1) structure so that information is borrowed from both time points that are immediately before and immediately after time t (except at the boundaries of the time series; see Section 12.3.4). W_{RW1} is the temporal weights matrix with all the diagonal elements equal to 0 and the off-diagonal element $w_{th} = 1$ if $|t - h| = 1$ and $w_{th} = 0$ otherwise.

To quantify the policy's impact, the assessment model uses the impact function, $f(t, b_k)$, to measure the departure of the trend of an NCC-COA from the control trend pattern estimated from the baseline model. The impact function is multiplied with the binary indicator function, $I(t \geq t_{k0})$, that takes the value 1 when time t is equal to or greater than the NCC

The baseline model (for $t = 1, \ldots, T$ and all i in the control group)	The assessment model (for $t = 1, \ldots, T$ and all k in the NCC group)
$y_{it}^{control} \sim Poisson\left(\mu_{it}^{control}\right)$	$y_{kt}^{treat} \sim Poisson(\mu_{kt}^{treat})$
$\mu_{it}^{control} = n_i^{control} \cdot \theta_{it}^{control}$	$\mu_{kt}^{treat} = n_k^{treat} \cdot \theta_{kt}^{treat}$
$\log\left(\theta_{it}^{control}\right) = \alpha^{control} + U_i^{control}$ $+ v_t + \varepsilon_{it}^{control}$	$\log(\theta_{kt}^{treat}) = \alpha^{control} + U_k^{treat} + v_t + I(t$ $\geq t_{k0}) \cdot f(t, b_k) + \varepsilon_{kt}^{treat}$
$\alpha^{control} \sim Uniform(-\infty, +\infty)$	$f(t, b_k) = b_k \cdot (t - t_{k0} + 1)$
$v_{1:T} \sim RW1(W_{RW1}, \sigma_v^2)$	$b_k \sim N(m, \sigma_b^2)$
$\sigma_v \sim N_{+\infty}(0, 10)$	$m \sim N(0, 1000000)$
$U_i^{control} \sim N(0, \sigma_U^2)$	$\sigma_b \sim N_{+\infty}(0, 10)$
$\varepsilon_{it}^{control} \sim N(0, \sigma_\varepsilon^2)$	$U_k^{treat} \sim N(0, \sigma_U^2)$
$\sigma_U \sim N_{+\infty}(0, 10)$	$\varepsilon_{kt}^{treat} \sim N(0, \sigma_\varepsilon^2)$
$\sigma_\varepsilon \sim N_{+\infty}(0, 10)$	

FIGURE 16.2
An interrupted time series model with a linear impact function to assess the overall and the local impacts of the NCC scheme in Peterborough.

starting year for the *kth* NCC area (e.g. $t_{k0} = 5$ if NCC was implemented in 2005 in that COA) and 0 before. Thus, the impact function only comes into play when $t \geq t_{k0}$. A linear impact function is used in the assessment model in Figure 16.2, representing a gradual departure. To assess impact locally, this linear impact function is specific to each NCC-COA so that for the *kth* NCC-COA ($k = 1,...,10$), $f(t,b_k) = b_k \cdot (t - t_{k0} + 1)$, where b_k quantifies the rate of departure of an NCC trend from the control trend on and after the implementation of the policy. So, a negative estimate of b_k would suggest a gradual reduction in the burglary rate in that NCC treated area compared to the areas in the control group, indicative of a positive NCC impact. The area-specific slopes are modelled hierarchically via a common normal distribution with unknown mean m and unknown variance σ_b^2. These area-specific slopes are spatially-unstructured because the NCC-COAs are scattered across the study region (see Figure 1.12(a)). Under this specification, both local and overall policy impacts, quantified by b_k and m respectively, can be simultaneously estimated. A possible extension would be to incorporate covariates into the model for b_k to investigate the effects that these factors might have on the programme's efficacy.

A number of terms are also added to the ITS model in Figure 16.2 to deal with the COA-level data. The baseline model includes the area-specific random effects, $U_i^{control}$, to account for the effects due to omitted COA-level characteristics. These random effects are modelled exchangeably, $U_i^{control} \sim N\left(0, \sigma_U^2\right)$, as opposed to being structured spatially because of the "holes" in the map created by the NCC areas. The term $\varepsilon_{it}^{control}$ is included in the baseline model to allow for overdispersion. Overdispersion is often encountered in the analysis of count data for small geographical areas. This extra variability may be due to unobserved risk factors and/or the non-independence of burglary events up to the scale of the areal unit possibly associated with repeat victimization (see for example Johnson and Bowers (2004), Tseloni and Pease (2004)). Each of the overdispersion terms, $\varepsilon_{it}^{control}$, follows a common normal distribution with mean 0 and an unknown (overdispersion) variance σ_ε^2. Similarly, the assessment model accounts for the place effects using U_k^{treat} as well as the effects of overdispersion through ε_{kt}^{treat}. The place effect (U_k^{treat}) and the overdispersion term (ε_{kt}^{treat}) are assumed to follow the corresponding distributions of $U_i^{control}$ and $\varepsilon_{it}^{control}$ in the baseline model. This specification implies that apart from the NCC implementation, the areas in both the NCC group and the control groups are assumed to be similar in all characteristics.

As highlighted in Section 12.3.5, the baseline model and the assessment model are jointly fitted in WinBUGS so that the full uncertainty associated with the estimation of the control time trend, $v_1,...,v_T$, is carried over into the estimation of the local and the global impacts ($b_1,...,b_{10}$ and m) in the assessment model. This uncertainty propagation is achieved using the cut function in WinBUGS – see Section 12.3.5.3 for more detail.

Finally, an advantage of placing inference in the Bayesian framework is the flexibility to transform some of the model parameters into quantities directly relevant to evaluation. As a measure of the policy impact, we calculate the quantity $\left[\exp\left(b_k \times t^*\right) - 1\right] \times 100$ that represents the post-policy percentage change in the burglary rate in the *kth* NCC-COA relative to the control group t^* years after the NCC implementation. A negative estimate corresponds to a relative reduction in the burglary rate in that NCC area, indicating a local success. In addition to providing the conventional point and interval estimates of the policy impact (using for example the posterior mean and the 95% credible interval), we can also calculate the posterior probability of a positive NCC impact in the treated area via $\Pr\left(b_k < 0 | \text{data}\right)$. Similarly, the quantity, $\left[\exp\left(m \times t^*\right) - 1\right] \times 100$, and its posterior summaries can be computed straightforwardly using the posterior distribution of m to measure the overall impact of the NCC policy. We will present the results of the policy's impacts after discussing the WinBUGS implementation of the model.

16.2.4 WinBUGS Implementation

Figure 16.3 gives the WinBUGS code for fitting the ITS model in Figure 16.2. The reader, by now, should be familiar with most of the code, but it is useful to comment on the following. Lines 45–46 specify the linear impact function specific to each NCC-COA. The area-specific slopes in the linear function are subsequently modelled hierarchically on Lines 70 and 73–75. Lines 55–57 "feed" the estimated control trend from the baseline model into the assessment model. Since this feeding of the trend estimate is done at each MCMC iteration, the uncertainty associated with the trend estimate is fully accounted for when estimating parameters in the assessment model. The cut function ensures that the data from the NCC group do not affect the estimation of the control trend so that the control trend is estimated using *only* control group data. The same procedure applies to feed the place effect precision (Line 60), the overdispersion precision (Line 61) and the intercept (Line 63) estimated from the baseline model into the assessment model. This reflects the assumption discussed in the previous section that areas in both the control and the treatment groups are similar apart from whether or not they had had the NCC treatment. Finally, Lines 80–87 calculate the local and the overall impacts of the scheme and the probability of a positive local/overall impact.

As discussed in Section 16.2.2, the pre-policy data from the NCC-COAs are also used to estimate the control trend. In Figure 16.3, y.control on Line 104 is a 449×8 matrix containing all the burglary counts used to estimate the control trend. The first eight entries of that matrix (Line 104) correspond to a NCC-COA with NCC implemented in 2005 so only the first four years of its data are kept, but the counts in the post-policy period are set to NA, i.e. they are removed. But the burglary counts of this area over the eight years are entered fully in y.treat (Line 115), a 10×8 matrix with burglary counts from the NCC areas.

It should also be noted that the matching criteria "RateMatch" and "RateMatch_MDI" carry out individual matching, meaning that each NCC-COA has its own set of control areas. The burglary rates of each NCC-COA are then compared against the burglary trend pattern estimated using its own set of control areas. Under these two matching criteria, we are essentially estimating 10 different control trends, one for each NCC-COA. However, the area-specific slopes in the assessment model are still hierarchically modelled in order to share information across the NCC-COAs regarding the policy's impact. Exercise 16.1 outlines the analysis under these two matching criteria.

16.2.5 Results

Figure 16.4 shows the estimates of the local impacts and the overall impact. The bottom part in Figure 16.4 provides support for the claim that, when taking the 10 local NCC schemes as a group, the NCC intervention has produced a relative reduction in burglary rates in the targeted areas when assessed against all five control groups using the linear impact function. Note this means a lower burglary rate than would be expected from the comparison areas – not necessarily an absolute reduction in burglary. On the other hand, when viewing the individual schemes, it is clear that there is some heterogeneity where several of the schemes (00JAPB0010, 00JANY0010, 00JANC0016, 00JANE0006, 00JANG0013, 00JANE0010) show various degrees of overlap between the 95% credible interval (referred

```
 1    #   implementation of the ITS model in Figure 16.2 with a
 2    #   linear impact function using only one control group
 3    #   (i.e. using matching criteria "All COAs", "NB" and "MDI")
 4    model {
 5    #################################################################
 6    #   the baseline model to estimate the control trend
 7    #################################################################
 8      for (i in 1:N.control) {
 9        for (t in 1:T) {
10          y.control[i,t] ~ dpois(mu.control[i,t])
11          mu.control[i,t] <- n.control[i] * theta.control[i,t]
12          log(theta.control[i,t]) <- alpha + U.control[i]
13                                   + gamma[t]
14                                   + epsilon.control[i,t]
15          #   overdispersion
16          epsilon.control[i,t] ~ dnorm(0,prec.epsilon.control)
17        }
18      }
19      #   priors for intercept and place effects
20      alpha ~ dflat()
21      for (i in 1:N.control) {
22        U.control[i] ~ dnorm(0,prec.U.control)
23      }
24      #   prior for the control trend (random walk of order 1)
25      gamma[1:T]
26        ~ car.normal(adj.tm[],weights.tm[],num.tm[],prec.gamma)
27      sigma.gamma ~ dnorm(0, 0.1)I(0,)
28      prec.gamma <- pow(sigma.gamma,-2)
29    #   priors on random effect and overdispersion variances
30      prec.U.control <- pow(sigma.U.control,-2)
31      sigma.U.control ~ dnorm(0, 0.1)I(0,)
32      prec.epsilon.control <- pow(sigma.epsilon.control,-2)
33      sigma.epsilon.control ~ dnorm(0, 0.1)I(0,)
34
35    #################################################################
36    #   the assessment model with the linear impact function
37    #################################################################
38      for (k in 1:all.NCC) {
39        for (t in 1:T) {
40          y.treat[k,t] ~ dpois(m.treat[k,t])
41          m.treat[k,t] <- n.treat[k] * theta.treat[k,t]
42          log(theta.treat[k,t]) <- alpha.treat + U.treat[k]
43                                 + gamma.treat[t]
44                                 + epsilon.treat[k,t]
45                                 + step(t-ncc.year[k])*b[k]
46                                 *(t-ncc.year[k] + 1)
47          #   Lines 45 and 46 (above) together define the linear
48          #   impact function specific to each NCC-COA
49          #   Line 50 (below) defines overdispersion
50          epsilon.treat[k,t] ~ dnorm(0,prec.epsilon.treat)
51        }
52      }
53      #   feed the control trend estimates into the
54      #   assessment model
55      for (t in 1:T) {
56        gamma.treat[t] <- cut(gamma[t])
57      }
58      #   feed the place and overdispersion precisions (variances)
59      #   into the assessment model
60      prec.U.treat <- cut(prec.U.control)
61      prec.epsilon.treat <- cut(prec.epsilon.control)
```

FIGURE 16.3
WinBUGS code to fit the ITS model presented in Figure 16.2 with a single control group (i.e. control groups constructed using the criteria "All COAs", "NB" and "MDI" in Table 16.2). Part of the input data is also given.

```
62        #  feed the grand mean into the assessment model
63        alpha.treat <- cut(alpha)
64        for (k in 1:all.NCC) {
65          #  place effects for the NCC-COAs
66          U.treat[k] ~ dnorm(0,prec.U.treat)
67          #  Line 70 (below) specifies a hierarchical model for
68          #  the local impacts (the area-specific slopes in the
69          #  area-specific linear impact functions)
70          b[k] ~ dnorm(m,prec.b)
71        }
72        #  priors on the mean and variance of the local slopes
73        m ~ dnorm(0,0.000001)
74        prec.b <- pow(sigma.b,-2)
75        sigma.b ~ dnorm(0, 0.1)I(0,)
76
77        ################################################################
78        ####   calculate the quantities relevant to evaluation
79        ################################################################
80        for (k in 1:all.NCC) {
81          local.impact[k] <- (exp(b[k])-1)*100
82          #  compute the posterior probability of b[k] less than 0
83          prob.local.success[k] <- 1-step(b[k])
84        }
85        overall.impact <- (exp(m)-1)*100
86        #  compute the posterior probability of m less than 0
87        prob.overall.success <- 1-step(m)
88      }
89
90      ##  part of the input data (based on the matching criterion
91      ##  All COAs in Table 16.2)
92      list(N.control=449 #  the number of selected areas in this
93                         #  control group (Table 16.2)
94        #  burglary counts from the COAs in the control group
95        #  e.g. Line 104 (below) shows counts from a 2005-NCC-COA
96        #  with all the post-NCC burglary counts (on and after
97        #  2005)removed (i.e. replaced by NA)
98        #  Line 105 (below) shows counts from a 2006-NCC-COA with
99        #  all the post-NCC burglary counts (on and after 2006)
100       #  removed (i.e. replaced by NA)
101       #  Line 106 (below) shows counts from a non-NCC-COA with
102       #  all its burglary counts used to estimate the
103       #  control trend
104       ,y.control=structure(.Data=c(3,3,1,0,NA,NA,NA,NA
105                                    ,6,9,5,2,1,NA,NA,NA
106                                    ,15,3,9,0,0,4,7,5
107                                    ,...)
108                            ,.Dim=c(449,8))
109       ,all.NCC=10  #  10 NCC-COAs
110       #  burglary counts from the 10 NCC COAs
111       #  e.g. Line 115 (below) shows the burglary counts from
112       #  the same 2005-NCC-COA in the control list above
113       #  but all its burglary counts are now used to estimate
114       #  the NCC trend
115       ,y.treat=structure(.Data=c(3,3,1,0,1,4,2,3
116                                  ,6,9,5,2,1,3,1,0
117                                  ,...)
118                            ,.Dim=c(10,8))
119       #  each element in ncc.year (below) indicates whether
120       #  a local NCC scheme was implemented in 2005 (=5)
121       #  or in 2006 (=6)
122       ,ncc.year=c(5,6,...)
123       ,... #  the rest of the data list
124     )
```

FIGURE 16.3 *(Continued)*

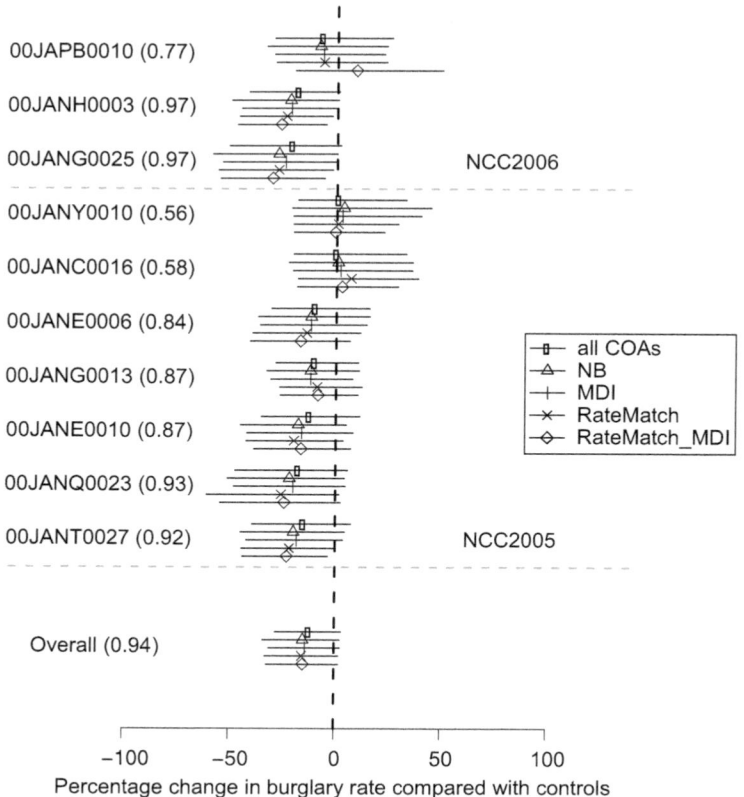

FIGURE 16.4

Percentage change in burglary rates in the NCC census output areas after the first year of NCC implementation compared with the different control groups using the linear impact function. A negative value indicates a reduction relative to the control group. The point symbols represent the posterior means and the horizontal bars are the 95% credible intervals. The number against the COA code is the posterior probability of local success ($\Pr(b_k < 0|$data$)$) and of overall success ($\Pr(m < 0|$data$)$) compared with the control group of all COAs in Peterborough (i.e. the matching criterion "All COAs" in Table 16.1).

to as 95% CI hereafter) and the vertical line at 0, which corresponds to no impact. The posterior probabilities of success for those areas are below 0.90. The four areas that seem to have benefited most from the scheme are: 00JANH0003, 00JANG0025, 00JANQ0023 and 00JANT0027. One possible explanation for this apparent heterogeneity, amongst many, may be the proportion of houses in a COA included by the police in any NCC scheme. Figure 16.5 shows a scatterplot of the posterior means of the local slopes (b_k) against the coverage percentage (see Table 16.1). This finding may be indicative of a "threshold effect" influencing the effectiveness of this spatially targeted scheme (Bowers et al., 2004).

16.2.6 Some Remarks

Whilst policy evaluation is best carried out using time series data collected through randomisation of subjects into treatment and control groups, this is often difficult to implement in practice, particularly in social sciences. Cases are not selected "at random" or even "as-if at random", and the scientist often has no control over whatever selection process is adopted. In such contexts, internal validity becomes a key concern. In this example, the

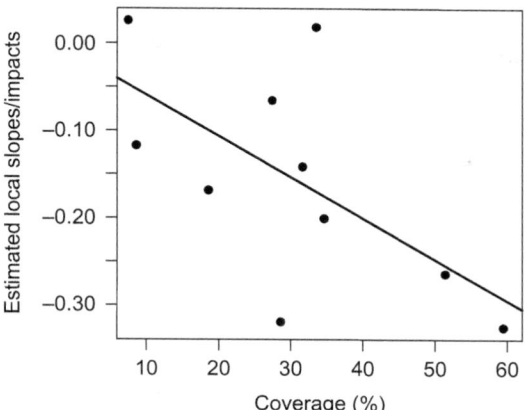

FIGURE 16.5
Initial assessment of the relationship between local impacts (posterior means of the slopes b_k) and coverage rates. The fitted least squares line is superimposed.

interrupted time series design embedded within a matched case-control framework was adopted with the aim of seeking to control for possible impacts linked not to the NCC policy but rather to extraneous and confounding variables. The choice of controls can also help counter the argument that any apparent effect is just a regression to the mean.[1] In addition, there is the issue of the generalizability of findings (external validity). Figure 16.6 shows the results of analysing all the NCC schemes implemented across Cambridgeshire, dividing them into those schemes in urban areas and those in rural areas. The evidence from this extension to the analysis is that the benefits of this initiative, in terms of burglary rates, are more apparent in urban than rural areas, a finding that may not be uncommon (see for example Mawby and Yarwood (2010)).

Finally, the findings presented in this section are based on the linear impact function, which assumes a gradual departure of the NCC trends from the control trend. Other forms of impact (such as step change or a non-linear departure) can be examined by considering different functional forms for the impact function $f(t,b)$ (Box and Tiao, 1975). However, one should bear in mind that the more complex the impact function becomes, the more data (e.g. longer time series and/or more treated areas) will be required. Li et al. (2013) considered a generalised nonlinear impact function but found that it is difficult to obtain reliable estimates because of the small number of NCC areas in Peterborough with a limited number of post-policy time points (see also the Appendix in Chapter 12).

16.3 Application 2: Assessing the Stability of Risk in Space and Time

16.3.1 Studying the Temporal Dynamics of Crime Hotspots and Coldspots: Background, Data and the Modelling Idea

Identifying crime hotspots has become a familiar activity undertaken by many police forces around the world drawing on their recorded crime databases. Such analyses may help in the proactive targeting of resources to tackle persistent areas of high crime (Groff and

---

[1] Regression to the mean is when, for example, a variable that is extreme at one time period moves back towards the average at the next.

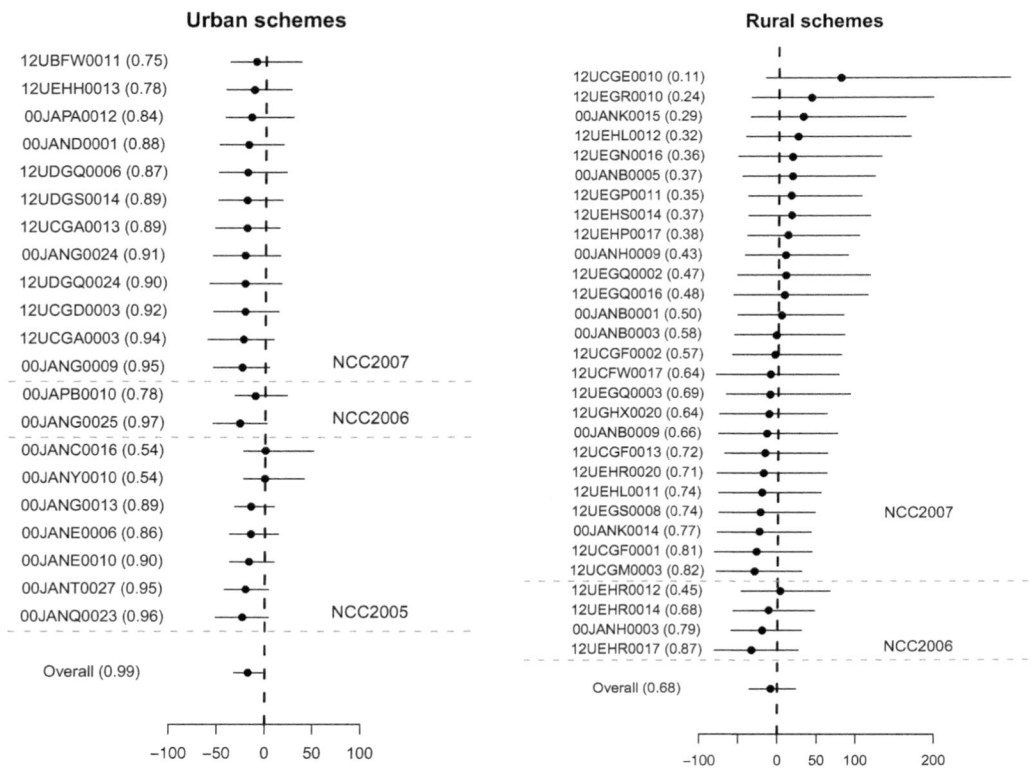

FIGURE 16.6
Results for urban and rural schemes in Cambridgeshire to assess external validity. See Figure 16.4 for interpretation of the graph. The evaluation presented in Figure 16.2 is run separately for the urban NCC-COAs and for the rural NCC-COAs due to their differences in impact. When assessing the impacts of the NCC schemes in the urban (rural) COAs, all urban (rural) non-NCC-COAs are used as controls. The Office for National Statistics (ONS) 2004 definition[2] is used to classify areas into urban (the "urban" type in the ONS definition) or rural (all other types).

LaVigne, 2001). Academic research in socio-spatial criminology and geography is typically interested in identifying high-risk areas and the different factors, demographic, socio-economic and environmental, associated with such areas.

Whilst analysing offence patterns within a specified time interval can be useful, for a deeper understanding, analysis needs to engage with their temporal dynamics (Grubesic and Mack, 2008). Poorer neighbourhoods tend to experience household burglary hotspots of long duration, whilst affluent neighbourhoods typically experience more short-term concentrations (Johnson and Bowers, 2004). Evidence gathered by criminologists and geographers suggests that there are two important components of variation in spatial-temporal counts of burglary offences in urban areas: (a) a stable component comprising a common spatial pattern modulated by an overall time trend pattern; (b) space-time interactions

---

[2] https://www.ons.gov.uk//methodology/geography/geographicalproducts/ruralurbanclassifications/2001ru ralurbanclassification/ruralurbandefinitionenglandandwales (last accessed: 31 January, 2019).

leading to departures from the stable pattern in which, for example, some areas show different trend patterns from the overall time trend whilst some others experience concentrations of cases that then dissipate. (Geographic scale is important when considering these types of questions, see for example Andresen and Malleson, 2011, Sherman et al. 1989.)

This application studies the Bayesian spatial-temporal model described in Li et al. (2014), an extension of the space-time model by Bernardinelli et al. (1995) proposed for studying disease risk. This space-time model is a space-time inseparable model (Knorr-Held, 2000 and Chapter 15 in this book), and it decomposes observed space-time variation into two main components, one that describes the stable spatial and temporal patterns and one that captures potential space-time interaction. Type IV interaction is used for modelling the space-time interaction component where a linear trend with a local slope is specified for each COA and the set of local slopes are further smoothed over space. From Section 15.7.1, this follows Strategy 1 to specify space-time interaction so that the emphasis of the modelling is on the individual time patterns (see also Section 15.7.4). The use of a simple linear specification for the local trend produces a number of benefits. It allows: (i) strengthening the estimation of change over time by imposing a spatial smoothing model on the local slopes; (ii) incorporating covariates to explain the variation in local time trend patterns; (iii) identifying and quantifying unusual local trends that depart from the common time trend; and (iv) enabling a classification of the identified unusual trends into stable/increasing/decreasing patterns, which helps in the interpretation of the detection results.

The utility of the model in the present context is enhanced further by introducing a two-stage method for classifying areas as well as identifying the covariates that might explain the space-time classification (Li et al., 2014). As an extension to the usual hotspot analysis, we are interested in (1) examining how hot and cold spots evolved over the time period; (2) identifying areas that, whilst not yet hotspots, show a tendency to become hotspots. Unlike Application 1, this is a non-experimental design. There is no subset of areas of special interest, rather our aim is to identify *any* areas where event rates differ markedly from the general common time trend.

In the study by Li et al. (2014), the model was applied to the burglary counts for 452 Census Output Areas (COAs) in Peterborough for the period 2005–08. The data are summarised in Table 16.3. The presence of small numbers of cases suggests a model-based approach will be needed to disentangle useful information from noise. The precision of space-time estimates of burglary rates can be improved by borrowing information not only spatially and temporally but also spatially-temporally. The type of exploratory analyses discussed in Chapter 13 is extremely helpful in informing and justifying how one can (and should) borrow information within a Bayesian space-time model – we will signpost the relevant exploratory analysis performed in Chapter 13 when discussing the model in the next section.

TABLE 16.3

Summary of Census Output Area (COA)-Level Burglary Counts in Peterborough, 2005–08

Year	Min	1st Quartile	Median	Mean	3rd Quartile	Max
2005	0	0	1	1.37	2	8
2006	0	0	1	1.43	2	9
2007	0	1	2	2.68	4	12
2008	0	1	2	2.82	4	16

16.3.2 Model Formulations

Let y_{it} denote the number of burglary cases in COA i $(=1,\ldots,452)$ in year t $(=1,\ldots,4)$ and let n_i denote the number of households in COA i. We assume the numbers of at-risk households remain constant throughout 2005–08. The number of burglary counts is modelled through a Poisson likelihood. That is, $y_{it} \sim Poisson(n_i \cdot \theta_{it})$, where θ_{it} represents the underlying unknown burglary rate. The process model describes the (log) burglary rate θ_{it} as follows:

$$\log(\theta_{it}) = \alpha + (S_i + U_i) + \left(b_0 t^* + v_t\right) + b_{1i} t^* + \varepsilon_{it},\qquad(16.1)$$

where $t^* = t - 2.5$ (i.e. centring time at the mid-observation period) and α is the overall log burglary rate over the four years in Peterborough. Eq. 16.1 partitions the observed space-time variation in burglary rates into the following components:

Component 1:	$(S_i + U_i)$ denotes the spatial component common across all four years, describing the overall spatial distribution of burglary risk.
Component 2:	$(b_0 t^* + v_t)$ describes the overall time trend common to all COAs. This common time trend is formulated as a linear trend $(b_0 t^*)$ – allowing for the overall-local trend comparison – with additional Gaussian noise (v_t) to allow for nonlinearity in the overall trend pattern (see also Section 12.3.3.1).
Component 3:	$b_{1i} t^*$ allows the time trend of each COA to deviate from the overall time trend pattern (as in Component 2). While b_0 represents the overall rate of change in (log) burglary rate, $b_{1i} t^*$ measures the *departure* from b_0 for each COA. For example, if b_0 is estimated to be positive (i.e. showing an overall increase in the burglary rate), then a negative estimate of b_{1i} would suggest a slower increase (or even a decline) in the burglary rate over time for that COA, while a positive estimate of b_{1i} would imply an even faster increase compared to the overall increasing trend.
Component 4:	ε_{it} captures additional variability in the data not captured by the other components. Overdispersion is a source of such additional variability which might be associated with any spatial clustering effects due, for example, to periods of repeat victimization.

In this model formulation, the first two components, the common spatial pattern and the common time trend, represent the stable components in burglary rate variation, whilst the other two capture space-time interaction. Component 3, in particular, defines a specific type of interaction that allows for trend detection.

For each model component, smoothing is carried out by assigning appropriate prior models to the random effects. For the overall spatial component, $(S_i + U_i)$, a BYM model is used. The random effects, $S = (S_1,\ldots,S_{452})$, are structured spatially through an ICAR model with a spatial weights matrix W of size 452×452. All diagonal entries in W are 0, and the off-diagonal entry, w_{ij} is specified as follows: $w_{ij} = 1$ if i and j share a common border and $w_{ij} = 0$ otherwise. The use of the ICAR model assumes that adjacent COAs have burglary rates that are more alike than those further apart – this is evident from the exploratory analysis of the burglary data in Figure 12.1. The reader is encouraged to calculate the global Moran's I using the COA-level raw burglary rates from each year as well as the COA-level raw burglary rates averaged over four years to assess spatial autocorrelation. To accommodate the part of the overall spatial pattern that does not show spatial structure, the random effects, $U_1,\ldots,U_{452}$, are modelled exchangeably, and each U_i follows a common normal distribution $N(0,\sigma_u^2)$. The same BYM prior model is assigned to the local slopes, b_{1i} for $i = 1,\ldots,452$, so that we are also assuming that adjacent COAs have *changes* in burglary rate that are more alike than those that are further apart. This is suggested by the apparent spatial structure in the map of the slope estimates in Figure 13.11. The temporal noises v_t (for $t = 1,\ldots,4$)

are modelled independently via a common normal distribution with mean 0 and variance σ_v^2. Similarly, the overdispersion terms, ε_{it}, are modelled independently using a common normal distribution with mean 0 and variance σ_ε^2. A strictly positive half normal prior, $N_{+\infty}(0,10)$, is independently assigned to each of the random effect standard deviations, i.e. σ_U, σ_S, σ_v, σ_{b1U}, σ_{b1S} and σ_e, where σ_{b1U} and σ_{b1S} are the standard deviations associated with the spatially-structured and the spatially-unstructured random effects in the BYM prior for the area-specific slopes. Finally, the prior for α and b_0 is taken to be the improper uniform prior defined on the whole real line, $Uniform(-\infty,+\infty)$ (see Section 8.2.1.1).

The model presented in Eq. 16.1 can be modified to include covariates that help to explain the spatial, temporal and spatial-temporal patterns of the data. Here, we include three COA-level variables from the 2011 UK census. The three variables are $x_{i,OWN}$, $x_{i,DET}$ and $x_{i,ETH}$, which measure respectively the percentage of home ownership, the percentage of detached and semi-detached houses[3] and Simpson's index of ethnic diversity at the COA level. Simpson's index ranges between 0 and 1. The larger the index, the less diverse the COA in terms of ethnic composition (Simpson, 1949). Also added in is a fourth covariate identifying whether the COA was part of the NCC scheme implemented by Cambridgeshire Constabulary (see Application 1 in Section 16.2), so that $x_{i,NCC}=1$ if COA i contained an NCC scheme and 0 otherwise. Since the NCC scheme was found to affect change in burglary risk (see Figure 16.4), this covariate was used to explain variation in the local slopes (b_{1i}). The resulting model now has the form:

$$\log(\theta_{it}) = \alpha + (S_i + U_i) + (b_0 t^* + v_t) + \beta_{OWN} x_{i,OWN} + \beta_{DET} x_{i,DET}$$

$$+ \beta_{ETH} x_{i,ETH} + (b_{1i} + \beta_{NCC} x_{i,NCC}) t^* + \varepsilon_{it}$$

(16.2)

A weakly-informative normal prior $N(0,1000000)$ is independently assigned to each of the four regression coefficients. The specification of all other model components is the same as those in Eq.16.1.

16.3.3 Classification of Areas

In addition to identifying areas with overall higher/lower burglary rates (i.e. hot/cold-spots) relative to the Peterborough average, as discussed in Section 16.3.1, the modelling also focuses on revealing how these identified hotspots (or coldspots) have changed over time, as well as identifying areas that are not yet showing higher/lower burglary rates but are showing a tendency to trend in one or the other direction. To do that, we adopt the two-step classification developed in Li et al. (2014), which is based on the overall spatial component, $(S_i + U_i)$, and the set of local slopes, b_{1i}, each quantifying the departure of a local trend from the overall trend. This two-step classification is given as follows.

Step 1 classifies a COA based on the posterior probability, $p_i = \Pr(\exp(S_i + U_i) > 1 \mid data)$, where $\exp(S_i + U_i)$ represents the average burglary risk over time in COA i compared to the overall Peterborough average (i.e. measured by α). Thus, if $p_i > 0.8$, COA i is classified as a burglary hotspot, suggesting that the overall burglary risk in this COA is considerably higher than that for Peterborough as a whole. On the other hand, if $p_i < 0.2$, then COA i is classified as a burglary coldspot. If p_i is not in either of those two extremes, i.e. $0.2 \leq p_i \leq 0.8$, then COA i is considered to have an overall burglary rate similar to the

---

[3] A semi-detached house (also referred to as a "twin home" or a "duplex") is a building with two attached homes, each of which is a distinct property.

Peterborough average, thus being considered neither a hotspot nor a coldspot. This is the classification rule developed in Richardson et al. (2004) in the context of detecting disease hotspots and coldspots. This classification rule results in the following set of indicators, $h = (h_1, \ldots, h_N)$, one for each COA where $h_i = 1$ if COA i is classified as a hotspot, $h_i = 2$ if it is a coldspot and $h_i = 3$ if the COA is neither a hotspot nor a coldspot.

Based on the classification results from Step 1, Step 2 further classifies every COA into one of the following three categories using the posterior probability $g_i = \Pr(b_{1i} > 0 \mid h_i, \text{data})$. So, COA i is classified to have an increasing trend that is steeper than the overall trend if g_i is greater than 0.8. A COA has a decreasing trend relative to the overall trend if g_i is less than 0.2. A COA has a trend that does not differ from the overall trend if g_i lies between 0.2 and 0.8. Note that these are all relative trends – relative to the overall trend – so they do not necessarily imply an absolute increase or decrease in the burglary rate.

This two-step classification gives rise to nine categories (three categories based on overall risk × three categories based on trend) of area, as can be seen in Table 16.4 in the next section. The decision rule probability cut-offs used at Step 1 have been shown (see Richardson et al., 2004) to yield a good balance between sensitivity (i.e. the power to detect a hot/cold spot when its overall risk is indeed above/below the average) and the false positive rate (i.e. the proportion of the declared hotspots and coldspots with risks that are *not* different from the average). The same probability thresholds (0.2 and 0.8) have been used at Step 2 to examine the nature of the local slopes/departures. This discussion is concerned with the multiple testing problem that any application involving classification needs to deal with. Richardson et al. (2004) select the classification thresholds based on a loss function approach (see Appendix B in their paper). We will return to this issue in Application 3 (Section 16.4), where the issue of multiple comparisons is addressed via controlling the false discovery (positive) rate, and again in Section 16.6.

16.3.4 Model Implementation and Area Classification

Figure 16.7 shows the WinBUGS code to implement the space-time model with four COA-level covariates, the more complex of the two models discussed. The hierarchically-centred version of the BYM model (see Section 8.5.1) is used for both the overall spatial component (Lines 24–28) and the local slopes (Lines 66–69 and 74) to achieve faster convergence and better mixing. A similar idea is used to centre the temporal noises at the overall linear trend (Lines 46–49) and to centre the overdispersion terms ε_{it} at the sum of all other terms in Eq. 16.2 (Lines 12–17). It should also be noted that Line 53 subtracts the overall time trend with temporal noise from the mean value of the trend (calculated on Line 55) so that the resulting overall trend has the value 0 at the midpoint of the study period. This is done to ensure that the overall intercept α is estimable (see Section 12.3.3.2.2). Lines 87–89 calculate

TABLE 16.4

Number of Areas Resulting from the Two-Step Classification of the 452 COAs in Peterborough

	Increasing Trend	Decreasing Trend	Not Differing from the Common Trend	Total
Hotspots	32	27	111	170 (38%)
Coldspots	21	21	84	126 (27%)
Neither hotspots nor coldspots	30	16	110	156 (35%)

```
 1    #  fitting the space-time detection model presented in
 2    #  Eq. 16.2 with three COA-level covariates:
 3    #  simpson (ethnicity) + p_owned + p_detached (on level)
 4    #  and the NCC indicator (on slope)
 5    model {
 6      for (i in 1:N) {
 7        for (t in 1:T) {
 8          y[i,t] ~ dpois(mu[i,t])
 9          log(mu[i,t]) <- log(n[i]) + log.rates[i,t]
10          #  Eq. 16.2 implemented in the hierarchically-centred
11          #  fashion - see main text for explanation
12          log.rates[i,t] ~ dnorm(theta[i,t],prec.epsilon)
13          theta[i,t] <- overall_spatial[i] + overall_temporal[t]
14                        + local_temporal[i,t]
15                        + beta_eth * simpson[i]
16                        + beta_owned * p_owned[i]
17                        + beta_detached * p_detached[i]
18        #  recovering the overdispersion terms
19          epsilon[i,t] <- log.rates[i,t] - theta[i,t]
20        }
21      #  BYM on the overall spatial pattern (implemented
22      #  under the hierarchically-centred version;
23      #  see Section 8.5.1)
24      overall_spatial[i] ~ dnorm(ms[i],prec.U)
25      ms[i] <- alpha + S[i]
26      #  recovering the spatially-unstructured random
27      #  effects U[i]
28      U[i] <- overall_spatial[i] - ms[i]
29      }
30    #  spatially-structured random effects using ICAR
31    S[1:N] ~ car.normal(adj[],weights[],num[],prec.S)
32    #  variances of the BYM model on the overall spatial pattern
33    sigma.S ~ dnorm(0,0.1)I(0,)
34    prec.S <- pow(sigma.S,-2)
35    sigma.U ~ dnorm(0,0.1)I(0,)
36    prec.U <- pow(sigma.U,-2)
37    #  variance for overdispersion
38    sigma.epsilon ~ dnorm(0,0.1)I(0,)
39    prec.epsilon <- pow(sigma.epsilon,-2)
40    #  prior for the overall intercept
41    alpha ~ dflat()
42    #  overall temporal component (an overall linear trend +
43    #  independent temporal noises)
44    b0 ~ dflat()    #  prior for the overall slope
45    for (t in 1:T) {
46      mu_tm[t] <- b0 * (t-mt)  #  mt=2.5, the midpoint of the
47                               #  study period
48      #  adding noise to the overall linear trend
49      overall_tm_temp[t] ~ dnorm(mu_tm[t],prec.v)
50      #  Line 53 (below) recentres the overall trend at the
51      #  overall mean (st) so that the resulting overall trend
52      #  is 0 at the midpoint of the study period
53      overall_temporal[t] <- overall_tm_temp[t] - st
54    }
55    st <- mean(overall_tm_temp[1:T]) #  the overall mean of the
56                                     #  uncentred overall trend
57    #  variance for the temporal noises
```

FIGURE 16.7
The WinBUGS implementation of the space-time model presented in Eq. 16.2 with four COA-level covariates.

```
58      prec.v <- pow(sigma.v,-2)
59      sigma.v ~ dnorm(0,0.1)I(0,)
60      #  departure of local time trend
61      for (i in 1:N) {
62        #  local slope = local departure (b1) + effect of NCC
63        local_slope[i] <- b1[i] + beta_ncc * ncc[i]
64        #  the hierarchical centred version of the BYM model
65        #  on the local departures (b1)
66        b1[i] ~ dnorm(b1S[i],prec.b1U)
67        #  recover the spatially-unstructured random effects
68        #  on the local departures (b1)
69        b1U[i] <- b1[i] - b1S[i]
70        for (t in 1:T) {
71          local_temporal[i,t] <- local_slope[i] * (t - mt)
72        }
73      }
74      b1S[1:N] ~ car.normal(adj[],weights[],num[],prec.b1S)
75      #  variances of the BYM model on the local linear departures
76      sigma.b1U ~ dnorm(0,0.1)I(0,)
77      prec.b1U <-pow(sigma.b1U,-2)
78      sigma.b1S ~ dnorm(0,0.1)I(0,)
79      prec.b1S <-pow(sigma.b1S,-2)
80      #  vague priors on the regression coefficients
81      beta_eth ~ dnorm(0,0.000001)
82      beta_owned ~ dnorm(0,0.000001)
83      beta_detached ~ dnorm(0,0.000001)
84      beta_ncc ~ dnorm(0,0.000001)
85
86      #  quantities for Step 1 classification
87      for (i in 1:N) {
88        spatial[i] <- exp(S[i] + U[i])
89      }
90      #  VPC to assess the importance of the stable space-time
91      #  component relative to the space-time interaction
92      #  component
93      for (i in 1:N) {
94        for (t in 1:T) {
95          stable[i,t] <- S[i] + U[i] + overall_temporal[t]
96          interaction[i,t] <- local_temporal[i,t] + epsilon[i,t]
97        }
98      }
99      #  empirical variance of the stable component
100     var.stable <- sd(stable[,]) * sd(stable[,])
101     #  empirical variance of the space-time interaction
102     #  component
103     var.interaction <- sd(interaction[,]) * sd(interaction[,])
104     vpc <- var.stable/(var.stable + var.interaction)*100
105     ##  covariate effects
106     #  change to burglary risk with a 0.1 increase in
107     #  Simpson's Index
108     RR.eth <- exp(beta_eth*0.1)
109     #  change to burglary risk with a 10% increase in
110     #  ownership percentage
111     RR.owned <- exp(beta_owned*10)
112     #  change to burglary risk with a 10% increase in
113     #  detached house percentage
114     RR.detached <- exp(beta_detached*10)
115   }
```

FIGURE 16.7 (*Continued*)

the quantities, spatial[i], for Step 1 classification, whilst the quantities, b1[i] on Line 66, are used in the Step 2 classification. Lastly, Lines 90–104 calculate the variance partition coefficient or VPC (see Section 15.3.2 and Section 15.6) in order to measure the importance of the stable space-time component relative to the space-time interaction component (after accounting for the covariates included). Fitting the model in Eq. 16.1 can be done easily by simply removing the lines in Figure 16.7 that are related to the four covariates.

The two-step classification cannot be done in WinBUGS but can be done in R. An easy way to fit the model and to obtain the classification results is through the use of the R2WinBUGS package in R. Exercise 16.4 outlines the procedure to carry out the whole analysis together with the two-step classification.

16.3.5 Interpreting the Statistical Results

Figure 16.8(a) and (b) show the stable spatial and stable temporal components of variation in the Peterborough burglary data using the model without covariates (i.e. Eq. 16.1). The general increase in burglary rates over the period 2005–08 is evident. The south-south-west of Peterborough shows a group of areas of high relative risk, as does the city centre and small pockets of COAs in the west. Figure 16.8(c) shows the estimated local slopes that measure the departure of local trends from the overall trend pattern. The map in Figure 16.8(c) suggests a north-south difference in the local departures from the overall trend for Peterborough, with areas in the north generally showing an even faster increase in their burglary rates whilst areas in the south display slower increases. There is some evidence, from a cursory inspection of Figures 16.8(a) and (c), of a convergence in burglary risk between the high relative risk in south-southwest Peterborough, where there is a

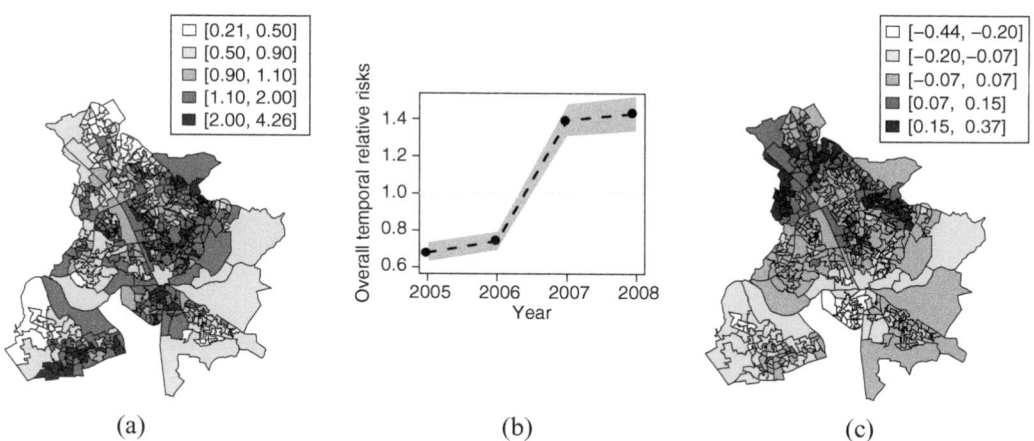

(a)   (b)   (c)

FIGURE 16.8

(a) A map of the estimated overall spatial pattern based on the posterior means of the spatial relative risks, $\exp(S_i + U_i)$; (b) the estimated overall temporal pattern (posterior means, the dots, with the corresponding 95% credible intervals, the grey band, of $\exp(b_0 t^* + v_t)$); and (c) a map of the posterior means of b_{1i} that measure the departures of each local trend from the overall time trend. All the estimates are obtained from fitting the Bayesian space-time model without covariates (Eq. 16.1) to the Peterborough COA-level burglary data from 2005 to 2008. For map (a) and plot (b), an estimated relative risk close to 1 suggests the corresponding COA (in (a)) or the corresponding time point (in (b)) has a risk of burglary close to the Peterborough average. For map (c), a COA with an estimate of b_{i1} close to 0 suggests that the burglary trend for that COA is similar to that for Peterborough as a whole; the Peterborough trend is shown in (b).

generally slow increase (even decrease) in burglary risk over the study period, and the low relative risk in north Peterborough but where there is a faster increase in risk over the study period. The VPC is estimated to be 80% (with 95% credible interval: 70%–87%), indicating that compared to the space-time interaction component, the overall spatial and the overall temporal components taken together explain the majority of the observed variability.

The classification of the 452 COAs into the nine groups is summarised in Table 16.4. A total of 38% of COAs are considered to be hotspots, whilst 27% are classified as coldspots. The remaining 35% of the 452 COAs are neither hotspots nor coldspots. The time trends associated with these areas within each of the three risk categories clearly differ. Although a majority of the areas in each category have time trends that do not differ from the overall Peterborough trend, some show trends with a faster/slower rate of increase. In particular, of the 170 identified hotspots, 32 showed a faster increase in their burglary rates over the period compared to the overall trend, whilst 27 hotspots tended to cool (Table 16.4). Figure 16.9 shows the locations of the identified hotspots with the inserted plots illustrating some of the local temporal dynamics that are different from the overall Peterborough trend. If we now consider Figure 16.10, which shows the COAs that are considered to be neither hotspots nor coldspots, together with Table 16.4, these areas do not all follow the same Peterborough trend: 30 of these COAs show a faster increasing trend compared to the overall trend, and inserted plots in Figure 16.10 illustrate this point for two such COAs. These areas may be trending towards becoming hotspots. Exercise 16.5 constructs the map of the identified coldspots and investigates their temporal trends.

We now turn to the model with four COA-level covariates (Eq. 16.2). The reason to include covariates was to explain (a) the overall spatial pattern, providing evidence as to why some COAs had persistently high or low risk; and (b) the departures of the local trends from the common trend. Table 16.5 reports the estimated relative risks (RRs) – exponential transformations of the regression coefficients – associated with the three covariates on the overall burglary rate level. The posterior means of the RRs are all below 1 (the value indicating no effect), suggesting that a COA with lower burglary rates tends to have a higher percentage of owner-occupied houses, a higher percentage of detached and semi-detached houses and a lower degree of ethnic heterogeneity. The effect of percentage of ownership is significant at the 5% level, with the 95% CI excluding 1. However, there is only weak evidence of effects from the other two covariates, as the upper bound of their 95% CIs is on the borderline of 1. The posterior mean of the coefficient associated with the implementation of a NCC scheme in a COA is negative ($\beta_{NCC} = -0.29$ with a 95% CI of −0.56, −0.03), suggesting a relatively slower increase in burglary risk over time for an NCC-COA compared to the Peterborough trend. For the trend estimates, the overall slope is estimated to have a positive posterior mean ($b_0 = 0.29$; 95% CI: −0.12, 0.73), reflecting the overall increasing trend in the burglary rate as noted earlier, although the 95% CI does include the value 0.

Inclusion of these covariates accounts for some of the hot/coldspots identified earlier in Table 16.4 from the model with no covariates included. Controlling for these covariates results in 48 of the 170 hotspots identified in Table 16.4 ceasing to be hotspots. These are the areas crosshatched in Figure 16.9. Of the 126 coldspots (Table 16.4), 26 ceased to be coldspots (see Exercise 16.6). These results imply that the persistently high (low) burglary risks in these identified hotspots (coldspots) can be explained by the included covariates. In other words, these 48 hotspots (or 26 coldspots) could arise because each has some combinations of the characteristics of a low (high) percentage of home ownership, a low (high) percentage of detached or semi-detached housing and/or a high (low) degree of ethnic heterogeneity. However, many identified hotspots and coldspots remain unaccounted for.

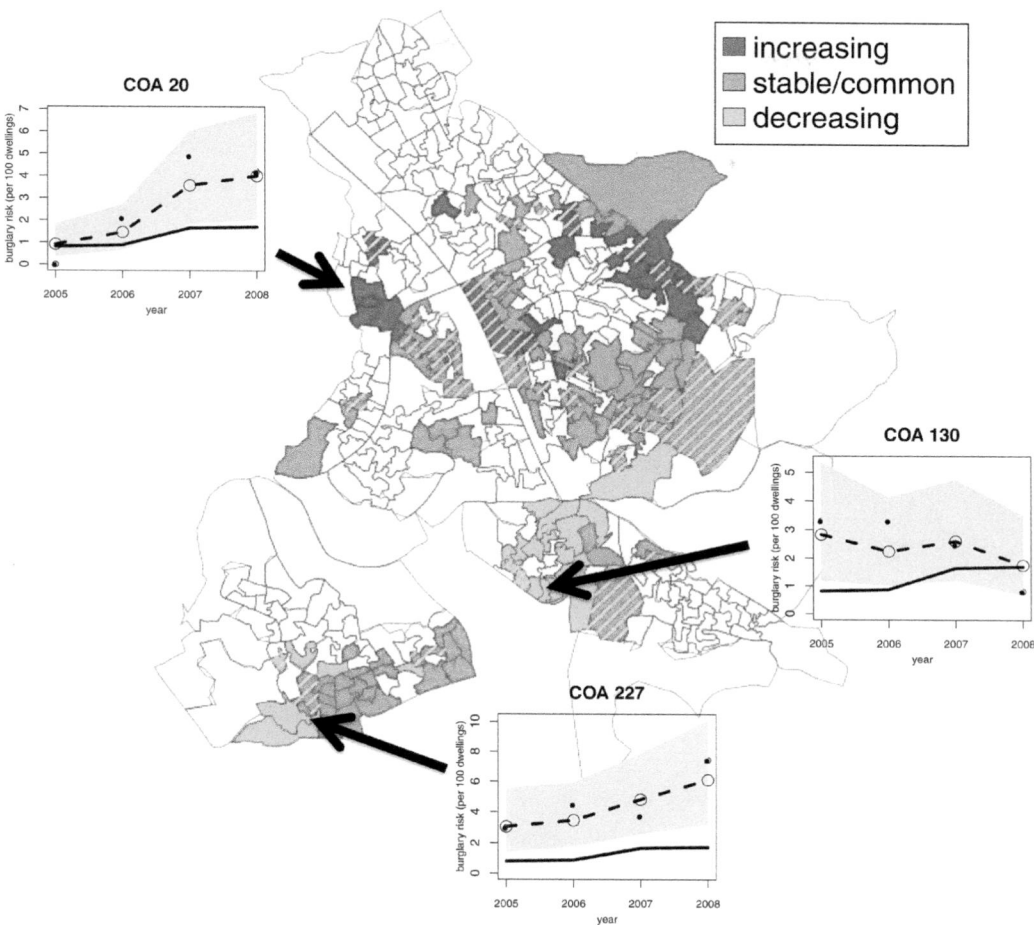

FIGURE 16.9
(See colour insert.) Persistently high-risk COAs (hotspots) in Peterborough, 2005–08. High-risk areas are further classified into those with an increasing or decreasing pattern relative to the common trend and those with a trend not differing from the common trend. For three COAs (one for each trend category), the inserted figures show the observed raw burglary rates (per 100 houses; black solid dots), the estimated burglary rates (open circles joined by dashed lines) with the 95% credible intervals (grey regions) and the estimated common trend (black solid line) over time. The crosshatched polygons are the COAs that cease to be hotspots after accounting for the effects of the included COA-level covariates. This figure is a revised version of Figure 2 in Li et al. (2014) where the inserted plot for COA 130 should be associated with the polygon indicated in this figure.

It is also worth noting that eight COAs became hotspots and 17 became coldspots which were not identified by the model without covariates (Eq. 16.1). This may be due to possible spatially-varying effects and/or interactions amongst some of the covariates. While Table 16.5 reports the "whole-map" effects of these covariates, the effects of some covariates may vary locally and/or be dependent on the values of other covariates. See Section 9.4 as to how one might address such issues.

One extension of the modelling presented here is to jointly analyse multiple related crime types to improve estimates, thus strengthening the detection results. Some crimes are considered to be "substitutable" – for example, an offender needs money and, opportunistically,

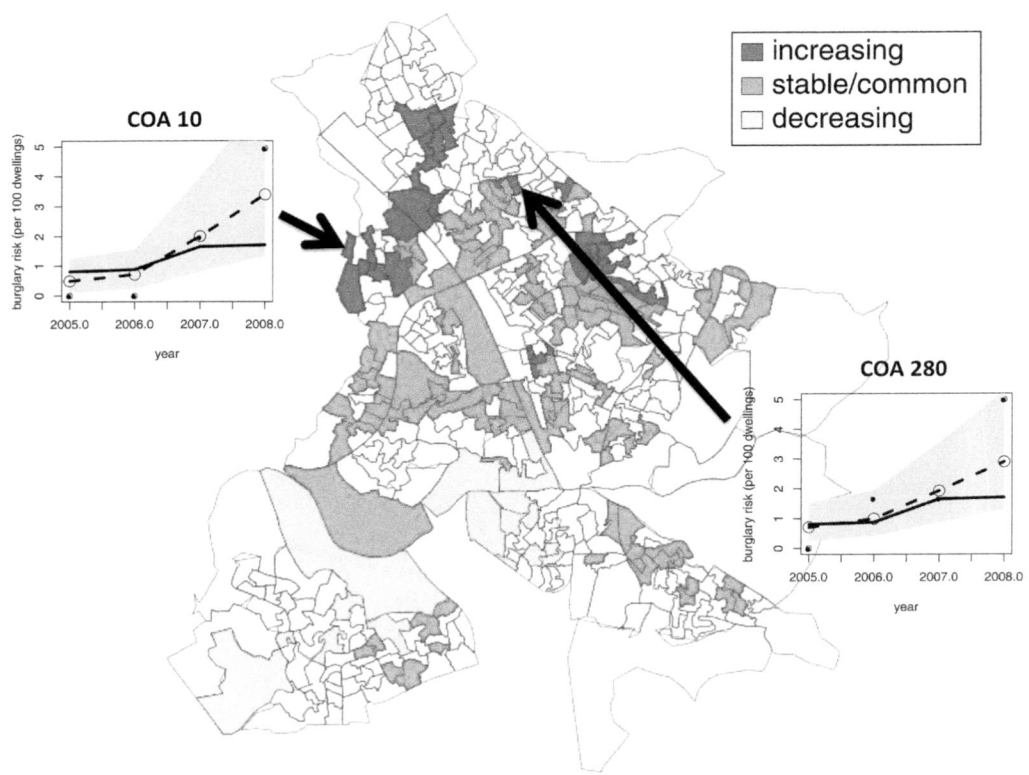

FIGURE 16.10
(See colour insert.) COAs in Peterborough with neither persistently high (hotspot) nor low (coldspot) risk of burglary. Two COAs are identified that are showing a tendency to becoming hotspots. (See Figure 16.9 for the explanation of the inserted figures.)

TABLE 16.5

Estimated Relative Risks (RRs), an Exponential Transformation of the Regression Coefficients for $x_{i,\text{OWN}}$, $x_{i,\text{DET}}$ and $x_{i,\text{ETH}}$ in Eq. 16.2

Relative Risk	Posterior Mean (95% CI)
RR_{OWN}	0.93 (0.90, 0.97)[a]
RR_{DET}	0.97 (0.94, 1.00)[a]
RR_{ETH}	0.94 (0.88, 1.00)[b]

Posterior means with 95% CIs are reported.
[a] with a 10% increase in the percentage of owner-occupied houses/percentage of detached or semi-detached houses in the COA. This corresponds to having about 13 more houses.
[b] with a 0.1 increase in Simpson's index.

may break into a car or burgle a house. Shared component modelling can be used to capture the shared pattern but also facilitates borrowing strength over multiple crime types (see, for example, Tzala and Best (2008) in the context of modelling spatial-temporal risks of multiple cancers and Quick et al. (2019) for modelling risks of multiple crime types over space and time). In addition to borrowing strength, a joint analysis of multiple crime types helps identify the crime-general (e.g. an identified hotspot showing high crime rates across all types) and the crime-specific (e.g. an identified hotspot showing high crime rates for some but not all crime types) patterns (Quick, M. et al., 2018).

Finally, the specification of the linear time trend is suitable for relatively short time series data with 2–5 time points, but it becomes restrictive when modelling longer time series data (>5 time points). In those circumstances, as we shall see in the next application, a different modelling approach based on mixture models is needed for detecting unusual local time patterns.

16.4 Application 3: Detecting Unusual Local Time Patterns in Small Area Data

16.4.1 Small Area Disease Surveillance: Background and Modelling Idea

Small area disease surveillance is an information-based activity in which large amounts of geographically referenced data are collected, analysed and interpreted with the aim of identifying those geographically defined populations at high risk from some disease and the timing of that raised risk. In Chapter 13 we described some of the exploratory methods that can be used for this purpose, and in particular the Knox test and Kulldorff's space-time permutation scan statistic. Here we describe a model-based space-time surveillance method that uses the Bayesian hierarchical modelling approach with a feature to allow information to be shared between areas and across time points, giving rise to more reliable estimates and in turn more accurate detection results.

Disease data collected for many small areas within a study region (e.g. a city) typically show a general temporal trend that most areas tend to follow. However, the time trends of some areas may depart from the common trend due to, for example, the emergence of localized risk factors or potential effects from targeted interventions (e.g. the NCC scheme in the case of deterring cold calls). Detecting areas with "unusual" temporal patterns provides a basis for further investigations in those areas. It should be noted that, in contrast to the case of policy *evaluation* (Application 1; Section 16.2), there is no subset of areas of interest a priori in the case of disease (or crime) *surveillance*. However, as in Application 2 (Section 16.3), our aim here is to identify any areas whose event rates differ markedly from the overall time trend. But whilst the detection model studied in 16.3 is suitable for two to five time points, the methodology that we shall discuss in this section deals with longer time series data with up to 10 time points. At the end of the section, we will briefly discuss models that have been developed to analyse longer time series data with 10 or more time points.

BaySTDetect, a space-time detection method proposed by Li et al. (2012), identifies unusual local time trends by comparing between two competing spatial-temporal models. Model 1 is a space-time separable model that decomposes the observed space-time variability into an overall spatial component and an overall temporal component. The implication of Model 1 is that all areas within the study region follow the same common time trend (see Chapter 14). Thus, Model 1 is referred to as the common trend model. Model 2, by contrast, estimates the time trends independently for each small area so that the trend estimate of each area can be different from those of other areas. Model 2 is referred to as the area-specific model. For each area, a Bayesian model comparison procedure is carried out to assess how likely it is that the observed time series data would have been generated under the common trend model (Model 1) or the area-specific trend model (Model 2). For example, if the data of an area are more likely to have come from the common trend model, then the time trend of this area is not unusual; otherwise, the time trend may be considered unusual. In Li et al. (2012), the proposed detection method was used for disease surveillance and also to explore the differential local impacts resulting from a *nationally* implemented government health policy. This contrasts with the geographically *targeted* implementation of NCC in Application 1 (Section 16.2). Due to data confidentiality, the data used in their application cannot be made available. Instead, we will reproduce some of the result of their simulation study to compare the detection performance of BaySTDetect with that from SaTScan.

A critical issue to address in this application is that of multiple comparisons (also referred to as the multiple testing problem): when a large number of areas are tested, some "unusual" areas are bound to arise just by chance. Therefore, as part of any detection method, it is important to represent the likely error (or uncertainty) in the detection results. The chosen procedure in BaySTDetect is the control of the false discovery rate (FDR), which is defined as the expected proportion of the areas declared to be "unusual" that are in fact *false positives* (i.e. an area with a truly usual trend but declared to be an unusual area by the method). The FDR controlling procedure will be discussed below.

16.4.2 Model Formulation

Here we present the formulation of BaySTDetect within the context of the empirical study by Li et al. (2012). In that study, they used annual national mortality data on chronic obstructive pulmonary disease (COPD; ICD9 490–496) for the 374 local authority districts (LADs; the spatial unit of the analysis) in England and Wales. The study period covered eight years from 1990 to 1997. The data analysed covered only men aged 45 years and over. The effect of age on COPD mortality was accounted for through the expected counts, which were standardised by five-year age bands with the age-specific reference rates calculated over the eight-year period in England and Wales (see Exercise 9.6 in Chapter 9 for age-sex standardisation).

Let y_{it} and E_{it} be the observed and expected numbers of COPD deaths respectively in LAD i ($i = 1, \ldots, N$) at time ($t = 1, \ldots T$). Because COPD is rare, the observed death count y_{it} is modelled using a Poisson likelihood, i.e. $y_{it} \sim \text{Poisson}(\mu_{it} \cdot E_{it})$, where μ_{it} is the (unknown) relative risk in LAD i in year t. With the aim of detecting LADs with unusual temporal trends, μ_{it} is described by the two competing models, where Model 1, the common trend model, assumes space-time separability for all LADs, so all LADs have the same time trend, whilst Model 2, the area-specific trend model, provides estimates for each LAD individually. Table 16.6 lays out the two models together with the full set of prior specifications.

The common trend model decomposes the log relative risk additively into an overall spatial component ($S_i + U_i$) and an overall temporal component (v_t). Since the temporal component is not indexed by area, all LADs have the same trend pattern represented by $v = (v_1, \ldots, v_T)$, a modelling feature that can oversmooth any local trends that display true departures from the common trend pattern. On the other hand, the area-specific trend model estimates the temporal trends individually for each LAD so it is able to accommodate any substantial departures of a local trend from the common trend. These two models are fitted separately to the observed space-time data so each LAD has two sets of posterior estimates of its temporal relative risks: one from each model. A third model, the selection model in Table 16.6, selects the posterior estimates from

TABLE 16.6

The Full Specification of the Common Trend Model, the
Area-Specific Model and the Selection Model in BaySTDetect

The common trend model	The area-specific trend model
$y_{it} \sim Poisson\left(E_{it} \cdot \mu_{it}^C\right)$	$y_{it} \sim Poisson\left(E_{it} \cdot \mu_{it}^{AS}\right)$
$\log\left(\mu_{it}^C\right) = \alpha + \left(S_i + U_i\right) + v_t$	$\log\left(\mu_{it}^{AS}\right) = H_i + \xi_{it}$
$\alpha \sim Uniform\left(-\infty, +\infty\right)$	$H_i \sim N\left(0, 10000\right)$
$S_{1:N} \sim ICAR\left(W_{sp}, \sigma_S^2\right)$	$\xi_{i,1:T} \sim ICAR\left(W_{RW1}, \sigma_{\xi,i}^2\right)$
$U_i \sim N\left(0, \sigma_U^2\right)$	$\log\left(\sigma_{\xi,i}^2\right) \sim N\left(a, b^2\right)$
$v_{1:T} \sim ICAR\left(W_{RW1}, \sigma_v^2\right)$	$a \sim N\left(0, 10000\right)$
	$b \sim N_{+\infty}\left(0, 2.5^2\right)$
σ_S, σ_U and σ_v each has the $N_{+\infty}\left(0, 10\right)$ prior	

The selection model

$$y_{it} \sim Poisson\left(E_{it} \cdot \mu_{it}\right)$$
$$\log\left(\mu_{it}\right) = z_i \cdot \log\left(\mu_{it}^C\right) + \left(1 - z_i\right) \cdot \log\left(\mu_{it}^{AS}\right)$$
$$z_i \sim Bernoulli\left(0.95\right)$$

either Model 1 or Model 2 to fit the observed count data for each area. Specifically, under this third selection model, a binary model indicator, z_i, is introduced for each LAD i so that the estimates from Model 1 are selected if $z_i = 1$ or the estimates from Model 2 are selected if $z_i = 0$. The posterior probability of selecting the common trend model, $f_i = \Pr(z_i = 1 | \text{data})$, is then calculated. A small value of f_i would indicate that the trend pattern of LAD i is unlikely to follow the pattern of the common trend represented by v.

We comment on the choice of prior models for the two models. Model 1 specifies the BYM model on the overall spatial component and the RW1 prior on the overall time trend, the same model formulation as specified in Eq. 14.5 in Chapter 14. The two weights matrices, W_{sp} and W_{RW1}, define the spatial neighbourhood structure and the temporal neighbourhood structure for information sharing across space and time, respectively. For W_{sp}, the first order rook's spatial contiguity is used so that the off-diagonal entry $w_{ij} = 1$ if LADs i and j share a common border and $w_{ij} = 0$ otherwise (see Section 4.2). See Section 12.3.4 for the specification of W_{RW1}. The focus of Model 2 is on capturing unusual time trends. Under Model 2, the disease rates of each area can behave differently from those of any other area. Thus, information borrowing across areas is used sparingly in Model 2. For example, a weakly informative prior $N(0, 10000)$ is independently assigned as the prior distribution for each of the area-specific intercepts (representing the overall log relative risk of each area), H_i. This prior choice ensures no information is shared spatially. Similarly, the time trend pattern of each LAD, $\xi_i = \left(\xi_{i1}, \ldots, \xi_{iT}\right)$, is modelled "almost independently" via the RW1 prior. The variance of the RW1 model, $\sigma_{\xi,i}^2$, is also specific to each LAD so that the amount of temporal smoothing in one area can be different from the amount of temporal smoothing in any other area. This flexibility allows the model to capture some abrupt changes (e.g. a sudden increase or decrease in disease rates) in a particular area. For that area, a large $\sigma_{\xi,i}^2$ – thus less temporal smoothing – is required (see also Section 15.4.1). However, with only eight time points, information sharing is needed in order to estimate these area-specific RW1 variances reliably. Thus, the log variances are modelled hierarchically, i.e. $\log(\sigma_{\xi,i}^2) \sim N(a,b^2)$, as opposed to modelling them independently. Modelling the *log* variances ensures that the variances are strictly positive. A weakly informative prior $N(0,10000)$ is assigned to a, the mean of the log variances, whilst a moderately informative prior $N_{+\infty}(0,2.5^2)$ is used for b, the standard deviation of the log variances. This prior choice on b assumes that roughly 10% of areas will display local temporal variability that is 10 times larger than the median of all local variances. That is, $\Pr(\sigma_{\xi,i}^2 > 10 \times \exp(a)) \approx 0.1$ under the prior that $b \sim N_{+\infty}(0,2.5^2)$ and for any value of a. (See Exercise 16.8 for the verification of this statement using WinBUGS.) Model 2 gives an example of modelling variances hierarchically. Typically, the mean of the hyperprior distribution (i.e. a in the case here) can be estimated with minimal prior information (thus a vague prior can be used). Additional information through the use of a moderately informative prior is typically needed to help the estimation of the variance of the log variances (i.e. b^2 here).

Under the selection model, the binary model indicator, z_i, has a Bernoulli prior distribution, namely, $z_i \sim \text{Bernoulli}(0.95)$. This prior specification suggests that, prior to seeing the data, there is a high probability of 0.95 that any one of the areas in the study region would select the common trend model (i.e. with $z_i = 1$), implying only a small number of areas would be expected to display unusual local time trends. We will return to this at the end of the section. Figure 16.11, adapted from Fig. 1 in Li et al. (2012), illustrates the modelling framework of BaySTDetect schematically.

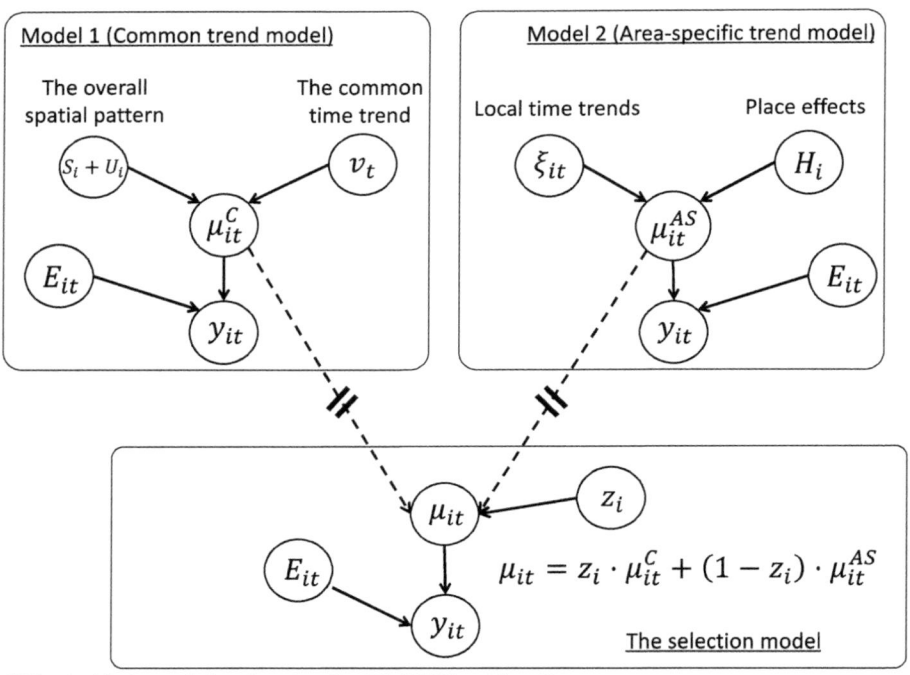

* The bolded equal sign denotes the WinBUGS *cut* function.

FIGURE 16.11
A schematic diagram of BaySTDetect adapted from Fig. 1 in Li et al. (2012).

16.4.3 Detecting Unusual Areas with a Control of the False Discovery Rate

The posterior model probability $f_i = \Pr(z_i = 1 | \text{data})$, calculated as the posterior mean of the model indicator z_i, indicates how likely LAD i is to follow the common trend. A small value of f_i would signal that the corresponding area shows an unusual time trend. With some chosen threshold value C, we can declare LAD i to have an unusual time trend if $f_i \leq C$. This quantity, f_i, also has another interpretation, where f_i can be thought of as the probability of false detection or what Wakefield (2007) terms the "Bayesian false discovery probability". So, in addition to assessing whether the time trend of LAD i is unusual or not, f_i can be used in conjunction with the method of Newton et al. (2004) and Ventrucci et al. (2011) to control the false discovery rate (FDR), the proportion of false positives within the set of areas detected to be unusual by the method. The FDR controlling procedure goes as follows. We choose the detection threshold C such that the average probability of false detection (the average value of f_i amongst all the areas declared to be unusual) is less than or equal to some predefined level η. Suppose k areas are so defined, this procedure aims to ensure that, on average, no more than $k \cdot \eta$ of these detected areas in fact follow the common trend, that is, are false positives. Li et al. (2012) investigated how well this procedure controls the number of false positives through a simulation study, part of which we will look at later on in the section. We will also illustrate how to apply this procedure through an example.

16.4.4 Fitting BaySTDetect in WinBUGS

The WinBUGS code for fitting the detection method BaySTDetect is given in Figure 16.12. We comment on two key features of this WinBUGS implementation. First, the three different models in BaySTDetect, the common trend model, the area-specific trend model and the selection model, are fitted jointly in WinBUGS. In particular, each model is fitted to the same set of outcome data so that each area has two sets of relative risk estimates, $\mu_{i,1:T}^{C} = \left(\mu_{i1}^{C}, \ldots, \mu_{iT}^{C} \right)$ and $\mu_{i,1:T}^{AS} = \left(\mu_{i1}^{AS}, \ldots, \mu_{iT}^{AS} \right)$, coming from the common trend model and the area-specific trend model, respectively. The selection model then selects either $\mu_{i,1:T}^{C}$ or $\mu_{i,1:T}^{AS}$ to fit the observed count data in that area. Fitting the three models jointly allows the uncertainty associated with the estimates of $\mu_{i,1:T}^{C}$ and $\mu_{i,1:T}^{AS}$ to be accounted for in the selection process. The observed count data are replicated three times so that the three models can be fitted simultaneously to the same set of outcome data. Thus, in Figure 16.12, the count data y1[i,t] (Line 17), y2[i,t] (Line 41) and y3[i,t] (Line 77) are all exactly the same. Similar to Application 1 (Section 16.2), the cut function is used to prevent the estimation of $\mu_{i,1:T}^{C}$ and $\mu_{i,1:T}^{AS}$ from being affected by the selection. That means the estimation of the common trend model and the estimation of the area-specific trend model are independent of whichever of the two models is selected. The cut function is used on Lines 27 and 30.

The second feature is the implementation of the hierarchical (prior) model on the log RW1 variances. Specifically, Line 100 specifies a common normal distribution to each of the log variances, which are then transformed back to precisions through Lines 91 and 93 so that they can be used in the car.normal function on Lines 85–86 for temporal smoothing. Lines 107 and 113 specify the hyperpriors on the mean and the standard deviation of the normal distribution on the log variances, respectively.

Line 103 calculates the unconditional variance of the local trend for each area under the area-specific model. Assessing the between-area variability of these variances helps justify the use of area-specific RW1 variances, as opposed to using a single common variance, in the area-specific trend model.

```
1     #  fitting the three models, the common trend model, the
2     #  area-specific trend model and the selection model under
3     #  BaySTDetect simultaneously
4     #
5     #  note that the three sets of space-time outcome values
6     #  y1, y2 and y3 are exactly the same so that the estimates
7     #  from the two models, the common trend model and the area-
8     #  specific trend model, are based on the same set of outcome
9     #  data, which are then used in the selection model to assess
10    #  which model fits the observed data in each area better
11    model {
12    ####################################################
13    #  the selection model
14    ####################################################
15      for (i in 1:N) {    #  loop through all the areas
16        for (t in 1:T) {  #  loop through all the time points
17          y1[i,t] ~ dpois(mu1[i,t])
18          log(mu1[i,t]) <- log.mu1[i,t] + log(E[i,t])
19          #  Lines 21-22 (below) use the binary indicator z[i] to
20          #  select between the two models
21          log.mu1[i,t] <- z[i] * common[i,t]
22                        + (1-z[i]) * specific[i,t]
23          ##  using the cut function to "feed" the estimates from
24          ##  both models into the (above) selection process
25          #  getting estimates from the common trend model
26          #  (see Lines 47-49 below)
27          common[i,t] <- cut(est.common[i,t])
28          #  getting estimates from the area-specific trend model
29          #  (see Line 80)
30          specific[i,t] <- cut(est.specific[i,t])
31        }
32        #  prior for each of the binary indicators
33        z[i] ~ dbern(0.95)
34      }
35
36    ####################################################
37    #  the common trend model (Model 1)
38    ####################################################
39      for (i in 1:N) {
40        for (t in 1:T) {
41          y2[i,t] ~ dpois(mu2[i,t])
42          #  Model 1 with the space-time separable structure
43          log(mu2[i,t]) <- log(E[i,t]) + alpha
44                        + overall.spatial[i]
45                        + overall.temporal[t]
46          #  the estimated log relative risk from Model 1
47          est.common[i,t] <- alpha
48                        + overall.spatial[i]
49                        + overall.temporal[t]
50        }
51        #  overall.spatial[i] is S[i] + U[i] in Table 16.6
52        overall.spatial[i] ~ dnorm(S[i],prec.U)
53        #  recover the spatially-unstructured random effect
54        #  in the BYM model
55        U[i] <- overall.spatial[i] - S[i]
56      }
57      #  prior specifications for Model 1
58      alpha ~ dflat()
```

FIGURE 16.12
The WinBUGS code for the joint fitting of the three models in BaySTDetect.

```
59      S[1:N] ~ car.normal(adj[],weights[],num[],prec.S)
60      overall.temporal[1:T]
61        ~ car.normal(adj.tm[],weights.tm[],num.tm[],prec.v)
62      #  priors on the random effect variances of the BYM model
63      prec.S <- pow(sigma.S,-2)
64      sigma.S ~ dnorm(0,0.1)I(0,)
65      prec.U <- pow(sigma.U,-2)
66      sigma.U ~ dnorm(0,0.1)I(0,)
67      #  prior on the random effect variance of the temporal
68      #  RW1 model
69      prec.v <- pow(sigma.v,-2)
70      sigma.v ~ dnorm(0,0.1)I(0,)
71
72      ##################################################
73      #  the area-specific trend model (Model 2)
74      ##################################################
75      for (i in 1:N) {
76        for (t in 1:T) {
77          y3[i,t] ~ dpois(mu3[i,t])
78          log(mu3[i,t]) <- log(E[i,t]) + H[i] + xi[i,t]
79          #  the estimated log relative risk from Model 2
80          est.specific[i,t] <- H[i] + xi[i,t]
81        }
82        #  modelling each local trend pattern using the
83        #  RW1 prior model
84        #  notice the RW1 precision (prec.xi) is area-specific
85        xi[i,1:T]
86          ~ car.normal(adj.tm[],weights.tm[],num.tm[],prec.xi[i])
87        #  priors on the area-specific intercepts
88        #  (no spatial smoothing)
89        H[i] ~ dnorm(0,0.0001)
90        #  hierarchical modelling of the RW1 log variances
91        prec.xi[i] <- pow(var.xi[i],-1) #  transforming variance
92                                        #  to precision
93        var.xi[i] <- exp(log.var.xi[i]) #  exponentiating a log
94                                        #  variance to obtain a
95                                        #  variance
96        #  Line 100 (below) model the log variances using a common
97        #  normal distribution with unknown mean a and unknown
98        #  precision prec, where prec=1/b^2 and b is the SD
99        #  of the log variances
100       log.var.xi[i] ~ dnorm(a,prec)
101       #  the unconditional variance of each local trend
102       #  pattern
103       uncon.var.xi[i] <- sd(xi[i,])* sd(xi[i,])
104     }
105     #  (hyper)priors on the common normal distribution for
106     #  the log variances
107     a ~ dnorm(0,0.0001)     #  a vague prior on the mean of the
108                             #  log variances
109     #  Line 113 (below) specifies a moderately informative
110     #  prior on the SD of the log variances; the variance of
111     #  the prior distribution is set to be 2.5*2.5, thus the
112     #  precision of this prior is 0.16
113     b ~ dnorm(0,0.16)I(0,)
114     prec <- pow(b,-2)       #  precision of the log variances
115   }
```

FIGURE 16.12 (*Continued*)

16.4.5 A Simulated Dataset to Illustrate the Use of BaySTDetect

To investigate the detection performance of BaySTDetect, Li et al. (2012) conducted a simulation study in which data were simulated based on three different departure patterns of local time trends as illustrated in Figure 16.13. Here, we consider a set of simulated space-time data that follows Pattern 1 (left-hand figure in Figure 16.13). Fifteen LADs (4% of the 354 LADs in England) were chosen to have an unusual trend pattern according to Pattern 1. Figure 13.7 in Chapter 13 shows where these 15 LADs are. All the other 339 LADs have the common trend (the grey line in Figure 16.13). This simulation scenario corresponds to Pattern 1 with original expected counts and a departure magnitude of $\theta = 1.5$ in Li et al. (2012). Exercise 16.9 carries out the data simulation, produces a map of the 15 chosen unusual areas as well as summarising the simulated death counts over the 354 areas across the eight time points.

16.4.6 Results from the Simulated Dataset

As discussed in Section 16.4.3, BaySTDetect detects an unusual local time pattern by applying a detection rule to the posterior probability $f_i = \Pr(z_i = 1 | \text{data})$, which indicates how likely it is that the common trend model is selected to describe the observed data for this area. The smaller f_i is, the more likely it is that the trend of area i differs from the common trend. Table 16.7 illustrates this detection procedure. The second column of Table 16.7 shows the estimated f_i ranked in an ascending order with the associated area IDs shown in the third column. A detection threshold C is sought so that the cumulative average of f_i (the values in the fourth column) does not exceed η, the FDR control level that is set by the analyst. When $\eta = 5\%$, C is chosen to be 0.11117, the f_i of area 99 with rank 7, so that the seven areas with f_i values that are less than or equal to this chosen threshold are declared to be unusual while the remaining 347 areas are considered to be usual, each following the same time trend (the fifth column in Table 16.7). In a simulation setting, since we know which areas have unusual time trends and which ones do not, we can classify each decision into one of the four types: a true positive, a true negative, a false positive or a false negative (see the note in Table 16.7 for description). All the seven areas detected to be unusual are those that have an unusual time trend, by construction, and thus they are true positives. This gives us eight false negatives, the eight unusual areas that are not detected as such by the model. None of the areas that followed the usual, common trend were assigned to the

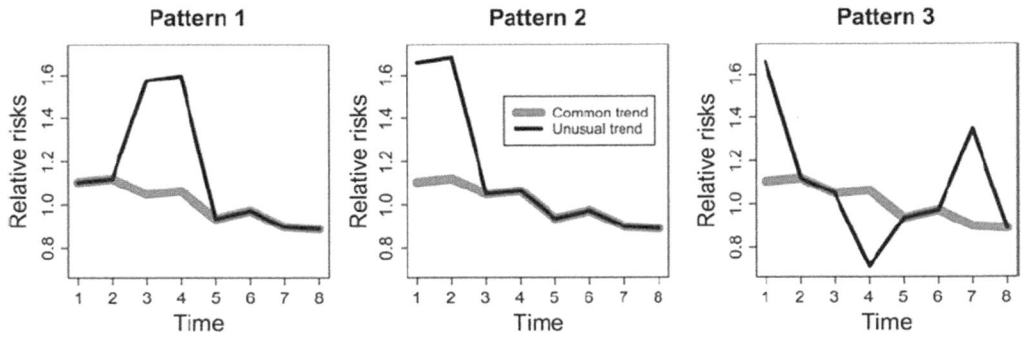

FIGURE 16.13
Three different departure patterns of the local trend (the black line) from the common trend (the grey line). The departure magnitude in each case is 1.5 – see Exercise 16.9 for more detail.

TABLE 16.7

The Detection Rule with a Control of FDR Chosen at the $\eta = 5\%$ Level Applied to the Simulated Data with 15 Areas Chosen to Have an Unusual Time Trend

Rank	Ordered f_i	Area ID	Cumulative Average of f_i	Model Decision	Decision Type[a]
1	0.00017	240	0.00017	Unusual	TP
2	0.01667	190	0.00842	Unusual	TP
3	0.02633	140	0.01439	Unusual	TP
4	0.03333	247	0.01913	Unusual	TP
5	0.06250	169	0.02780	Unusual	TP
6	0.06433	205	0.03389	Unusual	TP
7	0.11117	99	0.04493	Unusual	TP
8	0.11317	222	0.05346	Usual	FN
9	0.12183	186	0.06106	Usual	TN
10	0.15267	201	0.07022	Usual	FN
11	0.16433	136	0.07877	Usual	FN
...					

For illustration, only the first 11 areas with the smallest f_i are shown.

[a] TP = a true positive – an area with an unusual time trend that is correctly detected by the model; TN = a true negative – an area with a usual (common) trend that is correctly identified by the model; FP = a false positive – an area that follows a common time trend but is falsely declared by the model that it has an unusual time trend; FN = a false negative – an area that has an unusual time trend but is not detected by the model.

unusual trend category so that the number of false positives is 0, meaning that all the 339 areas with the same common trend are correctly identified as usual areas by the model. So, the number of true negatives is 339.

To assess the detection performance of the method, the above detection results can be summarised in terms of sensitivity (or power) – the percentage of the 15 unusual areas that are detected – and empirical FDR – the percentage of false positives amongst the detected areas. A good detection method would be expected to achieve a high sensitivity whilst keeping the empirical FDR as close as possible to the level set by the analyst. Table 16.8 shows the results with η set at different levels. Controlling FDR at the 5% level, just under

TABLE 16.8

Summary of the Detection Performance of BaySTDetect When the FDR Control is Set at Different Levels

Detection performance	Detection results with the control of FDR (η) set at			
	5%	10%	15%	20%
Sensitivity	47%	67%	67%	73%
Empirical FDR	0%	17%	29%	35%

With 15 areas set to have unusual time trends, sensitivity is calculated as $\left(\dfrac{TP}{15}\right) \times 100$, and the empirical FDR is calculated via $\left(\dfrac{FP}{TP + FP}\right) \times 100$.

half (seven) of the 15 selected unusual areas are detected, but none of the detected areas are false positives so the empirical FDR is 0%. An increased level of η yields a larger threshold for detection (e.g. $C = 0.11117$ with $\eta = 5\%$ and $C = 0.42017$ with $\eta = 15\%$), thus leading to higher power in detecting the unusual areas (Table 16.8). However, this also leads to more false positives. At the level of $\eta = 20\%$, BaySTDetect detected 17 areas to be unusual, amongst which 11 areas truly possess an unusual trend whilst six have the common trend. The latter gives the 35% empirical FDR in Table 16.8. Compared to the results from SaTScan in Section 13.4.2.1 (using the same set of data), BaySTDetect performs much better.

There are two points worth noting here. First, a detection method is liable to make two kinds of errors: (a) declaring an area as unusual when it actually follows the common time trend, i.e. a Type I error and (b) declaring an area as usual (normal) but its time trend is actually different from the common time trend, i.e. a Type II error. Increasing the detection threshold leads to an increase in detection power (that is, identifying an area as unusual when it is unusual) but increases the chance of a Type I error (that is, identifying an area as unusual when it is not). The detection rule discussed in Section 16.4.3 aims to allow the analyst to choose a detection threshold with the confidence that the proportion of false positives in the set of detected areas is, *on average*, controlled at the level set by the analyst. Here, "on average" means that this FDR control is a long-run behaviour, so that if we apply this detection method over many simulated datasets, then the mean of the set of empirical FDR values is close to the value that we set. Thus, a well-performing method should yield an empirical FDR as close as possible to the controlled level, but not 0. Second, the results presented above are only from a single set of simulated data. All the steps – data simulation, model fitting and result summary – need to be replicated in order to fully assess the detection performance (or long-run behaviour) of a detection method. Exercise 16.10 asks the reader to repeat the simulation over a number of times then to summarise the results. We will briefly summarise the results from Li et al. (2012) in the next section.

Finally, Table 16.9 shows that the variability of the local trends estimated from the area-specific trend model differs considerably between the unusual areas detected by the model and the ones that are considered to be usual. For the former, the variances of the estimated local trends tend to be higher, as we would expect (see Figure 16.13). Such variability supports the use of area-specific RW1 variances under the area-specific trend model.

16.4.7 General Results from Li et al. (2012) and an Extension of BaySTDetect

Li et al. (2012) compare BaySTDetect with Kulldorff's space-time scan test across the three different departure patterns illustrated in Figure 16.13, showing that the former method had greater power even when the data were sparse (i.e. cases are few in number). Their

TABLE 16.9

Comparing the Variances of the Estimated Local Trends Between the Seven Areas that Are Detected as Unusual (with the FDR Set at 5%) and the 347 Areas that Are Considered to Be Usual

	Summarising the Posterior Means of the Variances of the Estimated Local Trends
Within the seven areas detected as unusual	Mean = 0.18 with min = 0.16 and max = 0.21
Within the 347 areas considered as usual	Mean = 0.09 with min = 0.06 and max = 0.20; the posterior means of the variances of 80% of these areas are within 0.07 and 0.12

The reported numbers are based on the posterior means of the variances of the estimated local trends from the area-specific model (i.e. the posterior means of the quantities `uncon.var.xi[i]` on Line 103 in Figure 16.12).

results also show that the detection power of BaySTDetect is higher when the expected number of cases is large (so that the space-time data are less sparse) and/or the departure magnitude is higher (i.e. the unusual local trend pattern is further away from the common trend pattern). Regarding the type of departure pattern, Patterns 2 and 3 (as in Figure 16.13), with departures occurring at (or close to) the end(s) of the time series, were easier to detect compared to Pattern 1, where the departure was in the middle of the time series. This is because the bi-directional nature of information borrowing from the RW1 prior model (see Section 12.3.4 and also Section 15.6) imposes more temporal smoothing on departure Pattern 1.

BaySTDetect works better if the unusual areas are scattered across the study region without clustering since all areas are treated separately under the area-specific model and the model indicators, $z_1, \ldots, z_N$, are assumed to be uncorrelated from one area to another. However, the simulation results suggest that the spatial-temporal scan test in SaTScan, which places circular windows across the study region, tends to achieve higher power if the unusual areas are clustered (see Section 13.4.2).

When applied to the COPD mortality data from 1990 to 1997, the contrast between the two surveillance methodologies is striking (see Figure 16.14). Setting an FDR control at $\eta = 5\%$, BaySTDetect identified five LADs with unusual time trends. The uncertainty in this detection result is low as, on average, only 0.25 (=5%×5) of these five detected areas would be expected to be false detections. Of these five detected LADs, Tower Hamlets and Lewisham were of interest, as both display an increasing time trend of COPD mortality whilst the common trend is decreasing. Various enhanced services were introduced (from 2008) in Tower Hamlets to tackle such issues. The authors suggested the use of detection methods such as BaySTDetect would have identified such problems earlier. On the other hand, SaTScan declared a considerably larger number of areas to have unusual time trends.

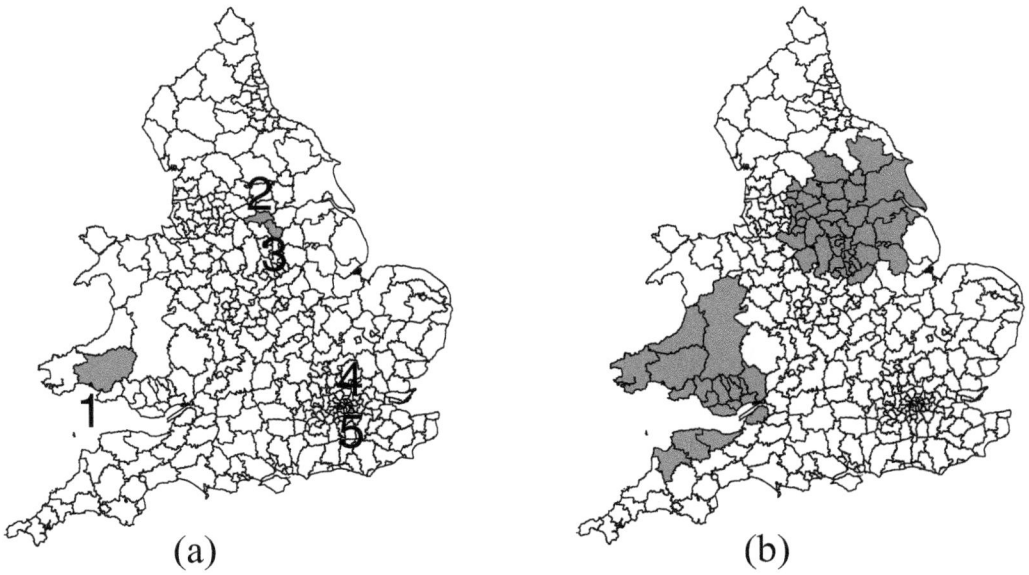

(a) (b)

FIGURE 16.14
Location of the identified unusual LADs. (a) BaySTDetect with a control of FDR at 5% (1 Carmarthenshire; 2 Barnsley; 3 Rotherham; 4 Tower Hamlets; 5 Lewisham). (b) The space-time permutation test in SaTScan with p-value threshold set at 0.05. Taken from Li et al., 2012.

In addition to the issue of having potentially large number of false positives, neither Tower Hamlets nor Lewisham were detected. The map of the detected areas by SaTScan also reflects the use of circular windows (Figure 16.14).

One possible extension to BaySTDetect would be to develop a sequential fitting so that the time series data are fed into the model one time point at a time. This would allow the analyst to both pinpoint when the departure occurs and estimate the magnitude of the departure. Such results would help in the prompt initiation of any public health response. Li et al. (2012) suggested that BaySTDetect is suitable for detecting unusual time trends using data with fewer than 10 time points. When dealing with longer time series data, changing the model indicator z_i to z_{it} would introduce more flexibility, as departure may occur at multiple time points. Boulieri et al. (2018) has extended BaySTDetect in a number of directions, one of which is to perform the model selection for each area at each time point, i.e. using z_{it}. These binary model indicators are modelled hierarchically so that each z_{it} comes from a Bernoulli distribution with parameter ϕ_{it}, which is modelled as

$$\text{logit}(\phi_{it}) = \text{logit}(\tau) + \pi_i + \delta_t \tag{16.3}$$

In Eq. 16.3, the two sets of random effects, $\pi = (\pi_1, \dots, \pi_N)$ and $\delta = (\delta_1, \dots, \delta_T)$, are structured spatially via the ICAR prior and temporally via the RW1 prior, respectively. In addition to borrowing information in estimating these space-time selection probabilities, imposing spatial and temporal structures on ϕ_{it} and in turn on z_{it} can achieve high detection power when departures tend to cluster over space and/or time. Boulieri et al. (2018) studied the detection performance of the proposed model through extensive simulation and applied their method to identify unusual local time trends of road traffic accidents across the 326 districts in England between 2005 and 2015. Their model implementation was carried out in OpenBUGS.

16.5 Application 4: Investigating the Presence of Spatial-Temporal Spillover Effects on Village-Level Malaria Risk in Kalaburagi, Karnataka, India

16.5.1 Background and Study Objective

In Section 9.4 we presented a study of malaria incidence in Kalaburagi taluk, a small administrative area in southern India. We reported results from a cross-sectional (purely spatial) analysis of incidence data for 139 small areas (villages) in June and July 2012. The aim was to estimate the spatial variation in malaria risk and examine the association with spatial variation in rainfall levels. Malaria is an infectious disease carried by the Anopheles mosquito, which acts as a vector in the transmission of the disease. Spatial variation in the incidence of malaria is therefore likely to be a function of not only village (place-specific) characteristics that are associated with the occurrence of the disease, as investigated in Section 9.4, but also spatial and temporal *spillover* effects. When data from other time intervals become available, one can investigate the presence (or absence) of such spillover effects, which is the aim of the analysis here.

In this analysis, *temporal* spillover effects refer to the size of the infected pool of people in a village at some past time point, say $t-1$, influencing the number of new cases in the current time period t in the same village. *Spatial* spillover effects are associated with population mixing (bringing infected and non-infected individuals, including those from different villages, into sufficiently close proximity) and the geographical range of mosquitoes that carry the parasite. A spatial-temporal analysis that incorporates some of the features of an econometric model, as described in Chapter 10, within a spatial-temporal hierarchical model, allows us to investigate the presence of such spillover effects, providing support for local health services to respond to new outbreaks in a timely fashion.

The objective of this spatial-temporal analysis is to investigate the presence of spillovers: whether the malaria risk in village i at time t is a function of the number of new cases that occurred:

1. In the same village in the previous time period $t-1$ (temporal spillover)
2. In the neighbouring villages in the previous time period $t-1$ (spatial-temporal spillover)

These numbers of new cases in the same village or in the neighbouring villages at $t-1$ can be taken as indicative of the size of the "effective" infected pool at the current time t in village i.

In this study, the time window defined for t is quite large (two months). At this temporal scale, given the speed with which infection can be transmitted between people, resulting in new reported cases of malaria (potentially within two weeks), there could be a "pure spatial spillover", in the spatial econometric sense, with the number of infected individuals at time t in areas that are neighbouring to area i contributing to the number of new cases in area i at the *same time*, t. The same applies to cases arising in the same village, that is cases arising early in the two-month time period inducing new cases in the village in the sense of an intra-village "contagious" or "local multiplier" process. These are not aspects of the process we investigate here, although part of the model does accommodate, to some extent, the effect of pure spatial spillovers (see the specification of the random effects below).

16.5.2 Data

The data analysed are the 1633 reported cases of malaria between February 2012 and January 2014 (inclusive) in Kalaburagi taluk, one of the seven administrative areas within Kalaburagi District in South India. The spatial units of analysis are the 139 small areas within Kalaburagi taluk, consisting of 138 small villages with an average population of approximately 2000 people, and Kalaburagi City, which has a population of over half a million. For each year, the malaria cases are aggregated to six two-monthly time periods, Feb–Mar, Apr–May and so on. This time aggregation is applied so that the number of cases within each time window is sufficiently large for analysis while the time window is reasonably short for informing timely responses to potential outbreaks.

In Sections 13.3 and 13.4.1.2, we carried out some exploratory analysis on this spatial-temporal dataset. Over the two-year period, the mean number of cases remains stable, around one per village per two months (Table 13.1). While there were some villages with large numbers of reported cases, most of the villages had zero cases throughout the two-year period (Table 13.1 and Figures 13.3 and 13.4). Thus, when modelling, we need to deal with two issues arising with this data: the heterogeneous risk distribution and the disproportionate number of zero observed values in the dataset. In Section 9.4, we described a Bayesian two-component mixture, zero-inflated Poisson model to handle these two issues in modelling a subset of the data (June–July 2012). As an extension of the spatial model, the next section presents a Bayesian two-component mixture, zero-inflated Poisson model with random effects for modelling the spatial-temporal malaria data with the aim of examining for the presence of spatial and temporal spillovers.

16.5.3 Modelling

To accommodate the excessive amount of zeros in the observed count data, y_{it}, the number of malaria cases observed in village i during time t is modelled using a zero-inflated Poisson (ZIP) model. The ZIP model describes an observed zero count by two distinct (underlying) processes. One process gives rise to a zero count because the village is risk-free, while the other generates a zero count by chance. Under the latter process, the underlying risk of malaria in that village is not exactly zero (disease-free), but there happened to be no observed cases during the specific time window. See Section 9.4.5 for a more detailed discussion of the two zero-generating processes. Eq. 16.4 shows the ZIP model applied to y_{it}:

$$\Pr\left(Y_{it} = y_{it} \mid g_i, \mu_{it}\right) = \begin{cases} g_i + \left(1 - g_i\right)e^{-\mu_{it}} & \text{when} & y_{it} = 0 \\ \left(1 - g_i\right)\dfrac{\mu_{it}{}^{y_{it}}}{y_{it}!}e^{-\mu_{it}} & \text{when} & y_{it} = 1, 2, \ldots \end{cases} \qquad (16.4)$$

where $g_i \sim$ Bernoulli(ϕ) is modelled as a binary 0/1 indicator. When $g_i = 1$, village i is modelled as a "disease-free" village, where the risk of malaria is equal to 0 for the entire study period. The parameter ϕ represents the proportion of villages that are considered to be disease-free. When $g_i = 0$, the observed case count y_{it} is modelled through a standard Poisson distribution with mean μ_{it}, which is the underlying malaria risk in village i in time window t. Eq. 16.5 shows how we model the (log) space-time risk:

$$\log\left(\mu_{it}\right) = \log\left(pop_i\right) + a_{z_i} + \beta_{\text{rain}} \cdot x_{it,\text{rain}} + \beta_{\text{prev}} \cdot y_{i,t-1}$$
$$+ \beta_{\text{NBprev}} \cdot \left(W_{sp}y\right)_{i,t-1} + U_i + v_t + \varepsilon_{it} \qquad (16.5)$$

where the definitions of various terms in Eq. 16.5 are given in Table 16.10.

We now explain some of these factors in turn. To accommodate the two distinctive groups of villages (see Figure 13.4 and the interpretation), each of the 139 villages is allocated to either a low-risk group (i.e. cluster of villages) or a high-risk group, with a_1 and

TABLE 16.10

Definitions/Interpretations of the Terms in Eq. 16.5

pop_i	The population size of village i (assumed constant over the study period)
a_{z_i}	One of the two intercepts chosen for this village (the two intercepts represent the overall residual risk levels of the two clusters of villages) – see main text for more information
$\beta_{\text{prev}} \cdot y_{i,t-1}$	The effect of the number of new cases that occurred in the same village in the previous time window
$\beta_{\text{NBprev}} \cdot (W_{sp}y)_{i,t-1}$	The effect of the total number of new cases that occurred in the neighbouring villages in the previous time window; and y is an $N \times T$ matrix containing all the observed case counts
$\beta_{\text{rain}} \cdot x_{it,\text{rain}}$	The effect of rainfall
$U = \left(U_1, \ldots, U_N\right)$ $v = \left(v_1, \ldots, v_T\right)$ $\varepsilon = \left(\varepsilon_{11}, \ldots, \varepsilon_{NT}\right)$	Three sets of random effects representing the place effects, the overall time trend and overdispersion; here $N = 139$ and $T = 12$

a_2 representing the overall residual risks of these two groups respectively. A constraint is imposed on these two intercepts where $a_1 \leq a_2$. The cluster allocation is done via the village-specific binary indicator z_i. Each of these indicators for cluster allocation follows a common discrete uniform distribution, with $\Pr(z_i = 1) = p$ and $\Pr(z_i = 2) = 1 - p$, and the parameter p represents the proportion of villages that are in the first cluster. A vague $\text{Beta}(1,1)$ prior is assigned to p. This is the part of the model that specifies a two-component mixture; see Section 9.4.4 for more information on this part and the interpretation of the two clusters.

The regression coefficient β_{prev} quantifies the effect of $y_{i,t-1}$, the number of new cases occurring at the previous time point in the same village, on this village's risk at the current time point. Here, $y_{i,t-1}$ acts as a proxy measure of the pool size. Similarly, β_{NBprev} measures the effect of $(W_{sp}y)_{i,t-1}$, which is the number of new cases in a village's spatial neighbours at the previous time point, on the current risk in that village. The term $(W_{sp}y)_{i,t-1}$ is taken to be the sum of all the new cases occurring in $t-1$ in the neighbouring villages of i but excluding those in i. The spatial weights matrix W_{sp} is defined through the first order rook's move contiguity (Section 4.2) but is not row-standardised (Section 4.7) because we are calculating the *total number* of cases in the neighbouring villages, as opposed to the average number. The two regression coefficients, β_{prev} and β_{NBprev}, quantify the effects of temporal spillover and spatial-temporal spillover respectively, as set out in Section 16.5.1. The parameter β_{rain} measures the effect of rainfall during the current time period. Rainfall level is included here as a continuous-valued covariate, as opposed to a categorical one as in Section 9.4, since the rainfall level varies a lot more when we have space-time data covering a two-year period. A vague prior $N(0,1000000)$ is assigned independently to the above regression coefficients.

The three sets of random effects, U, v and ε, are included to account for the additional spatial, temporal and spatial-temporal variability, respectively, that is not explained by other terms in the model. Such unexplained variability can come from, for example, unmeasured/unobserved covariates or malaria outbreaks due to spatial spillover effects operating within a two-month time window (see Section 16.5.1). Specifically, for $i = 1,\ldots,N$, $U_i \sim N(0,\sigma_U^2)$ forms a set of spatially-unstructured random effects. We do not consider imposing spatial structure on U because nearby villages do not tend to have similar malaria risks (see Figure 13.3). The temporal random effects, v, are modelled using a random walk prior of order 1. Finally, for all i and t, ε_{it} is modelled independently and identically by $N(0,\sigma_\varepsilon^2)$. The WinBUGS code for fitting the spatial model presented in Figures 9.12 and 9.13 can be modified to implement the spatial-temporal model in Eq. 16.5 (Exercise 16.11).

There is high correlation between $y_{1:N,t-1}$ and $(Wy)_{1:N,t-1}$. When covariates are correlated, the coefficient estimates may be sensitive to changes to the regression model, for example, by adding or removing a covariate from the regression model. To investigate the stability of the model estimates, we also fitted a version of the full model presented in Eq. 16.5 without $(Wy)_{1:N,t-1}$.

16.5.4 Results

As discussed in Section 9.4.6, to interpret the classification result, we derive a grouping indicator, $h_i = g_i + (1 - g_i) \cdot (z_i + 1)$, where g_i and z_i are the indicators in the zero-inflation part and the two-component mixture, respectively. This grouping indicator h_i assigns each village i to the "risk-free" ($h_i = 1$), low-risk ($h_i = 2$) or high-risk ($h_i = 3$) cluster (see Section 9.4.6 for full detail on h_i). Figure 16.15 shows the classification of the 139 villages. The proportion of villages with an underlying risk of zero (the white polygons) is estimated to be 47.7% (95% CI: 18.7%–64.0%). 46.0% (29.5%–74.1%) and 6.3% (5.8%–8.6%) of the villages are

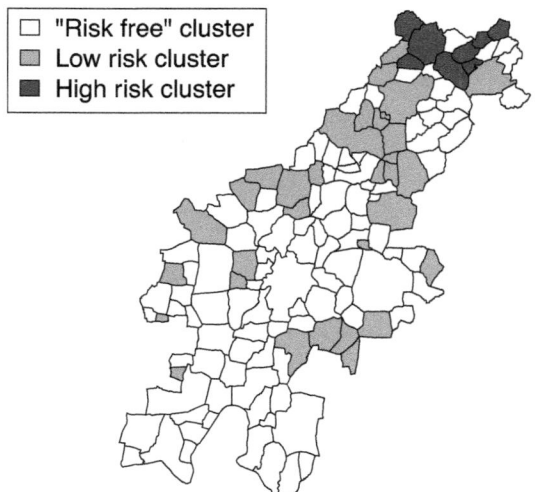

FIGURE 16.15
A classification of the villages based on the spatial-temporal model presented in Eq. 16.5.

allocated to the low (the polygons in light grey) and high (the polygons in dark grey) risk clusters, respectively. To present the uncertainty associated with the classification of each village, we calculate the posterior probability, p_{ik}, of assigning village i to each of the three clusters, and this is defined as

$$p_{ik} = \Pr\left(h_i = k \mid \text{data}\right)$$

for $k = 1, 2$ and 3. Figure 16.16 shows the resulting maps of p_{ik}. The model shows high certainty for all the village assignments to the high-risk cluster and all but four assignments to the low-risk cluster. A careful inspection of Figures 16.15 and 16.16(b) reveals three small polygons, villages with uncertain assignment to the low-risk cluster (shaded "low-risk" in Figure 16.15 but unshaded in Figure 16.16(b)). These three are to be seen in the north (very

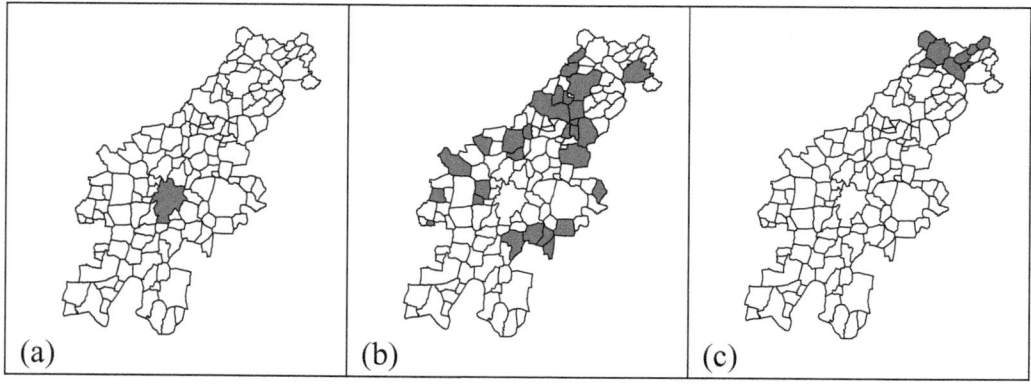

FIGURE 16.16
Three maps showing the posterior probabilities of assigning each village to (a) the "risk-free" cluster, (b) the low-risk cluster and (c) the high-risk cluster. The villages in dark grey are those with $p_{ik} > 0.8$, indicating that the model shows a high level of certainty in their classification.

small polygon in the centre of the low-risk cluster in the north), east and southwest of the map. All three villages had zero cases throughout the two years. The fourth uncertain assignment, village 49, is a relatively large polygon in the west. This village had 22 cases of malaria over two years. The model is uncertain whether to assign it to the low-risk or the high-risk cluster: for this village, $p_{i1} = 0$, $p_{i1} = 0.76$ and $p_{i3} = 0.24$. We will return to this village in the interpretation of Figure 16.18. Compared to the purely spatial analysis presented in Section 9.4.6 (Figure 9.14 in particular), the analysis of the spatial-temporal data yields a much more certain classification of villages into the low- and high-risk clusters.

Almost all assignments of villages to the "risk-free" cluster are uncertain, reflecting the difficulty in confidently stating "the extent to which the excess of zero cases is due to chance on the one hand or environmental factors on the other" (see Section 9.4.5). The model confidently classifies Kalaburagi City into the "risk-free" cluster because there were, apparently, zero reported cases over the study period in this large city with a population of over half a million.

Table 16.11 summarises the posterior estimates of the regression coefficients from both the full model and with the covariate $(Wy)_{1:N,t-1}$ removed. There is very little difference in the parameter estimates between the two models, suggesting that the correlations between $y_{1:N,t-1}$ and $(Wy)_{1:N,t-1}$ do not cause a problem. We will now look at the estimates from the full model in detail. As expected, the posterior estimate of β_{rain} shows the strong effect of rainfall on the risk of malaria. A one millimeter increase in the average rainfall during a two-month window, within the range of rainfall levels observed in this dataset, is found to be associated with a 5.3% (95% CI: 0.9%–9.6%) increase in malaria incidence during the same time window (see Section 5.4.2 in Chapter 5 on how this is calculated).

The posterior mean of $\exp(\beta_{prev})$ is 1.012, with a 95% CI of (1.003, 1.021), giving evidence of a within-village transmission over time. Within the same village, one additional case occurring during the previous time window increases the risk in the current time window by 1.2% (95% CI: 0.3%–2.1%). The estimated $\exp(\beta_{NBprev})$ does not differ from 1, as the posterior mean is close to 1 and the 95% CI contains 1. The analysis shows little evidence of an impact on a village's current risk arising from the number of new cases in its neighbouring villages in the previous time window.

To see how well the model fits the observations, we carry out the following posterior predictive checks. Here we summarise the results from the posterior predictive checks but refer the reader to Exercise 16.11 for detail. Figure 16.17 assesses how well the model accounts for the observed values of zero by comparing the total number of observed zero values to the posterior distribution of the predicted number from the model. The model reproduces the total number, 1513, of zero observations reasonably well.

For each of the 37 villages where there was at least one case occurring during the two years, 32 villages have their posterior predictive p-values (Section 5.4.2) lying between

TABLE 16.11

Posterior Summary (Posterior Mean and 95% CI) of the Regression Coefficients from Two Models: The Full Model Presented in Eq. 16.5 and the Full Model Without the Spatially- and Temporally-Lagged Covariate $(Wy)_{i,t-1}$

Parameter	Interpretation	The full model	Without $(Wy)_{i,t-1}$
$\exp(\beta_{rain})$	Rainfall effect	1.053 (1.009–1.096)	1.055 (1.008–1.106)
$\exp(\beta_{prev})$	Temporal risk spillover	1.012 (1.003–1.021)	1.012 (1.003–1.021)
$\exp(\beta_{NBprev})$	Spatial-temporal risk spillover	0.998 (0.992–1.004)	$\beta_{NBprev} = 0$

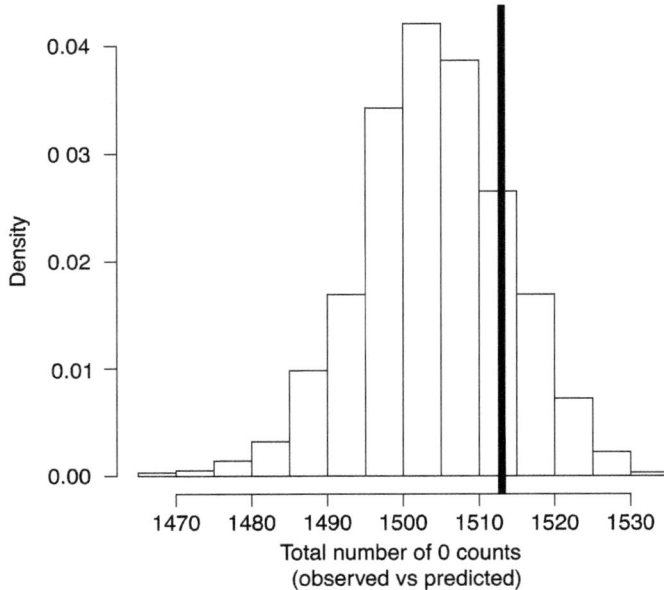

FIGURE 16.17
Comparing the total number of zero-value observations (the black vertical line) to the posterior distribution (the histogram) of this quantity predicted from the model.

0.05 and 0.95, indicating an adequate fit of the observed case counts from the model. However, there are five villages where, within each, some of the posterior predictive p-values are below 0.05, meaning that too few cases of malaria are generated from the model. Figure 16.18 shows the time sequence of the observed and predicted cases for these five villages. Specifically, these five villages are all classified into the lower risk cluster, and the poor fit mainly comes from the occasional occurrence of cases (see Figure 16.18). The malaria cases occur sporadically in these villages and the model does not deal with this feature particularly well. In particular, the poor fit to the case count of village 49 at time point 4 results in its uncertain assignment between the low-risk cluster and the high-risk cluster.

16.5.5 Concluding Remarks

From the purely spatial analysis described in Section 9.4, we noted how fitting a particular model (a mixture model with zero-inflation) provided a classification of the villages that could be used to help identify villages for targeted treatment as part of a strategy for disease control and prevention (Section 9.4.7). In this section, using space-time data, we have been able to refine the classification of high-risk villages (Figure 16.15) and make an assessment of the part played by temporal and spatial spillovers in the spread of malaria that may also inform any strategy for disease control.

We have developed and applied a Bayesian spatial-temporal model with a two-component mixture and zero-inflation to model the time sequence of observed counts of malaria cases at the village (small area) level. The analysis was motivated by the investigation of the presence of risk spillovers in time and/or space: whether the malaria risk in one village in the current time window is affected by the number of new cases that occurred (a) in

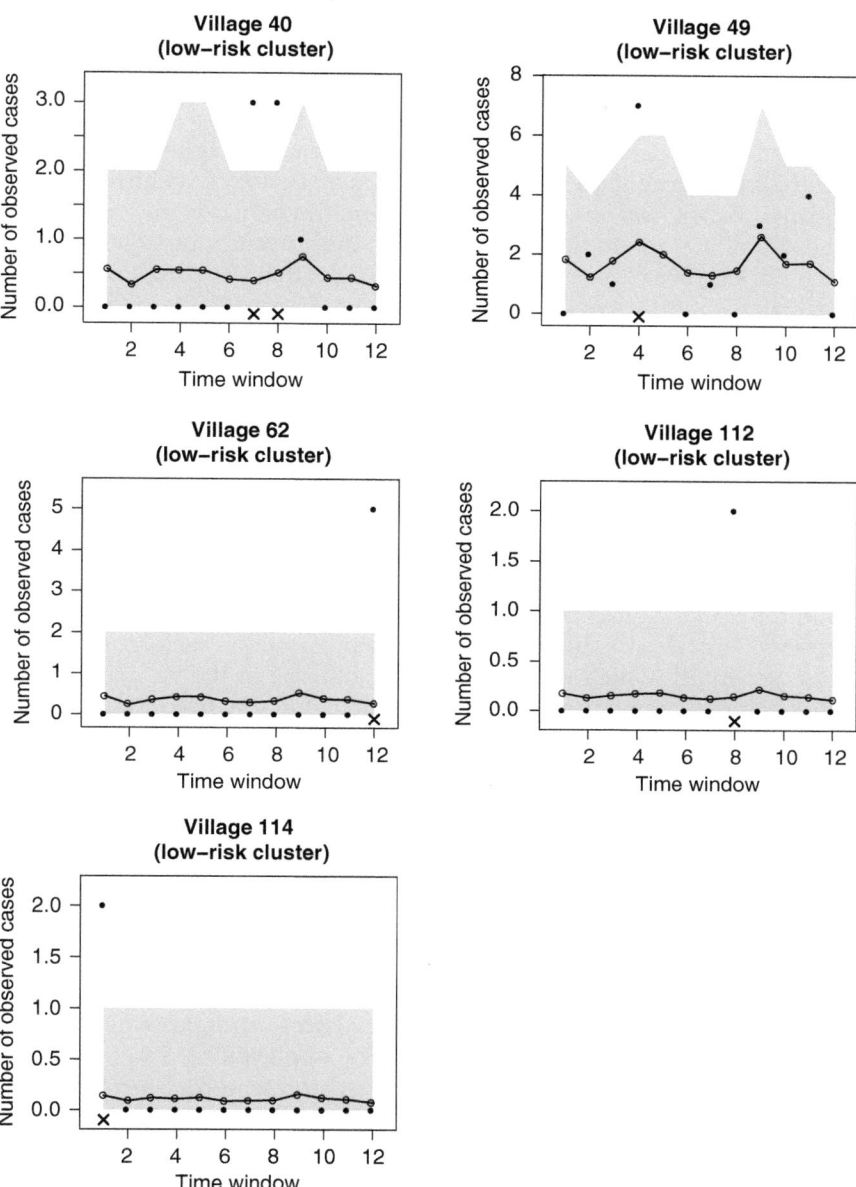

FIGURE 16.18
The observed and predicted time sequence of malaria cases for each of the five villages where at least one case occurred during 2012–2014 and the model shows signs of a poor fit to some of the data. The black dots denote the observed numbers of cases while the open circles and the black curve show the posterior means of the fitted values. The 95% uncertain region is shown in grey. For each village, the cluster assignment is given in the title. The lack of fit is indicated by the cross(es) at the bottom of each plot.

the same village in the previous time window and (b) in the neighbouring villages in the previous time window. We suggested that the inclusion of these covariates may provide a better indication of the size of the infected population to which individuals in area i may be exposed and which may increase their vulnerability to the disease.

While there is little evidence of a spillover of risk from the neighbouring villages in the previous time window, the modelling result shows evidence of within-village transmission.[4] A large number of new cases occurring during the previous time window is shown to elevate the risk of malaria in the same village in the current time window. This finding reinforces the importance of early targeting of villages that have high numbers of newly infected individuals but does not endorse the need to extend the targeting to nearby villages unless, of course, they too have a large number of new cases. The classification result (Figure 16.15) supplements such a targeting approach by identifying the high-risk villages, demonstrating the relevance of this modelling to informing/supporting the prevention and control of malaria.

We have considered spatial-temporal spillovers of risk by including temporally-lagged and spatially- and temporally-lagged *outcome* values as covariates. A similar idea can be applied to investigate the spatial-temporal spillovers of an *explanatory* (or exogenous) variable. For example, in this application, we have found that a high level of rainfall during the current time window increases the risk of malaria during the same time window. One can add a set of temporally-lagged (or a set of spatially- and temporally-lagged) rainfall values as covariates. This would allow us to investigate questions such as: does a prolonged period of heavy rainfall (which may result in an increase in the number of mosquitoes) increase a village's risk of malaria; is a village's risk of malaria affected by higher levels of rainfall occurring in the neighbouring villages (where, as a result, more mosquitoes may appear, which in turn may facilitate the transmission of malaria both within those villages and across the neighbouring villages). Exercise 16.12 investigates some of these ideas.

16.6 Conclusions

This chapter has presented four applications in spatial-temporal data analysis using Bayesian hierarchical models to obtain reliable small area estimates when data are sparse and to deal with various forms of uncertainty. The applications also presented various contrasting challenges in terms of how to handle spatial-temporal dependence and spatial-temporal heterogeneity. The first three applications all shared the common inferential goal of detecting "unusual" areas, which can be characterised as areas with local time trends that differ markedly from a "common" time trend. In Application 1, the unusual areas were defined as those that had been "treated" by the No Cold Calling policy of Cambridgeshire Police so that the process of detection was concerned with establishing whether the treatment had had a marked effect on crime risk relative to a group of control (untreated) areas. In Applications 2 and 3 we were concerned with surveillance with the aim of spotting unusual areas, wherever they might be – again, areas that differ markedly from the common trend. In all cases, the common trend was identified using the space-time separable structure, and the unusual areas were identified from the space-time interaction component as discussed in Chapters 14 and 15.

---

[4] Data at a finer temporal resolution, such as weekly data, might shed further light on the importance and timing of different types of spillovers in an outbreak of malaria.

Another common feature in the first three applications is that we are carrying out many comparisons simultaneously. In Application 1, the NCC scheme was evaluated locally by comparing the data from each of the NCC-COAs against the data from the control group. In Applications 2 and 3, trend detection is carried out by comparing each of the local trends to the common trend. The main concern of multiple comparisons is that some of the areas that are declared as "successful" (in the context of policy evaluation) or "unusual" (in the context of trend detection) are in fact areas in which the policy did not have any beneficial effects or areas with time trends that followed the common trend closely.[5] These wrongly declared areas are referred to as false positives and are associated with a Type I error, an incorrect rejection of a null hypothesis when the null hypothesis is true. As N, the number of tests (or areas as in the case of spatial-temporal modelling) increases, the chance of making a Type I error (of classifying an area as "unusual"/"successful" when it is not) increases. To address the multiple comparisons problem, the Bonferroni correction, a popular frequentist method, lowers the conventional 5% p-value threshold (used to reject/retain the null hypothesis for each comparison) to $5\%/N$, where N is the total number of tests performed. Lowering the p-value threshold is equivalent to widening the confidence interval and thus making each test more likely to retain the null hypothesis (Gelman et al., 2012). This leads to the criticism that the Bonferroni correction is too conservative (Nakagawa, 2004), lacking the power to declare a truly unusual area. Controlling the false discovery rate is a less conservative approach compared to the Bonferroni correction (Benjamini and Hochberg, 1995). The application of this method is illustrated within the disease surveillance context in Section 16.4.3.

More recently, Gelman et al. (2012) presents an interesting argument that the problem of multiple comparisons can be seen as a modelling/estimation problem. If we can model the problem in hand appropriately, model parameters can then be estimated reliably, thus providing a sound basis for decision-making. For example, in the context of trend detection, if we can estimate the difference between each local time trend and the common trend reliably, we can then make a more accurate decision about whether a local time trend is unusual or not. They advocate the Bayesian multilevel/hierarchical modelling framework, the framework that this book promotes. It is the idea of information sharing (shrinkage or partial pooling) that enables us to estimate model parameters more accurately. Moreover, in the spatial and spatial-temporal context, the many comparisons that we carry out are far from independent (as assumed in the Bonferroni correction); rather they are interdependent and correlated due to the properties of the data. The hierarchical modelling framework, as we have demonstrated in this book through both theory and applications, offers us the flexibility to incorporate a wide range of spatial, temporal and spatial-temporal dependence structures that arise from different applied problems.

In policy evaluation, a benefit of adopting the *Bayesian* approach is that evidence of the policy impact is communicated through probability statements (e.g. the posterior probabilities in Figure 16.4 and Figure 16.6). However, whenever a "binary" statement is made (e.g. declaring a local trend as unusual or not), we need to consider the uncertainty associated with the results (see Section 16.4.6). The FDR procedure in BaySTDetect illustrates a way to represent such uncertainty. Finally, a well-thought-out simulation study is often required when examining the performance of a newly developed detection/classification method.

---

[5] The converse can happen too. For example, the policy is estimated to have little/no impact in an NCC-COA, but in fact that COA experienced a reduction of burglary rates in the post-policy period. An area is declared to be usual (following the common trend closely), but in fact it has a different time trend. This is termed a Type II error.

The fourth application was concerned with a very different challenge, which was to identify the presence of spatial and temporal spillover effects in the spread of malaria, an infectious disease. This final application made an attempt to identify elements of a diffusion process unfolding over time both within and between a collection of villages. Similar challenges presented themselves in terms of data sparsity, various forms of uncertainty, spatial-temporal dependence and heterogeneity. But here, unlike the first three applications, we were concerned with using space-time data to get a better understanding of some of the important drivers of a spatial-temporal *process*. The availability of space-time data becomes essential if we are to study space-time processes and to have confidence in the estimation of spillover and feedback effects, as critics of spatial econometric models have drawn attention to (see Section 10.5).

In this application the Bayesian hierarchical model constructed, aside from the differences we have already drawn attention to in Section 10.5.2, shows some of the characteristics that can be seen in a spatial-temporal econometric model that takes the form:

$$Y_t = \alpha \mathbf{1} + \beta_{\text{rain}}\, X_t + \beta_{\text{prev}}Y_{t-1} + \beta_{NB\text{prev}}WY_{t-1} + e_t \tag{16.6}$$

where the vectors Y_t, Y_{t-1}, X_t and e_t are indexed on spatial units and W is a spatial weights matrix. As we saw in Chapters 10 and 11, spatial econometric models extend the standard regression model in ways that (a) recognize the special nature of spatial data and (b) enable the analyst to study the effects of covariates in "other places" on the outcome (spatial spillovers). These covariates can include outcomes in other places (e.g. the SLM model), which leads to the study of both spillover *and* feedback effects. But in Eq. 16.6, there is no simultaneity – no dependence of Y_t on WY_t, for example as encountered in many spatial econometric models – so that we do not encounter any of the problems posed by the need to specify a complicated multivariate distribution jointly for the observed data (see Section 10.5.2).[6] Thus, we can address features of the data (e.g. zero inflation and clustering of villages) through the flexible modelling available in the Bayesian hierarchical framework. Ripley (1981, p.101) refers to models of this type (e.g. in Eq. 16.6) as *multivariate time series*.

In conclusion, when engaging with these sorts of problems drawing on fine-grained spatial-temporal data, there may well be, in the future, convergence in application between space-time econometric models and Bayesian hierarchical space-time models. They become complementary approaches to tackle the same or similar problems. However, it needs to be remarked that using fine grained space-time data to study interaction effects presents unique challenges in terms of data sparsity and various forms of uncertainty, challenges that, as we have seen, Bayesian hierarchical modelling is particularly well-placed to handle.

16.7 Exercises

Exercise 16.1. Evaluate the NCC policy using each of the five control criteria given in Table 16.2. Note that for each of the first three control criteria (ALL COAs, NB and MDI), there is only one control group and hence one control trend for all the COA-level and group-level evaluations. For RateMatch and RateMatch_MDI,

---

[6] In several of the models we presented in Chapter 10, such as the SLM and SEM, this included the presence of a determinant term that was a function of one or more of the spatial parameters to be estimated.

each NCC-COA has its own control group (hence its own control trend). As a result, the WinBUGS code in Figure 16.3 needs to be modified accordingly. (Hint: It helps to first write down the mathematical form of the model when using either RateMatch or RateMatch_MDI as the control criterion. For the WinBUGS implementation, changes need to be made in the baseline model so that the control trend is estimated for each NCC-COA. The control trend in the assessment model (i.e. `gamma.treat` in Figure 16.3) also needs to be linked to the associated NCC-COA.)

Exercise 16.2. Replace the linear impact function in the NCC evaluation model in Figure 16.2 by the step change impact function (see Appendix in Chapter 12) and carry out the model fitting. Interpret the modelling results. Use all the non-NCC-COAs to form the control group in this exercise.

Exercise 16.3. Fit the model presented in Figure 16.2 to the urban NCC areas and the rural NCC areas in Cambridgeshire separately. For this analysis, use the linear impact function and use all the urban (rural) non-NCC areas to form the control trend for the urban (rural) NCC areas.

Exercise 16.4. The R script, `fit_two_step_BUGS.R` (available on the book's website), fits the space-time model presented in Eq. 16.1 to the COA-level burglary count data in Peterborough between 2001 and 2008. That R script calls upon another R script (`two_step_classification.R`, also available on the book's website) to carry out the classification of each COA into one of the nine categories in Table 16.4. Go through the two R scripts to ensure all the steps taken there can be clearly mapped onto the description of the model (Section 16.3.2) and that of the classification (Section 16.3.3). Explain why the two-step classification cannot be carried out together with the fitting of the model in WinBUGS.

Exercise 16.5. Based on the results from Exercise 16.4, construct a map of the identified coldspots and investigate their estimated time trends. Are there any coldspots showing an increasing time trend?

Exercise 16.6. Carry out the same analysis as in Exercise 16.4 but now include the four COA-level covariates into the model (i.e. the model presented in Eq. 16.2). Compare the classification results from this model to those obtained from Exercise 16.4 (where no covariates were included). Comment on and explain any difference between the two sets of classification results.

Exercise 16.7. Apply the methodology presented in Section 16.3 to a space-time dataset that you may have.

Exercise 16.8. Under the BaySTDetect model, the variance of the RW1 prior for each area in the area-specific trend model is modelled hierarchically as $\log(\sigma_{\xi,i}^2) \sim N(a,b^2)$. The prior for a is $N(0,10000)$ and the prior for b is $N_{+\infty}(0,2.5^2)$. Use WinBUGS to carry out forward sampling to show that the prior choice for b assumes that roughly 10% of areas will display local temporal variability that is 10 times larger than the median of all local variances, i.e. $\Pr\left(\sigma_{\xi,i}^2 > 10 \times \exp(a)\right) \approx 0.1$ based on the prior that $b \sim N_{+\infty}(0,2.5^2)$ and for any value of a. (Hint: the idea of forward sampling is to sample values for a and b from their respective prior distributions so that we can understand/visualise the corresponding (prior) distribution for $\sigma_{\xi,i}^2$ where $\sigma_{\xi,i}^2$ is some complicated function of a and b. See also Exercise 9.3 in Chapter 9 for the idea of forward sampling.)

Exercise 16.9. The R script online, `simulate_data_for_BaySTDetect.R`, simulates a set of annual mortality count data at the local authority district (LAD) level over eight years. As described in Section 16.4.5, 15 areas are chosen to follow an unusual trend (the black line in Figure 16.13 under Pattern 1), and all the other 339 LADs have the common trend (the grey line in Figure 16.13). Simulate a few sets of data using this script with different departure patterns (e.g. Pattern 2 or 3 in Figure 16.13) and different departure magnitudes. Explore each set of your simulated data to see how difficult (or easy) it is to identify the unusual trends by just looking at the raw data (without any form of modelling as discussed in Section 16.4).

Exercise 16.10. Follow Exercise 16.9 to simulate 50 sets of space-time count data from one departure pattern (say Pattern 1 in Figure 16.13) with departure magnitude fixed at, say, 1.5. For each of the simulated datasets, fit the BaySTDetect model and summarise the detection results in terms of sensitivity and empirical FDR (Table 16.8). How well does BaySTDetect perform over these 50 simulated datasets? For comparison, you may want to apply Kulldorff's space-time scan statistics (using SaTScan) to each of those simulated datasets. (Hint: you may want to automate the model fitting in R; see Section 13.4.2 for the detail on Kulldorff's space-time scan statistic.)

Exercise 16.11. Make necessary modifications to the WinBUGS code in Figures 9.12 and 9.13 that fits a spatial model with two-component mixture and zero-inflation to fit the space-time extension of that model (Section 16.5.3). Then carry out the analysis of the space-time malaria incidence data. (Hint: the covariate $(W_{sp}y)_{i,t-1}$, the total number of new cases that occurred in the neighbouring villages in the previous time window, can be computed in R (see for example Section 6.2.2).) Add the following features to the model fitting: (a) carry out a posterior predictive check to see how well the model describes the non-zero case counts in the observed dataset; and (b) carry out a posterior predictive check on how well the model accounts for the observed values of zero.

Exercise 16.12. Incorporate appropriate forms of the rainfall variable in order to investigate (a) whether a prolonged period of heavy rainfall would increase a village's risk of malaria; and (b) whether a village's risk of malaria would be affected by higher levels of rainfall occurring in the neighbouring villages. When investigating these problems, the issue of multicollinearity may arise due to the (strong) correlation between a temporally-lagged version of a covariate and the unmodified covariate itself. One needs to address that issue in the modelling.

Part IV

Addendum

17

Modelling Spatial and Spatial-Temporal Data: Future Agendas?

Throughout this book, the treatment of spatial and spatial-temporal data modelling has been grounded by four of the challenges such data present: spatial dependence, spatial heterogeneity, data sparsity and uncertainty. We have addressed these challenges through hierarchical modelling and spatial econometric modelling using Bayesian inference. Both modelling methodologies are, to varying degrees, widely encountered in the social, economic and public health science literatures, and we have attempted to present a balanced view of what they bring to the table when we seek methodologically rigorous answers to the questions such sciences tackle. Whilst hierarchical modelling uses a "process-based" approach to inference (making inferences about the underlying process generating the observed data), spatial econometric modelling uses a "likelihood-based" approach to inference (making inferences from the observed data through the likelihood function for the data). In this final chapter we discuss a number of topics that extend the models covered in the book, which we believe are of importance and which the reader is now in a position to pursue further.

The first set of topics (17.1 to 17.5) describe extensions to the hierarchical modelling we have covered here. At different times in the book we have stressed the complementarities between hierarchical and econometric modelling. In 17.6 we make reference to another important branch of spatial and spatial-temporal data modelling – geostatistics. We note there that this has been a branch of spatial statistics which to a large extent has mainly been of interest to the earth and environmental sciences. However, we briefly discuss three areas of geostatistics that we think should be of interest to the target audience of this book: modelling spatial structure; reducing visual bias in maps; modelling scale effects, especially in social and public health data. Geostatistics shares common ground with hierarchical modelling in the sense that it too is based on "borrowing strength" in space and space-time.

In Section 17.7 we discuss developments in spatial econometrics in order to model spatial count data. We have seen in earlier chapters the challenges that arise when building likelihood-based models of counts. We briefly refer to possible directions but also highlight the parameter estimation problems that continue to lie in wait. Finally, in 17.8, in a time of "big" and ever "bigger" data reported at "fine" and ever "finer" scales of resolution, we briefly draw the reader's attention to some of the computational challenges that arise in spatial and spatial-temporal data modelling.

17.1 Topic 1: Modelling Multiple Related Outcomes Over Space and Time

In this book, we have focused attention primarily on modelling variation in a single outcome – a single crime type, a single disease type. But criminologists often view apparently different crimes as related, and some crimes are considered to be substitutable (Hakim et al., 1984). For example, a motivated offender needs money to support a drug habit and is indifferent to whether it is obtained through burglary or breaking into a car. Their choice of crime may be strongly influenced by opportunity, which in turn may be linked to certain neighbourhood social and economic characteristics. Many diseases share common risk factors. Knorr-Held and Best (2001) report the joint spatial analysis of oral cavity and oesophageal cancer mortality cases across 544 districts of Germany (1986–1990). The two cancer sites are anatomically related, and both are thought to be associated with two well-established risk factors: smoking and alcohol consumption.

A shared components model allows the joint spatial analysis of two such outcomes (Knorr-Held and Best, 2001). The central idea of a shared component model is to separate the underlying risk surfaces of the two supposedly related outcomes into a shared component and two outcome-specific components. This can bring two substantive benefits. First, when estimating the risk of one type of crime or disease in an area, information can be borrowed not only from the outcome values observed in other neighbouring areas of *the same type* (of crime or disease), but also from the outcome values observed in other neighbouring areas of *the other type*. Modelling the two outcomes jointly should therefore help strengthen inference by sharing information spatially between the two outcomes. If similar patterns of geographical variation of related diseases, or related crimes, can be identified, this may help strengthen the detection of real clusters in the underlying risk surface (through the process model). Second, in the case of oral cavity and oesophageal cancer, it is possible there are differences in their pathogenesis, and they should not be treated as a single group of cancers. Fitting a shared component model allows that hypothesis to be explored. The analysis by Knorr-Held and Best (2001) reveals joint clusters of the two cancers, which they argue are associated with the two risk factors they have in common, but in addition they identify a north-south trend specific to oral cavity cancer. However, Knorr-Held and Best (2001, p.84) remark that their interpretation of results is built on unobserved covariates (no data on alcohol consumption or smoking are included in the modelling), which are assumed to have pronounced spatial structure and *different* spatial patterns. The shared and specific components are assumed to be independent, which ignores the possibility of interactions between the true covariates. Of course, as we have seen previously in Chapter 9, if data are available, observed area-level covariates can be added to a shared component model. The original shared component model has been extended to model more than two outcomes (Held et al., 2005 and Quick, M., et al., 2018). In Chapter 9, we discussed the use of the shared component modelling idea for a joint modelling of the two sets of related binary outcomes on the police-defined high intensity crime areas (PHIAs) and the empirically-defined high intensity crime areas (EHIAs). Slightly different from the context above, PHIAs and EHIAs are not two types of crime per se, but they are related, so a joint analysis of both helps to identify the underlying (shared) pattern of criminogenic properties of the areas. At the same time, the complementary nature of PHIAs and EHIAs is addressed through the outcome-specific components.

When spatial-temporal data are available over multiple related outcomes, an interest will be to reveal the sets of spatial, temporal and spatial-temporal patterns that are shared amongst related outcomes and the sets of spatial, temporal and spatial-temporal patterns that are outcome-specific. Again, extending the idea of shared component modelling to space and time allows these similarities and differences in the spatial, the temporal and the spatial-temporal patterns across multiple related outcomes to be explored. Tzala and Best (2008) model the spatial-temporal risks of multiple cancers. More recently, a number of modelling strategies have been proposed in modelling health data (Baker et al., 2017;

Carroll et al., 2017; Quick, H. et al., 2018), economic data (Bradley et al., 2015) and crime data (Quick et al., 2019).

17.2 Topic 2: Joint Modelling of Georeferenced Longitudinal and Time-to-Event Data

A longitudinal study follows a set of individuals over a period of time. Over the duration of the study, repeated measurements of one (or more) characteristic are recorded for each of the individuals. These repeated measurements are referred to as longitudinal data. Also recorded in a longitudinal study are time-to-event data, the time (typically measured from the beginning of the study) until an event of interest occurs to an individual. Quite often, longitudinal and time-to-event data are related to clinical research, in which measurements are taken each time a participant visits (the longitudinal data), and the time when the participant develops a disease of interest is recorded (the time-to-event data, which are also referred to as survival data). However, these two types of data can arise in social science research more generally. For example, in an educational setting, regular assessments of educational progression give rise to longitudinal data where, for example, school dropout could be an event of interest. There is an established literature on the joint modelling of longitudinal and time-to-event data (see, for example, Tsiatis and Davidian, 2004 and Rizopoulos, 2012). The aim of joint modelling is to better understand the relationship between such regular measurements and the onset of the event of interest, thereby improving accuracy in predicting event occurrence. More recently, in modelling a set of georeferenced longitudinal and survival data on AIDS (acquired immune deficiency syndrome), Martins et al. (2016) incorporate the spatial dimension into the joint modelling in order to account "for the unobserved heterogeneity amongst individuals living in the same region" (p.3368). Included in their model formulation is a set of spatially-structured random effects, and those random effects are modelled using the ICAR prior that we discussed in Chapter 8. Model fitting in Martins et al. (2016) is performed in OpenBUGS, so the introduction to WinBUGS throughout this book will facilitate the reader's understanding of their method.

17.3 Topic 3: Multiscale Modelling

Data relevant to the social sciences are being made available online at finer spatial and spatial-temporal scales. However, as we discussed in Section 3.3.3, properties of the spatial (spatial-temporal) data depend on the spatial (spatial-temporal) scale at which the data are represented. For example, any map of data values aggregated from the census output area (COA) level to the lower super output area (LSOA) level (or from the census tract level to the county level in the US) tends to be smoother compared to the map of COA level data (see Figure 3.5, for example). Aggregation of an attribute involves an element of spatial smoothing and so is likely to result in stronger spatial dependence in that attribute's spatial distribution. Whilst this may help to better identify global patterns in the data, we lose local variability (Section 3.3.1). In addition to affecting properties of the outcome data, a statistic such as a correlation coefficient that measures the association between a covariate and an outcome will also depend on the spatial (spatial-temporal) scale to which the data have been aggregated. For example, a covariate that is found to be associated with the outcome at one spatial aggregation level may not be associated with the same outcome at a different spatial level. It is often found that the correlation between two variables is greater at higher levels of aggregation.[1] The effect of scaling on the modelling results of a spatial dataset is referred to as the modifiable areal unit problem (see Chapter 3 and Gehlke and Biehl, 1934 and Openshaw, 1983). Whilst one attempt to address this problem would be to carry out the analysis at different spatial aggregation levels, Aregay et al. (2015) shows that this naïve approach performs poorly in terms of describing the observed data (based on a posterior predictive check between the outcome values predicted from the model and the actual observations). In a series of papers, Aregay et al. (2015, 2016 and 2017) develop a set of multiscale models to: (a) explain variation in disease count data aggregated at two spatial scales (159 counties nested within 18 public health districts in Georgia, USA) and (b) examine the effect of a risk factor on the health outcome at each of the two spatial scales. The key feature of their multiscale models is to link the outcome data at the two spatial scales using a set of spatially-structured random effects that share between the two scales. That is, each of these shared random effects is specific to a public health district, and the same random effect term is assigned to all the counties within that district. The inclusion of these shared random effects acknowledges the hierarchical (nested) structure of the two sets of count data. In addition to providing better fits to the data, the multiscale models are simpler (measured by pD in DIC) than the model from the naïve approach of modelling the data from the two spatial scales separately (Aregay et al., 2015). The method presented in Aregay et al. (2016) also addresses the issue associated with aggregating covariate values from one scale to another. This multiscale modelling idea is novel, but again the reader should be familiar with the modelling components as well as their WinBUGS implementation. The interested reader is also referred to the papers by Kolaczyk and Huang (2001) and Louie and Kolaczyk (2006) on specifying the likelihood (the data model) for the multiscale method. The idea of multiscale modelling is also relevant to analysing crime data. As Sampson argues (2013, p.7), "crime varies within societies, states, counties, cities, and certainly within neighborhoods, whether measured at the level of census tracts or even at the level of block groups". Multiscale modelling allows us to reveal variation in crime risk at different spatial scales and to evaluate the contributions coming from different covariates at different geographical levels.

17.4 Topic 4: Using Survey Data for Small Area Estimation

In Chapters 7 and 8 we discussed a number of challenges arising from the use of survey data to estimate small area estimates of average income. All the models considered in

---

[1] This is related to the regression attenuation (or regression dilution) phenomenon in regression modelling. Aggregating (disaggregating) the values of a covariate reduces (increases) the variability in the resulting values. Greater variability in a covariate's values tends to attenuate its association with the outcome (i.e. its regression coefficient moves closer towards zero). On the other hand, reducing the variability in the covariate's values tends to move the regression coefficient away from zero. Changing the variability of the outcome's values does not tend to change the value of the regression coefficient but does have an impact on the uncertainty of the coefficient estimate.

Chapters 7 and 8 are referred to as *unit-level models* because they deal with household-level outcome data and they are in the form:

$$y_{ij} = a + R_i + e_{ij}, \tag{17.1}$$

where y_{ij} is the outcome value for individual j in area i; a is the overall intercept; R_i is the area-specific random effect; and $e_{ij} \sim N(0, \sigma_y^2)$ is the independent error term. The inclusion of area-specific random effects is to share information amongst the areas in order to improve the quality of the estimates of the average income per household across the 109 middle super output areas (MSOAs) in Newcastle. Let N_i be the total number of households in MSOA i. The estimated average income for that MSOA, $\tilde{\bar{Y}}_i$, is obtained by

$$\tilde{\bar{Y}}_i = \frac{\sum_{j=1}^{N_i} \tilde{y}_{ij}}{N_i} = \frac{\sum_{j=1}^{N_i} \tilde{a} + \tilde{R}_i + \tilde{e}_{ij}}{N_i} = \tilde{a} + \tilde{R}_i, \tag{17.2}$$

where $\tilde{a}$ and $\tilde{R}_i$ are the posterior estimates of the intercept and the random effect term, and $\tilde{y}_{ij}$ is the predicted income value for (every) household j in MSOA i. Since the population size in each MSOA is large, $\left(\sum_{j=1}^{N_i} \tilde{e}_{ij}\right)/N_i$ is close to zero. Under Eq.17.1, each household in the same MSOA has the same income level, since only an area-level random effect is included but no household-level covariates. Note that the sum in Eq. 17.2 is taken over *all* households in the MSOA, not just the households in the survey. The size of the former is typically much larger than the size of the latter, i.e. $N_i \gg n_i$. When n_i is small, the sample mean, $\bar{y}_i = \left(\sum_{j=1}^{n_i} y_{ij}\right)/n_i$, is a rather poor estimate of the population mean, $\bar{Y}_i$ (Chapter 7). This is because we have a large number of households in the area ($N_i - n_i$) for which we do not have data from the survey.

The above unit-level model can be easily extended to include covariates (also referred to as auxiliary variables in the small area estimation context) at the unit-level (e.g. household-level characteristics, such as number of vehicles that a household possesses) and/or at the area-level (e.g. MSOA-level characteristics, such as a measure of area-level deprivation). The aim is to build a regression model using covariates so that we can obtain better predictions for the income of households not in the survey. Of course, how well we can achieve that aim depends on factors such as the usefulness of the chosen covariates in predicting the outcome and the robustness of the estimation results to model misspecification. We will return to the latter point later. A unit-level model with covariates takes the form:

$$y_{ij} = \alpha + \beta \cdot x_{ij} + \gamma \cdot H_i + R_i + e_{ij}, \tag{17.3}$$

where x_{ij} is a unit-level covariate, which often comes from the same survey that contains the data on y_{ij}, and H_i is an area-level covariate, which typically comes from a census. Here we use capital letter to denote a population-based quantity (so H_i denotes a covariate value obtained from a census). For simplicity, we only include one covariate at each level. Then, following Eq. 17.2, the area-level average income is given by:

$$\tilde{\bar{Y}}_i = \frac{\sum_{j=1}^{N_i} \tilde{y}_{ij}}{N_i} = \frac{\sum_{j=1}^{N_i} \tilde{\alpha} + \tilde{\beta} \cdot x_{ij} + \tilde{\gamma} \cdot H_i + \tilde{R}_i + \tilde{e}_{ij}}{N_i} \tag{17.4}$$

$$= \tilde{\alpha} + \tilde{\beta} \cdot \bar{X}_i + \tilde{\gamma} \cdot H_i + \tilde{R}_i$$

From Eq. 17.4, when producing the average income at the small area level, we need to have the *population mean* for the unit-level covariate included in the model. For example, if x_{ij} is the number of vehicles that a household possesses, then we would need $\bar{X}_i$, the population mean number of vehicles per household for that area, to produce the corresponding average income estimate. $\bar{X}_i$ is typically obtained from census data as opposed to plugging in $\bar{x}_i$, the sample mean calculated from the survey data. Intuitively, when n_i is small, using $\bar{x}_i$, which is a poor estimate for $\bar{X}_i$, in Eq. 17.4 would yield a poor estimate for $\bar{Y}_i$. A mathematical justification is that the best linear unbiased prediction (BLUP) estimator for the small area mean $\bar{Y}_i$ under the unit-level model in Eq. 17.3 depends on $\bar{X}_i$, in addition to other terms (Hidiroglou and You, 2016 and the references therein). When the population mean of a unit-level covariate is not available, then it is not advisable to include that covariate in the unit-level model.

When the outcome of interest at the unit (individual) level is binary (e.g. whether a person has a disease or not), the discussion of modelling the individual level outcome values for the purpose of deriving small area estimates (e.g. disease prevalence) is more complicated, mainly because $\tilde{y}_{ij}$ in the calculation of $\dfrac{\sum_{j=1}^{N_i}\tilde{y}_{ij}}{N_i}$ is a nonlinear function of the terms on the right-hand side of Eq. 17.1 or Eq. 17.3. We refer the interested reader to Rao and Molina (2015) for the methodological discussion and Moon et al. (2017), Yu et al. (2018) and Wakefield et al. (2018) for applications.

Unit-level models as discussed above cannot be used if outcome values at the individual level are not available, and the only available data from the survey that we do have access to are in the form of area-level summaries (e.g. $\bar{y}_i$, the sample mean income of each MSOA for which we have data in the income example in Chapters 7 and 8). In that situation, the Fay-Herriot model (Fay and Herriot, 1979) is a widely used model that builds on the sample mean values obtained directly from the survey to provide better small area estimates. The Fay-Herriot model is an area-level model in the sense that it takes the sample means as the outcome values, which are then linked to a set of area-level covariates. Note that area is the unit of analysis for both the area-level models for small area estimation discussed here and the models discussed in Chapter 9, but these two types of models are different. The former focus on producing reliable estimates of the outcome of interest at the area level (e.g. average income at the MSOA-level), whilst the latter (apart from the smoking model in the NO_x-stroke example) focus on estimating the outcome-covariate relationship. The smoking model in the NO_x-stroke example in Chapter 9 is an area-level model for small area estimation, as we model the counts of smokers (O_i^{smoke}), an ED-level summary of the smoking survey data, to provide ED-level estimates of smoking prevalence.

However, these two modelling purposes may not be unconnected. The paper by Namazi-Rad and Steel (2015) makes an interesting link between small area estimation and multi-scale modelling (that we discussed in Section 17.3). They argue that the use of unit-level models may run the risk of omitting the *contextual effect* of a covariate whereby the effect estimate of a covariate when included as a unit-level covariate (e.g. the number of vehicles that each household in the survey has) is different from the effect estimate when this covariate is included as an area-level covariate (the population mean number of vehicles in an area). The latter is called the contextual effect of a covariate. Ignoring such contextual effects in unit-level models may lead to biased estimates of the regression coefficients and in turn yield biased small area estimates[2]. By contrast, area-level models, as Namazi-Rad

---

[2] Material deprivation provides another example. When measured at the area level (rather than at the individual level), material deprivation is a contextual effect. Ignoring the possible impact of contextual effects is referred to as the atomistic or individualistic fallacy (see Section 3.3.3).

and Steel (2015) show, account for the contextual effects by construction. This discussion is interesting because both ecological bias (making individual-level inference using area-level data; Section 17.5) and contextual effect (learning about how the outcome at the individual level is affected by area-level covariates) are closely related to the modifiable areal unit problem. One line of future research will be to develop multiscale models so that the covariate effects at different scales (e.g. from individual to area or from area to individual) can be better estimated.

17.5 Topic 5: Combining Data at Both Aggregate and Individual Levels to Improve Ecological Inference

In many situations in the social, economic and public health sciences, the underlying process is at the individual level – it is individuals who contract a disease, houses that are burgled, families who have a weekly income. The process generating the data is an *individual level process*. Whilst the process that is hypothesized to be generating the data is at the individual level, our models are typically constructed at the aggregate level, usually because, for confidentiality reasons, that is how data both on outcomes and covariates are recorded. Ecological studies analyse area-level data in order to learn about the behaviours at the individual level and, as King et al. (2004, p.1) argue, "ecological inference is the best and often the only hope of making progress". There is, however, the problem of ecological bias. The process of data aggregation leads to loss of information about within-area variability. For example, whilst an area is considered to be affluent because its average income estimate is higher than other areas in the city, that does not mean every single household within that area has a high income. The same argument applies to the estimated outcome-covariate relationship from an ecological study. Take the example in Freedman (1999) as an illustration. Using the 1995 Current Population Survey in the US, the state-level proportion of families whose income is considered to be high is found to be positively correlated with the state-level proportion of people who are foreign-born, a result that may lead to a suggestion that the foreign-born may have higher income than the native-born. However, the individual data suggest the opposite: more native-born have higher income than foreign-born (see Freedman, 1999 for more information). The analysis of the Stockholm rape data gives another example where an area with more female residents is found to be associated with an increased risk of rape from the whole map analysis, but such an increase is found to be more pronounced when certain local characteristics are also present (e.g. they live in a deprived area with a large number of robbery cases).

A local analysis using Bayesian profile regression (Section 9.5) places the analysis into its local context so that the estimate of a risk factor's effect on outcome is local as opposed to being global (or whole map). However, these estimates remain at the area-level, not at the individual-level. To get closer to the individual-level estimates, we would need individual-level data. The good news is that more and more individual-level data georeferenced at fine spatial scales are becoming available. For example, in the UK, data about crime and policing are made publicly available via www.police.uk at the street level and updated on a monthly basis. In health, the Demographic and Health Surveys (DHS; https://dhsprogram.com/) conduct regular surveys to collect data on various demographic and health indicators in many developing countries. For many of these surveys, the locations (i.e. GPS coordinates) of the groupings of the participating households in the survey are often collected. Such fine scale spatial-temporal data can be combined with the often routinely available census area-level data to alleviate the problem of ecological bias. Methodologically, combining aggregate data with individual-level data is an actively developing area of research. Here we shall point the interested reader to Wilson and Wakefield (2018) and the references therein for the latest methodological development, and to Jackson et al. (2006 and 2008) for discussion of so-called hierarchical related regression models for combining individual and aggregate data in a non-spatial context. Even when the exact locations of the individual-level data are not available, Taylor et al. (2018) have introduced a method for delivering what they call "continuous inference" – inference which is at a spatial scale finer than the scale at which the data values are reported (a form of "downscaling" which we will refer to again in Section 17.6). Their approach is interesting because aggregations (spatial units) are allowed to overlap, the degree of overlap can be uncertain and if the modelling is in space-time, aggregations can change over time. Social scientists who analyse spatial and spatial-temporal data will be rather familiar with these sorts of problems when working with administrative data or when analysing such spatial units as hospital catchment areas or labour market regions. When modelling data recorded at point locations, we need to look at geostatistical modelling, a class of models that are different from those discussed in this book and which we now discuss.

17.6 Topic 6: Geostatistical Modelling

One of the shortcomings of the models we have encountered, irrespective of which modelling methodology we use, is the limited range of models for representing spatial (and spatial-temporal) structure. In the case of hierarchical modelling, spatial dependence, handled through the random effect, is typically defined using some form of *conditional* autoregressive model. In the case of spatial econometric modelling, spatial dependence, whether modelled through the error term or through spatially lagged functions of the outcome variable included as an explanatory variable, is modelled using a *simultaneous* autoregressive model. As we have seen throughout the book and looked at in particular detail in Chapter 4, both approaches depend on the specification of a connectivity or weights matrix (W), the form of which is often assumed rather than obtained from the data. Although consistent with the fundamental property (Section 3.3.2) that "values from observation sites close together in space tend to be more alike than values from observation sites that are further apart", the structure of spatial dependence that such models allow can be quite restrictive. One way to avoid the specification of neighbours and hence broaden the range of spatial dependence structures that can be represented is through the spatial dependence models encountered in the field of geostatistics.

Geostatistics is a distinctive methodology within the field of spatial statistics. Traditionally, this branch of spatial statistics has been linked to particular problems (such as spatial interpolation by kriging, see Section 2.3.2) and particular types of data (sampled data points or small regular blocks from a continuous surface of environmental or geological variables such as soil properties or sea surface temperature). There have been a number of recent developments in geostatistics that have a great deal to offer to the sorts of problems in this book. In addition to combining individual and aggregate spatial and spatial-temporal data, as discussed in Section 17.5, readers who are interested in the topics of this book will find geostatistics interesting in (a) its approach to modelling spatial dependence; (b) through relatively recent developments in binomial and Poisson kriging; and (c) through its application of factorial kriging to identify different scale components of variation. We shall comment on these only very briefly here, the details being beyond the scope of this book. However, the interested reader is referred to the special issue of *Geographical Analysis* (Kerry, Haining and Oliver, 2010) on Geostatistical Methods in Geography Part I: Applications in Human Geography, and more generally, Chun and Griffith (2013) and Diggle (2013).

17.6.1 Spatial Dependence

In Section 4.9 we described the implications of the choice of the spatial weights matrix W for the spatial dependence structure of the spatial parameters, either under the hierarchical approach or the spatial econometric approach, which W induces. How the W matrix is specified has important implications for that structure and, as we have frequently noted, W tends to assume a particular form based on the configuration of the set of areas. By contrast, geostatistics deal with a set of values, $z(x_1),...,z(x_N)$, observed over a set of points, $x_1,...,x_N$, in space. Typically, x_i takes the form of the longitude and the latitude of the ith location. The spatial dependence structure in the observed data is *estimated* via the spatial autocovariance in the case of weak stationarity or, more commonly, the variogram, when weak stationarity does not hold (for example, when the mean varies over the study region). The empirical variogram is defined as the variance of the differences in the observed value across all the pairs of points falling within each one of a sequence of distance separation bands (see for example Haining et al., 2010, p.12).[3] Essentially, the empirical variogram is an estimation of the (global or "whole map") spatial dependence in the observed data as a function of distance separation (between pairs of points). Providing that a "good" empirical variogram can be obtained, the next step is to find the best fitting permissible function to describe it (for some examples and references to a fuller discussion of permissible models see Haining et al., 2010, p.13 and, more comprehensively, Journel and Huijbregts, 1978, p.161–195 and Cressie, 1991 p.58–67). The empirical variogram is a useful exploratory tool that provides a great deal of information about the form of spatial dependence in the observed data. Rather than making a modelling assumption drawing on a limited range of spatial models, the long-term aim should be to infer spatial dependence structures using the observed data or model residuals. Recently Ver Hoff et al. (2018) have shown that a covariance matrix for a CAR-type of model can be derived from a geostatistical model (a spherical model in their case) the parameters of which can be estimated using observed area level data.

When estimating the empirical variogram for rate data on spatial units that differ in shape and size and where population counts are generally small and vary from area to area (e.g. disease or crime rates calculated for small census areas), how do we take these factors into account to obtain a "good" empirical variogram? First, since we are dealing with irregular areas, not points, we need to define the "location" of an area so that inter-area distances can be calculated. One approach is to represent each small area by its population weighted centroid so that distance separation between any two areas is calculated from their inter-centroid distance. Second, to adjust for population differences and to counter the effect of the small number problem (see Section 3.3.3), the differences between observations used in the variance calculation are population weighted. The difference of a data pair based on two small populations is given a smaller weight so that it has a

---

[3] Distance separation is defined in terms of "bands" (e.g., $h \pm \Delta$, where Δ denotes some small increment to a distance separation h to ensure that there are sufficient pairs of data values for estimating at each distance band. The distance measure between two points, x_i and x_j, can take the form of Euclidean distance. Other definitions of distance can also be used.

smaller impact on the variogram calculation. By contrast, the rate differences calculated from larger populations are considered to be more reliable than those based on smaller populations, so they are given larger weights (see for example Goovaerts et al., 2005, p.157). This adjustment is particularly important the greater the heterogeneity that is present in population sizes across the set of areal units.

17.6.2 Mapping to Reduce Visual Bias

If a goal of analysis is the detection of local clusters of high rates or to obtain a map of disease or crime rates that reduces the visual bias arising from different sizes of spatial units across the map where physically large areas dominate, then a procedure known as area to point (ATP) Poisson kriging can be used to filter, across the spatial units, the "spatially varying noise and account for the heterogeneity in shape, size, and population distribution" (Goovaerts, 2010, p.36). Since the empirical variogram has been calculated at the area-level, only an area-level variogram is available. ATP kriging (also referred to as "downscaling") requires knowledge of the point support variogram.[4] By a process known as deconvolution (Goovaerts, 2010, p.38), the variogram obtained from the area-level data is used to estimate a point support variogram. One way to undertake ATP kriging when areas are irregular in size and shape is to lay down a dense grid or mesh of points over the region, thereby discretising each area. The above procedure is not without its critics. For example, Journel and Huijbregts (1978, p.231) criticised the "supplementary and unverifiable hypotheses" involved in downscaling, and hence ATP kriging. Constructing a cartogram is another way to reduce visual bias in choropleth maps (see Figure 11.2).

17.6.3 Modelling Scale Effects

Another area of geostatistics relevant to the types of questions in this book is that of explanatory modelling, such as fitting regression models to outcome data in order to evaluate the contribution different covariates make to accounting for the observed spatial or spatial-temporal variation in the outcome variable. An area where geostatistical methods may offer particular insights is where different covariates impact on the outcome *at different geographical scales* (see also the discussion on multiscale modelling in Section 17.3). If the geographical scales at which these different covariates operate are different from each other, then this should be apparent from the parameters of the model fitted to the variogram of the outcome. Kerry et al. (2010), in a study of car-related thefts in the Baltic states, found that the variogram model for the risk estimate could be decomposed into the sum of a local and a regional component, plus a trend component. The decomposition they employed is a generalization of factorial kriging (Wakernagel, 1998) to Poisson kriging. Fitting a spatial error model (Section 10.2.4) to each of the spatial components from a factorial kriging analysis can help to indicate which covariates are most important at each scale (Kerry et al., 2010, p.62 and p.70–73).

In Part III of the book we drew attention, at the end of Chapter 16, to an apparent convergence in application between hierarchical modelling and spatial econometric modelling when analysing space-time data. As noted in the previous paragraph, there is complementarity between some forms of spatial econometric modelling and identifying scale effects (in terms of the impact of different covariates on an outcome) using factorial kriging. However, the links between hierarchical modelling and kriging are, in one key respect, even stronger. Both are based on a borrowing strength methodology.

---

[4] This is referred to as a change of support problem (in this case, from areas to points).

17.7 Topic 7: Modelling Count Data in Spatial Econometrics

As we commented at the end of Section 11.1.6, one active research area in spatial econometrics is to develop models for non-normal data, in particular, count data. To satisfy the normality assumption of the spatial econometric models we discussed earlier, log transforming the positive count data is one option. However, there is the question of what to do with zero values. Moreover, as we have seen in some of our applications, zero *inflation* is one of the issues that we often need to deal with in practice, so log transforming count data is not going to yield a satisfactory solution. There are other options. For example, whilst an auto-Poisson model only allows for negative spatial dependence (Section 10.5.2), it can be replaced by an auto-binomial model, which can allow for positive spatial dependence (Griffith, 2006) providing the at-risk population is known and the outcome of interest is rare. However, estimating the parameters of an auto-binomial model is not straightforward due to an intractable normalizing term in the likelihood (see Cressie, 1991, p.458–477 for an extended discussion of different approaches to estimation including pseudo-likelihood and coding methods; see also Section 10.5.2).

Another option is the spatial autoregressive Poisson model proposed by Lambert et al. (2010). In their model, each observed count is modelled by a Poisson distribution. The simultaneous autoregressive structure is applied to the log of the Poisson means. That is,

$$\log(\mu_i) = \delta \sum_{j=1}^{N} w_{ij}^* \log(\mu_j) + X_i \beta$$

The major drawback of this approach is that the reduced form of the model only depends on the covariates but not on the outcome variables, meaning that this model does not feature the feedback effects that we discussed in Section 10.3.3 and which constitute one of the special features of spatial econometric models. Essentially, this model can be considered as a hierarchical model where the equation above specifies the process model, and given the process model, the outcome values are modelled through a set of *independent* Poisson distributions. Then the properties of the above model follow the comparison we made in Section 10.5.2 between hierarchical modelling on the one hand and spatial econometric modelling on the other. Glaser (2017) provides a thorough review of the current development of spatial econometrics for count data. We also refer the interested reader to Griffith and Paelinck (2011) for discussion of what they term non-standard spatial statistics and spatial econometrics.

As we have seen in Chapters 10 and 11 as well as in this section, spatial econometrics is an important branch of spatial statistics. Being able to utilise the features in spatial econometrics (spatial spillovers and spatial feedback) for modelling non-normal outcome data will benefit many areas of research in the social and public health sciences. So, the search goes on.

17.8 Topic 8: Computation

As more and more data become available at finer spatial-temporal scales, and as the methodologies become more complex in order to deal with the dependence structures in the data, computation is another challenge that we face. As we have seen throughout this book, WinBUGS is a flexible program for fitting various spatial, temporal and spatial-temporal models. Fitting a model in WinBUGS is relatively straightforward (once you have fitted a few models successfully). But it has its limitations (see Section 8.4.3 and Section 15.7.4). WinBUGS is also not efficient at handling a large amount of data, owing to its value-by-value calculation of the likelihood function. In that situation, because of its flexibility, WinBUGS can still be used for developing models on a subset of the data. There are many other options that one can consider when WinBUGS becomes too slow or simply cannot fit the specific model in hand. Integrated Nested Laplace Approximations (INLA) is a fast alternative to WinBUGS, as it performs posterior inference through approximation (Rue et al., 2009). The book on spatial and spatial-temporal Bayesian models with R-INLA (Blangiardo and Cameletti, 2015) provides a practical guide to the use of INLA for spatial and spatial-temporal data analysis. Stan (Carpenter et al., 2017) and PyMC3 (Salvatier et al., 2016) both use MCMC for fitting Bayesian models, but both include efficient samplers for fast computation. Whilst running models in Stan can be done through R (via the R package `rstan` (Stan Development Team, 2018)), model implementation in PyMC3 requires coding in Python.

References

Abadir, K.M. and J.R. Magnus. 2005. *Matrix Algebra*. Cambridge:Cambridge University Press.

Abellan, J.J., S. Richardson and N. Best. 2008. The use of space-time models to investigate the stability of patterns of disease. *Environmental Health Perspectives*, 116(8):1111–1119.

Agresti, A. 2013. *Categorical Data Analysis*. New York:Wiley.

Alexander, M.S. and H. Maschner. 1996. *Anthropology, Space and Geographic Information Systems*. Oxford:Oxford University Press.

Anderson,C., D. Lee and N. Dean. 2014. Identifying clusters in Bayesian disease mapping. *Biostatistics*, 15(3):457–469.

Anderson, C., D. Lee and N. Dean. 2016. Bayesian cluster detection via adjacency modelling. *Spatial and Spatio-Temporal Epidemiology*, 16:11–20.

Andresen, M.A. and N. Malleson. 2011. Testing the stability of crime patterns: Implications for theory and policy. *Journal of Research in Crime and Delinquency*, 48(1):58–82.

Andrienko, N. and G. Andrienko. 2006. *Exploratory Analysis of Spatial and Temporal Data*. New York:Springer.

Anselin, L. 1988. *Spatial Econometrics: Methods and Models*. Dordrecht:Kluwer Academic Publishers.

Anselin, L. 1995. Local indicators of spatial association—LISA. *Geographical Analysis*, 27(2):93–115.

Anselin, L. 2001. Spatial econometrics. In Badi H. Baltagi (ed) *A Companion to Theoretical Econometrics*. Hoboken, NJ:Blackwell.

Anselin, L. 2003. GeoDa™ 0.9 User's guide. Centre for Spatially Integrated Social Sciences. http://www.unc.edu/~emch/gisph/geoda093.pdf

Anselin, L. 2005. Exploring spatial data with GeoDa: A workbook. Centre for Spatially Integrated Social Science. https://spatial.uchicago.edu/software

Anselin, L. 2010. Thirty years of spatial econometrics. *Papers in Regional Science*, 89(1):3–25.

Anselin, L. 2018. https://geodacenter.github.io/workbook/6a_local_auto/lab6a.html#significance

Anselin, L. and O. Smirnov. 1996. Efficient algorithms for constructing proper higher order spatial lag operators. *Journal of Regional Science*, 36(1):67–89.

Anselin, L., I. Syabri and Y. Kho. 2006. GeoDa: An introduction to spatial data analysis. *Geographical Analysis*, 38(1):5–22.

Anselin, L., I. Syabri and O. Smirnov. 2002. Visualizing multivariate spatial correlation with dynamically linked windows. In L. Anselin and S. Rey (eds) *New Tools for Spatial Data Analysis: Proceedings of the Specialist Meeting*. Santa Barbara, CA:Center for Spatially Integrated Social Science (CSISS), University of California.

Arbia, G. 2014. *A Primer for Spatial Econometrics: With Applications in R*. Basingstoke:Palgrave Macmillan.

Aregay, M., A.B. Lawson, C. Faes, R.S. Kirby, R. Carroll and K. Watjou. 2015. Impact of income on small area low birth weight incidence using multiscale models. *AIMS Public Health*, 2(4):667–680.

Aregay, M., A.B. Lawson, C. Faes, R.S. Kirby, R. Carroll and K. Watjou. 2016. Multiscale measurement error models for aggregated small area health data. *Statistical Methods in Medical Research*, 25(4):1201–1223.

Aregay, M., A.B. Lawson, C. Faes and R.S. Kirby. 2017. Bayesian multi-scale modeling for aggregated disease mapping data. *Statistical Methods in Medical Research*, 26(6):2726–2742.

Armstrong, B. 2001. Comments on the papers by Guthrie and Sheppard, Best et al., Chambers and Steele and Darby et al. *Journal of the Royal Statistical Society: Series A (Statistics in Society)*, 164(1):205–207.

Assunção, R.M., I.A. Reis and C.D. Oliveira. 2001. Diffusion and prediction of leishmaniasis in a large metropolitan area in Brazil with a Bayesian space-time model. *Statistics in Medicine*, 20(15):2319–2335.

Bailey, M.A. and M.C. Rom. 2004. A wider race? Interstate competition across health and welfare programs. *Journal of Politics*, 66(2):326–347.

Bailey, N. 1967. The simulation of stochastic epidemics in two-dimensions. *Proceedings, Fifth Berkeley Symposium on Mathematics and Statistics*, 4:237–257. University of California, Berkeley and Los Angeles, CA.

Baker, J., N. White, K. Mengersen, M. Rolfe and G.G. Morgan. 2017. Joint modelling of potentially avoidable hospitalisation for five diseases accounting for spatiotemporal effects: A case study in New South Wales, Australia. *PLoS ONE*, 12(8):e0183653.

Baldwin, R.E. and P. Krugman. 2004. Agglomeration, integration and tax harmonisation. *European Economic Review*, 48(1):1–23.

Banerjee, S., B. Carlin and A. Gelfand. 2004. *Hierarchical Modelling and Analysis for Spatial Data*. Boca Raton, FL:Chapman and Hall.

Batini, C. and M. Scannapieco. 2006. *Data Quality: Concepts, Methodology and Techniques*. New York:Springer-Verlag.

Bauer, C., J. Wakefield, H. Rue, S. Self, Z. Feng and Y. Wang. 2016. Bayesian penalized spline models for the analysis of spatio-temporal count data. *Statistics in Medicine*, 35(11):1848–1865.

Bavaud, F. 1998. Models for spatial weights: A systematic look. *Geographical Analysis*, 30(2):153–171.

Bayarri, M.J. and J.O. Berger. 2004. The interplay of Bayesian and frequentist analysis. *Statistical Science*, pp.58–80.

Bayes, T. 1763. An essay towards solving a problem in the doctrine of chances. https://royalsociety-publishing.org/doi/pdf/10.1098/rstl.1763.0053

Beck, N., K.S. Gleditsch and K. Beardsley. 2006. Space is more than geography: Using spatial econometrics in the study of political economy. *International Studies Quarterly*, 50(1):27–44.

Benjamini, Y. and Y. Hochberg. 1995. Controlling the false discovery rate: A practical and powerful approach to multiple testing. *Journal of the Royal Statistical Society: Series B (Methodological)*, 57(1):289–300.

Bennett, J.E., G. Li, K. Foreman, N. Best, V. Kontis, C. Pearson, P. Hambly and M. Ezzati. 2015. The future of life expectancy inequalities in England and Wales: Bayesian spatiotemporal forecasting. *The Lancet*, 386(9989), 11–17July: 163–170.

Benth, J.S., F.E. Benth and P. Jalinskas. 2007. A spatial-temporal model for temperature with seasonal variance. *Journal of Applied Statistics*, 34(7):823–841.

Bernal, J.L., S. Cummins and A. Gasparrini. 2017. Interrupted time series regression for the evaluation of public health interventions: A tutorial. *International Journal of Epidemiology*, 46(1):348–355.

Bernardinelli, L. and C. Montomoli. 1992. Empirical Bayes versus fully Bayesian analysis of geographical variation in disease risk. *Statistics in Medicine*, 11(8):983–1007.

Bernardinelli, L., D. Clayton, C. Pascutto, C. Montomoli, M. Ghislandi and M. Songini. 1995. Bayesian analysis of space-time variation in disease risk. *Statistics in Medicine*, 14(21–22):2433–2443.

Bernasco, W. and F. Luykx. 2003. Effects of attractiveness, opportunity and accessibility to burglars on residential burglary rates of urban neighbourhoods. *Criminology*, 41(3):981–1002.

Besag, J. 1974. Spatial interaction and the statistical analysis of lattice systems. *Journal of the Royal Statistical Society: Series B* (Methodological), 36(2):192–225.

Besag, J., J. York and A. Mollie. 1991. Bayesian image restoration, with two applications in spatial statistics. *Annals of the Institute of Statistical Mathematics*, 43(1):1–20.

Besag, J. and C. Kooperberg. 1995. On conditional and intrinsic autoregressions. *Biometrika*, 82(4):733–746.

Belsley, D.A. 1991. *Conditioning Diagnostics: Collinearity and Weak Data in Regression*. New York:John Wiley.

Besley, T., I. Preston and M. Ridge. 1997. Fiscal anarchy in the UK: Modelling poll tax non-compliance. *Journal of Public Economics*, 64(2):137–152.

Besley, T. and A. Case. 1995. Does electoral accountability affect economic policy choices? Evidence from gubanatorial term limits. *The Quarterly Journal of Economics*, 110(3):769–798.

Best, N., S. Richardson and A. Thomson. 2005. A comparison of Bayesian spatial models for disease mapping. *Statistical Methods in Medical Research*, 14(1):35–59.

Bishop, C.M. 2006. *Pattern Recognition and Machine Learning (Information Science and Statistics)*. Berlin, Heidelberg:Springer-Verlag.

Bijak, J. 2010. *Forecasting International Migration in Europe: A Bayesian View*. Dordrecht:Springer.

Bivand, R., E. Pebesma and V. Gómez-Rubio. 2014. *Applied Spatial Data Analysis with R* (2nd edition). New York:Springer.

Blangiardo, M. and M. Cameletti. 2015. *Spatial and Spatio-Temporal Bayesian Models with R-INLA*. New York:Wiley.

Blangiardo, M., F. Finazzi and M. Cameletti. 2016. Two stage Bayesian model to evaluate the effects of air pollution on chronic respiratory diseases using drug prescriptions. *Spatial and Spatio-Temporal Epidemiology*, 18:1–12.

Block, R. 1979. Community, environment and violent crime. *Criminology*, 17(1):46–57.

Bloom, H.S. 2003. Using "short" interrupted time-series analysis to measure the impacts of whole-school reforms: With applications to a study of accelerated schools. *Evaluation Review*, 27(1):3–49.

Boulieri, A., S. Liverani, K. de Hoogh and M. Blangiardo. 2017. A space–time multivariate Bayesian model to analyse road traffic accidents by severity. *Journal of the Royal Statistical Society: Series A (Statistics in Society)*, 180(1):119–139.

Boulieri, A., J.E. Bennett and M. Blangiardo. 2018. A Bayesian mixture modeling approach for public health surveillance. *Biostatistics* (published online). https://doi.org/10.1093/biostatistics/kxy038

Bowers, K. and A. Hirschfield. 1999. Exploring links between crime and disadvantage in N.W.England: An analysis using Geographical Information Systems. *International Journal of Geographical Information Science*, 13(2):159–184.

Bowers, K.J., S.D. Johnson and A. Hirschfield. 2004. The measurement of crime prevention intensity and its impact on levels of crime. *British Journal of Criminology*, 44(3):419–440.

Bowers, K.J. and S.D. Johnson. 2005. Domestic burglary repeats and space-time clusters. *European Journal of Criminology*, 2(1):67–92.

Box, G.E.P. 1976. Science and statistics. *Journal of the American Statistical Association*, 71(356):791–799.

Box, G.E.P., W.G. Hunter and J.A. Hunter. 1978. *Statistics for Experimenters*. New York:Wiley.

Box, G.E.P., G.M. Jenkins, G.C. Reinsel and G.M. Ljung. 2015. *Time Series Analysis: Forecasting and Control* (5th edition). New York:Wiley.

Box, G.E.P. and G.C. Tiao. 1975. Intervention analysis with applications to economic and environmental problems. *Journal of the American Statistical Association*, 70(349):70–79.

Bradley, J.R., S.H. Holan and C.K. Wikle. 2015. Multivariate spatio-temporal models for high-dimensional areal data with application to longitudinal employer-household dynamics. *The Annals of Applied Statistics*, 9(4):1761–1791.

Bradshaw, C.P., J.H. Zmuda, S.G. Kellam and N.S. Ialongo. 2009. Longitudinal impact of two universal preventive interventions in first grade on educational outcomes in high school. *Journal of Educational Psychology*, 101(4):926–937.

Braga, A.A. and D.L. Weisburd (eds). 2010. Special issue on empirical evidence of place in criminology. *Journal of Quantitative Criminology*, 26(1):1–6.

Brett, C. and J. Pinkse. 2000. The determinants of municipal tax rates in British Columbia. *Canadian Journal of Economics/Revue Canadienne d`Economique*, 33(3):695–714.

Brindley, P., R. Maheswaran, T. Pearson, S. Wise and R. Haining. 2004. Using modelled outdoor air pollution data for health surveillance. In R. Maheswaran and M. Craglia (eds) *GIS in Public Health Practice*, 125–149. Boca Raton, FL:CRC Press.

Brindley, P., S.M. Wise, R. Maheswaran and R.P. Haining. 2005. The effect of alternative representations of population location on the areal interpolation of air pollution exposure. *Computers, Environment and Urban Systems*, 29(4):455–469.

Brooks, S.P. 1998. Markov chain Monte Carlo method and its application. *Journal of the Royal Statistical Society: Series D* (the Statistician), 47(1):69–100.

Brooks, S.P. and A. Gelman. 1997. General methods for monitoring convergence of iterative simulations. *Journal of Computational and Graphical Statistics*, 7:434–455.

Brown, R.P. and J.C. Rork. 2005. Copycat gaming: A spatial analysis of state lottery structure. *Regional Science and Urban Economics*, 35(6):795–807.

Browne, W.J. 2004. An illustration of the use of reparameterisation methods for improving MCMC efficiency in crossed random effect models. *Multilevel Modelling Newsletter*, 16:13–25.

Browne, W.J., F. Steele, M. Golalizadeh and M.J. Green. 2009. The use of simple reparameterizations to improve the efficiency of Markov chain Monte Carlo estimation for multilevel models with applications to discrete time survival models. *Journal of the Royal Statistical Society: Series A (Statistics in Society)*, 172(3):579–598.

Brueckner, J.K. 2003. Strategic interaction among local governments. An overview of empirical studies. *International Regional Science Review*, 26(2):175–188.

Brunsdon, C., M. Charlton and P. Harris. 2012. Living with collinearity in local regression models. International Spatial Accuracy and Research Association. http://www.spatial-accuracy.org/BrunsdonAccuracy2012

Brunsdon, C. and L. Comber. 2015. *An Introduction to R for Spatial Analysis and Mapping* (1st edition). London:SAGE.

Brunsdon, C., A.S. Fotheringham and M. Charlton. 1996. Geographically weighted regression: A method for exploring spatial non-stationarity. *Geographical Analysis*, 28(4):281–298.

Burrough, P.A. and R.A. McDonnell. 2000. *Principles of Geographical Information Systems*. Oxford:Oxford University Press.

Bursik, R.J. and H.G. Grasmick. 1993. *Neighborhoods and Crime*. New York:Lexington Books.

Bursztyn, L. and D. Cantoni. 2016. A tear in the Iron Curtain: The impact of Western television on consumption behaviour. *Review of Economics and Statistics*, 98(1):25–41.

Cacoullos, T. 1965. A relation between t and F-distributions. *Journal of the American Statistical Association*, 60(310):528–531.

Cameron, A.C. and P.K. Trivedi. 1986. Econometric models based on Count Data: Comparisons and applications of some estimators and tests. *Journal of Applied Econometrics*, 1(1):29–53.

Cameron, A.C. and P.K. Trivedi. 1990. Regression-based tests for overdispersion in the Poisson model. *Journal of Econometrics*, 46(3):347–364.

Campbell, D. and H. Ross. 1970. The Connecticut crackdown on speeding: Time series data in quasi-experimental analysis. In E.R. Tufts (ed) *The Quantitative Analysis of Social Problems*. Reading, MA:Addison Wesley.

Campbell, D., J. Stanley and N. Gage. 1963. *Experimental and Quasiexperimental Designs for Research*. Chicago, IL:Rand McNally.

Cao, G., P.C. Kyriakidis and M.F. Goodchild. 2011. A multinomial logistic mixed model for the prediction of categorical spatial data. *International Journal of Geographical Information Science*, 25(12):2071–2086.

Carpenter, B., A. Gelman, M.D. Hoffman, D. Lee, B. Goodrich, M. Betancourt, M. Brubaker, J. Guo, P. Li and A. Riddell. 2017. Stan: A probabilistic programming language. *Journal of Statistical Software*, 76(1):1–32.

Carroll, R., A.B. Lawson, C. Faes, R.S. Kirby, M. Aregay and K. Watjou. 2017. Extensions to multivariate space time mixture modeling of small area cancer data. *International Journal of Environmental Research in Public Health*. May 9, 14(5). pii:E503. doi: 10.3390/ijerph14050503

Carroll, R.J., D. Ruppert, L.A. Stefanski and C.M. Crainiceanu. 2006. *Measurement Error in Nonlinear Models: A Modern Perspective* (2nd edition). Boca Raton, FL:Chapman & Hall/CRC.

Case, A.C., H.S. Rosen and J.R. Hines. 1993. Budget spillovers and fiscal policy interdependence: Evidence from the states. *Journal of Public Economics*, 52(3):285–307.

Ceccato, V. 2013. *Moving Safely: Crime and Perceived Safety in Stockholm's Subway Stations*. Lanham, MD:Lexington Books.

Ceccato, V. 2014. The nature of rape places. *Journal of Environmental Psychology*, 40:97–107.

Ceccato, V., G. Li and R. Haining. 2019. The ecology of outdoor rape: The case of Stockholm, Sweden. *European Journal of Criminology*, 16(2):210–236.

Celeux, G., F. Forbes, C.P. Robert and D.M. Titterington. 2006. Deviance Information criteria for missing data models. *Bayesian Analysis*, 1(4):651–673.

Cho, W. 2003. Contagion effects and ethnic contribution networks. *American Journal of Political Science*, 47(2):368–387.

Chou, Y.H. 1991. Map resolution and spatial autocorrelation. *Geographical Analysis*, 23(3):228–246.

Chun, Y. and D.A. Griffith. 2013. *Spatial Statistics and Geostatistics: Theory and Applications for Geographic Information Science and Technology*. Thousand Oaks, CA:Sage Publishing.

Clarke, R.V. 1997. *Situational Crime Prevention: Successful Case Studies* (2nd edition). New York:Harrow and Heston.

Clayton, D.G. 1996. Generalized linear mixed models. In W.R. Gilks, S. Richardon and D. Spiegelhalter (eds) *Markov Chain Monte Carlo in Practice*, 275–302. Boca Raton, FL:Chapman and Hall.

Clayton, D. and J. Kaldor. 1987. Empirical Bayes estimates of age-standardized relative risks for use in disease mapping. *Biometrics*, 43(3):671–681.

Cliff, A.D. and J.K. Ord. 1973. *Spatial Autocorrelation*. London:Pion.

Cliff, A. and J.K. Ord. 1981. *Spatial Processes: Models and Applications*. London:Pion.

Clifford, N.J., S.L. Holloway, S.P. Rice and G. Valentine. 2009. *Key Concepts in Geography* (2nd edition). Los Angeles, CA:Sage.

Clifford, P. and S. Richardson. 1985. Testing the association between two spatial processes. *Statistics and Decisions* Suppl. No. 2:155–160.

Clifford, P., S. Richardson and D. Hémon. 1989. Assessing the significance of the correlation between two spatial processes. *Biometrics*, 45(1):123–134.

Cockings, S., P.F. Fisher and M. Longford. 1997. Parameterization and visualization of the errors in areal interpolation. *Geographical Analysis*, 29(4):314–328.

Conway, K.S. and J.C. Rork. 2004. Diagnosis murder. The death of state taxes. *Economic Inquiry*, 42(4):537–559.

Corberan-Vallet, A. and A. Lawson. 2016. Spatial health surveillance. In A. Lawson, S. Banerjee, R. Haining and M. Ugarte (eds) *Handbook of Spatial Epidemiology*, 501–519. Boca Raton, FL:CRC Press.

Corrado, L. and B. Fingleton. 2012. Where is the economics in spatial econometrics? *Journal of Regional Science*, 52(2):210–239.

Correa, T.R., R.M. Assunção and M.A. Costa. 2015. A critical look at prospective surveillance using a scan statistic. *Statistics in Medicine*, 34(7):1081–1093.

Cowen, D.J. 1988. GIS versus CAD and DBMS: What are the differences? *Photogrammetric Engineering and Remote Sensing*, 54:1551–1554.

Craglia, M., R. Haining and P. Signoretta. 2001. Modelling high intensity crime areas in English cities. *Urban Studies*, 38(11):1921–1941.

Craglia, M., R. Haining and P. Signoretta. 2005. Modelling high-intensity crime areas: Comparing police perceptions with offence/offender data in Sheffield. *Environment and Planning A*, 37(3):503–524.

Cressie, N. 1991. *Statistics for Spatial Data*. New York:Wiley.

Cressie, N. 1992. Smoothing regional maps using empirical Bayes predictors. *Geographical Analysis*, 24(1):75–95.

Cressie, N. and C.K. Wikle. 2011. *Statistics for Spatio-Temporal Data*. New York:Wiley.

Cromley, E.K. and S.L. McLafferty. 2012. *GIS and Public Health* (2nd edition). London:Guildford Press.

Cutler, D.M. and E.L. Glaeser. 1997. Are ghettos good or bad? *The Quarterly Journal of Economics*, 112(3):827–872.

Cutts, D. and D.J. Webber. 2010. Voting patterns, party spending and space in England and Wales. *Regional Studies*, 44(6):735–760.

Cutts, D., D. Webber, P. Widdop, R. Johnston and C. Pattie. 2014. With a little help from my neighbours: A spatial analysis of the impact of local campaigns at the 2010 British general election. *Electoral Studies*, 34:216–231.

Cuzick, J. and P. Elliott. 1992. Small area studies: Purpose and methods. In P. Elliott, J. Cuzick, D. English and R. Stern (eds) *Geographical and Environmental Epidemiology: Methods for Small Area Studies*, 14–21. Oxford:Oxford University Press.

David, F.N. and D.E. Barton. 1966. Two space-time interaction tests for epidemicity. *British Journal for Preventative and Social Medicine*, 20(1):44–48.

Davies, R. and J. Voget. 2008. Tax competition in an expanding European Union. Oxford Centre for Business Taxation, Working Paper 08/30. https://core.ac.uk/download/pdf/28877951.pdf?repositoryId=662

Davison, A.C. and D.V. Hinkley. 1997. *Bootstrap Methods and Their Application*. Cambridge:Cambridge University Press.

Dean, C.B. 1992. Testing for overdispersion in Poisson and binomial regression models. *Journal of the American Statistical Association*, 87(418):451–457.

Dean, C.B. and J.F. Lawless. 1989. Tests for detecting overdispersion in Poisson regression models. *Journal of the American Statistical Association*, 84(406):467–472.

Dean, C.B. and E.R. Lundy. 2016. *Overdispersion*. Wiley StatsRef: Statistics Reference Online. New York:Wiley.

Diggle, P. 2013. *Statistical Analysis of Spatial and Spatio-Temporal Point Patterns* (3rd edition). Boca Raton, FL:CRC Press.

Diggle, P.J., A.G. Chetwynd, R. Häggkvist and S.E. Morris. 1995. Second order analysis of space-time clustering. *Statistical Methods in Medical Research*, 4(2):124–136. http://onlinelibrary.wiley.com/doi/10.1002/9781118445112.stat06788.pub2/pdf

Diggle, P.J. and E. Giorgi. 2016. Model-based geostatistics for prevalence mapping in low-resource settings. *Journal of the American Statistical Association*, 111(515):1096–1120.

Diggle, P.J., J.A. Tawn and R.A. Moyeed. 1998. Model-based geostatistics. *Journal of the Royal Statistical Society: Series C (Applied Statistics)*, 47(3):299–350.

DiNardo, J. 2008. Natural experiments and quasi-natural experiments. In S.N. Durlauf and L.E. Blume (eds) *The New Palgrave Dictionary of Economics* (2nd edition), 856–859. London: Palgrave Macmillan.

DiNardo, J. and D.S. Lee. 2004. Economic impacts of new unionization on private sector employees: 1984–2001. *Quarterly Journal of Economics*, 119(4):1383–1441.

DiNardo, J. and T. Lemieux. 2001. Alcohol, marijuana, and American youth: The unintended consequences of government regulation. *Journal of Health Economics*, 20(6):991–1010.

Dow, M.M., M.L. Burton and D.R. White. 1982. Network autocorrelation: A simulation study of a foundational problem in regression and survey research. *Social Networks* 4(2):169–200.

Dubin, R. 2009. Spatial weights. In A.S. Fotheringham and P.A. Rogerson (eds) *The SAGE Handbook of Spatial Analysis*, 125–157. Los Angeles, CA:SAGE.

Dunning, T. 2012. *Natural Experiments in the Social Sciences: A Design-Based Approach*. Cambridge:Cambridge University Press.

Eckley, D.C. and K.M. Curtin. 2013. Evaluating the spatio-temporal clustering of traffic accidents. *Computers, Environment and Urban Systems*, 37:70–81.

Elhorst, J.P. 2010. Applied spatial econometrics: Raising the bar. *Spatial Economic Analysis*, 5(1):9–28.

Elliott, P., S. Richardson, J.J. Abellan, A. Thomson, C. deHoogh, L. Jarup and D.J. Briggs. 2009. Geographic density of landfill sites and risk of congenital anomalies in England. *Occupational and Environmental Medicine*, 66(2):81–89.

Fahrmeir, L. and S. Lang. 2001. Bayesian inference for generalized additive mixed models based on Markov random fields. *Journal of the Royal Statistical Society: Series C* (Applied Statistics), 50(2):201–220.

Faraway, J. 2002. Practical regression and ANOVA using R. https://cran.r-project.org/doc/contrib/Faraway-PRA.pdf; last accessed February, 2019.

Farrington, D.P. and B.C. Welsh. 2005. Randomized experiments in criminology: What have we learned in the last two decades? *Journal of Experimental Criminology*, 1(1):9–38.

Fay, R.E. and R.A. Herriot. 1979. Estimation of income for small places: An application of James-Stein procedures to census data. *Journal of the American Statistical Association*, 74:268–277.

Fayet, A.L., J.A. Tobias, R.E. Hintzen and N. Seddon. 2014. Immigration and dispersal are key determinants of cultural diversity in a songbird population. *Behavioral Ecology*, 25(4):744–753.

Felson, M. and L.E. Cohen. 1980. Human ecology and crime: A routine activity approach. *Human Ecology*, 8(4):389–406.

Fernández, C. and P.J. Green. 2002. Modelling spatially correlated data via mixtures: a Bayesian approach. *Journal of the Royal Statistical Society: Series B (Statistical methodology)*, 64(4):805–826.

Fisher, P., C. Farrelly, A. Maddocks and C. Ruggles. 1997. Spatial analysis of visible areas from the Bronze Age cairns of Mull. *Journal of Archaeological Science*, 24(7):581–592.

Fisher, R. 1935. *The Design of Experiments*. Edinburgh:Oliver and Boyd.

Fisher, W.H., S.W. Hartwell and X. Deng. 2017. Managing inflation: On the use and potential misuse of zero-inflated count regression models. *Crime & Delinquency*, 63(1):77–87.

Florkowski, W.J. and C. Sarmiento. 2005. The examination of pecan price differences using spatial correlation estimation. *Applied Economics*, 37(3):271–278.

Forster, B.C. 1980. Urban residential ground cover using LANDSAT digital data. *Photogrammetric Engineering and Remote Sensing*, 46:547–558.

Fotheringham, A.S., C., Brunsdon and M., Charlton. 2002. *Geographically Weighted Regression: the analysis of Spatially Varying Relationships*. New York: Wiley.

Fotheringham, A.S. and T.M. Oshan. 2016. Geographically weighted regression and multicollinearity: Dispelling the myth. *Journal of Geographical Systems*, 18(4):303–329.

Fox, J. 1997. *Applied Regression Analysis, Linear Models and Related Methods*. Thousand Oaks, CA:SAGE.

Freedman, D.A. 1999. Ecological inference and the ecological fallacy. *International Encyclopedia of the Social & Behavioral Sciences*, 6:4027–4030.

Gallup, J.L., J.D. Sachs and A.D. Mellinger. 1999. Geography and economic development. *International Regional Science Review*, 22(2):179–232.

Gamarnikow, E. and A.G. Green. 1999. The third way and social capital: Education action zones and a new agenda for education, parents and communities. *International Studies in Sociology of Education*, 9(1):3–22.

Gao, S., Y. Liu, Y. Wang and X. Ma. 2013. Discovering spatial interaction communities from mobile phone data. *Transactions in GIS*, 17(3):463–481.

Garrett, T.A. and T.L. Marsh. 2002. The revenue impacts of cross border lottery shopping in the presence of spatial autocorrelation. *Regional Science and Urban Economics*, 32(4):501–519.

Gatrell, A.C. 1997. Structures of geographical and social space and their consequences for human health. *Geografiska Annaler: Series B, Human Geography*, 79(3):141–154.

Gehlke, C.E. and K. Biehl. 1934. Certain effects of grouping upon the size of the correlation coefficient in census tract material. *Journal of the American Statistical Association*, 29(185A):169–170.

Gelfand, A.E. and S. Banerjee. 2015. Bayesian wombling: Finding rapid change in spatial maps. *Wiley Interdisciplinary Reviews: Computational Statistics*, 7(5):307–315.

Gelfand, A.E., H.-J. Kim, C.F. Sirmans and S. Banerjee. 2003. Spatial modelling with spatially varying coefficient processes. *Journal of the American Statistical Association*, 98(462):387–396.

Gelfand, A.E., A. Kottas and S.N. MacEachern. 2005. Bayesian nonparametric spatial modelling with Dirichlet process mixing. *Journal of the American Statistical Association*, 100(471):1021–1035.

Gelfand, A.E., S.K. Sahu and B.P. Carlin. 1995. Efficient parameterisations for normal linear mixed models. *Biometrika*, 83:479–488.

Gelfand, A.E. and P. Vounatsou. 2003. Proper multivariate conditional autoregressive models for spatial data analysis. *Biostatistics*, 4(1):11–25.

Gelfand, A.E., L. Zhu and B.P. Carlin. 2001. On the change of support problem for spatio-temporal data. *Biostatistics*, 2(1):31–45.

Gelman, A. 2006. Prior distributions of variance parameters in hierarchical analysis. *Bayesian Analysis*, 1(3):515–534.

Gelman, A., J.B. Carlin, H.S. Stern, D.B. Dunson, A. Vehtari and D.B. Rubin. 2014. *Bayesian Data Analysis* (3rd edition). Boca Raton, FL:Chapman and Hall.

Gelman, A. and J. Hill. 2007. *Data Analysis Using Regression and Multilevel/Hierarchical Models.* Cambridge:Cambridge University Press.

Gelman, A., J. Hill and M. Yajima. 2012. Why we (usually) don't have to worry About multiple comparisons. *Journal of Research on Educational Effectiveness*, 5(2):189–211.

Gelman, A. and P.N. Price. 1999. All maps of parameter estimates are misleading. *Statistics in Medicine*, 18(23):3221–3234.

Gelman, A. and D.B. Rubin. 1992. Inference from iterative simulation using multiple sequences. *Statistical Science*, 7(4):457–472.

GeoBUGS User Manual, Version 1.2 September 2004 by Thomas, A., N. Best, D. Lunn, R. Arnold and D. Spiegelhalter. https://www.mrc-bsu.cam.ac.uk/wp-content/uploads/geobugs12manual.pdf; last accessed 18/03/2019.

Gershman, S.J. and D.M. Blei. 2012. A tutorial on Bayesian nonparametric models. *Journal of Mathematical Psychology*, 56(1):1–12.

Getis, A. and J.K. Ord. 1992. The analysis of spatial association by use of distance statistics. *Geographical Analysis*, 24:189–206 (with correction. 1993, 25:276).

Getis, A. and J.K. Ord. 1995. Local spatial autocorrelation statistics: Distributional issues and an application. *Geographical Analysis*, 27:286–306.

Gibbons, S. and H.G. Overman. 2012. Mostly pointless spatial econometrics? *Journal of Regional Science*, 52(2):172–191.

Gilks, W.R., S. Richardson and D.J. Spiegelhalter (eds). 1996. *Markov Chain Monte Carlo in Practice.* Boca Raton, FL:Chapman & Hall/CRC.

Gilman, E.A. and E.G. Knox. 1995. Childhood cancers: Space-time distribution in Britain. *Journal of Epidemiology and Community Health*, 49(2):158–163.

Glaser, S. 2017. A review of spatial econometric models for count data. Hohenheim Discussion Papers in Business, Economics and Social Sciences 19-2017, University of Hohenheim, Faculty of Business, Economics and Social Sciences.

Gleditsch, K.S. and M.D. Ward. 2006. Diffusion and the international context of democratization. *International Organization*, 60(4):911–933.

Gollini, I., B. Lu, M. Charlton, C. Brunsdon and P. Harris. 2015. GWmodel: An R package for exploring spatial heterogeneity using geographically weighted models. *Journal of Statistical Software*, 63(17):1–50.

Gómez-Rubio, V., N. Best, S. Richardson, G. Li and P. Clarke. 2008a. Bayesian statistics for small area estimation. http://www.bias-project.org.uk/papers/BayesianSAE.pdf; last accessed 17/03/2019.

Gómez-Rubio, V., N. Best and S. Richardson. 2008b. A comparison of different methods for small area estimation. http://www.bias-project.org.uk/papers/ComparisonSAE.pdf; last accessed 17/03/2019.

Gómez-Rubio, V. and H. Rue. 2018. Markov chain Monte Carlo with the integrated nested Laplace approximation. *Statistics and Computing*, 28(5):1033–1051.

Goodchild, M. 1989. Modelling error in objects and fields. In M. Goodchild and S. Gopal (eds) *Accuracy of Spatial Databases*, 107–113. London:Taylor and Francis.

Goodchild, M. and R.P. Haining. 2004. GIS and spatial data analysis: Converging perspectives. *Papers in Regional Science*, 83:363–385.

Goodwin, M.J. and O. Heath. 2016. The 2016 referendum, Brexit and the left behind: An aggregate-level analysis of the result. *The Political Quarterly*, 87(3):324–333.

Goovaerts, P. 2010. Geostatistical analysis of county-level lung cancer mortality rates in the southeastern United States. *Geographical Analysis*, 42(1):32–52.

Goovaerts, P., G.M. Jacquez and D. Greiling. 2005. Exploring scale-dependent correlations between cancer mortality rates using factorial kriging and population weighted semivariograms. *Geographical Analysis*, 37(2):152–182.

Gorman, D.M., L. Zhu and S. Horel. 2005. Drug 'hotspots', alcohol availability and violence. *Drug and Alcohol Review*, 24(6):507–513.

Gotway, C.A. and L.J. Young. 2002. Combining incompatible spatial data. *Journal of the American Statistical Association*, 97(458):632–648.

Green, P.J. 1995. Reversible jump Markov chain Monte Carlo computation and Bayesian model deter-mination. *Biometrika*, 82(4):711–732.

Green, P.J. and S. Richardson. 2002. Hidden Markov models and disease mapping. *Journal of the American Statistical Association*, 97(460):1055–1070.

Greene, S., E.R. Peterson, D. Kapell, A.D. Fine, M. Kulldorff and M. Daily. 2016. Reportable disease spatiotemporal cluster detection, New York City, New York, 2014–2015. *Emerging Infectious Diseases*, 22:1808–1812.

Greenland, S. and J. Robins. 1994. Invited commentary: Ecologic studies—Biases, misconceptions, and counter examples. *American Journal of Epidemiology*, 139(8):747–760.

Griffith, D.A. 2006. Assessing spatial dependence in count data: Winsorized and spatial filter specifi-cation alternatives to the auto-Poisson model. *Geographical Analysis*, 38(2):160–179.

Griffith, D.A. and J.H.P. Paelinck. 2011. *Non-Standard Spatial Statistics and Spatial Econometrics*. New York:Springer.

Groff, E.R. and N.G. LaVigne. 2001. Mapping an opportunity surface of residential burglary. *Journal of Research in Crime and Delinquency*, 38(3):257–278.

Grubesic, T.H. and E.A. Mack. 2008. Spatio-temporal interaction of urban crime. *Journal of Quantitative Criminology*, 24(3):285–306.

Grzebyk, M. and H. Wackernagel. 1994. Multivariate analysis and spatial/temporal scales: Real and complex models. *Proceedings of XVIIth International Biometric Conference*, 8–12 August 1994 at Hamilton, Ontario, Canada. Centre de Geostatistique, Ecoledes Mines de Paris 35 Rue Saint Honore, 77305 Fontainebleau, France. Volume 1:19–33.

Gschlößl, S. and C. Czado. 2008. Modelling count data with overdispersion and spatial effects. *Statistical Papers*, 49(3):531–552.

Guptill, S.C. and J.L. Morrison. 1995. *Elements of Spatial Data Quality*. Oxford:Elsevier Science.

Hagerstrand, T. 1967. *Innovation Diffusion as a Spatial Process*. Chicago, IL:Chicago University Press.

Haining, R.P. 1978. A spatial model for High Plains agriculture. *Annals, Association of American Geographers*, 68:593–604.

Haining, R.P. 1983. Modelling intra-urban price competition: An example of gasoline pricing. *Journal of Regional Science*, 23(4):517–528.

Haining, R.P. 1984. Testing a spatial interacting-markets hypothesis. *Review of Economics and Statistics*, 66(4):475–483.

Haining, R.P. 1986. Intra-Urban retail price competition: Corporate and neighbourhood aspects of spatial price variation In G. Norman (ed) *Spatial Pricing and Differentiated* Markets. London Papers in Regional Sciences Vol. 16, 144–164. London:Pion.

Haining, R.P. 1987. Small area aggregate income models: Theory and methods with an application to urban and rural income data for Pennsylvania. *Regional Studies*, 21(6):519–529.

Haining, R.P. 1988. Estimating spatial means with an application to remotely sensed data. *Communications in Statistics-Theory and Methods*, 17(2):573–597.

Haining, R.P. 1990a. The use of added variable plots in regression modelling with spatial data. *Professional Geographer*, 42(3):336–344.

Haining, R.P. 1990. *Spatial Data Analysis in the Social and Environmental Sciences*. Cambridge:Cambridge University Press.

Haining, R.P. 1991. Bivariate correlation with spatial data. *Geographical Analysis*, 23(3):210–227.

Haining, R.P. 2003. *Spatial Data Analysis: Theory and Practice*. Cambridge:Cambridge University Press.

Haining, R.P. 2014. Thinking spatially, thinking statistically. In E. Silva, P. Healey, N. Harris and P. Vander Boeck (eds) *The Routledge Handbook of Planning Research Methods*, 255–267. London:Routledge.

Haining, R.P., R. Kerry and M.A. Oliver. 2010. Geography, spatial data analysis and geostatistics. *Geographical Analysis*, 42(1):7–31.

Haining, R.P. and J. Law. 2007. Combining police perceptions with police records of serious crime areas: A modelling approach. *Journal of the Royal Statistical Society: Series A (Statistics in Society)*, 170(4):1019–1034.

Haining, R.P., G. Li, R. Maheswaran, M. Blangiardo, J. Law, N. Best and S. Richardson. 2010. Inference from ecological models: Estimating the relative risk of stroke from air pollution exposure using small area data. *Spatial and Spatio-Temporal Epidemiology*, 1(2–3):123–131. doi: 10.1016/j.sste.2010.03.006

Haining, R.P., S.M. Wise and J. Ma. 1998. Exploratory spatial data analysis. *Journal of the Royal Statistical Society: Series D (The Statistician)*, 47(3):457–469.

Hakim, S., U. Spiegel and J. Weinblatt. 1984. Substitution, size effects, and the composition of property crime. *Social Science Quarterly*, 65(3):719–734.

Hanes, N. 2002. Spatial spillover effects in the Swedish rescue services. *Regional Studies*, 36(5):531–539.

Haneuse, S.J. and J.C. Wakefield. 2007. Hierarchical models for combining ecological and case-control data. *Biometrics*, 63(1):128–136.

Haneuse, S.J. and J.C. Wakefield. 2008. The combination of ecological case-control data. *Journal of the Royal Statistical Society: Series B (Statistical Methodology)*, 70(1):73–93.

Harris, R. and M. Charlton. 2016. Voting out of the European Union: Exploring the geography of Leave. *Environment and Planning A*, 48(11):2116–2128.

Harris, R., J. Moffat and V. Kravtsova. 2011. In search of 'W'. *Spatial Economic Analysis*, 6(3):249–270.

Hassett, K.A. and A. Mathur. 2015. A spatial model of corporate tax incidence. *Applied Economics*, 47(13):1350–1365.

Hastie, T., R. Tibshirani and J. Friedman. 2009. *The Elements of Statistical Learning: Data Mining, Inference and Prediction* (2nd edition). New York:Springer.

Hearst, N., T.B. Newman and S.B. Hulley. 1986. Delayed effects of the military draft on mortality: A randomized natural experiment. *New England Journal of Medicine*, 314(10):620–624.

Held, L., I. Natàrio, S.E. Fenton, H. Rue and N. Becker. 2005. Towards joint disease mapping. *Statistical Methods in Medical Research*, 14(1):61–82.

Held, L. and H. Rue. 2010. Conditional and intrinsic autoregressions. In A.E. Gelfand, P.J. Diggle, M. Fuentes and P. Guttorp (eds) *Handbook of Spatial Statistics*, 201–216. Boca Raton, FL:Chapman and Hall, CRC Handbooks of Modern Statistical Methods. Chapter 13.

Hidiroglou, M.A. and Y. You. 2016. Comparison of unit level and area level small area estimators. *Survey Methodology*, 42(1):41–62.

Hilbe, J.M. 2011. *Negative Binomial Regression* (2nd edition). Cambridge:Cambridge University Press.

Hill, B. and R. Paynich. 2014. *Fundamentals of Crime Mapping* (2nd edition). Burlington, MA:Jones and Bartlett.

Hjort, N., C. Holmes, P. Müller and S.G. Walker (eds). 2010. *Bayesian Nonparametrics: Principles and Practice*. Cambridge:Cambridge University Press.

Hoaglin, D.C., F. Mosteller and J.W. Tukey. 1983. *Understanding Robust and Exploratory Data Analysis*. New York:Wiley.

Hodder, I. and C. Orton. 1976. *Spatial Analysis in Archaeology*. Cambridge:Cambridge University Press.

Hodges, J.S. and B.J. Reich. 2010. Adding spatially-correlated errors can mess up the fixed effect you love. *The American Statistician*, 64(4):325–334.

Holt, D., D.G. Steele and M. Tranmer. 1996. Area homogeneity and the modifiable areal unit problem. *Geographical Systems*, 3:181–200.

Home Office. 1997. *Formula for Police Specific Grant and Police Spending Assessment in 1998/9*. London:Government Statistical Service.

Hope, A.C.A. 1968. A simplified Monte Carlo significance test procedure. *Journal of the Royal Statistical Society: Series B (Methodological)*, 30(3):582–598.

Hoxby, C.M. 2000. Does competition among public schools benefit students and taxpayers? *American Economic Review*, 90(5):1209–1238.

Hu, M.C., M. Pavlicova and E.V. Nunes. 2011. Zero-inflated and hurdle models of count data with extra zeros: Examples from an HIV-risk reduction intervention trial. *The American Journal of Drug and Alcohol Abuse*, 37(5):367–375.

Huffer, F.W. and H. Wu. 1998. Markov chain Monte Carlo for autologistic regression models with application to the distribution of plant species. *Biometrics*, 54(2):509–524.

Hughes, G.J. and R. Gorton 2013. An evaluation of SaTScan for the prospective detection of space-time Campylobacter clusters in the North East of England. *Epidemiology and Infection*, 141(11):2354–2364.

Isaaks, E.H. and R.M. Srivastava. 1989. *An Introduction to Applied Geostatistics*. Oxford:Oxford University Press.

Isard, W. 1960. *Methods of Regional Analysis: An Introduction to Regional Science*. Cambridge:published jointly Technology Press of MIT and Wiley.

Jackman, S. 2009. *Bayesian Analysis for the Social Sciences*. New York:Wiley.

Jackson, C., N. Best and S. Richardson. 2006. Improving ecological inference using individual level data. *Statistics in Medicine*, 25(12):2136–2159.

Jackson, C., N. Best and S. Richardson. 2008. Hierarchical related regression for combining aggregate and individual data in studies of socio-economic disease risk factors. *Journal of the Royal Statistical Society: Series A (Statistics in Society)*, 171(1):159–178.

Jacobs, J. 1961. *The Death and Life of Great American Cities*. New York:Random House.

Jensen, C.D., D.J. Lacombe and S.G. McIntyre. 2013. A Bayesian spatial econometric analysis of the 2010 UK General Election. *Papers in Regional Science*, 92(3):651–667.

Johnson, N.L., S. Kotz and A.W. Kemp. 1992. *Univariate Discrete Distributions* (2nd edition). John Wiley & Sons.

Johnson, S.D., W. Bernasco, K.J. Bowers, H. Elffers, J. Ratcliffe, G. Rengert and M. Townsley. 2007. Space-time patterns of risk: A cross national assessment of residential burglary victimization. *Journal of Quantitative Criminology*, 23(3):201–219.

Johnson, S.D. and K. Bowers. 2004. The stability of space-time clusters of burglary. *British Journal of Criminology*, 44(1):55–65.

Journel, A.G. and C.J. Huijbregts. 1978. *Mining Geostatistics*. London:Academic press.

Kalnins, A. 2003. Hamburger prices and spatial econometrics. *Journal of Economics and Management Strategy*, 12(4):591–616.

Karlan, D. and J. Zinman. 2010. Expanding credit access: Using randomized supply decisions to estimate the impacts. *Review of Financial Studies*, 23(1):433–464.

Keele, L.J. and R. Titiunik. 2015. Geographic boundaries as regression discontinuities. *Political Analysis*, 23(1):127–155.

Kerry, R., P. Goovaerts, R.P. Haining and V. Ceccato. 2010. Applying geostatistical analysis to crime data: Car-related thefts in the Baltic states. *Geographical Analysis*, 42(1):53–77.

Kerry, R., R. Haining and M.A. Oliver. 2010. Geostatistical methods in geography Part I: Applications in human geography. (Introduction to special issue.) *Geographical Analysis* 42(1):5–6.

Khan, M., D. Hotchkiss, A. Berruti and P. Hutchinson. 2006. Geographic aspects of poverty and health in Tanzania: Does living in a poor area matter? *Health Policy and Planning*, 21(2):110–122.

King, G., O. Rosen and M.A. Tanner. 2004. *Ecological Inference: New Methodological Strategies*. New York:Cambridge University Press.

Klassen, A.C., M. Kulldorff and F. Curriero. 2005. Geographical clustering of prostate cancer grade and stage at diagnosis, before and after adjustment for risk factors. *International Journal of Health Geographics*, 4(1):1. https://doi.org/10.1186/1476-072X-4-1

Kneib, T. and L. Fahrmeir. 2006. Structural additive regression for categorical space-time data: A mixed model approach. *Biometrics*, 62(1):109–118.

Knorr-Held, L. 2000. Bayesian modelling of inseparable space-time variation in disease risk. *Statistics in Medicine*, 19(17–18):2555–2567.

Knorr-Held, L. and J. Besag. 1998. Modelling risk from a disease in space and time. *Statistics in Medicine*, 17(18):2045–2060.

Knorr-Held, L. and N.G. Best. 2001. A shared component model for detecting joint and selective clustering of two diseases. *Journal of the Royal Statistical Society: Series A (Statistics in Society)*, 164(1):73–85.

Knorr-Held, L. and G. Rasser. 2000. Bayesian detection of clusters and discontinuities in disease maps. *Biometrics*, 56(1):13–21.

Knorr-Held, L. and H. Rue. 2002. On block updating in markov random field models for disease mapping. *Scandinavian Journal of Statistics*, 29(4):597–614.

Knox, G. 1963. Detection of low intensity epidemicity: Application to cleft lip and palate. *British Journal of Preventive and Social Medicine*, 17:121–127.

Knox, G. 1964. Epidemiology of childhood leukaemia in Northumberland and Durham. *British Journal of Preventive and Social Medicine*, 18(1):17–24.

Knox, E.G. and M.S. Bartlett. 1964a. The detection of space-time interactions. *Journal of the Royal Statistical Society: Series C (Applied Statistics)*, 13(1):25–30.

Kolaczyk, E.D. and H. Huang. 2001. Multiscale statistical models for hierarchical spatial aggregation. *Geographical Analysis*, 33(2):95–118.

Kolko, J. and D. Neumark. 2010. Do enterprise zones create jobs? Evidence from California's Enterprise Zone Program. *Journal of Urban Economics*, 68(1):1–19.

Konrad, J. 2010. Who has the oil? A map of world oil reserves. https://gcaptain.com/who-has-the-oil-a-map-of-world-oil-reserves/

Kontis, V., J.E. Bennett, C.D. Mathers, G. Li, K. Foreman and M. Ezzati. 2017. Future life expectancy in 35 industrialised countries: Projections with a Bayesian model ensemble. *The Lancet*, 389(10076):1323–1335.

Kontopantelis, E., T. Doran and I. Buchan. 2015. Regression based quasi-experimental approach when randomisation is not an option. *British Medical Journal*, 350:h2750.

Kottas, A., J.A. Duan and A.E. Gelfand. 2008. Modeling disease incidence data with spatial and spatio temporal dirichlet process mixtures. *Biometrical Journal*, 50(1):29–42.

Krainski, E.T., V. Gómez-Rubio, H. Bakka, A. Lenzi, D. Castro-Camilo, D. Simpson, F. Lindgren and H. Rue. 2019. *Advanced Spatial Modeling with Stochastic Partial Differential Equations Using R and INLA*. Boca Raton, FL:CRC Press.

Krishna Iyer, P.V. 1949. The first and second moments of some probability distributions arising from points on a lattice and their applications. *Biometrika*, 36(1–2):135–141.

Krugman, P. 1996. Urban concentration: The role of increasing returns and transport costs. *International Regional Science Review*, 19(1–2):5–30.

Krugman, P. 1998. What's new about the New Economic geography? *Oxford Review of Economic Policy*, 14(2):7–17.

Kulldorff, M. 1997. A spatial scan statistic. *Communications in Statistics-Theory and Methods*, 26(6):1481–1496.

Kulldorff, M. 2001. Prospective time periodic geographical disease surveillance using a scan statistic. *Journal of the Royal Statistical Society: Series A (Statistics in Society)*, 164(1):61–72.

Kulldorff, M., R. Heffernan, J. Hartman, R. Assunção and F. Mostashari. 2005. A space-time permutation scan statistic for disease outbreak detection. *PLoS: Medicine*. https://doi.org/10.1371/journal.pmed.0020059

Kulldorff, M. and U. Hjalmars. 1999. The Knox method and other tests for space-time interaction. *Biometrics*, 55(2):544–552.

Kulldorff, M., L. Huang, L. Pickle and L. Duczmal. 2006. An elliptic spatial scan statistic. *Statistics in Medicine*, 25(22):3929–3943.

Kulldorff, M. and N. Nagarwalla. 1995. Spatial disease clusters: Detection and inference. *Statistics in Medicine*, 14(8):799–810.

Künsch, H.R. 1987. Intrinsic autoregressions and related models on the two dimensional lattice. *Biometrika*, 74(3):517–524.

Lambert, D. 1992. Zero-inflated Poisson regression, with an application to defects in manufacturing. *Technometrics*, 34(1):1–14.

Lambert, D.M., J.P. Brown and R.J. Florax. 2010. A two-step estimator for a spatial lag model of counts: Theory, small sample performance and an application. *Regional Science and Urban Economics*, 40(4):241–252.

Land, K.C., G. Deane and J.R. Blau. 1991. Religious pluralism and church membership: A spatial diffusion model. *American Sociological Review*, 56(2):237–249.

Law, J. 2016. Exploring the specification of spatial adjacencies and weights in Bayesian spatial modelling with intrinsic conditional autoregressive priors in a small area study of fall injuries. *AIMS Public Health*, 3(1):65–82. doi: 10.3934/publichealth.2016.1.65

Lawson, A. 2013. *Statistical Methods in Spatial Epidemiology* (2nd edition). New York:Wiley.

Lawson, A. 2018. *Bayesian Disease Mapping: Hierarchical Modelling in Spatial Epidemiology* (3rd edition). Boca Raton, FL:CRC Press.

Lawson, A., S. Banerjee, R. Haining and M. Ugarte. 2016. *Handbook of Spatial Epidemiology*. Boca Raton, FL:CRC Press.

Lawson, A.B. and A. Clark. 2002. Spatial mixture relative risk models applied to disease mapping. *Statistics in Medicine*, 21(3):359–370.

Leckie, G., R. French, C. Charlton and W. Browne. 2014. Modeling heterogeneous variance–Covariance components in two-level models. *Journal of Educational and Behavioral Statistics*, 39(5):307–332.

Lee, A.H., K. Wang and K.K.W. Yau. 2001. Analysis of zero-Inflated Poisson data incorporating extent of exposure. *Biometrical Journal*, 43(8):963–975.

Lee, D. 2011. A comparison of conditional autoregressive models used in Bayesian disease mapping. *Spatial and Spatio-Temporal Epidemiology*, 2(2):79–89.

Lee, D. and R. Mitchell. 2012. Boundary detection in disease mapping studies. *Biostatistics*, 13(3):415–426.

Lee, D. and R. Mitchell. 2013. Locally adaptive spatial smoothing using conditional auto-regressive models. *Journal of the Royal Statistical Society: Series C (Applied Statistics)*, 62(4):593–608.

Lee, D., A. Rushworth and G. Napier. 2018. Spatio-temporal areal unit modelling in R with conditional autoregressive priors using the CARBayes ST package. *Journal of Statistical Software*, 84(9):1–39.

Lee, D., A. Rushworth and S.K. Sahu. 2014. A Bayesian localized conditional autoregressive model for estimating the health effects of air pollution. *Biometrics*, 70(2):419–429.

Lee, S. 2001. Developing a bivariate spatial association measure: An integration of Pearson's r and Moran's I. *Journal of Geographical Systems*, 3(4):369–385.

Leroux, B., X. Lei and N. Breslow. 2000. Estimation of disease rates in small areas: A new mixed model for spatial dependence. In M. Halloran and D. Berry (eds) *Statistical Models in Epidemiology, the Environment and Clinical Trials*, 179–191. New York:Springer-Verlag.

LeSage, J.P. and M.M. Fischer. 2008. Spatial growth regressions: Model specification, estimation and interpretation. *Spatial Economic Analysis*, 3(3):275–304.

LeSage, J.P. and R.K. Pace. 2009. *Introduction to Spatial Econometrics*. Boca Raton, FL:CRC Press.

LeSage, J.P., R. K. Pace, N. Lam, R. Campanella and X. Liu. 2011. New Orleans business recovery in the aftermath of Hurricane Katrina. *Journal of the Royal Statistical Society: Series A (Statistics in Society)*, 174(4):1007–1027.

Li, P., S. Banerjee and A. McBean. 2011. Mining boundary effects in areally referenced spatial data using the Bayesian information criterion. *Geoinformatica*, 15(3):435–454.

Li, G., N. Best, A.L. Hansell, I. Ahmed and S. Richardson. 2012. BaySTDetect: Detecting unusual temporal patterns in small area data via Bayesian model choice. *Biostatistics*, 13(4):695–710.

Li, G., R.P. Haining, S. Richardson and N. Best. 2013. Evaluating the No Cold Calling zones in Peterborough, England: Application of a novel statistical method for evaluating neighbourhood policing methods. *Environment and Planning A*, 45(8):2012–2026.

Li, G., R.P. Haining, S. Richardson and N. Best. 2014. Space-time variability in burglary risk: A Bayesian spatio-temporal modelling approach. *Spatial Statistics*, 9:180–191.

Liesenfeld, R., J. Richard and J. Vogler. 2017. Likelihood evaluation of high-dimensional spatial latent Gaussian models with non-Gaussian response variables. In Badi H. Baltagi, J.P. LeSage and R.K. Pace (eds) *Spatial Econometrics: Qualitative and Limited Dependent Variables (Advances in Econometrics Vol. 37)*, 35–77. Bingley, UK: Emerald Publishing.

Liverani, S., D.I. Hastie, L. Azizi, M. Papathomas and S. Richardson. 2015. PReMiuM: An R package for profile regression mixture models using Dirichlet processes. *Journal of Statistical Software*, 64(7):1.

Lloyd, C. 2011. *Local Models for Spatial Analysis*. Boca Raton, FL:CRC Press.

Longley, P.A., M.F. Goodchild, D.J. Maguire and D.W. Rhind. 2001. *Geographical Information Systems and Science*. Chichester:Wiley.

Losch, A. 1957. *The Economics of Location* (English translation from 1939 original). New York:John Wiley and Sons.

Louie, M.M. and E.D. Kolaczyk. 2006. 2006. A multiscale method for disease mapping in spatial epidemiology. *Statistics in Medicine*, 25(8):1287–1306.

Lu, H., C. Reilly, S. Banerjee and B. Carlin. 2007. Bayesian areal wombling via adjacency modelling. *Environmental and Ecological Statistics*, 14(4):433–452.

Luechinger, S. 2009. Valuing air quality using the life satisfaction approach. *The Economic Journal*, 119(536):482–515.

Lunn, D., C. Jackson, N. Best, A. Thomas and D. Spiegelhalter. 2012. *The BUGS Book: A Practical Introduction to Bayesian Analysis*. Chapman & Hall/CRC.

Machin, S. and K. Salvanes. 2010. Valuing school quality via school choice reform. *London School of Economics, Spatial Economics Research Centre Discussion Paper*, SERCDP0082. https://papers.ssrn.com/sol3/papers.cfm?abstract_id=1545146

MacLeod, M., E. Graham, M. Johnston, C. Dibben and I. Morgan. 1999. How does relative deprivation affect health? *Health Variations: Official Newsletter of the ESRC Health Variations Programme*, No 3, 12–13. http://www.lancaster.ac.uk/fass/projects/hvp/newsletters/macleod3.htm

MacNab, Y.C. 2003. Hierarchical Bayesian Modeling of spatially correlated health service outcome and utilization rates. *Biometrics*, 59(2):305–316.

MacNab, Y.C. 2007. Mapping disability adjusted life years: A Bayesian hierarchical model framework for burden of disease and injury assessment. *Statistics in Medicine*, 26(26):4746–4769.

MacNab, Y.C. 2010. On Bayesian shared component disease mapping and ecological regression with errors in covariates. *Statistics in Medicine*, 29(11):1239–1249.

MacNab, Y.C. and C.B. Dean. 2001. Autoregressive spatial smoothing and temporal spline smoothing for mapping rates. *Biometrics*, 57(3):949–956.

MacNab, Y.C. and P. Gustafson. 2007. Regression B-spline smoothing in Bayesian disease mapping: With an application to patient safety surveillance. *Statistics in Medicine*, 26(24):4455–4474.

Maheswaran, R., P. Elliott and D.P. Strachan. 1997. Socio-economic deprivation, ethnicity and stroke mortality in Greater London and South-East England. *Journal of Epidemiology and Community Health*, 51(2):127–131.

Maheswaren, R. and M. Craglia (Eds). 2004. *GIS in Public Health Practice*. Boca Raton,FL: CRC Press.

Maheswaran, R., R.P. Haining, T. Pearson, J. Law, P. Brindley and N.G. Best. 2006. Outdoor NO_x and stroke mortality—Adjusting for small area level smoking prevalence using a Bayesian approach. *Statistical Methods in Medical Research*, 15(5):499–516.

Manley, D., K. Jones and R. Johnston. 2017. The geography of Brexit—What geography? Modelling and predicting the outcome across 380 local authorities. *Local Economy: The Journal of the Local Economy Policy Unit*, 32(3):183–203.

Mantel, N. 1967. The detection of disease clustering and a generalized regression approach. *Cancer Research*, 27(2):209–220.

Martin, D.J. 1998. Optimizing Census geography: The separation of collection and output geographies. *International Journal of Geographical Information Science*, 12(7):673–685.

Martin, D.J. 1999. Spatial representation: The social scientists' perspective. In P. Longley, M. Goodchild, D. Maguire and D. Rhind (eds) *Geographical Information Systems: Volume 1. Principles and Technical Issues* (2nd edition), 171–189. New York:Wiley.

Martin, R., A. Pike, P. Tyler and B. Gardiner. 2015. Spatially rebalancing the UK economy: The need for a new policy model. London:Regional Studies Association. http://www.regionalstudies.org/uploads/documents/SRTUKE_v16_PRINT.pdf

Martins, R., G.L. Silva and V. Andreozzi. 2016. Bayesian modeling of longitudinal and spatial survival AIDS data. *Statistics in Medicine*, 35(19):3368–3384.

Mason, M. 2009. Findings from the second year of the national neighbourhood policing programme evaluation. RR14. Research, Development and Statistics Directorate. London:Home Office.

Matheron, G. 1973. The intrinsic random functions and their applications. *Advances in Applied Probability*, 5(3):439–468.

Matula, D.W. and R.R. Sokal. 1980. Properties of Gabriel graphs relevant to geographic variation and the clustering of points in the plane. *Geographical Analysis*, 12(3):205–222.

Mawby, R. and R. Yarwood. 2010. *Rural Policing and Policing the Rural*. Farnham:Ashgate.

McCullagh, P. and J.A. Nelder. 1989. *Generalized Linear Models* (2nd edition). Boca Raton, FL:Chapman and Hall/CRC Press.

Mead, R. 1967. A mathematical model for the estimation of interplant competition. *Biometrics*, 23(2):189–205.

Melia, S. 2014. Do randomised controlled trials offer a solution to 'low quality' transport research. Working paper, University of the West of England, Bristol.

Meyer, S., L. Held and M. Höhle. 2017. Spatio-temporal analysis of epidemic phenomena using the R package surveillance. *Journal of Statistical Software*, 77(11):1–55.

Meyer, S., I. Warnke, W. Rössler and L. Held. 2016. Model-based testing for space-time interaction using point processes: An application to psychiatric hospital admissions in an urban area. *Spatial and Spatio-Temporal Epidemiology*, 17:15–25.

Michaels, G. 2008. The effect of trade on the demand for skill: Evidence from the interstate highway system. *Review of Economics and Statistics*, 90(4):683–701.

Min, Y. and A. Agresti. 2005. Random effect models for repeated measures of zero-inflated count data. *Statistical Modelling*, 5(1):1–19.

Molitor, J., M. Papathomas, M. Jerrett and S. Richardson. 2010. Bayesian profile regression with an application to the National survey of children's health. *Biostatistics*, 11(3):484–498.

Mollie, A. 1996. Bayesian mapping of disease. In W.R. Gilks, S. Richardson and D.J. Spiegelhalter (eds) *Markov Chain Monte Carlo in Practice*, 359–380. Boca Raton, FL:Chapman and Hall.

Monmonier, M. 1996. *How to Lie with Maps* (2nd edition). Chicago, IL:University of Chicago Press.

Moon, G., G. Aitken, J. Taylor and L. Twigg. 2017. Integrating national surveys to estimate small area variations in poor health and limiting long-term illness in Great Britain. *BMJ Open*, 7(8):e016936.

Moran, P.A.P. 1950. Notes on continuous stochastic phenomena. *Biometrika*, 37(1–2):17–23.

Morenoff, J.D. 2003. Neighbourhood mechanisms and the spatial dynamics of birth weight. *American Journal of Sociology*, 108(5):976–1017.

Morgan, C.J., M.F. Lenzenweger, D.B. Rubin and D.L. Levy. 2014. A hierarchical finite mixture model that accommodates zero-inflated counts, non-independence, and heterogeneity. *Statistics in Medicine*, 33(13):2238–2250.

Morgenstern, H. 1995. Ecologic studies in epidemiology: Concepts, principles and methods. *Annual Review of Public Health*, 16:61–81.

Mullahy, J. 1986. Specification and testing of some modified count data models. *Journal of Econometrics*, 33(3):341–365.

Müller, P. and R. Mitra. 2013. Bayesian nonparametric inference—Why and how. *Bayesian Analysis*, 8(2):323–356.

Nakagawa, S. 2004. A farewell to Bonferroni: The problems of low statistical power and publication bias. *Behavioral Ecology*, 15(6):1044–1045.

Namazi-Rad, M.-R.,. and D.G. Steel. 2015. What level of statistical model should we use in small area estimation? *Australian and New Zealand Journal of Statistics*, 57(2):275–298.

Neelon, B.H., A.E. Gelfand and M.L. Miranda. 2014. A multivariate spatial mixture model for areal data: Examining regional differences in standardised test scores. *Journal of the Royal Statistical Society: Series C (Applied Statistics)*, 63(5):737–761.

Neelon, B.H., A.J. O'Malley and S.L. Normand. 2010. A Bayesian model for repeated measures zero-inflated count data with application to outpatient psychiatric service use. *Statistical Modelling*, 10(4):421–439.

Nelder, J.A. 1977. A reformulation of linear models. *Journal of the Royal Statistical Society: Series A (General)*, 140:48–77.

Nelder, J.A. and R.W.M. Wedderburn. 1972. Generalized linear models. *Journal of the Royal Statistical Society: Series A (General)*, 135(3):370–384.

Nelsen, R.B. 2007. *An Introduction to Copulas*. New York:Springer.

Neprash, J.A. 1934. Some problems in the correlation of spatially distributed variables. *Journal of the American Statistical Association*, 29 (suppl.):167–168.

Neter, J., M.H. Kutner, C.J. Nachtsheim and W. Wasserman. 1996. *Applied Linear Regression Models*. New York:McGraw Hill.

Newton, A. and M. Felson. 2015. Editorial: Crime patterns in time and space: The dynamics of crime opportunities in urban areas. *Crime Science*, 4(1):11. doi: 10.1186/s40163-015-0025-6

Newton, M.A., A. Noueiry, D. Sarkar and P. Ahlquist. 2004. Detecting differential gene expression with a semi-parametric hierarchical mixture model. *Biostatistics*, 5(2):155–176.

Ning, X. and R.P. Haining. 2003. Spatial pricing in interdependent markets: A case study of petrol retailing in Sheffield. *Environment and Planning A*, 35(12):2131–2159.

O'Keeffe, A.G., L.D. Sharples and I. Petersen. 2014. Regression discontinuity design: An approach to the evaluation of treatment efficacy in primary care using observational data. *British Medical Journal*, 349:g5293.

Office for National Statistics. 2016. Model-based Estimates of households in poverty for Middle Layer Super Output Areas, 2011/12, Technical Report.

Openshaw, S. 1983. *The Modifiable Areal Unit Problem*. Norwick:Geo Books.

Openshaw, S., M. Charlton, C. Wymer and A. Craft. 1987. A mark 1 geographical analysis machine for the automated analysis of point data sets. *International Journal of Geographical Information Systems*, 1(4):335–358.

Ord, K. 1975. Estimation methods for models of spatial interaction. *Journal of the American Statistical Association*, 70(349):120–126.

Paciorek, C.J. 2010. The importance of scale for spatial-confounding bias and precision of spatial regression estimators. *Statistical Science*, 25(1):107–125.

Paelinck, J. and L. Klaassen. 1979. *Spatial Econometrics*. Farnborough:Saxon House.

Papachristos, A.V., D.M. Hureau and A.A. Braga. 2013. The corner and the crew: The influence of geography and social networks on gang violence. *American Sociological Review*, 78(3):417–447.

Patton, M. and S. McErlean. 2003. Spatial effects within the agricultural land market in Northern Ireland. *Journal of Agricultural Economics*, 54(1):35–54.

Pearson, R.K. 2018. *Exploratory Data Analysis Using R*. Boca Raton, FL:Chapman and Hall/CRC.

Porter, M.E. 1998. *The Competitive Advantage of Nations*. London:MacMillan.

Prado, R. and M. West. 2010. *Time Series: Modelling, Computation and Inference*. Boca Raton, FL:Chapman and Hall.

Pukelsheim, F. 1994. The three sigma rule. *American Statistician*, 48(2):88–91.

Qi, F. and F. Du. 2013. Tracking and visualization of space-time activities for a micro-scale flu transmission study. *International Journal of Health Geographics*, 12:6. https://dx.doi.org/10.1186%2F1476-072X-12-6

Quick, H., L.A. Waller and M. Casper. 2018. A multivariate space–time model for analysing county level heart disease death rates by race and sex. *Journal of the Royal Statistical Society: Series C (Applied Statistics)*, 67(1):291–304.

Quick, M., G. Li and I. Brunton-Smith. 2018. Crime-general and crime-specific spatial patterns: A multivariate spatial analysis of four crime types at the small-area scale. *Journal of Criminal Justice*, 58:22–32.

Quick, M., G. Li and J. Law. 2019. Spatiotemporal modelling of correlated small-area outcomes: Analyzing the shared and type-specific patterns of crime and disorder. *Geographical Analysis*, 51(2), pp.221–248.

R Core Team. 2019. R: A language and environment for statistical computing. R Foundation for Statistical Computing, Vienna, Austria. URL https://www.R-project.org/.

Raftery, A.E. and S.M. Lewis. 1992. [Practical Markov Chain Monte Carlo]: Comment: One long run with diagnostics: Implementation strategies for Markov Chain Monte Carlo. *Statistical Science*, 7(4):493–497.

Ratcliffe, J.H. 2002. Aoristic signatures and the temporal analysis of high volume crime patterns. *Journal of Quantitative Criminology*, 18(1):23–43.

Ratcliffe, J.H. and M.J. McCullagh. 1998. Aoristic crime analysis. *International Journal of Geographical Information Science*, 12(7):751–764.

Rao, J.N.K. and I. Molina. 2015. *Small Area Estimation* (2nd edition). New York:Wiley.

Redding, S.J. and D.M. Sturm. 2008. The costs of remoteness: Evidence from German division and reunification. *American Economic Review*, 98(5):1766–1797.

Relethford, J.H. 2008. Geostatistics and spatial analysis in biological anthropology. *American Journal of Physical Anthropology*, 136(1):1–10.

Revelli, F. 2002. Testing the taxmimicking versus expenditure spillover hypothesis using English data. *Applied Economics*, 34(14):1723–1731.

Richardson, S. 1992. Statistical methods for geographical correlation studies. In P. Elliot, J. Cuzich, D. English and R. Stern (eds) *Geographical and Environmental Epidemiology: Methods for Small Area Studies*, 181–204. Oxford:Oxford University Press.

Richardson, S. and C. Montfort. 2000. Ecological correlation studies. In P. Elliott, J. Wakefield, N. Best and D. Briggs (eds) *Spatial Emidemiology: Methods and Applications*, 205–220. Oxford:Oxford University Press.

Richardson, S., A. Thomson, N. Best and P. Elliott. 2004. Interpreting posterior relative risk estimates in disease mapping studies. *Environmental Health Perspectives*, 112(9):1016–1025.

Ripley, B. 1981. *Spatial Statistics*. New York:Wiley.

Rizopoulos, D. 2012. *Joint Models for Longitudinal and Time-To-Event Data: With Applications in R*. Boca Raton, FL:Chapman & Hall/CRC Biostatistics Series.

Rose, C.E., S.W. Martin, K.A. Wannemuehler and B.D. Plikaytis. 2006. On the use of zero-inflated and hurdle models for modeling vaccine adverse event count data. *Journal of Biopharmaceutical Statistics*, 16(4):463–481.

Rosenzweig, M.R. and K.I. Wolpin. 2000. Natural "natural experiments" in economics. *Journal of Economic Literature*, 38(4):827–874.

Rue, H. and L. Held. 2005. *Gaussian Markov Random Fields: Theory and Applications*. Boca Raton, FL:Chapman and Hall.

Rue, H., S. Martino and N. Chopin. 2009. Approximate Bayesian inference for latent Gaussian models by using integrated nested Laplace approximations. *Journal of the Royal Statistical Society: Series B (Statistical Methodology)*, 71(2):319–392.

Rushworth, A., D. Lee and R. Mitchell. 2014. A spatio-temporal model for estimating the long-term effects of air pollution on respiratory hospital admissions in Greater London. *Spatial and Spatio-Temporal Epidemiology*, 10:29–38.

Rushworth, A., D. Lee and C. Sarran. 2017. An adaptive spatiotemporal smoothing model for estimating trends and step changes in disease risk. *Journal of the Royal Statistical Society: Series C (Applied Statistics)*, 66(1):141–157.

Salehyan, I. and K.S. Gleditsch. 2006. Refugees and the spread of civil war. *International Organization*, 60(2):335–366.

Salvatier, J., T.V. Wiecki and C. Fonnesbeck. 2016. Probabilistic programming in Python using PyMC3. *PeerJ Computer Science*, 2:e55.

Sampson, R.J. 2013. The place of context: A theory and strategy for criminology's hard problems. *Criminology*, 51(1):1–31.

Sampson, R.J., S.W. Raudenbush and F. Earls. 1997. Neighbourhoods and violent crime: A multi-level study of collective efficacy. *Science*, 277(5328):918–924.

SaTScanTM 2018a. Kulldorff M and Information Management Services, Inc. SaTScan v.9.6: Software for the spatial and space-time scan statistics.

SaTScanTM User Guide. 2018b (March) https://www.satscan.org/cgi-bin/satscan/register.pl/SaTScan_Users_Guide.pdf?todo=process_userguide_download

Schmertmann, C.P. 2015. Adjusting for population shifts and covariates in space-time interaqction tests. *Biometrics*, 71(3):714–720.

Seya, H., Y. Yamagata and M. Tsutsumi. 2013. Automatic selection of a spatial weight matrix in spatial econometrics: Application to a spatial hedonic approach. *Regional Science and Urban Economics*, 43(3):429–444.

Shand, L.B., B. Li, T. Park and D. Albarracín. 2018. Spatially varying auto-regressive models for prediction of new human immunodeficiency virus diagnoses. *Journal of the Royal Statistical Society: Series C (Applied Statistics)*, 67(4):1003–1022.

Shekhar, S., E.H. Yoo, S.A. Ahmed, R. Haining and S. Kadannolly. 2017. Analysing malaria incidence at the small area level for developing a spatial decision support system: A case study in Kalaburagi, Karnataka, India. *Spatial and Spatio-Temporal Epidemiology*, 20:9–25.

Sheppard, E., R.P. Haining and P. Plummer. 1992. Spatial pricing in interdependent markets. *Journal of Regional Science*, 32(1):55–75.

Sherman, L.W., P.R. Gartin and M.E. Buerger. 1989. Hotspots of predatory crime: Routine activities and the criminology of place. *Criminology*, 27(1):27–56.

da Silva, A.R. and A.S. Fotheringham. 2016. The multiple testing issue in geographically weighted regression. *Geographical Analysis*, 48(3):233–247.

Simpson, E.H. 1949. Measurement of diversity. *Nature*, 163(4148):688.

Skogan, W.G. 1990. *Disorder and Decline*. New York:Free Press.

Snow, D.A. and D.M. Moss. 2014. Protest on the fly: Toward a theory of spontaneity in the dynamics of protest and social movement. *American Sociological Review*, 79(6):1122–1143.

Sparks, A., N. Bania and L. Leete. 2009. *Finding Food Deserts: Methodology and Measurement of Food Access in Portland, Oregon*, 1–26. Washington, DC:National Poverty Center/USDA Economic Research Service.

Spiegelhalter, D.J., K.R. Abrams and J.P. Myles. 2004. *Bayesian Approaches to Clinical Trials and Health-Care Evaluation*. New York:Wiley.

Spiegelhalter, D.J., N.G. Best, B.P. Carlin and A. Van der Linde. 2002. Bayesian measures of model complexity and fit (with discussion). *Journal of the Royal Statistical Society: Series B (Statistical Methodology)*, 64(4):583–639.

Stan Development Team. 2018. RStan: The R interface to Stan. R package version 2.18.2. http://mc-stan.org/

Stephan, F.F. 1934. Sampling errors and interpretations of social data ordered in time and space. *Journal of the American Statistical Association*, 29 (suppl.):165–166.

Stern, H.S. and N. Cressie. 2000. Posterior predictive model checks for disease mapping models. *Statistics in Medicine*, 19(17–18):2377–2397.

Szwarcwald, C., F. Bastos, C. Barcellos, M. de Pina and M. Esteves. 2000. Health conditions and residential concentration: A study in Rio de Janiero, Brazil. *Journal of Epidemiology and Community Health*, 54(7):530–536.

Tango, T. 1995. A class of tests for detecting general and focused clustering of rare diseases. *Statistics in Medicine*, 14(21–22):2323–2334.

Taylor, B.M., R. Andrade-Pacheco and H.J.W. Sturrock. 2018. Continuous inference for aggregated point process data. *Journal of the Royal Statistical Society: Series A (Statistics in Society)*, 181(4):1125–1150.

Tita, G.E. and S.M. Radil. 2010. Making space for theory: The challenges of theorizing space and place for spatial analysis in criminology. *Journal of Quantitative Criminology*, 26(4):467–479.

Tjøstheim, D. 1978. A measure of association for spatial variables. *Biometrika*, 65(1):109–114.

Tobler, W.R. 1970. A computer movie simulating urban growth in the Detroit region. *Economic Geography*, 46 (Suppl.):234–240.

Tolnay, S.E. 1995. The spatial diffusion of fertility: A cross sectional analysis of counties in the American South, 1940. *American Sociological Review*, 60(2):299–308.

Tolnay, S.E., G. Deane and E.M. Beck. 1996. Vicarious violence: Spatial effects on Southern lynchings, 1890–1919. *American Journal of Sociology*, 102(3):788–815.

Tango, T. and K. Takahashi. 2005. A flexibly shaped spatial scan statistic for detecting clusters. *International Journal of Health Geographics*, 4:11. https://dx.doi.org/10.1186%2F1476-072X-4-11

Tango, T. and K. Takahashi. 2012. A flexible spatial scan statistic with a restricted likelihood ratio for detecting disease clusters. *Statistics in Medicine*, 31(30):4207–4218.

Townsend, P., P. Phillimore and A. Beattie 1988. *Health and Deprivation: Inequality and the North*. London:Croom Helm.

Tseloni, A. and K. Pease. 2004. Repeat personal victimization: Random effects, event dependence and unexplained heterogeneity. *British Journal of Criminology*, 44(6):931–945.

Tsiatis, A.A. and M. Davidian. 2004. Joint modelling of longitudinal and time-to-event data: An overview. *Statistica Sinica*, 14(3):809–834.

Tuffin, R., J. Morris and A. Poole. 2006. An evaluation of the impact of the national reassurance policing programme. Home Office Research Study No. 296. Home Office Research, Development and Statistics Directorate, London.

Tukey, J. 1977. *Exploratory Data Analysis*. Reading, MA:Addison-Wesley.

Tzala, E. and N. Best. 2008. Bayesian latent variable modelling of multivariate spatio-temporal variation in cancer mortality. *Statistical Methods in Medical Research*, 17(1):97–118.

Ugarte, M.D., A. Adin and T. Goicoa. 2017. One-dimensional, two-dimensional and three-dimensional B-splines to model space-time interactions in Bayesian disease mapping: Model fitting and model identifiability. *Spatial Statistics*, 22:451–468.

Ugarte, M.D., T. Goicoa and A.F. Militino. 2010. Spatio-temporal modelling of mortality risks using penalized splines. *Environmetrics*, 21(3–4):270–289.

van den Broek, J. 1995. A score test for zero inflation in a Poisson distribution. *Biometrics*, 51(2):738–743.

van Ravenzwaaij, D., P. Cassey and S.D. Brown. 2018. A simple introduction to Markov chain Monte–Carlo sampling. *Psychonomic Bulletin and Review*, 25(1):143–154.

Vega, S. and J.P. Elhorst. 2015. The SLX model. *Journal of Regional Science*, 55(3):339–363.

Venables, A.J. 1999. But why does geography matter, and which geography matters? *International Regional Science Review*, 22(2):238–241.

Venables, W.N., D.M. Smith and the R Core Team. 2018. *An Introduction to R Notes on R: A Programming Environment for Data Analysis and Graphics*. Version 3.5.2 (2018-12-20). Bristol, UK:Network Theory.

Ventrucci, M., E.M. Scott and D. Cocchi. 2011. Multiple testing on standardized mortality ratios: A Bayesian hierarchical model for FDR estimation. *Biostatistics*, 12(1):51–67.

Ver Hoef, J.M., E.M. Hanks and M.B. Hooten. 2018. On the relationship between conditional (CAR) and simultaneous (SAR) autoregressive models. *Spatial Statistics*, 25:68–65.

Wacholder, S., J.K. McLaughlin and D. Silverman. 1991a. Selection of controls in case-control studies. I. Principles. *American Journal of Epidemiology*, 135(9):1019–1028.

Wacholder, S., J.K. McLaughlin and D. Silverman. 1991b. Selection of controls in case-control studies. II Types of controls. *American Journal of Epidemiology*, 135:1029–1041.

Wackernagel, H. 2003. Multivariate nested variogram. In H. Wackernagel (ed) *Multivariate Geostatistics: An Introduction with Applications* (3rd edition), 175–182. Berlin:Springer-Verlag.

Wall, M.M. 2004. A close look at the spatial structure implied by the CAR and SAR models. *Journal of Statistical Planning and Inference*, 121(2):311–324.

Wall, M.M. and X. Liu. 2009. Spatial latent class analysis model for spatially distributed multivariate binary data. *Computational Statistics & Data Analysis*, 53(8):3057–3069.

Wakefield, J. 2003. Sensitivity analyses for ecological regression. *Biometrics*, 59(1):9–17.

Wakefield, J. 2007. Disease mapping and spatial regression with count data. *Biostatistics*, 8(2):158–183.

Waller, L.A. 2014. Putting spatial statistics (back) on the map. *Spatial Statistics*, 9:4–19.

Wang, X., Y.R. Yue and J.J. Faraway. 2018. *Bayesian Regression Modeling with INLA*. Boca Raton, FL:CRC Press.

Wakefield, J. 2003. Sensitivity analyses for ecological regression. *Biometrics*, 59(1):9–17.

Wakefield, J. 2007. A Bayesian measure of the probability of false discovery in genetic epidemiology studies. *The American Journal of Human Genetics*, 81(2):208–227.

Wakefield, J. 2013. *Bayesian and Frequentist Regression Methods*. New York:Springer.

Wakefield, J., G.A. Fuglstad, A. Riebler, J. Godwin, K. Wilson and S.J. Clark. 2018. Estimating under-five mortality in space and time in a developing world context. *Statistical Methods in Medical Research*. https://doi.org/10.1177/0962280218767988

Wakernagel, H. 1998. *Multivariate Geostatistics: An Introduction with Applications*. Berlin:Springer.

Wall, M.M. 2004. A close look at the spatial structure implied by the CAR and SAR models. *Journal of Statistical Planning and Inference*, 121(2):311–324.

Waller, L.A. 2014. Putting spatial statistics (back) on the map. *Spatial Statistics*, 9:4–19.

Waller, L.A., B.P. Carlin, H.A. Xia and A.E. Gelfand. 1997. Hierarchical spatio-temporal mapping of disease rates. *Journal of the American Statistical Association*, 92(438):607–617.

Warboys, M. and K. Hornsby. 2004. From objects to events – GEM, the geospatial event model. In M. Egenhofer, C. Freksa and H. Miller (eds) *Proceedings of the 3rd International Conference on Geographic Information Science*, 327–343. New York:Springer.

Ward, M.P. and T.E. Carpenter. 2000. Analysis of space-time clustering in veterinary medicine. *Preventive Veterinary Medicine*, 43(4):225–237.

Webster, R. 1985. Quantitative analysis of soil in the field. In B.A. Stewart (ed) *Advances in Soil Science*, 1–70. New York:Springer-Verlag.

Weisburd, D., E. Groff and S.-M. Yang. 2012. *The Criminology of Place: Street Segments and Our Understanding of the Crime Problem*. Oxford:Oxford University Press.

Wheeler, D. 2007. Diagnostic tools and a remedial method for collinearity in geographically weighted regression. *Environment and Planning. A*, 39(10):2461–2481.

Wheeler, D. 2009. Simultaneous coefficient penalization and model selection in geographically weighted regression: The geographically weighted lasso. *Environment and Planning A*, 41:722–742.

Wheeler, D. and M. Tiefelsdorf. 2005. Multicollinearity and correlation among local regression coefficients in geographically weighted regression. *Journal of Geographical Systems*, 7(2):161–187.

Whittle, P. 1954. On stationary processes in the plane. *Biometrika*, 41(3–4):434–449.

Williams, L.K. and G.D. Whitten. 2014. Don't stand so close to me: Spatial contagion effects and party competition. *American Journal of Political Science*, 59(2):309–325.

Wilson, K. and J. Wakefield. 2018. Pointless spatial modeling. *Biostatistics*, kxy041. https://doi.org/10.1093/biostatistics/kxy041

Wilson, W. 1997. *When Work Disappears: The World of the New Urban Poor*. New York:Alfred Knopf.

WinBUGS User Manual Version 1.4 January 2003 by Spiegelhalter, D., A. Thomas, N. Best and D. Lunn. https://www.mrc-bsu.cam.ac.uk/wp-content/uploads/manual14.pdf; last accessed 18/03/2019.

Womble, W.H. 1951. Differential systematics. *Science*, 114(2961):315–322.

Woodall, W.H., J.B. Marshall, M.D. Joner, Jr, S.E. Fraker and A.-S.G. Abdel-Salam. 2008. On the use and evaluation of prospective scan methods for health-related surveillance. *Journal of the Royal Statistical Society: Series A (Statistics in Society)*, 171:223–237.

Xia, H. and B.P. Carlin. 1998. Spatio-temporal models with errors in covariates: Mapping Ohio lung cancer mortality. *Statistics in Medicine*, 17(18):2025–2043.

Ye, X. and S.J. Rey. 2013. A framework for exploratory space-time analysis of economic data. *The Annals of Regional Science*, 50(1):315–339.

Young, L.J., C.A. Gotway, J. Yang, G. Kearney and C. DuClos. 2009. Linking health and environmental data in geographical analysis: It's so much more than centroids. *Spatial and Spatio-Temporal Epidemiology*, 1(1):73–84.

Yu, H., Y. Wang, J. Opsomer, P. Wang and N.A. Ponce. 2018. A design based approach to small area estimation using a semiparametric generalized linear mixed model. *Journal of the Royal Statistical Society: Series A (Statistics in Society)*, 181(4):1151–1167.

Yule, G.U. 1912. On the methods of measuring the association between two attributes. *Journal of the Royal Statistical Society*, 75(6):579–652.

Zhang, Z., R. Assunção and M. Kulldorff. 2010. Spatial scan statistics adjusted for multiple clusters. *Journal of Probability and Statistics*. http://dx.doi.org/10.1155/2010/642379, Article ID 642379.

Index

Accessibility hypothesis, 21
Acquired immune deficiency syndrome (AIDS), 567
Action and reaction, 58
Adaptive local smoothing, 39
Adaptive spatial smoothing, 102, 503
AER R package, 204
Aggregate and individual levels data, 571–572
AIC, *see* Akaike's Information Criterion
AIDS, *see* Acquired immune deficiency syndrome
Air pollution and stroke mortality, 296–307
   data, 296–300
   modelling, 300–302
   statistical results interpretation, 302–305
Akaike's Information Criterion (AIC), 148
Analysis of variance (ANOVA), 170, 172, 201
Anisotropic structure, 75
ANOVA, *see* Analysis of variance
Applications of spatial-temporal data modelling, 517–560
   assessing stability of risk, 527–539
      classification of areas, 531–532
      formulations, 530–531
      implementation and area classification, 532–535
      interpreting statistical results, 535–539
      temporal dynamics of crime hotspots and coldspots, 527–529
   detecting unusual local time patterns, 539–550
      extension of BaySTDetect, 548–550
      with false discovery rate control, 543
      fitting BaySTDetect in WinBUGS, 543
      formulation, 540–542
      results from simulated dataset, 546–548
      small area disease surveillance, 539–540
      use of BaySTDetect, 546
   overview, 517
   spillover effects on village-level malaria risk, 550–558
      data, 551–552
      modelling, 552–553
      overview, 550–551
      results, 553–556
   targeted crime reduction intervention, 518–527
      constructing control groups, 520–521
      data, 518–520
      evaluation using ITS, 521–523
      results, 523–526
      WinBUGS implementation, 523
Area-specific random effects, 36, 37, 277, 280–282, 382, 522, 569
Area-specific regression coefficients, 198, 200
Area-specific trend model, 539–540, 543
Area to point (ATP) kriging, 574
Assessment model, 25, 428–429, 432, 521, 522, 523, 561
Asymptotic chi-squared distribution, 207
Atmospheric nitrogen oxides, 18, 281, 297–298
Atomistic fallacy, 80
ATP, *see* Area to point kriging
Attribute data, 63–85
   collection processes in social sciences, 63–70
      natural experiments, 64–67
      non-experimental observational studies, 68–70
      quasi-experiments, 67–68
   fundamental properties of spatial and spatial-temporal, 74–76
      dependence, 74–75
      heterogeneity, 75–76
   geographical reality and spatial database, 71–73
   overview, 63
   properties induced by measurement processes, 83–84
   properties induced by representational choices, 76, 78–83
Attribute values variation, 52–53
Auto-binomial model, 575
Auto-Poisson model, 575
Average direct effect, 349
Average indirect effect, 349
Average total effect, 349

Backtracking, 96, 103
Baseline model, 25, 428–429, 432, 521–523, 561
Bayes, Thomas, 118
Bayesian analysis, 34, 116–121, 149, 153, 266
   high-intensity crime areas modelling, 120–121
   likelihood function, 116–120

posterior distribution, 116–120
prior distribution, 116–120
Bayesian approach, 31, 32, 33, 35, 115, 121, 151, 224
Bayesian computation, 121–137
    integration and Monte Carlo integration, 123–127
    Markov chain Monte Carlo with Gibbs sampling, 127–128
    posterior distribution, 121–123
    WinBUGS
        fitting models in, 133–137
        overview, 129–133
Bayesian hierarchical models (BHMs), 1, 15, 17, 33–39, 41, 56, 213, 223, 274, 278, 296, 300, 301
    data, 35
    incorporating spatial and spatial-temporal dependence structures using random effects, 36–37
    information sharing through random effects, 37–39
    parameter, 36
    process, 35–36
    for spatial data, 281–331
        association between air pollution and stroke mortality, 296–307
        distribution of high intensity crime areas, 282–296
        modelling cases of rape, 321–330
        overview, 281–282
        village-level incidence of malaria, 308–321
Bayesian inference, 30–31, 33, 41, 115, 129, 150, 151, 153, 512, 565
Bayesian information criterion, 258
Bayesian latent variable model, 295
Bayesian model, 30, 32, 115, 118, 120, 129, 147–148
Bayesian multilevel/hierarchical modelling framework, 559, 560
Bayesian nonparametric methods, 320
Bayesian Poisson regression modelling, 138
Bayesian profile regression (BPR) model, 22, 323–326, 329, 571
Bayesian regression models, 121, 137–147, 152
    annual burglary rates, 143–147
    household-level income, 138–143
Bayesian space-time model, 503, 514
Bayesian spatial econometrics, 1, 39–40
Bayesian spatial models
    comparing fits, 274–277
        based on MSOA-level average income estimates, 276–277
        DIC comparison, 274–276

Bayesian two-component mixture, 551
Bayes' theorem, 34, 115, 118, 120
BaySTDetect model, 27, 539–540, 542, 559
    extension, 548–550
    fitting in WinBUGS, 543
    use, 546
Bernoulli random variables, 285, 306
Besag, York and Mollie (BYM) model, 234, 266–273, 470–471, 503, 506, 530, 532
    applying to Newcastle income data, 269–273
    remarks, 268–269
Best linear unbiased prediction (BLUP) estimator, 570
Beta distribution, 119
BGR, *see* Brooks-Gelman-Rubin (BGR) diagnostic
BHMs, *see* Bayesian hierarchical models
Big data algorithms, 57
Bimodal risk distribution, 316
Binary weights matrix, 89, 95, 245
Binomial distribution, 118, 138, 152
Binomial likelihood, 119, 120, 469
Binomial variability, 382
Bivariate association, 193–195
    Clifford-Richardson test, 193–195
    testing and global bivariate Moran's *I*, 195
Bi-variate correlation coefficient, 174
Bivariate Moran scatterplot, 191–193, 192
BLUP, *see* Best linear unbiased prediction estimator
Bonferroni adjustment, 185
Bootlegging effect, 343
Borrowing strength, 97
Box, George, 13
BPR, *see* Bayesian profile regression model
Brooks-Gelman-Rubin (BGR) diagnostic, 134, 141
Burglary case counts modelling, 398–400
Burglary rates, 4, 8, 9, 68, 166, 168, 181, 398, 403, 407, 408, 466–467, 475, 489, 491, 530, 536
BYM, *see* Besag, York and Mollie model

C++, 503
Cambridgeshire Constabulary, 68, 518, 531
CAR, *see* Conditional autoregressive structure
Cartogram, 81–82
Census output areas (COAs), 6, 9, 11, 25, 72, 80, 120–121, 123, 143–144, 146–147, 179, 283–284, 297, 398, 404, 406, 411, 412–414, 422, 424, 427, 434, 460, 465, 474–475, 498, 518, 521–522, 529, 531–532, 567

Challenges, 3–14, 41–46
    data sparsity, 10–12
    dependency, 4, 6–8
    heterogeneity, 8–10
    and opportunities, 27–40
        Bayesian hierarchical models, 33–39
        Bayesian spatial econometrics, 39–40
        Bayesian thinking in statistical analysis, 30–33
        statistical thinking, 27, 29–30
    uncertainty, 12–14
        data uncertainty, 12–13
        model/process uncertainty, 13
        parameter uncertainty, 13–14
Chi-square test, 168
Choropleth map, 83, 112, 161–162
Chronic obstructive pulmonary disease (COPD), 27, 540, 549
Clayton's rule, 498, 504–511
    exploring induced space-time dependence structure, 508–511
    Kronecker product, 506–508
    structure matrices and Gaussian Markov random fields, 505–506
Clifford-Richardson test, 193–195
Cluster detection methods, 181, 190
Cluster downweighting, 55
Cluster-specific risk factor profiles, 327
COAs, *see* Census output areas
Complete pooling, 217
Compound treatments, 66
Computation, 575–576
Conditional autoregressive (CAR) structure, 36, 98, 100, 105, 233, 369, 386, 503, 504, 572
Conditional permutation test, 183
Conditional probability, 34
Confounding bias, 66
Conjugate prior, 126
Continuous inference, 572
Continuous-valued data, 170–172
Continuous-valued risk factors, 324, 326
Convergence checking, 134–136
Convolution model, 37
COPD, *see* Chronic obstructive pulmonary disease
Core geographical concepts, 47
Count data, 168
Covariance matrix, 246, 250, 251, 289, 338, 368, 502, 506
Covariate effects interpretation, 346–356
    direct, indirect and total effects, 346–347
    direct and indirect effects without SAR, 347–349

LM and SEM models, 347
SLX model, 347–349
direct and indirect effects with SAR, 350–355
    average effects under SLM model, 354
    calculating in spatial feedback presence, 351
    properties, 351–354
    understanding in spatial feedback presence, 350–351
    estimated effects from cigarette sales data, 355–356
Covariate-outcome relationship, 18, 41, 321
Current Population Survey (1995), 571

Data-driven analysis, 29
Data generating process, 40
Data incompleteness, 83
Data inconsistency, 83–84
Data mapping, 48–52
Data model, 34, 35, 40, 41
Data sparsity, 1, 10–12, 11, 33, 208, 214
Data uncertainty, 12–13, 208
Delaunay triangulation, 91
Demographic and Health Surveys (DHS), 571
Dependency, data, 1, 4, 6–8
Deviance Information Criterion (DIC), 19, 148, 274–276, 290, 314, 362, 378, 381, 384, 474, 493
DHS, *see* Demographic and Health Surveys
DIC, *see* Deviance Information Criterion
Difference in differences (DID) regression model, 520
Diffuse prior, 148
Diffusion processes, 57
Digital scanners, 62
Direct spatial spillover effects, 333–334, 346
Dirichlet cells, 91
Dirichlet tessellation, 91
Discrete-valued covariates, 324
Dispersal processes, 57–58
Distributional outlier, 161
Domain-specific theory, 27

Ecological analyses, 18
Ecological bias, 79, 296, 307
Ecological fallacy, 80, 296
Ecological inference, 18, 79, 296, 571–572
Ecological modelling, 296–297, 321
Ecological studies, 571
ED, *see* Enumeration district
EDA, *see* Exploratory data analysis
EHIA, *see* Empirically-defined HIA

Empirical FDR, 459, 547–548

Empirically-defined HIA, 179, 180–181, 191, 283, 284, 285–290, 293, 294, 566

England COPD mortality data, 460

Enumeration district (ED), 18–19, 297–300, 305–307

Environmental exposure-disease relationships, 18

Error analysis, 83

Error variance, 7

ESDA, *see* Exploratory spatial data analysis

ESTDA, *see* Exploratory spatial-temporal data analysis

Euclidean distance, 90

Exchangeable hierarchical model, 223–230
 global smoothing, 225–226
 logic of information borrowing and shrinkage, 224–225
 to Newcastle income data, 228–230
 variance partition coefficient (VPC), 226–228

Exchange and transfer, 58

Exploratory data analysis (EDA), 159, 191, 311, 439

Exploratory spatial data analysis (ESDA), 159–211
 detecting local spatial clusters, 181–190
  Kulldorff's spatial scan statistic, 186–190
  local Moran's I, 182–185
  multiple testing problem, 185–186
 overview, 159–160
 relationships between variables, 191–203
  bivariate association, 193–195
  geographically weighted regression (GWR), 195–203
  scatterplots and bivariate Moran scatterplot, 191–193
 spatial autocorrelation, 172–181
  global Moran's I applied to regression residuals, 178–179
  global Moran's I statistic, 174–177
  join-count test for categorical data, 179–181
  Moran scatterplot, 173–174
  tests for assessing, 177–178
 techniques, 160–172
  checking for spatial trend, 165–167
  continuous-valued data, 170–172
  count data, 168
  mapping, 161–165
  Monte Carlo test, 169–170
 testing for overdispersion, 203–206
 testing for zero-inflation, 203–204, 206–207

Exploratory spatial-temporal data analysis (ESTDA), 439–464
 overview, 439–440
 patterns, 440–443
 tests of space-time interaction, 448–462
  assessing in form of varying local time trend patterns, 459–462
  Knox test, 449–454
  Kulldorff's space-time scan statistic, 454–459
 visualising, 443–448

Exposure-outcome relationships, 18–20

False discovery rate (FDR), 543, 559

Fay-Herriot model, 570

FDR, *see* False discovery rate

Feedback effects, 40

Feedback mechanism, 103

Finite mixture model, 320

First law of geostatistics, 278

Fitting regression models, 574

Fixed seed points, 91

Frequentist approach, 30, 31, 32, 33

Full conditional distributions, 245

Gabriel graph, 91

Gamma distribution, 138

Gaussian exchangeable model, 503

Gaussian Markov random field (GMRF), 498, 505–506, 509–510, 512, 513

GDR, *see* German Democratic Republic (GDR)

Generalized linear models (GLM), 137

General log-linear formulation, 500

General nesting spatial (GNS) model, 104, 366, 381

General weights matrix, ICAR model and, 245–249
 applying to Newcastle income data, 247–248
 as joint distribution and implied restriction on W, 245–246
 results, 248–249
 sum-to-zero constraint, 246–247

GeoBUGS, 238, 247, 250

GeoDa, 82, 159

Geodatabase in R, 107–110

Geographical features, 107

Geographical Information Systems (GISs), 47, 51, 61–62, 159

Geographically weighted regression (GWR), 195–203, 208, 321, 329

Geographical transport studies, 69

Geographic regression discontinuity (GRD) design, 66

Georeferenced longitudinal and time-to-event data, 567
Geostatistical modelling, 572–574
    mapping to reduce visual bias, 574
    scale effects, 574
    spatial dependence, 573–574
German Democratic Republic (GDR), 65, 66
Getis-Ord G statistic, 177, 181
Gibbs sampling, 127–128, 129
GISs, *see* Geographical Information Systems
GLM, *see* Generalized linear models
Global Moran's *I* statistic, 174–179, 181, 195, 247, 252, 344, 345, 375, 462
Global positioning system (GPS), 62, 81
Global regression model, 200, 202
Global smoothing, 39, 235
Global spatial autocorrelation, 174, 178, 203
Global spatial dependence, 160–161
Global whole map effect, 196
GMRF, *see* Gaussian Markov random field
GNS, *see* General nesting spatial model
GPS, *see* Global positioning system
GRD, *see* Geographic regression discontinuity design
G* statistics, 181
GWmodel package, 203
GWR, *see* Geographically weighted regression

Heterogeneity, data, 1, 8–10
HIAs, *see* High intensity crime areas
Hierarchical models, 30, 217, 226
Higher orders weights matrices, 95–97
High intensity crime areas (HIAs), 179, 191, 281, 282–296
    data and exploratory analysis, 283–285
    modelling, 120–121
    MVCAR model and limitation, 293
    overview, 282–283
    PHIA and EHIA
        combining maps, 285–286
        joint analysis of data, 286–290
        results, 290–292
HIV, *see* Human immunodeficiency virus
"Hotspots," 58
Human immunodeficiency virus (HIV), 503
Human visceral Leishmaniasis (HVL), 500
Hyperparameters, 223, 224
Hyperpriors, 223, 224

ICAR, *see* Intrinsic conditional autoregressive model
ICD9, *see* International Classification of Diseases, 9th revision

Identical parameter model, 218–220
Impact function, specifying, 436–437
Independent parameters model, 220–223
Independent Poisson distributions, 575
India, spillover effects on malaria risk in, 550–558; *see also* Malaria, modelling village-level incidence of
    data, 551–552
    modelling, 552–553
    overview, 550–551
    results, 553–556
Indirect spatial spillover effects, 333–334, 346
"Infinite hypothetical replications," 30
Information bias, 66
Information borrowing, 101–102, 213
Information sharing, 14, 32
Informative priors, 118, 151
Integrated nested Laplace approximation (INLA), 262, 266, 512, 576
International Classification of Diseases, 9th revision (ICD9), 297
Interrupted time series (ITS) models, 404, 406, 426–434, 519
    formulations, 428–429
    quasi-experimental designs and purpose, 426–428
    results, 432–434
    WinBUGS implementation, 429, 432
Intrinsic conditional autoregressive (ICAR) model, 111, 233–234, 275, 311, 387, 409, 418, 419, 434, 491–492, 502, 503, 505, 511, 530
    with general weights matrix, 245–249
        applying to Newcastle income data, 247–248
        as joint distribution and implied restriction on *W*, 245–246
        results, 248–249
        sum-to-zero constraint, 246–247
    using binary spatial weights matrix, 234–243
        properties, 244–245
        results, 240–242
        spatial contiguity to Newcastle income data, 238–239
        WinBUGS implementation, 236–238
Intrinsic partition, 71, 78
Iterative process, 29–30
ITS, *see* Interrupted time series models

Jacobian matrix, 360
Join-count test, 179–181
Joint probability distribution, 34, 151, 358, 360, 504

Knox test, 440, 449–454
    applying to malaria data, 453–454
    instructive example and methods to derive
        p-value, 451–453
Krishna Iyer join-count test, *see* Join-count test
Kronecker product, 506–508
Kulldorff's space-time scan statistic, 440,
        454–459
    simulated small area COPD mortality data,
        456–459
Kulldorff's spatial scan statistic, 186–190

LAD, *see* Local authority district
Lag model (LM), 347
Lagrange multiplier tests, 204
Latent class model, 296
Leave-one-out cross validation (LOOCV)
        method, 198
Leroux model, 503
Likelihood-based approach, 565
Likelihood-based model, 40
Likelihood-focused modelling, 8
Likelihood function, 103, 115, 116–120, 137, 368
Likelihood ratio, 186
Linear trend model, 6, 404, 500, 539
LISA, *see* Local indicator of spatial association
LM, *see* Lag model
Local authority district (LAD), 373–375, 378,
        456, 460, 540–541, 543, 546
Local indicator of spatial association (LISA),
        181, 182
Local information sharing, 98, 99, 101
Localised spatial-temporal patterns, 442
Locally adaptive spatial models, 256–266
    applying to Newcastle income data,
        262–266
    choosing optimal *W* matrix, 258
    modelling elements in *W* matrix, 258–262
Locally adaptive spatial smoothing, 105, 279
Locally-compensated bandwidth GWR, 202
Locally-compensated ridge GWR, 202
Local Moran's *I*, 182–186
Local regression models, 9
Local smoothing, 39, 76, 105, 226, 235–236,
        240–242, 244–245, 247, 252, 264, 272, 274,
        277, 281
Local spatial clusters detection, 181–190
    Kulldorff's spatial scan statistic, 186–190
    local Moran's *I*, 182–185
    multiple testing problem, 185–186
LOOCV, *see* Leave-one-out cross validation
        method
Lower super output area (LSOA), 72, 567

Macro-spatial configuration, 53
Malaria, modelling village-level incidence of,
        308–321
    data and exploratory analysis, 308–309
    overview, 308
    Poisson model, 310–311
        regression with random effects, 310–311
        two-component mixture, 311–312
        two-component mixture with zero-
            inflation, 313–314
    results, 314–318
Maptools, 109
Marginal posterior distribution, 105, 127, 132
Markov chain Monte Carlo (MCMC), 120, 121,
        127–128, 130, 133, 136, 140, 141, 146, 152,
        219–220, 269, 273, 302, 324, 326–327, 408,
        432, 503, 511–512
Maximum likelihood estimates, 149, 168
MCMC, *see* Markov chain Monte Carlo
Members of Parliament (MPs) election, 373
Meso-scale anonymity, 21, 326
Micro-scale anonymity, 21, 326
Middle super output area (MSOA), 12, 15, 17,
        31, 38–39, 72, 170, 172, 213, 214–218,
        220–226, 228–231, 239, 241–242, 248–249,
        253, 256, 262–263, 272, 276–277, 396, 569
Mixture modelling, 485
Model errors, 6
Model/process uncertainty, 13
Modifiable areal unit problem, 70, 78, 85, 397,
        568, 571, 586, 592
Modified *t*-test, 194
Monte Carlo error, 136, 141
Monte Carlo integration, 122, 123–127, 124,
        126, 127
Monte Carlo methods, 208
Monte Carlo procedure, 186, 201, 449, 450
Monte Carlo simulation, 61, 208
Monte Carlo test, 169–170, 186, 451, 453
Moran scatterplot, 173–174, 181, 184–185, 192–193
MPs election, *see* Members of Parliament election
MSOA, *see* Middle super output area
Multicollinearity, 202, 324
Multilevel models, *see* Hierarchical models
Multinomial logistic regression, 295
Multiple linear regression model, 24
Multiple related outcomes modelling, 565–567
Multiple time series design, 426
Multiscale modelling, 567–568
Multivariate conditional autoregressive
        (MVCAR), 283, 294–295
    limitation, 293
    model, 286–290, 293

Multivariate normal (MVN) distribution, 31, 40, 246, 357, 368, 505
Multivariate prior probability distribution, 31–32
Multivariate probability distribution, 36, 40
Multivariate time series, 560
MVCAR, *see* Multivariate conditional autoregressive
MVN, *see* Multivariate normal distribution

National census data, 76
Natural experiments, 64–67
NCC, *see* No cold calling
Neighbouring probability, 259–260
Newcastle income data, 238–239
    BYM model, 269, 271–273
    ICAR model, 238–239, 247–248
    locally adaptive spatial models, 262–266
    pCAR model, 253
    using non-hierarchical models, 218–223
        identical parameter model, 218–220
        independent parameters model, 220–223
No-cold calling (NCC), 24–26, 30–31, 32, 68, 427–429, 432–434, 518–523, 536, 558–559
Nomenclature of Territorial Units for Statistics (NUTS), 72
Non-binary weights matrix, 99
Non-conjugate priors, 119–120, 126
Non-experimental observational studies, 68–70
Non-geographic two-dimensional regression discontinuity design, 66
Non-HIA COAs, 286
Non-hierarchical models and Newcastle income data, 218–223
    identical parameter model, 218–220
    independent parameters model, 220–223
Non-intrinsic partition, 71, 78
Non-spatial multinomial logistic regression, 296
Null hypothesis, 168, 179, 185–186, 202, 205, 207, 449
NUTS, *see* Nomenclature of Territorial Units for Statistics

Office for National Statistics (ONS), 231, 375
OpenBUGS, 129, 550, 567
Opportunities, 14–27, 41–46
    and challenges, 27–40
        Bayesian hierarchical models, 33–39
        Bayesian spatial econometrics, 39–40
        Bayesian thinking in statistical analysis, 30–33
        statistical thinking, 27, 29–30

hypothesis, 21, 326
space-time dynamics, 27
statistical precision, 14–15, 17
variation in space and time, 18–27
    effects of intervention, 24–27
    exposure-outcome relationships, 18–20
    testing conceptual model at small area level, 20–22
    testing for spatial spillover effects, 22–24
Outcome-covariate relationship, 195–203
Overdispersion, 203–206, 281
Oversmoothing, 101, 166

Parameter estimation, 30, 31, 32
Parameter model, 34, 36, 41
Parameter uncertainty, 7, 13–14
pCAR, *see* Proper CAR model
Pearson's correlation coefficient, 194, 208
Permutation test, 176, 179
Peterborough burglary data, 493–498
    modelling trends in, 411–416
    using space-time separable model, 470–471, 474
Petrol retail price data, 390–392
PHIAs, *see* Police-defined high intensity crime areas
"Piecewise linear" effect, 192
Poisson approximation, 451
Poisson distribution, 78, 144, 152, 203, 204, 205, 314, 407
Poisson likelihood, 144, 152, 469, 500, 504
Poisson model, 78–79, 310–311, 317
    regression, 144, 203, 204–207, 309, 314, 316, 318, 326
        with random effects, 310–311
        "whole-map" analysis using, 322–323
    two-component mixture, 311–312
    two-component mixture with zero-inflation, 313–314
Police-defined high intensity crime areas (PHIAs), 123, 283, 284, 285–290, 293, 294, 566
Policy evaluation, 27
Population forecasting model, 396
Positive autocorrelation, 6, 7, 14, 15
Positive spatial autocorrelation, 174, 177, 462
Posterior distribution, 32, 115, 116–120, 126, 152, 318
Posterior predictive check, 144
Posterior predictive probabilities, 144, 318
Posterior predictive p-value, 146–147, 409, 471, 477–479, 489, 515, 555–556
Prior distribution, 115, 116–120, 137, 149, 235–236

Prior neighbouring probability, 259
Prior specifications, 148–151
    for modelling spatial and spatial-temporal
        data, 150–151
    vague/diffuse/weakly-informative, 148–150
Probability density function, 119
Probability distribution, 61, 149
Probability model, 12, 78, 120
Process-based approach, 565
Process-focused modelling, 8
Process model, 34, 35–36, 41, 138, 151, 403, 575
Proper CAR (pCAR) model, 98, 106, 111, 234,
        249–256, 269, 275, 338
    applying to Newcastle income data, 253
    vs. ICAR, 251–252
    prior choice for ρ, 251
    results, 253–256
PyMC3, 576
Python, 576

Quasi-experiments, 67–68
"Queen's move" definition for contiguity, 89

R2WinBUGS package, 535
Random effects, 13, 19–20, 35, 36, 37, 38, 98, 99,
        278, 288, 314, 320
Randomized controlled experiments, 64
Random walk (RW) model, 404–406, 416–426
    formulations, 417–418
    full conditionals and its properties, 418–421
    modelling burglary trends, 421–424
    model of order 2, 424–426
    WinBUGS implementation, 421
Rape cases in Stockholm, 321–330
    data, 321–322
    modelling, 322–326
        "localised" analysis using Bayesian
            profile regression, 323–326
        "whole-map" analysis using Poisson
            regression, 322–323
    risk factors
        "local" associations for, 326–329
        "whole map" associations for, 326
R codes, 108–109, 110–112, 124, 153, 163, 167, 182,
        183, 188, 205, 207, 443, 445, 446, 460
Reassurance policing programme, 68
Registry-based data, 11
Regression coefficients, 36, 152, 197, 201, 261, 353,
        387, 553, 555
Regression discontinuity design, 65
Regression equation, 24
Regression model, 18, 20, 41, 504
Replication and reproducibility, 69–70

Residual variance, 36
Reversible jump MCMC, 320
"Ripple effect," 99
RMSE, *see* Root mean square error
"Rook's move" definition for contiguity, 88
Rook's spatial contiguity, 253, 460, 541
Root mean square error (RMSE), 276–277
Routine Activity Theory, 58
Row standardisation, 93–95
Rural-urban difference, 8–9
RW, *see* Random walk model

Sampling error (SE), 80–81
SAR, *see* Simultaneous autoregressive
        structure
Satellite data collection systems, 62
SaTScan™, 190, 455, 457, 540
Scan-based method, 181–182
SCM, *see* Shared component model
SD, *see* Standard deviation
SDM, *see* Spatial Durbin model
SE, *see* Sampling error
Second-stage local analysis, 22
Selection bias, 66
SEM, *see* Spatial error model
Sexual assault, 21, 54, 281
Shared component model (SCM), 283, 293–295,
        538, 566
Simulated household income data, 231
Simultaneous autoregressive (SAR) structure,
        36, 233, 333, 336, 338, 354, 369, 375, 572
    direct and indirect effects with, 350–355
        average effects under SLM model, 354
        calculating in spatial feedback
            presence, 351
        properties, 351–354
        understanding in spatial feedback
            presence, 350–351
    direct and indirect effects without, 347–349
        LM and SEM models, 347
        SLX model, 347–349
SLM, *see* Spatial lag model
SLX, *see* Spatially-lagged covariates model
Small area estimation, 54–56, 97–102, 568–571
Small area income estimation, 214–218
    hierarchically, modelling parameters,
        217–218
    non-hierarchically, modelling parameters,
        216–217
Small area temporal data modelling, 401–435
    interrupted time series (ITS) models,
        426–434
        formulations, 428–429

quasi-experimental designs and purpose, 426–428
  results, 432–434
  WinBUGS implementation, 429, 432
issues when modelling temporal patterns in small area setting, 403–406
  flexibility of temporal model, 404–406
  temporal dependence, 403–404
  temporal heterogeneity and spatial heterogeneity, 404
linear time trend model, 407–416
  formulations, 407–411
  in Peterborough burglary data, 411–416
random walk (RW) models, 416–426
  formulations, 417–418
  full conditionals and its properties, 418–421
  modelling burglary trends, 421–424
  model of order 2, 424–426
  WinBUGS implementation, 421
setting scene, 406–407
Smoking model, 302–303
Social cohesion, 71
Space-time analysis, 50
Space-time dependence structure, 4, 499
Space-time detection model, 27
Space-time inseparable models, 400, 481–515, 517
  analysing Peterborough burglary data, 493–498
  overview, 481
  space-time separability and, 481–484
  type I space-time interaction, 484–486
    example, 484–486
    WinBUGS Implementation, 486
  type II space-time interaction, 486–490
    example, 489–490
    WinBUGS Implementation, 490
  type III space-time interaction, 490–493
    example, 491–492
    WinBUGS Implementation, 492–493
  type IV space-time interaction, 498–513
    Clayton's rule, 504–511
    extending type II to type IV, 499–501
    extending type III to type IV, 501–504
Space-time interaction tests, 51, 57, 448–462, 481, 495, 500, 513–514
  assessing in form of varying local time trend patterns, 459–462
  exploratory analysis in England COPD mortality data, 460–462
  exploratory analysis in Peterborough burglary data, 460

Knox test, 449–454
  instructive example and methods to derive *p*-value, 451–453
Kulldorff's space-time scan statistic, 454–459
  simulated small area COPD mortality data, 456–459
Space-time separable models, 449, 450, 465–479, 481, 485, 505, 513–514, 539
  analysing Peterborough burglary data, 470–474
  combining space and time components, 469–470
  estimating small area burglary rates, 465–467
  formulations, 467–469
  overview, 465
  results, 474–478
Spatial autocorrelation, 172–181
  global Moran's *I* applied to regression residuals, 178–179
  global Moran's *I* statistic, 174–177
  join-count test for categorical data, 179–181
  Moran scatterplot, 173–174
  structure, 334
  tests for assessing, 177–178
Spatial autoregressive Poisson model, 575
Spatial clustering, 441
Spatial confounding problem, 104
Spatial convergence, 58
Spatial correlation function, 55
Spatial dependence, 166, 208
Spatial Durbin model (SDM), 333, 335, 341–342, 377–378, 381, 384
  formulating, 341–342
  relating to other models, 342
Spatial econometric models, 23, 30, 41, 97, 102–104, 333–372, 388–389
  forms of spatial spillover, 334–335
  hierarchical modelling approach *vs.*, 367–369
  interpreting covariate effects, 346–356
    direct, indirect and total effects, 346–347
    direct and indirect effects without SAR, 347–349
    direct and indirect effects with SAR, 350–355
    estimated effects from cigarette sales data, 355–356
  model fitting in WinBUGS, 356–365
    calculating covariate effects in, 362–365
    derivation of likelihood function, 357–361
    simplifications to the likelihood computation, 361
    zeros-trick, 361–362

modelling cigarette sales, 343–346
  data description, exploratory analysis
      and model specifications, 343–345
  results, 345–346
overview, 333–334
price competition between petrol retail
      outlets, 382–388
  data, 383
  exploratory data analysis, 383–384
  hierarchical model with t_4 likelihood,
      385–387
  modelling and results, 384–385
  overview, 382–383
prior specifications, 342
problems of identifiability, 365–367
spatial Durbin model (SDM), 341–342
  formulating, 341–342
  relating to other models, 342
spatial error model (SEM), 340–341
spatial lag model (SLM), 335–339
  example, 336–337
  formulating, 335–336
  model fitting and interpreting
      coefficients, 339
  reduced form and constraint on δ, 337–338
  spatial weights matrix specification, 338
spatially-lagged covariates model (SLX),
      339–340
  example, 340
  formulating, 339–340
voting outcomes in England, 373–382
  data, 374–375
  exploratory data analysis, 375
  modelling, 375, 377–378
  overview, 373–374
  results, 378, 380–381
Spatial error model (SEM), 333, 335, 340–341,
      347, 377–378, 381, 384, 388–389
Spatial gradients in attribute values, 53
Spatial heterogeneity, 160, 161, 208
Spatial interpolation methods, 55
Spatial lag model (SLM), 23, 40, 102–103, 333,
      334, 335–339, 354, 377, 384
  example, 336–337
  formulating, 335–336
  model fitting and interpreting coefficients, 339
  reduced form and constraint on δ, 337–338
  spatial weights matrix specification, 338
Spatially-lagged covariates model (SLX), 19, 333,
      334, 339–340, 356
  example, 340
  formulating, 339–340

Spatially-lagged effect, 339
Spatially-structured model, 37
Spatially-structured patterns, 266–267
Spatially-structured random effects, 13, 296, 368
Spatially-unstructured patterns, 267, 272
Spatially-unstructured random effects, 13, 314
Spatially varying autoregressive model, 503
Spatial neighbours, 4, 59
Spatial precision, 81
Spatial prior probability distribution, 278
Spatial proximity, 450
Spatial regression models, 40
Spatial smoothing, 96, 160–161, 165–166
Spatial/spatially-referenced database, 47
Spatial spillover effects, 22–24, 40, 53–54, 550–551
Spatial-temporal interaction, 47, 57, 441–442, 449
Spatial-temporal models, 500
Spatial-temporal processes, 58
Spatial-temporal units, 59
Spatial thinking, 47–62
  data mapping, 48–52
  interpolation and small area estimation, 54–56
  overview, 47–48
  and temporal thinking, 56–59
      parameters for spatial-temporal units, 59
      variation, 56–59
  variation, 52–54
Spatial treatment effects, 66
Spatial trend, 160
Spatial weights matrix, 24, 37, 39, 87–113, 95, 100,
      233, 365; *see also* Geodatabase in R
  based on attribute values, 92
  based on contiguity, 88–89
  based on evidence, 92–93
  based on geographical distance, 90
  based on graph structure, 90–91
  choice and statistical implications, 97–104
      effects of observable covariates, 104
      small area estimation, 97–102
      spatial econometric modelling, 102–104
  construction and data stored in shapefile,
      110–113
  estimation, 105
  higher orders, 95–97
  ICAR model and binary, 234–243
      properties, 244–245
      results, 240–242
      spatial contiguity to Newcastle income
          data, 238–239
      WinBUGS implementation, 236–238
  overview, 87–88
  row standardisation, 93–95

Standard deviation (SD), 242
Standard linear regression model, 196, 344
Standard normal regression model, 40
Standard Poisson regression, 206
Standard probability theory, 360
Standard regression model, 6, 8, 103, 202, 347, 356, 358
Statistical analysis, 13, 27, 29, 31, 152
Statistical methods, 55
Statistical model, 20, 24, 29, 30–31, 34, 41, 321
Statistical precision, 11, 14–15, 17, 81
Statistical reasoning process, 29, 30
Stochastic nodes, 129
Stochastic process modelling, 75
Stroke deaths, 18–20, 306
Stroke mortality, 281
Stroke risk model, 302
Student-*t* distribution, 142
Subject matter theory, 69
Surveillance technique, 27
Sweden, rape cases in, *see* Rape cases in Stockholm

Temporal adaptive process, 417
Temporal and spatial thinking, 56–59
    parameters for spatial-temporal units, 59
    variation, 56–59
Temporal clustering, 441
Temporal neighbours, 4, 59
Temporal noise
    linear trend model with, 414–416
    linear trend model without, 412–414
Temporal smoothing, 417, 541
Temporal spillover effect, 550
Temporal weights matrix, 37
Theory-driven analysis, 27, 29
Thiessen polygons, 91
Time series modelling, 25, 441, 501–502
Tobler's First Law of Geography, 17, 74, 88, 278
Torus, 98–99
Two-component mixture, 316, 320, 321, 553, 556
Two-component Poisson mixture model, 311–314, 486
Two-stage rational choice model, 20, 21, 321
Type I space-time interaction, 484–486
    example, 484–486
    WinBUGS implementation, 486
Type II space-time interaction, 486–490
    example, 489–490
    WinBUGS implementation, 490
Type III space-time interaction, 490–493
    example, 491–492
    WinBUGS implementation, 492–493

Type IV space-time interaction, 498–513
    Clayton's rule, 504–511
        exploring induced space-time
            dependence structure, 508–511
        Kronecker product, 506–508
        structure matrices and Gaussian Markov
            random fields, 505–506
    extending type II to type IV, 499–501
    extending type III to type IV, 501–504
    examples, 502–504

UK Census Geography (2011), 214
UK Data Service, 108
Uncertainty, 12–14
    data, 12–13
    model/process, 13
    parameter, 13–14
Undersmoothing, 102
Unit-by-unit modelling, 462
Unit-level models, 569, 570
Univariate normal distribution, 368
Univariate probability distributions, 35
Unmeasured/unobserved covariates, 19
U.S. Census Bureau, 108
US states, modelling cigarette sales in, 343–346
    data description, exploratory analysis and
        model specifications, 343–345
    results, 345–346

Vague priors, 118, 138, 140, 149
VAR, *see* Vector autoregressive model
Variance-covariance matrix, 31, 98
Variance-covariance properties, 76
Variance partition coefficient (VPC), 226–228, 239, 486, 495, 515, 535, 536
Variogram, 55, 574
Vector autoregressive (VAR) model, 501, 503
Voronoi partition/decomposition, *see* Dirichlet tessellation
Voronoi polygons, 91
VPC, *see* Variance partition coefficient

Weakly-informative priors, *see* Vague priors
Weighted least squares approach, 197
WHO, *see* World Health Organization
WinBUGS, 41, 111–112, 116, 120, 127, 144, 146–147, 219, 227, 253, 262, 266, 268–269, 286, 288, 296, 302, 312, 344, 398, 409, 412, 418, 471, 484, 499, 511, 515–516, 532, 576
    fitting BaySTDetect in, 543
    fitting models in, 133–137
        checking convergence, 134–136

checking efficiency, 136–137
setting initial values, 133–134
implementation, 421, 429, 432, 486, 490,
    492–493
and ICAR model, 236–238
model fitting in, 356–365
    calculating covariate effects in, 362–365
    derivation of likelihood function, 357–361
    simplifications to the likelihood
        computation, 361
    zeros-trick, 361–362

modelling, 110
overview, 129–133
Wishart distribution, 289
W matrix, *see* Spatial weights matrix
World Health Organization (WHO), 308

Zero-inflated Poisson (ZIP) model, 206, 208, 314,
    316, 320, 551–552
Zero-inflation, 203–204, 206–207, 210–211, 281,
    320–321, 556, 575
ZIP, *see* Zero-inflated Poisson model